RIBOSOMES

RIBOSOMES

Edited by

M. Nomura
Institute for Enzyme Research
University of Wisconsin

A. Tissières
Department of Molecular Biology
University of Geneva

P. Lengyel
Departments of Molecular Biophysics and Biochemistry
Yale University

Cold Spring Harbor Laboratory
1974

**COLD SPRING HARBOR
MONOGRAPH SERIES**

The Lactose Operon
The Bacteriophage Lambda
The Molecular Biology of Tumour Viruses
Ribosomes

RIBOSOMES
© 1974 by Cold Spring Harbor Laboratory
All rights reserved
International Standard Book Number 0-87969-110-7
Library of Congress Catalog Card Number 74-83791
Book design by Emily Harste
Printed in the United States of America

Contents

Preface

Ribosomes present a unique challenge. The elucidation of their structure, assembled from nearly sixty different molecules, proteins and RNAs, is a problem of immense difficulty. Similarly, their function in the process of translation, still poorly understood, has appeared increasingly complex. Thus when Jim Watson suggested that we have a meeting at Cold Spring Harbor in the fall of 1973, followed by the publication of a book, a serious question came to our minds. Was it the right time?

In the last decade much work has been done on ribosomes regarding their structure, function, genetics and assembly. While the field of ribosome research is currently dynamic and fast progressing, the information we have obtained so far is very large. Thus we felt that reviewing the state of research in its various aspects, in addition to presenting the latest developments, was worthwhile at this time. This book should be useful to established investigators in presenting an overall view of the field. For newcomers, it will serve as a basic introduction to the field of ribosome research, and hopefully will stimulate fresh scientific interest. We also believe that some of the methods presently developed for the study of ribosomes will find their use in other problems of biology where specific aggregates of proteins, or of proteins and nucleic acids, are important.

The meeting was attended by a number of people far larger than we had first anticipated. This turned out to be fortunate as it allowed a more complete representation of the many facets of the research. The response and collaboration of each author in the book was very gratifying. To insure a certain balance and avoid overlaps between chapters, we asked some of the authors to modify their manuscripts, sometimes drastically and on short notice. We are very grateful for their help.

Jim Watson did much for many years to promote work on ribosomes and protein synthesis. Two of us, as well as several contributors to the book, started work on ribosomes with him and have fond memories of their time at Harvard. We would like to thank him for the idea of the meeting and the book, for his encouragement and for his help. Helen Parker took charge of all the organization at Cold Spring Harbor. Her task was difficult as we kept introducing new participants and new speakers up to the last moment. Her help was invaluable. We also wish to thank Nancy Ford who has efficiently dealt with the editing of the book.

Masayasu Nomura
Alfred Tissières
Peter Lengyel

RIBOSOMES

PART I
General
Reviews

Ribosome Research: Historical Background

Alfred Tissières
University of Geneva
Department of Molecular Biology
Geneva, Switzerland

Here I shall present a short, necessarily incomplete, and somewhat personal account of the history of ribosome research. I shall emphasize the steps which took place in the early period and only briefly mention some of the more recent developments leading to the present state of the work, for, in most cases, these will be discussed later in this book.

EARLY OBSERVATIONS

The discovery of ribosomes was preceded by two related observations made in the thirties. First, it was found that the bulk of the RNA is located in the cytoplasm, and second, a relation was established between the amount of RNA and the ability of a cell to synthesize proteins. I shall start by discussing these two points.

Bulk of RNA Is Present in Cytoplasm

Based on his own observations and those by Jorpes (1928), that in the course of spermatogenesis the size of the cytoplasm decreases simultaneously with the amount of RNA in the cell, Brachet (1933) came to the conclusion that very likely RNA was mainly located in the cytoplasm. But the first direct support of this hypothesis was made on plant material

3

several years later. A method had been worked out with rye embryos to achieve the separation of the nuclei from the cytoplasm, and it was shown that DNA was found exclusively in nuclei (Feulgen, Behrens and Mahidiassen 1937) and that a fraction consisting of purified cytoplasm contained most of the RNA (Behrens 1938). Until that time, the belief had been that RNA was present only in plants and DNA only in animal cells. It remained therefore to be shown that RNA was located in the cytoplasm of animal cells.

In the late thirties extensive studies on the location of nucleic acids in the cells of different tissues and in different conditions were made with the help of cytochemical techniques independently by Caspersson and Brachet. Brachet (1940, 1941, 1944) used basic stains (such as toluidine blue) for RNA, the Feulgen reaction for DNA, and introduced the use of a crucial control, that is, exposure of the section of tissue to purified nucleases. Thus RNA was no longer stained after treatment of the slide with pancreatic ribonuclease. Mixture of suitable stains allowed the selective staining of RNA and DNA independently.

Caspersson (1940) had developed a technique by which he could measure the absorption spectrum on a thin section of a tissue on a microscope slide at a point 0.1 μ or less in diameter. The apparatus, rather complicated and expensive, consisted of a quartz microscope coupled with a monochromatic source of light and a spectrophotometer. The location in cells of a number of substances with specific absorption spectra could therefore be determined.

Both authors thus showed that RNA is located in the cytoplasm of animal cells (Brachet 1941; Caspersson 1941, 1950).

Relation between Amount of RNA and Rate of Protein Synthesis

The tips of roots of *Allium cepa* and the imaginal discs of Drosophila larvae were found to be particularly rich in RNA, and it was concluded that a relationship existed between the amount of RNA and the ability of a tissue to grow (Caspersson and Schultz 1938, 1939). However Caspersson (1940) and Brachet (1940) pointed out that nonproliferating tissues that are active in protein synthesis, for instance, the cells secreting proteins in the pancreas and particularly the glands of the silk worm which are specialized in the abundant synthesis of one kind of protein, are rich in RNA, whereas other tissues that are physiologically very active, such as the heart muscle, but are not active in protein synthesis, contain little RNA. Thus both Brachet (1941) and Caspersson (1941) came to the conclusion that there is a direct relationship between the amount of RNA in a cell and the rate of protein synthesis, whereas the relation between growth and RNA seemed rather indirect.

DISCOVERY OF RIBOSOMES

In his studies of Rous sarcoma, Claude (1938, 1940, 1941) isolated tumor-inducing fractions from homogenates of infected tissue by differential centrifugation and found that the infectivity was associated with small particles, visible in the dark field microscope, consisting of ribonucleo-protein and lipids. But he noticed that almost identical, nonvirulent particles could be isolated from uninfected chick or mammal embryos and also from a variety of adult tissues. The difference in size between the Rous virus and the ribosomes was not noted. Claude was convinced that these particles in normal cells consisted of preexisting structures of widespread occurrence and were not artifacts of grinding the tissues, a point not proven until many years later. He called this particulate fraction "small granules" and later "microsomes," a term used to describe the fraction consisting of bits and pieces of endoplasmic reticulum and ribosomes bound and free. Jeener and Brachet (1941, 1942) isolated microsomes from various tissues, showed that they always contain RNA, and put forward the hypothesis of their crucial role in protein synthesis. The electron microscope work of Palade (1955) and Palade and Siekevitz (1956) confirmed the view of Claude that the particulate fraction represents structures present as such in the cell. They were able to show that the cytoplasmic basophilia was due to ribonucleoprotein particles, a component rich in RNA, of high density and about 200 Å in diameter seen in all the tissues examined from mammalian and avian origin. These particles, known today as ribosomes, were found mainly bound to the endoplasmic reticulum in some cells where this system is well developed but free in the cytoplasm of other cells, as in undifferentiated rapidly growing epithelia.

Extensive ultracentrifugal analysis of ribosomes from animal tissues started in the early fifties when Petermann and Hamilton (1952) observed discrete peaks in the analytical centrifuge having sedimentation coefficients ranging from 40S to 76S in extracts from mice spleen. In the case of leukemia, the same nucleoprotein components were extracted from the spleen, though much more abundantly. Such particles, purified by centrifugation and electrophoresis, were shown to contain about 50% RNA (Petermann, Hamilton and Mizen 1954). Clearly these components were the same as those studied by Claude (1941) and Palade (1955).

RIBOSOMES ARE THE SITES OF PROTEIN SYNTHESIS

From about 1950 a number of observations had indicated that the microsome contained the site of protein synthesis, since that fraction was the first where amino acids could be found incorporated into proteins. Thus Borsook et al. (1950) observed that 30 min after injection of labeled amino acids into mice, the microsome had the highest specific activity of the various cell fractions. Similar results were obtained by Hultin (1950).

But the clear demonstration that ribosomes were the site where amino acids were assembled to form proteins came from the laboratory of Zamecnik. Littlefield et al. (1955) showed that after injection of radioactive amino acids into rats, ribosomes extracted from the liver and detached from the reticulum with sodium deoxycholate were labeled maximally within a few minutes after the intravenous injection, whereas the bulk of microsomal proteins were labeled more slowly and progressively. They calculated that only a very small fraction (estimated to be 1%) of the amino acids in ribosomal proteins is turning over rapidly, suggesting a steady state in which the rates of formation and release of polypeptide chains were equal. The same type of observation was made in in vitro experiments using a previously developed cell-free system from rat liver (Zamecnik and Keller 1954). Similar results were also obtained using a cell-free system from mouse ascites tumor cells, a tissue relatively deficient in the endoplasmic reticulum, in which ribosomes are free and can be isolated without sodium deoxycholate. Amino acids were observed to be incorporated into proteins first on ribosomes and could be chased from there to the soluble protein fraction (Littlefield and Keller 1957).

BACTERIAL RIBOSOMES

When did the work on bacterial ribosomes start? Luria, Delbrück and Anderson (1943) in studying by electron microscopy the lysis of *E. coli* by phage noted: "The pictures of lyzed bacteria show, besides the particles of virus, also granular material of very regular units, 10–15 mμ in diameter. If these are to be interpreted as molecules, their size corresponds to a molecular weight of the order of 10^6. These particles are liberated from the cell in great abundance, and seem to constitute the bulk of the cytoplasm." Nine years later an important paper was published by Schachman, Pardee and Stanier (1952). They showed that when bacterial extracts prepared in 0.02–0.05 M NaCl are examined in the analytical centrifuge, three principal components forming very sharp boundaries are found with uncorrected sedimentation coefficients of about 40S, 29S and 5S. It should be noted that no magnesium was added to these extracts. The same type of pattern was obtained from a number of bacterial species and from extracts made by breaking the cells in different ways. These components were shown to comprise the bulk of the cellular RNA, and the purified 40S peak contained 40% RNA and appeared in the electron microscope as spherical particles (Luria et al. 1943). That same year similar centrifugal patterns were observed in *E. coli* extracts by Siegel, Singer and Wildman (1952).

Chao and Schachman (1956) and Tso, Bonner and Vinograd (1956) studied the effect of Mg^{++} ions on yeast and pea seedling ribosomes, respectively. But Chao (1957) was the first to report that this ion was necessary for the stability of the yeast 80S particles, which dissociated in its absence into two components with sedimentation coefficients of 60S and 40S, respectively, an observation which became basic thereafter for work on ribosomes from any source.

Thus by the mid- to late fifties it had been established that somewhat spherical particles from a variety of tissues in higher organisms, with a diameter of 200–300 Å and formed of roughly equal proportions of RNA and proteins, were the site of protein synthesis. Such particles were also found in plants, yeast and bacteria, though in the latter case of somewhat smaller dimension.

Properties and Function

I had been interested in bacterial ribosomes since about 1950, when Roy Markham and I in Cambridge had observed in an analytical centrifuge two sharp boundaries (which we now know to be the 30S and 50S components) in a bacterial preparation (unpublished observation). Jim Watson was naturally interested in these problems, and I joined him at Harvard in 1957 in order to study ribosomes from *E. coli* and, to start with, their gross chemical and physical properties (Tissières and Watson 1958). At that time there were few laboratories working on ribosomes, of which the group at the Carnegie Institution of Washington was one of the most active. Under the guidance of Dick Roberts, then the head of this group, a whole session of the Biophysical Society meeting at M.I.T. in 1958 was devoted to discussions of ribonucleoprotein particles. A book containing the papers presented at the meeting was edited by Roberts (1958) in which he introduced the name RIBOSOME. Charles Kurland joined our group at Harvard that same year to start his work on ribosomal RNA (Kurland 1960), and David Schlessinger joined Jim Watson and myself for further work on the particles (Tissières et al. 1959). We were impressed by the studies of the mechanism of protein synthesis in systems from animal cells, but there was then no direct indication that ribosomes in bacteria were the site of protein synthesis. In a number of laboratories attempts at amino acid incorporation into proteins, as well as trials to synthesize enzymes in vitro, were made with bacterial extracts. The reports were that the most active fractions in protein synthesis were the ones sedimenting at low speed, consisting of pieces of membranes or other large cell fractions. There was no indication in these experiments that ribosomes were involved, and altogether the field of protein synthesis in bacteria was in a depressive state. There were even rumors that perhaps ribosomes in bacteria served an entirely different function than that of protein synthesis. The experiments of McQuillen, Roberts and Britten (1959), showing conclusively that bacterial ribosomes were the site of protein synthesis, were therefore very welcome. They reported that radioactive sulfur or amino acids added to a culture of *E. coli* were first incorporated into proteins on ribosomes, before they could be traced or chased to other parts of the cell. These experiments made use of sucrose gradients, to my knowledge for the first time, to study the distribution and function of macromolecular components in cell extracts.

It had become very reasonable to try obtaining from bacteria and in vitro system similar to that used by Zamecnik and Keller (1954), and therefore we visited Paul Zamecnik to discuss these matters and get his advice. It turned out that he and Marvin Lamborg were just starting such work (Lamborg and Zamecnik 1960). From our first experiments it was clear that the system from *E. coli,* which was easy to handle, had the same general properties as the liver system (Tissières, Schlessinger and Gros 1960). We failed to know at the time that similar experiments had already been described by Schachtschabel and Zillig (1959) about a year earlier. Such systems turned out to be crucial for the study of both the coding problem and the mechanism of protein synthesis.

Ribosomal Proteins and Ribosomal RNAs

Though a working hypothesis had been that the structure of ribosomes was similar to that of the small spherical plant viruses, with a core of RNA surrounded by a simple spherical shell of protein, it soon became obvious that ribosomes had very different properties from these viruses. That the situation was likely to be complex was shown by Waller and Harris (1961), and the very thorough study of Waller (1964) left little doubt that the protein part of the ribosome was formed of a large number of different molecules. Both on starch gel electrophoresis and on carboxymethyl-cellulose chromatography, he obtained a large number of components. Several experiments were done to rule out the possibility of aggregation or chemical modification which would give the illusion of a complex picture. Every evidence in this work was that there were a large number of different proteins and that the proteins from the 30S subunit were different from those of the 50S.

In spite of this, and for over two years thereafter, there were rumors from some laboratories that ribosomal proteins were much simpler than could be deduced from Waller's work, and it was not until the proteins were purified in several laboratories and peptide maps and other types of analysis done on them that the situation became really clear (Kaltschmidt et al. 1967; Traut et al. 1967; Hardy et al. 1969). As is known today, the 30S *E. coli* subunit contains 21 and the 50S 34 different proteins (Witt-mann et al. 1971). Sequence work on these ribosomal proteins is already well on the way (Wittmann and Wittmann-Liebold, this volume).

The last RNA component to be discovered in bacterial ribosomes, the 5S RNA (Rosset and Monier 1963), was also the first one for which the nucleotide sequence became known (Brownlee, Sanger and Barrell 1968). The work from the Strasbourg group has led to the almost complete knowledge of the sequence of the 16S RNA from *E. coli* (see Fellner, this volume), and rapid progress is already being made on the 23S RNA. Thus a great deal of chemical information is already available on the numerous components of the ribosomes from *E. coli*.

RIBOSOMES FROM HIGHER ORGANISMS

Although most of the early work was done with material from animal cells, information on ribosomes from higher organisms lagged behind until a few years ago, when the conditions were worked out to dissociate and reassociate the subunits without losing their properties and functions (Martin and Wool 1968). Because of their size, which is almost twice that of the *E. coli* particles, and because of some of their properties, it was thought at first that the ribosomes from eukaryotes differed basically from those of prokaryotes. Now, however, the studies on the function of the two kinds of particles suggest that the differences are less important. Work on ribosomes from higher organisms is rapidly catching up and will, I believe, before long reach a level comparable in many ways to that of the *E. coli* particles, since in several instances the experience gained with the prokaryote ribosomes can be applied to those from animal cells (see Wool and Stöffler, this volume).

COMPLEXITY OF RIBOSOMES AND PROSPECTS

Why are ribosomes so complex? The progress made in recent years on protein synthesis, on the mechanisms of initiation, elongation, and termination of the polypeptide chain, has indicated the very active and complex role of the ribosome. A number of highly specific sites, formed of ribosomal proteins and possibly RNA and serving particular functions, such as interactions with tRNA, mRNA, and the various protein factors, are being investigated. Taking this into account, the great complexity of the ribosome is not as surprising as it might have been a number of years ago, when, in the absence of our present knowledge, it was thought that ribosomes played a rather passive role.

The experiments on the reconstruction of ribosomes from RNA and proteins (Traub and Nomura 1968; Nomura and Erdmann 1970) have led to new concepts on assembly and to new ways of studying the function of particular components. We are now reaching the questions of the topography and architecture of the particles in relation to function, a problem of still formidable dimension.

Nevertheless, it is pleasing to see that the state of research has reached a very high level, both by the sophistication of the tools available and by the importance of the expected results. It appears to me that research on ribosomes has entered a most interesting phase, various aspects of which are treated in the following chapters. Indeed, a variety of tools, such as cross-linking agents, affinity labeling, immunological techniques, genetics, and the study of the interaction between proteins and RNAs, are just starting to be exploited and should yield a great deal of information concerning structure and function. Knowledge of the constituents of the specific sites on ribosomes, as those interacting with tRNA, mRNA, and the various protein factors, should result from such studies. In addition, the new physical approaches presented here seem very promising. However, it still remains to be seen how far these techniques will lead us in the understanding of the mechanism of protein synthesis.

References

Behrens, M. 1938. Ueber die Lokalisation der Hefenucleinsäure in pflanzlichen Zellen. *Hoppe-Seylers Z. Physiol. Chem.* **253**:185.

Borsook, H., C. L. Deasy, A. J. Haagen-Smit, G. Keighly and P. H. Lowy. 1950. Metabolism of C^{14}-labeled glycine, l-histidine, l-leucine and l-lysine. *J. Biol. Chem.* **187**:829.

Brachet, J. 1933. Recherches sur la synthèse de l'acide thymonucléique pendant le développement de l'oeuf d'oursin. *Arch. Biol.* **44**:519.

———. 1940. La détection histochimique des acides pentosenucléiques. *C. R. Soc. Biol.* **133**:88.

———. 1941. La détection histochimique et le microdosage des acides pentose-nucléiques. *Enzymologia* **10**:87.

———. 1944. *Embryologie chimique.* Masson & Cie, Paris.

Brownlee, G. G., F. Sanger and B. G. Barrell. 1968. The sequence of 5S rRNA. *J. Mol. Biol.* **34**:379.

Caspersson, T. 1940. II. Methods for the determination of the absorption spectra of cell structures. *J. Roy. Microscop. Soc.* **60**:8.

———. 1941. The protein metabolism of the cell. *Naturwissenschaften* **29**:33.

———. 1950. *Cell growth and cell function: A cyto-chemical study.* W. W. Norton, New York.

Caspersson, T. and J. Schultz. 1938. Nucleic acid metabolism of the chromosomes in relation to gene reproduction. *Nature* **142**:294.

———. 1939. Pentose nucleotides in the cytoplasm of growing tissues. *Nature* **143**:602.

Chao, F. C. 1957. Dissociation of macromolecular ribonucleoprotein of yeast. *Arch. Biochem. Biophys.* **70**:426.

Chao, F. C. and H. K. Schachman. 1956. The isolation and characterization of a macromolecular ribonucleoprotein from yeast. *Arch. Biochem. Biophys.* **61**:220.

Claude, A. 1938. A fraction from normal chick embryo similar to the tumor producing fraction of chicken tumor. I. *Proc. Soc. Exp. Biol. Med.* **39**:1.

———. 1940. Particulate components of normal and tumor cells. *Science* **91**:77.

———. 1941. Particulate components of cytoplasm. *Cold Spring Harbor Symp. Quant. Biol.* **9**:263.

Feulgen, R., M. Behrens and S. Mahdihassan. 1937. Darstellung und Identifizierung der in den pflanzlichen Zellkernen vorkommenden Nucleinsäure. *Hoppe-Seyler's Z. Physiol. Chem.* **246**:203.

Hardy, S. J. S., C. G. Kurland, P. Voynow and G. Mora. 1969. The ribosomal proteins of *E. coli*. I. Purification of the 30S proteins. *Biochemistry* **8**:2897.

Hultin, T. 1950. Incorporation in vivo of ^{15}N-labeled glycine into liver fractions of newly hatched chicks. *Exp. Cell Res.* **1**:376.

Jeener, R. and J. Brachet. 1941. Association, dans le même granule, de ferments et des pentosenucléoprotéides cytoplasmiques. *Acta Biol. Belg.* **1**:476.

————. 1942. Sur la présence d'hormones protéiques et d'hémoglobine dans les granules à pentosenucléoprotéides. *Acta Biol. Belg.* **2**:447.

Jorpes, E. 1928. Zur Analyse der Pankreasnucleinsäuren. *Acta Med. Scand.* **68**:503.

Kaltschmidt, E., M. Dzionara, D. Donner and H. G. Wittmann. 1967. Ribosomal proteins. I. Isolation, amino acid composition, molecular weights and peptide mapping of proteins from *E. coli* ribosomes. *Mol. Gen. Genet.* **100**:364.

Kurland, C. G. 1960. Molecular characterization of ribonucleic acid from *E. coli* ribosomes. I. Isolation and molecular weights. *J. Mol. Biol.* **2**:83.

Lamborg, M. R. and P. C. Zamecnik. 1960. Amino acid incorporation into protein by extracts of *E. coli*. *Biochim. Biophys. Acta* **42**:206.

Littlefield, J. W. and E. B. Keller. 1957. Incorporation of C^{14} amino acids into ribonucleoprotein particles from the Ehrlich mouse ascites tumor. *J. Biol. Chem.* **224**:13.

Littlefield, J. W., E. B. Keller, J. Gros and P. C. Zamecnik. 1955. Studies on cytoplasmic ribonucleoprotein particles from the liver of the rat. *J. Biol. Chem.* **217**:111.

Luria, S. E., M. Delbrück and T. F. Anderson. 1943. Electron microscope studies of bacterial viruses. *J. Bacteriol.* **46**:57.

McQuillen, K., R. B. Roberts and R. J. Britten. 1959. Synthesis of nascent protein by ribosomes in *Escherichia coli*. *Proc. Nat. Acad. Sci.* **45**:1437.

Martin, T. E. and I. G. Wool. 1968. Formation of active hybrids from subunits of muscle ribosomes from normal and diabetic rats. *Proc. Nat. Acad. Sci.* **60**:569.

Nomura, M. and V. Erdmann. 1970. Reconstruction of 50S ribosomal subunits from dissociated molecular components. *Nature* **228**:744.

Palade, G. E. 1955. A small particulate component of the cytoplasm. *J. Biophys. Biochem. Cytol.* **1**:59.

Palade, G. E. and P. Siekevitz. 1956. Liver microsomes, an integrated morphological and biochemical study. *J. Biophys. Biochem. Cytol.* **2**:171.

Petermann, M. L. and M. G. Hamilton. 1952. An ultracentrifugal analysis of the macromolecular particles of normal and leukemic mouse spleen. *Cancer Res.* **12**:373.

Petermann, M. L., M. G. Hamilton and N. A. Mizen. 1954. Electrophoretic analysis of the macromolecular nucleoprotein particles of mammalian cytoplasm. *Cancer Res.* **14**:360.

Roberts, R. B. 1958. *Microsomal particles and protein synthesis* (ed. R. B. Roberts), p. viii. Pergamon Press, New York.

Rosset, R. and R. Monier. 1963. A propos de la présence d'acide ribonucléique de faible poids moléculaire dans les ribosomes d'*E. coli*. *Biochim. Biophys. Acta* **68**:653.

Schachman, H. K., A. B. Pardee and R. Y. Stanier. 1952. Studies on the macromolecular organization of microbial cells. *Arch. Biochem. Biophys.* **38**:245.

Schachtschabel, D. and W. Zillig. 1959. Untersuchungen zur Biosynthese der Proteine. I. Ueber den Einbau ^{14}C-markierter Aminosäuren ins Protein zellfreier Nucleoproteid-Enzym-Systeme aus *Escherichia coli B*. *Hoppe-Seyler's Z. Physiol. Chem.* **314**:262.

Siegel, A., S. J. Singer and S. G. Wildman. 1952. A preliminary study of the high molecular weight component of normal and virus infected *Escherichia coli*. *Arch. Biochem. Biophys.* **41**:278.

Tissières, A. and J. D. Watson. 1958. Ribonucleoprotein particles from *Escherichia coli*. *Nature* **182**:778.

Tissières, A., D. Schlessinger and F. Gros. 1960. Amino acid incorporation into proteins by *Escherichia coli* ribosomes. *Proc. Nat. Acad. Sci.* **46**:1450.

Tissières, A., J. D. Watson, D. Schlessinger and B. R. Hollingworth. 1959. Ribonucleoprotein particles from *Escherichia coli*. *J. Mol. Biol.* **1**:221.

Traub, P. and M. Nomura. 1968. Structure and function of *E. coli* ribosomes. V. Reconstitution of functionally active 30S ribosomal particles from RNA and proteins. *Proc. Nat. Acad. Sci.* **59**:777.

Traut, R. R., P. B. Moore, H. Delius, H. Noller and A. Tissières. 1967. Ribosomal proteins of *E. coli*. I. Demonstration of different primary structures. *Proc. Nat. Acad. Sci.* **57**:1294.

Tso, P.O.P., J. Bonner and J. Vinograd. 1956. Microsomal nucleoprotein particles from pea seedlings. *J. Biophys. Biochem. Cytol.* **2**:451.

Waller, J. P. 1964. Fractionation of the ribosomal protein from *Escherichia coli*. *J. Mol. Biol.* **10**:319.

Waller, J. P. and J. H. Harris. 1961. Studies on the composition of the protein from *Escherichia coli* ribosomes. *Proc. Nat. Acad. Sci.* **47**:18.

Wittmann, H. G., G. Stöffler, I. Hindennach, C. G. Kurland, L. Randall-Hazelbauer, E. A. Birge, M. Nomura, E. Kaltschmidt, S. Mizushima, R. R. Traut and T. A. Bickle. 1971. Correlation of 30S ribosomal proteins of *Escherichia coli* isolated in different laboratories. *Mol. Gen. Genet.* **111**:327.

Zamecnik, P. C. and E. B. Keller. 1954. Relation between phosphate energy donors and incorporation of labeled amino acids into proteins. *J. Biol. Chem.* **209**:337.

The Process of Translation: A Bird's-eye View

Peter Lengyel
Department of Molecular Biophysics and Biochemistry
Yale University
New Haven, Connecticut 06520

INTRODUCTION: SURVEY OF PROTEIN SYNTHESIS

Ribosomes function in protein synthesis and the complexity of ribosome structure can be accounted for by the complexity of this process. This is why a short survey of protein synthesis is presented here. Aspects of ribosome structure and function discussed in detail in other chapters of this book are deemphasized. To make the reading smoother and to save space, with some exceptions, only recent references are provided. The missing references can be obtained from the more detailed reviews devoted to this topic in recent years (e.g., Lengyel and Soll 1969; Lucas-Lenard and Lipmann 1971; Kozak and Nathans 1972; Haselkorn and Rothman-Denes 1973). A collection of papers on protein synthesis can be found in the Cold Spring Harbor Symposium, Volume 34 (1969). A collection of reviews on various aspects of this process was edited by L. Bosch (1972).

Messenger RNA and Cell-Free Systems

The ribosome is the most complex component of the protein synthesizing machinery of the cell. This machinery translates the information provided in the form of nucleotide sequences in mRNAs into polypeptides. The mRNAs are read in the 5' to 3' direction and the polypeptides are synthesized from the amino-terminal towards the carboxy-terminal amino acid. Messenger RNAs contain (a) sequences which are not translated and (b) sequences specifying initiation of translation (initiation signals); these are followed by (c) sequences that are actually translated and which in turn are followed by (d) sequences specifying the termination of translation (termination signals), which are not translated (Weissmann et al. 1973; Steitz 1975; Bronson, Squires and Yanofsky 1973; Gaskill and Kabat 1971; Brownlee et al. 1973).

Each initiation signal contains an initiator codon. This specifies the first (N-terminal) amino acid of a polypeptide. Each termination signal begins with a terminator codon. This is not translated and is adjacent to the 3'-terminal side of the codon specifying the last (C-terminal) amino acid of the polypeptide. Each codon, including initiator codons, other codons specifying amino acids, and terminator codons, consists of three adjacent nucleotides.

Much of our knowledge about protein synthesis was gained by studying the process first in crude and later in fractionated extracts from broken cells. The most often used extracts are prepared from (a) bacteria, e.g., *Escherichia coli* (Nirenberg and Matthaei 1961; Lengyel, Speyer and Ochoa 1961; Nathans et al. 1962); (b) mammalian cells, e.g., rabbit reticulocytes (Lockard and Lingrel 1969) and tumor cells growing either in ascitic form in the peritoneal cavity of mice or in cell culture (Smith, Marcker and Mathews 1970); (c) an invertebrate, the brine shrimp (McCroskey, Zasloff and Ochoa 1972); or (d) a plant, wheat germ (Marcus 1972; Roberts and Paterson 1973).

The endogenous protein-synthesizing activity of the extracts is usually decreased by incubating them under conditions of protein synthesis to allow for the inactivation of the endogenous mRNA ("preincubation"). Preincubated extracts are capable of translating added mRNA into polypeptides. The mRNAs used include enzymatically or chemically synthesized polynucleotides (homopolymers and copolymers of random or defined sequence) and natural mRNAs. The natural mRNAs include species extracted from phages or viruses, carrying their hereditary information in the form of an mRNA, and mRNAs transcribed from phage or viral genomes in vitro either by enzymes present in the virion (e.g., in reovirus, vaccinia virus and vesicular stomatitis virus [Kates and McAuslan 1967; Borsa and Graham 1968; Shatkin and Sipe 1968; Baltimore, Huang and Stampfer 1970]) or by cellular RNA polymerases. The isolation of mixtures of eukaryotic cellular mRNAs is facilitated by the fact that a majority of these carry a polyadenylate segment attached to their 3'-terminal end. This serves as a handle for their selective purification on affinity columns with polyuridylic acid (Perry et al. 1973). An RNA fraction greatly enriched in a specific mRNA species can be isolated by precipitation of the polysomes translating the mRNA with an antiserum against the protein into which the required mRNA is translated and subsequent extraction of the mRNA from the immunoprecipitate (Shapiro et al. 1974).

The microinjection of an RNA species into intact cells, especially large cells like toad eggs or oocytes makes it possible to test whether this RNA species can serve as a messenger in the cell in vivo (Gurdon 1974).

At present it seems that there are several structural and functional differences between prokaryotic and eukaryotic mRNAs. Thus as noted earlier, a majority of eukaryotic cellular and viral mRNAs (though not those, e.g., specifying histones [Adesnik and Darnell 1972] or reovirus proteins [Stoltzfus, Shatkin and Banerjee 1973]) have a polyadenylate sequence as their 3'-terminal segment; this can be 200 nucleotides long (Mendecki, Lee and Brawerman 1972). The function of this polyadenylate segment has not been established. Its removal does not seem to impair the messenger activity as tested in a cell-free system (Bard et al. 1974) but does decrease it as tested in toad oocytes (Huez et al. 1974). No polyadenylate segments have been found in prokaryotic mRNAs. Eukaryotic (but not prokaryotic) mRNA was reported to contain methylated nucleotides (Perry and Kelley 1974). Eukaryotic mRNAs have a long half-life, e.g., those from HeLa cells have a half-life of 7 to 24 hours (Singer and Penman 1973). Prokaryotic mRNAs have a much shorter half-life, e.g., those of *Bacillus subtilis* of 2 minutes (Levinthal, Keynan and Higa 1962). The structural features determining the length of the half-life of a messenger are not known. The long half-life of eukaryotic mRNAs raises the question of whether the translation of specific mRNAs is slowed down or accelerated according to the metabolic necessities of eukaryotic cells. Some prokaryotic mRNAs specify the synthesis of several polypeptides (polycistronic or polygenic mRNA); i.e., they contain several signals for initiation and termination of translation (Weissmann et al. 1973; Blasi et al. 1973). In eukaryotic cells, there is no evidence as yet for the translation of any mRNA into more than one polypeptide. However, the translating machinery from eukaryotic cells may be able to translate polycistronic messengers: It was reported that a eukaryotic cell extract initiated the translation of a prokaryotic polycistronic mRNA from RNA phage at internal initiation sites (Schreier et al. 1973; Morrison and Lodish 1973; Aviv et al. 1972). Moreover, some eukaryotic mRNAs (e.g., those in poliovirus) are translated into a very long polypeptide, which in turn is cleaved into several active proteins (Baltimore et al. 1971).

Many eukaryotic mRNAs may occur in the cytoplasm in a complex with proteins. Such complexes are designated as messenger ribonucleoproteins (Spirin 1969; Williamson 1973). Some of the proteins in these are associated with the polyadenylate segments (Kwan and Brawerman 1972; Blobel 1973). It was claimed that messenger ribonucleoproteins contain proteins that are required for the translation of the messenger in fractionated, but not in crude, protein-synthesizing systems (Cashion and Stanley 1974).

Splicing of Two Polypeptides into a Single Polypeptide

One of the unexpected recent findings is the splicing of two segments of the β-galactosidase protein into one, apparently full-size and biologically fully active β-galactosidase protein in vivo (Apte and Zipser 1973; see also Hendrix and Casjens 1974). The protein segments that are joined are apparently specified by two unconnected, independently controlled copies of the gene specifying β-galactosidase. One of these seems to specify the N-terminal one-third of the protein, the other the C-terminal two-thirds. The authors propose as possible explanations of this phenomenon that the splicing occurs either at the level of polypeptide synthesis, i.e., on the ribosomes, or is catalyzed by enzymes between polypeptide segments. The tertiary structure of the native proteins would supply the information for the specific joining. Splicing in vitro has not yet been demonstrated.

Types of Protein-Synthesizing Machineries

Prokaryotic cells have a single kind of protein-synthesizing machinery. Eukaryotic cells have several types: one in the cytoplasm, one in mitochondria and one in chloroplasts (in those cells harboring chloroplasts). Remarkably, the protein-synthesizing machinery in mitochondria and chloroplasts is more similar to that in prokaryotes than to that in the eukaryotic cytoplasm (Sager 1972; Dawid 1972; see also Chua and Luck, this volume). Subsequently in this chapter eukaryotic protein synthesis refers to eukaryotic cytoplasmic protein synthesis.

Translation: Phases and Components

In the course of the first phase, *peptide chain initiation,* the components of the translating machinery are assembled on an initiation signal of mRNA. In the second phase, *peptide chain elongation,* the translation proper occurs. The translating machinery moves along the mRNA in steps of three nucleotides, and one aminoacyl residue is added to the growing peptide chain per step. *Peptide chain termination,* the third phase, is triggered when, in the course of its movement along the mRNA, the translating machinery reaches a chain terminating signal. At that time the completed polypeptide chain is released from the translating machinery and the translating machinery is released from the mRNA and is disassembled.

The mRNA translating machinery in prokaryotes, which is simpler than that in eukaryotes, consists of over 130 macromolecules. Some of these, e.g., tRNAs, aminoacyl-tRNA synthetases and ribosomes, are involved in all phases of translation. Other macromolecules, designated as factors, are specifically involved in only one of the three phases of the process: thus initiation factors (IF) function in initiation, elongation factors (EF) in elongation, and release factors (RF) in termination. The term "factor" is supposed to indicate that a macromolecule so designated is attached to the ribosome only during a particular phase of the translation process. It is possible that some of the proteins which are designated at present as ribosomal may also turn out to be factors.

Transfer RNAs and Aminoacyl-tRNA Synthetases

In the course of translation the codons in mRNA are read by aminoacyl-tRNAs and the growing polypeptide chain is attached to tRNA. The known tRNAs are 73 to 93 nucleotide-long copolyribonucleotides (Schafer and Soll 1974). It was estimated that there are about 60 different species of tRNAs in bacterial cells and 100 to 110 in mammalian cells. The aminoacyl residues are esterified in the tRNAs to the 3'-terminal adenosine residues. These adenosine residues are preceded in all tRNAs by two cytidylate residues. Transfer RNAs contain various modified nucleotides, formed by modification of nucleotides in the polynucleotide. About 50 different modified nucleotides are known. The proportion of these in tRNAs increases with evolutionary complexity. It can be lower than 3% in a bacterial tRNA species and as high as 20% in a mammalian tRNA species (Soll 1971; Nishimura 1972). The nucleotide sequences of all the tRNAs (from phages, prokaryotes and eukaryotes) that have been sequenced so far can be arranged in a cloverleaf structure (Holley et al. 1965), consisting of four double-stranded helical regions and four single-stranded loops. The *anticodon loop,* which is thought to interact with the codon in the mRNA, includes the anticodon, i.e., three nucleotides essentially complementary in sequence to the codon specifying the aminoacyl residue attached to the tRNA. The *TψC-containing loop* is believed to interact with the ribosome. No specific function could be attributed so far to the *dihydro-U-containing loop*. The size of the anticodon loop and the TψC loop is essentially independent of the size of the tRNA, that of the dihydro-U loop varies slightly. The additional nucleotides in larger tRNAs are part of a so-called *extra loop*.

The three-dimensional structure of a tRNA, phenylalanyl-tRNA from yeast, was determined recently (Suddath et al. 1974; for details, see Rich, this volume). The molecule is L-shaped, the L consisting of two complex helical regions perpendicular to each other. The anticodon loop is at one end of the molecule and the CCA moiety at the other end. The distance between the two ends is close to 80 Å. The thickness of the L is that of a double helix, i.e., only 20 Å. The fact that a mixture of tRNAs can be crystallized together, along with the common cloverleaf structure, may indicate that there is a profound similarity in three-dimensional structure between the various tRNAs.

It is conceivable that tRNAs do exist in different conformations during the different steps in protein synthesis in which they participate. Consequently, the determination of the three-dimensional structure of a tRNA molecule in a complex with aminoacyl-tRNA synthetase on the one hand, and with one of the peptide chain elongation factors (EF-Tu) and GTP on the other hand, would be of great interest (Arai et al. 1973b).

Transfer RNAs are synthesized from longer precursor polyribonucleotides (Schafer and Soll 1974). Some of these precursor molecules contain the sequences of one tRNA (Altman and Smith 1971), others of two, or possibly even more, different tRNAs (Guthrie et al. 1973). Both endo- and exonucleases, as well as a set of modifying enzymes, are involved in the processing of the precursors into tRNAs. Surprisingly, the precursors for some tRNAs contain the CCA sequence at the 3′ end of the mature tRNA, but others do not.

In prokaryotes there is apparently one aminoacyl-tRNA synthetase for each amino acid (Soll and Schimmel 1974). In eukaryotes there is usually more than one. This is due in part to the fact that subcellular organelles contain some aminoacyl-tRNA synthetases, as well as tRNAs, that are different from those in the cytoplasm (Barnett and Epler 1966; Buck and Nass 1969).

The first step in the aminoacylation of tRNA, as catalyzed by most aminoacyl-tRNA synthetases, is the formation from ATP and an amino acid of an aminoacyl-adenylate which remains enzyme bound. The second step is the transfer of the aminoacyl residue from the aminoacyl-adenylate to the 2′-hydroxyl of the 3′-terminal adenosine residue of the tRNA, to form an aminoacyl-tRNA (Sprinzl and Cramer 1973). However, some of the aminoacyl-tRNA synthetases apparently catalyze the formation of aminoacyl-tRNA in a concerted reaction between ATP, amino acid and tRNA. The rate-limiting step in the formation of aminoacyl-tRNA, at least in the only case in which this was established, is apparently the release of the aminoacyl-tRNA from the synthetase (Eldred and Schimmel 1973). There is a rapid and reversible aminoacyl migration between the 2'-hydroxyl and the 3'-hydroxyl groups of the 3'-terminal adenosine moiety of the tRNA (Wolfenden, Rammler and Lipmann 1964). It appears that the function of aminoacyl-tRNA as an acceptor in translation requires that the aminoacyl residue should be esterified to the 3'-hydroxyl group (Sprinzl and Cramer 1973).

Although all aminoacyl-tRNA synthetases catalyze the same type of reaction with the same type of substrate, they differ greatly from each other in both size (with MW from 50,000 to 200,000) and subunit structure. Some of them are composed of a single subunit, others of two or four, the subunits being identical in some cases and not identical in others. Fingerprint analyses revealed that some of the larger single subunit aminoacyl-tRNA synthetases (MW 100,000 to 135,000) may consist of two similar polypeptides linked to each other by a peptide bond. This type of structure could be accounted for by gene duplication and fusion of the duplicated genes (Waterson and Konigsberg 1974; Koch, Boulanger and Hartley 1974).

The mode of recognition between tRNA and the synthetase charging it is not known. The interaction involves in some cases, directly or indirectly, the anticodon. Thus the mutational replacement of a single base in the anticodon or in the aminoacyl acceptor stem region of the tRNA was found in some cases to change the recognition, i.e., resulted in the attachment of a different aminoacyl residue to the tRNA (Smith and Celis 1973; Yaniv et al. 1974). The fidelity of aminoacyl-tRNA synthesis (i.e., the avoidance of attaching the wrong amino acid to the tRNA) may be increased by the ability of at least some of the aminoacyl-tRNA synthetases to deacylate tRNA acylated with the wrong amino acid (Yarus 1972). The physiological significance of this error-correcting mechanism is, however, doubtful.

Reading of Codons by Aminoacyl-tRNA

One aminoacyl-tRNA may donate aminoacyl residues into polypeptides in response to several related codons. The relationship between such codons is the following. There is an apparent requirement for a Watson-Crick type complementarity between the central nucleotide of the codon and the central nucleotide of the anticodon, as well as between one of the outside nucleotides (i.e., the first or the third) of the codon and the corresponding nucleotide of the anticodon. (Note that the codon in mRNA and the anticodon in tRNA are thought to be lined up in an antiparallel fashion on the ribosome during translation.) The relationship between the third (the other outside) nucleotide of the codon and the corresponding nucleotide in the anticodon seems to obey less strict rules of complementarity (Crick 1966). Such relaxed rules appear to prevail between the third nucleotide of the codon and the first nucleotide of the anticodon in the case of aminoacyl-tRNAs involved in elongation, and between the first nucleotide of the codon and the third nucleotide of the anticodon in the case of the chain initiator tRNA. Thus a yeast alanyl-tRNA species (anticodon IGC) recognizes as codons GCU, GCC and GCA, and methionyl-tRNA$_f$, the initiator tRNA (anticodon CAU), seems to recognize initiator codons AUG, GUG and probably also UUG or CUG (see also the section on initiator codons) (Files, Weber and Miller 1974). The specificity of codon recognition by aminoacyl-tRNA is affected by those modified nucleotides which in some tRNAs are part of the anticodon and in other tRNAs are located adjacent to the anticodon (Nishimura 1972).

Suppression of Chain Terminator Codons and
Frameshifts by Mutated tRNAs

Chain terminator codons (see the section on chain termination) can be mistakenly translated as amino acids (nonsense suppression) by certain mutated tRNAs (Garen 1968; Berg 1973). Such nonsense suppressor tRNAs may arise by the mutational replacement of a nucleotide in the anticodon, resulting in a new anticodon complementary to a nonsense codon (Goodman et al. 1968). In some cases this nucleotide replacement allows the mutant tRNA to be charged by the same aminoacyl-tRNA synthetase and with the same amino acid as the wild-type tRNA, but in other cases by a different aminoacyl-tRNA synthetase with a different amino acid (Yaniv et al. 1974). It is probable that an *E. coli* strain could not survive after losing the ability to translate a particular amino acid-specifying codon. Thus it may be presumed that the strain in which such a nonsense suppressor tRNA was obtained by mutation must carry a second tRNA species not affected by the mutation which can read the particular amino acid-specifying codon. At least one nonsense suppressor tRNA differs from the wild-type tRNA from which it arose by mutation in a nucleotide substitution in the stem of the dihydro-U loop and not in the anticodon (Hirsch 1971). Certain aminoacyl-tRNAs, which have an anticodon loop containing an extra nucleotide, read four nucleotides in an mRNA as a codon. Such mutant aminoacyl-tRNAs can correct for frameshift mutations in which an additional nucleotide is inserted into a translated portion of a gene (Riddle and Carbon 1973).

Involvement of tRNAs in Other Processes

In addition to serving as carriers of aminoacyl residues for protein synthesis, tRNAs also function in several other roles. Particular species provide aminoacyl residues for the biosynthesis of bacterial cell-wall oligopeptides (Matsuhashi, Dietrich and Strominger 1965). Others are the source of aminoacyl residues which become attached to the N termini of certain proteins. This attachment is catalyzed by so-called aminoacyl-tRNA protein transferases. One such enzyme, apparently specific for arginyl-tRNA, is present in several mammalian tissues. Another one specific for leucyl and phenylalanyl-tRNAs is found in enteric bacteria (Soffer 1973).

Transfer RNAs are involved in adjusting the rate of various metabolic processes (including stable RNA synthesis, etc.) according to the supply of amino acids. This regulatory process in *E. coli* is designated as stringent control. Thus if an aminoacyl-tRNA required for addition to a growing polypeptide chain on ribosomes is not available, then the corresponding uncharged tRNA will become attached to the ribosome. This triggers the synthesis of unusual nucleotides (ppGpp and pppGpp). It is thought that these unusual nucleotides are involved in the control of the rate of those metabolic processes which are affected by the lack of a required amino acid (see Cashel and Gallant, and Block and Haseltine, this volume).

Aminoacyl-tRNAs are also involved in the control of the biosynthesis of certain amino acids, e.g., histidine, leucine, valine and isoleucine in *E. coli* and in *Salmonella typhimurium* (Singer et al. 1972; Allaudeen, Yang and Soll 1972). Histidyl-tRNA seems to have a high affinity for the first enzyme of the histidine biosynthetic pathway. It was proposed that in a complex with this enzyme, histidyl-tRNA may specifically repress the synthesis of the mRNA specifying the enzymes involved in histidine biosynthesis (Vogel et al. 1972; Blasi et al. 1973). It is noteworthy that mutant histidine-specific tRNAs which are lacking particular pseudouridylate residues are inactive in regulating the histidine biosynthetic pathway but still functional apparently in protein synthesis. This finding may indicate that many of the modifications of tRNA are required for functions other than protein synthesis (Singer et al. 1972; Allaudeen, Yang and Soll 1972).

Aminoacyl-tRNAs are intermediates in the biosynthesis of glutamine in *Bacillus megaterium*. In this microorganism, a specific glutamyl-tRNA is amidated to glutaminyl-tRNA (Wilcox and Nirenberg 1968).

A tRNA species was shown to be involved in viral nucleic acid replication. One of the several species of tRNAs carried in the virion of an RNA tumor virus (Rous virus) which has a tryptophan-specific anticodon was reported to serve as a primer for the transcription of the tumor virus RNA into DNA by reverse transcriptase (J. Dahlberg, personal communication).

The possible role of aminoacyl-tRNAs in the control of translation will be discussed in a later section.

It is a curious fact that various plant RNA viruses (turnip yellow mosaic virus, tobacco mosaic virus, etc.) and at least one animal RNA virus (mengo) appear to have a tRNA-like segment covalently linked at the 3' end of the genomic RNA. Thus in turnip yellow mosaic virus RNA, it was observed that tRNA nucleotidyl transferase, the enzyme attaching the CCA sequence to tRNA, can attach an A residue to the CC sequence at the 3' end of the viral RNA. The resulting molecule is charged with valine by valyl-tRNA synthetase. The resulting turnip yellow mosaic virus valyl RNA can form a complex with an elongation factor (EF1, see the section on chain elongation) and GTP. The turnip yellow mosaic virus valyl RNA can also donate its valine residue into peptide linkage in a cell-free system from *E. coli,* with turnip yellow mosaic virus RNA serving as the messenger. It is likely that before serving as a donor of valine residues, the 3'-terminal tRNA-like segment is cleaved off from the rest of the viral RNA (Yot et al. 1970; Litvak et al. 1973; Haenni et al. 1973; Salomon and Littauer 1974). The physiological significance of these observations remains to be elucidated.

TRANSLATION PHASE I: PEPTIDE CHAIN INITIATION

In this phase of translation, the components of the translating machinery are assembled on an initiation signal in mRNA. The components known to be involved in the process are an initiator tRNA, several peptide chain initiation factors, the small and the large ribosomal subunit, and GTP (Grunberg-Manago et al. 1973).

Initiator tRNA

The peptide chain initiator tRNAs are apparently unique Met-tRNA species (designated as Met-tRNA$_f^{Met}$ or Met-tRNA$_f$) (Rudland and Clark 1972). These do not serve in peptide chain elongation. An N-formylated derivative of Met-tRNA$_f$ (fMet-tRNA$_f$) is the initiator in prokaryotic cells and in subcellular organelles, i.e., mitochondria and chloroplasts of eukaryotic cells. fMet-tRNA$_f$ is formed by the transfer of a formyl residue from formyltetrahydrofolate to Met-tRNA$_f$ as catalyzed by a specific transformylase. Met-tRNA$_f$ (nonformylated) is apparently the initiator in eukaryotic cytoplasmic (i.e., nonmitochondrial and nonchloroplastic) protein synthesis. It is a curious fact that at least one prokaryotic microorganism, *Streptococcus faecalis,* which normally initiates with fMet-tRNA$_f$ can also initiate with (unformylated) Met-tRNA$_f$. This occurs when *Streptococcus faecalis,* which cannot synthesize folate, is grown in a medium deficient in folate. Remarkably, the tRNA$_f$ from cells grown in a folate-deficient medium has an unmodified uridylate residue in place of a ribothymidylate residue in the tRNA$_f$ from cells grown in a folate-containing medium (Samuel and Rabinowitz 1974).

Peptide Chain Initiation Factors

The best described prokaryotic initiation system is that from *E. coli.* From this organism three types of initiation factors have been characterized: IF1 (MW around 9400), two species of IF2 (MW 80,000 and 91,200), and two or more species of IF3 (MW 21,500 and 23,500). These three types can be obtained from native small ribosomal subunits by washing with solutions of high ionic strength (e.g., 0.5 to 1 M ammonium chloride or potassium chloride) and fractionating the proteins present in the washing solutions. The large ribosomal subunit and couples of small and large subunits, as well as ribosomes translating mRNA, do not appear to have initiation factors attached. It is conceivable that the smaller IF2 factor is derived from the larger one by proteolysis. The relationship of the various IF3 species to each other is unknown (Revel et al. 1972; Revel 1972).

Eukaryotic initiation factors were isolated from various sources (Schreier and Staehelin 1973; Cashion and Stanley 1973), including rabbit reticulocytes (Merrick, Lubsen and Anderson 1973; Gupta et al. 1973), calf liver, calf brain, tumor cells in cell culture, including L cells (Levin, Kyner and Acs 1973), ascites cells, brine shrimp (McCroskey, Zasloff and Ochoa 1972) and wheat germ (Seal and Marcus 1973). Most of these factors were less well-characterized than those from *E. coli*. We will discuss some of them in a later section.

Outline of the Initiation Process

This process is usually studied in cell-free systems under ionic conditions (low Mg^{++}) in which the initiation on natural mRNA strictly depends on initiation factors and initiator tRNA. (The ionic conditions of chain initiation in intact cells are not known.) In these conditions an early intermediate in chain initiation is a complex consisting of a small ribosomal subunit, mRNA, GTP, initiation factors, and initiator tRNA.

We will designate this as small subunit initiation complex. This complex is converted into the initiation complex by attachment of the large ribosomal subunit and the concomitant cleavage of GTP to GDP and inorganic phosphate; release from the ribosome of the cleavage products of GTP and the initiation factors then follows.

The initiation complex can donate its Met-tRNA$_f$ or fMet-tRNA$_f$ residue into peptide linkage with aminoacyl-tRNA (see the section on chain elongation) or with the aminoacyl-tRNA analog puromycin. These and other data indicate that the initiator tRNA is located in the so-called P site of the ribosomes.

Assembly of the Small Subunit Initiation Complex

Despite much effort, our knowledge of the temporal sequence in which the components of the small subunit initiation complex are assembled is still sketchy. This is due in part to the following conceptual and technical difficulties. It is likely, but not certain, that there is a unique temporal sequence for this assembly of components in vivo. In vitro studies revealed that the complex is cooperatively stabilized; most, if not all, of the components are required for the formation of a stable complex that can be isolated from a reaction mixture, e.g., by centrifugation through a sucrose gradient. Incomplete complexes are formed in reaction mixtures from which any one of a number of components of the complex (e.g., mRNA or initiator tRNA) is omitted; however, these are generally unstable (Vermeer et al. 1973).

The following is a possible outline of the order of assembly. The first intermediate in initiation is apparently a small subunit·initiation factor complex. The initiation factors in this complex may include IF1, IF2 and IF3. In the case of eukaryotic initiation, the binding of Met-tRNA$_f$ and GTP apparently precedes that of mRNA (eukaryotic cell extracts contain small subunit·Met-tRNA$_f$ complexes not associated with mRNA) (Schreier and Staehelin 1973; Levin, Kyner and Acs 1973; Darnbrough et al. 1973). In the case of prokaryotic initiation, it is unclear whether the initiator tRNA or the mRNA is bound first in the small subunit initiation complex. The significance of determining the temporal sequence of this assembly is the following: If the initiator tRNA is bound first, then it may be involved, together with the ribosome and initiation factors, in the selection of the initiation signal. It should be noted that there are some reports implicating the involvement of ATP in eukaryotic chain initiation (Marcus 1970).

Role of the Various Initiation Factors

The elucidation of the function of the various initiation factors is difficult. This is because, as noted, the small subunit initiation complex is highly cooperative; i.e., some, if not all, of its components, including initiation factors, affect more than one function. This difficulty accounts in part for the many gaps and uncertainties in our knowledge of this problem.

IF3 is involved in promoting the binding of natural mRNA in the small subunit initiation complex. It was proposed that various species of IF3 may preferentially recognize different initiation signals (Lee-Huang and Ochoa 1973). (For a discussion see a subsequent section.) Moreover, IF3 added to a solution containing small and large subunits, as well as couples of small and large subunits, shifts the equilibrium towards dissociation by binding apparently to free small subunits. For this reason IF3 is also designated as a dissociation factor (Subramanian, Davis and Beller 1969). IF2 bound to the small subunit stabilizes the complex of the small subunit with IF3 and IF1. IF2 is also involved in binding the initiator tRNA in the small subunit initiation complex. Free IF2 can bind GTP and fMet-tRNA$_f$ loosely. It is uncertain whether this loose complex is an intermediate in peptide chain initiation. It should be noted that, unlike in the case of prokaryotic systems, the complex between one of the eukaryotic initiation factors (presumably the one corresponding to IF2) and the initiator Met-tRNA$_f$ and GTP is stable (Gupta et al. 1973).

The binding of the large subunit to the small subunit initiation complex triggers the release of all initiation factors. (According to some reports IF3 is released even earlier, already upon completion of the small subunit initiation complex [Vermeer et al. 1973].) The release of IF2 depends on GTP cleavage and on the presence of IF1; IF2 remains bound in an initiation complex formed in the absence of IF1. It was proposed that the major role of IF1 is to promote the release of IF2 from the initiation complex (Benne et al. 1973).

Additional factors are apparently also involved in prokaryotic and eukaryotic initiation. Since these seem to affect the initiation signal specificity of the process, we will discuss them in the section devoted to this topic.

Initiator Codons

Studies with Synthetic Oligo- and Polynucleotides

Of all 64 ribotrinucleotides tested, only AUG, GUG and CUG promote the binding of fMet-tRNA$_f$ to *E. coli* ribosomes significantly. In cell-free systems from *E. coli,* synthetic polynucleotides with AUG or GUG at or near the 5' end promote incorporation of fMet-tRNA$_f$ into peptide linkage. In cell-free systems from a eukaryote, only those synthetic polynucleotides having AUG or GUG at the 5' end, promote polypeptide chain synthesis at low Mg^{++} concentration. In these conditions the N-terminal residue of the polypeptides synthesized is methionine, originating from the initiator methionyl-tRNA$_f$ (Brown and Smith 1970).

Studies with Natural mRNAs

The characterization of AUG and GUG as initiator codons, as well as the elucidation of some other features of initiation signals in natural mRNAs, resulted mainly from two approaches:

Isolation of Ribosome-Binding Segments In Vitro.   The procedure (Steitz 1975) consists of four steps: (a) formation of an initiation complex by incubating mRNA with all components required for chain initiation (but no aminoacyl-tRNA other than the chain initiator to make it impossible for the ribosome to be attached to the mRNA anywhere but on initiation signals) (Kondo et al. 1968); (b) digestion of the initiation complex with ribonuclease at such a low concentration that it should digest all parts of the mRNA except the segment bound to the ribosomes (Takanami, Yan and Jukes 1965); (c) reisolation and sequencing of the ribosome-bound mRNA segment; and (d) verification that it contains the sequence of codons which specifies the N-terminal portion of the protein translated from this mRNA (Steitz 1969). The studies were performed mainly with *E. coli* phage RNAs. In each of the ribosome-binding initiation segments, AUG and GUG were found to serve as initiator codons.

Identification of "Reinitiation" Signals In Vivo (Miller 1974). A mutational replacement of an amino acid-specifying codon in a gene by a chain terminator codon results in premature chain termination, i.e., formation of only the N-terminal fragment of the wild-type protein. However in some cases, e.g., those of β-galactosidase and lac repressor (both from *E. coli*), the translation of a new peptide chain is initiated on the same mRNA nearby the site of chain termination. This results in the formation of C-terminal fragments of the wild-type protein. These fragments can be isolated by immunoprecipitation with the antiserum prepared against the wild-type protein. A comparison of the N-terminal amino acid sequence of a reinitiation fragment with that of the appropriate internal portion of the wild-type protein allows the identification of the codon specifying reinitiation. The codon is the one assigned to the aminoacyl residue in the wild-type protein corresponding in position to the N-terminal methionine with which the C-terminal fragment is "reinitiated." In a reinitiation fragment lacking N-terminal methionine, it is assumed that methionine was cleaved off after synthesis as is true of most wild-type proteins. In this case that amino acid in the wild-type protein is taken to correspond to the chain "reinitiator" which precedes the aminoacyl residue corresponding to the N-terminal of the reinitiation fragment. These particular internal amino acids included methionine (codon AUG), valine (the valine codon which could specify a chain initiator methionine is GUG), and leucine (leucine codons which conceivably could specify a chain initiator methionine are UUG or CUG). On this basis it was proposed that AUG, GUG and CUG or UUG can serve as "reinitiator" codons and presumably also as initiator codons in *E. coli* in vivo (Files, Weber and Miller 1974). In one eukaryotic gene, that specifying cytochrome *c* in yeast, a mutational substitution of GUG for an initiator AUG resulted, however, in inactivation of the gene (Stewart et al. 1971).

Initiation Signals

Codons of the same nucleotide sequence as initiator codons also specify internal aminoacyl residues of the peptide chain. Thus further characteristics may exist in mRNA to distinguish initiator codons from codons of the same sequence not specifying initiation. To learn about these, the nucleotide sequence of several ribosome-binding segments was compared. Most of the segments were obtained from *E. coli* RNA bacteriophage RNAs (including Qβ, R17, f2 and MS2 (Steitz 1975). These RNAs if intact are translated in vivo and in vitro into at least three proteins: the maturation or A protein, the coat protein, and the replicase protein. Intact phage RNA added to the ribosome-binding reaction binds only a single ribosome at the coat protein initiation signal. However, phage RNA fragmented into large segments binds ribosomes to each of the three initiation signals: those for A protein, coat protein and replicase protein. The lengths of the segments protected by the attached ribosome against cleavage by pancreatic ribonuclease are as short as about 30 nucleotides; those protected against cleavage by ribonuclease T1 can be as long as about 80 nucleotides. The initiator codon is usually nearer to the 3′ end than to the 5′ end of the segments.

A comparison of the nucleotide sequences of the various initiation signals from the various RNA phage RNAs, a DNA phage T7 mRNA (Arrand and Hindley 1974), a DNA phage ϕX174 mRNA (Robertson et al. 1973), and E. coli lac operon mRNA (Maizels 1974) permits the following conclusions:

(1) Each signal contains an initiator codon, either AUG or GUG.
(2) Neither of the initiator codons is preceded by an adjacent chain terminator codon. Each ribosome-binding segment contains, however, at least one chain terminator codon in or out of phase with the initiator codon. The distance between the initiator codon and the chain terminator codon specifying the C-terminal residue of the adjacent 5′-proximal gene varied between 26 and several hundred nucleotides. This indicates the existence of untranslated segments in the mRNAs.
(3) No nucleotide sequences other than initiator codons were found to occur in all initiation signals. Several of them did share, however, some sequences. Thus one set of initiation signals contained an AGGA sequence, and a different, but overlapping set, a purine-purine-UUU-purine-purine sequence. The shared sequences did not occur, however, at the same distance from the initiator codons.
(4) The nucleotide sequences of the various initiation signals did not allow the construction of an obvious common secondary structure which would distinguish these segments from other segments in the RNA.

The fact that only one initiation signal (that for the coat protein site) is available for ribosome binding in intact phage RNA, but all three are available in fragmented mRNA, indicates that secondary structure may block the initiation at potential initiation signals. Sequence studies on MS2 phage RNA revealed that a larger segment of the coat cistron is complementary to the replicase initiation signal (Min Jou et al. 1972). Thus the block of initiation at (and of ribosome binding to) the replicase initiation signal in intact MS2 RNA may be due to hydrogen bonding between these two complementary segments. The replicase initiation signal becomes available for initiation in intact RNA in vivo and in vitro (Engelhardt, Webster and Zinder 1967) only after a ribosome has translated at least part of the coat protein cistron. This ribosome may cover up the replicase initiation signal by removing from it the segment of the coat gene complementary to it in sequence. Certain initiation signals, for example, in the *E. coli* DNA bacteriophages f_1 and ϕX174, contain true palindromic sequences including or near to the chain initiator codon (e.g., AUGA*UUAAAGUUGAA-AUU*). Such sequences, which read the same forwards and backwards, may not allow the formation of a double-stranded region from segments adjacent to the initiator codon. Their function in initiation signals, if any, remains to be established (H. Robertson, personal communication).

Segments of the phage RNA beyond the isolatable ribosome binding sites seem to be involved in recognition. Thus fragmentation of the mRNA to small size increases ribosome binding at the A initiation site but decreases it greatly at the coat and the replicase sites, so much so that the ribosome-binding coat segment, reisolated after RNase treatment from the initiation complex, cannot be rebound at all to the ribosomes (Steitz 1975).

The second type of approach to the study of initiation, or more accurately "reinitiation," signals resulted in the following observations (Files, Weber and Miller 1974; Miller 1974). Reinitiation after chain termination occurs in the case of the *lac* repressor gene at three in-phase codons out of the first seventy. These reinitiation sites do not seem to function in wild-type genes as initiation signals either because of limitations by the secondary structure or because ribosomes translating the mRNA cover them up. Upon nearby chain termination, however, the ribosomes may uncover these mRNA segments allowing initiation to occur at appropriate codons.

Further data will be needed to reconcile the apparent discrepancy between (a) initiation on the phage RNA, which is restricted to very few sites even after opening of the secondary structure (e.g., by formaldehyde treatment) and (b) "reinitiation" on the *lac* repressor RNA, on which numerous "reinitiation" sites can be created by nearby chain termination. It remains to be seen if the discrepancy is only apparent and due to the different conditions of the experiments or if there is a difference in the secondary structure between the bacterial and the phage mRNA which could account for the discrepancy.

Components Involved in Recognition of Initiation Signals

Ribosomal Components

As noted above, RNA phage RNA (e.g., from R17), fragmented into large segments, allows the initiation of the translation of three proteins in the cell free system from *E. coli*. In a cell-free system from *Bacillus stearothermophilus*, however, the translation of only one, the A protein, is initiated. The addition of small ribosomal subunits from *E. coli* to a *B. stearothermo-*

philus system allows the initiation of the translation of all three proteins. This indicates that components determining initiation specificity are parts of the small subunit (Lodish 1970). At least two of these were identified by dissociating small subunits from *E. coli* and from *B. stearothermophilus* into individual components and reassembling various sets of small subunits, each containing one or two components from *B. stearothermophilus* and all others from *E. coli,* and comparing the capacity of the heterologous subunits obtained to translate intact R17 phage RNA. (In an *E. coli* system, 90% of the translation products from this messenger are coat protein, less than 10% replicase protein, and even less A protein.)

Heterologous subunits with the small ribosomal protein S12 from *B. stearothermophilus* were only 50% as active as *E. coli* subunits in these tests. No other *B. stearothermophilus* small subunit protein impaired the translation significantly. Heterologous subunits containing S12 protein as well as the small subunit RNA from *B. stearothermophilus* (and all other components from *E. coli*) were 85% less active than the *E. coli* small subunit. These results seem to indicate that both protein S12 and the small subunit RNA are involved in initiation signal selection (Held, Gette and Nomura 1974; see also Goldberg and Steitz 1974). The involvement of S12 may be related to the finding that this protein on the small subunit can be crosslinked to IF3, indicating that it may interact directly with IF3 (R. Traut, personal communication). The involvement of the RNA may conceivably be related to a short sequence complementarity between a purine-rich segment preceding the initiator codon in some initiation signals (e.g., a hexanucleotide in the initiation signal for A protein) and a pyrimidine-rich segment located at the 3'-terminal region of the *E. coli* small subunit RNA (Shine and Dalgarno 1974).

Initiation Factors

As noted earlier, there are reports indicating that *E. coli* contains more than one species of IF3 factor. Two of these, designated as IF3α and IF3β, were isolated. The initiation signal specificity of the two was reported to be different. IF3α preferentially translates MS2 RNA over late phage T4 mRNA, whereas IF3β has the reverse preference. It is not known if the two IF3 species are specified by two different genes or, if one is produced by modification from the other (Lee-Huang and Ochoa, 1973).

Other Agents Affecting Specificity of Chain Initiation

Factors Promoting Translation of Particular mRNAs. A protein isolated from Krebs ascites cells promotes in vitro the initiation of translation of encephalomyocarditis viral RNA without affecting that of various other natural mRNAs tested (Wigle and Smith 1973). Another protein isolated from rabbit reticulocytes promotes in an ascites cell extract the translation of alpha-globin mRNA several-fold, of beta-globin mRNA slightly, and has no effect on that of mengo viral mRNA (Nudel, Lebleu and Revel 1973). One agent from chick embryo red muscle promotes the translation in a rabbit reticulocyte extract of chick myoglobin mRNA, and a second agent, that of chick myosine mRNA (Heywood, Kennedy and Bester 1974).

Factors Interfering with Translation of Particular mRNAs. The *E. coli* RNA phage coat protein and the replicase serve as "translational repressors" controlling the translation of the phage RNA (Weissmann et al. 1973). In the late phase of infection, the synthesis of replicase is reduced. This is due to the binding of coat protein, accumulated in the cell, to the initiation site for the translation of replicase interfering with ribosome attachment to this site. The replicase can bind tightly to the initiation site for coat protein translation. If bound, it allows ribosomes already attached to mRNA to continue translation and be released, but thereafter blocks the binding of further ribosomes. Thereby it can free the RNA to serve as the template for the synthesis of complementary RNA that is catalyzed by the replicase. The latter proceeds along the RNA template in the opposite direction than do ribosomes, avoiding collision with them (Weissmann et al. 1973).

In mouse myeloma tumor cells, an increased intracellular concentration of myeloma protein, composed of two heavy (H) and two light (L) chains, results in the reduction of the synthesis of H-chain protein (Stevens and Williamson 1973a). This is presumably a consequence of the binding of the myeloma protein to H-chain mRNA (Stevens and Williamson 1973b).

In *E. coli* cells infected with phage T7, the translation of T7 mRNA proceeds while that of host mRNA is blocked. This selective inhibition is apparently mediated by a T7-specified inhibitor. The mechanism of action of this inhibitor remains to be established (Herrlich et al. 1974).

A series of factors from *E. coli* were reported to inhibit the initiation of translation on certain initiation signals and stimulate it on others. These factors were designated as "interference factors." The first one discovered (designated as f_i) blocked the translation of RNA-phage RNA (e.g., MS2 or $Q\beta$ RNA) but not of late T4 phage mRNA in a cell extract (Revel et al. 1973). This factor was identified also as part of the phage $Q\beta$·replicase complex (Groner et al. 1972; Kamen et al. 1972). The complex consists of one phage-specified protein and three host proteins, one of which is f_i (Blumenthal, Landers and Weber 1972). Moreover, f_i is also indistinguishable from one of the small ribosomal subunit proteins (designated as S_1) by N-terminal amino acid sequence, gel electrophoresis under various conditions, immunological studies and functional experiments (Inouye, Pollack and Petre 1974; Wahba et al. 1974).

Ribosomes deficient in S1 protein are impaired in their capacity to bind and to translate various mRNAs (van Duin and van Knippenberg 1974; van Knippenberg, Hooykaas and van Duin 1974; Szer and Leffler 1974). The physiological relevance of the data on the effect on initiation of f_i (S_1) added to cell extracts remains to be clarified.

General Inhibitors of Peptide Chain Initiation.   Double-stranded RNA remarkably at low concentration (0.01 μg/ml) but not at higher concentration (10 μg/ml) blocks the formation of the Met-tRNA$_f$·small subunit initiation complex in extracts from eukaryotic cells by a mechanism which has not been determined (Darnbrough et al. 1973; Ehrenfeld and Hunt 1971). The same process is blocked by an inhibitor that is accumulated in reticulocyte extracts in the absence of hemin (Gross and Rabinowitz 1973). This block can be overcome by adding to the system a eukaryotic initiation factor (Mizuno, Fischer and Rabinowitz 1972; Kaempfer and Kaufman 1973), which forms a tight complex with Met-tRNA$_f$ and GTP (Beuzard and London 1974). Agents like the above two inhibitors decrease the rate of peptide chain initiation by decreasing the concentration of a component (in the above case, the Met-tRNA$_f$·small subunit initiation complex) required in the initiation of all proteins. Nevertheless, the net effect of their action may be specific (Beuzard and London 1974; Giglioni et al. 1973). By increasing the competition for a component needed in peptide chain initiation, the ratio of initiation on efficient messengers to that on less efficient messengers may increase.

A word of caution: It is dangerous to draw conclusions about the physiological role of an agent whose effect on peptide chain initiation was only established in cell extracts. This is in part because these extracts translate mRNA much slower than intact cells do. Thus in a Krebs ascites cell extract, the rate of polypeptide chain elongation is probably only 5–10% as fast as in vivo (Mathews and Osborn 1974). Consequently, the rate-limiting step in translation is not necessarily the same in vivo and in vitro. It should also be noted that the fact that a factor affects peptide chain initiation discriminately (i.e., promotes the translation of certain mRNAs selectively) is no safe basis to conclude that it has a regulatory role in vivo. Unless otherwise documented, it may just be that there is no single factor capable of initiating translation on all kinds of initiation signals, but several factors are needed.

TRANSLATION PHASE II: PEPTIDE CHAIN ELONGATION

During this phase the translating machinery is moving along the mRNA towards the 3′-terminal nucleotide in steps of three nucleotides. It attaches one aminoacyl residue at each step to the peptide chain. The peptide chain is propagated from the amino-terminal towards the carboxy-terminal amino acid. It is linked to tRNA and through this to the messenger·ribosome complex. At least three peptide chain elongation factors, aminoacyl-tRNA species and GTP are involved in this phase (Leder 1973).

Peptide Chain Elongation Factors

The best-characterized prokaryotic factors are again from *E. coli* (Kaziro, Inoue-Yokosawa and Kawakita 1972; Arai et al. 1973a) and from the thermophilic bacterium *Bacillus stearothermophilus* (Beaud and Lengyel 1971). They are designated as EF-Tu, EF-Ts, and EF-G. The molecular weights of the factors from *E. coli* are, in the above order, about 47,000, 34,000 and 83,000. EF-Tu and EF-Ts also occur as an EF-Tu·EF-Ts complex. This is designated as EF-T. The factors are isolated from the high-speed supernatant fraction of the cell extract and can be crystallized. They comprise 2–3% of the total protein in the supernatant fraction from rapidly growing *E. coli* cells. Mutants of EF-Ts (Kuwano et al. 1973) and EF-G (see Haselkorn and Rothman-Denes 1973) were obtained.

Factors in eukaryotes corresponding to EF-T (designated EF1) and EF-G (designated EF2) were described (Haselkorn and Rothman-Denes 1973). EF1 from some sources (e.g., calf brain) exists in heavy and light forms. The heavy forms contain free and esterified cholesterol (Moon et al. 1973). EF1 from rabbit reticulocytes was shown to harbor two activities: a heat-stable activity corresponding to EF-Ts and a heat-labile activity corresponding to EF-Tu (Prather et al. 1974). EF1 from pig liver was recently resolved into two complementary fractions: $EF1_\alpha$ and $EF1_\beta$, which may correspond to EF-Tu and EF-Ts, respectively (Iwasaki et al. 1974).

The study of peptide chain elongation is facilitated by the fact that, at least under presumably nonphysiological conditions, peptide chains can be initiated in cell extracts in the absence of the agents required for chain initiation. Thus at high Mg^{++} concentration poly(U) can direct the formation of polyphenylalanyl-tRNA from phenylalanyl-tRNA in the presence of salts, ribosomes, GTP and the three above-listed elongation factors. According to one report, the translation of mRNA other than poly(U) may require a further factor. The factor is proposed to function in a step taking place after the formation of the first dipeptide (Ganoza and Fox 1974).

Outline of the Elongation Process in Prokaryotes

The addition of each aminoacyl residue to the growing peptide chain is a cyclic process (chain elongation cycle). As noted earlier, the end product in the initiation phase is an initiation complex in which fMet-tRNA$_f$ is bound in the P site. In the first part of the elongation cycle (*aminoacyl-tRNA binding*), an aminoacyl-tRNA is bound to the A site. The aminoacyl-tRNA is the one specified by the codon adjacent to the 3' side of the initiator codon. It is attached to the mRNA·ribosome complex in the form of an aminoacyl-tRNA·EF-Tu·GTP complex (designated as ternary complex). In the second part of the cycle (*peptide bond formation*), a GTP in the ternary complex is cleaved to GDP and P_i and an EF-Tu·GDP complex and P_i are released from the ribosome; the fMet residue of the fMet-tRNA in the P site is then released from its linkage to tRNA, forming a peptide bond with the α-amino group of the aminoacyl-tRNA at the A site. The enzyme(s) catalyzing peptide bond formation is called peptidyl transferase and is part of the large ribosomal subunit. In the third part of the cycle (*translocation*), EF-G and GTP are bound to the ribosome. Thereafter the discharged tRNA is released from the P site, the dipeptidyl-tRNA is shifted from the A site to the P site, the ribosome moves along the mRNA three nucleotides further towards the 3′ end, the ribosome-bound GTP is cleaved to GDP and P_i, and the cleavage products, together with EF-G, are released from the ribosome. This finishes the process and the stage is set for attaching another aminoacyl-tRNA by repeating the cycle.

What remains to be done is the regeneration of the aminoacyl-tRNA·EF-Tu·GTP complex from EF-Tu·GDP. The process is catalyzed by EF-Ts, promoting the release of GDP from EF-Tu·GDP and resulting in the formation of EF-Tu·EF-Ts. This in turn interacts with GTP, resulting in the formation of EF-Tu·GTP and free EF-Ts. Finally the ternary complex is regenerated by binding aminoacyl-tRNA to EF-Tu·GTP.

We know less about peptide chain elongation in eukaryotes than in prokaryotes. So far no significant difference has been found in the characteristics of the process between the two cell types.

Further Characteristics of Factors, Intermediates and Processes in Chain Elongation

Aminoacyl-tRNA Binding: Role of EF-Tu and EF-Ts

EF-Tu can bind either GDP or GTP; however, binding of aminoacyl-tRNA by EF-Tu requires GTP. The GTP apparently affects the conformation of that region of EF-Tu which interacts with aminoacyl-tRNA in the EF-Tu·GTP·aminoacyl-tRNA complex (Arai et al. 1974). The molar ratio of the components in this complex is 1:1:1. EF-Tu can form a ternary complex will all aminoacyl-tRNAs except the initiator Met-tRNA (whether formylated or not), uncharged tRNA or N-acyl-aminoacyl-tRNA. The substrate specificity of EF-Tu assures that, as long as available, only charged tRNAs specified by the appropriate codons are bound to the ribosome. (In the absence of a required charged tRNA as already noted, the corresponding uncharged tRNA is bound, triggering the stringent control mechanism; see Block and Haseltine, this volume.) The binding of one molecule of aminoacyl-tRNA to the ribosome is accompanied by the cleavage of one GTP molecule, the one present in the ternary complex which carried the aminoacyl-tRNA to the ribosome.

EF-Ts is apparently not involved in either the cleavage or the formation of any covalent linkage. Its only known function in peptide chain elongation is to promote dissociation of the EF-Tu·GDP complex (Weissbach, Redfield and Hachman 1970; Beaud and Lengyel 1971). The need for a separate protein for this task may be the consequence of the high affinity of EF-Tu for GDP. Remarkably, EF-Tu and EF-Ts also function in RNA replication: The two factors are components of the $Q\beta$ bacteriophage RNA replicase enzyme complex and may serve as a nucleotide-binding moiety (Blumenthal, Landers and Weber 1972; Hori 1974).

Translocation: Role of EF-G

In previous sections the terms tRNA binding sites A and P were used repeatedly. These are operational terms: The ribosomal site to which peptidyl-tRNA not reactive with puromycin (an aminoacyl-tRNA analog) is bound is defined as the A site. The ribosomal site to which peptidyl-tRNA reactive with puromycin is bound is defined as the P site. The reaction promoted by EF-G and GTP is stated to be the shifting of peptidyl-tRNA from site A to site P. It is conceivable that the two sites are distinct; i.e., that the A site consists always of one and the same set of ribosomal components and the P site constantly of a different set. It is also conceivable, however, that the two tRNA binding sites are functionally equivalent; i.e., at any time one of the two is the A site and the other the P site, but after one translocation of peptidyl-tRNA the (former) A site becomes the (new) P site and the (former) P site the (new) A site (Woese 1973).

The movement of the ribosome along the mRNA during translocation can be experimentally demonstrated. It can be shown that the mRNA fragment protected against ribonuclease cleavage by the attached ribosome extends three nucleotides (one codon) further towards the 3′ end in the mRNA·ribosome complex that was treated with EF-G and GTP (post-translocation complex) than in the complex prior to this treatment (pre-translocation complex) (Gupta et al. 1971; Thach and Thach 1971). Apparently a single GTP molecule is cleaved during the translocation of one peptidyl-tRNA molecule (Modolell, Cabrer and Vazquez 1973b).

There is a competition between EF-Tu and EF-G for binding to an mRNA·ribosome complex not carrying peptidyl-tRNA. This seems to indicate that the two factors interact with the ribosome at a common site or at least at overlapping sites. During the chain elongation cycle, the two factors have to function at different stages: EF-G before translocation, EF-Tu after translocation. Competition for interaction with the ribosome between the two factors is apparently diminished during chain elongation. This is reflected in the following observations: Pretranslocation complexes in vitro support EF-G-dependent GTPase activity readily but EF-Tu-dependent GTPase activity sluggishly, whereas the opposite is the case for post-translocation complexes (Modolell, Cabrer and Vazquez 1973a; Nombela and Ochoa 1973; see also Haselkorn and Rothman-Denes 1973).

Translational Control during Chain Elongation

The rate of translation of different mRNAs could in principle be selectively controlled. Such a control could be based on the occurrence in mRNAs of codons whose translation depended on aminoacyl-tRNA present in rate-limiting concentrations (Wilcox 1971; Scherberg and Weiss 1972; Wilson 1973; Gupta, Sopori and Lengyel 1974; Content et al. 1974).

TRANSLATION PHASE III: PEPTIDE CHAIN TERMINATION

Peptide chain termination is triggered when, in the course of the movement of the ribosome along the mRNA, a chain terminator codon (UAA, UAG or UGA) is reached at the ribosomal A site. Thereupon the completed polypeptide chain is released from its linkage to tRNA and to the ribosome and the mRNA·ribosome·tRNA complex falls apart. Three release factors and GTP are involved in chain termination in prokaryotes; one (complex) release factor and GTP in rabbit reticulocytes (Beaudet and Caskey 1972).

Termination Signals and "Readthrough" Proteins

In *E. coli* strains not carrying nonsense suppressors UAA (also called ochre codon), UAG (also called amber codon) and UGA do not seem to specify amino acids but trigger chain termination. Thus, e.g., RNA from a mutant bacteriophage f2 or R17, with a UAG codon within the genome specifying the coat protein, promotes the synthesis in vitro of a coat protein fragment, whereas RNA from a corresponding wild-type protein directs the formation of the entire coat protein. This and other results with synthetic and natural messengers reveal that any of the three above codons can alone specify chain termination. This was established both for *E. coli* and for rabbit reticulocytes (Beaudet and Caskey 1972).

Apparently chain termination in vivo is not always 100% efficient, even in *E. coli* strains not known to harbor nonsense suppressors. Thus "read-through" occurs across the chain terminator codon at the 3' end of the coat protein gene in bacteriophage Qβ RNA. The product of the readthrough is a protein consisting of the phage coat protein as its amino-terminal portion, to which a protein that had been translated from the intercistronic region between the coat protein gene and the replicase gene is covalently linked (Horiuchi, Webster and Matsuhashi 1971; Weiner and Weber 1971; Moore et al. 1971; Steitz 1972).

Remarkably, the readthrough protein is required for the assembly of the virion and is present in the virion (Hofstetter, Monstein and Weissmann

1974). There is severalfold more readthrough protein in strains harboring UGA suppressors than in wild-type strains. The amino acid translated (mistakenly) from the readthrough chain terminator codon in such suppressor strains is tryptophan (codon UGG). These observations reveal that, in this case at least, the inefficient chain terminator codon is UGA (Weiner and Weber 1973).

Interestingly, in the case of the f2 and R17 phages, two chain terminator codons adjacent to each other, a UAAUAG sequence, serves as chain termination signal at the end of the coat protein genes (Nichols 1970). Such a tandem terminator sequence may have evolved to eliminate translation through inefficient terminator codons. It was proposed that about 13% of all chain termination signals in *E. coli* may be tandem terminator codons. The proposal is based on the increase in the proportion of C-terminal tyrosine in total protein in *E. coli* cells carrying tyrosine-inserting nonsense suppressor tRNAs compared to that in wild-type strains (Lu and Rich 1971).

The existence of efficient nonsense suppressors in strains growing at the same rate as those without suppressors is a puzzling phenomenon. Its most likely explanation is that in such strains nonsense mutations are more efficiently suppressed than natural chain termination signals. This difference in suppressibility can be accounted for in the case of the natural chain termination signals consisting of tandem terminator sequences. These, however, comprise only the minority of all chain termination signals. The majority of these is presumably more efficient in termination than nonsense mutations because of the sequence of nucleotides adjacent to the signal (Salser, Fluck and Epstein 1969). In support of this view, it was found that in a bacteriophage T4 lysozyme gene, a nonsense mutation followed by the amino acid-specifying codon CAA (resembling the UAA terminator codon) is less efficiently suppressed than the same nonsense codon in a different context (Yahata, Ocada and Tsugita 1970). Thus the chain-terminating machinery may conceivably recognize a sequence longer than a single chain terminator codon.

Release Factors

Prokaryotic

RF1 (MW about 44,000) is a protein active in chain termination with the codons UAA and UAG, RF2 (MW about 47,000) with the codons UAA or UGA. One *E. coli* cell contains about 500 molecules of RF1 and about 700 molecules of RF2. RF3 is a protein designated earlier α by one group of investigators, who stated it to be identical with EF-Tu (Capecchi and Klein 1969), and also a factor designated earlier S_1 by a different group of investigators, who found it not to be identical with EF-Tu (Caskey et al. 1969). All the above factors were obtained from the high-speed supernatant fraction of an *E. coli* extract and purified to apparent homogeneity (Beaudet and Caskey 1972).

Eukaryotic

RF (MW about 255,000), purified from rabbit reticulocytes several hundredfold, apparently is a single factor consisting of subunits active in chain termination with any one of the three terminator codons (Beaudet and Caskey 1971).

Outline of the Chain Termination Process

A simple model system for studying chain termination in *E. coli* is obtained by incubating washed ribosomes with AUG codons and fMet-tRNA. This results in the formation of a fMet-tRNA·AUG·ribosome complex. Release of formylmethionine from this complex is considered to be a process analogous to chain termination. The release is promoted by RF1 in the presence of the chain terminator codons UAA or UAG and by RF2 in the presence of UAA or UGA. The rate of release, in the presence of RF1 or RF2 and the appropriate terminator codon at the proper concentration, is further increased by RF3 and GTP or GDP (Beaudet and Caskey 1972).

Chain termination in eukaryotes has been studied with ribosomes and factors from rabbit reticulocytes. An fMet-tRNA·ribosome complex was formed without the addition of an AUG codon. Release of formyl-methionine from this complex is promoted by RF in the presence of tetranucleotides of defined sequences, containing any one of the codons UAA, UAG or UGA, and GTP. The latter is hydrolyzed to GDP in the process (Tate, Beaudet and Caskey 1973).

Further Characteristics of Factors, Intermediates, and Processes in Chain Termination

During chain termination, the terminator codons promote the binding of the release factors to the ribosome. In the reticulocyte system, the binding is further promoted by GTP. After the cleavage of GTP to GDP, the RF can be released from the ribosome and reutilized.

In the *E. coli* system RF3 increases the rate of chain termination by increasing the binding of RF1 and RF2 to the ribosomes. GTP stimulation of *E. coli* RF1 and RF2 binding in the presence or absence of RF3 has not been observed. The presence of GDP or GTP and RF3 labilizes the RF1 (or RF2) ribosome·terminator codon complex. Thus the role of guanine nucleotides in *E. coli* chain termination is still unclear (Tate, Beaudet and Caskey 1973).

Release factors and nonsense suppressor tRNA species recognize terminator codons. Thus it is not unexpected that by varying the ratio of suppressor tRNA to release factors in a cell-free system containing mRNA with a nonsense codon, the ratio of readthrough translation to chain termination is varied (Beaudet and Caskey 1970).

It is conceivable that the ribosomal components involved in peptide bond formation, the peptidyl transferases, are also involved in chain termination: Several antibiotics blocking peptide bond formation also block chain termination. Moreover, the peptidyl-tRNA must be located in the P site in order to undergo chain termination.

It is a curious fact that the 3'-terminal sequence of the RNA from the small ribosomal subunit in *E. coli,* yeast, Drosophila, and rabbit reticulocytes is UUA. This triplet is complementary to the UAA terminator codon. It remains to be seen if this complementarity reflects an interaction between the complementary sequences during chain termination (Dalgarno and Shine 1973).

Fate of the Termination Complex after Release of the Polypeptide

The disassembly of the termination complex in *E. coli* extracts seems to result in the formation of 70S ribosomes (i.e., couples of a small and a large subunit) which dissociate into subunits (Subramanian and Davis 1973). Disassembly of the termination complex in rabbit reticulocyte extracts also results in free subunits. However, in this case there is no evidence so far for the existence of 80S ribosomes as intermediates in the process. The equilibrium between free subunits and 70 or 80S ribosomes is affected by dissociation factors in both cases. Moreover 70 or 80S ribosomes accumulate in vivo (in both cases) if the rate of protein synthesis, and thus the number of ribosomes in polysomes, decreases. These ribosomes can be mobilized for protein synthesis when the rate of the latter increases (see Kaempfer, and Davis, this volume).

The involvement of one further agent, RR factor, in chain termination has been proposed. In the presence of EF-G and GTP, RR factor promotes the dissociation of mRNA·ribosome·uncharged tRNA complexes into their components. It is conceivable that such mRNA·ribosome·discharged tRNA complexes are intermediates in chain termination that are formed upon release of the completed polypeptide from the chain termination complex (Hirashima and Kaji 1973).

FURTHER AGENTS INVOLVED IN PROTEIN SYNTHESIS

There are indications that peptidyl-tRNA may fall off ribosomes during protein synthesis. Apparently to avoid the accumulation of such toxic fallen-off molecules, the cells contain a peptidyl-tRNA hydrolase that hydrolyzes them into free peptides and tRNA (Menninger 1974).

Many *E. coli* proteins do not have a formyl residue and many not even a methionyl residue at their N termini. This is because, after the synthesis of their peptide chain has been initiated, the formyl residue is removed by a specific deformylase. Subsequently the methionyl residue or even further aminoacyl residues may be cleaved off by aminopeptidases (Livingston and Leder 1969).

ROLE OF GTP IN THE TRANSLATION PROCESS

No covalent linkages are known to be synthesized during either chain initiation or chain termination. Only one covalent linkage, a peptide bond, is formed during each turn of the chain elongation cycle. Energetically there seems to be no need for the free energy of GTP in peptide bond formation since the ΔF of the peptidyl donor peptidyl tRNA (estimated to be -7 Kcal/mole) is ample for the synthesis of the peptide bond (ΔF -3 Kcal/mole) (Lipmann 1967).

Nevertheless as noted in earlier sections, GTP is apparently cleaved to GDP and P_i at four stages in translation: (1) in initiation, presumably after the joining of the large subunit to the small subunit initiation complex, carrying among others GTP and IF2; (2) in elongation, after binding of the EF-Tu·GTP·aminoacyl-tRNA complex to the ribosome; (3) in elongation, after binding of EF-G and GTP to the ribosome, and (4) in termination, after binding of RF and presumably GTP to the ribosome. As noted earlier the stoichiometry of GTP cleavage has been determined only in (2) and (3). Moreover, the involvement of GTP cleavage in chain termination was established only in the reticulocyte system, not as yet in the *E. coli* system. There is no evidence in any of the four stages for the transfer of any cleavage product of GTP to form a high energy intermediate with a component of the translating machinery (i.e., of the $X \sim P$, $X \sim$ GMP or GDP $\sim X$ type).

In spite of the profound differences in the transformations taking place at the four stages of translation, there seems to be a remarkable similarity in the pattern in which GTP is involved in each:

(1) GTP (not GTP cleavage) seems to be required for the binding of a factor, or a factor in complex with another component such as aminoacyl-tRNA, to the ribosome.
(2) GTP cleavage seems to be required for the release of the factor attached to the ribosome.
(3) GTP seems to function by causing a conformational change in one or two components (e.g., factors) of the translating machinery. This change enables them to perform a function, including attachment to the ribosome. The cleavage of GTP reverses this change, resulting in, among other things, their release from the ribosome. Thus GTP acts apparently as an allosteric effector, whose action is modified by its cleavage to GDP.

Support for the three parts of this hypothesis (see also Lengyel and Soll 1969; Ono et al. 1969; Dubnoff, Lockwood and Maitra 1972; Tate, Beaudet and Caskey 1973; Kaziro 1973) comes from the following observations and considerations: (1) GMPPCP, an analog of GTP that cannot be cleaved to GDP and P_i, can substitute for GTP in allowing the binding to the ribosomes of the same components as GTP does. Thus GMPPCP was reported to allow the formation of initiation complexes (Anderson et al. 1967) and the attachment to the ribosomes of aminoacyl-tRNA in complex with EF-Tu (Ertel et al. 1968), of EF-G (Brot and Spears 1970) and of RF (Tate, Beaudet and Caskey 1973). Moreover EF-G bound to a pre-translocation complex in the presence of GMPPCP was reported to trigger translocation (Inoue-Yokosawa, Ishikawa and Kaziro 1974). It should be noted that GMPPCP is no perfect substitute for GTP; its interaction with EF-Tu and GTP is much weaker than that of GTP (Skoultchi et al. 1970).

(2) Ribosome complexes in which GMPPCP is substituted for GTP cannot undergo all the transformations that ribosome complexes with GTP can. Such complexes retain the factors whose binding was made possible by the analog. Complexes formed in the presence of GTP do not retain the factors after they have performed their function on the ribosome. This difference in retention is presumably because GMPPCP is not cleaved, whereas GTP is; and the cleavage product GDP does not substitute for GTP in allowing the binding of the factors to the ribosome. The retention of factors on ribosomal complexes formed in the presence of GMPPCP was noted for EF-Tu (Ono et al. 1969; Lucas-Lenard, Tao and Haenni 1969; Skoultchi et al. 1970) and subsequently for IF2 (Dubnoff, Lockwood and Maitra 1972; Benne et al. 1973) and EF-G (Brot and Spears 1970; Bodley, Zieve and Lin 1970) and seems to be the case for RF too (Tate, Beaudet and Caskey 1973). The release of factors is needed because a factor retained on the ribosome after it has performed its function blocks some of the further transformations. Although as noted, GMPPCP allows the binding of EF-Tu and aminoacyl-tRNA to an initiation or post-translocation complex, the resulting complex does not undergo peptide bond formation (Haenni and Lucas-Lenard 1968). Peptide bond formation can take place only after GMPPCP and EF-Tu are removed from the complex. This removal can be obtained, e.g., by sedimenting an incubated complex through a sucrose gradient (Yokosawa et al. 1973). Similar observations have been reported in the case of chain initiation complexes containing GMPPCP. These do not donate their fMet residue into peptide linkage with puromycin unless IF2 and GMPPCP are first removed from the complex (Dubnoff, Lockwood and Maitra 1972).

The release of factors is also needed to allow the ribosome to bind other factors required for the subsequent transformations. Remarkably, the four factors known to be involved in GTP cleavage together with the ribosome, i.e., IF2, EF-Tu, EF-G and RF, cannot be attached to the same ribosome simultaneously. In some or all cases this might be due to shared or at least overlapping binding sites (Haselkorn and Rothman-Denes 1973; Tate, Beaudet and Caskey 1973; Lee-Huang 1974). Last but not least, factors have to be released because a factor stuck on a ribosome cannot be re-utilized.

(3) The electron spin resonance spectrum of spin-labeled EF-Tu·GDP is markedly different from that of spin-labeled EF-Tu·GTP. This difference may reflect upon a difference in conformation between the two complexes (Yokosawa et al. 1973). The latter may account for the fact that EF-Tu·GTP binds aminoacyl-tRNA, whereas EF-Tu·GDP does not.

It has not been established which ribosomal components are directly involved in the cleavage of GTP that is triggered by the four factors. A potential candidate for this role is a complex formed between the 5S RNA and several proteins from the large ribosomal subunit. (Such a complex was obtained from both *E. coli* and *B. stearothermophilus*.) The complex was reported to cleave GTP. The GTPase activity is not affected by EF-G. The same complex also hydrolyzes ATP. GTP and ATP noncompetitively inhibit each other's cleavage. This may indicate that the two nucleoside triphosphates are cleaved at different sites (Horne and Erdman 1973). The significance of this ATP cleavage remains to be seen. The only ribosomal function known to involve ATP in bacteria is the formation of ppGpp and pppGpp during stringent control. In the synthesis of these nucleotides the β-γ pyrophosphoryl moiety of ATP is transferred to the 3'-OH of GDP or GTP (see Block and Haseltine, this volume). ATP was, however, implicated in peptide chain initiation in a wheat germ extract (Marcus 1970).

IN CONCLUSION

Though the outlines of mRNA translation are familiar, much remains to be learned about this process and its evolution. At present the mode of discrimination between initiation signals by the translating machinery and the significance and mechanism of translational control are unclear. The elucidation of further details of the translation process at the molecular level will have to be based largely on progress in our understanding of the structure and function of the most complex participant in this process—the ribosome.

Acknowledgment

Studies in the author's laboratory were supported by a grant from the National Science Foundation.

References

Adesnik, M. and J. E. Darnell. 1972. Biogenesis and characterization of histone messenger RNA in HeLa cells. *J. Mol. Biol.* **67**:397.

Allaudeen, H. S., S. K. Yang and D. Soll. 1972. Leucine tRNA from His T mutant of *Salmonella typhimurium* lacks two pseudouridines. *FEBS Letters* **28**:285.

Altman, S. and J. D. Smith. 1971. Tyrosine tRNA precursor molecule polynucleotide sequence. *Nature New Biol.* **233**:35.

Anderson, J. S., M. S. Bretscher, B. F. C. Clark and K. A. Marcker. 1967. A GTP requirement for binding initiation tRNA to ribosomes. *Nature* **215**:490.

Apte, B. N. and D. Zipser. 1973. In vivo splicing of proteins: One continuous polypeptide from two independently functioning operons. *Proc. Nat. Acad. Sci.* **70**:2969.

Arai, K., M. Kawakita, Y. Kaziro, T. Kondo and N. Ui. 1973a. Studies on the polypeptide elongation factors from *E. coli*. III. Molecular characteristics of EF Tu·guanosine diphosphate, EF Ts and EF Tu·Ts complex. *J. Biochem.* **73**:1095.

Arai, K. I., M. Kawakita, S. Nishimura and Y. Kaziro. 1973b. Crystallization of phenylalanyl tRNA^{Phe}-EF Tu·GTP complex. *Biochim. Biophys. Acta* **324**:440.

Arai, K., M. Kawakita, Y. Kaziro, T. Maeda and S. Ohnishi. 1974. Conformational transition in polypeptide elongation factor Tu as revealed by electron spin resonance. *J. Biol. Chem.* **249**:3311.

Arrand, J. R. and J. Hindley. 1974. Nucleotide sequence of a ribosome binding site on RNA synthesized in vitro from coliphage T7. *Nature New Biol.* **244**:10.

Aviv, H., I. Boime, B. Loyd and P. Leder. 1972. Translation of bacteriophage Qβ messenger RNA in a murine Krebs 2 ascites tumor cell-free system. *Science* **178**:1293.

Baltimore, D., A. S. Huang and M. Stampfer. 1970. Ribonucleic acid synthesis of vesicular stomatitis virus. II. An RNA polymerase in the virion. *Proc. Nat. Acad. Sci.* **66**:572.

Baltimore, D., A. Huang, K. F. Manley, D. Rekosh and M. Stampfer. 1971. The synthesis of proteins by mammalian RNA viruses. In *Strategy of the viral genome*. Ciba Foundation Symp. (ed. G. E. W. Wolstenholme and M. O'Connor) p. 101.

Bard, E., D. Efron, A. Marcus and R. P. Perry. 1974. Translational capacity of deadenylated messenger RNA. *Cell* **1**:101.

Barnett, W. E. and J. L. Epler. 1966. Fractionation and specificities of two aspartyl ribonucleic acid and two phenylalanyl ribonucleic acid synthetases. *Proc. Nat. Acad. Sci.* **55**:184.

Beaud, G. and P. Lengyel. 1971. Peptide chain elongation. Role of the S_1 factor in the pathway from S_3-guanosine diphosphate complex to aminoacyl transfer ribonucleic acid-S_3-guanosine triphosphate complex. *Biochemistry* **10**:4899.

Beaudet, A. L. and C. T. Caskey. 1970. Release factor translation of RNA phage terminator codons. *Nature* **227**:38.

————. 1971. Mammalian peptide chain termination. II. Codon specificity and GTPase activity of release factor. *Proc. Nat. Acad. Sci.* **68**:619.

————. 1972. Polypeptide chain termination. In *The mechanism of protein synthesis and its regulation* (ed. L. Bosch) p. 133. North-Holland, Amsterdam.

Benne, R., N. Naaktgeboren, J. Gubbens aid H. O. Voorma. 1973. Recycling of initiation factors IF 1, IF 2 and IF 3. *Eur. J. Biochem.* **32**:372.

Berg, P. 1973. Suppression: A subversion of genetic decoding. In *Harvey Lectures 1971–72*, p. 243. Academic Press, New York.

Beuzard, Y. and I. M. London. 1974. The effects of hemin, initiation factors and inhibitors on the synthesis of globin in reticulocyte and ascites tumor lysates. *Fed. Proc.* **33**:1262.

Blasi, F., C. B. Bruni, A. Avitabile, R. G. Deeley, R. F. Goldberger and M. M. Meyers. 1973. Inhibition of transcription of the histidine operon in vitro by the first enzyme of the histidine pathway. *Proc. Nat. Acad. Sci.* **70**:2692.

Blobel, G. 1973. A protein of molecular weight 78,000 bound to the poly-adenylate region of eukaryotic messenger RNAs. *Proc. Nat. Acad. Sci.* **70**:924.

Blumenthal, T., T. A. Landers and K. Weber. 1972. Bacteriophage Qβ replicase contains the protein biosynthesis elongation factors EF Tu and EF Ts. *Proc. Nat. Acad. Sci.* **69**:1313.

Bodley, J. W., F. J. Zieve and L. Lin. 1970. Studies on translocation. IV. The hydrolysis of a single round of guanosine triphosphate in the presence of fusidic acid. *J. Biol. Chem.* **245**:5662.

Borsa, J. and A. F. Graham. 1968. Reovirus RNA polymerase activity in puri-fied virions. *Biochem. Biophys. Res. Comm.* **33**:896.

Bosch, L., ed. 1972. *The mechanism of protein synthesis and its regulation*. North-Holland, Amsterdam.

Bronson, M. J., C. Squires and C. Yanofsky. 1973. Nucleotide sequences from tryptophan messenger RNA of *Escherichia coli:* The sequence corresponding to the amino terminal region of the first polypeptide specified by the operon. *Proc. Nat. Acad. Sci.* **70**:2335.

Brot, N. and C. Spears. 1970. The interaction of transfer factor G, ribosomes and various guanosine nucleotides in the presence of fusidic acid. *Fed. Proc.* **29**:862.

Brown, J. C. and A. E. Smith. 1970. Initiator codons in eukaryotes. *Nature* **226**:610.

Brownlee, G. G., E. M. Cartwright, N. J. Cowan, J. M. Jarvis and C. Milstein. 1973. Purification and sequence of messenger RNA for immunoglobulin light chains. *Nature New Biol.* **244**:236.

Buck, C. A. and M. M. K. Nass. 1969. Studies on mitochondrial tRNA from animal cells. *J. Mol. Biol.* **41**:67.

Capecchi, M. R. and H. A. Klein. 1969. Characterization of three proteins in-volved in polypeptide chain termination. *Cold Spring Harbor Symp. Quant. Biol.* **34**:469.

Cashion, L. M. and W. M. Stanley, Jr. 1973. Comparative studies on the prop-erties of the eukaryotic protein synthesis initiation factor IF 1 from several sources. *Biochim. Biophys. Acta* **324**:410.

————. 1974. The eukaryotic initiation factors (IF I and IF II) of protein synthesis that are required to form an initiation complex with rabbit reticulocyte ribosomes. *Proc. Nat. Acad. Sci.* **71**:436.

Caskey, T., E. Scolnick, R. Tompkins, J. Goldstein and G. Milman. 1969. Peptide chain termination, codon, protein factor and ribosomal requirements. *Cold Spring Harbor Symp. Quant. Biol.* **34**:479.

Content, J., B. Lebleu, A. Zilberstein, H. Berissi and M. Revel. 1974. Mechanism of the interferon induced block of mRNA translation in mouse L cells: Reversal of the block by transfer RNA. *FEBS Letters* **41**:125.

Crick, F. H. C. 1966. Codon-anticodon pairing: The wobble hypothesis. *J. Mol. Biol.* **19**:548.

Dalgarno, L. and J. Shine. 1973. Conserved terminal sequence in 18S rRNA may represent terminator anticodons. *Nature New Biol.* **245**:261.

Darnbrough, C., S. Legon, T. Hunt and R. J. Jackson. 1973. Initiation of protein synthesis: Evidence for messenger RNA-independent binding of methionyl-transfer RNA to the 40S ribosomal subunit. *J. Mol. Biol.* **76**:379.

Dawid, I. B. 1972. Mitochondrial protein synthesis. In *Mitochondria: biogenesis and bioenergetics. FEBS Proc.* **28**:35.

Dubnoff, J. S., A. H. Lockwood and U. Maitra. 1972. Studies on the role of GTP in polypeptide chain initiation in *Escherichia coli. J. Biol. Chem.* **247**:2884.

Ehrenfeld E. and T. Hunt. 1971. Double stranded poliovirus RNA inhibits initiation in protein synthesis by reticulocyte lysates. *Proc. Nat. Acad. Sci.* **68**:1075.

Eldred, E. W. and P. R. Schimmel. 1973. Release of aminoacyl transfer ribonucleic acid from transfer ribonucleic acid synthetase. *Biochemistry* **51**:229.

Engelhardt, D. L., R. E. Webster and N. D. Zinder. 1967. Amber mutants and polarity in vitro. *J. Mol. Biol.* **29**:45.

Ertel, R., N. Brot, B. Redfield, J. E. Allende and H. Weissbach. 1968. Binding of guanosine 5′ triphosphate by soluble factors required for polypeptide synthesis. *Proc. Nat. Acad. Sci.* **59**:861.

Files, J. G., K. Weber and J. H. Miller. 1974. Translational reinitiation: reinitiation of lac repressor fragments at three internal sites early in the *lac i* gene of *Escherichia coli. Proc. Nat. Acad. Sci.* **71**:667.

Ganoza, M. C. and J. L. Fox. 1974. Isolation of a soluble factor needed for protein synthesis with various messenger ribonucleic acids other than poly(U). *J. Biol. Chem.* **249**:1037.

Garen, A. 1968. Sense and nonsense in the genetic code. *Science* **160**:149.

Gaskill, P. and D. Kabat, 1971. Unexpectedly large size of globin messenger ribonucleic acid. *Proc. Nat. Acad. Sci.* **68**:72.

Giglioni, B., A. M. Gianni, P. Comi, S. Ottolenghi and D. Rungger. 1973. Translation control of globin synthesis by haemin in Xenopus ococytes. *Nature New Biol.* **246**:99.

Goldberg, M. L. and J. A. Steitz. 1974. Cistron specificity of 30S heterologously reconstituted with components from *Escherichia coli* and *Bacillus stearothermophilus. Biochemistry* **13**:2123.

Goodman, H. M., J. Abelson, A. Landy, S. Brenner and J. D. Smith. 1968.

Amber suppression: A nucleotide change in the anticodon of a tyrosine transfer RNA. *Nature* **217**:1019.

Groner, Y., R. Scheps, R. Kamen, D. Kolakofsky and M. Revel. 1972. Host subunit of Qβ replicase is translation control factor i. *Nature New Biol.* **239**:19.

Gross, M. and M. Rabinowitz. 1973. Partial purification of a translational repressor mediating hemin control of globin synthesis and implication of results on the site of inhibition. *Biochem. Biophys. Res. Comm.* **50**:832.

Grunberg-Manago, M., Th. Godefroy-Colburn, A. D. Wolfe, P. Dessen, D. Pantaloni, M. Springer, M. Graffe, J. Dondon and A. Kay. 1973. Initiation of protein synthesis in prokaryotes. In *Regulation of transcription and translation in eukaryotes* (ed. E. K. F. Bautz) pp. 213. Springer-Verlag, Berlin.

Gupta, S. L., M. L. Sopori and P. Lengyel. 1974. Release of the inhibition of messenger RNA translation in extracts of interferon-treated Ehrlich ascites tumor cells by added transfer RNA. *Biochem. Biophys. Res. Comm.* **57**:763.

Gupta, N. K., C. L. Woodley, Y. C. Chen and K. K. Bose. 1973. Protein synthesis in rabbit reticulocytes. Assays, purification and properties of different ribosomal factors and their role in peptide chain initiation. *J. Biol. Chem.* **248**:2500.

Gupta, S. L., J. Waterson, M. L. Sopori, S. M. Weissman and P. Lengyel. 1971. Movement of the ribosome along the messenger ribonucleic acid during protein synthesis. *Biochemistry* **10**:4410.

Gurdon, J. B. 1974. Molecular biology in a living cell. *Nature* **248**:772.

Guthrie, C., J. G. Seidman, B. G. Barrell, J. D. Smith and W. H. McClain. 1973. Identification of tRNA precursor molecules made by phage T4. *Nature New Biol.* **246**:6.

Haenni, A. and J. Lucas-Lenard. 1968. Stepwise synthesis of a tripeptide. *Proc. Nat. Acad. Sci.* **61**:1363.

Haenni, A. L., A. Prochiantz, O. Bernhard and F. Chapeville. 1973. TYMV valyl-RNA as an amino acid donor in protein biosynthesis. *Nature New Biol.* **241**:166.

Haselkorn, R. and L. B. Rothman-Denes. 1973. Protein synthesis. *Ann. Rev. Biochem.* **42**:397.

Held, W. A., W. R. Gette and M. Nomura. 1974. Role of 16S ribosomal ribonucleic acid and the 30S ribosomal protein S12 in the initiation of natural messenger ribonucleic acid translation. *Biochemistry* **13**:2115.

Hendrix, R. W. and S. R. Casjens. 1974. Protein fusion: A novel reaction in bacteriophage λ head assembly. *Proc. Nat. Acad. Sci.* **71**:1451.

Herrlich, P., H. J. Rahmsdorf, S. H. Pai and M. Schweiger. 1974. Translation control induced by bacteriophage T7. *Proc. Nat. Acad. Sci.* **71**:1088.

Heywood, S. M., D. S. Kennedy and A. J. Bester. 1974. Separation of specific initiation factors involved in the translation of myosin and myoglobin mRNAs and the isolation of a new RNA involved in translation. *Proc. Nat. Acad. Sci.* **71**:2428.

Hirashima, A. and A. Kaji. 1973. Role of elongation factor G and a protein factor in the release of ribosomes from messenger ribonucleic acid. *J. Biol. Chem.* **248**:7580.

Hirsch, D. 1971. Tryptophan transfer RNA as the UGA suppressor. *J. Mol. Biol.* **58**:439.

Hofstetter, H., H. J. Monstein and C. Weissmann. 1974. The readthrough protein A_1 is required for in vitro reconstitution of infectious $Q\beta$ particles. *Experientia* **30**:687.

Holley, R. W., J. Apgar, G. A. Everett, J. T. Madison, M. Marquisee, S. H. Merrill, J. R. Penswick and A. Zamir. 1965. Structure of a ribonucleic acid. *Science* **147**:1462.

Hori, K. 1974. Subunit complex III-IV as the nucleotide binding site of $Q\beta$ replicase. *Nature* **250**:659.

Horiuchi, K., R. E. Webster and S. Matsuhashi. 1971. Gene products of bacteriophage $Q\beta$. *Virology* **45**:429.

Horne, J. R. and V. A. Erdmann. 1973. ATPase and GTPase activities associated with a specific 5S-RNA-protein complex. *Proc. Nat. Acad. Sci.* **70**:2870.

Huez, G., G. Marbaix, B. Lebleu, U. Nudel, M. Revel and U. Littauer. 1974. Role of the polyadenylic segment in the translation of globin messenger RNA in Xenopus oocytes. *Proc. Nat. Acad. Sci.* **71**: (in press).

Inouye, H., Y. Pollack and J. Petre. 1974. Physical and functional homology between ribosomal protein S1 and interference factor i. *Eur. J. Biochem.* (in press).

Inoue-Yokosawa, N., C. Ishikawa and Y. Kaziro. 1974. The role of guanosine triphosphate in translocation reaction catalyzed by elongation factor G. *J. Biol. Chem.* **249**:4321.

Iwasaki, K., K. Mizumoto, M. Tanaka and Y. Kaziro. 1974. A new protein factor required for polypeptide elongation in mammalian tissues. *J. Biochem.* **74**:849.

Kaempfer, R. and J. Kaufman. 1973. Inhibition of cellular protein synthesis by double stranded RNA: Inactivation of an initiation factor. *Proc. Nat. Acad. Sci.* **70**:1222.

Kamen, R., M. Kondo, W. Romer and C. Weissmann. 1972. Reconstitution of $Q\beta$ replicase lacking subunit α with protein synthesis interference factor i. *Eur. J. Biochem.* **31**:44.

Kates, J. R. and B. R. McAuslan. 1967. Poxvirus DNA-dependent RNA polymerase. *Proc. Nat. Acad. Sci.* **58**:134.

Kaziro, Y. 1973. The role of guanosine triphosphate in the polypeptide elongation reaction in *Escherichia coli*. In *Organization of energy-transducing membranes* (ed. M. Nakas and L. Packer) p. 187. Univ. Tokyo Press, Tokyo.

Kaziro, Y., N. Inoue-Yokosawa and M. Kawakita. 1972. Studies on polypeptide elongation factor from *E. coli:* I. Crystalline factor G. *J. Biochem.* **72**:853.

Koch, G. L. E., Y. Boulanger and B. S. Hartley. 1974. Repeating sequences in aminoacyl-tRNA synthetases. *Nature* **249**:316.

Kondo, M., G. Eggertsson, J. Eisenstadt and P. Lengyel. 1968. Ribosome formation from subunits: Dependence on formylmethionyl-transfer RNA in extracts from *E. coli. Nature* **220**:368.

Kozak, M. and D. Nathans. 1972. Translation of the genome of a ribonucleic acid bacteriophage. *Bact. Rev.* **36**:109.

Kuwano, M., M. Ono, M. Yamamoto, H. Endo and T. Kamiya. 1973. Elongation factor T altered in a temperature-sensitive *Escherichia coli* mutant. *Nature New Biol.* **244**:107.

Kwan, S. and G. Brawerman. 1972. A particle associated with the polyadenylate segment in mammalian messenger RNA. *Proc. Nat. Acad. Sci.* **69**:3247.

Leder, P. 1973. The elongation reactions in protein synthesis. *Adv. Protein Chem.* **27**:213.

Lee-Huang, S. 1974. Inhibition of polypeptide chain initiation by elongation factor G. *Fed. Proc.* **33**:1261.

Lee-Huang, S. and S. Ochoa. 1973. Purification and properties of two messenger discriminating species of *E. coli* initiation factor 3. *Arch. Biochem. Biophys.* **156**:84.

Lengyel, P. and D. Soll. 1969. Mechanism of protein synthesis. *Bact. Rev.* **33**:264.

Lengyel, P., J. F. Speyer and S. Ochoa. 1961. Synthetic polynucleotides and the amino acid code. *Proc. Nat. Acad. Sci.* **47**:1936.

Levin, D. H., D. Kyner and G. Acs. 1973. Protein initiation in eukaryotes. Formation and function of a ternary complex composed of partially purified ribosome factors, methionyl transfer RNA and guanosine triphosphate. *Proc. Nat. Acad. Sci.* **70**:41.

Levinthal, C., A. Keynan and A. Higa. 1962. Messenger RNA turnover and protein synthesis in *B. subtilis* inhibited by actinomycin D. *Proc. Nat. Acad. Sci.* **48**:1631.

Lipmann, F. 1967. Peptide bond formation in protein biosynthesis. In *Regulation of nucleic acid and protein biosynthesis* (ed. V. V. Koningsberger and L. Bosch). Biochim. Biophys. Acta. Library, vol. 10, p. 117. Elsevier, New York.

Litvak, S., A. Tarrago, L. Tarrago-Litvak and J. E. Allende. 1973. Elongation factor-viral genome interaction dependent on the aminoacylation of TYMV and TMV RNAs. *Nature New Biol.* **241**:88.

Livingston, D. M. and P. Leder. 1969. Deformylation and protein synthesis. *Biochemistry* **8**:435.

Lockard, R. E. and J. B. Lingrel. 1969. The synthesis of mouse hemoglobin β chains in a rabbit reticulocyte cell-free system programmed with mouse reticulocyte 9S RNA. *Biochem. Biophys. Res. Comm.* **37**:204.

Lodish, H. F. 1970. Specificity in bacterial protein synthesis: Role of initiation factors and ribosomal subunits. *Nature* **226**:705.

Lu, P. and A. Rich. 1971. The nature of the polypeptide chain termination signal. *J. Mol. Biol.* **58**:513.

Lucas-Lenard, J. and F. Lipmann. 1971. Protein biosynthesis. *Ann. Rev. Biochem.* **40**:409.

Lucas-Lenard, J., P. Tao and A. Haenni. 1969. Further studies on bacterial polypeptide elongation. *Cold Spring Harbor Symp. Quant. Biol.* **34**:455.

Maizels, N. 1974. *E. coli* lactose operon ribosome binding site. *Nature* **249**:647.

Marcus, A. 1970. Tobacco mosaic virus ribonucleic acid-dependent amino acid incorporation in a wheat embryo system in vitro. *J. Biol. Chem.* **245**:955.

————. 1972. Protein synthesis in extracts of wheat embryo. In *Protein biosynthesis in nonbacterial systems, Methods in molecular biology* (ed. J. A. Last and A. I. Laskin) vol. 2, p. 127. M. Dekker, New York.

Mathews, M. B. and M. Osborn. 1974. The rate of polypeptide chain elongation in a cell-free system from Krebs II ascites cells. *Biochim. Biophys. Acta* **340**:147.

Matsuhashi, M., C. P. Dietrich, and J. L. Strominger. 1965. Incorporation of glycine into the cell wall glycopeptide in *Staphylococcus aureus:* Role of sRNA and lipid intermediates. *Proc. Nat. Acad. Sci.* **54**:587.

McCroskey, R. P., M. Zasloff and S. Ochoa. 1972. Polypeptide chain initiation in eukaryotes. III. Initiation and stepwise elongation with Artemia ribosomes and factors. *Proc. Nat. Acad. Sci.* **69**:2451.

Mendecki, J., S. Y. Lee and G. Brawerman. 1972. Characteristics of the polyadenylic segment associated with messenger ribonucleic acid in mouse sarcoma 180 ascites cells. *Biochemistry* **11**:792.

Menninger, J. R. 1974. Peptidyl-tRNA can fall off ribosomes during protein synthesis. *Fed. Proc.* **33**:1335.

Merrick, W. C., N. H. Lubsen and W. F. Anderson. 1973. A ribosome dissociation factor from rabbit reticulocytes distinct from initiation factor M3. *Proc. Nat. Acad. Sci.* **70**:2220.

Miller, J. H. 1974. GUG and UUG are initiation codons in vivo. *Cell* **1**:73.

Min Jou, W., G. Haegeman, M. Ysebaert and W. Fiers. 1972. Nucleotide sequence of the gene coding for the bacteriophage MS2 coat protein. *Nature* **237**:82.

Mizuno, S., J. M. Fischer and M. Rabinowitz. 1972. Hemin control of globin synthesis: Action of an inhibitor formed in the absence of hemin on the reticulocyte cell-free system and its reversal by a ribosomal factor. *Biochim. Biophys. Acta* **272**:638.

Modolell, J., B. Cabrer and D. Vazquez. 1973a. The interaction of elongation factor G with N-acetylphenylalanyl transfer RNA ribosome complexes. *Proc. Nat. Acad. Sci.* **70**:3561.

————. 1973b. The stoichiometry of ribosomal translocation. *Proc. Nat. Acad. Sci.* **248**:8356.

Moon, H., B. Redfield, S. Millard, F. Vane and H. Weissbach. 1973. Multiple forms of elongation factor 1 from calf brain. *Proc. Nat. Acad. Sci.* **70**:3282.

Moore, C. H., F. Farron, D. Bohnert and C. Weissmann. 1971. Possible origin of a minor virus specific protein (A₁) in Qβ particles. *Nature New Biol.* **234**:204.

Morrison, T. C. and H. F. Lodish. 1973. Translation of bacteriophage Qβ RNA by cytoplasmic extracts of mammalian cells. *Proc. Nat. Acad. Sci.* **70**:315.

Nathans, D., G. Notani, J. H. Schwartz and N. D. Zinder. 1962. Biosynthesis of the coat protein of coliphage f₂ by *E. coli* extracts. *Proc. Nat. Acad. Sci.* **48**:1424.

Nichols, J. L. 1970. Nucleotide sequence from the polypeptide chain termination region of the coat protein cistron in bacteriophage R 17 RNA. *Nature* **225**:147.

Nirenberg, M. W. and J. H. Matthaei. 1961. The dependence of cell-free protein synthesis in *E. coli* upon naturally occurring or synthetic polynucleotides. *Proc. Nat. Acad. Sci.* **47**:1588.

Nishimura, S. 1972. Minor components in transfer RNA: Their characterization, location and function. *Progr. Nucleic Acid Res. Mol. Biol.* **12**:49.

Nombela, C. and S. Ochoa. 1973. Conformational control of the interaction of eukaryotic elongation factors EF 1 and EF 2 with ribosomes. *Proc. Nat. Acad. Sci.* **70**:3556.

Nudel, N., B. Lebleu and M. Revel. 1973. Discrimination between messenger ribonucleic acids by a mammalian translation initiation factor. *Proc. Nat. Acad. Sci.* **70**:2139.

Ono, Y., A. Skoultchi, J. Waterson and P. Lengyel. 1969. Peptide chain elongation: GTP cleavage catalyzed by factors binding aminoacyl-transfer RNA to the ribosome. *Nature* **222**:645.

Perry, R. P. and D. E. Kelley. 1974. Existence of methylated messenger RNA in mouse L cells. *Cell* **1**:37.

Perry, R. P., J. R. Greenberg, D. E. Kelley, J. La Torre and G. Schochetman. 1973. Messenger RNA: Its origin and fate in mammalian cells. In *Gene expression and its regulation* (ed. F. T. Kenney et al.) p. 137. Plenum Press, New York.

Prather, N., J. M. Ravel, B. Hardesty and W. Shive. 1974. Evidence for activities of rabbit reticulocyte elongation factor 1 analogous to bacterial factors EF Ts and EF tu. *Biochem. Biophys. Res. Comm.* **57**:578.

Revel, M. 1972. Polypeptide chain initiation: The role of ribosomal protein factors and ribosomal subunits. In *The mechanism of protein synthesis and its regulation* (ed. L. Bosch) p. 87. North-Holland, Amsterdam.

Revel, M., Y. Pollack, Y. Groner, R. Scheps, H. Inouye, H. Berissi and H. Zeller. 1972. Protein factors in *Escherichia coli* controlling initiation of mRNA translation. In *Ribosomes: structure, function and biogenesis,* 8th FEBS Proc. (ed. H. Bloemendal et al.) vol. 27, p. 261. North-Holland/-American Elsevier.

————. 1973. IF 3-interference factors: Protein factors in *Escherichia coli* controlling initiation of mRNA translation. *Biochimie* **55**:41.

Riddle, D. L. and J. Carbon. 1973. Frameshift suppression: a nucleotide addition in the anticodon of a glycine transfer RNA. *Nature New Biol.* **242**:230.

Roberts, B. E. and B. M. Paterson. 1973. Efficient translation of tobacco mosaic virus RNA and rabbit globin 9S RNA in a cell-free system from commercial wheat germ. *Proc. Nat. Acad. Sci.* **70**:2330.

Robertson, H. D., B. G. Barrell, H. L. Weith and J. E. Donelson. 1973. Isolation and sequence analysis of a ribosome-protected fragment from bacteriophage ϕX174 DNA. *Nature New Biol.* **241**:38.

Rudland, P. S. and B. F. C. Clark. 1972. Polypeptide chain initiation and the role of a methionine tRNA. In *The mechanism of protein synthesis and its regulation* (ed. L. Bosch) p. 55. North-Holland, Amsterdam.

Sager, R. 1972. *Cytoplasmic genes and organelles.* Academic Press, New York.

Salomon, R. and U. Z. Littauer. 1974. Enzymatic acylation of histidine to mengo virus RNA. *Nature* **249**:32.

Salser, W., M. Fluck and R. Epstein. 1969. The influence of the reading context upon the suppression of nonsense codons III. *Cold Spring Harbor Symp. Quant. Biol.* **34**:513.

Samuel, C. E. and J. R. Rabinowitz. 1974. Initiation of protein synthesis by folate sufficient and folate deficient *Streptococcus faecalis* R. *J. Biol. Chem.* **249**:1198.

Schafer, K. P. and D. Soll. 1974. New aspects in tRNA biosynthesis. *Biochimie* (in press).

Scherberg, N. H. and S. B. Weiss. 1972. T4 transfer RNAs: Codon recognition and translational properties. *Proc. Nat. Acad. Sci.* **69**:1114.

Schreier, M. H. and T. Staehelin. 1973. Initiation of eukaryote protein synthesis: (Met-tRNA$_f$·40S ribosome) initiation complex catalyzed by purified initiation factors in the absence of mRNA. *Nature New Biol.* **242**:35.

Schreier, M. H., T. Staehelin, R. F. Gesteland and P. F. Spahr. 1973. Translation of bacteriophage R17 and Qβ RNA in a mammalian cell-free system. *J. Mol. Biol.* **75**:575.

Seal, S. N. and A. Marcus. 1973. Translation of the initial codons of satellite tobacco necrosis virus ribonucleic acid in a cell-free system from wheat embryo. *J. Biol. Chem.* **248**:6577.

Shapiro, D. J., J. M. Taylor, G. S. McKnight, R. Palacios, C. Gonzales, M. L. Kiely and R. T. Schimke. 1974. Isolation of hen oviduct ovalbumin and rat liver albumin polysomes by indirect immunoprecipitation. *J. Biol. Chem.* **249**:3665.

Shatkin, A. J. and J. D. Sipe. 1968. RNA polymerase activity in purified reoviruses. *Proc. Nat. Acad. Sci.* **61**:1462.

Shine, J. and L. Dalgarno. 1974. The 3'-terminal sequence of *Escherichia coli* 16S ribosomal RNA: Complementarity to nonsense triplets and ribosome binding sites. *Proc. Nat. Acad. Sci.* **71**:1342.

Singer, R. H. and S. Penman. 1973. Messenger RNA in HeLa cells: Kinetics of formation and decay. *J. Mol. Biol.* **78**:321.

Singer, C. E., G. R. Smith, R. Cortese and B. N. Ames. 1972. Mutant tRNA[His] ineffective in repression and lacking two pseudouridine modifications. *Nature New Biol.* **238**:72.

Skoultchi, A., Y. Ono, J. Waterson and P. Lengyel. 1970. Peptide chain elongation: Indications for the binding of an amino acid polymerization factor, guanosine 5'-triphosphate·aminoacyl transfer ribonucleic acid complex to the messenger·ribosome complex. *Biochemistry* **9**:508.

Smith, J. D. and J. E. Celis. 1973. Mutant tyrosine transfer RNA that can be charged with glutamine. *Nature New Biol.* **243**:66.

Smith, A. E., K. A. Marcker and M. B. Mathews. 1970. Translation of RNA from encephalomyocarditis virus in a mammalian cell-free system. *Nature* **225**:184.

Soffer, R. L. 1973. Posttranslational modification of proteins catalyzed by aminoacyl-tRNA protein transferases. *Mol. Cell Biochem.* **2**:3.

Soll, D. 1971. Enzymatic modification of transfer RNA. *Science* **173**:293.

Soll, D. and P. Schimmel. 1974. Aminoacyl-tRNA synthetases. In *The enzymes* (ed. P. Boyer) vol. 10, p. 489. Academic Press, New York.

Spirin, A. 1969. Informosomes. *Eur. J. Biochem.* **10**:70.

Sprinzl, M. and F. Cramer. 1973. Accepting site for aminoacylation of tRNA[Phe] from yeast. *Nature New Biol.* **245**:3.

Steitz, J. A. 1969. Nucleotide sequences of the ribosomal binding sites of bacteriophage R17 RNA. *Cold Spring Harbor Symp. Quant. Biol.* **34**:621.

―――. 1972. Oligonucleotide sequence of replicase initiation site in $Q\beta$ RNA. *Nature New Biol.* **236**:71.

―――. 1975. Ribosome recognition of initiator regions in the RNA bacteriophage genome. In *RNA phages* (ed. N. D. Zinder). Cold Spring Harbor Laboratory (in press).

Stevens, H. R. and A. R. Williamson. 1973a. Isolation of messenger RNA coding for mouse heavy-chain immunoglobulin. *Proc. Nat. Acad. Sci.* **70**:1127.

―――. 1973b. Translational control of immunoglobulin synthesis: I, II. *J. Mol. Biol.* **78**:505, 517.

Stewart, J. W., F. Sherman, N. A. Slipman and M. Jackson. 1971. Identification and mutational relocation of the AUG codon initiating translation of iso-1-cytochrome c in yeast. *J. Biol. Chem.* **246**:7429.

Stolzfus, C. M., A. J. Shatkin and A. K. Banerjee. 1973. Absence of polyadenylic acid from reovirus messenger ribonucleic acid. *J. Biol. Chem.* **248**:7993.

Subramanian, A. R. and B. D. Davis. 1973. Release of 70S ribosomes from polysomes in *Escherichia coli*. *J. Mol. Biol.* **74**:45.

Subramanian, A. R., B. D. Davis and R. J. Beller. 1969. The ribosome dissociation factor and the ribosome-polysome cycle. *Cold Spring Harbor Symp. Quant. Biol.* **34**:223.

Suddath, F. L., G. J. Quigley, A. McPherson, D. Sneden, J. J. Kim, S. H. Kim and A. Rich. 1974. Three-dimensional structure of yeast phenylalanine transfer RNA at 3.0 Å resolution. *Nature* **248**:20.

Szer, W. and S. Leffler. 1974. Interaction of *E. coli* 30S ribosomal subunits with MS2 phage RNA in the absence of initiation factors. *Proc. Nat. Acad. Sci.* (in press).

Tate, W. P., A. L. Beaudet and C. T. Caskey. 1973. Influence of guanine nucleotides and elongation factors on interaction of release factors with the ribosome. *Proc. Nat. Acad. Sci.* **70**:2350.

Takanami, M., Y. Yan and T. H. Jukes. 1965. Studies on the site of ribosomal binding of f2 bacteriophage RNA. *J. Mol. Biol.* **12**:761.

Thach, S. S. and R. E. Thach. 1971. Translocation of messenger RNA and "accommodation" of fMet-tRNA. *Proc. Nat. Acad. Sci.* **68**:1791.

van Duin, J. and P. H. van Knippenberg. 1974. Functional heterogeneity of the 30S ribosomal subunit of *Escherichia coli*. *J. Mol. Biol.* **84**:185.

van Knippenberg, P. H., P. J. J. Hooykaas and J. van Duin. 1974. The stoichiometry of *E. coli* 30S ribosomal protein S_1 on in vivo and in vitro polyribosomes. *FEBS Letters* **41**:323.

Vermeer, C., W. van Alphen, P. van Knippenberg and L. Bosch. 1973. Initiation factor-dependent binding of MS2 RNA to 30S ribosomes and the recycling of IF 3. *Eur. J. Biochem.* **40**:295.

Vogel, T., M. Meyers, J. S. Kovach and R. F. Goldberger. 1972. Specificity of interaction between the first enzyme for histidine biosynthesis and aminoacylated histidine transfer ribonucleic acid. *J. Bact.* **112**:126.

Wahba, A. J., M. J. Miller, A. Niveleau, T. A. Landers, G. C. Carmichael, K. Weber, D. A. Hawley and L. I. Slobin. 1974. Subunit 1 of Qβ replicase and 30S ribosomal protein S₁ of *Escherichia coli*. *J. Biol. Chem.* **249**:3314.

Waterson, R. M. and W. H. Konigsberg. 1974. Peptide mapping of aminoacyl-tRNA synthetases: Evidence for internal sequence homology in *Escherichia coli* leucyl-tRNA synthetase. *Proc. Nat. Acad. Sci.* **71**:376.

Weiner, A. M. and K. Weber. 1971. Natural readthrough at the UGA termination signal of Qβ coat protein cistron. *Nature New Biol.* **234**:206.

―――. 1973. A single UGA codon functions as a natural termination signal in the coli phage Qβ coat protein cistron. *J. Mol. Biol.* **80**:837.

Weissbach, H., B. Redfield and J. Hachman. 1970. Studies on the role of factor Ts in aminoacyl-tRNA binding to ribosomes. *Arch. Biochem. Biophys.* **141**:384.

Weissmann, C., M. A. Billeter, H. M. Goodman, J. Hindley and H. Weber. 1973. Structure and function of phage RNA. *Ann. Rev. Biochem.* **42**:303.

Wigle, D. T. and A. E. Smith. 1973. Specificity in initiation of protein synthesis in a fractionated mammalian cell-free system. *Nature New Biol.* **242**:136.

Wilcox, M. 1971. Transfer RNA and regulation at the translational level. In *Metabolic pathways*, 3rd Ed. vol. 5, *Metabolic regulation*, p. 143.

Wilcox, M. and N. Nirenberg. 1968. Transfer RNA as a cofactor coupling amino acid synthesis with that of protein. *Proc. Nat. Acad. Sci.* **61**:229.

Williamson, R. 1973. The protein moieties of animal messenger ribonucleoproteins. *FEBS Letters* **37**:1.

Wilson, J. H. 1973. Function of the bacteriophage T4 transfer RNAs. *J. Mol. Biol.* **74**:753.

Woese, C. R. 1973. Evolution of the genetic code. *Naturwissenschaften* **60**:447.

Wolfenden, R., D. H. Rammler and F. Lipmann. 1964. On the site of esterification of amino acids to soluble RNA. *Biochemistry* **3**:329.

Yahata, H., Y. Ocada and A. Tsugita. 1970. Adjacent effect on suppression efficiency. II. *Mol. Gen. Genet.* **106**:208.

Yaniv, M., W. R. Folk, P. Berg and L. Soll. 1974. A single mutational modification of a tryptophan-specific transfer RNA permits aminoacylation by plutamine and translation of the codon UAG. *J. Mol. Biol.* **86**:245.

Yarus, M. 1972. Phenylalanyl-tRNA synthetase and isoleucyl-tRNA^{Phe}: A possible verification mechanism for aminoacyl-tRNA. *Proc. Nat. Acad. Sci.* **69**:1915.

Yokosawa, H., N. Inoue-Yokosawa, K. Arai, M. Kawakita and Y. Kaziro. 1973. The role of guanosine triphosphate hydrolysis in elongation factor Tu-promoted binding of aminoacyl transfer ribonucleic acid to ribosomes. *J. Biol. Chem.* **248**:375.

Yot, P., M. Pinck, A. L. Haenni, H. M. Duranton and F. Chappeville. 1970. Valine-specific tRNA-like structure in turnip yellow mosaic virus RNA. *Proc. Nat. Acad. Sci.* **67**:1345.

General Physical Properties of Ribosomes

K. E. Van Holde
Department of Biochemistry and Biophysics
Oregon State University
Corvallis, Oregon 97331

W. E. Hill
Department of Chemistry
University of Montana
Missoula, Montana 59801

INTRODUCTION

This chapter is concerned with the morphology and structure of ribosomes, as revealed by physicochemical techniques. To attempt a comprehensive review of all reported measurements would be a formidable and unrewarding task. Rather, we feel that what is needed at this point is a critical review, which selects those observations that appear to be best documented. We shall attempt to correlate and interrelate these observations in such a manner as to provide a comprehensive picture of what we now know, or think we know, about ribosome structure.

In doing this, we are almost forced into proposing tentative models. These models may be useful, not because our knowledge of ribosome structure is complete, but because it is incomplete. Models provide a means of summarizing existing data in a way that provokes further questions and of focusing on the areas of our ignorance. Finally, they allow concentration on the interrelation between structure and function. We believe that our knowledge of ribosome structure is now at the stage where such questions can be seriously considered.

STRUCTURE OF *E. COLI* RIBOSOMES

Scope and Difficulties

In this section, we shall limit discussion to the 70S ribosomes of *E. coli* and the 50S and 30S subunits thereof. While we recognize that the ribosomes of other prokaryotes and eukaryotes may be of equal or greater interest to many, it is only with the *E. coli* particles that enough precise information has been gathered to present a coherent picture of physical properties. We presume, with some evidence in support, that these results will prove to be typical of prokaryotic ribosomes, and perhaps those of mitrochondria and chloroplasts as well.

Information concerning the quite different ribosomes of eukaryotic cytoplasm is much more limited; it will be reviewed in the next section.

At the outset, we must recognize that there are fundamental difficulties in definition of a ribosome or ribosomal subunit. These difficulties are of several kinds:

1. Different methods of ribosome preparation appear to yield different kinds of particles, with somewhat different physical properties. Many of the early studies (and some more recent ones) have employed variants of the method of preparation used by Tissières et al. (1959) in their pioneering work. This method involved grinding the bacteria in dilute Tris or phosphate buffer, containing 0.01 M $MgCl_2$, followed by repeated centrifugation in the same buffer. Later a number of workers utilized techniques in which the ribosomes, after separation as 70S particles from other cellular constitutents, were "washed" in 0.5 or 1 M NH_4Cl (Spirin et al. 1963; Salas et al. 1965; Hill et al. 1969b). This procedure was originally intended to remove ribonuclease; it has since been shown to remove other proteins as well (Stanley et al. 1966) and is thought to completely remove ribosomal protein S1 (Nomura, personal communication). Particles so treated will be referred to as "NH_4Cl-washed" ribosomes, to distinguish them from the "unwashed" ribosomes prepared by the method of Tissières et al. (1959). A third procedure, which has been widely employed by Kurland and coworkers and other groups (Kurland 1966), involves precipitation of the ribosomes and their subunits in ammonium sulfate. Such particles will be referred to as "precipitated" ribosomes or subunits.

There has been considerable controversy as to whether such washed or precipitated particles are truly ribosomes, or ribosomal subunits, or whether *essential* proteins have been removed. Such particles have been shown to be active in protein synthesis (Kurland 1966; Salas et al. 1965; Kurland et al. 1969; Voynow and Kurland 1971). While Kurland et al. (1969) showed that addition of protein factors increased the protein synthetic activity of such washed particles, exactly the same degree of increase was found upon addition of the same factors to unwashed ribosomes. However, more recent studies have shown that the total activity in each case could be accounted for by the number of subunits containing about 20 proteins (Nomura, personal communication). On the basis of this evidence, it would seem that most physical studies have been made on somewhat heterogeneous samples since the totally active subunit which contains 20 proteins has not been purified homogeneously.

2. Ribosomal subunits may exist in more than one conformation. There is now strong evidence that the 30S subunit exists in a different form at $[Mg^{++}] < 2$ mM than it does at $[Mg^{++}] > 2$ mM, and that the former structure is "inactive" in binding tRNA and in associating with the 50S particle (Zamir et al. 1971; Spitnik-Elson et al. 1972). Somewhat similar effects have been seen with 50S particles as well (Miskin et al. 1970). Recent gel electrophoresis studies by Talens et al. (1973) have shown that both the washed 50S and washed 30S subunits show electrophoretic heterogeneity, which they suggest is probably due to differences in size and/or shape in the various subclasses of each subunit. In any event, it is clear that we must be very cautious in interpreting results that purport to describe "the conformation" of a subunit.

3. The 70S particles are especially poorly defined. The 70S ribosomes, as obtained from cell extracts, may contain varying amounts of mRNA, nascent polypeptide and protein cofactors, depending upon the accidents of cell breakage and the metabolic state of the cell. In addition, as will be discussed below, some evidence has developed that the 70S particle may be capable of two or more conformational states, perhaps reflecting different states in the protein synthesis process (Spirin 1969; Talens et al. 1970, 1973; Waterson et al. 1972).

Finally, under some circumstances the 70S particle may be in reversible equilibrium with its 50S and 30S subunits or may undergo dimerization to a 100S particle. With all of these complications, it is hardly surprising that physical data on 70S particles appear to be less reproducible than those for the subunits.

With these caveats, we turn to consideration of data on the individual particles.

The 30S Subunit

Hydrodynamic Studies. The results shown in Table 1, part A, indicate general agreement that the unwashed 30S subunits have a molecular weight of about 1.0×10^6, whereas the NH_4Cl-washed or $(NH_4)_2SO_4$-precipitated particles give values of 0.90×10^6 daltons. The difference in weight of 100,000 daltons would correspond to about six proteins of the average size found in 30S subunits (17,000 daltons). Voynow and Kurland (1971) have compared the protein compositions of washed and unwashed 30S particles, and their results indicate a protein loss of just about this amount. Evidence from sedimentation equilibrium experiments (Figure 1) shows that the washed 30S particles are quite homogeneous in molecular weight,

Table 1 Molecular weights of *E. coli* ribosomal particles

Reference	Preparation[a]	Physical[b] method	$(M \times 10^{-6})$[c]
A. 30S subunits			
Tissières et al. (1959)	unwashed	S & [η]	1.00 ± .16
		S & D	0.70 ± .05
Hill et al. (1969a, 1970)	unwashed	SE	1.00 ± .05
	NH₄Cl wash	SE	0.90 ± .03
	(NH₄)₂SO₄ ppt	SE	0.90 ± .03
Scafati et al. (1971)	unwashed[d]	LS	1.00 ± .10
B. 50S subunits			
Tissières et al. (1959)	unwashed	S & D	1.80 ± .07
		S & [η]	1.80 ± .29
Hill et al. (1969a, 1970)	unwashed	SE	1.65 ± .07
	NH₄Cl washed	SE	1.55 ± .05
	(NH₄)₂SO₄ ppt	SE	1.55 ± .05
Scafati et al. (1971)	unwashed	LS	1.70 ± .05
C. 70S particles			
Tissières et al. (1959)	unwashed	S & D	2.60 ± 0.2
		S & [η]	3.10 ± 0.5
Hill et al. (1969a)	NH₄Cl washed	SE	2.65 ± 0.2
Scafati et al. (1971)	unwashed[d]	LS	2.90 ± 0.3

[a] "Unwashed" ribosomes are those prepared by the method of Tissières et al. (1959), or a modification thereof, which does not involve a high-salt washing step. "NH₄Cl-washed" ribosomes are prepared with a 0.5 M NH₄Cl wash, as described by Hill et al. (1969b). "(NH₄)₂SO₄ ppt" refers to the ammonium sulfate precipitation method of purification of Kurland (1966).

[b] The code is as follows: S & D refers to the combination of sedimentation and diffusion measurements, using the Svedberg equation; S & [η] refers to the combination of sedimentation and intrinsic viscosity measurements using the Scheraga-Mandelkern (1953) equation; SE means sedimentation equilibrium; LS means light scattering.

[c] The estimated uncertainties in these values are those given by the various researchers.

[d] Prepared by "a method derived from Tissières et al."

although heterogeneity resulting from *exchange* of a few proteins would not be detected by this method.

Estimates of the molecular weight of the 16S RNA of the 30S subunit range from 0.53×10^6 (end group analysis, Midgely 1965) and 0.56×10^6 (light scattering, sedimentation and viscosity, Kurland 1960) to a recent value of 0.64×10^6 (sedimentation equilibrium, Ortega and Hill 1973). With this most recent value, the percent RNA is predicted to be 64% for the unwashed particles and 71% for the washed subunits. These results are in good agreement with published composition data (Tissières et al. 1959; Kurland et al. 1969) for particles prepared in these two ways.

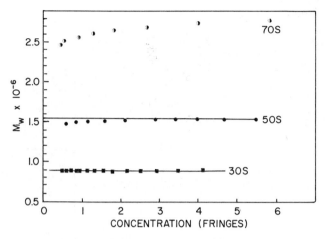

Figure 1 M_w vs. c plots for 70S, 50S and 30S ribosomes taken from sedimentation equilibrium data of Hill et al. (1969a). The data have been selected to demonstrate that homogeneous preparations of 30S and 50S subunits can be obtained.

However such comparisons must be treated with some caution, for the molecular weight of the RNA (and the ribosome) given by physical methods will be that of the polyelectrolyte salt and will thus depend upon the ionic milieu and the nature of counter-ion binding. The kind of differences which may result are illustrated below.

If we assume 16S rRNA contains 1650 nucleotides with nucleotide percentages as given by Spirin and Gavrilova (1969), then:

(a) 16S rRNA—564,000 daltons

(b) 16S rRNA (Na^+ salt)—602,000 daltons

(c) 16S rRNA (K^+ salt)—628,000 daltons

assuming sufficient salt concentration to neutralize charge groups in cases (b) and (c).

Because of the polyelectrolyte nature of both ribosomes and RNA, one must be careful to compare molecular weights determined in similar salts. If we assume that the ribosomes studied by Hill et al. (1969a) were predominantly K^+ salts and subtract the values of M for the K^+ salt of the RNA (Ortega and Hill 1973), we find that the 30S ribosome contains about 360,000 daltons of protein before NH_4Cl washing or $(NH_4)_2SO_4$ precipitation and about 260,000 daltons after such treatment. These numbers are close to those used by Traut et al. (1969) and Kurland et al. (1969) in enumeration of proteins present. It should be remembered that ribosome composition values determined by chemical determinations of RNA and protein content are subject to other uncertainties.

It should be noted from the data in Table 2 that the "30S" particle is

Table 2 Other physical data on *E. coli* ribosomal particles

$s^\circ_{20,w}$ (S)	$\bar{v}\left(\dfrac{ml}{g}\right)$ (25°)	$[\eta]\left(\dfrac{ml}{g}\right)$	R_G(Å)	Max. dim. (Å)[a]	Prep.[b]	Reference
A. 30S subunits						
30.6 ± 1.0	0.64	8.0 ± 0.6	—	—	U	Tissières et al. (1959)
31.8 ± 0.3		—	69 ± 2.0	220	W	Hill et al. (1969b)
—	0.601 ± .006	8.1 ± 0.15	—	—	W	Hill et al. (1969a)
31.1		—	72 ± 1.6	215	W	Smith (1971)
—		—	72[c]	—	U	Scafati et al. (1971)
B. 50S subunits						
50.0 ± 1.0	0.64	5.4 ± 0.4	—	—	U	Tissières et al. (1959)
50.2 ± 0.5		—	77 ± 2.0	—	W	Hill et al. (1969b)
—	0.592 ± .006	5.6 ± 0.2	—	—	W	Hill et al. (1969a)
50.1 ± 0.5		5.5 ± 0.3	75.5 ± 1.1	233	W	Tolbert (1971)
—		—	74.3	—	—	Serdyuk et al. (1970)
—		—	66.5[c]	—	U	Scafati et al. (1971)
C. 70S ribosomes						
69.1 ± 1.0	0.64	6.1 ± 0.6	—	—	U	Tissières et al. (1959)
70.5 ± 0.7		—	125	400	W	Hill et al. (1969b)
—	0.606 ± .006	6.8 ± 0.2	—	—	W	Hill et al. (1969a)
—		—	88[c]	—	U	Scafati et al. (1971)

Uncertainties in values, as estimated by the various researchers, are given after each number when available.

[a] The maximum particle dimension, as estimated from the pair-distribution function obtained from low-angle X-ray scattering.

[b] U means unwashed, W means washed in 0.5 M NH₄Cl.

[c] From light scattering. All other R_G values from X-ray scattering.

actually a "31S" particle, with a value of $s^\circ_{20, w}$, as obtained in several laboratories, close to 31.0 svedbergs.

A number of experimental studies indicate that the 30S subunit is a rather asymmetric and highly hydrated particle. It has the largest intrinsic viscosity of all the particles, and its radius of gyration (from either low-angle X-ray scattering or light scattering) is nearly as great as that of the much more massive 50S subunit.

Since hydrodynamic methods, like sedimentation and viscometry, do not distinguish unambiguously between hydration and asymmetry, we must turn to other techniques to learn something about the shape and volume of the particles. Two techniques have been widely used; each has its limitations.

Electron Microscope Studies. At first glance, it would seem that the problem could be solved easily by electron microscopy, but it must be remembered that many electron microscope techniques subject highly hydrated particles like ribosomes to rather damaging conditions during sample preparation. If the particles are dried from the solution state on the electron microscope grid, they may be expected to shrink and may change not only in volume but even in relative dimensions and morphology. A considerable volume decrease on air-drying has been demonstrated in several cases (see, for example, Hart 1962; Vasiliev 1971; Spiess 1973). The best preparative technique for such particles would seem to be the freeze-drying method (Hart 1962; Vasiliev 1971) in which the preparation is quickly frozen on the EM grid and the water removed under vacuum at low temperature. As we shall show below, this method gives results in best agreement with other techniques. However, it is still difficult to guarantee that even the most careful sample preparation will not damage highly hydrated particles like ribosomes.

Small-angle X-ray Scattering. A quite different approach to the problem of deducing particle shape and dimensions, a technique applicable to particles in solution, is small-angle X-ray scattering. The very low-angle region of the scattering curve yields quite unambiguously the radius of gyration (R_G), whereas the region at slightly higher angles is sensitive to the shape and dimensions of the particles. It is also possible to obtain with some reliability a largest dimension by computing the pair distribution function (see Beeman et al. 1957).

The analysis of the "shape region" of the scattering curves is a curve-fitting procedure; one tries various models and finds which one yields a predicted scattering curve in best agreement with the experimental result. Dimensions are to some extent fixed by R_G and the maximum dimension, and quite different shapes do give quite different scattering curves. Furthermore, good small-angle data extend over several decades of scattering intensity; therefore a really good fit can be quite convincing. The residual ambiguity is evidenced, however, by the fact that for some ribosomal subunits, a number of slightly different models have been shown to give equally good representation of the data. The analyses by Tolbert (1971) of the 50S subunit scattering, as listed in Table 3, give an indication of the degree of ambiguity involved. While a number of shapes which had been proposed were clearly inconsistent with Tolbert's data, a number of somewhat different structures could not be rejected. Note, however, that these models all correspond to roughly equal volumes.

With this aside, we can return to the question of the shape and dimensions of the 30S subunits, attempting to utilize both EM and X-ray results, and steer a course between the Scylla of shrinkage and the Charybdis of ambiguity.

Shape and Dimensions of the 30S Subunit.   Small-angle X-ray scattering by two workers (Table 3) yields in each case an oblate ellipsoid of about 55 Å × 220 Å × 220 Å as a best model. The fit of the curve calculated for the ellipsoid to the experimental curve is exceptionally good in this case, giving nearly exact agreement over a factor of almost three decades in scattered intensity (see Figure 2). Furthermore, direct analysis of the small-angle data yields values for the largest dimension (220 Å) and thickness (50 Å) in good agreement with the model (Smith 1971). Thus there is strong reason to place confidence in these results.

Unfortunately, most of the early electron microscope studies on the 30S subunit have used air-dried preparations and do not agree well with one another nor with the X-ray results (see Table 3). The only feature these early EM studies have in common is the particle volume, which comes out in every case to be close to the *anhydrous* volume calculated from M and $\bar{v}$, about 0.90×10^6 Å^3. This is much smaller than the volume obtained from the X-ray data and is inconsistent with hydrodynamic results. It is clear that particle shrinkage and perhaps deformation have occurred. More recent studies of 30S ribosomes, using either positive staining (Amelunxen 1971) or negative staining (Wabl et al. 1973a,b) show somewhat different particles. Amelunxen observed two different kinds of objects: prolate ellipsoids similar to those reported by Spirin et al. (1963) and elongated particles (see Table 3). The fraction of each present in a preparation depended upon the way in which the ribosomal subunits were deposited onto the electron microscope carbon film. Wabl et al. (1973a,b) find only the elongated particles. Although reporting only a length (220 Å), they depict a model roughly resembling a 2:1 prolate ellipsoid, which exhibits a cleft or notch on one side dividing the particle into two unequal parts. The observation of different forms is confusing, and the confusion is increased by the fact that quite different conditions have been used in preparing the ribosome suspensions. Amelunxen (1971) used a very low Mg^{++} concentration (0.25 mM); this is close to the value (0.1 mM) that has been found sufficient to cause unfolding of 31S to 23S particles (Blair and Hill, unpublished results). In the papers by Wabl and co-workers, a higher Mg^{++} concentration was used (1 mM), but the ribosome preparations were heated to 37°C to "accomplish stable conformation." Neither study was made on the active form of the subunit as defined by Spitnik-Elson et al. (1972).

Comparison of Various Models of the 30S Subunit.   It is possible to compare various models with the experimental data by computing a scattering curve for each and fitting this to that obtained from the 30S subunit. This has been done, as shown in Figure 2, for a uniform density ellipsoid of axial ratios 1:3.6:3.6, for the model proposed by Wabl et al. (1973a) and for a model adapted from that given by Cox and Bonanau (1969) for the 40S rabbit reticulocyte ribosomal subunit.

The best overall fit is provided by the curve calculated for the uniform density ellipsoid, although in the > 25-mrad region the intensity falls off faster than that observed experimentally due to a lack of internal structure in the ellipsoidal model. Nonetheless, there is excellent agreement over almost three decades of scattered intensity, giving strong evidence that the

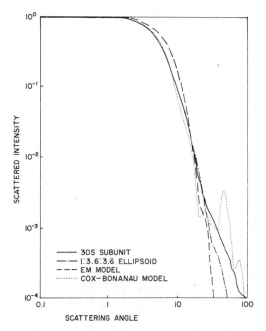

Figure 2   A comparison of the experimental scattering curve of the 30S subunit and the theoretical scattering curves of a 1:3.6:3.6 oblate ellipsoid, the Cox-Bonanau model, and a model derived from electron microscope studies. Scattering angle was measured in radians $\times$ 10^{-3}.

Table 3 Reported dimensions of *E. coli* ribosomal particles

Reference	Method[a]	Shape[b]	Dimension (Å)	$V(\text{Å}^3) \times 10^{-6}$
A. 30S subunits				[0.90][f]
Hill et al. (1969b)	X-ray	OE	55 × 220 × 220	1.39
Smith (1971)	X-ray	OE	56 × 224 × 224	1.47
Hall and Slayter (1959)	EM (AD)	PE	95 × 95 × 170	(0.80)[g]
Huxley and Zubay (1960)	EM (AD)	OE	70 × 180 × 180	1.18
Spirin (1963)	EM (AD)	TE	80 × 120 × 140	(0.70)
Wabl et al. (1973)	EM (AD)	irreg	? × ? × 220	—
Amelunxen (1971)	EM (AD)	TE	75 × 100 × 200	(0.79)
	EM (AD)	TE	100 × 130 × 145	0.99
B. 50S subunits				[1.51]
Hill et al. (1969b)	X-ray	OE	115 × 230 × 230	3.19
		TE	130 × 173 × 260	3.03
		EC	121 × 162 × 194	2.99
		OE	113 × 225 × 225	3.00
Tolbert (1971)	X-ray	irreg	(156 × 184 × 216)[c, h]	3.00
		irreg	(101 × 189 × 158)[d, h]	3.00

Reference	Method	Shape	Dimensions (Å)	Volume (× 10⁶ ų)...
Hall and Slayter (1959)	EM (AD)	OE	140 × 170 × 170	2.18
Huxley and Zubay (1960)	EM (AD)	S	150 × 150 × 150	1.77
Hart (1962)	EM (AD)	OE	130 × 225 × 225	3.44
	EM (FD)	OE	160 × 230 × 230	4.43
Spirin and Gavrilova (1969)	EM	HC	140 × 160 × 160	1.88
Spiess (1973)	EM (AD)	OE	138 × 164 × 164	1.94
	EM (CP)	OE	166 × 186 × 186	3.01

C. 70S particles

Reference	Method	Shape	Dimensions (Å)	Volume
Hill et al. (1969b)	X-ray	TE	135 × 200 × (400)[e]	[2.41]
Hall and Slayter (1959)	EM (AD)	OE	170 × 200 × 200	5.65
Huxley and Zubay (1960)	EM (AD)	TE	160 × 180 × 200	3.56
Spirin et al. (1963)	EM (AD)	PE	150 × 150 × 190	3.01 (2.23)
Vasiliev (1971)	EM (FD)	TE	170 × 230 × 250	5.1

[a] Code: X-ray = small-angle X-ray scattering; EM = electron microscope; AD = air-dried specimen; FD = freeze-dried specimen; CP = critical-point method.

[b] Code: OE = oblate ellipsoid; PE = prolate ellipsoid; TE = triaxial ellipsoid; EC = elliptical cylinder; S = sphere; HC = hemispherical cap; irreg = irregular shape.

[c] This is an irregular object with a "notch." Maximum dimensions are given. Lubin (1968).

[d] Roughly half cylinder. Bruskov and Kiselev (1968a).

[e] Probably too large, since some end-to-end dimerization is likely.

[f] Values in square brackets are anhydrous volumes.

[g] Values in parentheses are less than anhydrous value.

[h] Dimensions used by Tolbert to fit the x-ray data. The shapes were suggested by the EM studies cited, which did not give dimensions.

30S subunit in solution must have a shape approximating that of a 55–60 Å × 220 Å × 220 Å ellipsoid.

The curve calculated for the Cox-Bonanau model gives a good fit with that of the experimental curve over the first two decades of scattered intensity, although there are differences, especially in the 8–20-mrad region. In this model most of the RNA is placed on top of a two-dimensional array of proteins, which gives a structure having approximate axial ratios of 1:3:3. This difference in shape causes the calculated scattering curve to fall off somewhat faster than the experimental curve, thus accounting for the discrepancy in the 8–20-mrad region. The various peaks in the 20–100-mrad region arise largely due to the regularity in the positioning of the proteins and RNA strands within the model. It is apparent that the proposed structure is too regular, giving rather large peaks instead of the observed slight shoulders. It is also apparent that some of the maxima found (for instance that seen at 25 mrad) do not correspond to any of the experimentally observed shoulders. Nonetheless, the basic concept of a two-dimensional array of proteins seems correct (see also Kurland 1972 and this volume).

The scattering curve calculated for the model derived from the electron microscope studies of Wabl et al. (1973a) does not provide a good fit with the experimental curve. If the radius of gyration were increased from the 60 Å calculated for their model to that of the 30S subunit (72 Å), the curve for their model would fit rather well out to about 5 mrad and then it would fall below the experimental curve in increasing increments. This is a fundamental difference typical of prolate versus oblate ellipsoidal comparisons in general. Thus the EM model is quite different from the 30S subunit in solution. The difference may result in part from the different conditions used in preparing samples for electron microscope studies (for instance, heating to 37°C) and/or in part from the drying of the sample on the grids.

In any case, although the X-ray scattering results (which were all obtained at 4°C and 0.7 mM or 1 mM Mg^{++}) seem so definite and the electron microscopy results seem inconclusive, we must still raise the question as to whether more than one stable conformation of the 30S ribosome can exist. Evidence for conformational motility is indicated in the studies of Zamir et al. (1971) and Spitnik-Elson et al. (1972), who found that 30S subunits at $[Mg^{++}] < 2$ mM became "inactive"; they could not bind tRNA nor associate with 50S subunits. They could be activated, however, by heating in $[Mg^{++}] > 2$ mM. It should be noted, however, that Wabl et al. (1973a,b) used a *lower* $[Mg^{++}]$, although they heat-treated the particles. Since the $[Mg^{++}]$ requirement is strict, the state of the particles studied by Wabl et al. is uncertain. The small-angle X-ray studies were also carried out at 1 mM Mg^{++}, suggesting that these particles were also in an inactive configuration.

It should also be reemphasized, as mentioned at the outset, that various modes of preparation can alter the resultant particle preparation. Almost all of the physical studies on the 30S subunit, as well as on the 50S subunit, have been carried out on "washed" ribosomal subunits. Since it is now felt that the fully active 30S subunit contains 20 proteins (Nomura, personal communication), with the presence of S1 being questioned, and since most physical studies have been made on subunits which may contain somewhat fewer proteins, the structure of the fully active 30S subunit could be somewhat different than any of those studied.

Studies on Altered 30S Subunits.   In an effort to discern the fine structure of the 30S subunit, studies have been made on altered forms of the subunits. One method is to cause the subunit to unfold in some fashion. This has been done in various ways, including magnesium ion depletion (Cammack and Wade 1965; Gavrilova et al. 1966; Gesteland 1966; Ghysen et al. 1970; Miall and Walker 1969; Weller et al. 1968), thermal denaturation (Bodley 1969; Cox et al. 1973; Tal 1969a,b), urea denaturation (Roberts and Walker 1970), and the presence of phosphate ions together with oligonucleotides (Natori et al. 1968). The unfolding was monitored either by sedimentation velocity measurements, viscosity measurements, or by means of hyperchromic effects. Using the last three methods of unfolding, there is reason to believe that much, if not all, of the secondary structure of the RNA is disrupted. However in the case of magnesium ion depletion, it has been shown that the unfolding does not initially disrupt the secondary structure of the RNA (Eilam and Elson 1971), but rather, appears to be merely an extension of the ribonucleoprotein strand in some fashion.

From sedimentation velocity studies of the 30S subunit ($s°_{20,w} = 31S$), it has been shown that the unfolding probably occurs as follows:

$$31S \longrightarrow (27S) \longrightarrow 23S$$

If the subunits are treated with EDTA in the absence of other salts, other forms with lower sedimentation coefficients appear (Ghysen et al. 1970), but it is likely that in most of these there is extensive disruption of the secondary structure of the RNA (Eilam and Elson 1971).

Although the 27S particle mentioned above is not always seen, it can be observed in time by monitoring the sample as dialysis against 0.01 M EDTA continues. As the magnesium concentration is reduced, the 31S peak first shows a slight shoulder and later a more pronounced 27S peak, which is never fully separated from the 31S peak. This shoulder could well be due to a slow or Mg^{++}-dependent conformational equilibrium between the 23S and 31S particles and thus might not represent a separate entity. Concurrently the 23S peak appears and increases until all of the 31S sample is transformed to the 23S form. This peak has been assigned sedimentation coefficients that vary from 14S to 23S, depending upon the conditions used for the study. In a buffer containing 0.1 mM Mg^{++}, 0.07 M KCl, 0.01 M Tris pH 7.4, we have found it to have a sedimentation coefficient of $s^{\circ}_{20,w} = (23.2 - 1.2c)$ where c is measured in mg/ml Blair and Hill, unpublished results). The variation of the sedimentation coefficients reported is probably due to ionic conditions of the samples (Ghysen et al. 1970).

The fact that this is an unfolded particle has been concluded from results of viscometry and small-angle X-ray scattering. We have found (Blair and Hill, unpublished results) that the intrinsic viscosity was 11.0 ml/g, rather than the 0.1 ml/g observed for the 30S subunit. This indicates that the unfolding is limited, since extension to an extended nucleoprotein strand should yield a much larger value. Small-angle X-ray scattering by Smith (1971) has shown this 23S particle (his 16S particle) to have a radius of gyration of 123 Å and a maximum dimension of about 415 Å. This is considerably larger than the 72 Å radius of gyration and 215 Å maximum dimension reported for the 30S subunit. It is interesting to note that the scattering curve from the unfolded particle is best matched by that from a parallelpiped of dimensions 50 Å × 190 Å × 380 Å, which would indicate that the particle unfolds primarily in one dimension (see Table 4).

Another method by which the 30S subunit has been physically altered is by removing various proteins or groups of proteins (Traub and Nomura 1968; Spitnik-Elson and Atsmon 1969; Atsmon et al. 1969; Itoh et al. 1968). However, the physical studies on these particles have been few, other than routine sucrose gradient or sedimentation velocity studies. Smith (1971) has made a small-angle X-ray scattering study of the protein-deficient 23S *core* particles (these are not to be confused with the 23S *unfolded* particles). In this study, the radius of gyration was found to be 89 Å, somewhat larger than the 72 Å of the intact subunit. This would indicate some unfolding, again mainly in one dimension since the maximum dimension was about 295 Å and the scattering curve was best approximated

Table 4   Comparative physical data on *E. coli* ribosomal subunits and unfolded particles

$s^{\circ}_{20,w}$(S)	R_G(Å)	Long dim. (Å)	Shape[a]	Dimensions (Å)	Reference
50.1	75	240	TE	130 × 170 × 260	Tolbert (1971)
40.3	117	370	EC	125 × 185 × 370	Hill et al. (1970)
31.2	72	215	OE	55 × 220 × 220	Smith (1971)
23.3	123	415	pp	48 × 191 × 381	Smith (1971)

[a] Code: TE = triaxial ellipsoid; EC = elliptical cylinder; OE = oblate ellipsoid; pp = parallelpiped.

by that of an ellipsoid of dimensions 45 Å × 180 Å × 350 Å. These results indicate that the proteins do contribute to the stability of the subunit morphology.

The putative rate-limiting step in the assembly of the 30S subunit—the RI to RI* transformation—also appears to be a conformation change (Traub and Nomura 1969; Held and Nomura 1973). In this case, about 15 proteins are attached to the 16S rRNA, after which the resulting RI particle is heated to transform it to the RI* conformation. That this involves a conformation change has been indicated by the sedimentation velocity experiments of Held and Nomura (1973).

From these studies on altered ribosomal subunits, some features of the fine structure of the 30S subunit are evident. First, a loosening or unfolding can occur, in which process the total protein complement and the secondary structure of the RNA remains essentially intact (Eilam and Elson 1971) but the particle unfolds mainly in one direction. Second, as proteins are removed from the subunit, the structure is loosened (Smith 1971), suggesting that the proteins themselves stabilize the structure significantly.

Hydration of the 30S Subunit. That the structure of the intact 30S subunit itself must be quite "loose" in solution is suggested by the calculations of Smith (1971) (see Table 5) which indicate a very high degree of hydration. These X-ray studies imply about 0.4 g/g of internal water; another 0.9 g/g of additional hydration is indicated by the hydrodynamic results. It is noteworthy that this large amount is required to explain the sedimentation coefficient and intrinsic viscosity even though a very asymmetric model has been used. Assumption of a more nearly spherical structure would necessitate even greater hydration.

Table 5   Estimates of hydration of *E. coli* ribosomal subunits

Particle	V_a	V_x	V_i	h_i	V_s	h_s	V_n	h_t	h_e
30S	0.90	1.47	0.57	0.39	1.37	0.34	2.79	1.27	0.88
(Smith 1971)							2.96	1.39	1.00
50S	1.51	3.00	1.49	0.58	2.55	0.40	5.00	1.38	0.80
(Tolbert 1971)							5.00	1.35	0.77

Quantities indicated by V's are volumes per particle, all in units of 10^6 Å^3. Code: V_a = anhydrous volume from M and $\bar{v}$; V_x = volume of particle observed in low-angle and x-ray scattering; $V_i = V_x - V_a$ = internal hydrated volume; V_s = total volume per particle excluded to sucrose; V_n = hydrodynamic volume. In each case the upper figure is from S and M, the lower from [η] and M, using the shape as determined by low-angle scattering.

The quantities denoted by h's are hydration values, in g water/g dry ribosomes. Cole: h_i = internal hydration, from V_i; h_s = internal hydration that excludes sucrose; h_t = total hydration, from V_n; $h_e = h_t - h_i$ = external hydration.

The 50S Subunit

Hydrodynamic Studies. As in the case of the 30S subunit, molecular weight studies indicate that the process of NH_4Cl washing or $(NH_4)_2SO_4$ precipitation removes an appreciable amount of protein. The average of results (by several different methods) for the molecular weight of un- washed particles is about 1.7×10^6, whereas sedimentation equilibrium studies on 50S particles prepared from washed or precipitated 70S ribo- somes give reproducibly a value of 1.55×10^6 (see Figure 1, Table 1). As in the case of the 30S subunit, the loss of mass does not lead to a decrease in the sedimentation coefficient (see Table 2). It would appear that changes in $\bar{v}$ and the frictional coefficient accompanying the protein loss compen- sates for the mass loss, so that the sedimentation coefficients are approxi- mately the same for both particles. The difference in $\bar{v}$ values reported by Tissières et al. (1959) and Hill et al. (1969a) is somewhat surprising, for it is considerably larger than would be expected on the basis of the com- positional change alone.

The problem of obtaining accurate values of $\bar{v}$ continually plagues these studies. It is necessary to make very accurate measurements of density and concentration on substances which are difficult to obtain and purify in large quantities. Furthermore, since ribosomes appear to exhibit some poly- electrolyte properties and interact strongly with solvent, the calculation of $\bar{v}$ from composition data is sure to be unreliable. Similarly, use of buoyant density measurements to estimate $\bar{v}$ for such complex, strongly interacting particles is fraught with hazard.

Values for the molecular weight of the 23S rRNA have been reported as 1.0×10^6 (Midgley 1965, end-group analysis), 1.07×10^6 (Stanley and Bock 1965, sedimentation equilibrium), and 1.1×10^6 (Kurland 1960, light scattering, sedimentation and viscosity). Again it is somewhat difficult to compare these values, since those obtained by physical techniques presumably correspond to a salt of the polymer, whereas Midgley's end- group analysis has been calculated to a molecular weight on the basis of the average nucleotide residue weight. Therefore the differences may not be so great as they appear to be. In any event, it would seem that the 23S rRNA weight should be about 1.05×10^6 daltons. Adding to this the weight of the 5S rRNA (0.04×10^6), we have a value of about 1.1×10^6 daltons of rRNA. This corresponds to about 65% RNA in the unwashed particles, 71% in the washed particles. As in the case of the 30S subunits, these values are in reasonable agreement with the analytical results; they indicate that the unwashed particles contain about 600,000 daltons of protein, the washed particles about 450,000 daltons. The former value is somewhat greater, the latter slightly less than the sum of molecular weights of the known 50S proteins (about 520,000 daltons: Traut et al. 1969; Garrett and Wittmann 1973).

The value of the intrinsic viscosity of the 50S ribosomal subunit has been reproduced in a number of laboratories (see Table 2) and is consistently found to be much smaller than that of the 30S subunit. This indicates that the 50S particle has either a lower net hydration or is a more symmetrical structure. That the latter is probably the case is indicated by the radius of gyration measurements (see Table 2). Even excluding the R_G value obtained by Scafati et al. (1971), which seems unreasonably low, the average of the other values is little larger than that found for the 30S subunit.

Size and Shape of the 50S Subunit.   The general shape of the 50S subunit is indicated by data in Table 3. While a number of different models have been found to fit the small-angle X-ray data, all are roughly similar and share the common feature that the particle is not highly asymmetric. There is gratifying agreement in this case between the X-ray scattering data and electron microscopy, at least when samples are prepared for the latter technique by the freeze-drying method. The results of Hart's (1962) study are particularly revealing, for with freeze-dried specimens he observes dimensions close to some proposed by the X-ray scattering workers, although the particle proposed by Hart is thicker and the volume larger. Hart finds that air-drying reduces ribosomal volumes by a factor of about 1.3. Similar results have recently demonstrated that critical-point drying of 50S subunits yields an apparent particle volume about 50% greater than air-drying.

A number of more recent electron microscope studies of 50S subunits have indicated that the particles are of somewhat uneven topography; frequently a notch or crevice is observed (Lubin 1968; Brushov and Kiselev 1968a). While there exists the danger that some such features may be artifacts of sample preparation, it is of interest that Tolbert (1971) found that his small-angle X-ray scattering curve could be fitted quite well by a scattering curve calculated for a model of the shape proposed by Lubin. That all is still not simple is illustrated by the fact that an equally good fit was provided by a half-cylinder model like that proposed by Bruskov and Kiselev, but only if the deep crevice were removed! Furthermore, a simple oblate ellipsoid provided the best fit of all.

An analysis of the hydration of the 50S subunit, as deduced by Tolbert (1971) from low-angle X-ray scattering and hydrodynamic studies, is given in Table 5. The hydration, both internal and external, seems to be somewhat larger than that of the 30S particle.

Unfolded 50S Subunits.   The earliest observation of the loosening or unfolding of a ribosomal subunit was made by Elson (1961) when he observed that upon placing the 70S ribosome into buffers containing increasing amounts of NaCl, the sedimentation coefficient of the 50S subunit decreased with a concomitant release of a slow moving component. He concluded that the decrease in the sedimentation coefficient was due to the release of RNA (now known to be 5S RNA). This observation has more recently been substantiated by others (Gormly et al. 1971; Siddiqui and Hosokawa 1969). However the release of the 5S RNA alone is not sufficient to account for the entire decrease in the sedimentation coefficient. Calculations show that the release of a mass equal to that of the 5S RNA would only account for a decrease of about 2 Svedbergs, not the 5–10 Svedbergs generally observed. Thus there must also be other material lost, a loosening of the particle itself, or a dramatic change in the partial specific volume.

Further studies by many workers (Gesteland 1966; Gavrilova et al. 1966; Weller et al. 1968; Ghysen et al. 1970) showed that the 50S subunit does unfold as the magnesium concentration decreases in rather discrete steps, approximately as follows:

$$50S \longrightarrow 40S \longrightarrow 30S \longrightarrow 20S$$

This is illustrated in the Schlieren pattern shown in Figure 3.

Viscosity studies, which generally have been carried out on mixtures of the various unfolded stages, confirmed that the subunits are definitely being unfolded (Gesteland 1966; Weller et al. 1968; Tal 1969a,b). However the isolation of homogeneous unfolded particles is somewhat difficult and thus far has only been accomplished in the presence of 0.5 M NH_4Cl at low concentrations of magnesium ($\leq$ 0.1 mM).

Not all changes in sedimentation coefficients reflect conformational changes. For instance, upon placing the 50S subunit in 0.5 M NH_4Cl and 0.5−1 mM magnesium, the sedimentation coefficient is immediately decreased to 45S. At first this was thought to be due to unfolding, but small-angle X-ray scattering studies (Hill et al. 1970) showed the 45S particle to be almost identical—perhaps slightly smaller—to the 50S subunit. The molecular weights were also identical, and it was ultimately found that the

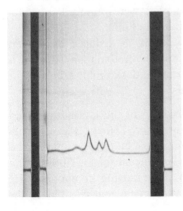

Figure 3 Schlieren pattern showing unfolded 50S particles. Sample of 50S ribosomal subunits was dialyzed 8 hours against 0.07 M KCl, 0.01 M Tris pH 7.4. Sedimentation is from left to right. Approximate sedimentation coefficients from left to right are 20S, 30S, 40S and 50S.

entire decrease in the sedimentation coefficient from 50S to 45S was due to an increase in $\bar{v}$ from 0.59 to 0.64 ml/g (Hill et al. 1970). This conclusion has recently been supported by an electron microscopic study on loosened *B. subtilis* ribosomes by Nanninga (1970), in which he observed no shape change between the 50S and his 43S particle (isolated in 0.5 M NH₄Cl). Thus a decrease in sedimentation coefficient alone is not sufficient to characterize unfolding and/or a loss of molecular weight.

The first true stage of unfolding—the 40S particle—has been isolated and partially characterized by Hill et al. (1970). The sedimentation coefficient was found to be 39S, in the presence of the released 5S RNA (Hill et al. 1970). The intrinsic viscosity is 8.7 ml/g and the partial specific volume 0.639 ml/g. Small-angle X-ray scattering results give a radius of gyration of 117 Å as compared with the 75-Å radius of gyration of the 50S subunit. The scattering curve was best fit by one calculated for an elliptical cylinder of dimensions 125 Å × 185 Å × 370 Å (see Table 4). This suggests that the unfolding of the 50S subunit is essentially in one dimension, as is the case with the 30S subunit.

There have been some electron microscopic studies made on various unfolded forms of the 50S subunit. Spirin et al. (1963) made some early observations on the "23S" particle, which appeared to be in the form of a "beaded chain." More recent studies by Nanninga (1970) also show quite extended particles. Similar results were obtained by Matsuura et al. (1970) from precursor particles. These studies are not quantitative, but rather serve to show that unfolding does occur to various degrees.

Protein-deficient 50S Subunits. While several groups have obtained protein-deficient 50S subunits (Maruta et al. 1971; Atsmon et al. 1969; Ghysen et al. 1970), to our knowledge there have been no physical studies in this area, except for an electron microscopic study by Spiess and Amelunxen (1972) of the α, β, γ core particles (50S subunits containing decreasing amounts of protein). They found the α and β particles to be rather globular in shape, but the γ particle appeared to be an extended RNP strand of about 30 Å diameter. This suggests, as in the case of the 30S subunit, that the proteins themselves are responsible for some of the conformation of the 50S subunit.

Internal Structure of the 50S Subunit. There is one final question which is often asked. Does the 50S subunit have a dense RNA "core" or perhaps a hole? Serdyuk et al. (1970) suggested the presence of a dense RNA "core" after studying radius of gyration and diffusion measurements. However no attempt was made to account for the effects of hydration as they compared the two dissimilar measurements. Venable et al. (1970) have also indicated that a nonuniform density may well account for some of the scattering maxima they observe, but they have not yet suggested models to fit their data. However Tolbert (1971) reported that X-ray scattering studies on the 50S subunit in various sucrose concentrations showed that there was no evidence for a more dense internal "core." Recently Moore et al. (this volume) have shown through neutron scattering that the protein and RNA radii of gyration are apparently displaced in the 50S subunit. These results, coupled with the lack of an exceptionally good fit of the 50S X-ray scattering curve with those of rather standard geometrical structures, tend to lend credence to the view that the 50S subunit may have a nonuniform structure of some nature.

The 70S Particle

As mentioned above, the definition of the 70S ribosomal particles studied by various workers is somewhat ambiguous, and it is not surprising that results with these are not as consistent as those found for the 30S and 50S subunits. It may be expected that 70S particles prepared under different conditions will contain varying amounts of mRNA, tRNA, protein co-factors, etc., and thus that their weight may vary from one preparation to another and will generally exceed the sum of purified 50S and 30S subunits. This is reflected in the results shown in Table 1. Another complication has been clear in the sedimentation equilibrium experiments of Hill et al. (1969a, 1970). These studies reveal a strong tendency for the 70S particles to associate at magnesium concentrations which stabilize the 70S ribosomes. This can be seen in Figure 1 in the dependence of M_w on concentration shown by the 70S particles. While the tendency to associate appears to be decreased by NH_4Cl washing (Hill et al. 1970), it is still appreciable and can cause serious complications at the concentration used in X-ray scattering.

The evidence concerning the conformation and dimensions of the 70S particle in solution is also much less definite than the corresponding data for subunits. For example, small-angle X-ray scattering (Hill et al. 1969b) and light scattering (Scafati et al. 1971) yield very different values for R_G (see Table 2). It should be pointed out, however, that the latter data show wide variation, with values of R_G obtained ranging from 64 Å to 110 Å. The model proposed by Hill et al. (1969b) (see Table 3) exhibits a long dimension considerably larger than any suggested by electron microscopy. It is quite likely that this dimension is somewhat high because of the presence of some aggregated particles, since such are now known to occur under the conditions of these X-ray studies.

Most of the early electron microscope studies of 70S particles, which employed air-drying, may be criticized on the same grounds as were discussed in the preceding sections; the volumes obtained are close to the anhydrous volume and cannot reflect the state of the hydrated particles in solution. Vasiliev (1971) has applied the freeze-drying technique to the study of 70S particles. The dimensions obtained (See Table 3) are somewhat larger than those obtained from electron microscopy of air-dried samples, as might be expected. The 30S subunit is observed as a thin, curved cap on the more symmetrical 50S subunit. In some images this cap has been released at one end and projects from the particle. We shall consider this point later.

It is of interest to attempt to compare our best estimates of 70S dimensions, both from X-ray scattering and the electron microscope studies of Vasiliev, with the best estimates of subunit dimensions, combined so as to yield a reasonable model. We shall make the following assumptions concerning the subunits:

1. The maximum dimension of the 30S subunit is about 220 Å. This is indicated by the small-angle X-ray scattering data and the more recent electron micrographs and corresponds roughly to the length of the "cap" as seen on the 70S particle in Vasiliev's micrographs. The other dimensions are less certain for two reasons: First, as we have described above, there appears to be evidence for multiple conformations of the 30S subunit in solution. Second, there exists the possibility that the 30S subunit is deformed upon interaction with the 50S subunit (Huang and Cantor 1972). Electron micrographs of the 70S particles give little information about this point, since most ribosomes seem to bind to the grid in one preferential orientation and thus yield but one view of the attached 30S subunit.

2. The 50S subunit is approximated by the Lubin model, roughly a prolate ellipsoid with a shallow notch on one edge. An oblate ellipsoid with dimensions around 150 Å × 200 Å × 200 Å would give approximately the same picture. Both Hart's (1962) and Spiess' (1973) electron microscope results and many of the models used to fit the X-ray data are roughly comparable to this.

3. We assume that the 30S subunit fits over the end of the 50S subunit on which the "notch" occurs. The "notch," together with deformation of the 30S particle, allows contact of the subunits in two regions, with an open "tunnel" between. The significance of these features will be discussed later.

A somewhat romanticized view of the model is presented in Figure 4. We have arbitrarily assumed that the 30S subunit, at least in its active form, takes a shape similar to that suggested by the electron microscope studies of Wabl et al. (1973a). This is also similar to the form suggested for the small subunit of eukaryotic ribosomes by the work of Nonomura et al. (1971). Somewhat different shapes for the 30S subunit would not materially affect the model.

Such a model agrees quite well with the dimensions given by Vasiliev (1971), but the long axis will be somewhat smaller than the 400 Å proposed by Hill et al. (1969b). While we feel that 400 Å is probably too large (because of partial dimerization), it seems possible that in solution the "tunnel" is more opened, and the particle longer, than when dried on electron microscope grids. In many respects, this is very similar to the

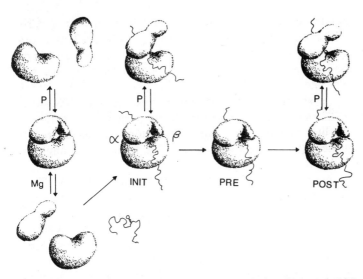

Figure 4 Somewhat speculative pictures of the 50S and 30S subunits and their association to form a 70S ribosome. On the left is shown the formation of 50S and 30S "couples" upon the addition of Mg^{++}. These couples can be dissociated by high pressures, as exist in the ultracentrifuge cells at high rotor velocity. The structure labeled "INIT" is the initiation complex, involving the mRNA, N-formyl-aminoacyl-tRNA, and other appropriate factors. Two sites of interaction between the 30S and 50S subunits are indicated by α and β. Interactions at the β site are presumed to involve most of the volume change and thus are released at high pressure (*top*). The α site is stabilized by mRNA and/or tRNA. The pretranslocation complex ("PRE") is stable against such pressure effects, but the post-translocation complex ("POST") lacks this additional stabilization at the β site.

model suggested a number of years ago by Hill et al. (1969b). In general the model seems to satisfy the available data fairly well. It has two additional features that will be of significance in our further discussions:

1. The tunnel between 50S and 30S surfaces could accommodate mRNA, the tRNAs, and protein synthesis factors. There are three kinds of evidence to support this idea: First, such entrapment of mRNA is strongly suggested in electron microscope studies of eukaryotic polysomes (Nonomura et al. 1971). Second, Castles and Singer (1969) have noted that poly (U) is protected from ribonuclease attack when associated with *E. coli* 70S ribosomes. The length of the protected segments varied from 70–120 nucleotides, depending upon the initial size of the poly (U). Third, it has been shown by Malkin and Rich (1967) that nascent polypeptide is protected against proteolysis while on the ribosome.

2. The model indicates *two* separated binding sites for the 30S subunit on the 50S subunit. Release at one of these sites would give a partially open structure of a kind seen by Vasiliev (1971).

It must be emphasized that this model is still probably oversimplified, both in terms of the detailed shapes of the subunits (which are still inaccessible) and in the possible changes of conformation that may occur on binding.

Association Equilibria

In this section we shall discuss briefly some recent results on the reversible Mg^{++}-mediated association of 30S and 50S subunits to form 70S particles (the so-called "30–50 couples"). We shall *not* describe the much more complex process that accompanies the functional association of subunits in the initiation of protein synthesis; that will be discussed in detail in other chapters. Thus the process we describe here is to an extent artificial; nevertheless, it provides important information concerning the interaction of the subunits.

While there have been many attempts to study this process, most have been complicated either by artifacts of preparation of the ribosomal subunits or by the effects of hydrostatic pressure in the sedimentation methods that have been most commonly used. These pressure effects will be considered in more detail below. It will suffice to say here that there is now abundant evidence that the association equilibrium can be strongly perturbed by the hydrostatic pressures existing in ultracentrifuge cells during high-speed sedimentation.

For this reason, the most revealing study to date is that of Zitomer and Flaks (1972), who used the light scattering method. In brief, they find that the reaction

$$30S + 50S + nMg^{++} \rightleftharpoons (70S) \cdot Mg_n^{++}$$

has the following characteristics:

(1) It is fully reversible over a wide range of conditions.
(2) It behaves as if $n \cong 8$, with some evidence that other ions can compete with Mg^{++}.
(3) It yields the thermodynamic parameters $\Delta H^\circ = -70$ kcal/mole, $\Delta S^\circ = -120$ cal/$^\circ$mole.

When the reaction is studied as a function of $[Mg^{++}]$, the value of $[Mg^{++}]$ required for half-association varies with temperature and the concentration of other ions. However, in no case is appreciable association observed below $[Mg^{++}] = 2$ mM, except when spermidine is also present. This correlates well with the observation of Zamir et al. (1971), that the 30S subunit is inactive in association at $[Mg^{++}] < 2$ mM.

The signs and magnitudes of ΔH° and ΔS° are of interest. Anywhere in the vicinity of room temperature the large negative enthalpy term dominates the entropy term; consequently, we must say that the association of ribosomal subunits is *energetically,* rather than *entropically,* favored. The

$\Delta S°$ term is large, however, and much larger than one would expect simply from the losses in translational and rotational entropy. Thus both terms suggest that there may be significant changes in conformation and/or hydration accompanying association. Independent evidence for hydration changes comes from the work of White et al. (1972). These workers used nuclear magnetic resonance (NMR) to measure "site-bound" water of hydration, water presumably bound to charged groups on the surface. A decrease of 15–20% in such water was observed upon association of subunits.

Whatever the conformational change may be, however, it apparently does not involve drastic changes in the RNA secondary structure, for circular dichroism measurements (Miall and Walker 1968) show that there is only a small difference in this respect between subunits and 70S ribosomes. Furthermore, recent studies by Kabasheva et al. (1971) indicate that such changes as occur may be due to the effects of divalent ions, rather than to association per se.

More recently, the equilibrium studies of Zitomer and Flaks (1972) have been continued and extended by kinetic investigation of the association and dissociation reaction (Wolfe et al. 1973; Grunberg-Manago, personal communication).

There is evidently a substantial volume increase accompanying the Mg^{++}-dependent association of subunits. The partial specific volume values obtained by 30S, 50S, and 70S preparations by Hill et al. (1969a) would suggest an increase of the order of 10^4 ml/mole. While this is probably too large, it would represent only a small fractional increase in the 1.5×10^6 ml molar volume of the 70S particle. However such estimates (which depend on small difference in $\bar{v}$) are hazardous. Much more direct evidence comes from studies in which the association equilibrium is shown to be influenced by hydrostatic pressure. If there is a molar volume change (ΔV) in a reaction, the equilibrium constant for the reaction will be a function of pressure (P);

$$K(P) = K(P_o)e^{-\Delta V (P-P_o)/RT}, \tag{1}$$

where P_o is a reference pressure, usually taken as 1 atm. If ΔV is large, pressures that exist within an ultracentrifuge cell at high rotor speeds can cause enormous changes in $K(P)$ and, consequently, in the composition of the equilibrating mixture at various points in the ultracentrifuge cell. In accordance with Le Chatlier's principle, Equation (1) shows that if there is a volume increase in the reaction ($\Delta V > 0$), the effect of increased pressure will be to drive the reaction back toward reactants. This is exactly what has been observed by a number of workers for the Mg^{++}-dependent ribosome subunit association (Spirin 1971; Infante and Baierlein 1971; Hauge 1971).

Infante and Baierlein have carried out an especially careful study of the effect of pressure on sucrose gradient centrifugation of sea urchin ribo-

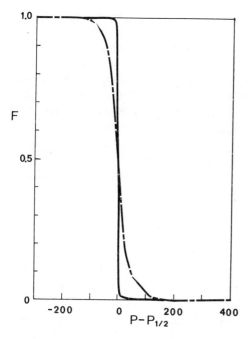

Figure 5 Effect of pressure on a conformational equilibrium. It is assumed that the equilibrium constant at 1 atm is very large and that ΔV is either 10^3 ml/mole (———) or 10^4 ml/mole (——— — ———). In either case, the conformational change will be observed over a relatively narrow pressure range around $P_{1/2}$, the pressure of half-change. Pressure is in atmospheres.

somes. They deduce a value of $\Delta V = 500$ ml/mole. This is more likely a reasonable estimate for the order of magnitude of ΔV than that given above.

These pressure effects give rise to sedimentation patterns which are liable to serious misinterpretation. Two factors are principally responsible:

1. If K(1 atm) is very large and ΔV is large, the observed pressure effects will occur over a relatively narrow pressure range. Figure 5 shows calculation of the effect in an isomerization reaction for $\Delta V = 10^3$ and 10^4 ml/mole. Note that if the experiment is conducted under conditions where $P_{max} < P_{1/2}$, very little effect will be seen; thus in a sucrose gradient experiment at low rotor speeds or in the early part of an experiment at high speed, the effect will not be observed. If the sedimenting zone passes the point where $P = P_{1/2}$, the equilibrium will rather sharply shift. An association-dissociation equilibrium will behave qualitatively in much the same fashion, with the very important difference that the equilibrium composition will also depend upon the local concentration. This concentration dependence can be used to distinguish between conformational and association equilibria (Infante and Baierlein 1971; Hauge 1971).

2. If the equilibrium is shifted, but forward and reverse reactions are fast enough to maintain local equilibrium, the result may not be the appearance of peaks corresponding to dissociation products, but rather a decrease in the apparent sedimentation coefficient of the associated component. This can be (and has been) misinterpreted as indicating a *conformational* change (see Infante and Krauss 1971 for a more detailed discussion). Again the distinction can be made on the basis of the fact that a conformational equilibrium is concentration independent. Thus Spirin (1971) has shown that the so-called "60S ribosomes" observed in sedimentation velocity experiments with uncharged 70S couples are artifactual; the boundary is actually a reaction boundary involving 30S, 50S, and 70S species.

Does the 70S Ribosome "Pulsate"?

The model of ribosome function proposed by Spirin (1969) requires a major change in the spatial relationship of the 30S and 50S subunits accompanying the steps in protein synthesis. A few years ago evidence emerged to indicate that there was in fact some such conformational change, based largely on work by Noll and his associates (Schreier and Noll 1971) and Chuang and Simpson (1971). These workers, utilizing polypeptide-synthesizing 70S particles blocked in the initiation, pre-translocation, and post-translocation steps, observed differences in sedimentation coefficients between such structures. While the numerical values for the changes in sedimentation coefficient varied, both groups agreed that the pre-translocation complex exhibited a larger sedimentation coefficient than either the post-translocation complex or the initiation complex. Since the difference could not be explained on the basis of expected gains or losses of material, the concept of the "pulsating ribosome" developed. It was suggested that upon formation of the pre-translocation complex a tightening of the structure occurred, which was then released following translocation.

Such behavior makes sense with respect to both models of the synthesis process (Spirin 1969) and the structural picture that is beginning to emerge. However a serious complication has appeared. As described above, a number of workers have pointed out that 70S ribosomes can be pressure-dependent in their stability and might tend to dissociate at the high pressure existing in the bottom of the ultracentrifuge cell. If the 70S ribosomes could be in equilibrium with subunits, and if this equilibrium were sensitive to the state of the ribosome cycle, the observations of a more slowly sedimenting boundary might correspond to small differences in degree of dissociation rather than a conformational change. In other words, the 60S peak observed by Schreier and Noll, for example, might result simply from displacement of an association equilibrium. It must be pointed out, however, that 70S ribosomes complexed to mRNA do *not* appear to dissociate in response to pressure effects (Infante and Baierlein 1971).

Some very careful recent studies by Waterson et al. (1972) have somewhat clarified the problem. They demonstrate that a pressure effect is clearly involved, for while the shift in sedimentation coefficient is observed in experiments at high rotor speed, the difference disappears at low rotor speed. On the other hand, it does *not* appear to involve association-dissociation equilibrium, for the effect is insensitive to changes in ribosome concentration. These workers also point out that there clearly is pressure inhibition of protein synthesis in vivo (Landau 1967) and in vitro (Arnold and Albright 1971).

We believe that the results may be reconciled by the following model: Suppose that the 30S subunit has two regions of attachment to the 50S subunit (α and β) and that the tunnel lies between these (see Figure 4). If most of the volume change which occurs on association is a consequence of binding at the β site, and if the α site is stabilized by mRNA and/or tRNA, we have the following consequences:

1. In the absence of mRNA and tRNA, high pressure promotes dissociation, since the β site is released and binding at the α site is weak. Hence 30–50 couples should exhibit the observed behavior.

2. In the presence of mRNA and tRNA, high pressure favors the release at the β site alone, since the α site is stabilized. This must *not* be able to occur when the ribosome is in the pre-translocation conformation, implying an especially tight conformation at this step. This unfolding will lead to (a) a lower sedimentation coefficient value for the initiation and post-translocation complexes at high pressure, with little or no effect at low pressure, and (b) blocking of protein synthesis at high pressure. It should be noted that Landau (1967) and Arnold and Albright (1971) showed that synthesis resumes *immediately* upon pressure release, suggesting that no ribosome reattachment or reinitiation was required. Furthermore, the pressure range in which protein synthesis is inhibited (ca. 500 atm) corresponds well to that in which pressure effects are noted in the ultracentrifuge.

It is a corollary of the model that there must be conformational changes *at the β site* during the steps of protein synthesis, if the results of Waterson et al. are to be explained. We would also suggest that the electron micrographs of Vasiliev (1971), some of which reveal the 30S cap half-released, are in fact illustrations of this unfolded conformation.

EUKARYOTIC RIBOSOMES

While eukaryotic ribosomes have not been studied in as great depth as those from *E. coli,* there have been several rather broad physical studies on such ribosomes from a variety of sources. The technical problems of mem-

Table 6 Physical data on eukaryotic ribosomes

Source	Sed. coef.(S)	$\bar{v} \frac{ml}{g}$	$[\eta] \frac{ml}{g}$	$D \frac{cm^2}{sec}$	$M[\times 10^{-6}]$	% Protein	$\rho \frac{g}{ml}$	References
Yeast	80	0.67	—	—	4.1(c,$\bar{v}$)[a]	58	—	Chao (1957)
	60	0.67	—	—	2.2(c,$\bar{v}$)	58	—	Chao (1957)
	40	0.67	—	—	1.48(c,$\bar{v}$)	58	—	Chao (1957)
	82.0	0.630	—	—	3.60(SE)	47	—	Mazelis & Petermann (1973)
	60.7	0.635	—	—	2.57(SE)	50	—	Mazelis & Petermann (1973)
	37.8	0.620	—	—	1.01(SE)	42.5	—	Mazelis & Petermann (1973)
Yeast (S. pombe)	83	0.65	4.5	—	3.8(S,$[\eta]$)	59	—	Lederberg & Mitchison (1962)
Pea seedlings	80	—	11–12	—	4–4.5(S,$[\eta]$)	60–65	—	Ts'o et al. (1958)
Dry pea seeds	79	—	—	—	—	50	1.55	Gumilevskaya et al. (1971)
Rat liver (urea-treated)	59.1	0.623	—	—	3.0(SE)	43	1.614	Hamilton et al. (1971)
	40.9	0.654	—	—	1.5(SE)	55	1.551	Hamilton et al. (1971)
Rat liver (EDTA-treated)	49.9	0.623	—	—	3.1(SE)	44.7	1.602	Hamilton & Ruth (1969)
	28.6	0.633	—	—	1.2(SE)	49.5	1.593	Hamilton & Ruth (1969)

Paramecium	82.6	0.61	—	1.52×10^{-7}	3.33(S,D)	—	—	Reisner et al. (1968)
Guinea pig	80	0.66	—	—	5(SE)	—	—	Tashiro & Yphantis (1965)
	47	0.66	—	—	3.2(SE)	—	—	Tashiro & Yphantis (1965)
	32	0.66	—	—	—	—	—	Tashiro & Yphantis (1965)
Chick embryo								
(free)	78.1	—	—	1.04×10^{-7}	5.2(S,D)	—	1.61	Vournakis & Rich (1971)
(polysomal)	85.1	—	—	1.14×10^{-7}	5.2(S,D)	—	1.60	Vournakis & Rich (1971)
Beef liver (mito)	56.3	0.674	—	—	2.83(SE)	69	1.42	Hamilton & O'Brien (1973)
Rat liver	80	0.637	—	—	4.6(c,ρ)	47	1.590	Cammarano et al. (1972)
Chick liver	80	—	—	—	4.3(c,ρ)	46	1.593	Cammarano et al. (1972)
Sea urchin	80	—	—	—	4.1(c,ρ)	49	1.581	Cammarano et al. (1972)
Plants	80	0.635	—	—	3.9(c,ρ)	49	1.581	Cammarano et al. (1972)
Rabbit retic	80	—	—	—	4.0(x)	—	—	Dibble & Dintzis (1960)

[a] (c,$\bar{v}$)—calculated from $\bar{v}$ and the diameter; (S,[η])—sedimentation and intrinsic viscosity; (SE)—sedimentation equilibrium; (S,D)—sedimentation and diffusion; (c,ρ)—calculated using RNA molecular weight and particle densities; (x)—absolute intensity from small-angle x-ray diffraction.

braneous contaminants, as well as the tenacious binding together of the subunits, make physical studies in this area much more difficult.

Hydrodynamic Studies

Before the first physical studies on *E. coli* ribosomes were published, Chao and Schachman (1956) had obtained some physical data on yeast ribosomes. Later Chao (1957) completed this study and reported data as shown in Table 6. Ts'o et al. (1958) shortly thereafter published some physical studies on the ribosomes from pea seedlings in which they reported the sedimentation coefficients, intrinsic viscosity, RNA composition, and molecular weights of the large and small subunits, as well as the ribosome as a unit. Since that time there have been a substantial number of physical studies made on ribosomes from a wide variety of sources (see Table 6).

It should be noted that there have been considerable variations in techniques and conditions used. It is therefore difficult to compare the results systematically and concisely. For instance, molecular weights have been obtained by sedimentation and diffusion studies (Reisner et al. 1968; Vournakis and Rich 1971), sedimentation and viscosity studies (Ts'o et al. 1958; Lederberg and Mitchison 1962), calculations assuming spherical molecules (Chao 1957), sedimentation equilibrium (Tashiro and Yphantis 1965; Hamilton et al. 1969, 1971, 1973), and by small-angle X-ray scattering (Dibble and Dintzis 1960). Cammarano et al. (1972) used Loening's (1968) values for the molecular weights of various RNA molecules from different sources, and the percentage of protein and RNA calculated from the buoyant density of formaldehyde-fixed ribosomes, to find the molecular weight of the protein and, therefore, that of the total particle. This method involves a large number of assumptions of dubious validity, not the least of which is that $1/\rho = \bar{v}$.

One of the serious problems in most of these studies was that of the partial specific volume, $\bar{v}$, which has been previously discussed. Often it was calculated from the protein to RNA ratio. Calculating a partial specific volume ignores any conformational or salt effects on the particle volume (Van Holde 1967). In other cases the inverse of the buoyant density has been used. As has been abundantly demonstrated in nucleic acid studies, the observed buoyant density may depend upon the nature of the salt used for the gradient and may bear little relation to the anhydrous specific volume. In some instances the densities of the particles were determined using 3-ml or 5-ml pycnometers, which are quite inaccurate. An exception was the study of Lederberg and Mitchison (1962), in which they used 50-ml pycnometers.

Therefore, while studies have been made, and in some cases with much detail (see especially Hamilton et al. 1969, 1971, 1973; Mazelis and Petermann 1973; Tashiro and Yphantis 1965), there are still some intrinsic sources of error, in addition to those of sample heterogeneity, which could prove to be quite serious. However where relative values under the same conditions were obtained, the results may be quite reliable. An example is the study of Vournakis and Rich (1971), whose results indicated a size change of approximately 20 Å between cytoplasmic and mRNA-bound chick embryo muscle ribosomes.

Small-angle X-ray scattering studies on eukaryotic ribosomes have only been attempted in two cases. Dibble and Dintzis (1960) scattered from rabbit reticulocyte ribosomes and obtained a 108 Å radius of gyration. A later effort by Bohn et al. (1967) on beef pancreas ribosomes gave a radius of gyration for these of 91 Å. Full scattering curves were not given in either case.

Electron Microscope Studies

There have been a great number of electron microscope studies on the ribosomes from many different eukaryotic cells. We have tabulated these below (Table 7). The conditions and staining methods vary considerably from study to study, so no effort will be made here to compare the various values. It could be assumed, however, that the same problems with dehydration upon drying occur here as with bacterial ribosomes. Therefore the dimensions of these eukaryotic ribosomes should be used with caution since hydrated dimensions are certain to be larger.

There are two recent studies that deserve special mention—those of Nonomura et al. (1971) and Amelunxen and Spiess (1971). Using negative staining techniques, Nonomura et al. made a careful study of ribosomes from rat liver. They concluded that the small subunit is an elongated and slightly bent prolate-shaped particle of approximate dimensions 230 Å × 120–140 Å. It is divided transversely into two regions of unequal size. The large subunits are often rounded structures of about 230 Å having a flattened face containing a notch of about 40 Å. The small subunit binds to this flattened face across the notch, in somewhat the manner suggested in Figure 4.

Amelunxen and Spiess (1971) observed that the small subunit had a rodlike shape and appeared either with or without a groove around it. The large subunit was observed to be arc-shaped, with a long groove or channel apparent. In the monosome form, the small subunit appeared to be located within this channel.

While these two studies show common features of ribosomes from these two sources, there are also some perplexing contradictions, especially in light of the studies made on prokaryotic ribosomes. These contradictions emphasize the problem of comparing results from studies using different methods and various types of ribosomes. It is apparent that many further, careful studies will yet be necessary with eukaryotic ribosomes.

Table 7 Electron microscopic studies of eukaryotic ribosomes

Source	Approx. sed. coef. (S)	Length (Å)	Width (Å)	References
Yeast	80	250	206	Mazelis & Petermann (1973)
	60	202	211	Mazelis & Petermann (1973)
	40	52	185	Mazelis & Petermann (1973)
Pea seedlings	80	350	160	Ts'o et al. (1958)
	60	420–500	180 (calc.)	Bayley (1964)
	40	210–250	80	Bayley (1964)
	26	210–250	90	Bayley (1964)
	60	290–330	250–300	Amelunxen & Spiess (1971)
	40	250–300	90–130	Amelunxen & Spiess (1971)
Pea and bean (cytoplasm)	80	260 ± 10	$190\text{–}200 \pm 10$	Odintsova et al. (1967) Bruskov & Kiselev (1968a) Bruskov & Odintsova (1968)
(chloroplast)	70	220 ± 10	170 ± 10	Odintsova et al. (1967) Bruskov & Kiselev (1968a) Bruskov & Odintsova (1968)

Tobacco leaves				
(cytoplasm)	80	286 ± 28	222 ± 25	Miller et al. (1966)
(chloroplast)	70	268 ± 24	214 ± 20	Miller et al. (1966)
Mouse Plasma tumor	80	300	280 (calc.)	Sheton & Kuff (1966)
	large	220	180	Sheton & Kuff (1966)
	small	260	100	Sheton & Kuff (1966)
Novikoff hepatoma	80	240	180	Kuff & Zeigel (1960)
Rat liver	82	303 ± 18	256 ± 23	Haga et al. (1970)
	59	244 ± 17	207 ± 18	Haga et al. (1970)
	60	230	230	Nonomura et al. (1971)
	40	230	140	Nonomura et al. (1971)
Mouse liver	60	175–200	150–175	Florendo (1969)
	40	100–200	75–100	Florendo (1969)

Acknowledgments

We especially thank Dr. John W. Anderegg for critical reading of the manuscript. We are especially indebted to him since it is mainly the efforts coming from his laboratory that have supplied so many of the substantive small-angle X-ray results.

References

Amelunxen, F. 1971. Untersuchungen zur Struktur der Ribosomen. Ein Beitrag zur Konformation der Ribosomen, in besondere der 30S Untereinheit. *Cytobiologie* **3**:111.

Amelunxen, F. and E. Spiess. 1971. Untersuchungen zur Struktur der Ribosomen. Ein Beitrag zur Konformation der 80S-Ribosomen von *Pisum sativum. Cytobiologie* **4**:293.

Arnold, R. M. and L. J. Albright. 1971. Hydrostatic pressure effects on the translation stages of protein synthesis in a cell-free system from *Escherichia coli. Biochim. Biophys. Acta* **238**:347.

Atsmon, A., P. Spitnik-Elson and D. Elson. 1969. Detachment of ribosomal proteins by salt. II. Some properties of protein-deficient particles formed by the detachment of ribosomal proteins. *J. Mol. Biol.* **45**:125.

Bayley, S. T. 1964. Physical studies on ribosomes from pea seedlings. *J. Mol. Biol.* **8**:231.

Beeman, W. W., P. Kaesberg, J. W. Anderegg and M. B. Webb. 1957. Size and shape of particles from small-angle X-ray scattering. *Handbuch der Physik* (ed. S. Flugge) vol. 32, p. 321. Springer-Verlag, Berlin.

Bodley, James W. 1969. Irreversible thermal denaturation of *Escherichia coli* ribosomes. *Biochemistry* **8**:465.

Bohn, T. S., R. K. Farnsworth and W. E. Dibble. 1967. Small-angle X-ray scattering studies of beef pancreas ribosomes. *Biochim. Biophys. Acta* **138**:212.

Bruskov, V. I. and N. A. Kiselev. 1968a. Electron microscope study of the structure of *Escherichia coli* ribosomes and CM-like particles. *J. Mol. Biol.* **37**:367.

———. 1968b. Electron microscopy investigation of the structure of cytoplasmic ribosomes of bean leaves. *J. Mol. Biol.* **38**:443.

Bruskov, V. I. and M. S. Odintsova. 1968. Comparative electron microscopic studies of chloroplast and cytoplasmic ribosomes. *J. Mol. Biol.* **32**:471.

Cammack, K. A. and H. E. Wade. 1965. The sedimentation behavior of ribonuclease-active and -inactive ribosomes from bacteria. *Biochem. J.* **96**:671.

Cammarano, P., A. Romeo, M. Gentile, A. Felsani and C. Gualerzi. 1972. Size heterogeneity of the large ribosomal subunits and conservation of the small subunits in eukaryote evolution. *Biochim. Biophys. Acta* **281**:597.

Castles, J. J. and M. F. Singer. 1969. Degradation of poly (U) by ribonuclease. II. Protection by ribosomes. *J. Mol. Biol.* **40**:1.

Chao, F. C. 1957. Dissociation of macromolecular ribonucleoprotein of yeast. *Arch. Biochem. Biophys.* **70**:426.

Chao, F. C. and H. Schachman. 1956. The isolation and characterization of a macromolecular ribonucleoprotein from yeast. *Arch. Biochem. Biophys.* **61**:220.

Chuang, D. M. and M. V. Simpson. 1971. A translocation-associated ribosomal conformational change detected by hydrogen exchange and sedimentation velocity. *Proc. Nat. Acad. Sci.* **68**:1474.

Cox, R. A. and S. A. Bonanau. 1969. A possible structure of the rabbit reticulocyte ribosome. *Biochem. J.* **114**:769.

Cox, R. A., H. Pratt, P. Huvos, B. Higginson and W. Hirst. 1973. A study of the thermal stability of ribosomes and biologically active subribosomal particles. *Biochem. J.* **134**:775.

Dibble, W. E. and H. M. Dintzis. 1960. The size and hydration of rabbit reticulocyte ribosomes. *Biochim. Biophys. Acta* **37**:152.

Eilam, Y. and D. Elson. 1971. Unfolding of the 30S ribosomal subunit of *Escherichia coli* and the conformation of the unfolded subunit. Parallel sedimentation and optical rotatory dispersion studies. *Biochemistry* **10**:1489.

Elson, D. 1961. A ribonucleic acid particle released from ribosomes by salt. *Biochim. Biophys. Acta* **53**:232.

Florendo, N. T. 1969. Ribosome substructure in intact mouse liver cells. *J. Cell Biol.* **41**:335.

Garrett, R. A. and H. G. Wittmann. 1973. Structure and function of the ribosome. *Endeavor* **32**:8.

Gavrilova, L. P., D. A. Ivanov and A. S. Spirin. 1966. Studies on the structure of ribosomes. III. Stepwise unfolding of the 50S particles without loss of ribosomal protein. *J. Mol. Biol.* **16**:473.

Gesteland, R. F. 1966. Unfolding of *Escherichia coli* ribosomes by removal of magnesium. *J. Mol. Biol.* **18**:356.

Ghysen, A., A. Bollen and A. Herzog. 1970. Ionic effects on the ribosomal quaternary structure. *Eur. J. Biochem.* **13**:132.

Gormly, J. R., C. H. Yang and J. Horowitz. 1971. Further studies on ribosome unfolding. The reversible release of 5S RNA. *Biochim. Biophys. Acta* **247**:80.

Gumilevskaya, M. A., E. B. Kuvaeva, L. V. Chumikina, and V. L. Kretovich. 1971. Ribosomes of dry pea seeds. *Biochimiya* (Eng.) **36**:228.

Haga, J. Y., M. G. Hamilton and M. L. Petermann. 1970. Electron microscope observations on the large subunit of the rat liver ribosome. *J. Cell Biol.* **47**:211.

Hall, C. E. and H. S. Slayter. 1959. Electron microscopy of ribonucleoprotein particles from *Escherichia coli*. *J. Mol. Biol.* **1**:329.

Hamilton, M. G. and T. W. O'Brien. 1973. The size of the 55S mitochondrial ribosome from beef liver. *Fed. Proc.* **32**:583 (Abs.)

Hamilton, M. G. and M. E. Ruth. 1969. The dissociation of rat liver ribosomes by ethylenediaminetetraacetic acid; molecular weight, chemical composition, and buoyant densities of the subunits. *Biochemistry* **8**:851.

Hamilton, M. G., A. Pavlovec and M. L. Petermann. 1971. Molecular weight, buoyant density, and composition of active subunits of rat liver ribosomes. *Biochemistry* **10**:3424.

Hart, R. G. 1962. Electron microscopy of the 50S ribosomes of *Escherichia coli*. *Biochim. Biophys. Acta* **60**:629.

Hauge, J. G. 1971. Pressure-induced dissociation of ribosomes during ultracentrifugation. *FEBS Letters* **17**:168.

Held, W. A. and M. Nomura. 1973. Rate-determining steps in the reconstitution of *Escherichia coli* 30S ribosomal subunits. *Biochemistry* **12**:3273.

Hill, W. E., J. W. Anderegg and K. E. Van Holde. 1970. Effects of solvent environment and mode of preparation on the physical properties of ribosomes from *Escherichia coli*. *J. Mol. Biol.* **53**:107.

Hill, W. E., G. P. Rossetti and K. E. Van Holde. 1969a. Physical studies of ribosomes from *Escherichia coli*. *J. Mol. Biol.* **44**:263.

Hill, W. E., J. D. Thompson and J. W. Anderegg. 1969b. X-ray scattering study of ribosomes from *Escherichia coli*. *J. Mol. Biol.* **44**:89.

Huang, K. and C. R. Cantor. 1972. Surface topography of the 30S *Escherichia coli* ribosomal subunit: Reactivity towards fluorescein isothiocyanate. *J. Mol. Biol.* **67**:265.

Huxley, H. E. and G. Zubay. 1960. Electron microscope observations of the structure of microsomal particles from *Escherichia coli*. *J. Mol. Biol.* **2**:10.

Infante, A. A. and R. Baierlein. 1971. Pressure-induced dissociation of sedimenting ribosomes: Effect on sedimentation patterns. *Proc. Nat. Acad. Sci.* **68**:1780.

Infante, A. A. and M. Krauss. 1971. Dissociation of ribosomes induced by centrifugation: Evidence for doubting conformational changes in ribosomes. *Biochim. Biophys. Acta* **246**:81.

Itoh, T., E. Otaka and S. Osawa. 1968. Release of ribosomal proteins from *Escherichia coli* ribosomes with high concentration of lithium chloride. *J. Mol. Biol.* **33**:109.

Kabasheva, G. N., L. S. Sandakhehiev and A. P. Sevastyanov. 1971. Circular dichroism and association-dissociation of ribosomes. *FEBS Letters* **14**:161.

Kuff, E. L. and R. F. Zeigel. 1960. Cytoplasmic ribonucleoprotein components of the Novikoff hepatoma. *J. Biophys. Biochem. Cytol.* **7**:465.

Kurland, C. G. 1960. Molecular characterization of ribonucleic acid from *Escherichia coli* ribosomes. I. Isolation and molecular weight. *J. Mol. Biol.* **2**:83.

―――. 1966. The requirements for specific sRNA binding by ribosomes. *J. Mol. Biol.* **18**:90.

―――. 1972. Structure and function of the bacterial ribosome. *Ann. Rev. Biochem.* **41**:377.

Kurland, C. G., P. Voynow, S. J. S. Hardy, L. Randall and L. Lutter. 1969. Physical and functional heterogeneity of *E. coli* ribosomes. *Cold Spring Harbor Symp. Quant. Biol.* **34**:17.

Landau, J. V. 1967. Induction, transcription and translation in *Escherichia coli:* A hydrostatic pressure study. *Biochim. Biophys. Acta* **149**:506.

Lederberg, S. and J. M. Mitchison. 1962. Interaction of the ribosomes of *Schizosaccaromyces pombe* and *Escherichia coli*. *Biochim. Biophys. Acta* **55**:104.

Loening, U. E. 1968. Molecular weights of ribosomal RNA in relation to evolution. *J. Mol. Biol.* **38**:355.

Lubin, M. 1968. Observations on the shape of the 30S ribosomal subunit. *Proc. Nat. Acad. Sci.* **61**:1454.

Malkin, L. I. and A. Rich. 1967. Partial resistance of nascent polypeptide chain to proteolytic digestion due to ribosomal shielding. *J. Mol. Biol.* **26**:329.

Maruta, H., T. Tsuchiya and D. Mizuno. 1971. *In vitro* reassembly of functionally active 50S ribosomal particles from ribosomal proteins and RNA's of *Escherichia coli. J. Mol. Biol.* **61**:123.

Matsuura, S., Y. Tashiro, S. Osawa and E. Otaka. 1970. Electron microscopic studies on the biosynthesis of the 50S ribosomal subunit in *Escherichia coli. J. Mol. Biol.* **47**:383.

Mazelis, A. G. and M. L. Petermann. 1973. Physical-chemical properties of stable yeast ribosomes and ribosomal subunits. *Biochim. Biophys. Acta* **312**:111.

Miall, S. H. and I. O. Walker. 1968. Circular dichroism of *Escherichia coli* ribosomes and tobacco mosaic virus. *Biochim. Biophys. Acta* **166**:711.

————. 1969. Structural studies on ribosomes. II. Denaturation and sedimentation of ribosomal subunits unfolded in EDTA. *Biochim. Biophys. Acta* **174**:551.

Midgley, J. E. M. 1965. The estimation of polynucleotide chain length by a chemical method. *Biochim. Biophys. Acta* **108**:340.

Miller, A., U. Karlsson and N. K. Boardman. 1966. Electron microscopy of ribosomes isolated from tobacco leaves. *J. Mol. Biol.* **17**:487.

Miskin, R., A. Zamir and D. Elson. 1970. Inactivation and reactivation of ribosomal subunits: The peptidyl transferase activity of the 50S subunit of *Escherichia coli. J. Mol. Biol.* **54**:355.

Nanninga, N. 1970. Electron microscopy of loosened 50S ribosomal subunits of *Bacillus subtilis. J. Mol. Biol.* **48**:367.

Natori, S., H. Maruta and D. Mizuno. 1968. Unfolding of *Escherichia coli* ribosomes by phosphate ion in the presence of oligonucleotides. *J. Mol. Biol.* **38**:109.

Nonomura, Y., G. Blobel and D. Sabatini. 1971. Structure of liver ribosomes studies by negative staining. *J. Mol. Biol.* **60**:303.

Odintsova, M. S., V. I. Bruskov and E. V. Golubeva. 1967. Comparative study of ribosomes of chloroplasts and cytoplasm of certain plant species. *Biochimiya* **32**:1047 (Eng.).

Ortega, J. P. and W. E. Hill. 1973. A molecular weight determination of the 16S ribosomal ribonucleic acid from *Escherichia coli. Biochemistry* **12**:3241.

Reisner, A. H., J. Rowe and H. M. Macindoe. 1968. Structural studies on the ribosomes of paramecium: Evidence for a "primitive animal ribosome." *J. Mol. Biol.* **32**:587.

Roberts, M. E. and I. O. Walker. 1970. Structural studies on *Escherichia coli* ribosomes. III. Denaturation and sedimentation of ribosomal subunits unfolded in urea. *Biochim. Biophys. Acta* **199**:184.

Salas, M., M. A. Smith, W. M. Stanley, Jr., A. J. Wahba and S. Ochoa. 1965. Direction of reading of the genetic message. *J. Biol. Chem.* **240**:3988.

Scafati, A. R., M. R. Stornaivolo and P. Novaro. 1971. Physicochemical and light scattering studies on ribosome particles. *Biophys. J.* **11**:370.

Schreier, M. H. and H. Noll. 1971. Conformational changes in ribosomes during protein synthesis. *Proc. Nat. Acad. Sci.* **68**:805.

Serdyuk, I. N., N. I. Smirnov, O. B. Ptitsyn and B. A. Fedorov. 1970. On the presence of a dense internal region in the 50S subparticle of *E. coli* ribosomes. *FEBS Letters* **9**:324.

Shelton, E. and E. L. Kuff. 1966. Substructure and configuration of ribosomes isolated from mammalian cells. *J. Mol. Biol.* **22**:23.

Scheraga, H. A. and L. Mandelkern. 1953. Consideration of the hydrodynamic properties of proteins. *J. Amer. Chem. Soc.* **75**:179.

Siddiqui, M. A. Q. and K. Hosokawa. 1969. Role of 5S ribosomal RNA in polypeptide synthesis. *Biochem. Biophys. Res. Comm.* **36**:711.

Smith, W. S. 1971. An X-ray scattering study of the smaller ribosomal subunit of *Escherichia coli*. Ph. D. thesis, University of Wisconsin, Madison.

Spiess, E. 1973. Untersuchungen zur Struktur der Ribosomen. Preparation der 50S Untereinheit von *Escherichia coli* nach der Kritischepunkt-Methode. *Cytobiologie* **7**:28.

Spiess, E. and F. Amelunxen. 1972. Untersuchungen zur Struktur der Ribosomen. Elektronenmikroskopishe Analyse CsCl-behandelter 50S Unterein-heiten von *Escherichia coli. Cytobiologie* **5**:190.

Spirin, A. S. 1969. A model of the functioning ribosome: Locking and unlock-ing of the ribosomal subparticles. *Cold Spring Harbor Symp. Quant. Biol.* **34**:197.

————. 1971. On the equilibrium of the association-dissociation reaction of ribosomal subparticles and on the existence of the so-called "60S intermedi-ate" (swollen 70S) during centrifugation of the equilibrium mixture. *FEBS Letters* **14**:349.

Spirin, A. S. and L. P. Gavrilova. 1969. *The ribosome.* Springer-Verlag, New York.

Spirin, A. S., N. A. Kiselev, R. S. Shakulov and A. A. Bogdanov. 1963. Study of the structure of the ribosomes: Reversible unfolding of the ribosome parti-cles in ribonucleoprotein strands and a model of the packing. *Biochimiya* **28**:920 (Eng.).

Spitnik-Elson, P. and A. Atsmon. 1969. Detachment of ribosomal proteins by salt. I. Effect of conditions on the amount of protein detached. *J. Mol. Biol.* **45**:113.

Spitnik-Elson, P., A. Zamir, M. Miskin, Y. Kaufmann, B. Greenman, A. Breiman and D. Elson. 1972. Experiments on ribosome structure and function. *FEBS Symp.* **23**:175. Academic Press, New York.

Stanley, W. M., Jr. and R. Bock. 1965. Isolation and physical properties of the ribosomal ribonucleic acid of *Escherichia coli. Biochemistry* **4**:1302.

Stanley, W. M., Jr., M. Salas, A. Wahba and S. Ochoa. 1966. Translation of the genetic message: Factors involved in the initiation of protein synthesis. *Proc. Nat. Acad. Sci.* **58**:290.

Tal, M. 1969a. Thermal denaturation of ribosomes. *Biochemistry* **8**:424.

————. 1969b. Metal ions and ribosomal conformation. *Biochim. Biophys. Acta* **195**:76.

Talens, J., F. Kalousek and L. Bosch. Dissociation of *E. coli* ribosomes induced by a ribosomal fraction (DF). Electrophoretic studies of the ribosomal par-ticles. *FEBS Letters* **12**:4.

Talens, A., O. P. Van Diggelen, M. Brougers, L. M. Popa and L. Bosch. 1973. Electrophoretic separation of *Escherichia coli* ribosomal particles on poly-acrylamide gels. *Eur. J. Biochem.* **37**:121.

Tashiro, Y. and D. A. Yphantis. 1965. Molecular weights of hepatic ribosomes and their subunits. *J. Mol. Biol.* **11**:174.

Tissières, A., J. D. Watson, D. Schlessinger and B. R. Hollingworth. 1959. Ribo-nucleoprotein particles from *Escherichia coli. J. Mol. Biol.* **1**:221.

Tolbert, W. R. 1971. A small-angle X-ray scattering study of 50S ribosomal subunits from *Escherichia coli*. Ph. D. thesis, University of Wisconsin, Madison.

Traub, P. and M. Nomura. 1968. Structure and function of *Escherichia coli* ribosomes. I. Partial fractionation of the functionally active ribosomal proteins and reconstitution of artificial subribosomal particles. *J. Mol. Biol.* **34**:575.

————. 1969. Structure and function of *Escherichia coli* ribosomes. VI. Mechanism of assembly of 30S ribosomes studied *in vivo*. *J. Mol. Biol.* **40**:391.

Traut, R. R., H. Delius, C. Ahmad-Zadeh, T. A. Bickle, P. Pearson and A. Tissières. 1969. Ribosomal proteins of *E. coli:* Stoichiometry and implication for ribosome structures. *Cold Spring Harbor Symp. Quant. Biol.* **34**:25.

Ts'o, P. O. P., J. Bonner and J. Vinograd. 1958. Structure and properties of microsomal nucleoprotein particles from pea seedlings. *Biochim. Biophys. Acta* **30**:570.

Van Holde, K. E. 1967. Sedimentation equilibrium. *Fractions* **1**:1. Beckman Instruments, Spinco Div., Palo Alto, Ca.

Vasiliev, V. D. 1971. Electron microscopy study of 70S ribosomes of *Escherichia coli*. *FEBS Letters* **14**:203.

Venable, J. H., Jr., M. Spencer and E. Ward. 1970. Low-angle X-ray diffraction maxima from ribosomes. *Biochim. Biophys. Acta* **209**:493.

Vournakis, J. and A. Rich. 1971. Size change in eukaryotic ribosomes. *Proc. Nat. Acad. Sci.* **68**:3021.

Voynow, P. and C. G. Kurland. 1971. Stoichiometry of the 30S ribosomal proteins of *Escherichia coli*. *Biochemistry* **10**:517.

Wabl, M. R., P. J. Barends and N. Nanninga. 1973a. Tilting experiments with negatively stained *E. coli* ribosomal subunits. An electron microscopic study. *Cytobiologie* **7**:1.

Wabl, M. R., H. G. Doberer, S. Höglund and L. Ljung. 1973b. Electron microscopic study of isolated 30S ribosomal subunits of *Escherichia coli*. *Cytobiologie* **7**:111.

Waterson, J., M. L. Sopori, S. L. Gupta and P. Lengyel. 1972. Apparent changes in ribosomes conformation during protein synthesis. Centrifugation at high speed distorts initiation, pre-translocation, and post-translocation complexes to a different extent. *Biochemistry* **11**:1377.

Weller, D. L., Y. Schechter, D. Musgrave, M. Rougvie and J. Horowitz, 1968. Conformational changes in *Escherichia coli* ribosomes at low magnesium ion concentrations. *Biochemistry* **7**:3668.

White, J. P., I. D. Kuntz and C. R. Cantor. 1972. Studies on the hydration of *Escherichia coli* ribosomes by nuclear magnetic resonance. *J. Mol. Biol.* **64**:511.

Wolfe, A. D., P. Dessen and P. Pantaloni. 1973. Direct measurement and kinetic analysis of the association of *E. coli* ribosomal subunits. *FEBS Letters* **37**:112. (See *Erratum, FEBS Letters* **41**:179.)

Zamir, A., R. Miskin and D. Elson. 1971. Inactivation and reactivation of ribosomal subunits. Amino-acyl tRNA binding activity of the 30S subunit of *Escherichia coli*. *J. Mol. Biol.* **60**:347.

Zitomer, R. S. and E. G. Flaks. 1972. Magnesium dependence and equilibrium of the *Escherichia coli* ribosomal subunit association. *J. Mol. Biol.* **71**:203.

Purification and Identification of *Escherichia coli* Ribosomal Proteins

H. G. Wittmann
Max-Planck-Institut für Molekulare Genetik
Berlin-Dahlem, Germany

INTRODUCTION

The complex nature of the protein moiety of *E. coli* ribosomes was at first revealed by Waller and Harris (1961). Starch gel electrophoresis resulted in the separation of more than twenty bands. Although these and later studies (Waller 1964) supported the hypothesis that there are many different proteins in *E. coli* ribosomes, uncertainty still existed for several years about the number of ribosomal proteins in each of the two subunits. This uncertainty was mainly due to the observation in other systems that the number of bands in gel electrophoresis was often much higher than the number of individual proteins. Aggregation and modification of proteins, e.g., by oxidation or by carbamylation in urea-containing buffers, can give rise to additional bands in polyacrylamide gel electrophoresis.

To solve this problem, it became necessary to isolate all ribosomal proteins and, by chemical, physical and immunological characterization, to determine the exact number of individual proteins with a unique structure. Furthermore, purification and characterization of the ribosomal proteins are important prerequisites for an understanding at the molecular level of the structure and function of ribosomes and their components.

This chapter presents an outline of the different methods used in the extraction and purification of ribosomal proteins. Based on the characterization of the proteins, the question of how many of the isolated proteins have unique and individual structures is answered. Furthermore, it is discussed whether all proteins which can be isolated from a population of ribosomes are "true" ribosomal proteins, and finally, a correlation of ribosomal proteins isolated in different laboratories is presented.

EXTRACTION

Two methods are mainly applied for the extraction of proteins from the ribosomal subunits: (1) extraction with 67% acetic acid; this procedure was originally used for the extraction of proteins from plant viruses (Fraenkel-Conrat 1957) and was efficiently modified for ribosomal proteins by addition of a high concentration of Mg^{++} (Hardy et al. 1969), and (2) treatment of the ribosomes with 2 M LiCl in the presence of 4 M urea (Spitnik-Elson 1965; Marcot-Queiroz and Monier 1966; Lerman et al. 1966). Both methods are very efficient and extract all proteins from *E. coli* ribosomes (Kaltschmidt and Wittmann 1972). On the other hand, one protein, L3, from ribosomes of *Bacillus stearothermophilus,* is not extracted by LiCl-urea but only by the acetic acid procedure (Erdmann, Fahnestock and Nomura 1971).

Another method resulting in extraction of all *E. coli* ribosomes is treatment of unfolded ribosomes with RNase. It can be used if contact of proteins with urea or high H^+ concentration is to be avoided. In addition to these three methods (acetic acid, LiCl-urea and RNase), Kaltschmidt and Wittmann (1972) studied extraction with phenol and with acetic acid in the presence of mono- and bivalent cations, e.g., Li^+, Na^+, K^+, Mg^{++}, Ca^{++} and Mn^{++}, at various concentrations. Treatment of ribosomes with phenol extracts only a portion of the proteins. This is also true when acetic acid is applied at low concentrations of salts. The rest of the proteins remaining bound to the RNA can be obtained by repeating the extraction with actic acid at high salt concentration. Therefore successive extraction at low and at high salt concentrations can be used for prefractionation of ribosomal proteins prior to isolation with cellulose exchangers.

PURIFICATION

30S Proteins

After their extraction, the proteins from 30S subunits are subjected to column chromatography with cellulose cation exchangers. Either cellulose phosphate or carboxymethyl-cellulose were extensively used by various groups, as illustrated in Figures 1–5. The size of the columns used in these runs varied very widely (from 30 × 0.5 cm to 120 × 3 cm) according to the load of the proteins applied (from 3–2000 mg). Since only a portion of the proteins can be purified to homogeneity by a single column run, the rest must be further purified by other methods. This is done by gel filtration of the unseparated proteins on Sephadex or by rechromatography under

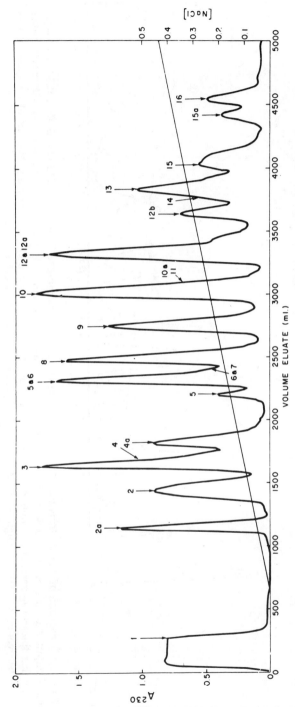

Figure 1 Separation of 30S proteins according to Hardy et al. (1969). Cellulose phosphate column: 60 × 2.8 cm. Flow rate: 45 ml/hr. Buffer: 0.05 M NaH₂PO₄, 0.012 M methylamine, pH 6.5; 6 M urea; 50 μl mercaptoethanol per liter. Linear gradient: 0–0.6 M NaCl. Volume: 6 liter. Load: 400 mg 30S proteins.

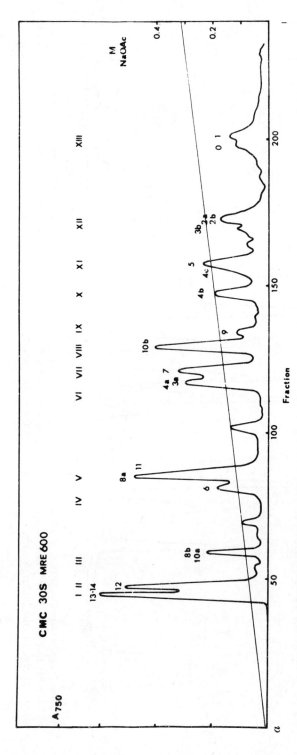

Figure 2 Separation of 30S proteins by CM-cellulose chromatography according to Traut et al. (1969). Buffer: 0.03 M methylamine adjusted to pH 5.6 by acetic acid, 6 M urea, 0.003 M mercaptoethanol. Elution was done with the described buffer containing linearly increasing amounts of sodium acetate (pH 5.6) up to a molarity of 0.5 M Na+. Volume of the gradient was 25–40 bed volumes. Ratio of height to diameter of the columns: between 20 and 50 to one.

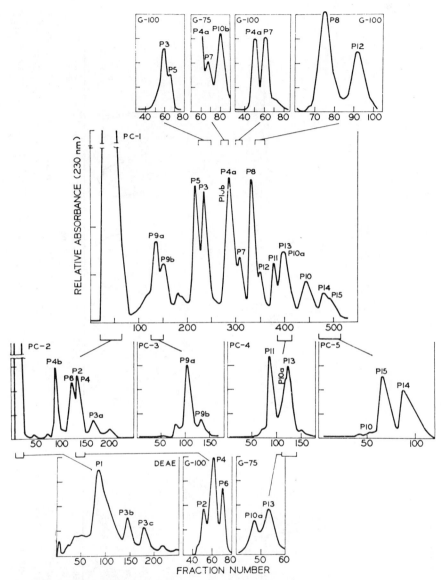

Figure 3 Separation of 30S proteins according to Held, Mizushima and Nomura (1973). Cellulose phosphate column (PC-1): 90 × 2.5 cm. Flow rate: 80–90 ml/hr. Buffer: 0.01 M phosphoric acid, pH adjusted to 8.0 with methylamine, 0.003 M mercaptoethanol, 6 M urea. Linear gradient: 0.15–0.6 M LiCl in the buffer. Volume of gradient: 8800 ml. Load: 1000 mg 30S proteins. Appropriate fractions (15 ml each) were pooled and applied to other columns as shown in the figure.

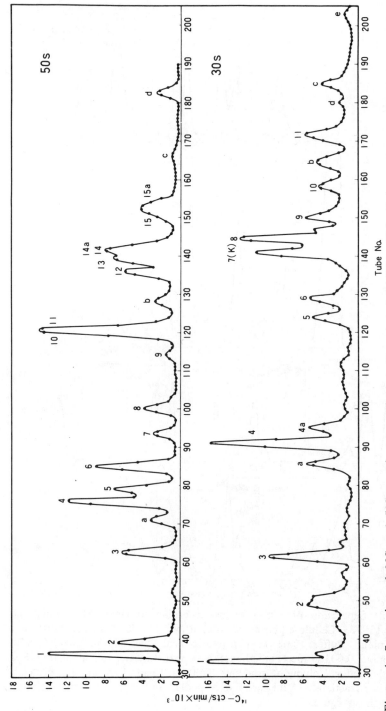

Figure 4 Separation of 30S and 50S proteins according to Osawa et al. (1972). CM-cellulose column: 30 × 0.5 cm. Flow rate: 4.5 ml/hr. Buffer: 0.005 M sodium acetate pH 5.6. Linear gradient: 0.05 M to 0.5 M sodium acetate pH 5.6 in 6 M urea. Volume of gradient: 400 ml. Load: 3 mg proteins. Conditions for chromatography are described by Otaka, Itoh and Osawa (1968).

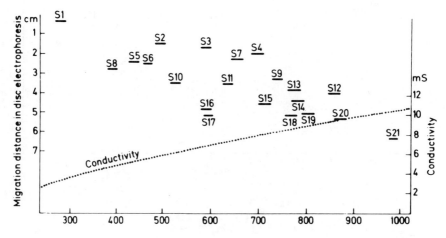

Figure 5 Separation of 30S proteins according to Hindennach, Stöffler and Wittmann (1971). CM-cellulose column: 120 × 3 cm. Flow rate: 50 ml/hr. Equilibration buffer: 0.8 ml pyridine, 4.8 ml formic acid, 1.0 ml mercaptoethanol, 360 g urea per 1 liter. Linear gradient: 7500 ml equilibration buffer plus 7500 ml of the following buffer: 48 ml pyridine, 19.2 ml formic acid, 1.0 ml mercaptoethanol, 360 g urea per 1 liter. Fractions: 14 ml. Load: 2000 mg 30S proteins. Aliquots of every fifth fraction were run in disc electrophoresis (ordinate).

modified conditions, e.g., with a different exchanger, buffer or gradient. Acidic proteins not retained by cellulose phosphate are rechromatographed on DEAE-cellulose as illustrated in Figure 3. The full details of the purification procedures used in the different laboratories cannot be given here, but they are described in the papers cited in the legends of the figures.

50S Proteins

50S subunits contain about 60% more proteins than 30S. Therefore when the 50S proteins were to be isolated on a large scale, they were prefractionated before they were applied to cellulose exchangers. The prefractionation was done either by stepwise treatment of 50S with LiCl (Hindennach, Kaltschmidt and Wittmann 1971) or by differential precipitation of extracted 50S proteins with ammonium sulfate (Mora et al. 1971). After their prefractionation, the proteins in each group are further separated on CM-cellulose (Figure 6) or cellulose phosphate (Figure 7). Purification of 50S proteins can also be achieved by omitting the prefractionation step. In this case, total 50S proteins are applied onto a cellulose phosphate column and unseparated proteins are rechromatographed on CM-cellulose columns (Figure 8).

The elution profiles of the proteins can be determined by several

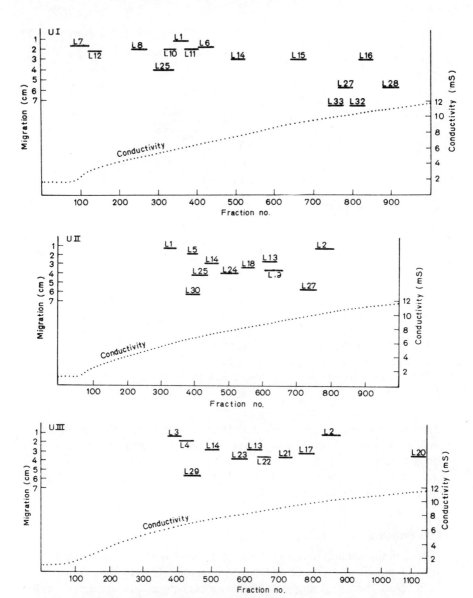

Figure 6 Separation of 50S proteins according to Hindennach, Kaltschmidt and Wittmann. 1971. 50S particles were subjected to stepwise treatment with LiCl and urea, resolving them into three protein groups (UI, II, III) which were applied to CM-cellulose columns. Chromatography was done as described in Figure 5. Load for each of the columns: 1200–1600 mg.

Figure 7 (*facing page*)   Separation of 50S proteins according to Mora et al. (1971). Proteins were extracted from 50S subunits and fractionated into five groups by precipitation with increasing amounts of ammonium sulfate. Each of the five protein groups (A, B, C, D, E) were applied to cellulose phosphate chromatography as described in Figure 1.

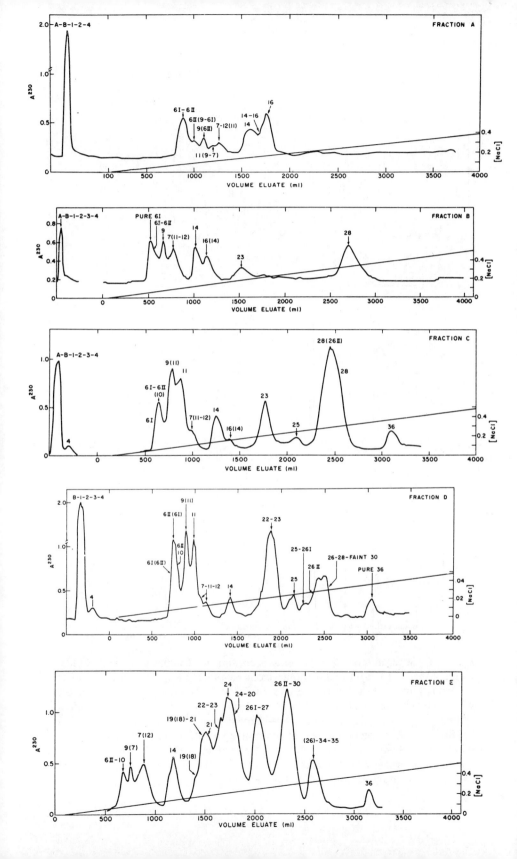

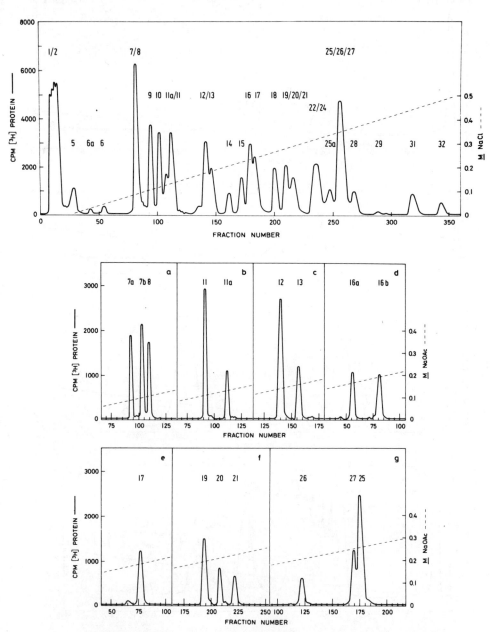

Figure 8 Separation of 50S proteins according to Zimmermann and Stöffler (unpublished). (Top) Elution profile from cellulose phosphate column (40 × 0.5 cm). Flow rate: 4.5 ml/hr. Buffer: 0.05 M NaH_2PO_4 adjusted to pH 6.5 with methylamine; 6 M urea. Linear gradient: 0–0.5 M NaCl in the buffer. Volume of gradient: 300 ml. Load: 10–15 mg of total 50S proteins.

(Bottom) Rechromatography of unresolved protein mixtures on CM-cellulose columns (40 × 0.3 cm). Buffer: 0.5 M sodium acetate pH 5.6. Flow rate: 1.8 ml/hr. Linear gradient of sodium acetate (pH 5.6) in which the salt concentration increased by 0.1 M per 40 ml buffer.

methods, e.g., UV absorption at 230 nm, radioactivity, colorimetric protein determination, or disc electrophoresis. The purity of the isolated proteins is tested either by one-dimensional polyacrylamide electrophoresis with different extent of cross-linking (Hardy et al. 1969; Mora et al. 1971; Pearson, Delius and Traut 1972) or by two-dimensional polyacrylamide gel electrophoresis (Hindennach, Kaltschmidt and Wittmann 1971).

70S Proteins

Purification of proteins on a large scale is limited by the amounts of subunits which can be separated by centrifugation in zonal rotors. To avoid this, 70S ribosomes were treated with 2 M LiCl, at first in the absence and then in the presence of urea. The proteins in each group were further fractionated by isoelectric precipitation at two pH values, by zonal electrophoresis, by column chromatography with CM-cellulose, and, if necessary, by gel filtration on Sephadex G100. In this way pure proteins in amounts up to 150 mg have been obtained (Kaltschmidt et al. 1971).

PROPERTIES

The chemical, physical and immunological properties of the proteins isolated from *E. coli* ribosomes have been extensively studied. As the chemical and immunological characterization is described elsewhere in this volume (see contributions by Wittmann and Wittmann-Liebold and by Stöffler), only the studies on molecular weights and secondary structure of the proteins remain to be discussed here.

Three methods were used for determination of the molecular weights: sedimentation equilibrium, SDS-gel electrophoresis, and a calculation based on the number of tryptic peptides and the content of arginine and lysine. The molecular weights of the proteins from the 30S and 50S subunits are listed in Table 1 and 2, respectively. They range between 9000 and 35,000 daltons for all proteins with the exception of protein S1, which has a much higher molecular weight, namely, 65,000 daltons. The average molecular weight of the 30S proteins is 19,000 and of the 50S proteins 16,300 daltons. The sum of the molecular weights is about 400,000 daltons for the 30S and 555,000 daltons for the 50S proteins. Comparison of the mass of protein in the 30S subunit with the sum of the molecular weights of the isolated 30S proteins was the first hint for a possible heterogeneity of the ribosome population.

The α-helix content of isolated proteins is in the range of 20–40% (Dzionara 1970). Only two 50S proteins, L7 and L12, which are very rich in alanine (24%), have a relatively high α-helix content of 50–60% (Möller, Castleman and Terhorst 1970; Dzionara 1970).

Table 1 Comparison of molecular weights of 30S proteins

	Berlin[a]		Madison[b]			Geneva[c]	
	SDS gel	Sed. equil.	Protein	Sed. equil.	Chemical[d]	Protein	SDS gel
S1	65,000	n.d.	1	65,000	31,000	13	68,000
S2	28,300	24,000	4a	30,000	27,300	11	29,800
S3	28,200	23,000	9	33,000	14,200	10b	29,900
S4	26,700	23,000	10	26,700	19,300	9	26,600
S5	19,600	18,500	3	24,000	16,200	8a	20,200
S6	15,600	15,500	2	18,000	18,000	10a	13,500
S7	22,700	26,000	8	21,500	14,500	7	19,600
S8	15,500	15,500	2a	17,600	19,000	8b	14,100
S9	16,200	14,500	12	21,000	13,500	5	16,200
S10	12,400	18,000	4	16,000	14,800	6	10,500
S11	15,500	n.d.	11	18,300	16,300	4c	14,500
S12	17,200	15,000	15	19,000	16,000		
S13	14,900	14,000					
S14	14,000	14,000	12b	15,600	14,200		
S15	12,500	13,000	14	13,200	15,800	4b	10,700
S16	11,700	13,000	6	13,500	11,700	4a	9,600
S17	10,900	n.d.	7	10,700	15,600	3a	9,800
S18	12,200	10,500	12a	14,600	11,000	2b	11,000
S19	13,100	14,000	13	15,000	13,000	2a	11,400
S20	12,000	12,500	16	14,000	12,000	1	10,800
S21	12,200	13,500	15a	13,000	15,700	0	19,700

n.d.: not determined. [a]Dzionara et al. (1970); [b]Craven et al. (1969); [c]Traut et al. (1969); [d]For calculation of "chemical" molecular weight see Craven et al. (1969).

NUMBER

30S Proteins

There is general agreement that ammonium chloride-washed 30S ribosomes contain 21 proteins, although in different stoichiometries. The extensive characterization of the isolated proteins with chemical, physical and immunological methods clearly demonstrated that all of them have unique structures different from each other (see Wittmann and Wittmann-Liebold and Stöffler, this volume).

However the question arises whether all of the 21 proteins can be regarded as "true" ribosomal proteins. Some criteria relevant to this question are as follows:

(1) Several proteins bind specifically and at a molar ratio of 1:1 to ribosomal but not to other RNAs.

(2) Addition of "fractional" ribosomal proteins (which do not occur in all ribosomal particles) to ribosomes stimulates ribosomal functions.

(3) Reconstitution with a mixture of ribsomal proteins from which one at the time is omitted leads in most cases to biologically inactive or even physically incomplete particles.

(4) Salt treatment of subunits results in split proteins and cores. Both fractions separately are often functionally inactive, but activity can be regained by adding purified single proteins from the split fraction to the core particles.

(5) Amino acid replacements in certain ribosomal proteins cause changes in the ribosomal phenotype, e.g., resistance to or dependence on antibiotics.

At least one of these criteria can be applied for most of the 30S proteins. For instance, protein S4 is a binding protein, is altered in ribosomal mutants, and is necessary for reconstitution of complete 30S particles. Therefore there is no doubt that S4 is a "true" ribosomal protein. The only ambiguity exists for protein S1. It is unusual not only because it has by far the greatest molecular weight of the ribosomal proteins (Table 1), but also because it occurs in only 10–30% of the purified ribosomes. This protein stimulates the binding of poly(U) to the 30S subunits and, as a consequence, also stimulates the binding of Phe-tRNA (Van Duin and Kurland 1970; Tal et al. 1972; Randall-Hazelbauer and Kurland 1972). However, ribosomes reconstituted in the absence of S1 are active in polyphenylalanine synthesis when the ribosomes are supplied with supernatant proteins (Held, Mizushima and Nomura 1973). Therefore it is possible that S1 is found in the supernatant fraction, and it has been suggested that S1 might be considered a factor (Held, Mizushima and Nomura 1973). Finally, it remains to be seen whether all laboratories designate the same protein as S1.

50S Proteins

When proteins from 50S subunits washed with ammonium chloride are extracted and subjected to two-dimensional polyacrylamide gel electrophoresis, 34 protein spots are found (Kaltschmidt and Wittmann 1970b; Figure 9). Although this result is in good agreement with the number of bands obtained by separation of 50S proteins by CM-cellulose chromatography followed by SDS gel electrophoresis (Traut et al. 1969), it cannot be concluded that there are 34 proteins in the 50S subunits because a given protein might give rise to more than one spot due to its aggregation, proteolytic cleavage or chemical modification, e.g., oxidation, carbamylation, etc. This is illustrated by the observation that a few proteins can give rise to two spots in two-dimensional gel electrophoresis. Thus proteins S12 and S17 (Kaltschmidt and Wittmann 1972) as well as S11 (Wittmann et al. 1971) can appear as double spots, possibly reflecting oxidized and reduced forms of the same protein. Therefore the proteins must be isolated

Table 2 Comparison of molecular weights of 50S proteins

	Berlin[a]		Uppsala[b]			Geneva[c,d]			
	SDS gel	Sed. equil.	Protein	Sed. equil.	Chemical	Protein (c)	(d)	SDS gel[c]	Chemical[d]
L1	26,700	22,000	7	26,900	18,000–20,100	11	17 III	27,600	31,300
L2	31,500	28,000	28	28,800	17,500–19,500	25	13–14 X	32,400	22,100
L3	27,000	23,000	11	24,500	21,800–24,800	10		25,800	
L4	25,800	28,500	14	25,600	19,300–21,400	12	16 V	26,200	26,800
L5	22,000	17,500	6 II	23,700	17,600–23,400	7b	13–14 III	21,700	19,000
L6	22,200	21,000	6 I	17,300	17,700–24,100	9	12 III	21,200	21,600
L7	13,400	15,500	1 (2)	14,600	16,100–21,200	1a(1b)		10,200	
L8	17,300	19,000	3	17,200	15,300–17,900	2b			
L9	17,300	n.d.	3	17,200	15,300–17,900	2a		16,900	
L10	19,000	21,000	4	19,400	16,300–19,200	5		17,800	
L11	19,600	19,000	9	21,200	15,000–20,700	7a		15,300	
L12	13,200	15,500	2 (1)	14,600	16,100–21,200	1b(1a)		10,100	
L13	17,800	20,000	22	20,100	17,800–19,600	17		17,700	
L14	16,200	18,500	16	17,600	13,500–15,800	13	10 VI	13,800	17,200
L15	17,500	17,000	25	18,200	11,400–14,600	20	9 VIII	16,500	14,500

L16	17,900	22,000	26 II	20,200	15,900–18,800	27	8 X	17,400	16,800
L17	16,700	15,000	26 I	17,700	14,000–15,600	22	8 IX	14,000	19,900
L18	14,300	17,000	24	17,100	16,100–18,900	19		11,300	
L19	14,900	17,500	23	17,300	14,700–19,400	16b		12,800	
L20	17,200	16,000	36	17,400	13,700–16,200	31	4b XII	16,700	12,400
L21	13,900	14,000				26		11,800	
L22	14,800	17,000	20	12,000	11,200–14,400	18		12,600	
L23	12,700	12,500				16a		11,200	
L24	14,300	17,500	21	14,500	11,800–15,900	15		12,000	
L25	12,000	12,500	12	13,800	11,100–15,500	11a		9,800	
L26	12,000	12,500				(25a)		9,500	
L27	12,700	12,000	30	12,800	12,000–14,500	24		10,200	
L28	12,300	15,000	34	10,200	6,100– 7,900	(25a)		9,500	
L29	12,000	12,000	10	8,900	11,600–14,200	8		7,100	
L30	11,200	10,000	19	10,000	9,000–12,400	14	3 VI	8,800	11,000
L31	10,000	n.d.							
L32	10,500	n.d.							
L33	10,500	9,000	27	10,800	800–10,800	21	1 IX	8,100	7,300
L34	9,600	n.d.							

[a]Dzionara et al. (1970); [b]Mora et al. (1971); [c]R. A. Zimmermann and G. Stöffler (unpublished); [d]Pearson et al. (1972).

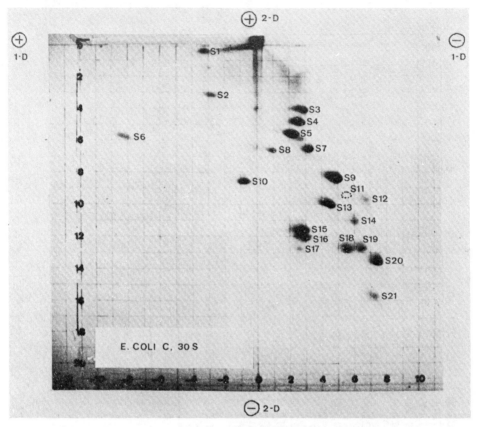

Figure 9 Separation of 30S and 50S proteins by two-dimensional polyacrylamide gel electrophoresis (Kaltschmidt and Wittmann 1970b). Electrophoresis in the first dimension is done in an 8% acrylamide gel at pH 8.6 (Tris-boric acid buffer) and that in the second dimension in a 18% acrylamide gel at pH

and shown by chemical and immunological methods to have unique structures.

The most direct evidence would be the comparison of the amino acid sequences of all proteins. This is, of course, a very laborious and time-consuming task. Therefore other methods are necessary for a quicker although more indirect answer. Fingerprints and especially immunological studies with antibodies against single proteins have proved useful for this purpose.

Fingerprints of the tryptic peptides from most single 50S proteins (Mora et al. 1971; Rombauts, Peeters and Wittmann 1971; Pearson, Delius and Traut 1972) differed from each other with one exception; only those

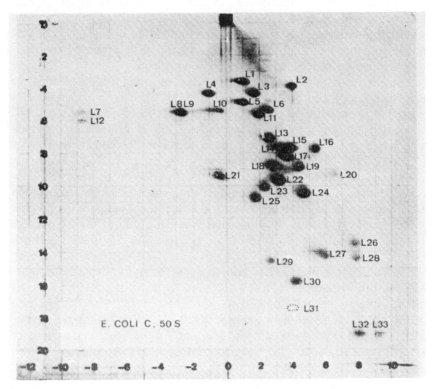

4.6 (acetate buffer). For details of buffers and gels see Kaltschmidt and Wittmann (1970a). Separation of proteins L8 and L9 is achieved by using 4%
instead of 8% acrylamide in the first dimension. Protein L34 can be detected by
shortening the time for the runs in the first and second dimension.

of proteins L7 and L12 were almost identical (Rombauts et al. 1971;
Terhorst et al. 1972a), indicating structural homology between these two
proteins. This result is in very good agreement with immunological studies
that showed the complete immunological identity of proteins L7 and L12
(Stöffler and Wittmann 1971) and with studies on the amino acid sequences, which were shown to be almost identical for both proteins
(Terhorst et al. 1972b).

Furthermore, antisera against 24 of the 50S proteins were prepared and
tested for immunological homologies among these proteins. No cross-
reaction was detected (Stöffler and Wittmann 1971). The studies were
recently extended by Stöffler et al. (1974), who tested antisera against each

of the 34 electrophoretically identified 50S proteins for cross-reaction with any other 50S protein. With the exception of L7 and L12, no such cross-reaction was found.

These immunological results demonstrate that each of the 34 spots (with the exception of L7 and L12) revealed by two-dimensional polyacrylamide gel electrophoresis of 50S proteins corresponds to an individual protein of unique structure and that none of these proteins have arisen by modification, aggregation or degradation of other 50S proteins.

It still remains to be seen whether there are proteins with identical or homologous structure when 30S and 50S proteins are compared. This question was investigated by Tischendorf, Stöffler and Wittmann (1974), who tested antisera directed against each of the 21 proteins of the small subunit and the 34 proteins of the large in all possible combinations. Besides L7 and L12, only two other proteins gave complete cross-reaction, namely, S20 from 30S subunit and L26 from 50S. In addition to these immunological studies, protein-chemical and genetic evidences support the notion that proteins S20 and L26 have identical structures. Therefore the number of individual proteins in 70S ribosomes is not the sum of the 21 proteins from the small subunit and the 34 from the large, i.e., 55 proteins, but one less.

The fact that a protein has been isolated from saltwashed ribosomes does not prove that it is a "true" ribosomal element. It could be a supernatant protein strongly absorbed to ribosomes during the isolation procedure and not removed by washing with salt. As mentioned above, several criteria can be applied to settle this question. Of the five criteria discussed for 30S proteins, three (i.e., specific binding to rRNA, addition of split proteins to cores, and altered proteins in ribosomal mutants) can so far also be used to determine that a protein isolated from 50S subunits is a "true" ribosomal protein and not a supernatant protein strongly absorbed to the particles. Two further criteria are the identification by affinity labeling or antibodies of proteins involved in binding of antibiotics, aminoacyl-tRNA or factors and the assignment of ribosome-specific enzymatic activity (e.g., peptidyltransferase or GTPase) to a certain protein.

About 20 proteins from the 50S subunits can so far be confirmed as "true" ribosomal proteins by at least one of the criteria mentioned above. Evidence for more proteins will probably become available with further studies on the role of 50S ribosomal proteins.

CORRELATION

When several groups started the isolation and characterization of *E. coli* ribosomal proteins, they of course used their own nomenclature. For a common nomenclature to be adopted by all groups, it was necessary to correlate the proteins unambiguously by several methods. This was done by comparison of their molecular weights, amino acid compositions, mobilities in one- and two-dimensional polyacrylamide electrophoresis, elution from cellulose exchangers, and immunological cross-reaction.

In this way all of the 30S proteins (Table 3) and most of the 50S proteins (Table 4) could be correlated among several laboratories. The standard nomenclature is based on the mobility of the proteins in two-dimensional polyacrylamide gel electrophoresis (Kaltschmidt and Wittmann 1970a). The correlation is necessary for comparison of the studies done with ribosomal proteins in different laboratories. However, it is possible that proteins isolated in different laboratories with nonidentical procedures for extraction, purification and storage may have somewhat different properties. This possibility has to be kept in mind when functional or other properties (e.g., specific binding to rRNA) of a given protein studied in different laboratories are compared.

Table 3   Correlation of 30S ribosomal proteins studied in different laboratories

Berlin	Uppsala	Madison	Geneva	Hiroshima
S1	1	P1	13	30-1
S2	4a	P2	11	30-4a
S3	9(+5)	P3	10b	30-8
S4	10	P4a	9	30-9
S5	3	P4	8a	30-4
S6	2	P3b + P3c	10a	30-2
S7	8	P5	7	30-7
S8	2a	P4b	8b	30-3
S9	12	P8	5	30-11
S10	4	P6	6	30-a
S11	11	P7	4c	30-b
S12	15	P10		30-c
S13	15b	P10a		30-c
S14	12b	P11		30-c
S15	14	P10b	4b	30-10
S16	6	P9a*	4a	30-5
S17	7	P9b*	3a	30-6
S18	12a	P12	2b	30-d
S19	13	P13	2a	30-c
S20	16	P14	1	30-e
S21	15a	P15	0	30-f

* Correlated by Held et al. (1973).

The correlation of the proteins from Berlin, Uppsala, Madison and Geneva is described by Wittmann et al. (1971). The proteins separated in Hiroshima are correlated with those in Berlin by Muto et al. (1974).

Table 4   Correlation of 50S proteins isolated in different laboratories

Berlin	Uppsala[a]	Geneva[b]	Hiroshima[c]
L1	7	11	50-4
L2	28	25	
L3	11	10	50-6
L4	14	12	50-8
L5	6 II	7b	
L6	6 I	9	50-5
L7	1 (2)	1a (1b)	50-1
L8	3	2b	
L9	3	2a	
L10	4	5	
L11	9	7a	50-3
L12	2 (1)	1b (1a)	50-1
L13	22	17	50-12
L14	16	13	
L15	25	20	
L16	26 II	27	50-d
L17	26 I	22	
L18	24	19	
L19	23	16b	
L20	36	31	
L21		26	
L22	20	18	50-13
L23	18	16a	
L24	21	15	
L25	12	11a	50-7
L26		(25a)	
L27	30	24	
L28	34	(25a)	
L29	10	8	50-5
L30	19	14	
L31			
L32			
L33	27	21	
L34	35		

The correlations with the Berlin nomenclature were done by: [a]C. G. Kurland, D. Donner, G. Tischendorf, G. Stöffler and H. G. Wittmann (unpublished results); [b]R. A. Zimmermann and G. Stöffler (unpublished results); [c]T. Itoh et al. (unpublished results).

References

Craven, G. R., P. Voynow, S. J. S. Hardy and C. G. Kurland. 1969. Ribosomal proteins of *E. coli. II.* Chemical and physical characterization of 30S proteins. *Biochemistry* **8**:2906.

Dzionara, M. 1970. Ribosomal proteins. Secondary structure of individual ribosomal proteins of *E. coli* studied by circular dichroism. *FEBS Letters* **8**:197.

Dzionara, M., E. Kaltschmidt and H. G. Wittmann. 1970. Ribosomal proteins. XIII. Molecular weights of isolated ribosomal proteins of *Escherichia coli*. *Proc. Nat. Acad. Sci.* **67**:1909.

Erdmann, V. A., S. Fahnestock and M. Nomura. 1971. Reconstitution of 50S ribosomes from 23S and 5S RNA and protein components. *Fed. Proc.* **30**:1203.

Fraenkel-Conrat, H. 1957. Degradation of tobacco mosaic virus with acetic acid. *Virology* **4**:1.

Hardy, S. J. S., C. G. Kurland, P. Voynow and G. Mora. 1969. The ribosomal proteins of *Escherichia coli*. I. Purification of the 30S ribosomal proteins. *Biochemistry* **8**:2897.

Held, W. A., S. Mizushima and M. Nomura. 1973. Reconstitution of *Escherichia coli* 30S ribosomal subunits from purified molecular components. *J. Biol. Chem.* **248**:5720.

Hindennach, I., E. Kaltschmidt and H. G. Wittmann. 1971. Ribosomal proteins. Isolation of proteins from 50S ribosomal subunits of *Escherichia coli*. *Eur. J. Biochem.* **23**:12.

Hindennach, I., G. Stöffler and H. G. Wittmann. 1971. Ribosomal proteins. Isolation of the proteins from 30S ribosomal subunits of *Escherichia coli* *Eur. J. Biochem.* **23**:7.

Kaltschmidt, E. and H. G. Wittmann. 1970a. Ribosomal proteins. VII. Two-dimensional polyacrylamide gel electrophoresis for fingerprinting of ribosomal proteins. *Ann. Biochem.* **36**:401.

————. 1970b. Ribosomal proteins. XII. Number of proteins in small and large ribosomal subunits of *Escherichia coli* as determined by two-dimensional gel electrophoresis. *Proc. Nat. Acad. Sci.* **67**:1276.

————. 1972. Ribosomal proteins. XXXII. Comparison of several extraction methods for proteins from *Escherichia coli* ribosomes. *Biochimie* **54**:167.

Kaltschmidt, E., V. Rudloff, H. G. Janda, M. Cech, K. Nierhaus and H. G. Wittmann. 1971. Isolation of proteins from 70S ribosomes of *Escherichia coli*. *Hoppe-Seyler's Z. Physiol. Chem.* **352**:1545.

Lerman, M. I., A. S. Spirin, L. P. Gavrilova and V. F. Colon. 1966. Studies on the structure of ribosomes. II. Stepwise dissociation of protein from ribosomes by caesium chloride and the reassembly of ribosome-like particles. *J. Mol. Biol.* **15**:268.

Marcot-Queiroz, J. and R. Monier. 1966. Préparation de particules 18S et 25S a partir des ribosomes d'*Escherichia coli*. *Bull. Soc. Chim. Biol.* **48**:446.

Möller, W., H. Castleman and C. Terhorst. 1970. Characterization of an acidic protein in 50S ribosomes of *E. coli*. *FEBS Letters* **8**:192.

Mora, G., D. Donner, P. Thammana, L. Lutter and C. G. Kurland. 1971. Purification and characterization of 50S ribosomal proteins of *Escherichia coli*. *Mol. Gen. Genet.* **112**:229.

Muto, A., E. Otaka, T. Itoh, S. Osawa and H. G. Wittmann. 1974. Correspondence of 30S ribosomal proteins of *Escherichia coli* fractionated on carboxymethyl-cellulose chromatography to the standard nomenclature. *Mol. Gen. Genet.* (in press).

Osawa, S., E. Otaka, R. Takata, S. Dekio, M. Matsubara, T. Itoh, A. Muto,

K. Tanaka, H. Teraoka and M. Tamaki. 1972. Ribosomal protein genes in bacteria. *FEBS Symp.* **23**:313.

Otaka, E., T. Itoh and S. Osawa. 1968. Ribosomal proteins of bacterial cells: Strain and species-specificity. *J. Mol. Biol.* **33**:93.

Pearson, P., H. Delius and R. R. Traut. 1972. Purification and characterization of 50S ribosomal proteins of *Escherichia coli. Eur. J. Biochem.* **27**:482.

Randall-Hazelbauer, L. and C. G. Kurland. 1972. Identification of three 30S proteins contributing to the ribosomal A-site. *Mol. Gen. Genet.* **115**:234.

Rombauts, W., B. Peeters and H. G. Wittmann. 1971. Comparison of peptide patterns from isolated 30S and 50S ribosomal proteins of *Escherichia coli* by column chromatography. *FEBS Letters* **18**:164.

Spitnik-Elson, P. 1965. The preparation of ribosomal protein from *Escherichia coli* with lithium chloride and urea. *Biochem. Biophys. Res. Comm.* **18**:557.

Stöffler, G. and H. G. Wittmann. 1971. Ribosomal proteins. XXV. Immunological studies on *Escherichia coli* ribosomal proteins. *J. Mol. Biol.* **62**:407.

Stöffler, G., G. W. Tischendorf, R. Hasenbank and H. G. Wittmann. 1974. The number of individual proteins of the 50S ribosomal subunit from *Escherichia coli* as determined by immunochemical methods. *Eur. J. Biochem.* (in press).

Tal, M., M. Aviram, A. Kanarek and A. Weiss. 1972. Polyuridylic acid binding and translating by *Escherichia coli* ribosomes: Stimulation by protein 1, inhibition by aurintricarboxylic acid. *Biochim. Biophys. Acta* **281**:381.

Terhorst, C. P., B. Wittmann-Liebold and W. Möller. 1972a. 50S ribosomal proteins. Peptide studies on two acidic proteins, A_1 and A_2, isolated from 50S ribosomes of *Escherichia coli. Eur. J. Biochem.* **25**:13.

Terhorst, C. P., W. Möller, R. Laursen and B. Wittmann-Liebold. 1972b. Amino acid sequence of a 50S ribosomal protein involved in both EFG and EFT dependent GTP-hydrolysis. *FEBS Letters* **28**:325.

Tischendorf, G. W., G. Stöffler and H. G. Wittmann. 1974. The comparative characterization of the proteins of *Escherichia coli* 30S and 50S ribosomal subunits. The identity of S20 and L26. *Eur. J. Biochem.* (in press).

Traut, R. A., H. Delius, C. Ahmad-Zadeh, T. A. Bickle, P. Pearson and A. Tissières. 1969. Ribosomal protein of *E. coli:* Stoichiometry and implication for ribosome structure. *Cold Spring Harbor Symp. Quant. Biol.* **34**:25.

Van Duin, J. and C. G. Kurland. 1970. Functional heterogeneity of the 30S ribosomal subunit of *E. coli. Mol. Gen. Genet.* **109**:169.

Waller, J. P. 1964. Fractionation of the ribosomal protein from *Escherichia coli. J. Mol. Biol.* **10**:319.

Waller, J. P. and J. I. Harris. 1961. Studies on the composition of the protein from *E. coli* ribosomes. *Proc. Nat. Acad. Sci.* **47**:18.

Wittmann, H. G., G. Stöffler, I. Hindennach, C. G. Kurland, L. Randall-Hazelbauer, E. A. Birge, M. Nomura, E. Kaltschmidt, S. Mizushima, R. R. Traut and T. A. Bickle. 1971. Correlation of 30S ribosomal proteins of *Escherichia coli* isolated in different laboratories. *Mol. Gen. Genet.* **111**:327.

Chemical Structure of Bacterial Ribosomal Proteins

H. G. Wittmann and B. Wittmann-Liebold
Max-Planck-Institut für Molekulare Genetik
Berlin-Dahlem, Germany

INTRODUCTION

The isolation of ribosomal proteins which is described in the preceding contribution has enabled chemical, physical and immunological studies to be made on the structure of these proteins. The chemical investigations to be summarized here include studies on the amino acid composition, isoelectric points, N-termini, peptide maps, isolation and analysis of tryptic peptides, and determination of the complete sequence of some proteins. By far the majority of these investigations were done on ribosomal proteins from *E. coli* and its mutants, but recently similar studies have begun on proteins from *Bacillus stearothermophilus* ribosomes. Furthermore, conclusions about the chemical structure of ribosomal proteins from numerous bacterial species can be drawn by comparative studies with electrophoretic and immunological methods.

ESCHERICHIA COLI WILD TYPE

Isoelectric Points

It became evident from early electrophoretic studies on *E. coli* ribosomal proteins (Waller 1964; Leboy, Cox and Flaks 1964) that many of the proteins must be very basic. Determination of the isoelectric point for each of the proteins from *E. coli* ribosomes was possible by two-dimensional polyacrylamide gel electrophoresis at different pH values (Kaltschmidt

1971). About 70% of the proteins have isoelectric points of pH 10 and higher, whereas those of only three proteins, namely S6, L7 and L12, are at about pH 5. There is no significant difference between 30S and 50S proteins with respect to their isoelectric points.

Amino Acid Compositions

The basic nature of most *E. coli* ribosomal proteins is also reflected by their amino acid compositions. However, as it is not possible to differentiate by amino acid analysis between glutamic acid or glutamine and between aspartic acid or aspargine, there remains a rather large ambiguity in trying to determine the net charge of the proteins from their amino acid compositions.

Many proteins from both *E. coli* 30S (Moore et al. 1968; Fogel and Sypherd 1968; Craven et al. 1969; Kaltschmidt et al. 1970a) and 50S subunits (Kaltschmidt et al. 1970a; Mora et al. 1971) are very rich in basic amino acids and poor in aromatic amino acids. With the exception of proteins L7 and L12, whose compositions (Möller, Castleman and Terhorst 1970; Kaltschmidt et al. 1970a) and sequences (Terhorst et al. 1972b, 1973) are (almost) identical, all other proteins differ from each other in their amino acid compositions. Some proteins contain high percentages of a given amino acid, e.g., proteins L7 and L12 about 24% alanine, S20 about 19% alanine, S21 about 20% arginine, L10 and L20 about 18% alanine, L15 about 16% glycine, and L33 about 21% lysine. The cysteine content was determined by labeling the proteins with cysteine-specific radioactive reagents, such as iodoacetamide, *N*-ethylmaleimide or 5,5'-dithio-bis-(2-nitrobenzoic acid), and separation by chromatographic or electrophoretic methods (Moore 1971; Acharya and Moore 1973; Bakardjieva and Crichton 1974).

Modified lysine residues have been found in proteins L7 and L12 (details are given in the section on the amino acid sequence of these proteins) as well as in protein L11. In contrast to ε-*N*-monomethyllysine in L7 and L12, the modified residue in L11 is ε-*N*-trimethyllysine. Alix and Hayes (1974) found that the ribosomal proteins of *E. coli* cells grown in the presence of ethionine are submethylated and that after in vivo or in vitro methylation of the ribosomes, protein L11 contains approximately one ε-*N*-trimethyllysine residue.

Peptide Maps

A relatively quick and efficient method for deciding whether ribosomal proteins are similar or different in their primary structures is digestion with trypsin and separation of the tryptic peptides by column chromatography (Craven et al. 1969; Rombauts, Peeters and Wittmann 1971; Mora et al.

1971) or by two-dimensional paper electrophoresis and chromatography (Traut et al. 1967; Kaltschmidt et al. 1967; Fogel and Sypherd 1968; Pearson, Delius and Traut 1972). Comparison of the elution profiles and of the "fingerprints" showed that (almost) all proteins studied have different structures. Only proteins L7 and L12 were found to be similar in their peptide maps (Rombauts, Peeters and Wittmann 1971; Terhorst et al. 1972a), a result which is in complete agreement with the amino acid sequences of the two proteins (Terhorst et al. 1972b, 1973).

Isolation and Analysis of Tryptic Peptides

The first step towards the determination of the complete amino acid sequence of a protein is to isolate the tryptic peptides on a preparative scale by column and paper chromatography and to analyze their amino acid compositions. In analyses made in this way on more than 20 ribosomal proteins (Wittmann-Liebold 1971, 1973b and unpublished results), no identical tryptic peptides longer than four amino acids were found among the proteins studied. This result is in good agreement with the conclusion drawn from peptide maps and from immunological studies, namely, if there are homologous structures among ribosomal proteins, the homology must be very low. Knowledge of the amino acid sequences of the ribosomal proteins is necessary to settle finally the question of homology.

N-Terminal Regions

There is a rather unequal distribution of amino acids as N-terminal groups among *E. coli* ribosomal proteins. Methionine and alanine were found to be the most frequently occurring amino acids at the N-termini, both when unfractionated 70S (Waller 1964) or purified ribosomal proteins (Wittmann et al. 1970) were analyzed for end groups.

The availability of sequenators for efficient sequence analysis from the N-terminus (Edman and Begg 1967) enabled amino acid sequence determination of the N-terminal regions of individual ribosomal proteins. There are so far three reports on *E. coli* proteins: Yaguchi et al. (1973) determined the sequences of the first 12 residues from 15 proteins of the small subunit, Wittmann-Liebold (1973a) the first 32–60 residues from 20 proteins (Figure 1), and Higo and Loertscher (1974) the first 25–30 residues from 5 proteins of the small subunit. As far as sequence data for a given protein are at present available from more than one group, there is a good agreement among them. It remains to be seen whether the few points of disagreement are due to errors in the identification of the amino acids or to differences in the proteins which were isolated from different *E. coli* strains. The amino acid sequences for many of the proteins listed in Figure 1 are independently supported by isolation of the tryptic peptides

```
                          1                    5                   10                  15                  20                  25                  30   32
```

S2 Ala-Thr-Val-Ser-Met-Arg-Asp-Met-Leu-Lys-Ala-Gly-Val-His-Phe-Gly-His-Gln-Thr-Arg-Tyr-Trp-Asn-Pro-Lys-Met-Lys-Pro-Phe-Leu-Phe-Gly-

S3 Gly-Gln-Lys-Val-His-Pro-Asn-Gly-Ile-Arg-Leu-Gly-Ile-Val-Lys-Pro-Gly-His-Gly-Asn-Ser-Thr-Gly-Phe-Ala-Asn-Thr-Lys-Glu-Asp-Asn-Leu-

S4 Ala-Arg-Tyr-Leu-Gly-Pro-Lys-Leu-Lys-Leu-Ser-Arg-Arg-Glu-Gly-Thr-Asp-Leu-Phe-Leu-Lys-Ser-Gly-Val-Arg-Ala-Ile-Asp-Thr-Lys-Cys-Lys-

S6 Met-Arg-His-Tyr-Glu-Ile-Val-Phe-Met-Val-His-Pro-Asp-Gln-Ser-Glu-Gln-Val-Pro-Gly-Met-Ile-Glu-Arg-Tyr-Thr-Ala-Ile-Ile-Thr-Gly-Ala-

S7 Pro-Lys-Phe-(Ser)-Val-Lol-Gly-Gln-(Ala)-Lys-(Phe-Asn)-Pro-Asp-Pro-Met-Phe-Gly-Ser-Glu-Lol-Lol-Ala-Lys-Ala-Val-Lol-Met-Val-Ala-

S8 Ser-Met-Gln-Asp-Pro-Ile-Ala-Asp-Met-Leu-Thr-Arg-Ile-Arg-Asn-Gly-Gln-Ala-Ala-Asn-Lys-Ala-Ala-Val-Thr-Met-Pro-Ser-Lys-Leu-Lys-

S9 Ala-Glu-Asn-Gln-Tyr-Gly-Thr-Gly-Arg-Arg-Lys-Ser-Ser-Ala-Ala-Arg-Val-Phe-Ile-Lys-Pro-Gly-Asn-Gly-Lys-Ile-Val-Ile-Asn-Gln-Arg-

S10 Met-Gln-Asn-Gln-Arg-Ile-(Val-Phe)-Met-Val-Leu)-Ala-Phe(Phe)-His-Arg-(Leu(Asp-Ile)-Gln-Ala-Thr-Glu-Lol-Val-Glu-Thr-Ala-(Ala)-Glu-(Thr)-

S11 Phe-Lys-Ala-Pro-Lol-Arg-Ala-Arg-Lys-Arg-Val-Arg-Lys-Gln-Val-Ser-(Arg)-Gly-Val-Ala-His-Lol-(His)-Ala-Ser-Phe-Asn-Asn-Thr-Lol-Val-(Thr)-

S12 Ala-Thr-Val-Asn-Gln-Leu-Val-Arg-Lys-Pro-Arg-Ala-Arg-Lys-Val-Ala-Lys-(Lys)-Asn-Val-Pro-Ala-Leu-Glu-Ala-(Cys)-Pro-Glx-Ser-Arg-Gly-Val-

S13 Ala-Arg-Ile-Ala-Gly-Ile-Asn-Ile-Pro-Asp-His-Lys-His-Ala-Val-Ile-Ala-Leu-Thr-Ser-Ile-Tyr-Gly-Val-Gly-Lys-Ala-Lol-(Lol)-Lys-Ala-Lol-(Ser)-Arg-

S14 Ala-Lys-Gln-Ser-Met-Lys-Ala-Arg-Glu-Val-Lys-Arg-Val-Ala-Leu-Ala-Asp-Lys-Tyr-Phe-Ala-Lys-Arg-Ala-Glu-Lol-(Lol)-Lys-Ala-Lol-(Ser)-Arg-

S15 Ser-Leu-Ser-Thr-Glu-Ala-Thr-Ala-Lys-Ile-Val-Ser-Glu-Phe-Gly-Arg-Asp-Ala-Asn-Asp-Thr-Gly-Ser-Thr-Glu-Val-Gln-Val-Ala-Leu-Leu-Thr-

S16 Met-Val-Thr-Ile-Arg-Leu-Ala-Arg-His-Gly-Ala-Lys-Lys-Arg-Pro-Phe-Tyr-Gln-Val-Val-Val-Ala-Asp-Ser-Arg-Asn-Ala-Arg-Asn-Gly-Arg-Phe-

S17 Thr-Asp-Lys-Ile-Arg-Thr-Leu-Gln-Gly-Arg-Val-Val-Ser-Asp-Lys-Met-Glu-Lys-(Ser)-Lol-Val-Val-Ala-Lol-Val-Glu-Arg-Phe-Val-Lys-His-Pro-Lol-

S19 Pro-Arg-Ser-Lol-Lys-Lys-Gly-Pro-Phe-Lol-Asp-Lol-His-Lol-Leu-Lys-Lys-Val-Glu-Lys-Ala-Val-Glu-Ser-Gly-Asp-Lys-Lys-Pro-Lol-(Pro)-Arg-

S20 Ala-Asn-Ile-Lys-Ser-Ala-Lys-Lys-Arg-Ala-Ile-Gln-Ser-Glu-Lys-Ala-Arg-Lys-His-Asn-Ala-Ser-(Ala)-Arg-Ser-Met-Met-Arg-Thr-Phe-Ile-Lys-

S21 Pro-Val-Ile-Lys-Val-Arg-Glu-Asn-Glu-Pro-Phe-Asp-Val-Ala-Leu-Arg-Arg-Phe-Lys-Arg-Ser-Cys-Glu-Lys-Ala-Gly-Val-Leu-Ala-Glu-Val-Arg-

L12 Ser-Ile-Thr-Lys-Asp-Gln-Ile-Ile-Glu-Ala-Val-Ala-Ala-Met-Ser-Val-Met-Asp-Val-Val-Glu-Leu-Ile-Ser-Ala-Met-Glu-Glu-Lys-Phe-Gly-Val-

L24 Ala-Ala-Lys-Ile-Arg-Arg-Asp-Asp-Glu-Val-Leu-Val-Ile-Thr-Gly-Lys-Asp-Lys-Gly-Lys-Val-Gly-Lys-Val-Lys-Asn-Val-Leu-Ser-Gly-Lys-

118

	33	35		40		45		50		55		60

S2: Ala-Ala-Asn-

S3: Asp-Ser-Asp-Phe-Lys-Val-Arg-Gln-Tyr-Leu-Thr-Lys-Glu-Leu-Ala-Lys-Ala-

S4: Ile-Gln-Ala-Pro-Gly-Gln-His-Gly-Ala-Arg-Lys-Pro-Arg-Leu-Ser-Asp-Tyr-Gly-

S6: Glu-Gly-Lys-Lol-(Lol)-His-Lol-Glu-Asp-Gly-

S8: Val-Ala-

S9: Ser-Leu-Glu-Gln-Tyr-Phe-Gly-Arg-Glu-Ala-Ala-Met-Arg-Val-Val-Gln-Pro-(Ser-Glu)-Leu-Val-Met-(Arg)-Arg-

S10: Gly-Ala-Gln-Leu-Val-Gly-Pro-Phe(Leu-Leu)-

S12: Cys-Thr-Arg-Val-Tyr-Thr-Thr-Thr-Pro-Lys-Pro-Asn-Ser-Ala-Leu-Arg-

S13: Leu-Ala-Ala-Ala-Gly-Lol-Ala-Arg-(Glu)-Val-Lys-Leu-Thr-Leu-Thr-Leu-Glu-Ser-Gly-Gln-Lol-Arg-Thr-Lol-Leu-Arg-Glu-Val-Ala-

S14: Asn-Val-(Asn)-Ala-

S15: Ala-Gln-Ile-Asn-His-Leu-Gln-Gly-Gly-Phe-Ala-Gla-Glu-(Arg-Lys-Arg-Ser-Arg-Ser)-Phe-(Gly-Lol)-Lol-Met-Val-

S16: Ile-Glu-Arg-Val-Gly-Phe-Phe-Asn-Pro-Ile-Ala-Ser-Glu-Lys-Glu-Gly-Thr-Arg-Leu-Asp-Leu-Asp-Arg-

S17: Tyr-Gly-

S19: Val-Asn-Ala-Ser-

S20: Lys-Val-Tyr-Ala-Ala-Ile-Glu-Ala-Glu-Asp-Lys-Arg-Ala-Ala-Gln-Lys-Ala-

S21: Arg-Arg-Glu-Phe-Tyr-Glu-Lys-Pro-Arg-Thr-Glu-Thr-Arg-Lys-Ala-Lys-Ala-Ser-Ala-Val-Lys-

L12: Ser-Ala-Ala-Ala-Val-Ala-Ala-Ala-Gly-Pro-Val-Ala-Ala-Glu-Glu-Gln-Lys-

L24: Val-Ile-Val-Glu-Gly-Ile-Asn-Leu-Val-Lys-Lys-

Figure 1 Amino acid sequences of 20 *E. coli* ribosomal proteins (Wittmann-Liebold 1973a). LoI means leucine or isoleucine. Identification of amino acids in brackets is not ambiguous. Proteins S5 and S18 did not give PTH-amino acids by Edman degradation. Protein S7: Although only one amino acid was found in many positions, two amino acids, namely, one in high concentration and one in low concentration (not shown), were obtained in some positions. Further studies are in progress to clarify these findings.

Sequence analysis of the N-terminal region of protein S10 from different batches resulted in two different sequences for positions 7–11 and 14, namely, one given here and the other corresponding to that determined by Higo and Loertscher (1974) and by M. Yaguchi, L. P. Visentin and A. T. Matheson (personal communication). The sequences for the other positions (1–6, 12–13, 15–29) were identical.

119

from these proteins and determination of their amino acid compositions, which agrees very well with the reported sequences (for details see Wittmann-Liebold 1973b).

The most important finding from the sequence analysis of *E. coli* ribosomal proteins is the direct and unambiguous evidence that all 20 proteins so far studied (Figure 1) have primary structures completely different from each other. This result is in excellent agreement with immunological studies (Stöffler and Wittmann 1971a,b; Stöffler et al. 1974) in which no cross-reaction among the individual ribosomal proteins (with the exception of L7/L12 and S20/L26) could be detected. It was concluded from these studies that no extensive identical regions exist in different ribosomal proteins of *E. coli*. Since an antigenic determinant

Table 1   Comparison of identical regions in ribosomal proteins

	Sequence	Positions
S4	Arg-Lys-Pro-Arg	43–46
S12	Arg-Lys-Pro-Arg	8–11
S21	Lys-Pro-Arg	39–41
S16	Ala-Lys-Lys-Arg	11–14
S20	Ala-Lys-Lys-Arg	6–9
S3	Lys-Pro-Gly-Asn	15–18
S9	Lys-Pro-Gly-Asn	21–24
S13	Ala-Ala-Ala	34–36
L12	Ala-Ala-Ala	34–36
S9	Ala-Ala-Ala	42–44
S2	Ala-Thr-Val	1–3
S12	Ala-Thr-Val	1–3
S17	Lys-Ile-Arg	3–5
L24	Lys-Ile-Arg	3–5
S19	Lys-Lys-Val	16–18
S20	Lys-Lys-Val	32–34
S16	Lys-Lys-Arg	12–14
S20	Lys-Lys-Arg	7–9
S15	Lys-Lys-Arg	48–50
S14	Val-Ala-Leu	13–15
S21	Val-Ala-Leu	13–15
S15	Val-Ala-Leu	28–30
S3	Val-His-Pro	4–6
S6	Val-His-Pro	10–12

consists of at least 4–5 amino acids (Sela et al. 1967; Arnon 1971; Benjamin, Michaeli and Young 1972), identical regions with a lower number of amino acids would not have been detected by immunological techniques (see Stöffler, this volume).

Comparison of the amino acid sequences summarized in Figure 1 shows that some proteins have identical tetra- and tripeptides (Table 1). Furthermore, there exist regions of homologous structure among different proteins. These regions consist of very short identical sequences interrupted by nonidentical amino acids, as illustrated in Figure 2. Certain homologous patterns are recognizable.

As ribosomal proteins are rich in basic amino acids (up to about 35 moles percent), a clustering of lysines and/or arginines can be expected. In agreement with this expectation, long regions with basic amino acids occur in several proteins (Table 2). Similarly, apolar amino acids are clustered in some regions (Table 3), and it is possible that these regions are located in the interior of the molecule. One of the apolar regions contains four consecutive alanine residues, namely, positions 34–37 in protein L12. Three alanines in a row occur in positions 42–44 of S9, three threonines in positions 38–40 of S12, and three arginines in 32–34 of S21. Regions containing at least 50% acidic amino acids are listed in Table 4.

```
        7                                                         26
S3   Asn-Gly-Ile -Arg-Leu-Gly-Ile -Val- Lys-Pro-Gly-Asn-Ser-Thr-Gly- Phe-Ala-Asn-Thr-Lys
        13                                                        32
S9   Ser-Ser-Ala-Ala-Arg-Val-Phe-Ile - Lys-Pro-Gly-Asn-Gly-Lys- Ile -Val- Ile -Asn-Gln-Arg
        35                                         49
S12  Arg-Val-Tyr-Thr-Thr-Thr-Pro-Lys-Lys-Pro- Asn-Ser- Ala-Leu-Arg


        1                       9
S19  Pro-Arg-Ser-Lol-Lys-Lys-Gly-Pro-Phe
        1                            11
S21  Pro-Val- Ile -Lys-Val- Arg-Glu-Asn-Glu-Pro-Phe


        41                           52
S4   Gly-Ala-Arg-Lys-Pro-Arg-Leu-Ser-Asp-Tyr- Gly-Val
        6                            17
S12  Leu-Val-Arg-Lys-Pro-Arg-Ala- Arg-Lys-Val-Ala-Lys
        36                                48
S21  Phe-Tyr-Glu-Lys-Pro-Arg -Thr-Glu-Thr- Arg-Lys-Ala-Lys


        16                                28
S19  Lys-Lys-Val-Glu-Lys-Ala-Val-Glu-Ser-Gly-Asp-Lys-Lys
        32                                44
S20  Lys-Lys-Val-Tyr-Ala-Ala- Ile -Glu-Ala-Gly-Asp-Lys-Arg


        9                                                                  30
L12  Glu-Ala-Val-Ala-Ala-Met-Ser-Val-Met -Asp-Val-Val-Glu-Leu-Ile-Ser-Ala-Met-Glu-Glu-Lys-Phe
        5                                                             26
S6   Glu-Ile -Val-Phe-Met-Val-His-Pro-Asp-Gln-Ser-Glu-Gln-Val-Pro-Gly-Met-Ile-Glu-Arg-Tyr-Thr
```

Figure 2 Homologous regions in *E. coli* ribosomal proteins. The sequence regions of S19 and S20 are taken from Higo and Loertscher (1974). All other sequences are from Figure 1.

Table 2 Regions with at least 50% basic amino acids

	Positions	Total no. amino acids	No. basic amino acids
S4	7–13	7	4
S4	40–46	7	4
S4	55–62	8	5
S4	143–153	11	6
S4	180–185	6	4
S9	10–12	3	3
S11	6–13	8	6
S12	8–18	11	7
S13	11–13	3	3
S13	26–30	5	3
S14	6–12	7	4
S15	47–52	6	5
S16	5–14	10	6
S17	26–30	5	3
S19	2–6	5	3
S19	13–20	8	4
S19	27–32	6	3
S20	4–9	6	4
S20	15–19	5	4
S20	43–48	6	3
S21	16–24	9	5
S21	32–34	3	3
S21	39–48	10	5
L24	3–6	4	3
L24	16–25	10	6

Table 3 Regions with apolar amino acids

	Positions
S2	28–35
S3	6– 9, 11–14, 21–24
S4	35–39, 51–54, 88–94, 104–107, 120–125, 127–131
S6	6–10, 17–22
S7	25–32
S8	15–20
S9	27–31, 42–45, 47–50
S12	3– 7, 19–23
S13	3– 9, 14–18, 31–39
S14	13–16
S15	26–31, 33–36, 38–43, 54–59
S16	18–22, 36–43
S17	20–24
S19	7–10
S21	25–29
L12	10–14, 34–45, 60–64, 66–69, 76–79, 90–94
L24	10–13, 37–41

Table 4   Regions with at least 50% acidic amino acids

	Positions	Total no. amino acids	No. acidic amino acids
S16	45–48	4	3
S20	39–42	4	3
S21	7–12	6	3
L12	46–50	5	3
L12	101–104	4	3
L12	111–118	8	4
L24	7–9	3	3

Complete Amino Acid Sequences

Three proteins, two from the large subunit and one from the small, have so far been completely sequenced. The determination of the complete amino acid sequences for several other proteins can be expected in the near future.

Proteins L7 and L12.   The sequence determination (Terhorst et al. 1972b, 1973) was carried out with a Beckman protein sequencer as improved by Wittmann-Liebold (1973b), with a solid phase peptide sequenator as developed by Laursen (1971), and with conventional methods. It was already suggested from immunological studies (Stöffler and Wittmann 1971b), as well as from comparison of the amino acid compositions (Kaltschmidt et al. 1970a; Möller et al. 1972) and of the tryptic peptides (Terhorst et al. 1972a; Rombauts, Peeters and Wittmann 1971), that proteins L7 and L12 are very similar. The complete sequence analyses of the two proteins (Terhorst et al. 1972b, 1973) demonstrated unambiguously the almost complete identity of L7 and L12. The only difference between them is the N-terminus: L7 begins with *N*-acetyl-Ser and L12 with Ser. All other amino acid positions are identical, even the occurrence of ε-*N*-monomethyllysine at position 81 (Figure 3). It is notable that the modified lysine residue is present in only 50% of the protein chains of both L7 and L12.

Hydrophobic and charged amino acids are distributed along the protein chain in three regions (Terhorst et al. 1973). The first region (from the N-terminus up to position 55) is negatively charged and contains many hydrophobic amino acids, e.g., only alanine and valine residues in positions 34–42 (Figure 3). In the second region, comprising positions 56–81, no negatively charged residues occur, whereas there is a cluster of polar and of negatively charged amino acids in the third region (positions 82–120). It is likely that the first region contributes strongly to the relatively high α-helix content of proteins L7 and L12. Ptitsyn et al. (1973) predicted

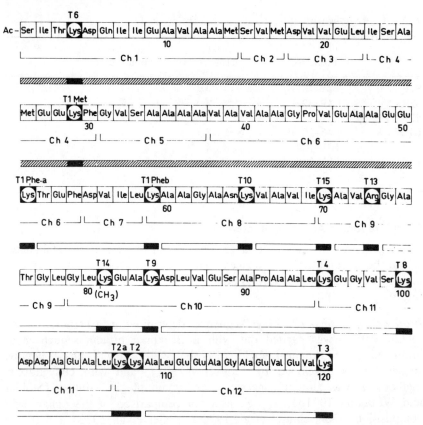

Figure 3 Amino acid sequence of *E. coli* protein L7. (Reprinted with permission from Terhorst et al. 1973.)

that there are five fixed helical regions (namely, positions 4–30, 32–40, 65–73, 82–88 and 104–112) in the compact protein globule and four fluctuating helical regions (7–42, 44–73, 80–88 and 91–119) in the unfolded chain.

The functions of protein L7 and L12 in the translation process, especially in the translocation step, and the possible analogy with motile proteins is discussed elsewhere in this volume by Möller.

Protein S4. The sequence of this protein (Figure 4) was determined by Reinbolt and Schiltz (1973) using the methods mentioned in the previous chapter. With 203 amino acid residues and a molecular weight of 22,550, it is one of the largest proteins in the 30S subunit. In contrast to proteins L7 and L12, whose isoelectric points are at about pH 5, protein S4 is a basic protein with 42 lysine and arginine residues. Basic amino acids are clustered in the following regions: positions 7–13, 43–46, 55–62, 143–153 and 180–185. As protein S4 is one of the strongest RNA-binding proteins, it would be interesting to see which regions and amino acids are involved in the RNA protein interaction. Another reason for determining the sequence of S4 is the fact that many mutants with altered S4 protein have been isolated and studied by genetic and electrophoretic methods. Among them are mutants with S4 chains shorter or longer than that of the wild type and with drastic changes at the C-termini (see section on S4 mutants).

ESCHERICHIA COLI MUTANTS

Alterations in at least ten ribosomal proteins have so far been found in *E. coli* mutants and/or strains. Most of these mutants have an altered response towards antibiotics or temperature.

Protein S4

This protein was found to be altered in revertants from streptomycin dependence to independence (Deusser et al. 1970; Birge and Kurland 1970; Kreider and Brownstein 1971; Hasenbank et al. 1973) and in Ram (ribosomal ambiguity) mutants (Zimmermann, Garvin and Gorini 1971; Hasenbank et al. 1973).

The differences between proteins S4 from the mutants and wild type are rather drastic (Donner and Kurland 1972; Funatsu et al. 1972c). That more than one amino acid replacement is involved is shown by the following results (Funatsu et al. 1972c): The molecular weights of mutant S4 proteins can be considerably (up to 30 amino acids) shorter or longer than S4 from the wild type. The amino acid sequence at the C-terminal region of S4 is completely different among the various mutants as well as between the mutants and the wild type. The molecular mechanism for a shortening of the mutant S4 protein chain is illustrated in Figure 5. It consists of a combination of a deletion and frame shift leading to a new termination codon (E. Schiltz, unpublished results). The changes at the C-terminal region of the mutant S4 protein lead to a suppression of the dependence on streptomycin which is caused by an alteration in protein S12 (see section on protein S12).

A difference in the chromatographic mobility has been found between protein S4 from a so-called Ram B mutant and its parental strain (Zimmermann, Ikeya and Sparling 1973). This mutation maps near the gene for kasugamycin resistance, which is located at 0.5 min. Since it is well established by the genetic and protein-chemical studies cited above that the structural gene for protein S4 is in the *str*A region, the alteration in protein S4 of the Ram B mutants is probably a secondary modification of the protein, e.g., by methylation, phosphorylation or acylation. Further work is necessary to test this hypothesis.

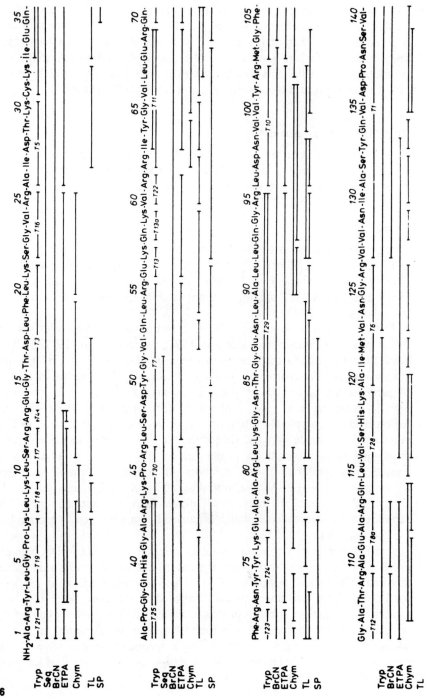

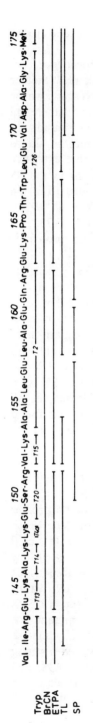

Figure 4 Amino acid sequence of *E. coli* protein S4. (Reprinted with permission from Reinbolt and Schiltz 1973.)

```
              180       185                 190                 195                 200      203
Wild type:.... Lys-Arg-Lys-Pro-Glu-Arg-Ser-Asp-Leu-Ser-Ala-Asp-Ile-Asn-Glu-His-Leu-Ile-Val-Glu-Leu-Tyr-Ser-Lys-
               AAA AGA AAA CCU GAA AGA UCU GAU CUG UCU GCG GAC AUU AAU GAA CAC UUA AUU GUU GAA UUA UAU UCU AAA

Mutant:....... AAA AGA AAA ┏━ deletion ━┓ UCU GUC UGC GGA CAU UAA UGA
               Lys-Arg-Lys             Ser-Val-Cys-Gly-His end end
```

Figure 5 Amino acid sequence of the C-termini of *E. coli* proteins S4 from wild type and from a mutant with a shorter S4 chain. The possible codons are assigned to the amino acids. The deletion of 14 nucleotides leads to a frame shift and to a new termination codon (E. Schiltz, unpublished).

Protein S5

Altered S5 proteins have so far been detected in the following four classes of mutants:

1. In revertants from streptomycin dependence to independence, not only protein S4 (see above) but also protein S5 can be altered (Stöffler et al. 1971; Kreider and Brownstein 1972). Among 100 revertants whose ribosomal proteins were analyzed by two-dimensional polyacrylamide gel electrophoresis (Hasenbank et al. 1973) three classes were found: 24 mutants had altered S4 proteins and 16 had altered S5, whereas no alteration could be detected among the rest. In contrast to the rather extensive differences in the S4 proteins of the revertants, the alterations in the S5 proteins are single amino acid replacements. This is concluded from the electrophoretic mobility of the mutant proteins (Hasenbank et al. 1973) and directly shown by sequence analysis (Itoh and Wittmann 1973): A leucine residue in the tryptic peptide T2 of the wild-type S5 protein is replaced by arginine in one of the revertants.

2. Spectinomycin resistance is conferred by an altered S5 protein, as shown by hybrid reconstitution with proteins from the wild type and a mutant (Bollen et al. 1969). This result agrees with the finding that S5 is the only chromatographically altered protein in spectinomycin-resistant ribosomes (Dekio and Takata 1969; Dekio et al. 1970). Comparison of the total amino acid composition of a spectinomycin-resistant mutant with that of its parental strain revealed a difference of one amino acid (Bollen and Herzog 1970). The amino acid replacements in protein S5 were determined for several spectinomycin-resistant mutants (Funatsu et al. 1971, 1972b; DeWilde and Wittmann-Liebold 1973), and it was found that they are located within a very narrow region, namely, three adjacent amino acids of the S5 protein chain. Clearly this region is very important: apparently only alterations in this region lead to spectinomycin resistance. Similar observations were made with mutants resistant to other antibiotics, e.g., streptomycin and erythromycin (see proteins S12 and L4).

3. Two types of mutants with unusual ribosomal sedimentation profiles and with altered S5 proteins have been described: (a) A mutant (*spc*-49-1) was isolated as SpcR and was found to have the *"sad"* phenotype. It is cold sensitive and accumulates precursors of 30S and 50S subunits at low temperature. Genetic studies showed that the same mutation is responsible for both the *spc*R and the *"sad"* phenotypes (Guthrie et al. 1969; Nashimoto and Nomura 1970). Further studies revealed that protein S5 is altered in this mutant (Nashimoto et al. 1971). (b) Another mutant was isolated as a suppressor of a defect in the gene for alanyl-tRNA synthetase (Buckel et al. 1972). Its ribosomal sedimentation profile showed a strong increase in amount of a 50S precursor particle sedimenting at 43S, and protein S5 was found to be altered (Wittmann et al. 1974). The alteration in S5 of this mutant differs from all other alterations so far known in S5

mutant proteins. Genetic studies with revertants and transductants provided evidence that the mutation in protein S5 leads to suppression of the defect in alanyl-tRNA synthetase.

4. Protein S5 is one of the two ribosomal proteins which differ among various *E. coli* strains (the other protein is S7). Protein S5 of strain B is different from S5 of strains C, K and MRE 600 as shown by column chromatography (Takata et al. 1969; Osawa et al. 1970) and by two-dimensional gel electrophoresis (Kaltschmidt et al. 1970b). From the different electrophoretic mobilities it can be concluded that the difference between the S5 proteins probably consists of a single amino acid replacement. This was directly demonstrated by sequence analysis: An alanine residue in peptide T1 of protein S5 from strain B is exchanged for glutamic acid at the corresponding position in S5 from strain K12 (Wittmann-Liebold and Wittmann 1971).

Protein S6

In mutants resistant to neo- and kanamycin, an altered S6 protein has been detected by two-dimensional polyacrylamide gel electrophoresis (H. G. Wittmann and D. Apirion, unpublished results). The altered protein is more basic or less acidic than the S6 protein from the parental strain.

Protein S7

Besides S5, another protein, namely S7, differs among *E. coli* strains. Protein S7 from K strains, the so-called K protein, is electrophoretically and chromatographically different from S7 of strains B, C and MRE 600 (Leboy, Cox and Flaks 1964; Birge et al. 1969; Sypherd 1969; Takata et al. 1969; Osawa et al. 1970; Kaltschmidt et al. 1970b). There are not only differences in charge and amino acid composition, but also in molecular weight (Kaltschmidt et al. 1970b): S7 from strain K is about 10% longer than S7 from the other strains. This drastic difference in size is remarkable in view of the almost identical sets of ribosomal proteins in all *E. coli* strains studied: None of the 50S and 30S proteins, except S5 and S7, have been found so far to differ among the various *E. coli* strains.

Protein S8

Temperature-sensitive mutants, which grow at 30°C but not at 42°C, were isolated by a special method (Phillips, Schlessinger and Apirion 1969). The ribosomal proteins from 12 of these mutants were studied by two-dimensional polyacrylamide gel electrophoresis and immunological techniques (G. Stöffler, H. G. Wittmann and D. Apirion, manuscript in preparation). Protein S8 in at least two of these mutants was found to be different from that of the parental strain. It is clear from the electrophoretic comparison that the alterations in the mutant S8 proteins are not identical. The complete primary structure of protein S8 was determined and the amino acid replacements in the mutants were localized (H. Stadler, manuscript in preparation).

Protein S12

Mutation in this protein confers resistance to (Ozaki, Mizushima and Nomura 1969) and dependence on streptomycin (Birge and Kurland 1969) as shown by hybrid reconstitution and in vitro tests. The amino acid replacements in protein S12 from streptomycin-resistant mutants are clustered in only two amino acid positions (Funatsu et al. 1972a; Funatsu and Wittmann 1972): The lysine residue at position 42 (in the tryptic peptide T6) of the S12 wild-type protein chain is replaced by asparagine, threonine and arginine in the mutants A1, A2 and A60, respectively. Another lysine residue at position 87 (in peptide T15) is exchanged for arginine in mutant A40. This result is in very good agreement with genetic studies (Breckenridge and Gorini 1970), according to which the mutation A1, A2 and A60 are clustered at one site on the bacterial chromosome, whereas A40 is located 0.3 map unit away from this site.

The four mutants A1, A2, A40 and A60 differ in their ability to restrict phenotypic and genotypic suppression in translation (Gorini 1969, 1971). For instance, addition of streptomycin to a cell-free system containing ribosomes from A60 mutants causes much more error in translation than is the case with ribosomes from A1 under the same conditions. As the difference between the mutants consists of single amino acid replacements, it follows that the exchange of only one out of the approximately 8000 amino acids in the ribosome can be sufficient to change drastically the error frequency in translation. It follows from the protein-chemical studies that both position and type of the amino acid replacements influence translational misreading. The importance of the position of the replacement is illustrated by the finding that two mutants (A40 and A60) with different degrees of translational misreading have the same type of amino acid exchange (lysine by arginine) but at different positions in S12. On the other hand, in three mutants (A1, A2 and A60), which strongly differ in translational misreading, different amino acid replacements occur at the same position. Further support for the importance of the type of replacement is the finding that the substitution of lysine by another basic amino acid (arginine in mutant A60) does not lead to such a drastic change in misreading frequency as the replacement of this lysine residue by neutral amino acids (asparagine and threonine in mutants A1 and A2, respectively).

Mutations to streptomycin dependence lead to amino acid replacements not only in the same ribosomal protein, namely S12, as mutations to streptomycin resistance, but also in the same regions of the S12 protein chain (Funatsu and Wittmann 1972; Itoh and Wittmann 1973; U. van Acken, unpublished results). These findings strengthen the conclusion drawn from the sequence analysis of streptomycin-resistant mutants that two small regions in protein S12 are very important. Amino acid replacements within these regions lead to an altered response (resistance or dependence instead of sensitivity) of the ribosome towards streptomycin.

Protein S17

E. coli strain AT2472 has an altered S17 protein as found by CM-cellulose chromatography (Dekio 1971) and by two-dimensional polyacrylamide gel electrophoresis (Muto et al. 1974). From the altered chromatographic and electrophoretic mobilities it follows that the altered protein is more basic or less acidic than protein S17 from other *E. coli* strains.

Protein S18

In an *E. coli* mutant originally isolated as a temperature-sensitive mutant, protein S18 is altered, as found by column chromatography and polyacrylamide gel electrophoresis (Bollen et al 1973). The alteration in the mutant protein consists of a replacement of arginine by cysteine (Kahan et al. 1973). Genetic analysis showed that the gene of protein S18 maps between 76–88 min (Bollen et al. 1973), i.e., outside the chromosomal region where all cistrons for ribosomal proteins so far identified are located.

Protein S20

Mutants with altered S20 proteins were found as follows: The first step was isolation of temperature-sensitive mutants with altered alanyl-tRNA synthetases. From these mutants "revertants" were isolated which had lost the temperature sensitivity and which showed an increased amount of stable RNA (Buckel et al. 1972). Some of them had altered S20 proteins as detected by two-dimensional gel electrophoresis and by immunological methods (Wittmann et al. 1974). It can also be concluded from these studies that the mutant S20 proteins are more acidic or less basic than S20 from the wild type and that the alterations in the S20 proteins from several mutants are not identical.

Protein L4

This protein, which corresponds to protein 50-8 in the nomenclature of the group in Hiroshima, has been found to be altered in erythromycin-resistant *E. coli* mutants (Otaka et al. 1970; Dekio et al. 1970). The ribosomes isolated from these mutants do not bind erythromycin, and the formation of N-acetylphenylalanyl puromycin is strongly impaired (Wittmann et al. 1973). The amino acid replacements in proteins L4 from several mutants are located within the same tryptic peptide, as concluded from comparison of peptide maps of the mutant proteins (Otaka et al. 1971). This finding points to a situation similar to that discussed for mutants resistant to spectinomycin (altered in protein S5) and to streptomycin (altered in S12) in which the amino acid replacements are clustered within narrow regions of the protein chain. Protein L4 is not only altered in mutants resistant to erythromycin, but also in those which are relatively sensitive to the action of this drug but resistant to other antibiotics belonging to the macrolides, such as spiramycin, leucomycin and tylosin (Tanaka et al. 1971).

Protein L22

Besides erythromycin-resistant mutants in which protein L4 is altered, there exists another class, namely, one with an alteration in protein L22 (Wittmann et al. 1973). Ribosomes from the latter mutants bind erythromycin as efficiently as those from wild type, and the formation of N-acetylphenylalanyl puromycin is not decreased. There are several explanations for the possible mechanism of erythromycin resistance in the mutants with altered L22 proteins (Wittmann et al. 1973), and it remains to be seen which of them can be demonstrated to be true.

ENTEROBACTERIACEAE

Ribosomal proteins of *E. coli* were compared with those of other species belonging to the family of Enterobacteriaceae by CM-cellulose column chromatography (Otaka et al. 1968; Osawa et al. 1972), by one-dimensional (Sun, Bickle and Traut 1972) and two-dimensional polyacrylamide gel electrophoresis (Geisser 1971; Geisser et al. 1973b), as well as by immunochemical methods (Wittmann et al. 1970; Geisser et al. 1973b). The studies allow conclusions about the taxonomical relationship of bacterial species and the degree of conservation in the structure of ribosomal proteins from numerous species belonging to Enterobacteriaceae, e.g., *E. coli, E. freundii, E. paraintermedia, E. adecarboxylata, Salmonella typhimurium, Paracolobactrum coliforme, Shigella dispar, Shigella dysenteriae, Aerobacter aerogenes, Serratia marcescens, Erwinia carotovorum* and *Proteus vulgaris*.

The comparative investigations gave several interesting results (Osawa et al. 1972; Geisser et al. 1973b): (a) Some proteins (e.g., S3, S4, S11, S14, S19, S21, L2, L14, L24 and L28) are more conserved than others (e.g., S10, S16 and L27) as shown by two-dimensional gel electrophoresis and by immunological methods. This finding probably reflects the evolutionary pressure for conservation of a given protein structure. (b) All Enterobacteriaceae ribosomes contain an S7 protein very similar to that from *E. coli* strain B and distinctly different from K strains. (c) Ribosomal proteins from species belonging to different genera, e.g., *E. coli* and *Shigella dispar,* can be much more similar than those from the same genus, e.g., *E. coli* and *E. freundii.* (d) There is not good agreement between the current taxonomy and the relationship among ribosomal proteins from various species. On the other hand, such an agreement does exist for the studies on ribosomes and other biochemical data, e.g., on hybridization and GC content.

BACILLACEAE

Comparison of ribosomal proteins from numerous species belonging to the Bacillaceae by column chromatography (Otaka et al. 1968), as well as by two-dimensional gel electrophoresis and immunological techniques (Geisser et al. 1973a), revealed that there is much less resemblance among ribosomes from different species of Bacillaceae than among those of Enterobacteriaceae. For instance, the ribosomal proteins of *Bacillus subtilis,* the type species of Bacilli (Bergey 1957), on the one hand and those of *B. circulans, B. coagulans, B. brevis* and *B. cereus* on the other hand are rather dissimilar, whereas a great similarity between *B. subtilis* and *B. licheniformis* was revealed by immunological and electrophoretic methods. The ribosomal proteins of *B. megaterium* and *B. pumilis* are related to those of *B. subtilis* to an extent which is in between the two (dissimilar and similar) groups (Geisser et al. 1973a). The classification of the various Bacillus species according to their ribosomal proteins agrees rather well with data on the GC content (Marmur, Falkow and Mandel 1963) and hybridization of DNA with RNA (Doi and Igarashi 1965). Extensive comparisons have been made between ribosomal proteins from *B. stearothermophilus* and *E. coli.* One of the reasons for this comparison is the successful reconstitution of *B. stearothermophilus* 50S subunits (Nomura and Erdmann 1970). The finding that *E. coli* 30S proteins and *B. stearothermophilus* 16S RNA can form active 30S particles (Nomura, Traub and Bechmann 1968) shows that ribosomal proteins of *E. coli* can replace those of *B. stearothermophilus* in reconstitution and that therefore homologies must exist among them. These homologies have been demonstrated for individual proteins by functional tests and by immunological and chemical comparisons.

The functional correspondence was revealed by reconstitution of 30S *E. coli* ribosomes in which one 30S protein at a time was replaced by the corresponding protein from *B. stearothermophilus* 30S subunits (Higo

134

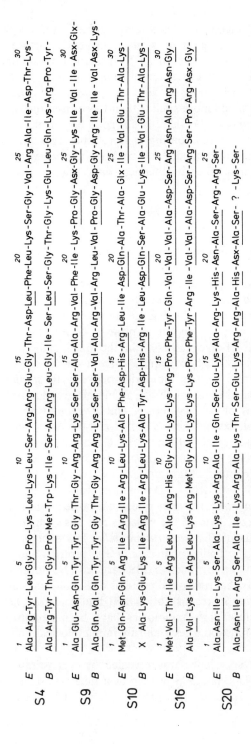

Figure 6 Comparison of the amino acid sequences of five corresponding ribosomal proteins from *E. coli* and *Bacillus stearothermophilus* (Higo and Loertscher 1974). The N-terminal regions of other ribosomal proteins from both bacteria are compared by Yaguchi et al. (1973).

et al. 1973). In this way a functional correspondence was found for many of the *E. coli* and *B. stearothermophilus* 30S proteins.

A structural homology between each of the 30S proteins from both bacterial species was demonstrated by studies on the amino acid sequence of the N-terminal region, on the molecular weight, on the amino acid composition, on the mobility in two-dimensional polyacrylamide gel electrophoresis, and on the immunological cross-reaction of the isolated 30S proteins (Isono et al. 1973). A few of these proteins have been independently correlated by immunological methods (Higo et al. 1973) and by amino acid sequence studies (Higo and Loertscher 1974). The degree of structural homology between corresponding proteins of *E. coli* and *B. stearothermophilus* is illustrated in Figure 6. The relatively high number (30–70%) of identical positions in the amino acid sequences reflects the rather strong conservation of the primary structure during evolution of the two bacterial species.

References

Acharya, A. S. and P. B. Moore. 1973. Reaction of ribosomal sulfhydryl groups with 5,5′-dithiobis-(2-nitrobenzoic acid). *J. Mol. Biol.* **76**:207.

Alix, J.-J. and D. Hayes. 1974. Properties of ribosomes and RNA synthesised by *Escherichia coli* grown in the presence of ethionine. III. Methylated protein in 50S ribosomes of *E. coli* EA2. *J. Mol. Biol.* (manuscript submitted).

Arnon, R. 1971. Antibodies to enzymes—a tool in the study of antigenic specificity determinants. *Current Topics Microbiol. Immunol.* **54**:47.

Bakardjieva, A. and R. R. Crichton. 1974. Topography of *E. coli* ribosomal protiens: The order of reactivity of sulfhydryl groups. *Biochem. J.* (in press).

Benjamin, E., D. Michaeli and J. D. Young. 1972. Antigenic determinants of proteins of defined sequences. *Current Topics Microbiol. Immunol.* **58**:85.

Bergey's Manual of determinative bacteriology. 1957. Williams & Wilkins, Baltimore.

Birge, E. A. and C. G. Kurland. 1969. Altered ribosomal protein in streptomycin-dependent *Escherichia coli. Science* **166**:1282.

————. 1970. Reversion of a streptomycin dependent strain of *Escherichia coli. Mol. Gen. Genet.* **109**:356.

Birge, E. A., G. R. Craven, S. J. S. Hardy, C. G. Kurland and P. Voynow. 1969. Structure determinant of a ribosomal protein: K locus. *Science* **164**:1285.

Bollen, A. and A. Herzog. 1970. The ribosomal protein altered in spectinomycin-resistant *Escherichia coli. FEBS Letters* **6**:69.

Bollen, A., J. Davies, M. Ozaki and S. Mizushima. 1969. Ribosomal protein conferring sensitivity to the antibiotic spectinomycin in *Escherichia coli. Science* **165**:85.

Bollen, A., M. Faelen, J. P. Lecocq, A. Herzog, J. Zengel, L. Kahan and M. Nomura. 1973. The structural gene for the ribosomal protein S18 in *Escherichia coli.* I. Genetic studies on a mutant having an alteration in the protein S18. *J. Mol. Biol.* **76**:463.

Breckenridge, L. and L. Gorini. 1970. Genetic analysis of streptomycin resistance in *Escherichia coli. Genetics* **65**:9.

Buckel, P., D. Ruffler, W. Piepersberg and A. Böck. 1972. RNA overproducing revertants of an alanyl-tRNA synthetase mutant of *Escherichia coli. Mol. Gen. Genet.* **119**:323.

Craven, G. R., P. Voynow, S. J. S. Hardy and C. G. Kurland. 1969. Ribosomal proteins of *E. coli.* II. Chemical and physical characterization of 30S proteins. *Biochemistry* **8**:2906.

Dekio, S. 1971. Genetic studies of the ribosomal proteins in *Escherichia coli.* VII. Mapping of several ribosomal protein components by transduction experiments between *Shigella dysenteriae* and *Escherichia coli,* and between different strains of *Escherichia coli. Mol. Gen. Genet.* **113**:20.

Dekio, S. and R. Takata. 1969. Genetic studies of the ribosomal proteins in *Escherichia coli.* II. Altered 30S ribosomal protein components specific to spectinomycin-resistant mutants. *Mol. Gen. Genet.* **105**:219.

Dekio, S., R. Takata, S. Osawa, K. Tanaka and M. Tamaki. 1970. Genetic studies of the ribosomal proteins in *Escherichia coli.* IV. Pattern of the alteration of ribosomal components in mutants resistant to spectinomycin or erythromycin in different strains of *Escherichia coli. Mol. Gen. Genet.* **107**:39.

Deusser, E., G. Stöffler, H. G. Wittmann and D. Apirion. 1970. Ribosomal proteins. XVI. Altered S4 proteins in *Escherichia coli* revertants from streptomycin dependence to independence. *Mol. Gen. Genet.* **109**:298.

DeWilde, M. and B. Wittmann-Liebold. 1973. Localization of the amino acid exchange in protein S5 from an *Escherichia coli* mutant resistant to spectinomycin. *Mol. Gen. Genet.* **127**:273.

Doi, R. H. and R. T. Igarashi. 1965. Conservation of ribosomal and messenger ribonucleic acid cistrons in Bacillus species. *J. Bact.* **90**:384.

Donner, D. and C. G. Kurland. 1972. Changes in the primary structure of a mutationally altered ribosomal protein S4 of *Escherichia coli. Mol. Gen. Genet.* **115**:49.

Edman, P. and G. Begg. 1967. A protein sequenator. *Eur. J. Biochem.* **1**:80.

Fogel, S. and P. S. Sypherd. 1968. Chemical basis for heterogeneity of ribosomal protein. *Proc. Nat. Acad. Sci.* **59**:1329.

Funatsu, G. and H. G. Wittmann. 1972. Ribosomal proteins. XXXIII. Location of amino acid replacements in protein S12 isolated from *Escherichia coli* mutants resistant to streptomycin. *J. Mol. Biol.* **68**:547.

Funatsu, G., K. H. Nierhaus and H. G. Wittmann. 1972a. Determination of allele types and amino acid exchanges in protein S12 of three streptomycin resistant mutants of *Escherichia coli. Biochim. Biophys. Acta* **287**:282.

Funatsu, G., K. Nierhaus and B. Wittmann-Liebold. 1972b. Ribosomal proteins. XXII. Studies on the altered protein S5 from a spectinomycin-resistant mutant of *Escherichia coli. J. Mol. Biol.* **64**:201.

Funatsu, G., E. Schiltz and H. G. Wittmann. 1971. Ribosomal proteins. XXVII. Localization of the amino acid exchanges in protein S5 from two *E. coli* mutants resistant to spectinomycin. *Mol. Gen. Genet.* **114**:106.

Funatsu, G., W. Puls, E. Schiltz, J. Reinbolt and H. G. Wittmann. 1972c. Ribosomal proteins. XXXI. Comparative studies on S4 proteins of six

Escherichia coli revertants from streptomycin dependence to independence. *Mol. Gen. Genet.* **115**:131.

Geisser, M. 1971. Vergleichende immunologische und elektrophoretische Untersuchungen an Ribosomen von Enterobacteriaceae. Thesis (Berlin).

Geisser, M., G. W. Tischendorf and G. Stöffler. 1973a. Comparative immunological and electrophoretical studies on ribosomal proteins of Bacillaceae. *Mol. Gen. Genet.* **127**:129.

Geisser, M., G. W. Tischendorf, G. Stöffler and H. G. Wittmann. 1973b. Immunological and electrophoretical comparison of ribosomal proteins from eight species belonging to Enterobacteriaceae. *Mol. Gen. Genet.* **127**:111.

Gorini, L. 1969. The contrasting role of *str*A and *ram* gene products in ribosomal functioning. *Cold Spring Harbor Symp. Quant. Biol.* **34**:101.

————. 1971. Ribosomal discrimination of tRNAs. *Nature* **234**:261.

Guthrie, C., H. Nashimoto and M. Nomura. 1969. Studies on the assembly of ribosomes in vivo. *Cold Spring Harbor Symp. Quant. Biol.* **34**:69.

Hardy, S. J. S., C. G. Kurland, P. Voynow and G. Mora. 1969. The ribosomal proteins of *Escherichia coli*. I. Purification of the 30S ribosomal proteins. *Biochemistry* **8**:2897.

Hasenbank, R., C. Guthrie, G. Stöffler, H. G. Wittmann, L. Rosen and D. Apirion. 1973. Electrophoretic and immunological studies on ribosomal proteins of 100 *Escherichia coli* revertants from streptomycin dependence. *Mol. Gen. Genet.* **127**:1.

Higo, K. J. and K. Loertscher. 1974. Amino terminal sequences of some *Escherichia coli* 30S ribosomal proteins and functionally corresponding *Bacillus stearothermophilus* ribosomal proteins. *J. Bact.* **118**:180.

Higo, K., W. Held, L. Kahan and M. Nomura. 1973. Functional correspondence between 30S ribosomal proteins of *Escherichia coli* and *Bacillus stearothermophilus*. *Proc. Nat. Acad. Sci.* **70**:944.

Isono, K., S. Isono, G. Stöffler, L. P. Visentin, M. Yaguchi and A. T. Matheson. 1973. Correlation between 30S ribosomal proteins of *Bacillus stearothermophilus* and *Escherichia coli*. *Mol. Gen. Genet.* **127**:191.

Itoh, T. and H. G. Wittmann. 1973. Amino acid replacements in proteins S5 and S12 of two *Escherichia coli* revertants from streptomycin dependence to independence. *Mol. Gen. Genet.* **127**:19.

Kahan, L., J. Zengel, M. Nomura, A. Bollen and A. Herzog. 1973. The structural gene for the ribosomal protein S18 in *Escherichia coli*. II. Chemical studies on the protein S18 having an altered electrophoretic mobility. *J. Mol. Biol.* **76**:473.

Kaltschmidt, E. 1971. Isoelectric points of ribosomal proteins of *E. coli* as determined by two-dimensional polyacrylamide gel electrophoresis. *Anal. Biochem.* **43**:25.

Kaltschmidt, E., M. Dzionara and H. G. Wittmann. 1970a. Ribosomal proteins. XV. Amino acid compositions of isolated ribosomal proteins from 30S and 50S subunits of *Escherichia coli*. *Mol. Gen. Genet.* **109**:292.

Kaltschmidt, E., M. Dzionara, D. Donner and H. G. Wittmann. 1967. Ribosomal proteins. I. Isolation, amino acid composition, molecular weights and peptide mapping of proteins from *E. coli* ribosomes. *Mol. Gen. Genet.* **100**:364.

Kaltschmidt, E., G. Stöffler, M. Dzionara and H. G. Wittmann. 1970b. Ribosomal proteins. XVII. Comparative studies on ribosomal proteins of four strains of *Escherichia coli. Mol. Gen. Genet.* **109**:303.

Kreider, G. and B. L. Brownstein. 1971. A mutation suppressing streptomycin dependence. II. An altered protein in the 30S ribosomal subunit. *J. Mol. Biol.* **61**:135.

————. 1972. Ribosomal proteins involved in the suppression of streptomycin dependence in *Escherichia coli. J. Bact.* **109**:780.

Laursen, R. A. 1971. Solid-phase Edman degradation. An automatic peptide sequencer. *Eur. J. Biochem.* **20**:89.

Leboy, P. S., E. C. Cox and J. G. Flaks. 1964. The chromosomal site specifying a ribosomal protein in *Escherichia coli. Proc. Nat. Acad. Sci.* **52**:1367.

Marmur, J., S. Falkow and M. Mandel. 1963. New approaches to bacterial taxonomy. *Ann. Rev. Microbiol.* **17**:329.

Möller, W., H. Castleman and C. Terhorst. 1970. Characterization of an acidic protein in 50S ribosomes of *E. coli. FEBS Letters* **8**:192.

Möller, W., A. Groene, C. Terhorst and R. Amons. 1972. 50S ribosomal proteins. Purification and partial characterization of two acidic proteins, A1 and A2, isolated from 50S ribosomes of *Escherichia coli. Eur. J. Biochem.* **25**:5.

Moore, P. B. 1971. Reaction of *N*-ethyl maleimide with the ribosomes of *Escherichia coli. J. Mol. Biol.* **60**:169.

Moore, P. B., R. R. Traut, M. Noller, P. Pearson and H. Delius. 1968. Ribosomal proteins of *Escherichia coli.* II. Protein from the 30S subunit. *J. Mol. Biol.* **31**:441.

Mora, G., D. Donner, P. Thammana, L. Lutter and C. G. Kurland. 1971. Purification and characterization of 50S ribosomal proteins of *Escherichia coli. Mol. Gen. Genet.* **112**:229.

Muto, A., E. Otaka, T. Itoh, S. Osawa and H. G. Wittmann. 1974. Correspondence of 30S ribosomal proteins of *Escherichia coli* fractionated on carboxymethyl-cellulose chromatography to the standard nomenclature. *Mol. Gen. Genet.* (in press).

Nashimoto, H. and M. Nomura. 1970. Structure and function of bacterial ribosomes. XI. Dependence of 50S ribosomal assembly on simultaneous assembly of 30S subunits. *Proc. Nat. Acad. Sci.* **67**:1440.

Nashimoto, H., W. Held, E. Kaltschmidt and M. Nomura. 1971. Structure and function of bacterial ribosomes. XII. Accumulation of 21S particles by some cold-sensitive mutants of *Escherichia coli. J. Mol. Biol.* **62**:121.

Nomura, M. and V. A. Erdmann. 1970. Reconstitution of 50S ribosomal subunits from dissociated molecular components. *Nature* **228**:744.

Nomura, M., P. Traub and H. Bechmann. 1968. Hybrid 30S ribosomal particles reconstituted from components of different bacterial origins. *Nature* **219**:793.

Osawa, S., R. Takata and S. Dekio. 1970. Genetic studies of the ribosomal proteins in *Escherichia coli.* III. Compositions of ribosomal proteins in various strains of *Escherichia coli. Mol. Gen. Genet.* **107**:32.

Osawa, S., E. Otaka, R. Takata, S. Dekio, M. Matsubara, T. Itoh and A. Muto. 1972. Ribosomal protein genes in bacteria. *FEBS Symp.* **23**:313.

Otaka, E., T. Itoh and S. Osawa. 1968. Ribosomal proteins of bacterial cells: Strain and species-specificity. *J. Mol. Biol.* **33**:93.

Otaka, E., T. Itoh, S. Osawa, K. Tanaka and M. Tamaki. 1971. Peptide analyses of a protein component, 50-8, of 50S ribosomal subunit from erythromycin resistant mutants of *Escherichia coli* and *Escherichia freundii*. *Mol. Gen Genet.* **114**:14.

Otaka, E., H. Teraoko, M. Tamaki, K. Tanaka and S. Osawa. 1970. Ribosomes from erythromycin-resistant mutants of *Escherichia coli*. *J. Mol. Biol.* **48**:499.

Ozaki, M., S. Mizushima and M. Nomura. 1969. Identification and functional characterization of the protein controlled by streptomycin-resistant locus in *E. coli. Nature* **222**:333.

Pearson, P., H. Delius and R. R. Traut. 1972. Purification and characterization of 50S ribosomal proteins of *Escherichia coli*. *Eur. J. Biochem.* **27**:482.

Phillipps, S. L., D. Schlessinger and D. Apirion. 1969. Mutants in *E. coli* ribosomes: A new selection. *Proc. Nat. Acad. Sci.* **62**:772.

Ptitsyn, O. B., A. J. Denesyuk, A. V. Finkelstein and V. J. Lim. 1973. Prediction of the secondary structure of the L7, L12 proteins of the *E. coli* ribosomes. *FEBS Letters* **34**:55.

Reinbolt, J. and E. Schiltz. 1973. The primary structure of ribosomal protein S4 from *Escherichia coli*. *FEBS Letters* **36**:250.

Rombauts, W., B. Peeters and H. G. Wittmann. 1971. Comparison of peptide patterns from isolated 30S and 50S ribosomal proteins of *Escherichia coli* by column chromatography. *FEBS Letters* **18**:164.

Sela, M., B. Schechter, J. Schechter and F. Borek. 1967. Antibodies to sequential and conformational determinants. *Cold Spring Harbor Symp. Quant Biol.* **32**:537.

Stöffler, G. and H. G. Wittmann. 1971a. Sequence differences of *Escherichia coli* 30S ribosomal proteins as determined by immunochemical methods. *Proc. Nat. Acad. Sci.* **68**:9.

Stöffler, G., E. Deusser, H. G. Wittmann and D. Apirion. 1971. Ribosomal proteins. XIX. Altered S5 ribosomal protein in an *Escherichia coli* revertant from streptomycin dependence to independence. *Mol. Gen. Genet.* **111**:334.

Stöffler, G., G. W. Tischendorf, R. Hasenbank and H. G. Wittmann. 1974. The number of individual proteins of the 50S ribosomal subunit from *Escherichia coli* as determined by immunochemical methods. *Eur. J. Biochem.* (in press).

―――. 1971b. Ribosomal proteins. XXV. Immunological studies on *Escherichia coli* ribosomal proteins. *J. Mol. Biol.* **62**:407.

Sun, T., T. A. Bickle and R. R. Traut. 1972. Similarity in size and number of ribosomal proteins from different prokaryotes. *J. Bact.* **111**:474.

Sypherd, P. S. 1969. Amino acid differences in a 30S ribosomal protein from two strains of *Escherichia coli*. *J. Bact.* **99**:379.

Takata, R., S. Dekio, E. Otaka and S. Osawa. 1969. Genetic studies on the ribosomal proteins in *Escherichia coli*. I. Mutants and strains having 30S ribosomal subunit with altered protein components. *Mol. Gen. Genet.* **105**:113.

Tanaka, K., M. Tamaki, T. Itoh, E. Otaka and S. Osawa. 1971. Ribosomes from spiramycin-resistant mutants of *Escherichia coli* Q13. *Mol. Gen. Genet.* **114**:23.

Terhorst, C., B. Wittmann-Liebold and W. Möller. 1972a. 50S ribosomal proteins. Peptide studies on two acidic proteins, A1 and A2, isolated from 50S ribosomes of *Escherichia coli. Eur. J. Biochem.* **25**:13.

Terhorst, C. P., W. Möller, R. Laursen and B. Wittmann-Liebold. 1972b. Amino acid sequence of a 50S ribosomal protein involved in both EFG- and EFT-dependent GTP hydrolysis. *FEBS Letters* **28**:325.

————. 1973. The primary structure of an acidic protein from 50S ribosomes of *Escherichia coli* which is involved in GTP hydrolysis dependent on elongation factors G and T. *Eur. J. Biochem.* **34**:138.

Traut, R. R., P. B. Moore, H. Delius, H. Noller and A. Tissières. 1967. Ribosomal proteins of *E. coli.* Demonstration of different primary structures. *Proc. Nat. Acad. Sci.* **57**:1294.

Waller, J. P. 1964. Fractionation of the ribosomal protein from *Escherichia coli. J. Mol. Biol.* **10**:319.

Wittmann, H. G., G. Stöffler, W. Piepersberg, P. Buckel, D. Ruffler and A. Böck. 1974. Altered S5 and S20 ribosomal proteins in revertants of an alanyl-tRNA synthetase mutant of *Escherichia coli. Mol. Gen. Genet.* (in press).

Wittmann, H. G., G. Stöffler, D. Apirion, L. Rosen, K. Tanaka, M. Tamaki, R. Takata, S. Dekio, E. Otaka and S. Osawa. 1973. Biochemical and genetic studies on two different types or erythromycin-resistant mutants of *Escherichia coli* with altered ribosomal proteins. *Mol. Gen. Genet.* **127**:175.

Wittmann, H. G., G. Stöffler, E. Kaltschmidt, V. Rudloff, H. G. Janda, M. Dzionara, D. Donner, K. Nierhaus, M. Cech, I. Hindennach and B. Wittmann-Liebold. 1970. Proteinchemical and serological studies on ribosomes of bacteria, yeast and plants. *FEBS Symp.* **21**:33.

Wittmann-Liebold, B. 1971. Ribosomal proteins. XXI. Amino acid composition of the tryptic peptides isolated from proteins S4, S18 and S20 of *Escherichia coli* ribosomes. *Hoppe-Seyler's Z. Physiol. Chem.* **352**:1705.

————. 1973a. Studies on the primary structure of 20 proteins from *Escherichia coli* ribosomes by means of an improved sequenator. *FEBS Letters* **36**:247.

————. 1973b. Amino acid sequence studies on ten ribosomal proteins of *Escherichia coli* with an improved sequenator equipped with an automatic conversion device. *Hoppe-Seyler's Z. Physiol. Chem.* **354**:1415

Wittmann-Liebold, B. and H. G. Wittmann. 1971. Ribosomal proteins. XX. Isolation and analyses of the tryptic peptides of proteins S5 from strains K and B of *Escherichia coli. Biochim. Biophys. Acta.* **251**:44.

Yaguchi, M., C. Roy, A. T. Matheson and L. P. Visentin. 1973. The amino sequence of the N-terminal region of some 30S ribosomal proteins from *Escherichia coli* and *Bacillus stearothermophilus:* Homologies in ribosomal proteins. *Canad. J. Biochem.* **51**:1215.

Zimmerman, R. A., R. T. Garvin and L. Gorini. 1971. Alteration of a 30S ribosomal protein accompanying *ram* mutation in *Escherichia coli. Proc. Nat. Acad. Sci.* **68**:2263.

Zimmermann, R. A., Y. Ikeya and P. F. Sparling. 1973. Alteration of ribosomal protein S4 by mutation linked to kasugamycin resistance in *Escherichia coli. Proc. Nat. Acad. Sci.* **70**:71.

5S RNA

Roger Monier
Institut de Recherches Scientifiques sur le Cancer
Villejuif, France

INTRODUCTION

5S RNA was identified for the first time as a 50S subunit component in *E. coli* by Rosset and Monier (1963). It was later found in all cytoplasmic and chloroplastic ribosomes (for references see the review of Monier 1972) but not in mitochondrial ribosomes either from fungi (Lizardi and Luck 1971) or from higher organisms (Zylber and Penman 1969; Attardi et al. 1970). It remains to be seen whether a 5'-terminal section of the larger ribosomal mitochondrial RNA plays the same role as 5S RNA in other ribosomes (Lizardi and Luck 1971) or whether an RNA with a sedimentation constant of 4S replaces the usual 5S component (Wu et al. 1972).

In prokaryotes, 5S RNA is the only ribosomal RNA of low molecular weight so far identified. In eukaryotes, the larger cytoplasmic subunit also contains the so-called 7S RNA (Forget and Weissman 1967; Pene, Knight and Darnell 1968), which becomes detached from the 28S RNA after exposure to denaturing conditions. The 7S RNA in eukaryotes is part of the same transcription unit as 18S and 28S RNA (Pene, Knight and Darnell 1968), whereas 5S RNA has an entirely different genetic origin (Brown and Weber 1968). Therefore the 7S RNA, which is restricted to eukaryotic cytoplasmic ribosomes, probably has not the same significance as 5S RNA as regards the structure and function of the ribosome.

The presence in Rous sarcoma virions of a low molecular weight RNA, which has the same electrophoretic mobility and the same fingerprint as the cytoplasmic 5S RNA from uninfected chicken embryo fibroblasts, has been reported (Faras et al. 1973). One molecule of this RNA is noncovalently bound to each 35S subunit of the virus RNA. It is released upon heating, in parallel with the dissociation of the 70S RNA into 35S subunits.

PRIMARY STRUCTURE OF 5S RNA

5S RNA from all organisms is a chain of about 120 nucleotides, which does not contain any methylated or otherwise modified nucleotide. Because of its ease of purification after heavy labeling with ^{32}P, E. coli 5S RNA was the first RNA whose sequence was entirely determined by Sanger and his group (Brownlee, Sanger and Barrell 1968).

The major sequences of all the 5S RNAs which have been sequenced at the present time are shown in Figure 1. The sequence shown for mammalian 5S RNA is the one established for the 5S RNA from human KB cells by Forget and Weissman (1969). Except for a minor difference in HeLa cell 5S RNA sequence (Hatlen, Amaldi and Attardi 1969), the sequences of all the mammalian 5S RNAs that have been studied are considered to be identical. They include those of a marsupial (Averner and Pace 1972), of two mouse cell lines, of rat pituitary and of rabbit reticulocytes (Williamson and Brownlee 1969).

The chain lengths of 5S RNA preparations from particular prokaryotic species are constant. By contrast, 5S RNAs prepared from the cytoplasmic ribosomes of vertebrates usually are mixtures of molecules with either 120 or 121 nucleotides. They are also heterogeneous at their 5' end, where various proportions of tri-, di- and monophosphate groups are found (Hatlen, Amaldi and Attardi 1969). This particularity actually extends to all eukaryotic 5S RNAs (Soave, Galante and Torti 1970; Soave et al. 1973).

Brownlee, Sanger and Barrell (1968) have pointed out that the two halves of the 5S RNA from E. coli display a large extent of homology. Sections 10–60 and 61–110 have identical nucleotide residues at 34 sites and section 1–9 is also very similar to section 111–120. This observation, together with comparisons between E. coli 5S RNA and several tRNAs has led Mullins, Lacey and Hearn (1973) to propose a hypothetical model for the origin of tRNA and 5S RNA from a common ancestral gene.

In all organisms, the ribosomal RNAs, including 5S RNAs, are transcribed from structural genes which exist in multiple copies (Birnstiel, Chipchase and Speirs 1971). It is therefore of importance to ascertain whether the rRNAs from one particular organism have unique sequences or whether some sequence heterogeneity can be related to the number of genes. The sequence heterogeneity of high molecular weight RNAs is difficult to study exhaustively. The 5S RNAs offer a simpler case; from the very first determination of a 5S RNA sequence it became clear that limited sequence heterogeneity does exist in *E. coli* K12 and MRE600 (Brownlee, Sanger and Barrell 1968). A careful investigation by Jarry and Rosset (1971) of 5S RNA extracted from various *E. coli* strains has shown that single base substitutions can occur at least at five different sites along the 5S RNA molecule. They are all localized in the vicinity of either one or the other end of the polynucleotide chain. The distribution of some of these substitutions differs among various strains, and these strain-specific differences have been exploited to map four of the six cistrons for 5S RNA in *E. coli* (Jarry and Rosset 1973).

Sequence heterogeneity has also been found in *Ps. fluorescens* 5S RNA (DuBuy and Weissman 1971). It has been claimed, on the other hand, that no such sequence variability is evident in eukaryotic 5S RNAs from a single cell type (Forget and Weissman 1969; Williamson and Brownlee 1969). A clear-cut difference in the sequences of 5S RNAs prepared from different cells of the same organism has been disclosed by the study of 5S RNAs from *Xenopus laevis* (Wegnez, Monier and Denis 1972; Ford and Southern 1973). Previous observations on the physical chemical properties of 5S RNAs extracted either from maturing oocytes or from various somatic cells (Denis, Wegnez and Willem 1972) have been related to differences in primary sequence (Wegnez, Monier and Denis 1972). Since no evidence for 5S RNA gene amplification in oocytes has been found (Wegnez and Denis 1972), the results can only be explained on the basis of a differential control of 5S RNA gene classes in maturing oocytes as contrasted to somatic cells. Nevertheless, since the number of 5S RNA gene copies per haploid genome in *X. laevis* is about 25,000 (Brown and Weber 1968), the observed sequence heterogeneity bears no direct relation to the actual gene multiplicity. It can be assumed that 5S RNA genes in this organism are distributed among several classes, each of which contains many copies with identical sequences.

Provided deletions and insertions of a few nucleotides are assumed, a very large extent of homology is easily discernible between either prokaryotic or eukaryotic 5S RNAs primary structures (Figure 1). Although direct comparison of prokaryotic and eukaryotic sequences does not show at first sight a strong resemblance, Sankoff and Cedergren (1973) have concluded on the basis of a statistical test that a significant homology does exist between *E. coli* or *Ps. fluorescens* and KB 5S RNAs. Comparison of the extent of changes between 5S RNAs from various vertebrates has led Pace, Walker, Pace and Erikson (unpublished results) to conclude that 5S RNA structure has drifted to the same extent or even more rapidly than the cytochromes *c*.

Nevertheless, the differences between prokaryotic and eukaryotic 5S RNAs are sufficiently important to prevent any cross recognition between

	1									10										20					
E. coli	U	G	C	C	U	G	G	C	G	G	C	C	-	G	U	A	G	C	G	C	G	G	U	G	G	U

	1									10										20					
Ps. fluorescens	U	G	U	U	C	U	U	U	G	A	C	G	A	G	U	A	G	U	G	G	C	A	U	U	G	G

	1								10											
B. stearothermophilus	C	C	U	A	G	U	G	A	C	A	A	U	A	G	C	G (A G, A G, A G, G)	-	-	A

	1									10										20						
Yeast	G	-	G	U	U	G	C	G	G	C	C	A	U	A	C	C	A	U	C	U	-	A	G	-	A	A	A G

	1									10										20						
Chlorella	A	U	G	C	U	A	C	G	U	U	C	A	U	A	C	-	A	C	C	A	-	C	G	-	A	A	A G

	1									10										20						
Xenopus	G	C	-	C	U	A	C	G	G	C	C	A	C	A	C	C	A	C	C	C	-	U	G	-	A	A	A G

	1									10										20						
Chicken	G	C	-	C	U	A	C	G	G	C	C	A	U	C	C	C	A	C	C	C	C	U	G	U	A	A	C G

	1									10										20						
Human	G	U	-	C	U	A	C	G	G	C	C	A	U	A	C	C	A	C	C	C	-	U	G	-	A	A	C G

				70									80									
E. coli	U	A	G	C	-	G	C	C	G	A	U	G	G	U	A	G	U	G	U	G	G	G G U - - - C

				70									80									
Ps. fluorescens	C	A	U	C	-	G	C	C	G	A	U	G	G	U	A	G	U	G	U	G	G	G G U - - - U

								80											90
B. stearothermophilus	C	C	A	G C, G)	C	C	G	A	U	-	-	-	A	G	U	-	U	G	G G G C C A G C

				70									80									90
Yeast	A	G	A	G	C	C	U	G	A	C	C	G	A	G	U	A	G	U	G	U	A	G U G G G U G

				70								80									
Chlorella	-	G	G	G	C	U	C	G	A	C	U	-	A	G	U	A	C	U	G	G	G U U G G G A G

				70								80									
Xenopus	-	G	G	G	C	C	U	G	G	U	U	-	A	G	U	A	C	U	U	G	G A U G G G A G

				70								80									
Chicken	-	G	G	G	C	C	U	G	G	U	U	-	A	G	U	A	C	U	U	G	G A U G G G A G

				70								80								
Human	-	G	G	G	C	C	U	G	G	U	U	A	G	U	A	C	U	U	G	G A U G G G A G

Figure 1 Primary structure of various 5S RNAs. The structures presented here are taken from the following original reports: *Escherichia coli* (Brownlee, Sanger and Barrell 1968); *Pseudomonas fluorescens* (DuBuy and Weissman 1971); *Bacillus stearothermophilus* (Marotta et al. 1973); yeast (Hindley and

144

```
      30            40              50              60
C C C A C C U G A C C C C A U G C C G A A C U C A G A A G U G A A A C G C C G

      30            40              50              60
A A C A C C U G A U C C C A U C C C G A A C U C A G A G G U G A A A C G A U G

        30            40
A A C A C C C G U C U C C A U C C C G A A C A C G (G A A G) U U A A G (C U C U C


      30            40              50              60
C A C C G U U C U C C G U C C G A U A A C C U G U A G U U.A A G C - U G G U A

      30            40              50              60
C A C C C G A U C C C A U C A G A A C U C G G A - A G U.U A A A C G U G G U U

      30            40              50              60
U G C C C G A U C U C G U C U G A U C U C G G A - A G C C A A G C A G G G U C

      30            40              50              60
- - C C C G A U C U G G U C U G A U C U C G G A - A G C U A A G C A G G G U C

      30            40              50              60
C G C C C G A U C U C G U C U G A U C U C G G A - A G C U A A G C A G G G U C


      90          100             110               120
U C C C C A U G C G - A G A G U A G G G A A C U G C C A G G C A U OH

  90          100             110               120
U C C C C A U G U C A A G A U C U C G - - A C C A U A G A G C A U OH

            100             110           119
G C C C C - U G C - A A G A G U A G G U U G U C G C U A G G C OH


          100               110               120
- A C C A U A C G C - - - - G A A - A C C U A G G U G C U G C A A U C U OH

  90          100     105     110             120
G A U U A C C U G A G U G G G G A A - - C C C C G A C G U A G U G U OH

    90            100             110               120
- A C C G C C U G G - - - - G A A U A C C - A G G U G U C G U A G G C U U (U) OH

    90            100             110               120
- A C C G C C U G G - - - - G A A U A C C - G G G U G C U G U A G G C U U OH

    90            100             110               120
- A C C G C C U G G - - - - G A A U A C C - G G G U G C U G U A G G C U U (U) OH
```

Page 1972); *Chlorella* (Jordan and Galling 1973); *Xenopus laevis* (Brownlee et al. 1972); chicken (Pace, Walker, Pace and Erikson, unpublished data); man (KB cells) (Forget and Weissman 1969). The most frequent sequence of each prokaryotic 5S RNA is shown.

E. coli 50S subunit proteins and eukaryotic 5S RNAs (Bellemare, Vigne and Jordan 1973) or the functional substitution of prokaryotic 5S RNAs by eukaryotic ones in 50S subunit complete reconstitution tests, whereas 5S RNAs from *E. coli, Proteus vulgaris, B. subtilis, Micrococcus lysodeikticus, Staphylococcus aureus, Ps. fluorescens* and *Azotobacter vinelandii* can efficiently replace the homospecific 5S RNA in a *B. stearothermophilus* system (Wrede and Erdmann 1973).

SECONDARY STRUCTURE AND CONFORMATION

Physical Properties of Solutions

The 5S RNA molecule in solution of moderate ionic strength and containing Mg^{++} ions is a prolate ellipsoid with an axial ratio of $5/1$ and a radius of gyration of 34.5 ± 1.5 Å, according to the results of X-ray scattering experiments at small angles at room temperature (Connors and Beeman 1972). No significant differences were found between *E. coli* and yeast 5S RNAs. A high degree of asymmetry is also indicated by sedimentation studies (Boedtker and Kelling 1967; Comb and Zehavi-Willner 1967). The sedimentation coefficient of *E. coli* 5S RNA is 4.5–4.8S.

Many optical determinations of the extent of base pairing in 5S RNA have been performed using UV or IR absorption and ORD or CD measurements (Boedtker and Kelling 1967; Cantor 1968; Scott et al. 1968; Richards et al. 1972). Most of these observations were made on *E. coli* 5S RNA, although some comparisons between *E. coli* and sea urchin embryo (Bellemare, Cedergren and Cousineau 1972) or *B. stearothermophilus* and *B. subtilis* (Gray and Saunders 1973) have been performed. The number of base pairs which has been calculated varies from 34 to 45, the percent of GC pairs being 60–70%. The recent application of the NMR technique has led its originators to conclude that the extent of base pairing is likely to be much lower than previously assumed (Wong et al. 1972). No more than 25 ± 3 or 28 ± 3 base pairs in *E. coli* or yeast 5S RNAs, respectively, were detected.

Denaturation

5S RNA can be easily interconverted into different molecular conformations, which can be separated from each other by chromatography (Aubert et al. 1968) or polyacrylamide gel electrophoresis (Hindley 1967; Weinberg and Penman 1968; Phillips and Timko 1972). The UV absorption and ORD and CD properties of these molecular forms are distinct (Aubert et al. 1968; Richards, Lecanidou and Geroch 1973). The conformation having the highest affinity for ribosomal proteins in 50S subunit partial in vitro assembly has been considered as native (Aubert et al. 1968; Aubert, Bellemare and Monier 1973). As judged by electrophoretic mobility, 5S

RNA previously exposed to denaturing conditions is converted by heating in the presence of Mg^{++} ions to a conformation identical with the native one, with an activation energy of 62 kcal/mole. The conversion proceeds to completion irrespective of temperature and Mg^{++} ion concentration above 1 mM (Richards, Lecanidou and Geroch 1973). The native form is therefore the thermodynamically stable conformation under these conditions. The existence of various conformations should, of course, be taken into consideration by experimenters who plan physical studies on 5S RNA, and preparations should be adequately tested by measuring their affinities for ribosomal proteins (Aubert et al. 1968, 1973).

Enzymatic Degradation and Chemical Modifications

Several attempts to provide information on the single-stranded regions of the 5S RNA molecule have been made. Various base-specific chemical reagents have provided consistent results with *E. coli* 5S RNA. Glyoxal and kethoxal have thus been shown to react preferentially with G_{41} and more slowly with G_{13} and G_{44} (Bellemare et al. 1972c; Litt 1973). Among the residues most easily accessible to methoxyamine are those belonging to sequences C_{35}–C_{38}, C_{42}–C_{43} and C_{47}–C_{49}. C residues at C_{28}–C_{31} react more slowly (Bellemare et al. 1972c). A fast reaction of carbodiimide reagents with U_{40} has also been demonstrated; another accessible U residue belongs to the trinucleotide sequence UAG, which is present four times at positions 14, 65, 77 and 103 (Lee and Ingram 1969; Bellemare, unpublished data). It therefore appears that residues located on both sides of position G_{41} are accessible to chemical reagents and probably belong to a single-stranded region.

Another very exposed region, including residues U_{87} C_{88} U_{89}, has also been demonstrated with the use of methoxyamine and a carbodiimide (Bellemare et al. 1972c, Bellemare, unpublished data). Cramer and Erdmann (1968) have estimated that 10 out of 23 A residues readily react with monoperphtallic acid and are probably unpaired, but no attempt at determining their localization in the sequence has been made.

The existence of a well-exposed, single-stranded region around residue G_{41} in *E. coli* 5S RNA was also demonstrated by partial enzymatic digestion with a variety of enzymes (Jordan 1971a; Vigne and Jordan 1971; Bellemare, Jordan and Monier 1972), in agreement with the results of chemical reactivity measurements. Nevertheless, it is probably of interest, with respect to the tertiary folding of the 5S RNA molecule in solution, that nucleotide bonds between residues U_{87} C_{88} U_{89}, which react with appropriate reagents at the same rate as residues located near G_{41}, are not as easily hydrolyzed either by pancreatic or T_2 RNases as bonds which belong to the U_{22}–C_{49} section of the molecule. The first long degradation products obtained with pancreatic RNase are lacking the sequence between U_{22} and

C_{49} but do not show any interruption in the U_{87} C_{88} U_{89} sequence (Jordan 1971a; Vigne and Jordan 1971). With T_2 RNase, a first split at A_{39} is followed by splits at A_{29} and C_{49}, again leaving the U_{87} C_{88} U_{89} sequence untouched (Vigne 1972). On the other hand, more extensive digestion of 5S RNA in the reconstitution buffer of Nomura and Erdmann (1970) with either pancreatic or T_2 RNase can lead to the production with a yield of 15–20% of a fragment lacking all the sequence between C_{11} and C_{68}. Although this fragment moves as a unit during polyacrylamide gel electrophoresis in nondenaturing conditions, it contains both extremities of the original molecule and is interrupted between U_{87} and U_{89} (Feunteun, unpublished data). The access of enzymes to the U_{87} U_{89} bonds is therefore hindered by other parts of the sequence and becomes possible only after these parts have been removed by previous degradations.

In order to ascertain the possible relevance of these data, obtained on free 5S RNA extracted from *E. coli* ribosomes, to the biological properties of 5S RNA in general, two types of approaches have been followed. 5S RNAs prepared from another prokaryote (*Ps. fluorescens*) and from eukaryotes (yeast and HeLa cells) have been submitted to partial enzymatic degradation (Vigne, Jordan and Monier 1973). When differences in primary structures are taken into account, it appears that appropriate enzymes do detect an easily accessible region around position 40 in all 5S RNAs tested, whether from prokaryotes or eukaryotes. Nevertheless, when T_1 RNase is used, the two bonds located at G_{37} and G_{89} in yeast and HeLa (Vigne, Jordan and Monier 1973) as well as in *X. laevis* 5S RNAs (Wegnez, Monier and Denis 1972) are split with equal probabilities. Therefore, although all these eukaryotic RNAs possess in common with prokaryotic ones a well accessible region located at the first third of the molecule, they also possess a second region of equal accessibility to T_1 RNase located at the last fourth of the sequence.

In the second approach, attempts have been made to elucidate the situation of 5S RNA inside an intact 50S subunit. It was thus demonstrated that this RNA is not freely exposed on the surface of the subunit, since nucleolytic enzymes have no access to any part of the molecule under conditions where the 23S RNA itself is profoundly degraded (Feunteun and Monier 1971). On the other hand, kethoxal and the carbodiimide reagent, 1-cyclohexyl-3-(2-morpholinyl)-4 ethylcarbodiimide metho-*p*-toluene sulfonate react very rapidly with G_{41}, U_{40}, U_{87} and/or U_{89}. G_{13} and/or G_{64} as well as U_{25} and/or U_{32} react to some extent, although more slowly (Bellemare, unpublished data). There is therefore a strong similarity between the results obtained in solution and in the ribosomal subunit.

All the results presented above concur in indicating the existence both in free and in ribosome-bound *E. coli* 5S RNA of a single-stranded accessible region located around G_{41}. There are nevertheless in the literature two reports which do not agree with this conclusion. Lewis and Doty (1970) have proposed that sequences 9–13, 25–32, 58–65 and 93–99 are single stranded because they efficiently bind complementary tri- and tetranucleotides. According to Mirzabekov and Griffin (1972), the internucleotide bond which is the most accessible to T_1 RNase digestion is located at G_{61}, other breaks occurring more slowly at G_{16}, G_{18}, G_{56}, G_{100} and G_{102}. Aubert, Bellemare and Monier (1973) have observed that in the 5S RNA conformer obtained after exposure to urea and EDTA, the G residue most accessible to glyoxal is G_{61}, followed by G_{100} and/or G_{102}. Upon renaturation in the presence of Mg^{++} ions, G_{61} ceases to be highly reactive, whereas the easy accessibility of G_{41} is recovered. Similarly, the results of Jordan (1971a) show that the first splits produced by T_1 RNase in denatured 5S RNA do not occur after residue G_{41} but somewhere between G_{44} and G_{61}. Further splits occur between G_{13} and G_{24} and between G_{76} and G_{79}.

Secondary Structure Models

Systematic explorations of all possible secondary structures of 5S RNA with the help of computer programs indicate that no unique pairing scheme can be proposed on the basis of the primary structure alone (Richards 1969; Jordan 1971b). All 5S RNA sequences nevertheless are compatible with the existence of a stem formed by pairing the 3' and 5' termini with each other (Figure 2). A number of secondary structure models of 5S RNA have actually been published (Boedtker and Kelling 1967; Cantor 1967; Brownlee, Sanger and Barrell 1968; Raacke 1968; Madison 1968; Du Buy and Weissman 1971), none of which are fully compatible with the properties of native *E. coli* 5S RNA. The only model in which a high degree of base pairing is assumed and in which the region surrounding residue G_{41} is single stranded is the model proposed by Madison (1968) (Figure 3A). Two other possible models are also suggested in Figure 3B,C).

5S RNA BIOSYNTHESIS

Control of Biosynthesis

In mature ribosomes, 5S RNA and the two high molecular weight RNAs occur in stoichiometric amounts. In prokaryotes, a stoichiometric production of all three rRNAs is ensured through the existence of a common transcription unit (Pato and von Meyenburg 1970; Pace, Peterson and Pace 1970; Ford Doolittle and Pace 1970, 1971). In accordance with this mode of transcription, no large pool of free 5S RNA accumulates in exponentially growing bacteria, although a small pool of precursor 5S RNA unlinked to ribosomes or ribosome precursors probably exists (Galibert and Sanfourche 1969; Jordan, Feunteun and Monier 1970; Hayes and Hayes 1971; Feunteun, Jordan and Monier 1972).

In eukaryotes, by contrast, there is no genetic linkage between high

E. coli

```
                  1                     10
E. coli        p  U  G  C  C  U  G  G  C  G  G  C
                  |  |  |  |  |  |  |  |  |     |
           H  O  U  A  C  G  G  A  C  C  G  U  C  A
                 120                    110
```

Ps. fluorescens

```
                  1                     10
Ps. fluorescens  p  U  G  U  U  C  U  U  U  G  A
                  |  |     |  |  |     |  |
           H  O  U  A  C  G  A  G  A  U  A  C  C
                 120                    110
```

B. stearothermophilus

```
                       1                     10
B. stearothermophilus  p  C  C  U  A  G  U  G  A  C  A  A  U  A
                          |  |  |  |  |     |  |  |  |  |
                 H  O  C  G  G  A  U  C  G  C  U  G  U  U  G  G
                      119                    110
```

Yeast

```
                 1                     10
Yeast         p  G  G  U  U  G  C  G  G  C  C
                 |     |  |  |  |  |     |  |
        H  O  U  C  U  A  A  C  G  U  C  G  U
                120                    111
```

Chlorella

```
                 1                     10
Chlorella     p  A  U  G  C  U  A  C  G  U  U  C
                 |        |  |  |  |  |  |  |
        H  O  U  G  U  G  A  U  G  C  A  G  C
                120                    110
```

Xenopus

```
                 1                     10
Xenopus       p  G  C  C  U  A  C  G  G  C  C
                 |  |  |  |  |  |  |  |     |
   H  O  (U) U  U  C  G  G  A  U  G  C  U  G  U
                120                    110
```

Chicken

```
                 1                     10
Chicken       p  G  C  C  U  A  C  G  G  C  C
                 |  |  |  |  |  |  |     |  |
   H  O  (U) U  U  C  G  G  A  U  G  U  C  G  U
                120                    110
```

Human

```
                                        10
Human         p  G  U  C  U  A  C  G  G  C  C
                 |     |  |  |  |  |     |  |
   H  O  (U) U  U  C  G  G  A  U  G  U  C  G  U
                120                    110
```

Figure 2 Suggested pairing schemes for the 3′- and 5′-terminal sequences of various 5S RNAs.

150

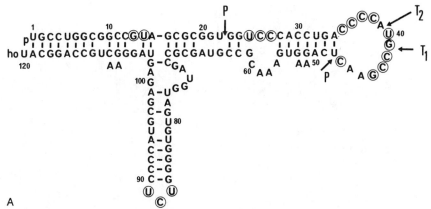

Figure 3 (*A, B, C*) Possible secondary structure models of *E. coli* 5S RNA. The base residues which are the most accessible to specific reagents have been circled and the bonds which are the most easily digested by RNases have been indicated by arrows.

(*A*) Madison's model (reprinted, with permission, from Madison 1968). The base residues (circled) belong to T_1 RNase oligonucleotides known to be affected by specific chemical reagents in T_1 RNase digests of chemically modified molecules. T_1 and T_2 arrows indicate the bonds that are the first to be split during limited enzyme degradation by either T_1 or T_2 RNases. During limited pancreatic RNase digestion, the first product which could be isolated extended from residue 1 to residue 22 and from residue 50 to residue 120. This product, which ran as a single band on polyacrylamide gels under nondenaturing conditions, completely lacked the sequence from 23 to 49 (Jordan 1971a). Arrows therefore indicate the limits of the first long fragment obtained during limited pancreatic RNase digestion and not necessarily the bonds at which the first splits occurred. (See following page for B and C.)

molecular weight rRNA genes and 5S RNA genes. The biosynthesis of 5S RNA can be dissociated from that of the giant precursor of large rRNAs in various situations, physiological or nonphysiological. It is much less sensitive to actinomycin D (Perry and Kelley 1968; Abe and Yamana 1971) and is not inhibited by α-amanitin (Price and Penman 1972; Reynolds and Penman, cited in Leibowitz, Weinberg and Penman 1973). 5S RNA synthesis still occurs during mitosis, when the synthesis of 18S and 28S RNA is blocked (Zylber and Penman 1971). In oocytes at early stages of maturation, it is produced in large excess (Ford 1971; Mairy and Denis 1971) and is accumulated in a 42S nucleoprotein particle, which also contains tRNAs (Denis and Mairy 1972). Even in exponentially growing cells in culture, the amount of 5S RNA transcribed exceeds about four times that of 18S and 28S RNA (Leibowitz, Weinberg and Penman 1973). The control of the final level of 5S RNA must therefore be regulated by a constant turnover rather than at the level of transcription.

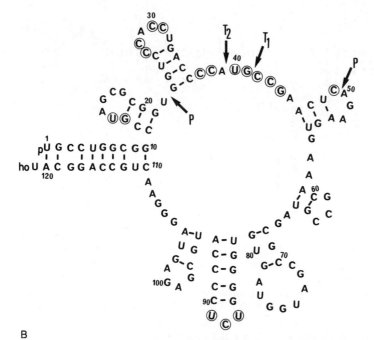

B

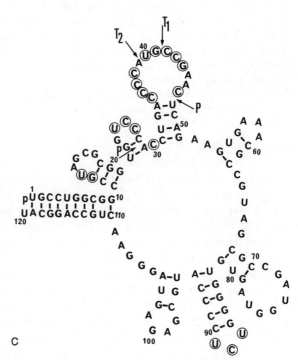

C

Figure 3 (*B, C*) Other possible models of 5S RNA.

Early after transcription, the newly synthesized 5S RNA is recovered from the cytoplasm of cell extracts (Leibowitz, Weinberg and Penman 1973), whereas at later times it enters a nuclear pool that represents 20–25% of the total 5S RNA of the cell (Knight and Darnell 1967). The presence of newly made 5S RNA in the cytoplasm probably means that it is not yet linked to a nuclear structure and leaks out of the nucleus during processing of the cell extracts.

Precursors

Pulse-labeled 5S RNA from HeLa cells has the same electrophoretic mobility on polyacrylamide gel as the steady-state labeled product (Leibowitz, Weinberg and Penman 1973). Since 5S RNA isolated from mature ribosomes often bears a triphosphate group at its 5′ end (Hatlen, Amaldi and Attardi 1969), it is likely that the transcription of each 5S RNA gene starts at the 5′ end of the molecule and stops at its 3′ end, or close to it, without stretching into the long spacer sequence that is intercalated between successive genes (Brown, Wensink and Jordan 1971). No precursor form of eukaryotic 5S RNA is therefore suggested by the data reported so far.

In prokaryotic organisms, 5S RNA precursors that have at their 5′ end the trinucleotide pAUU in excess of the mature sequence have been characterized in *E. coli* (Jordan, Feunteun and Monier 1970). More recently, another precursor has been isolated from a mutant which is temperature sensitive with respect to RNA synthesis. This precursor not only bears the trinucleotide pAUU at its 5′ end, but also a short run of oligo-A at its 3′ end (Griffin and Baillie 1973). In *B. subtilis,* two precursors of 180 and 150 nucleotides, respectively, with supplementary sequences at both extremities have been found (Pace et al. 1973; Pace, Sogin and Cuneo, unpublished observations).

The predominant 3′-terminal sequence of the longest precursor has recently been shown to be U_6Pu_{OH} (Pace, unpublished observation; Pace, Rosenberg and Weissman, unpublished observation). It remains to be seen whether the above 3′-terminal sequence is identical to that of the original transcript or is produced by nucleolytic cleavage of the transcript. However, U_6Pu_{OH} was found as the 3′-terminal sequence of transcripts synthesized in vitro by *E. coli* RNA polymerase (free of readily detectable nuclease activity) on λ bacteriophage DNA template (Larsen et al. 1970; Lebowitz, Weissman and Radding 1971; Dahlberg and Blattner 1973).

Precursors of 5S RNA accumulate in prokaryotes under conditions where protein synthesis is inhibited (Galibert, Lelong and Larsen 1967; Adesnik and Levinthal 1970; Jordan, Forget and Monier 1971; Pace et al. 1973) because at least some of the maturation steps occur after binding to 50S subunit precursors. However, as shown by observations on precursor particles from *E. coli sad* mutants and on active polysomes isolated from exponentially growing cells, there is no obligatory relation between the final maturation step and either the completion of a 50S subunit or its participation in protein synthesis (Feunteun, Jordan and Monier 1972).

Binding to Ribosome Precursors during Ribosome Assembly

Early observations suggested that the entry of 5S RNA into the 50S subunit structure took place very late during subunit assembly (Osawa et al. 1969; Morell and Marmur 1968). These observations were based on the measurement of the ratio between radioactive 5S RNA and radioactive 23S RNA isolated from the late 43S precursor. They were actually biased due to both the lability of the binding of 5S RNA to early precursors (Monier et al. 1969) and the existence of a small pool of free precursor 5S RNA, which delays the entry of radioactivity into bound 5S RNA. Under appropriate experimental conditions, stoichiometric amounts of 5S RNA precursors and 23S RNA are found in the two precursor particles sedimenting at about 32S and 43S (Hayes and Hayes 1971). These two incomplete particles also contain proteins L5, L18 and L25 (Nierhaus, Bordasch and Homann 1973).

RELEASE OF 5S RNA FROM THE LARGE SUBUNIT

5S RNA is a permanent constituent of the ribosome and is not exchanged during protein biosynthesis (Kaempfer and Meselson 1968). Its release from the larger ribosomal subunit can be obtained after either extended unfolding of the subunit structure or stripping of a large fraction of the proteins.

Prokaryotic Ribosomes

In prokaryotes, unfolding with complete release of 5S RNA is produced by prolonged dialysis against EDTA (Aubert et al. 1967; Morell and Marmur 1968; Sarkar and Comb 1969; Siddiqui and Hosokawa 1969). Protein L25 is simultaneously released (Garrett, unpublished data). Other unfolding processes based on the use of Mg^{++}-complexing agents have been proposed (Hosokawa 1970; Yogo, Fujimoto and Mizuno 1971). Replacement of Mg^{++} ions by NH_4^+ ions can also be used to produce both unfolding and solubilization of 5S RNA (Siddiqui and Hosokawa 1968; Reynier and Monier 1970). Several proteins are solubilized at the same time (Gormly, Yang and Horowitz 1971), among which proteins L18 and L25 have been identified (Garrett, unpublished data).

Unfolded particles can be refolded more or less extensively by further dialysis against Mg^{++}-containing buffers. In the case of EDTA-unfolding, the release of 5S RNA is not reversible (Aubert et al. 1967), whereas in the case of NH_4^+-treated particles, refolding is accompanied by rebinding of 5S RNA (Reynier and Monier 1970). The protein fraction, which is released with 5S RNA, is required (Gormly, Yang and Horowitz 1971). Maruta, Natori and Mizuno (1969) claimed that NH_4^+-unfolded ribosomes can be refolded into a biologically active form, but this observation has not been substantiated by Gormly, Yang and Horowitz (1971).

    E. coli 50S subunits can be converted into 25S cores after release of about 50% of the proteins and 85% of 5S RNA by exposure to 2M LiCl (Marcot-Queiroz and Monier 1967). Proteins L5, L18 and L25 are completely, and protein L2 partly, released in the 2M LiCl split (Homann and Nierhaus 1971). After dialysis against Mg^{++}- and K^+-containing buffers, the core and the split fraction can be reassembled into a 48S particle, which is devoid of biological activity but which incorporates 5S RNA with a high degree of specificity (Reynier and Monier 1968). The cores, obtained by sedimentation through CsCl gradients by the technique of Staehelin, Maglott and Monro (1969), still contain 5S RNA, although according to Nierhaus and Montejo (1973) they have lost proteins L18 and L25.

Eukaryotic Ribosomes

Early observations on eukaryotic ribosomes suggested that exposure to EDTA did not release 5S RNA (Knight and Darnell 1967) unless the treatment was performed at a slightly alkaline pH (Zehavi-Willner 1970). Later experiments by Lebleu et al. (1971) and Blobel (1971) have demonstrated that EDTA treatment near neutrality actually removes 5S RNA in the form of a protein complex. This complex, which is of low stability and which can be dissociated by moderate concentrations of KCl or NaCl, can also be obtained in the presence of formamide (Petermann, Hamilton and Pavlovec 1972). The molecular weight of the associated protein, whose isoelectric point is near neutrality (Petermann, Hamilton and Pavlovec 1972), has been estimated at 38,000–41,000 in the case of rat liver ribosomes (Blobel 1971; Petermann, Hamilton and Pavlovec 1972) or at 45,000 in the case of rabbit reticulocytes (Lebleu et al. 1971).

INTERACTIONS OF 5S RNA WITH RIBOSOMAL PROTEINS

Apart from the observations on the release of a 5S RNA–protein complex mentioned above, this subject has not yet been studied in detail in eukaryotic cytoplasmic ribosomes. The following discussion will therefore be restricted to bacterial ribosomes.

Identification of 5S RNA-specific Proteins from *E. coli*

Among the 35 proteins of the 50S subunit, two proteins, L18 and L25, can individually form stable complexes with 5S RNA in the ionic conditions which permit the in vitro assembly of *B. stearothermophilus* 50S subunits. Protein L18 also has a weak affinity for 23S RNA (Gray et al. 1973). Protein L2 and a fourth protein are required in addition to build up, under the same ionic conditions, a complex which contains equimolar amounts of 23S and 5S RNA and which can be retained on nitrocellulose filters (Gray et al. 1972). The fourth protein had previously been identified as protein L6 (Gray et al. 1972), but more recent experiments have shown that this identification was incorrect. The active protein actually is protein L5, which is very difficult to separate from protein L6 (Garrett and Stöffler, unpublished results).

Protein L2 belongs to the group of proteins that individually bind to 23S RNA (Stöffler et al. 1971a,b; Garrett et al. 1971). Protein L2 plays an indirect role, which is only detected by the filtration technique, in enhancing the retention of the complex on filters. When complexes are isolated by sucrose gradient centrifugation, the requirement for protein L2 completely vanishes. The binding of 5S RNA to 23S RNA, as measured by gradient sedimentation, is actually ensured with the same efficiency by a combination of either proteins L5 and L18 or proteins L18 and L25 (Gray et al. 1972). On the other hand, 23S RNA can be replaced by the 18S fragment derived from its 3′ end by partial RNase degradation, according to Allet and Spahr (1971) (Gray and Monier 1972).

The results of Gray et al. (1972) have been confirmed by Horne and Erdmann (1972), who have shown that upon dialysis of the total 50S subunit proteins plus 5S RNA against the reconstitution buffer, a complex of 5S RNA and a few proteins is formed. In the case of *E. coli* 50S subunits, the main protein components in the complex are proteins L18 and L25, accompanied by smaller amounts of proteins L5, L20 and L30. When *B. stearothermophilus* 50S subunits are used, the complex contains two major components, B-L5 and B-L22, plus some minor ones (B-L6, B-L10, B-L26 and B-L27). Proteins B-L5 and B-L22 from *B. stearothermophilus* are probably similar to proteins L6 and L18 from *E. coli,* as judged from their electrophoretic behavior.

Localization of Binding Sites in 5S Primary Structure

Partial degradation of various 5S RNA-protein complexes with pancreatic, T_1 and T_2 RNases, followed by isolation of the 5S RNA fragments which remain bound to proteins after polyacrylamide gel electrophoresis, has been used to identify the areas of the RNA molecule which are involved in protein–RNA interactions (Gray et al. 1973; Feunteun, unpublished data). The most important sites of interaction belong to sequence 70–103. The tetranucleotide G_{86} U_{87} C_{88} U_{89} is certainly not involved, because it is easily accessible to enzymes in the complexes (Figure 4).

These observations are in agreement with previous data on the binding

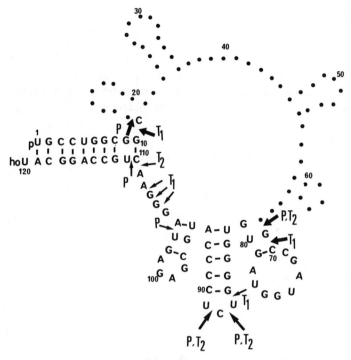

Figure 4   *E. coli* 5S RNA fragments isolated after RNase digestion of 5S RNA–protein complexes. The nucleotide residues that are eventually present in the fragments are shown. The residues which are always missing are indicated by dots. ➤ : Limits of the fragments when pancreatic (P),T₁ or T₂ RNases are used. ⟶ : Nucleotide binds where the fragments are frequently interrupted. ⟶ : Nucleotide bonds that are accessible to some extent and where interruptions in the fragments are eventually observed.

of modified 5S RNA molecules to 50S subunit proteins. Jordan and Monier (1971) had shown that molecules with a hidden T_1 RNase break at G_{41} bind to reconstituted particles. The binding is competitive with that of intact 5S RNA. Similarly, molecules in which G_{41} had been chemically modified either with glyoxal and periodate or with kethoxal have almost the same affinity for ribosomal proteins as unmodified molecules (Bellemare et al. 1972c; Aubert, Bellemare and Monier 1973). The results of Bellemare (unpublished data) on the chemical modification of 5S RNA inside the 50S subunit, which have already been mentioned above, also confirm the absence of interaction of the G_{41} region and the U_{87} C_{88} U_{89} loop with ribosomal proteins.

5S RNA in solution can form intercalation complexes with dyes such as ethidium bromide (Gray and Saunders 1971). As demonstrated by recent measurements by Feunteun, Le Bret and Le Pecq (unpublished data) in the reconstitution buffer used in the study of protein–5S RNA interactions, 5–6 binding sites, with an association constant of 4×10^4 M, are available. The addition of protein L25 to 5S RNA does not decrease the number of ethidium bromide binding sites, but the addition of a mixture of proteins containing both L25 and L18 reduces the number of binding sites to 2–3 (Figure 5). The specificity of these effects was demonstrated by replacing native 5S RNA with the denatured conformation obtained by exposure to

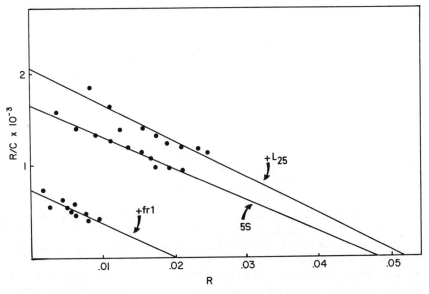

Figure 5 Scatchard plots of ethidium bromide binding to 5S RNA and 5S RNA–protein complexes.

5S RNA: Ethidium bromide (200 μg/ml in the standard reconstitution buffer) was progressively added to 1 ml of 5S RNA solution (5 μg/ml in the standard reconstitution buffer, 20 mM Tris-HCl pH 7.8; 300 mM KCl; 20 mM MgCl$_2$; 6 mM mercaptoethanol). The fluorescence of the dye excited at 520 nm was measured at 610 nm by photon counting in a Jobin-Yvon fluorimeter. A parallel experiment performed on a sample without RNA gave the fluorescence of free ethidium bromide. The results were plotted according to Scatchard. On abscissa, R = bound ethidium bromide concentration/amount of RNA expressed as the number of nucleotides. On ordinate, C = free ethidium bromide concentration.

5S RNA–protein complexes: 5S RNA (5 μg) was complexed with a stoichiometric amount of L25 protein or of fraction 1 proteins (Gray et al. 1972) by incubation in the standard reconstitution buffer for 15 min at 30°C. The same experimental procedure as above was applied to this complex. In a control experiment it was shown that the proteins did not bind ethidium bromide by themselves.

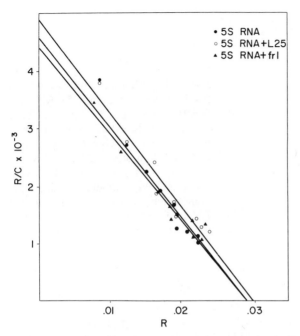

Figure 6 Scatchard plots of ethidium bromide binding to denatured 5S RNA in the absence or presence of proteins. 5S RNA was denatured by exposure to 8 M urea and 10 mM EDTA in 10 mM Tris-HCl buffer pH 8.9 for 45 min at 23°C. The RNA was recovered by ethanol precipitation and dissolved in the reconstitution buffer. The binding of ethidium bromide was studied as described in the legend to Figure 5.

EDTA and urea (Aubert et al. 1968) (Figure 6). The number of binding sites on the denatured conformation is 3–4, with an association constant of 1.7×10^5 M. The binding is completely unaffected by the addition of proteins, in agreement with the absence of affinity of 50S subunit proteins for the denatured RNA (Aubert et al. 1968). Protein L18 alone is able to chase ethidium bromide from its complexes with 5S RNA. When the chasing effects on free 5S RNA and on the L25–5S RNA complex are compared, it clearly appears from the shapes of the respective curves that the affinity of protein L18 for the L25–5S RNA complex is higher than for the free 5S RNA molecule (Figure 7). The binding of proteins L25 and L18 to 5S RNA is therefore highly cooperative. The resulting sharpening in the chasing curve enables one to calculate accurately the maximum amount of protein L18 that can bind to the L25–5S RNA complex. This amount corresponds to two moles of protein L18 (mol wt = 14,300) per mole of 5S RNA, in agreement with previous observations (Gray et al. 1973).

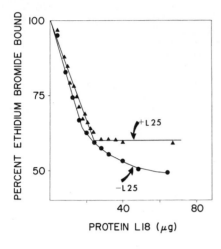

Figure 7 Ethidium bromide release from its complexes with 5S RNA or 5S RNA–protein L25 upon addition of protein L18. Ethidium bromide was bound to free 5S RNA (30 μg/ml) or to 5S RNA–protein L25 complex. Increasing amounts of protein L18 were added and the ethidium bromide fraction which remained bound to 5S RNA was estimated by fluorimetry as described in the legend to Figure 5.

FUNCTION OF 5S RNA

The in vitro assembly of biologically active *B. stearothermophilus* 50S subunit is dependent upon the presence of 5S RNA (Erdmann et al. 1971b). The omission of 5S RNA prevents the binding of at least protein B-L5 (Horne and Erdmann 1972), and possibly of three other proteins (Erdmann et al. 1971b), to the reconstituted particle. The (−5S RNA) particle displayed low activities in the following assays: polypeptide synthesis directed by natural mRNA; peptidyl transferase assay; peptide chain termination factor R_1-dependent [³H]UAA binding; G factor-dependent [³H]GTP binding; codon-directed tRNA binding in the presence of 30S subunits. Erdmann et al. (1971b) moreover point out that the extent of some residual activities (7% for peptidyl transferase, 25% for GTP binding, and 20% for Phe–tRNA binding) notably exceed the percentage of 5S RNA (1%) which was still present in the (−5S RNA) particles compared to the amount normally present in (+5S RNA) particle controls.

These experiments demonstrate that 5S RNA has an important structural role in the assembly of fully biologically active 50S subunits. They do not provide any evidence for the direct participation of 5S RNA in any particular function of the subunit. It is also likely that in the simple complexes, which contain 5S RNA and a few proteins and which display both a GTPase and an ATPase activity (Erdmann et al., unpublished observations), 5S RNA plays a structural role. Therefore the biological function of 5S RNA, as indicated by these various observations, would not be very different from that of any structural protein component of the 50S subunit. However, the conservation of the 5S RNA structure through evolution strongly suggests that it should have a more specific function of its own. Two hypotheses have been formulated regarding this function. According to the first one, 5S RNA might take part in the peptidyl transferase activity of the 50S subunit by serving as an intermediate acceptor of the growing peptide chain (Raacke 1971). The chemical modification of the 3' terminus of 5S RNA does not prevent in vitro assembly of the normally active 50S subunit (Erdmann, Doberer and Sprinzl 1971; Fahnestock and Nomura 1972). Therefore, unless one assumes that some other part of the molecule is used to accept the growing peptide chain, this hypothesis can no longer be contemplated.

According to the second hypothesis, 5S RNA could play a role in the binding of tRNA to the ribosome by providing a complementary sequence to the GTφC sequence which is found in most tRNAs (Pestka 1969; Chladek 1971; Jordan 1971a). As a matter of fact, two GAAC tetranucleotides are present in the sequence of *E. coli* 5S RNA, and one of them is also found in a similar position in *Ps. fluorescens* and *B. stearothermophilus* 5S RNAs (Figure 1). No such situation exists however in eukaryotic 5S RNAs, but the hypothesis could still be considered in that case if one assumed that the trinucleotide GAA was sufficient for the tRNA–5S RNA interaction to occur. A trinucleotide GAA indeed can be found in similar positions in all eukaryotic 5S RNAs whose sequences are known (Figure 1).

Several experimental evidences in favor of this hypothesis have actually been reported. It has been shown, for example, that the tetranucleotide TφCG actually binds to 5S RNA–protein complexes. The chemical modification with monoperphtallic acid of two unidentified A residues in 5S RNA reduces this binding and, at the same time, diminishes by about 50% the biological activities of reconstituted 50S subunits (Erdmann et al., unpublished observations). Although highly suggestive, these evidences are not yet conclusive, and the specific function, if any, of 5S RNA still remains an open question.

References

Abe, H. and K. Yamana. 1971. The synthesis of 5S RNA and its regulation during early embryogenesis of *Xenopus laevis. Biochim. Biophys. Acta* **240**:392.

Adesnik, M. and C. Levinthal. 1970. RNA metabolism in T4-infected *Escherichia coli. J. Mol. Biol.* **48**:187.

Allet, B. and P. F. Spahr. 1971. Binding sites of ribosomal proteins on two specific fragments derived from *E. coli* 50S ribosomes. *Eur. J. Biochem.* **19**:250.

Attardi, G., Y. Aloni, B. Attardi, D. Ojala, L. Pica-Mattocia, D. L. Robberson and B. Storrie. 1970. Transcription of mitochondrial DNA in HeLa cells. *Cold Spring Harbor Symp. Quant. Biol.* **35**:599.

Aubert, M., G. Bellemare and R. Monier. 1973. Selective reaction of glyoxal with guanine residues in native and denatured *Escherichia coli* 5S RNA. *Biochimie* **55**:135.

Aubert, M., R. Monier, M. Reynier and J. F. Scott. 1967. Structure and function of transfer RNA. Attachment of 5S RNA to *E. coli* ribosomes. *Proc. 4th FEBS Meeting* (Oslo: Universitet Folaget) **3**:151.

Aubert, M., J. F. Scott, M. Reynier and R. Monier. 1968. Rearrangement of the conformation of *Escherichia coli* 5S RNA. *Proc. Nat. Acad. Sci.* **61**:292.

Averner, M. J. and N. R. Pace. 1972. The nucleotide sequence of marsupial 5S ribosomal ribonucleic acid. *J. Biol. Chem.* **247**:4491.

Bellemare, G., R. J. Cedergren and G. H. Cousineau. 1972a. Comparison of the physical and optical properties of *Escherichia coli* and sea urchin 5S ribosomal RNAs. *J. Mol. Biol.* **68**:445.

Bellemare, G., B. R. Jordan and R. Monier. 1972b. Demonstration of a highly exposed region in *Escherichia coli* 5S RNA by partial hydrolysis with ribonuclease IV and sheep kidney nuclease. *J. Mol. Biol.* **71**:307.

Bellemare, G., R. Vigne and B. R. Jordan. 1973. Interaction between *Escherichia coli* ribosomal proteins and 5S RNA molecules: Recognition of prokaryotic 5S RNAs and rejection of eukaryotic 5S RNAs. *Biochimie* **55**:29.

Bellemare, G., B. R. Jordan, J. Rocca-Serra and R. Monier. 1972c. Accessibility of *Escherichia coli* 5S RNA base residues to chemical reagents. Influence of chemical alterations on the affinity of 5S RNA for the 50S subunit structure. *Biochimie* **54**:1453.

Birnstiel, M. L., M. Chipchase and J. Speirs. 1971. The ribosomal RNA cistrons. In *Progress in nucleic acid research and molecular biology* (ed. J. N. Davidson and W. E. Cohn) vol. 11, pp. 351–389. Academic Press, New York.

Blobel, G. 1971. Isolation of a 5S RNA–protein complex from mammalian ribosomes. *Proc. Nat. Acad. Sci.* **68**:1881.

Boedtker, H. and D. G. Kelling. 1967. The ordered structure of 5S RNA. *Biochem. Biophys. Res. Comm.* **29**:758.

Brown, D. D. and C. S. Weber. 1968. Gene linkage by RNA-DNA hybridization. I. Unique DNA sequences homologous to 4S, 5S and rRNA. *J. Mol. Biol.* **34**:661.

Brown, D. D., P. C. Wensink and E. Jordan. 1971. Purification and some characteristics of 5S DNA from *Xenopus laevis*. *Proc. Nat. Acad. Sci.* **68**:3175.

Brownlee, G. G., F. Sanger and B. G. Barrell. 1968. The sequence of 5S rRNA. *J. Mol. Biol.* **34**:379.

Brownlee, G. G., E. Cartwright, T. McShane and R. Williamson. 1972. The nucleotide sequence of somatic 5S RNA from *Xenopus laevis*. *FEBS Letters* **25**:8.

Cantor, C. R. 1967. Possible conformations of 5S rRNA. *Nature* **216**:513.

———. 1968. The extent of base pairing in 5S rRNA. *Proc. Nat. Acad. Sci.* **59**:478.

Chladek, S. 1971. Possible relationship of peptidyl transferase binding sites, 5S RNA and peptidyl-tRNA. *Biochem. Biophys. Res. Comm.* **45**:695.

Comb, D. G. and T. Zehavi-Willner. 1967. Isolation, purification and properties of 5S rRNA: A new species of cellular RNA. *J. Mol. Biol.* **23**:441.

Connors, P. G. and W. W. Beeman. 1972. Size and shape of 5S ribosomal RNA. *J. Mol. Biol.* **71**:31.

Cramer, F. and V. A. Erdmann. 1968. Amount of adenine and uracil base pairs in *Escherichia coli* 23S, 16S and 5S rRNA. *Nature* **218**:92.

Dahlberg, J. E. and F. R. Blattner. 1973. Sequence of a self-terminating RNA made near the origin of DNA replication of phage lambda. *Fed. Proc.* (Abs.) **32**:664.

Denis, H. and M. Mairy. 1972. Recherches biochimiques sur l'oogenèse. I. Distribution intracellulaire du RNA dans les petits oocytes de *Xenopus laevis*. *Eur. J. Biochem.* **25**:524.

Denis, H., M. Wegnez and R. Willem. 1972. Recherches biochimiques sur l'oogenèse. V. Comparison entre le RNA 5S somatique et le RNA 5S des oocytes de *Xenopus laevis*. *Biochimie* **54**:1189.

DuBuy, B. and S. M. Weissman. 1971. Nucleotide sequence of *Pseudomonas fluorescens* 5S ribonucleic acid. *J. Biol. Chem.* **246**:747.

Erdmann, V. A., H. G. Doberer and M. Sprinzl. 1971a. Structure and function of 5S RNA: The role of the 3′ terminus in 5S RNA function. *Mol. Gen. Genet.* **114**:89.

Erdmann, V. A., S. Fahnestock, K. Higo and M. Nomura. 1971b. Role of 5S RNA in the functions of 50S ribosomal subunits. *Proc. Nat. Acad. Sci.* **68**:2932.

Fahnestock, S. R. and M. Nomura. 1972. Activity of ribosomes containing 5S RNA with a chemically modified 3′ terminus. *Proc. Nat. Acad. Sci.* **69**:363.

Faras, A. J., A. C. Garapin, W. E. Levinson, J. M. Bishop and H. M. Goodman. 1973. Characterization of the low-molecular-weight RNAs associated with the 70S RNA of Rous sarcoma virus. *J. Virol.* **12**:334.

Feunteun, J. and R. Monier. 1971. Accessibility of 5S RNA to ribonucleases in *Escherichia coli* ribosomes. *Biochimie* **53**:657.

Feunteun, J., B. R. Jordan and R. Monier. 1972. Study of the maturation of 5S RNA precursors in *Escherichia coli*. *J. Mol. Biol.* **70**:465.

Ford, P. J. 1971. Non-coordinated accumulation and synthesis of 5S rRNA by ovaries of *Xenopus laevis*. *Nature* **233**:561.

Ford, P. J. and E. M. Southern. 1973. Different sequences for 5S RNA in kidney cells and ovaries of *Xenopus laevis*. *Nature New Biol.* **241**:7.

Ford Doolittle, W. and N. R. Pace. 1970. Synthesis of 5S ribosomal RNA in *E. coli* after rifampicin treatment. *Nature* **228**:125.

————. 1971. Transcriptional organization of the ribosomal RNA cistrons in *E. coli*. *Proc. Nat. Acad. Sci.* **68**:1786.

Forget, B. G. and S. M. Weissman. 1967. Low molecular weight RNA components from KB cells. *Nature* **213**:878.

————. 1969. The nucleotide sequence of ribosomal 5S ribonucleic acid from KB cells. *J. Biol. Chem.* **244**:3148.

Galibert, F. and F. Sanfourche. 1969. Mise en évidence dans le surnageant de sédimentation des ribosomes d'un "pool" de RNA 5S précurseur. *C. R. Acad. Sci.*, Ser. D, **269**:851.

Galibert, F., J. C. Lelong and C. J. Larsen. 1967. Absence du RNA 5S dans les particules ribosomales 25S accumulées sous l'effet du chloramphénicol ou de la puromycine chez *E. coli*. *C. R. Acad. Sci.*, Ser. D, **265**:279.

Garrett, R. A., K. H. Rak, L. Daya and G. Stöffler. 1971. Ribosomal proteins. XXIX. Specific protein binding sites on 16S rRNA of *Escherichia coli*. *Mol. Gen. Genet.* **114**:112.

Gormly, J. R., C. H. Yang and J. Horowitz. 1971. Further studies on ribosomes unfolding the reversible release of 5S RNA. *Biochim. Biophys. Acta* **247**:80.

Gray, P. N. and R. Monier. 1972. Partial localization of the 5S RNA binding site on 23S RNA. *Biochimie* **54**:41.

Gray, P. N. and G. F. Saunders. 1971. Binding of ethidium bromide to 5S rRNA. *Biochim. Biophys. Acta* **254**:60.

―――. 1973. Comparative studies on bacterial 5S ribosomal RNA. *Arch. Biochem. Biophys.* **156**:104.

Gray, P. N., R. A. Garrett, G. Stöffler and R. Monier. 1972. An attempt at the identification of the proteins involved in the incorporation of 5S RNA during 50S ribosomal subunit assembly. *Eur. J. Biochem.* **28**:412.

Gray, P. N., G. Bellemare, R. Monier, R. A. Garrett and G. Stöffler. 1973. Identification of the nucleotide sequences involved in the interaction between *Escherichia coli* 5S RNA and specific 50S subunit proteins. *J. Mol. Biol.* **77**:133.

Griffin, B. E. and D. L. Baillie. 1973. Precursors of stable RNA accumulated in a mutant of *E. coli*. *FEBS Letters* **34**:273.

Hatlen, L. E., F. Amaldi and G. Attardi. 1969. Oligonucleotide pattern after pancreatic ribonuclease digestion and the 3' and 5' termini of 5S ribonucleic acid from HeLa cells. *Biochemistry* **8**:4989.

Hayes, F. and D. H. Hayes. 1971. Biosynthesis of ribosomes in *E. coli*. I. Properties of ribosomal precursor particles and their RNA components. *Biochimie* **53**:369.

Hindley, J. 1967. Fractionation of ^{32}P-labeled RNA on polyacrylamide gel and their characterization by fingerprinting. *J. Mol. Biol.* **30**:125.

Hindley, J. and S. M. Page. 1972. Nucleotide sequence of yeast 5S ribosomal RNA. *FEBS Letters* **26**:157.

Homann, H. E. and K. H. Nierhaus. 1971. Ribosomal proteins. Protein composition of biosynthetic precursors and artificial subparticles from ribosomal subunits in *E. coli* K12. *Eur. J. Biochem.* **20**:249.

Horne, J. R. and V. A. Erdmann. 1972. Isolation and characterization of 5S RNA-protein complexes from *Bacillus stearothermophilus* and *Escherichia coli* ribosomes. *Mol. Gen. Genet.* **119**:337.

Hosokawa, K. 1970. Binding of 5S ribosomal ribonucleic acid to the unfolded 50S ribosomes of *E. coli*. II. *J. Biol. Chem.* **245**:5880.

Jarry, B. and R. Rosset. 1971. Heterogeneity of 5S RNA in *E. coli*. *Mol. Gen. Genet.* **113**:43.

―――. 1973. Localization of some 5S RNA cistrons on *E. coli* chromosome. *Mol. Gen. Genet.* **121**:151.

Jordan, B. R. 1971a. Studies on 5S RNA conformation by partial ribonuclease hydrolysis. *J. Mol. Biol.* **55**:423.

―――. 1971b. Computer generation of pairing schemes for RNA molecules. *J. Theoret. Biol.* **34**:363.

Jordan, B. R. and G. Galling. 1973. Nucleotide sequence of *Chlorella* cytoplasmic 5S RNA. *FEBS Letters* **32**:333.

Jordan, B. R. and R. Monier. 1971. 5S RNA molecules formed by reassociation of separated fragments. *J. Mol. Biol.* **59**:219.

Jordan, B. R., J. Feunteun and R. Monier. 1970. Identification of a 5S RNA precursor in exponentially growing *E. coli* cells. *J. Mol. Biol.* **50**:605.

Jordan, B. R., B. G. Forget and R. Monier. 1971. A low molecular weight ribonucleic acid synthesized by *E. coli* in the presence of chloramphenicol: Characterization and relation to normally synthesized 5S ribonucleic acid. *J. Mol. Biol.* **55**:407.

Kaempfer, R. and M. Meselson. 1968. Permanent association of 5S RNA molecules with 50S ribosomal subunits in growing bacteria. *J. Mol. Biol.* **34**:703.

Knight, E., Jr. and J. E. Darnell. 1967. Distribution of 5S RNA in HeLa cells. *J. Mol. Biol.* **28**:491.

Larsen, C. J., P. Lebowitz, S. M. Weissman and B. DuBuy. 1970. Studies of the primary structure of low molecular weight RNA other than tRNA. *Cold Spring Harbor Symp. Quant. Biol.* **35**:35.

Lebleu, B., G. Marbaix, G. Huez, J. Temmerman, A. Burny and H. Chantrenne. 1971. Characterization of the messenger ribonucleoprotein released from reticulocyte polyribosomes by EDTA treatment. *Eur. J. Biochem.* **19**:264.

Lebowitz, P., S. M. Weissman and C. M. Radding. 1971. Nucleotide sequence of an RNA transcribed *in vitro* from λ phage DNA. *J. Biol. Chem.* **246**: 5120.

Lee, J. C. and V. M. Ingram. 1969. Reaction of 5S RNA with a radioactive carbodiimide. *J. Mol. Biol.* **41**:431.

Leibowitz, R. D., R. A. Weinberg and S. Penman. 1973. Unusual metabolism of 5S RNA in HeLa cells. *J. Mol. Biol.* **73**:139.

Lewis, J. B. and P. Doty. 1970. Derivation of the secondary structure of 5S RNA from its binding of complementary oligonucleotides. *Nature* **225**:510.

Litt, M. 1973. Kethoxal as a chemical probe for the conformation of 5S RNA. *Fed. Proc.* **32**:586.

Lizardi, P. M. and D. J. L. Luck. 1971. Absence of a 5S RNA component in the mitochondrial ribosomes of *Neurospora crassa. Nature New Biol.* **229**:140.

Madison, J. T. 1968. Primary structure of RNA. *Annu. Rev. Biochem.* **37**:131.

Mairy, M. and H. Denis. 1971. Recherches biochimiques sur l'oogenèse. I. Synthèse et accumulation du RNA pendant l'oogenèse du crapaud sud-africain *Xenopus laevis. Develop. Biol.* **24**:143.

Marcot-Queiroz, J. and R. Monier. 1967. Les ribosomes d'*E. coli*. II. Préparation de particules 18S et 25S par traitement des ribosomes au chlorure de lithium. *Bull. Soc. Chim. Biol.* **49**:477.

Marotta, C. A., C. C. Levy, S. M. Weissman and F. Varrichio. 1973. Preferred sites of digestion of a ribonuclease from *Enterobacter sp.* in the sequence analysis of *B. stearothermophilus* 5S RNA. *Biochemistry* **12**:2901.

Maruta, H., S. Natori and D. Mizuno. 1969. Protein synthesis with *Escherichia coli* ribosomes altered in conformation by monovalent cations. *J. Mol. Biol.* **46**:513.

Mirzabekov, A. D. and B. E. Griffin. 1972. 5S RNA conformation. Studies of its partial T_1 RNase digestion by gel electrophoresis and two-dimensional thin-layer chromatography. *J. Mol. Biol.* **72**:633.

Monier, R. 1972. Structure and function of ribosomal RNA. In *The mechanism of protein synthesis and its regulation* (ed. L. Bosch), pp. 353–394. North-Holland, Amsterdam.

Monier, R., J. Feunteun, B. Forget, B. Jordan, M. Reynier and F. Varrichio. 1969. Presence of 5S RNA in 43S particle accumulated by *E. coli* K12 high growth rate lability of 5S RNA binding to precursor particles. *Cold Spring Harbor Symp. Quant. Biol.* **34**:139.

Morell, P. and J. Marmur. 1968. Association of 5S RNA to 50S ribosomal subunits of *E. coli* and *B. subtilis. Biochemistry* **7**:1141.

Mullins, D. W., Jr., J. C. Lacey, Jr. and R. A. Hearn. 1973. RNA—Common evolutionary origin of 5S rRNA and tRNA. *Nature New Biol.* **242**:80.

Nierhaus, K. H. and V. Montejo. 1973. A protein involved in the peptidyltransferase activity of *E. coli* ribosomes. *Proc. Nat. Acad. Sci.* **70**:1931.

Nierhaus, K. H., K. Bordasch and H. E. Homann. 1973. Ribosomal proteins. XLIII. *In vivo* assembly of *Escherichia coli* ribosomal proteins. *J. Mol. Biol.* **74**:587.

Nomura, M. and V. A. Erdmann. 1970. Reconstitution of 50S ribosomal subunits from dissociated molecular components. *Nature* **228**:744.

Osawa, S., E. Otaka, T. Itoh and T. Fukui. 1969. Biosynthesis of 50S ribosomal subunit in *Escherichia coli. J. Mol. Biol.* **40**:321.

Pace, B., R. L. Peterson and N. R. Pace. 1970. Formation of all stable RNA species in *E. coli* by post-transcriptional modification. *Proc. Nat. Acad. Sci.* **65**:1097.

Pace, N. R., M. L. Pato, J. McKibbin and C. W. Radcliffe. 1973. Precursors of 5S ribosomal RNA in *Bacillus subtilis. J. Mol. Biol.* **75**:619.

Pato, M. L. and K. von Meyenburg. 1970. Residual RNA synthesis in *Escherichia coli* after inhibition of initiation of transcription by rifampicin. *Cold Spring Harbor Symp. Quant. Biol.* **35**:497.

Pene, J. J., E. Knight, Jr. and J. E. Darnell, Jr. 1968. Characterization of a new low molecular weight RNA in HeLa cell ribosomes. *J. Mol. Biol.* **33**:609.

Perry, R. P. and D. E. Kelley. 1968. Persistent synthesis of 5S RNA when production of 28S and 18S ribosomal RNA is inhibited by low doses of actinomycin D. *J. Cell. Physiol.* **72**:235.

Pestka, S. 1969. Translocation, aminoacyl-oligonucleotides, and antibiotic action. *Cold Spring Harbor Symp. Quant. Biol.* **34**:395.

Petermann, M. L., M. G. Hamilton and A. Pavlovec. 1972. A 5S ribonucleic acid-protein complex extracted from rat liver ribosomes by formamide. *Biochemistry* **11**:2323.

Phillips, G. R. and J. L. Timko. 1972. Simple method for the characterization of 5S RNA. *Annu. Biochem.* **45**:319.

Price, R. and S. Penman. 1972. A distinct RNA polymerase activity, synthesizing 5.5S, 5S and 4S RNA in nuclei from adenovirus 2-infected HeLa cells. *J. Mol. Biol.* **70**:435.

Raacke, I. D. 1968. "Cloverleaf" conformation for 5S RNAs. *Biochem. Biophys. Res. Comm.* **31**:528.

———. 1971. A model for protein synthesis involving the intermediate formation of peptidyl-5S RNA. *Proc. Nat. Acad. Sci.* **68**:2357.

Reynier, M. and R. Monier. 1968. Les ribosomes d'*Escherichia coli.* III. Etude de la dissociation réversible du RNA 5S par le chlorure de lithium. *Bull. Soc. Chim. Biol.* **50**:1583.

———. 1970. Les ribosomes d'*E. coli.* IV. Etude de la dissociation réversible du RNA 5S par dépliement de la sous-unité 50S. *Bull. Soc. Chim. Biol.* **52**:607.

Richards, E. G. 1969. 5S RNA. An analysis of possible base pairing schemes. *Eur. J. Biochem.* **10**:36.

Richards, E. G., R. Lecanidou and M. E. Geroch. 1973. The kinetics of renaturation of 5S RNA from *Escherichia coli* in the presence of Mg^{2+} ions. *Eur. J. Biochem.* **34**:262.

Richards, E. G., M. G. Geroch, H. Simpkins and R. Lecanidou. 1972. Optical properties and base pairing of *E. coli* 5S RNA. *Biopolymers* **11**:1031.

Rosset, R. and R. Monier. 1963. A propos de la présence d'acide ribonucléique de faible poids moléculaire dans les ribosomes d'*E. coli*. *Biochim. Biophys. Acta* **68**:653.

Sankoff, D. and R. J. Cedergren. 1973. A test for nucleotide sequence homology. *J. Mol. Biol.* **77**:159.

Sarkar, N. and D. G. Comb. 1969. Studies on the attachment and release of 5S ribosomal RNA from the large ribosomal subunit. *J. Mol. Biol.* **39**:31.

Scott, J. F., R. Monier, M. Aubert and M. Reynier. 1968. Some optical properties of 5S RNA from *E. coli*. *Biochem. Biophys. Res. Comm.* **33**:794.

Siddiqui, M. A. Q. and K. Hosokawa. 1968. Role of 5S rRNA in polypeptide synthesis. II. Dissociation of 5S rRNA from 50S ribosomes in *E. coli*. *Biochem. Biophys. Res. Comm.* **32**:1.

———. 1969. Role of 5S rRNA in polypeptide synthesis. *Biochem. Biophys. Res. Comm.* **36**:711.

Soave, C., E. Galante and G. Torti. 1970. Hydroxyapatite chromatography of 5S RNA from wheat. *Bull. Soc. Chim. Biol.* **52**:857.

Soave, C., R. Nucca, E. Sala, A. Viotti and E. Galante. 1973. 5S RNA: Investigation of the different extent of phosphorylation at 5′ terminus. *Eur. J. Biochem.* **32**:392.

Staehelin, T., D. Maglott and R. E. Monro. 1969. On the catalytic center of peptidyl transfer: A part of the 50S ribosome structure. *Cold Spring Harbor Symp. Quant. Biol.* **34**:39.

Stöffler, G., L. Daya, K. H. Rak and R. A. Garrett. 1971a. Ribosomal proteins. XXVI. The number of specific protein binding sites on 16S and 23S RNA of *Escherichia coli*. *J. Mol. Biol.* **62**:411.

———. 1971b. Ribosomal proteins. XXX. Specific protein binding sites on 23S RNA of *Escherichia coli*. *Mol. Gen. Genet.* **114**:125.

Vigne, R. 1972. Etude comparée de la conformation de RNAs 5S procaryotes et eucaryotes. *Ph. D. thesis,* Université d'Aix-Marseille, France.

Vigne, R. and B. R. Jordan. 1971. Conformational analysis of RNA molecules by partial RNase digestion and two-dimensional acrylamide gel electrophoresis. Application to *E. coli* 5S RNA. *Biochimie* **53**:981.

Vigne, R., B. R. Jordan and R. Monier. 1973. A common conformational feature in several prokaryotic and eukaryotic 5S RNAs. A highly exposed, single-stranded loop around position 40. *J. Mol. Biol.* **76**:303.

Wegnez, M. and H. Denis. 1972. Biochemical research on oogenesis. IV. Absence of amplification of 5S RNA and tRNA organization genes in small oocytes of *Xenopus laevis*. *Biochimie* **54**:1069.

Wegnez, M., R. Monier and H. Denis. 1972. Sequence heterogeneity of 5S RNA in *Xenopus laevis*. *FEBS Letters* **25**:13.

Weinberg, R. A. and S. Penman. 1968. Small molecular weight monodisperse nuclear RNA. *J. Mol. Biol.* **38**:289.

Williamson, R. and G. G. Brownlee. 1969. The sequence of 5S rRNA from two mouse cell lines. *FEBS Letters* **3**:306.

Wong, Y. P., D. R. Kearns, B. R. Reid and R. G. Shulman. 1972. The extent of base pairing in 5S RNA yeast 5S RNA. *J. Mol. Biol.* **72**:741.

Wrede, P. and V. A. Erdmann. 1973. Activities of *B. stearothermophilus* 50S ribosomes reconstituted with prokaryotic and eukaryotic 5S RNA. *FEBS Letters* **33**:315.

Wu, M., N. Davidson, G. Attardi and Y. Aloni. 1972. Expression of the mitochondrial genome in HeLa cells. XIV. The relative positions of the 4S RNA genes and of the ribosomal RNA genes in mitochondrial DNA. *J. Mol. Biol.* **71**:81.

Yogo, Y., H. Fujimoto and D. Mizuno. 1971. Dissociation of 5S RNA from 50S ribosomal subunits of *Escherichia coli* by phosphate treatment. *Biochim. Biophys. Acta* **240**:564.

Zehavi-Willner, T. 1970. The release of 5S RNA from reticulocyte ribosomes. *Biochem. Biophys. Res. Comm.* **39**:161.

Zylber, E. A. and S. Penman. 1969. Mitochondrial-associated 4S RNA synthesis inhibition by ethidium bromide. *J. Mol. Biol.* **46**:201.

————. 1971. Synthesis of 5S and 4S RNA in metaphase-arrested HeLa cells. *Science* **172**:947.

Structure of the 16S and 23S Ribosomal RNAs

Peter Fellner*

Institut de Chimie Biologique
Université de Strasbourg
67000 Strasbourg, France

1. **Determination of Primary Sequences**
   A. Sequence Studies on 16S RNA
   B. Sequence Studies on 23S RNA
2. **A Possible Model of rRNA Secondary Structure**
3. **Heterogeneity of rRNA**
4. **Conservation and Repetition of rRNA Sequences in Prokaryotes**
   A. Conservation of rRNA Sequences among Different Strains
   B. Sequence Homologies and Repetitions among rRNAs of a Single Organism
5. **Post-transcriptional Modification of rRNA**
   A. Methylation
   B. Enzymatic Cleavage

INTRODUCTION

The important role of the rRNAs in establishing and maintaining the structure of ribosomes has been clearly revealed by extensive studies on the reconstitution of ribosomal particles (see Nomura and Held, this volume). The rRNAs appear to provide the framework to which the ribosomal proteins are attached during assembly. In the mature ribosome, they hold the proteins in a configuration which permits the particle to correctly carry out its functions in protein biosynthesis. The conformations of the RNA and protein molecules and the manner in which they are associated together result from a variety of interactions, both inter- and intramolecular. These interactions, which are cooperative, cause the RNA and protein components to assume specific secondary and tertiary structures and to become attached to each other in a precise way.

In view of the fundamental role of rRNAs, considerable efforts have been made in the last few years to determine their nucleotide sequences. Extensive information about the primary structures of these molecules is necessary in order to gain a more detailed understanding of their role in ribosome structure, function and assembly and might also provide a means

* Present address: Searle Research Laboratories, Lane End Road, High Wycombe, Bucks, England.

of resolving several questions concerning the genetics and evolution of ribosomes. In particular, such knowledge should prove helpful in clarifying the following areas:

(1) Primary sequence data is expected to give some indication of the secondary structures of the rRNA molecules, both from consideration of the sequences themselves and from the relative susceptibilities of specific bonds within them to enzymatic digestion. In any event, it would be difficult to suggest a plausible model for the secondary structure of rRNA in the absence of such information.

(2) Knowledge of the primary structures of these molecules is required in any attempt to determine which portions of their sequence serve as sites for the attachment of individual ribosomal proteins during the assembly of ribosomes. Such information is also essential for a detailed description of the specific RNA–protein interactions governing such associations, which of course form the basis of ribosomal assembly. This topic is treated in the article by Zimmerman elsewhere in this volume.

(3) It is known that multiple cistrons are used for the transcription of the 16S and 23S RNAs (see Jaskunas, Nomura and Davies, this volume). It would be of interest to learn the extent to which individual rRNA classes are homogeneous, that is, the degree to which the sequences of the multiple cistrons have been conserved. This information might in turn shed some light on mechanisms involved in restricting the variability of such sequences.

(4) The evolution of rRNAs and ribosomes in different species can be investigated by examining the extent to which sequence homologies are found among the rRNAs from different organisms. With regard to the evolution of rRNA molecules within a particular species, it might be profitable to inspect the sequences of each rRNA class for internal repetitions as well as to search for possible homologies among the various rRNA classes.

(5) During the maturation of ribosomal particles, the rRNAs undergo two post-transcriptional modifications: certain nucleotides in the rRNAs are modified, mainly by methylation, while larger precursor RNAs are trimmed by enzymatic cleavage. Studies of the primary structures of the rRNAs would permit determination of the positions and identities of the modified nucleotides and would contribute to a fuller understanding of the trimming process.

In this chapter, information presently available concerning the primary structures of 16S and 23S RNA is summarized. The topics listed above are then discussed in the light of the sequence data which has so far been obtained. It will be seen that our current knowledge of rRNA structure is sufficiently detailed to resolve at least some of these problems.

DETERMINATION OF PRIMARY SEQUENCES

Because of their very great size, complete sequence analysis of the large rRNAs presents considerable technical problems. The molecular weights of the 16S and 23S RNAs of *Escherichia coli* were estimated by physical methods to be 0.56×10^6 and 1.1×10^6 daltons, respectively (Kurland 1960; Stanley and Bock 1965), indicating their chain lengths to be about 1700 and 3500 nucleotides. End-group determinations (Midgley 1965) suggested similar values of 1500 and 3000 nucleotides, respectively. The earliest studies of the primary structures of these molecules were therefore limited to determining the sequences of a few nucleotides occurring at specific positions, either at the 3' or 5' terminus (Takanami 1967; McIlreavy and Midgley 1967) or in the neighborhood of the modified nucleotides (Fellner and Sanger 1968; Fellner 1969). However, the introduction of rapid methods for fractionating radioactive oligonucleotides (Sanger, Brownlee and Barrell 1965; Brownlee and Sanger 1969; Adams et al. 1969) made it feasible to attempt sequence analysis of these molecules on a much larger scale. With these techniques, extensive studies of the primary structures of the 16S and 23S RNAs have been carried out, mainly in Strasbourg.

Sequence Studies on 16S RNA

In the case of the 16S RNA, isolated from *E. coli* strain MRE 600, the classical protocol of sequence analysis was followed. Accordingly, the molecule was cleaved enzymatically, under a variety of conditions into a large number of overlapping fragments of different sizes. These fragments were then analyzed in an effort to deduce the complete sequence of the RNA molecule. In the first stage of the analysis, the sequences and frequencies of occurrence of all the oligonucleotide products obtained by fingerprinting complete T_1 RNase digests of the 16S RNA were determined (Fellner, Ehresmann and Ebel 1972). Many of the pancreatic RNase products were also similarly cataloged.

The more difficult task of arranging these oligonucleotides in a single defined sequence was approached by isolating and characterizing large defined fragments of the 16S RNA arising from partial enzymatic hydrolysis (Ehresmann et al. 1972). The products were, in general, fractionated by two gel electrophoresis steps, using polyacrylamide slabs as described by Ehresmann et al. (1972), and then characterized by fingerprinting both before and after further partial digestion. Two types of enzymatic hydrolysis have proved particularly helpful in this phase of the sequence analysis. (1) T_1 RNase hydrolysis of the 16S RNA within the 30S subunit (Fellner, Ehresmann and Ebel 1970a), under the conditions used, produces a fragment of about 600 nucleotides arising from the 5′ half, as well as several smaller fragments derived from other regions of the molecule. (2) Partial T_1 RNase digestion of the 16S RNA in the ribosome reconstitution buffer (Zimmermann et al. 1972a,b; Muto et al. 1974) yields fragments of roughly 12S and 8S, representing the 5′ 60% and the 3′ 35% of the molecule, respectively. This digestion procedure was initially employed in studies of the location of specific protein binding sites on the RNA (see Zimmermann, this volume) but has proved helpful in rRNA sequence determination as well. The 12S and 8S fragments each contain a number of "hidden breaks" and reproducibly provide a further series of useful subfragments upon polyacrylamide gel electrophoresis in the presence of 8 M urea. Several smaller fragments of the 16S RNA that are specifically protected from enzymatic digestion by the presence of bound ribosomal proteins (see Zimmermann, this volume; Zimmermann et al. 1974) have also been of value in the sequence analysis.

Ehresmann et al. (1972) have described in detail the sequence data relating to about 70% of the 16S RNA molecule. Since that time, much of the remaining sequence has also been characterized by means of the same procedures (C. Ehresmann, P. Stiegler and P. Fellner, unpublished results). In addition, a systematic check of earlier results has led to the reordering of several blocks in the sequence relative to one another, although no revision of the sequences within these blocks was necessary. Initial uncertainties in the order of a number of the larger fragments were caused by difficulties in obtaining and satisfactorily purifying overlapping sequences from several portions of the molecule. The primary structure of the 16S RNA, as it is known at the present time, is given in Figure 1. The sequences of the T_1 RNase oligonucleotides from a large part of the 16S RNA molecule, as well as the association of several blocks of sequences within these regions, have recently been confirmed by Santer and Santer (1973).

The nucleotide sequence shown in Figure 1 contains about 1520 base residues. The contiguity of a sequence of approximately 690 nucleotides, running from the 5′ terminus of the molecule to the end of section C_1' has been established, as has the contiguity of a segment containing about 760 nucleotides, which extends from the beginning of section D′ to the 3′ terminus of the molecule. As yet no formal overlaps between the extremities of the central 74- residue fragment $K'C_2'$ and the two long terminal fragments have been found, and the number of intervening bases is unknown. However all the characteristic T_1 RNase oligonucleotides in the 16S RNA are believed to be contained within the sequence presented. If any unidentified linking sequences exist, they are therefore likely to be rather short, probably encompassing no more than 20 or 30 residues in all. The length of the

16S RNA is thus 1520 to 1550 nucleotides, a value lying between the length of 1700 nucleotides suggested by earlier physicochemical studies (Kurland 1960; Stanley and Bock 1965) and that of 1500 nucleotides proposed on the basis of end-group analysis (Midgley 1965) and within the limits of accuracy of both types of determination. It is nonetheless substantially lower than the value of 1900 nucleotides suggested by the recent physical measurements of Ortega and Hill (1973). The reasons for this discrepancy are not clear since, as these authors point out, their method should be more precise than those employed in previous work. The overall base composition of the 16S RNA derived from the sequence data is 31.8% guanine, 24.8% adenine, 23.3% cytosine and 20.0% uracil, in close agreement with the values obtained by alkaline hydrolysis of the molecule (Spahr and Tissières 1959).

Sequence Studies on 23S RNA

Studies on the primary structure of the 23S RNA are also being carried out in Strasbourg (Branlant et al. 1973; C. Branlant and P. Fellner, unpublished results). The methods being used are similar to those employed in the sequence analysis of 16S RNA. A slightly different protocol has been adopted, however, in that for the most part work is being done on large fragments from defined portions of the molecule. Fragments from the 5'- and 3'-terminal regions were prepared in the following way: The RNA was digested with T_1 RNase in 0.05 Tris-HCl, pH 7.6 – 0.02 M $MgCl_2$ – 0.35 M KCl (TMK buffer), and the products were partially separated by sucrose gradient centrifugation. Fractions from the gradient were subsequently subjected to two successive cycles of gel electrophoresis in the presence of 8 M urea, which yielded a large number of pure fragments. A search among these fragments was first carried out to detect products containing the characteristic 5'-terminal (Takanami 1967) and 3'-terminal (Fellner and Ebel 1970a) oligonucleotides. A family of fragments from the 5'-terminal region that ranged up to 500 residues in length was found, and a fragment of about 350 residues from the 3'-terminal region was isolated as well. The sequences of all the oligonucleotides resulting from T_1 and pancreatic RNase digestion of these fragments have been cataloged, and some of the fragments have in addition been subjected to partial enzymatic hydrolysis. Analysis of the latter products has yielded information about the order of the oligonucleotides within the original fragments but was insufficient to permit deduction of their complete nucleotide sequences. Some smaller fragments from the interior of the 23S RNA have also been partially cataloged.

Fragments from the 5'-terminal portion of the 23S RNA can be isolated in good yield by RNase digestion of the complex between 23S RNA and 50S subunit protein L24 (Branlant et al. 1973), since the specific binding

L AAAUUGAGAG̲A̲GUUUGAUCAUGGCUCAGAUUGAACGCUGG̲CGGCAGGC(C. U)AACACAUGCAAGUCGAACGGUAACAGCA̲GGAAGGCUUG [C(UU.C̲)G. CUG. ACG] CUG [GACG. AG. UG. G̲CG] —

H' UAAUGCUCUG [GGAAACUG. CCUG. (G₅AG)AUG] A̲UAAACACUUGGAAA̲ACGGUAGCUAAUACCGUGC̲AUAAC̲U̲ GUCCGAAGACC(AAAG. AG)GGGACCUUCGGGCCUCUGCCAUCGGAUGCCAGAUG —

Q GGAUAUG [GGG. UAG. CUAG. UAACG] AGUGGCUCACCUACAGCCGACGAU(C₂U)AGCUGGUCUG(AG. AG. G₂)AUGACCAGCCACACUGGAACUGAGACACGGUCCAGACUCCUACG̲G̲GGGGCAG —

B CAGUGGGGAAUAUUGCACAAUGGGCGCAAGCCUGAUGCAGCCAUGCCGCGUGUAUGAAGAAGGCCUUCGGGUUGUAAAGUACUUUCAGCAGGGAAGAAGGGAGUAAGGUUAAUA̲ACCUUAUUCAUUGAUG —

R AAGC(C. A̲)CGUACUUCUAGGC̲G(AG. AG. AAG. G₆)UGGGUAACUCCCGUGCCAG CCGCGGUAAUACGGAGGGUGCAAGCGUUAAUCGGAAUUACUGGGCGUAAAGCGCACGCAGGCGGUUUGUUAAGUCAGAUG —

H UGAAAUCCCCGGGCUCAACCCUGGGAACUGCAUCUGAUACUGGCAAGCUUGAGUCUCGUAGAGGGGGGUAGAAUUCCAGGUGUAGCGGUGAAAUGCGUAGAGAUCUGGAGGAAUACCGGUGGCGAAGGCGG —

G (G. AAG)] GACUGACGAAAGCGUCAGGUG [UG. 1-2CG. GGGG]C. . .AUACUCCGGCCGUAAACGAUGUCGACUUGGAGGUUGUGCCCUUGAGGCGUGGCUUCCGGAGCUAACGCGUUAAGUCGACCGCCUGGGGAGUACGGCCGCAAGGUUAAAAACUCAAAUGAAUUGACGGGGGCCCGCACAAGCGGUGGAGCAUGUGGUUUAAUUCGAUG —

M ACGUACUGGUUCUG [A. GIG. UCCG. GAUUG] C(UU. AAC. C)G —

I' GGGGACUGACCAUCACG̲UUUUCAGAGAUG̲CAGGUCAAGUGCUCGGGGAACCGAGAAUGUGCCGGGCCGCACGGAUG —

I CAGUGGGGAAUAUUGCACAAUGGGCGCAAGC... (continued in B)

C''   ↓CCm⁷GCGGUAAUACGGAGGGUGCAAGCGUUAAUCGGAAUUACUGGGCGUAAAG UGGUUUGUUAAGUCAGAUG —

C AAGC(C. A̲)CGUACUUCUAGGC̲G [CAG. CACG. AG. CG. CG]

C'1 UGAAAUCCCCGGGCUCAACCCUGGGAACUGCAUCUGAUACUGGCAAGCUUGAGUCUCGUAGAGGGGGGUAGAAUUCCAGGUGUAGCGGUGAAAUGCGUAGAGAUCUGGAGGAAUACCGGUGGCGAAGGCGG —

C'2   ↓ ACUGGCCCUUGAGG(CG. UG. G. CCUUCG)A. G̲G(CG. .

K' UAGAGAUCUGGAGGAAUACC(GUG. GCG) [CCCCCUG. G. CG.]

K CUAAGGUUAAGUCGCGUCAGGUGCUCGGGGAACCGAGGAUGUGCCGGGCCGCACGGAUGACAGCAG [CAUG. UUG. UUG. UCAG. 1-2CUG. G. AG. UCCCG] AAAU d̲CGUCCCG̲ CUUAAGUUC̲ d̲CUGm⁵CAACGAG] CGCAACGCGAAGAACCUUACCUG —

D GUCUUGACCAUCACG̲UUUUCAGAGAUG̲ ⌐GAAGUG⌐ AGAAUGUGCCGGGCCGCACGGAUGACAGCAG —

O' UUAACGCGCGCGCAAGGUUAAAAACUCAAAUGAAUUGACGGGGGCCCGCACAAGCGG(A. AG)] ⌐CACAAGCGGUGGAGCAUGUGGUUUAAUUCGAUG —

O ⌐   A ↑²ᵐG
    GACUGACGAAAGCGUCAGGUG[UG. 1-2CG. GGGG]C...↑U

E' ⌐CAUG. UUG. UUG. UCAG. 1-2CUG. G. AG. UCCCG]

P AUUAAACGG(AG. AAG. G₅)UGGGGAUGACGUCAAGUCAUCAUGGdP²CCCCCUUCACG —

E* AUUAAACG(AG. AG. G₅)UGGGGAUGACGUCAUC ...

P' ACCAGGd̲GC(AC. AC. AC. U)GUG(CUACAAUG. G. CG)CAUCACAAAG̲Gd̲A̲G̲Cd̲ACUCCCG̲Gd̲ACAGCAAGGCAᵈCCUCAUAAAG(UCG. UG. CG.)UAG [UCUG. (A. G̲)G. UCCG. GAUUG] C(UU. AAC. C)G —

A ACUCCAUGAAGUCGGAAUCGCUAGUAAUCGUd̲AAUCG̲GA̲GUUAG̲ ⌐CG⌐ CUCACACCAUGGGAGUGGGUUGCACCAGAAGUAGCUAGUCUAACCUUCGGGAGGACGGGUGACCACGGUG —

J m⁴C U m⁵
   ᵐᵐ₂C CGCCCGᵐ²₆₂C CACACCAUGGGAGUGGGUUGCACCAGAAGUAGCUAGUCUAACCU...

   AUAUUCGUGCAAAAGAAGUAGGUAGᵐᵐ₆²CUUAAAACCUUUGᵈUCGUAACAAGGUAACCGUAGGGGA̲_A̲CCUGCGGUUGGAUCCUCACUCA_OH

Figure 1 A plan of the nucleotide sequence of the 16S RNA, representing the primary structure of the 16S RNA molecule insofar as it is known at the present time. Roughly 90% of the sequence has been completed. Overlapping fragments have permitted the establishment of contiguous sequences between the 5' terminus of the molecule and the 3' end of section C'_1, and between the 5' end of section D' and the 3' terminus of the molecule. The remaining region, $K'C'_2$, must lie between these blocks, but direct contiguity has not yet been formally established with pure fragments isolated from gels containing 8 M urea. Therefore a few nonunique oligonucleotides may be present at each end of $K'C'_2$, linking it to sections C'_1 and D'. Since publication of the previous plan (Ehresmann et al. 1972), several new sections have been characterized and others completed (C. Ehresmann, P. Stiegler, G. A. Mackie, R. A. Zimmermann, J.-P. Ebel and P. Fellner, unpublished results). A number of alterations in the relative order of certain blocks of sequence have also been made, based on the isolation and analysis of pure overlapping fragments, particularly from those parts of the molecule where some doubts or inconsistencies were apparent. Major revisions include the following: The order of H'H and QR around F has been reversed, and it is now known that the overlaps originally proposed were deduced from impure fragments. The isolation of overlapping sequences linking section I directly to section C" has led to the elimination of section N, which probably represented a contaminant arising from the 23S RNA. In addition, earlier conclusions regarding the partial order of the oligonucleotides within section C", predicted on the grounds of the putative overlap with section N, were found to be incorrect. Suggested overlaps between sections K' and K, sections D' and D, and sections O' and A, which were based on information derived from fragments that were purified in the absence of 8 M urea, have proved to be spurious. In fact, section K' is contiguous with section C'_2, and sections O and O' have been found to intervene between sections D' and D. In the previous plan, the precise arrangement of several blocks of sequence within the 3'-half of the 16S RNA was not established. Through the analysis of extensive overlapping fragments, the order of the sections in this part of the molecule has recently been resolved. In a few places, segments of the sequence are known to be contiguous, but the overlapping fragments obtained thus far have been too large to yield the precise primary structure at the point of contiguity. Such segments are underlined with a dashed line. Regions of the molecule in which the order of individual oligonucleotides remains unknown are placed in parentheses.

175

site for this protein appears to be located within the 5′-terminal 500 nucleotide residues of the molecule (see Zimmermann, this volume). A specific RNA fragment protected by 50S subunit protein L1, which probably arises from the 3′-terminal half of the 23S RNA, has also been partially characterized.

Recently two large fragments of the 23S RNA have been produced by T_1 RNase digestion of the 50S ribosomal subunit. These fragments arise from the 5′ and 3′ ends of the RNA molecule and contain 1300 and 1000 nucleotides, respectively. The oligonucleotide catalogs of these fragments, together with the information obtained from the smaller fragments, have permitted the assignment of many of the unique oligonucleotides to their approximate positions within the molecule.

Limited sequence analysis of rRNAs from other prokaryotes will be discussed later in this chapter.

A POSSIBLE MODEL OF rRNA SECONDARY STRUCTURE

It has been shown by studies of hyperchromicity during heat denaturation (Cox 1966) and by optical rotatory dispersion measurements (McPhie and Gratzer 1966) that 60–70% of the nucleotides in both the 16S and 23S RNAs participate in secondary structure under physiological buffer conditions. Moreover enzymatic hydrolysis of both RNAs in TMK buffer is restricted to a relatively small number of points, reproducibly giving rise to a limited set of specific RNA fragments (Ehresmann et al. 1972; C. Branlant and P. Fellner, unpublished results). This is presumably due to the insensitivity toward RNase of bases involved in the substantial amount of secondary, and possibly tertiary, structure expected to be present under the conditions of partial digestion used.

In the simplest model of rRNA secondary structure, which has been suggested by several different authors (Doty et al. 1959; Fresco, Alberts and Doty 1960; Cox 1966; Cotter, McPhie and Gratzer 1967; Cotter and Gratzer 1969; Cox and Bonanou 1969), the polynucleotide chains are folded back on themselves, forming a series of hairpin loops linked by short single-stranded regions. In this scheme, the predominant points of scission during partial digestion would be expected to occur in the single-stranded areas between the hairpins or in the unpaired loops. The resulting fragments should thus consist either of the hairpin loops themselves or of sequences corresponding to half loops which are capable of base-pairing with other fragments from adjacent regions of the primary structure. Inspection of the sequences of many of the fragments obtained indicated that they indeed contained self-complementary segments that would be able to form very stable hairpin loops. We have examined this question more systematically, using the graphical method proposed by Tinoco, Uhlenbeck and Levine (1971), to search for the most stable possible secondary structures for each of the fragments according to the thermodynamic criteria suggested. We have set out the hairpin structures obtained in this way in Figure 2, in which a possible secondary structure for the 16S RNA is proposed. In this model, the 16S RNA is folded up as a series of hairpins of various sizes linked by short single-stranded regions, as previously suggested. The points where enzymatic hydrolysis initially occurs almost all fall either between the hairpins or in the main unpaired loops. In some cases, scissions also occur in short unpaired sequences which are thought to loop out from the main helical regions (Ehresmann et al. 1972; C. Ehresmann, P. Stiegler and P. Fellner, unpublished results).

The secondary structure proposed in Figure 2 contains 40 or so loops, including 34 hairpins in regions where the primary structure has been mainly completed, and a further 9 hypothetical loops, shown by dashed lines, which might be present in areas where the sequence is still largely unfinished. These hypothetical loops have been drawn on the basis of the sizes of the incomplete sections and from indications of possible base-pairing inferred from partial sequences. In the secondary structure proposed, the amount of base-pairing is 63%, in good agreement with previous physicochemical estimates (Cox 1966; McPhie and Gratzer 1966). Since some of the unfinished sequences may be base-paired to adjacent segments of the molecule, which have been shown as single-stranded for the time being, it is possible that a slightly higher amount of secondary structure will be derived from the complete primary sequence.

The secondary structure of the 16S RNA proposed in Figure 2 bears considerable resemblance to that suggested for the RNA of the coat protein cistron of bacteriophage MS2 (Min Jou et al. 1972). In both cases, much of the RNA chain is thought to be folded into a series of hairpin loops, which exhibit a number of irregularities in base-pairing. Furthermore the two proposed structures have similar stabilities according to the criteria of Tinoco, Uhlenbeck and Levine (1971). Finally, the 66% base-pairing postulated for the MS2 RNA is very close to the estimate of 63% for 16S RNA.

Using infrared spectroscopy, Cotter and Gratzer (1969) estimated the proportion of the total nucleotides in *E. coli* rRNA that are present in G:C and A:U base pairs as 33–35% and 27–29%, respectively. Cramer and Erdmann (1968) studied the amount of A:U pairing in the 16S and 23S RNAs by oxidizing unpaired adenine residues with monoperphtallic acid. They estimated the A:U pairs to be about 20% of the total nucleotides in both rRNAs. In the secondary structure proposed in Figure 2, the composition of the helical regions of the 16S RNA agrees well with these findings, with 34% and 23% of the bases in G:C and A:U pairs, respectively, as well as 6% in G:U pairs. In the hairpins shown in Figure 2, there are

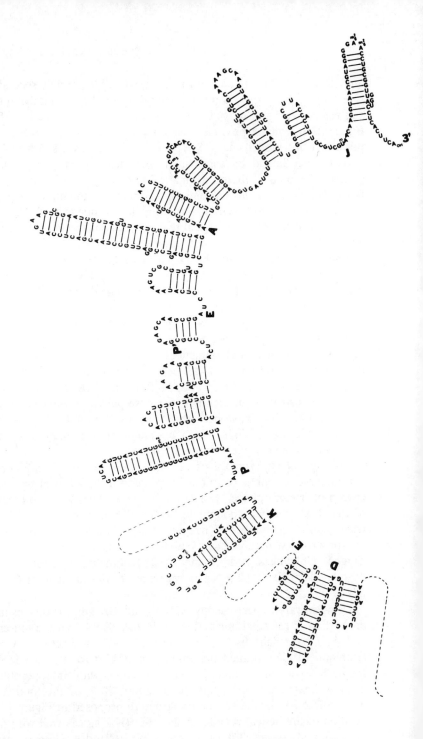

Figure 2 A possible secondary structure for the 16S RNA. The proposed
secondary structure, which is characterized by a series of hairpin loops connected by short single-stranded segments, has been derived from the primary
sequence according to the principles suggested by Tinoco, Uhlenbeck and Levine

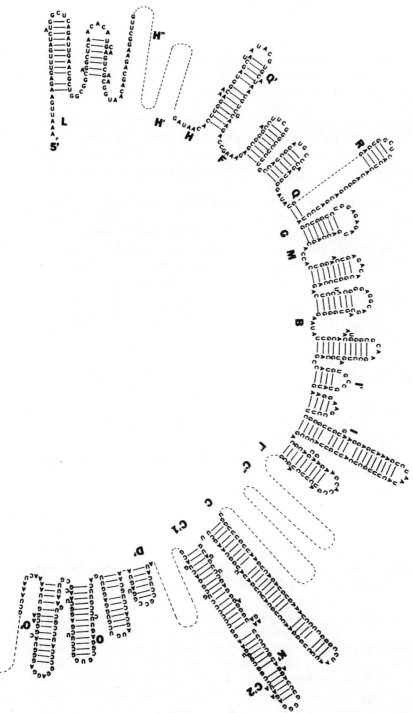

(1971). The base-pairing among sections C'_1, K' and C'_2 has been predicted upon incomplete sequence data. This portion of the secondary structure must consequently remain highly speculative, although it is consistent with all the analytical information presently available.

sometimes small looped-out or unpaired regions within the helices, involving from one to five nucleotides. It is noteworthy that these sequences are greatly enriched in adenine residues, containing 53% adenine, 19% guanine, 14% cytosine and 14% uracil.

The physicochemical studies cited above suggested that the secondary structure of the 23S RNA closely resembles that of the 16S RNA. However the sequence analysis of the 23S RNA is not sufficiently advanced to permit the postulation of a detailed model of its secondary structure at the present time.

It is clear that many other secondary structures of roughly equal stability can be proposed for the 16S RNA, involving base-pairing between more widely separated regions of the molecule. However recent studies of the specific interaction of individual ribosomal proteins with isolated fragments of the 16S RNA (see Zimmermann, this volume) impose certain limitations upon the number of possible alternative secondary structures. Individual protein binding sites do not require the integrity of the entire 16S RNA primary structure, and it seems probable that specific features of the secondary, and possibly tertiary, structures of such sites are recognized by the proteins. If parts of the molecule separated from each other by large distances in the primary structure were base-paired together to any significant extent, the disruption of the secondary structure during isolation of the comparatively small RNA fragments involved in protein binding would probably interfere with the specific recognition process. In other words, the characteristics of the protein-binding fragments isolated thus far imply that the secondary structure of the rRNA is "local," in the way proposed in Figure 2, rather than "dispersed" between widely separated regions of the molecule. Nonetheless additional interactions between distant points of the primary structure cannot be completely excluded at present.

Optical rotatory dispersion studies by McPhie and Gratzer (1966) have demonstrated that the amount of secondary structure possessed by the rRNAs within the ribosome is the same as that of the rRNAs in solution. This observation, together with the fact that many ribosomal proteins appear to bind to their specific sites on the rRNAs in vitro, implies, but does not prove, that the distribution of rRNA secondary structure in the ribosomal particle is similar or identical to that in solution.

Little information is available concerning the tertiary structure or conformation of the 16S RNA molecule either free in solution or within the 30S ribosomal particle. However several approaches to this problem, based on the results of primary sequence determination, have recently been made. Fellner, Ehresmann and Ebel (1970a) characterized an RNA fragment encompassing about 600 nucleotides from the 5' half of the 16S molecule that was resistant to hydrolysis upon partial T_1 RNase digestion of the intact 30S subunit. By contrast, the 3' half of the 16S RNA appeared to be much less resistant to T_1 RNase digestion within the 30S particle. Santer and Santer (1973) have recently examined the RNase susceptibility of the 16S RNA within the subunit structure in greater detail. They also found that the 3' half of the molecule is more accessible to RNase and reported that the 3'-terminal 150 nucleotides are especially labile. On the basis of these results, it was suggested that the latter sequence, which is also heavily methylated, may be particularly active in functions relating to protein synthesis. The resistant fragment from the 5' half, by contrast, contains the binding sites for many of the early assembly proteins (Zimmermann et al. 1972b, 1974).

HETEROGENEITY OF rRNA

In the *E. coli* genome, there are six cistrons that code for each of the rRNAs (Spadari and Ritossa 1970). Early studies of rRNA structure suggested that substantial sequence heterogeneity occurred within each class (Aronson and Holowczyk 1965; McIlreavy and Midgley 1967; Young 1968). However later investigations of the 5'-terminal sequences (Takanami 1967) and of the areas around the methylated nucleotides (Fellner and Sanger 1968; Fellner 1969) afforded no evidence of heterogeneity. Subsequent cataloging and quantitation of pancreatic RNase (Muto 1970) and T_1 RNase (Fellner, Ehresmann and Ebel 1970b) products of the 16S RNA and of T_1 RNase products from the 23S RNA (Fellner and Ebel 1970b) showed that the rRNAs must be predominantly homogeneous. This conclusion was in large measure confirmed by the sequence analysis of the 16S RNA. However a low level of heterogeneity was encountered, involving about 1% of the nucleotides. A similar level of heterogeneity was found in the sequence of 5S RNA as well (Jarry and Rosset 1971). This presumably arises from the existence of multiple genes for rRNA as noted above. Points of heterogeneity in the 16S RNA chain are indicated by asterisks in the linear diagram of the RNA sequence presented in Figure 1. It is striking that although heterogeneity is found at several places along the sequence, most of the positions at which variability has been detected occur in clusters, with two or more such points in close proximity. This finding lends support to the idea that hot spots of genetic variation exist within regions of the RNA where the conservation of primary sequence is not required for proper biological function. An alternative explanation of the observed clustering of the points of variability might be that a change in one nucleotide could necessitate a compensatory change in a neighboring nucleotide in order for the RNA molecule to retain its functional properties.

The studies of the 23S RNA primary structure so far carried out (Branlant et al. 1973; C. Branlant and P. Fellner unpublished results) indicate that there is a low degree of heterogeneity in the 23S RNA sequence, similar to that found in the 16S and 5S RNAs. Two points of heterogeneity have been detected in the 5'-terminal 500 nucleotides.

CONSERVATION AND REPETITION OF rRNA SEQUENCES IN PROKARYOTES

Conservation of rRNA Sequences among Different Strains

Hybridization studies initially demonstrated that substantial sequence homologies exist among the rRNAs from a variety of prokaryotes (Moore and McCarthy 1967; Brenner et al. 1969). Subsequently, sequencing methods were used to determine more precisely the extent to which rRNA primary structure has been conserved, both among different species and among different strains of the same species. These investigations have been addressed mainly to two problems: (1) Efforts have been made to map the positions of rRNA genes on the bacterial chromosome by examining the occurrence of specific marker sequences in the rRNAs synthesized by interspecies or interstrain hybrids. (2) Comparative sequence analysis has been used to study the evolutionary relationships among rRNAs from different organisms and to determine whether sequences in functionally important regions, such as those involved in protein binding, have been conserved to a greater extent than others across species lines.

Muto (1970) detected a single characteristic difference among the decanucleotide products arising upon pancreatic RNase digestion of the 16S RNA from *E. coli* strains K and B. This difference was subsequently used as a marker in the determination of the chromosomal location of 16S RNA cistrons (Muto, Takata and Osawa 1971; Matsubara, Takata and Osawa 1972). All other pancreatic RNase products appeared to be identical in the two strains. The 16S RNA sequence also appears to be very highly conserved in other Enterobacteriaceae. Thus, for example, de Graaf, Niekus and Klootwijk (1973) found that the sequence of a 50-nucleotide fragment from the 3′ terminus of the 16S RNA of *Enterobacter cloacae* is identical to that determined for the corresponding region of the *E. coli* 16S RNA (Ehresmann et al. 1972). The primary structure of the 16S RNA of *Proteus vulgaris* has also been compared to that of *E. coli* (J. L. Fischel and P. Fellner, unpublished results). A fragment of about 400 nucleotides from the 3′ terminus of *P. vulgaris* 16S RNA is found to be indistinguishable from the corresponding sequence in *E. coli* 16S RNA, at least in the products that arise from it upon T_1 RNase digestion. A 5′-terminal fragment of 90 nucleotides is also apparently conserved in the two species, with the exception of the 5′-terminal oligonucleotide itself, where an adenine residue in *E. coli* is changed to a uracil residue in *P. vulgaris*. By contrast, certain sequences from the interior of the molecule, including section C, are less well conserved in the two bacteria.

Pechman and Woese (1972) have investigated similarities between the 16S RNAs of *E. coli* and *Bacillus megaterium,* which represent two different bacterial genera. They found that the sequences of pentamers and hexamers arising upon T_1 RNase digestion are conserved to an extent greatly in excess of that expected on a random basis, but that larger products, containing eight or more nucleotides, are generally not the same in the two species. These results suggest that although substantial sequence homology persists across family lines, there is not sufficient evolutionary pressure to conserve long uninterrupted sequences within the rRNAs of these distantly related organisms. In the course of their study, Pechman and Woese stated that the T_1 RNase products obtained from the *E. coli* 16S RNA differ substantially from those in the sequence presented in Figure 1. The reasons for this assertion are unclear, however, since only three small differences are evident among the 60 or so products they describe, although several of the larger products are omitted from their list. Moreover Santer and Santer (1973) have independently confirmed many of the oligonucleotide sequences originally reported by Fellner, Ehresmann and Ebel (1972).

No systematic comparisons of 23S RNA sequences have yet been reported, although Sypherd, O'Neil and Taylor (1969) have used two differences detected among the pancreatic RNase products of the 23S RNAs from *E. coli* and *Salmonella typhosa* as an aid in genetic mapping experiments.

Sequence Homologies and Repetitions among rRNAs of a Single Organism

Only limited information is available concerning possible homologies between different rRNA classes in the same organism. Competition hybridization revealed that the 16S and 23S RNAs of *Bacillus subtilis* are unrelated in sequence (Smith et al. 1968). There are persistent reports, however, that partial sequence homology can be detected between the 16S and 23S RNAs of *E. coli* by the same method (Attardi, Huang and Kabat 1965; Mangiarotti et al. 1968; Avery, Midgley and Pigott 1969). Nonetheless, sequence studies carried out on the 16S and 23S RNAs of *E. coli* do not thus far provide any support for this notion. Fellner and Ebel (1970b) showed that fingerprints of the products of T_1 RNase digests of 23S RNA are completely different from those of the 16S RNA and that no noticeable homologies exist among the larger, characteristic oligonucleotides. Moreover the termini (Takanami 1967; Fellner and Ebel 1970a) and the methylated regions (Fellner and Sanger 1968; Fellner 1969) of the two *E. coli* rRNAs are entirely dissimilar. More extensive studies on the 23S RNA sequence (C. Branlant and P. Fellner, unpublished results) have also failed to reveal any striking homologies with 16S RNA. In the long run, only a systematic examination of large tracts of completed sequences in the two molecules will finally resolve this question.

Sequence analysis of the 5S RNA of *E. coli* has brought to light a substantial duplication of sequences present within the molecule (Brownlee, Sanger and Barrell 1967). From the T_1 RNase products arising upon complete hydrolysis of the *E. coli* 16S RNA, it was suggested that there might also be a greater degree of nucleotide sequence repetition within this molecule than would be expected if the nucleotides occurred in a random order (Fellner, Ehresmann and Ebel 1970b). More extensive primary structure analysis (Ehresmann et al. 1972) indicated that a significant amount of repetition occurs in the 16S RNA, but such sequences have neither been sought systematically nor analyzed statistically up to the present time. It will be necessary to have the complete nucleotide sequence of the 16S RNA in hand before this question can be thoroughly explored. In the 23S RNA, the methylated sequences were initially thought to be duplicated (Fellner and Sanger 1968; Fellner 1969), but as discussed below, it is likely that they each occur only once. No further information is presently available regarding sequence repetition in the 23S RNA.

POST-TRANSCRIPTIONAL MODIFICATION OF rRNA

Methylation

During the assembly of ribosomal particles, the 16S and 23S RNAs of *E. coli* are methylated at a number of specific positions by rRNA methylases, which employ *S*-adenosyl methionine as a donor of methyl groups (Starr and Fefferman 1964; Dubin and Gunalp 1967; Helser, Davies and Dahlberg 1972). Certain uridylic acid residues are also modified to pseudouridylic acid (Dubin and Gunalp 1967). The T_1 RNase digestion products containing these modified nucleotides were among the earliest rRNA sequences to be determined (Fellner and Sanger 1968; Fellner 1969). These oligonucleotides were subsequently encountered during the sequence analysis of the 16S RNA (Ehresmann et al. 1972; C. Ehresmann, P. Stiegler and P. Fellner, unpublished results), permitting the positions of the methylated bases within the molecule to be determined. The sequence analysis shows that the 16S RNA contains ten methylated nucleotides, with a total of 14 methyl groups. This value is lower than the estimate of 22 methyl groups obtained earlier by a double-labeling method (Fellner and Sanger 1968; Fellner 1969) and implies that the estimate of 28 methyl groups in the 23S RNA, which was obtained in the same way, is also too high. It is thus likely that the 14 methylated nucleotides in the 23S RNA, which contain 16 methyl groups, are each present once (or possibly in submolar amounts in one or two cases) rather than twice as previously suggested (Fellner and Sanger 1968; Fellner 1969).

Sequence analysis of the 16S RNA has revealed that most of the methylated nucleotides occur in the 3′-terminal 25% of the molecule, with the exceptions of an m^7G residue in section C′′ and an m^2G residue in section O′, which may occur in submolar amounts. Ten out of the 14 methyl groups in the molecule are clustered in two segments within the 3′-terminal 150 nucleotides. A similar clustering of modified bases occurs in the 23S RNA, where pseudouridylic acid is also encountered in the sequence close to the methylated nucleotides (Fellner and Sanger 1968; Fellner 1969), but the locations of these clusters within the molecule are not yet known. A further unidentified nucleotide, which is modified but not methylated, has been found in the 5′-terminal region of the 23S RNA, within the binding site of 50S ribosomal subunit protein L24 (Branlant et al. 1973). Dubin and Gunalp (1967) have reported that two pseudo-uridylic acid residues are present in the 16S RNA. These bases would be indistinguishable from uridylic acid in the course of sequence analysis, since no chromatographic procedures have been used to search for them. Hence the positions of these nucleotides within the sequence are unknown. The presence of thionucleotides in *E. coli* rRNA has been excluded (Champney and Sypherd 1970).

Both the wide variety of modified nucleotides occurring in the rRNAs and their presence in specific but dissimilar sequences suggest that a comparatively large number of different modifying enzymes are involved in their synthesis.

The function of the modified nucleotides is not known, but their presence clearly influences the properties of the ribosomal particles. This is illustrated by the recent discovery that the modified sequence $m_2^6Am_2^6A$, which occurs close to the 3′ terminus of the 16S RNA in wild-type strains (see Figure 2), is absent in a kasugamycin-resistant mutant, where the AA sequence remains unmethylated (Helser, Davies and Dahlberg 1971). This change was shown to be responsible for the property of kasugamycin resistance, and the absence of the methyl groups most likely provokes an alteration in the conformation of the 30S ribosomal particles which influences their susceptibility to kasugamycin. In a similar vein, resistance to erythromycin in *Staphylococcus aureus* has also been shown to result from alterations in ribosomal RNA methylation (Lai and Weisblum 1971). In this case however, antibiotic resistance depends upon the formation of m_2^6A via the *introduction* of methyl groups into a specific oligonucleotide of the 23S RNA (Lai, Dahlberg and Weisblum 1973).

The observations suggesting that alterations in the methyl group complement of rRNAs change the conformation of ribosomal particles raise puzzling questions about the functional role of the modified nucleotides. Thus, for example, ribosomes of the kasugamycin- and erythromycin-resistant bacteria discussed above are apparently fully active since cells containing the mutations grow at normal rates. In addition, Björk and Isaksson (1970) have isolated a mutant lacking the single m^1G residue usually present in the 23S RNA, in which normal ribosomal function also seems to be unaffected. Despite these results, the methylated nucleotides and the neighboring sequences have been found to be highly conserved in prokaryotic rRNAs (Sogin et al. 1972). In particular, fingerprinting studies have shown that the sequences of the methylated oligonucleotides in several prokaryotic 16S RNAs are far more highly conserved than other regions of the molecule. Furthermore, the sequence $Gm_2^6Am_2^6ACCUG$, which mediates kasugamycin susceptibility in *E. coli* 16S RNA (see above), must possess a special biological significance, since very similar sequences, containing the same methylated nucleotides, are found in the small ribosomal subunit rRNAs of both yeast (Klootwijk, van den Bos and Planta 1972) and HeLa cells (Maden, Salim and Summers 1972). In the latter cases, the introduction of the methyl groups appears to be an isolated late event in the maturation process, occurring after other methylation steps have been completed. The conservation of this sequence is particularly striking, since most of the methyl groups in eukaryotic rRNAs are attached to the ribose moiety of the nucleotide residues and not to the base moiety as in prokaryotic rRNAs.

Methylation of the 16S RNA precursor (see below) occurs late in the maturation of the 30S ribosomal subunit, since the RNA isolated from 21S precursor particles, which contain most of their protein complement (Nashimoto et al. 1971), appears to be entirely unmethylated (Lowry and Dahlberg 1971; Feunteun et al. 1974). The clustering of the methylated nucleotides close to the 3′ terminus of the 16S RNA may be related to the temporal sequence of events occurring during the assembly and maturation of the 30S particle, since the binding sites for most of the proteins which interact individually with the 16S RNA and which are believed to become attached early in the assembly process (Mizushima and Nomura 1970; Nashimoto et al. 1971) are located within the 5′ half of the molecule (Zimmermann et al. 1972b, 1974; see Zimmermann, this volume). The presence of these proteins might thus occlude any potential methylation sites in this region of the RNA.

Enzymatic Cleavage

The mature rRNAs in functional bacterial ribosomes arise by specific enzymatic cleavage of larger precursor molecules (Hecht and Woese 1968; Dahlberg, Dingman and Peacock 1969; Adesnik and Levinthal 1969). It was estimated that the precursor 16S RNA and precursor 23S RNA in *E. coli* are each about 50,000 daltons larger than the mature 16S RNA and 23S RNA, respectively (Dahlberg, Dingman and Peacock 1969).

Several authors have used fingerprinting methods to study the additional nucleotide sequences present in precursor 16S RNA isolated from pulse-labeled, exponentially growing cells (Sogin et al. 1971; Hayes et al. 1971; Brownlee and Cartwright 1971), chloramphenicol-treated cells (Lowry and Dahlberg 1971), and cold-sensitive (Lowry and Dahlberg 1971) or thermosensitive (Feunteun et al. 1974) mutants of *E. coli* defective in ribosomal assembly. The supplementary sequences were detected by the appearance of characteristic oligonucleotide products in fingerprints of the precursor RNAs that were not present in fingerprints of the mature 16S RNA. The precursor 16S RNAs isolated from the cold-sensitive and thermosensitive mutants, which could readily be prepared in pure form and in large amounts, have been characterized most thoroughly (Lowry and Dahlberg 1971; Feunteun et al. 1974). The results indicate that the precursor 16S RNAs accumulated by the different mutants are virtually identical. The other precursor 16S RNAs, obtained from exponentially growing or chloramphenicol-treated cells, are probably identical with those from the mutants as well. Both Lowry and Dahlberg (1971) and Feunteun et al. (1974) characterized eight large additional oligonucleotides in the precursor 16S RNA which, together with the new 5′-terminal oligonucleotide pUG, encompass about 90 nucleotides. These sequences represent only about half of the residues expected on the basis of the difference in size between the precursor 16S and mature 16S RNAs measured by polyacrylamide gel electrophoresis (Dahlberg, Dingman and Peacock 1969). The remaining residues are probably present in smaller, nonunique products and therefore cannot readily be detected in fingerprints of the entire precursor 16S RNA. Although additional nucleotide sequences occur at both the 3′ and 5′ termini of the RNA molecule, the way in which the extra nucleotides are distributed between the two termini is not known. However, the absence of a 5′-terminal triphosphate in the precursor 16S RNA indicates that this precursor is not the primary transcription product. Finally, the precursor RNAs lack most of the characteristic methylated nucleotides (Lowry and Dahlberg 1971; Feunteun et al. 1974) and are therefore believed to be entirely unmethylated. The complete absence of methyl groups, however, has not yet been confirmed.

The recognition system that governs the specificity with which the precursor 16S RNA is cleaved has not been defined. The novel thermosensitive mutant studied by Feunteun et al. (1974) contains an altered ribosomal protein S4 and accumulates precursor 16S RNA in 26S precursor particles at the nonpermissive temperature of 42°C. At 30°C, however, the mutant synthesizes apparently normal functional ribosomal particles, but the mature 16S RNA within them is incompletely trimmed, containing either one or two additional uracil residues at the 5′ terminus. This might be due to an accumulation of two late RNA intermediates in the maturation process or to "mis-trimming" of the RNA. In either event, it appears that the change in protein S4 responsible for the thermosensitive character of ribosomal assembly in this mutant also causes an alteration in the conformation of the cleavage site, affecting its recognition by the modification enzyme.

No work on the primary structure of the precursor 23S RNA has so far been reported.

References

Adams, J. M., P. G. N. Jeppesen, F. Sanger and B. G. Barrell. 1969. Nucleotide sequence from the coat protein cistron of R17 bacteriophage RNA. *Nature* **223**:1009.

Adesnik, M. and C. Levinthal. 1969. Synthesis and maturation of ribosomal RNA in *Escherichia coli. J. Mol. Biol.* **46**:281.

Aronson, A. I. and M. A. Holowczyk. 1965. Composition of bacterial ribosomal RNA. Heterogeneity within a given organism. *Biochim. Biophys. Acta* **95**:217.

Attardi, G., P. C. Huang and S. Kabat. 1965. Recognition of ribosomal RNA sites in DNA. I. Analysis of the *E. coli* system. *Proc. Nat. Acad. Sci.* **53**:1490.

Avery, R. J., J. E. M. Midgley and G. H. Pigott. 1969. An analysis of the ribosomal ribonucleic acids of *Escherichia coli* by hybridization techniques. *Biochem. J.* **115**:395.

Björk, G. R. and L. A. Isaksson. 1970. Isolation of mutants of *Escherichia coli* lacking 5-methyluracil in transfer ribonucleic acid or 1-methylguanine in ribosomal RNA. *J. Mol. Biol.* **51**:83.

Branlant, C., A. Krol, J. Sriwidada, P. Fellner and R. Crichton. 1973. The identification of the RNA binding site for a 50S ribosomal protein by a new technique. *FEBS Letters* **35**:265.

Brenner, D. J., G. R. Fanning, K. E. Johnson, R. V. Citarella and S. Falkow. 1969. Polynucleotide sequence relationships among members of *Enterobacteriaceae. J. Bact.* **98**:637.

Brownlee, G. G. and E. Cartwright. 1971. Sequence studies on precursor 16S ribosomal RNA of *Escherichia coli. Nature New Biol.* **232**:50.

Brownlee, G. G. and F. Sanger. 1969. Chromatography of ^{32}P-labelled oligonucleotides on thin layers of DEAE-cellulose. *Eur. J. Biochem.* **11**:395.

Brownlee, G. G., F. Sanger and B. G. Barrell. 1967. Nucleotide sequence of 5S ribosomal RNA from *Escherichia coli. Nature* **215**:735.

Champney, W. S. and P. Sypherd. 1970. Absence of thiolated nucleotides in ribosomal RNA of *E. coli. Biochem. Biophys. Res. Comm.* **41**:1328.

Cotter, R. I. and W. B. Gratzer. 1969. Conformation of ribosomal RNA of *E. coli:* An infrared analysis. *Nature* **221**:154.

Cotter, R. I., P. McPhie and W. B. Gratzer. 1967. Internal organization of the ribosome. *Nature* **216**:864.

Cramer, F. and V. A. Erdmann. 1968. Amount of adenine and uracil base pairs in *E. coli* 23S, 16S and 5S ribosomal RNA. *Nature* **218**:92.

Cox, R. A. 1966. The secondary structure of ribosomal ribonucleic acid in solution. *Biochem. J.* **98**:841.

Cox, R. A. and S. A. Bonanou. 1969. A possible structure of the rabbit reticulocyte ribosome. *Biochem. J.* **114**:769.

Dahlberg, A. E., C. W. Dingman and A. C. Peacock. 1969. Electrophoretic characterization of bacterial polyribosomes in agarose-acrylamide composite gels. *J. Mol. Biol.* **41**:139.

Doty, P., H. Boedtker, J. R. Fresco, R. Haselkorn and M. Litt. 1959. Secondary structure in ribonucleic acids. *Proc. Nat. Acad. Sci.* **45**:482.

Dubin, D. T. and A. Gunalp. 1967. Minor nucleotide composition of ribosomal, precursor, and ribosomal, ribonucleic acid in *Escherichia coli. Biochim. Biophys. Acta* **134**:106.

Ehresmann, C., P. Stiegler, P. Fellner and J.-P. Ebel. 1972. The determination of the primary structure of the 16S ribosomal RNA of *Escherichia coli.* (II).

Nucleotide sequences of products from partial enzymatic hydrolysis. *Biochimie* **54**:901.

Fellner, P. 1969. Nucleotide sequences from specific areas of the 16S and 23S ribosomal RNAs of *E. coli. Eur. J. Biochem.* **11**:12.

Fellner, P. and J.-P. Ebel. 1970a. The 3'-terminal nucleotide sequence of the 23S ribosomal RNA from *Escherichia coli. FEBS Letters* **6**:102.

―――. 1970b. Observations on the primary structure of the 23S ribosomal RNA from *E. coli. Nature* **225**:1131.

Fellner, P. and F. Sanger. 1968. Sequence analysis of specific areas of the 16S and 23S ribosomal RNAs. *Nature* **219**:236.

Fellner, P., C. Ehresmann and J.-P. Ebel. 1970a. Nucleotide sequences in a protected area of the 16S RNA within 30S ribosomal subunits from *Escherichia coli. Eur. J. Biochem.* **13**:583.

―――. 1970b. Nucleotide sequences present within the 16S ribosomal RNA of *Escherichia coli. Nature* **225**:26.

―――. 1972. The determination of the primary structure of the 16S ribosomal RNA of *Escherichia coli.* (I). Nucleotide sequence analysis of T_1 and pancreatic ribonuclease digestion products. *Biochimie* **54**:853.

Feunteun, J., R. Rosset, C. Ehresmann, P. Stiegler and P. Fellner. 1974. Abnormal maturation of precursor 16S RNA in a ribosomal assembly defective mutant of *E. coli. Nucleic Acid Res.* **1**:141.

Fresco, J. R., B. M. Alberts and P. Doty. 1960. Some molecular details of the secondary structure of ribonucleic acid. *Nature* **188**:98.

de Graaf, F. K., H. G. D. Niekus and J. Klootwijk. 1973. Inactivation of bacterial ribosomes *in vivo* and *in vitro* by cloacin DF13. *FEBS Letters* **35**:161.

Hayes, F., D. Hayes, P. Fellner and C. Ehresmann. 1971. Additional nucleotide sequences in precursor 16S ribosomal RNA from *Escherichia coli. Nature New Biol.* **232**:54.

Hecht, N. B. and C. R. Woese. 1968. Separation of bacterial ribosomal ribonucleic acid from its macromolecular precursors by polyacrylamide gel electrophoresis. *J. Bact.* **95**:986.

Helser, T. L., J. E. Davies and J. E. Dahlberg. 1971. Change in methylation of 16S ribosomal RNA associated with mutation to kasugamycin resistance in *Escherichia coli. Nature New Biol.* **233**:12.

―――. 1972. Mechanism of kasugamycin resistance in *Escherichia coli. Nature New Biol.* **235**:6.

Jarry, B. and R. Rosset. 1971. Heterogeneity of 5S RNA in *Escherichia coli. Mol. Gen. Genet.* **113**:43.

Klootwijk, J., R. C. van den Bos and R. J. Planta. 1972. Secondary methylation of yeast ribosomal RNA. *FEBS Letters* **27**:102.

Kurland, C. G. 1960. Molecular characterization of ribonucleic acid from *Escherichia coli* ribosomes. I. Isolation and molecular weights. *J. Mol. Biol.* **2**:83.

Lai, C. J. and B. Weisblum. 1971. Altered methylation of ribosomal RNA in an erythromycin-resistant strain of *Staphylococcus aureus. Proc. Nat. Acad. Sci.* **68**:856.

Lai, C.-J., J. E. Dahlberg and B. Weisblum. 1973. Structure of an inducibly methylatable nucleotide sequence in 23S ribosomal ribonucleic acid from erythromycin-resistant *Staphylococcus aureus. Biochemistry* **12**:457.

Lowry, C. V. and J. E. Dahlberg. 1971. Structural differences between the 16S ribosomal RNA of *E. coli* and its precursor. *Nature New Biol.* **232**:52.

Maden, B. E. H., M. Salim and D. F. Summers. 1972. Maturation pathway for ribosomal RNA in the Hela cell nucleolus. *Nature New Biol.* **237**:5.

Mangiarotti, G., D. Apirion, D. Schlessinger and L. Silengo. 1968. Biosynthetic precursors of 30S and 50S ribosomal particles in *Escherichia coli. Biochemistry* **7**:456.

Matsubara, M., R. Takata and S. Osawa. 1972. Chromosomal loci for 16S ribosomal RNA in *Escherichia coli. Mol. Gen. Genet.* **117**:311.

McIlreavy, D. J. and J. E. M. Midgley. 1967. The chemical structure of bacterial ribosomal RNA. I. The terminal nucleotide sequences of *Escherichia coli* ribosomal RNA. *Biochim. Biophys. Acta* **142**:47.

McPhie, P. and W. B. Gratzer. 1966. The optical rotary dispersion of ribosomes and their constituents. *Biochemistry* **5**:1310.

Midgley, J. E. M. 1965. The estimation of polynucleotide chain length by a chemical method. *Biochim. Biophys. Acta* **108**:340.

Min Jou, W., G. Haegemann, M. Ysebaert and W. Fiers. 1972. Nucleotide sequence of the gene coding for the bacteriophage MS2 coat protein. *Nature* **237**:82.

Mizushima, S. and M. Nomura. 1970. Assembly mapping of 30S ribosomal proteins from *E. coli. Nature* **226**:1214.

Moore, R. L. and B. J. McCarthy. 1967. Comparative study of ribosomal ribonucleic acid cistrons in Enterobacteria and Myxobacteria. *J. Bact.* **94**:1066.

Muto, A. 1970. Nucleotide distribution of *Escherichia coli* 16S ribosomal ribonucleic acid. *Biochemistry* **9**:3683.

Muto, A., R. Takata and S. Osawa. 1971. Chemical and genetic analysis of 16S ribosomal RNA in *Escherichia coli. Mol. Gen. Genet.* **111**:15.

Muto, A., C. Ehresmann, P. Fellner and R. A. Zimmermann. 1974. RNA-protein interactions in the ribosome. I. Characterization and RNase digestion of 16S RNA-ribosomal protein complexes. *J. Mol. Biol.* (in press).

Nashimoto, H., W. Held, E. Kaltschmidt and M. Nomura. 1971. Structure and function of bacterial ribosomes. XII. Accumulation of 21S particles by some cold-sensitive mutants of *Escherichia coli. J. Mol. Biol.* **62**:121.

Ortega, J. P. and W. E. Hill. 1973. A molecular weight determination of the 16S ribosomal ribonucleic acid from *Escherichia coli. Biochemistry* **12**:3241.

Pechman, K. J. and C. R. Woese. 1972. Characterization of the primary structural homology between the 16S ribosomal RNAs of *Escherichia coli* and *Bacillus megaterium* by oligomer cataloguing. *J. Mol. Evolution* **1**:230.

Sanger, F., G. G. Brownlee and B. G. Barrell. 1965. A two-dimensional fractionation procedure for radioactive nucleotides. *J. Mol. Biol.* **13**:373.

Santer, M. and U. Santer. 1973. Action of ribonuclease T1 on 30S ribosomes of *Escherichia coli* and its role in sequence studies on 16S ribonucleic acid. *J. Bact.* **116**:1304.

Smith, I., D. Dubnau, P. Morell and J. Marmur. 1968. Chromosomal location of DNA base sequences complementary to transfer RNA and to 5S, 16S and 23S ribosomal RNA in *Bacillus subtilis. J. Mol. Biol.* **33**:123.

Sogin, M., B. Pace, N. R. Pace and C. R. Woese. 1971. Primary structural relationship of p16S m16S ribosomal RNA. *Nature New Biol.* **232**:48.

Sogin, M. L., K. J. Pechman, L. Zablen, B. J. Lewis and C. R. Woese. 1972. Observations on the post-transcriptionally modified nucleotides in the 16S ribosomal ribonucleic acid. *J. Bact.* **112**:13.

Spadari, S. and F. Ritossa. 1970. Clustered genes for ribosomal ribonucleic acids in *Escherichia coli. J. Mol. Biol.* **53**:357.

Spahr, P. F. and A. Tissières. 1959. Nucleotide composition of ribonucleoprotein particles from *Escherichia coli. J. Mol. Biol.* **1**:237.

Stanley, W. M., Jr. and R. M. Bock. 1965. Isolation and physical properties of the ribosomal ribonucleic acid of *Escherichia coli. Biochemistry* **4**:1302.

Starr, J. L. and R. Fefferman. 1964. The occurrence of methylated bases in ribosomal ribonucleic acid of *Escherichia coli* K12 W-6. *J. Biol. Chem.* **239**:3457.

Sypherd, P. S., D. M. O'Neil and M. M. Taylor. 1969. The chemical and genetic structure of bacterial ribosomes. *Cold Spring Harbor Symp. Quant. Biol.* **34**:77.

Takanami, M. 1967. Analysis of the 5'-terminal nucleotide sequences of ribonucleic acids. I. The 5'-termini of *Escherichia coli* ribosomal RNA. *J. Mol. Biol.* **23**:135.

Tinoco, I., O. C. Uhlenbeck and M. D. Levine. 1971. Estimation of secondary structure in ribonucleic acids. *Nature* **230**:362.

Young, R. J. 1968. The fractionation of quaternary ammonium complexes of nucleic acids. Evidence for heterogeneity of ribosomal ribonucleic acid. *Biochemistry* **7**:2263.

Zimmermann, R. A., A. Muto, P. Fellner and C. Branlant. 1972a. The interaction of 30S ribosomal proteins with 16S RNA and RNA fragments. In *Functional units of protein biosynthesis,* 7th FEBS Symp. (Varna 1971) vol. 23, pp. 53–73. Academic Press, London.

Zimmermann, R. A., A. Muto, P. Fellner, C. Ehresmann and C. Branlant. 1972b. Location of ribosomal protein binding sites on 16S ribosomal RNA. *Proc. Nat. Acad. Sci.* **69**:1282.

Zimmermann, R. A., G. A. Mackie, A. Muto, R. A. Garrett, E. Ungewickell, C. Ehresmann, P. Stiegler, J.-P. Ebel and P. Fellner. 1974. Location and characteristics of ribosomal protein binding sites in the 16S RNA of *Escherichia coli. Nature* (in press).

Reconstitution of Ribosomes: Studies of Ribosome Structure, Function and Assembly

Masayasu Nomura and William A. Held
Institute for Enzyme Research and
Departments of Biochemistry and Genetics
University of Wisconsin
Madison, Wisconsin 53706

INTRODUCTION

The in vitro reconstitution of the bacterial 30S ribosomal subunit (Traub and Nomura 1968) followed by that of the 50S subunit (Nomura and Erdmann 1970; Fahnestock, Erdmann and Nomura 1973) has established that the information required for the assembly of this complex organelle is entirely contained in the structures of the RNA and protein components. The mechanism involved in this complex self-assembly process is now being elucidated, especially with respect to 30S ribosome assembly, by a variety of in vitro experiments. The in vitro assembly reaction, while differing in some details from in vivo assembly, very likely reflects the in vivo process in important ways. In addition, reconstitution techniques have facilitated the study of ribosome structure and structure-function relationships.

A detailed review on the in vitro reconstitution of 50S subunits (Fahnestock, Held and Nomura 1972), as well as reviews on ribosome assembly in general (Nomura 1970, 1973) have appeared previously. In this article, we summarize and discuss the work related to ribosome assembly, concentrating mainly on the 30S subunit, including several recent developments in this area.

PARTIAL RECONSTITUTION
OF RIBOSOMES

The first step in the analysis of the functional role of ribosomal components was the development of a system for reconstituting ribosomes that had been partially disassembled (Staehelin and Meselson 1966; Hosokawa, Fujimura and Nomura 1966). Selective removal of a portion of the ribosomal protein ("split proteins") was accomplished by centrifugation of the ribosomes in 5 M CsCl in the presence of 0.04 M Mg^{++}. The functionally inactive nucleoprotein particles ("core particles") that remain can then be mixed with the split proteins to form functionally active ribosomal particles. This partial reconstitution is rapid and relatively insensitive to variations in experimental conditions. The reaction is complete within a few minutes at 37°C (Nomura and Traub 1968). In these respects, the partial reconstitution reaction is very different from the total reconstitution that will be discussed below.

The conditions originally described for preparing 30S cores (Hosokawa, Fujimura and Nomura 1966) completely split off seven 30S proteins (S1, S2, S3, S5, S9, S10 and S14), and only minute amounts of other proteins are released (Traub et al. 1967; Cohlberg 1974). This system was used to demonstrate, for the first time, requirements for individual ribosomal proteins for the functional activity of ribosomes (Traub et al. 1967). In these experiments S3, S10 and S14 were absolutely required and S5 was partially required for various ribosomal functions. A requirement for S9 for activity in poly(U)-dependent polyphenylalanine synthesis was *not* clearly demonstrated (Traub et al. 1967). However in some later experiments, S9 was required for maximum activity (Cohlberg 1974). The reason for the discrepancy is not clear.

Since the development of the total reconstitution system for 30S subunits, which can examine all twenty-one 30S proteins, the partial reconstitution system has become less advantageous. However, since at the moment the complete 50S reconstitution system is available only for *Bacillus stearothermophilus* ribosomes, many experiments on the functional analysis of *E. coli* 50S ribosomal proteins are still being done using this initial partial reconstitution system or similar systems with some modifications. The particular proteins split off from the ribosomes vary depending on the divalent cation and salt concentrations during centrifugation. Thus various kinds of "core" particles were obtained, and the reconstitution of functional ribosomal particles from these various particles and the corresponding split proteins was demonstrated (Staehelin, Maglott and Monro 1969; Atsmon, Spitnik-Elson and Elson 1969). It should be noted that the protein composition of core particles varies considerably with slight variations in the conditions used for preparation. Thus experiments in which the protein composition of the various core particles is not precisely known are difficult to interpret. It is advisable, therefore, that each investigator examine the protein composition of his own preparations of core particles and split proteins.

TOTAL RECONSTITUTION OF 30S SUBUNITS

Protein-free 16S RNA, prepared either by phenol extraction or urea-LiCl precipitation, is mixed with 30S ribosomal proteins prepared by urea-LiCl extraction. The reconstitution requires 20 mM Mg^{++} ions, has a sharp ionic strength optimum of 0.37, and proceeds most rapidly at 40–50°C (Traub and Nomura 1969). The sensitivity to ionic strength is probably indicative of a narrow range of salt concentration at which specific RNA–protein or protein–protein interactions are permitted but nonspecific interactions are weak. This suggestion is supported by the finding that the binding of S8 to 16S RNA shows a plateau of one mole of S8 per mole 16S RNA in a narrow range of ionic strength centered at 0.37. At higher ionic strength the binding decreases, and at lower ionic strength more than one mole is bound, presumably in a nonspecific manner (Schulte and Garrett 1972). As mentioned previously, unlike the total reconstitution system, partial reconstitution from core particles and split proteins occurs at low temperatures and at low ionic strength, suggesting that the later stages of ribosome assembly are quite different from the initial stages.

The original reconstitution experiments were done using an unfractionated mixture of 30S proteins. Subsequently each of the twenty-one different 30S proteins has been purified by a combination of several column chromatographic methods (see Wittmann, this volume), and complete reconstitution using 16S RNA and purified 30S ribosomal proteins has been demonstrated (Held, Mizushima and Nomura 1973). The reconstituted 30S subunits are physically and functionally similar to the original particles, except that incorporation of the protein S1 is usually very weak (see below). The functional assays tested include the following: poly(U)-directed polyphenylalanine synthesis, natural messenger RNA-directed polypeptide synthesis, poly(U)-directed Phe-tRNA binding, AUG-directed fMet-puromycin formation, the binding of termination codon UAA in the presence of chain termination factors, and AUG-directed fMet-tRNA binding. All assays, except the last, were performed in the presence of 50S subunits. In all cases activities comparable to reference 30S subunits can be observed, except that the reconstituted particles usually have lower activity than the reference (non-salt washed) 30S subunits in AUG-directed fMet-tRNA binding assayed in the presence of purified initiation factor IF-2 or IF-2 and IF-1. This lower activity could be explained by the presence in the reference 30S subunits of some nonribosomal protein factors that stimulate fMet-tRNA binding (such as IF-1 or IF-3) and their absence in the reconstituted particles assembled from known purified components. Recently 30S subunits reconstituted from 16S RNA and unfractionated

30S proteins were shown to be active in the synthesis of S-adenosyl-methionine-cleaving enzyme in a cell-free system coupled with the transcription of phage T3 DNA (Egberts et al. 1972). Although such experiments have not been done with 30S subunits reconstituted from purified components, we expect that similar results would be obtained.

The kinetics of reconstitution using purified protein mixtures is also essentially identical to that obtained using unfractionated 30S proteins. These results strongly suggest that twenty-one purified 30S proteins together with 16S RNA are sufficient to reconstitute 30S subunits, and that no essential 30S components were lost during the fractionation and purification of the 30S proteins. However this does not necessarily mean that all the proteins are necessary. In fact, omission of S1 or S6 usually does not result in significant reduction in the activities of the reconstituted particles (Held, Mizushima and Nomura 1973; see below for further discussion).

MECHANISM OF ASSEMBLY OF *E. COLI* 30S SUBUNITS

The reconstitution of active 30S subunits follows first-order kinetics (Traub and Nomura 1969), indicating that the rate-determining step is a unimolecular reaction. The rate of reconstitution is strongly temperature dependent, and an Arrhenius activation energy of about 40 kcal/mole has been calculated from the dependence of the rate constant on the temperature of the reaction. At low temperatures (10°C or below) an inactive intermediate particle (RI) is assembled which contains 16S RNA and some, but not all, of the 30S ribosomal proteins. If these particles are isolated and heated (40°C) in the absence of free ribosomal proteins to form an activated RI particle (RI*) and then returned to 0°C, they are to a considerable extent capable of binding the remaining ribosomal proteins and forming active 30S subunits at low temperatures. The isolated RI particles sediment at 21S (Traub and Nomura 1969). Similar intermediate particles can also be observed when reconstitution is performed at 30°C. At this temperature, the rate of reconstitution of functionally active 30S subunits is slow. It was found that particles sedimenting at 21S appear first, followed by a gradual decrease in the amount of 21S particles and a corresponding increase in the amount of particles sedimenting at 30S (Held and Nomura 1973b). The degree of reconstitution judged by the activity assay is roughly proportional to the amount of 30S particles produced. Thus the following sequence can be observed:

$$\text{16S RNA + proteins} \xrightarrow{} \text{RI} \xrightarrow{\text{heat}} \text{RI*} \xrightarrow{\text{+ proteins}} \text{30S subunits.} \quad (1)$$

The step RI to RI* is the major rate-limiting step that requires the high activation energy and represents a unimolecular structural rearrangement of the intermediate particle (Traub and Nomura 1969).

Isolated RI particles appear to be heterogeneous with respect to protein composition (Traub and Nomura 1969). It is possible that RI particles are unstable and that some essential proteins are lost during purification. Conversely, other proteins may be bound to RI particles but might not actually be required for the heat-dependent structural rearrangement of the intermediate. Therefore experiments were done to determine, in a more precise manner, which 30S proteins must be present during the heat-dependent step to obtain efficient assembly of functionally active 30S particles.

The twenty-one purified proteins were divided into two groups in various ways. One group of proteins was incubated with 16S RNA at high temperature; after cooling, the remaining proteins were added and incubated at low temperature for a short time. By omitting or adding proteins during the first high-temperature incubation, the proteins whose presence is required during the heat step for efficient reconstitution have been identified ("proteins required for RI* formation," see Table 1).

From these results, the following tentative reaction scheme has been proposed for the in vitro 30S assembly reaction (Held and Nomura 1973b):

$$16S\ RNA + \begin{bmatrix} S4,\ S5,\ S6,\ S7,\ S8,\ S9, \\ S11,\ S12,\ S13,\ S15,\ S16, \\ S17,\ S18,\ S19,\ S20 \end{bmatrix} \longrightarrow RI\ particles \qquad (2)$$

$$RI\ particles \xrightarrow{\text{heating}} RI^*\ particles \qquad (3)$$

$$RI^*\ particles + \begin{bmatrix} (S1),\ S2,\ S3 \\ S10,\ S14,\ S21 \end{bmatrix} \longrightarrow 30S\ particles \qquad (4)$$

$$RI\ particles \rightleftharpoons 21S\ particles + [S5,\ S12,\ S19] \qquad (5)$$

In addition to the proteins required for the formation of RI* particles (Table 1), S13, S20 and S6 are included as proteins participating in the first step since these proteins are found in the 21S particles isolated from the reconstitution mixture without heating. The proposed RI particles which undergo the rate-determining unimolecular reaction (reaction 3) may contain all of those proteins described in reaction 2, but they are presumably unstable and lose some proteins during isolation (especially S5, S12 and possibly S19, reaction 5). Some discrepancy between the protein composition of the isolated RI particles obtained in the present authors' laboratory (Traub and Nomura 1969; Held and Nomura 1973b) and those of Homann and Nierhaus (1971) (see Table 1) could be explained by differences in the conditions used for isolation and purification of the particles. Because of the presumed instability of RI particles, it is difficult to study reaction 3 as a separate reaction. However particles obtained according to reaction 2 and those obtained according to reactions 2 and 3 showed a large difference in sedimentation coefficient. The former particles

Table 1 Comparison of proteins required for RI*
formation with those found in 21S particles

| | Required for RI* formation[a] | In vitro 21S | | In vivo 21S precursor | |
		1[b]	2[c]	cold-sensitive mutant[d]	wild-type strain[e]
S4	++	++	++	+	+
S7	++	++	−	+	−
S8	++	++	++	++	+
S16	++	++	++	+	+
S19	++	+	−	+	−
S15	+	++	++	++	+
S17	+	++	++	+	+
S5	±	+	++	−	+
S9	±	(++)[f]	±	±	−
S11	±	(++)[f]	−	−	−
S12	±	−	−	−	−
S18	±	++	+	±	−
S6	−	++	++	+	−
S13	−	++	++	+	+
S20	−	++	+	+	+
S1	−	−	+	−	+
S2	−	−	±	−	−
S3	−	−	−	−	−
S10	−	−	−	−	−
S14	−	−	−	−	−
S21	−	−	+	−	−

[a] Extent to which each protein is required for the temperature-dependent step during reconstitution (see text). ++, strongly required; +, moderately required; ±, weakly required; −, not required (Held and Nomura 1973b).

[b] Protein composition of the isolated "RI particles" (in vitro 21S particles). ++, present; +, present in reduced amounts; −, absent or almost absent (Erdmann, Kaltschmidt and Nomura, unpublished experiments; Traub and Nomura 1969; Held and Nomura 1973b).

[c] Protein composition of the isolated "RI particles" determined by Homann and Nierhaus 1971. ++, present in normal amounts; +, present in reduced amounts; ±, present in trace amounts; −, not detectable.

[d] Protein composition of 21S particles accumulated by a cold-sensitive mutant. ++, present in amounts comparable to those in 30S subunits; +, present in reduced amounts; ±, found only in some preparations; −, not detected (Nashimoto et al. 1971).

[e] Protein composition of in vivo 30S precursor isolated from wild-type cells (Nierhaus, Bordasch and Homann 1973). +, present in normal amounts; −, not detectable.

[f] Both S11 and S9 are probably present, but because of poor resolution of these two proteins in the polyacrylamide gel electrophoresis, one cannot make a definite conclusion as to which one (or both) is present.

(RI particles) sediment at about 21–22S, whereas the latter ("RI*" particles) sediment at about 25–26S (Held and Nomura 1973b). These results indicate the occurrence of a large conformational change in the particle during reaction 3. It is known that the effective hydrodynamic volume of isolated 16S RNA in solution (even under conditions in which the RNA shows its maximum degree of folding) is much larger than that of 30S subunits (cf. Miall and Walker 1969). Thus "folding" or "tightening" of RNA must take place during the reconstitution reaction. Reaction 3 may represent one such folding process.

Recently Cantor and his coworkers studied structural alterations during the assembly of 30S particles by first making a complex of 16S RNA with a 30S protein having a fluorescent dye attached, then adding total unfractionated 30S ribosomal proteins and following the changes in fluorescence (Cantor, Huang and Fairclough, this volume). Such studies may give more detailed information on structural alterations during the assembly reaction.

ASSEMBLY MAPPING OF 30S RIBOSOMAL PROTEINS

The "sequence" of addition of proteins to 16S RNA has been studied with purified 30S ribosomal proteins (Mizushima and Nomura 1970; Held et al. 1974). Under reconstitution conditions, only seven of the 30S proteins bind individually to 16S RNA. Certain other proteins bind only after some of the first seven proteins are bound. The binding of the remaining proteins requires the presence of several proteins in addition to some or all of the initial binding proteins. In this manner, the sequence of addition of proteins to 16S RNA has been analyzed and an assembly map has been constructed (Figure 1). Regarding the "initial binding proteins," some investigators have presented results suggesting that only five of the seven proteins shown in Figure 1 are specific initial binding proteins (Schaup, Green and Kurland 1970, 1971; Garrett et al. 1971). These workers failed to observe the direct binding of S13 to 16S RNA and also suggested that the binding of S17 to 16S RNA was nonspecific. Experimental results obtained in the present authors' laboratory, however, strongly indicate that S17 binds specifically and stoichiometrically to 16S RNA (Held et al. 1974). It should be noted that the extent of binding to 16S RNA of some of these initial binding proteins varies, depending on the preparation and storage time of the RNA and the proteins. It is possible that some discrepancies among the results obtained by various workers could be explained on this basis. (For further discussion on RNA–protein interactions, see chapters by Fellner and Zimmermann, this volume.)

Several different types of experiments support the validity of the assembly map. (1) The assembly reaction was "inhibited" by omitting a single protein from the mixture of all the proteins, and the resulting protein-deficient particles were purified and analyzed with respect to their protein composi-

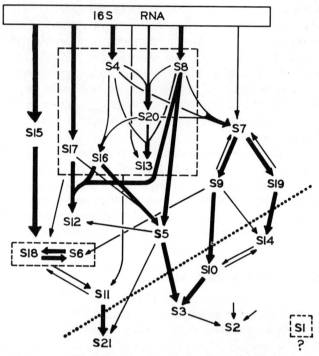

Figure 1 Revised assembly map of *E. coli* 30S ribosomal proteins. Arrows between proteins indicate the facilitating effect of one protein on the binding of another—a thick arrow indicates a major facilitating effect. The map may be used to indicate the following relationships: The thick arrow from 16S RNA to S4 indicates that S4 binds directly to 16S RNA in the absence of other proteins. The thin arrow from 16S RNA to S7 indicates that S7 binds weakly to 16S RNA in the absence of other ribosomal proteins. Thin arrows pointing toward S7 from S4, S8, S20, S9 and S19 indicate that the latter proteins all enhance the binding of S7 to RNA. The arrow to S11 from the large box with dashed outline indicates that S11 binding depends on some of the proteins enclosed in the box; it is not known exactly which proteins. The binding of S2 takes place at a later stage in the assembly sequence and is stimulated by the presence of S3 and probably by several other proteins; hence extra arrows are added toward S2 to emphasize the point. Proteins above the dotted line are those either required for the formation of RI* particles or found in the isolated 21S RI particles (see text). It should be noted that the arrows connecting initial binding proteins to 16S RNA have no relationship to the actual order of the binding sites along the 16S RNA molecule. Information on the order of the binding sites has come from other types of studies (see chapters by Fellner and Zimmermann, this volume).

tion. When the incubation time for the reconstitution was relatively short (40 min at 42°C), the results were in general consistent with the map. The most striking example is the omission of S7. In this case, the particles assembled are deficient in S1, S2, S3, S7, S9 (or S11), S10, S14 and S19, in agreement with the sequence shown in the map (Mizushima and Nomura 1970 and other unpublished data; concerning some unexpected results obtained after a longer incubation time, see below). (2) The assembly reaction was also "inhibited" by modifying 16S RNA with several reagents such as monoperphthallic acid. Incomplete particles formed during reconstitution with modified 16S RNA were also purified and analyzed to determine their protein composition (Held, Mansour and Nomura, unpublished experiments, cited in Mizushima and Nomura 1970). There was no inconsistency with the map; that is, no "distal" protein was bound when a "proximal" one was not. (3) The assembly reaction can also be studied in the presence of certain "inhibitors." Bollen and his coworkers performed reconstitution in the presence of ethidium bromide and found that particles accumulate which have a protein composition similar to that of isolated RI particles; that is, the proteins bound were the more proximal ones on the assembly map (Bollen et al. 1970). (4) The protein compositions of various 30S ribosomal fragments have been compared with the assembly map. In this case, the test assumes that the interactions in the assembly map are important not only for the assembly reaction itself, but also for the stability of the assembled particles. This assumption may not necessarily be true. However with many different kinds of fragments (mostly obtained by mild RNase digestion), no serious inconsistency has been found. One example is the 23S core particles obtained by the CsCl treatment mentioned earlier. The seven split proteins (S1, S2, S3, S5, S9, S10 and S14) are distal proteins, as indicated by the assembly map. Their removal does not affect the binding of proximal proteins to the 16S RNA. Results of the analysis of other fragmented particles are described in the chapter by Traut et al. (this volume).

It should be noted that the arrows connecting proteins in the assembly map represent only the interactions (direct or indirect) that have been detected in the experiments used to construct the map. Many other interactions that stabilize ribosomal structure probably occur. [Green and Kurland (1973) have reported that considerable binding of S13 takes place in the presence of S4, S8 and S20, in agreement with our earlier results, but a further stimulation of S13 binding can be observed by the presence of S7 and S16 in addition to the above three proteins.] It is also likely that proteins other than the initial binding proteins also interact with 16S RNA in the finished ribosome structure. It should also be emphasized that cooperative interaction between two proteins shown in the assembly map does not necessarily mean the direct interaction of the two proteins, as pointed out in our earlier papers (Mizushima and Nomura 1970; Nomura 1973). For example, protein A may help the binding of protein B by directly interacting with B; alternatively, protein A may interact with either 16S RNA or other proteins and may create a correct binding site for B indirectly through some conformational alteration. Nevertheless, experiments performed in several laboratories have given strong support for the original suggestion that the assembly map reflects topological relationships between ribosomal proteins in the finished ribosome structure (see Traut et al. and Craven, Rigby and Changchien, this volume).

It is also probable that the "sequence" described in the assembly map corresponds approximately to the temporal sequence of assembly. As described in the previous section, there is other evidence which shows that the assembly is sequential (see equations 2, 3 and 4 above). It can be seen that all of the proteins required for the formation of the RI* particles, as well as three proteins (S20, S13, S6) found in isolated RI particles (see equation 2), occupy an earlier part of the assembly map, which was constructed from completely independent binding studies. These proteins are shown above the dotted line in the assembly map in Figure 1. The incorporation of other proteins (under the dotted line) into the reconstituted 30S particles depends on the prior formation of RI* particles as discussed in the previous section.

However it is evident from the complexity of the map, as well as the way in which the map was constructed, that the precise temporal order during in vitro assembly would be difficult to determine. In fact, it is probable that there are several different routes for the assembly of 30S ribosomal subunits and that differences in the free energy of activation among them may be rather small. Such an inference is supported by the striking observation that 30S-like particles with nearly full activity are slowly produced even in the absence of S16, which plays a major role in the assembly process (Held and Nomura, unpublished experiments).

FUNCTIONAL ANALYSIS OF 30S RIBOSOMAL COMPONENTS USING RECONSTITUTION TECHNIQUES

Protein synthesis can be divided, conceptually as well as operationally, into distinct processes and interactions. These can be attributed, in general, mainly to one or the other of the two subunits (see Lengyel, this volume). The 30S subunit is mainly concerned with the interaction of messenger RNA codons with transfer RNA anticodons, contributing to the mRNA-directed binding of tRNA to the ribosome, and participates in the first stages of formation of the initiation complex, i.e., the selection of the proper initiation signal and binding of the initiator tRNA with the aid of several initiation factors. Related to 50S subunits, several distinct processes are known, including peptidyl transferase—the major catalytic activity of the ribosome (see below, and several other pertinent articles in this volume).

If one extends the correlation of specific functions with ribosomal components to the level of individual molecular components, a model can be envisioned wherein each component is responsible for some specific function and the overall process results from the summation of the co-ordinated contributions of all of the individual components. Alternatively, it may be that each function is due to a structure created by the cooperative interaction of many components and that no function can be attributed exclusively to one component of the ribosome. These two models are extreme views. Very likely the ribosome has some features of both models, and the different steps of protein synthesis mentioned above may involve different degrees of cooperativity among ribosome components. In addition, it is conceivable that some ribosomal components are required only for ribosome assembly but not for ribosomal function. Thus four possible roles can be considered for the components of the ribosome:

(1) A given component may be essential for ribosome assembly but not required for any function.

(2) A given component may be required indirectly for some ribosomal function because its presence maintains an active center in a proper configuration in the ribosome structure.

(3) A given component may be a part of the active center, playing a direct role in a given ribosomal function.

(4) A given component may be required for both assembly and function of the ribosome.

Several different types of approaches have been taken to obtain information on the role of the ribosomal components in the assembly and/or function of 30S subunits. We will discuss these approaches, giving some typical experiments as examples.

Single Component Omission Experiments

With this approach, reconstitution experiments are performed with one component omitted. The resulting particles are then analyzed to determine whether physically intact 30S particles are formed, and if so, whether they are functionally active. One can examine activities in various functional assays related to protein synthesis, such as chain initiation or chain elongation. If the loss of activity occurs only in one function, then one could conclude that the given component is probably specifically involved in that function. On the other hand, if all ribosomal functions are impaired, then the omission of the component may result in faulty assembly or an incorrect ribosomal conformation, causing a general loss of ribosomal activity rather than a specific lesion.

The first major conclusion obtained from such experiments is the absolute requirement of 16S RNA for 30S subunit reconstitution (Traub and Nomura 1968). In the absence of 16S RNA, no activity is found after incubation; in fact no aggregate particle with a definite structure is produced. Thus 16S RNA is essential for the assembly of 30S subunits. Information pertinent to the functional role of 16S RNA in the assembled 30S subunits will be described below.

Regarding the role of individual proteins, earlier work used both the partial reconstitution system (Traub et al. 1967) and the complete reconstitution system (Nomura et al. 1969). In the latter case, experiments were done using a combination of several purified proteins and protein fractions containing the remaining proteins. At that time, not all the proteins had been identified, and only one incubation time was used. Therefore the results with respect to the role of individual proteins should still be considered preliminary. More recently, similar experiments have been performed using only purified proteins (Held, Mizushima and Nomura 1973; Held and Nomura, unpublished experiments).

In general, it has proved difficult to assign a particular role to any one protein component by single component omission experiments; omission of a single protein usually affects several functions, and a given function can be shown to require the presence of each of several proteins. Nevertheless, some useful information has been obtained. For example, omission of S1 or S6 singly from the reconstitution mixture did not result in any decrease in the ribosomal functions tested. The requirement of S6 for the initiation function observed in earlier studies (Nomura et al. 1969) has not been observed in the more recent experiments (Held and Nomura, unpublished experiments). However S6 is clearly a ribosomal protein. Although the stoichiometry of this protein in the ribosome is unclear (there is a large discrepancy between two published results: Voynow and Kurland 1971; Weber 1972), S6 has an important role in the binding of S18 during ribosome assembly (Mizushima and Nomura 1970; see Figure 1).

The role of S1 in ribosome assembly and function, however, is not clear. This protein is not incorporated into reconstituted ribosomes under reconstitution conditions. Purified reconstituted particles lack S1 almost completely (Mizushima and Nomura 1970; Held, Mizushima and Nomura 1973). Isolated 30S ribosomal particles contain only about 0.1–0.14 copy of this protein per 30S particle (Voynow and Kurland 1971; Weber 1972). Also S1 has a very high molecular weight (65,000), whereas the molecular weights of most of the other 30S ribosomal proteins range between 10,000 and 30,000 (see Wittmann, this volume). Thus S1 might not be a "true" ribosomal protein, although the presence of S1 in assay mixtures has been reported to stimulate poly(U)-directed Phe-tRNA binding. (Van Duin and Kurland 1970). On the basis of such arguments, it was suggested that S1 might be better considered a protein "factor" associated with 30S subunits (Held, Mizushima and Nomura 1973). Recently S1 has been found to be identical to "i-factor" (H. Inouye, Y. Pollack and J. Petre, and A. J. Wahba et al., manuscripts in preparation), which in turn is identical to Qβ replicase subunit I (Groner et al. 1972).

As mentioned previously, omission of S16 decreased the rate of assembly very strongly, but functionally active 30S subunits were produced slowly. These 30S particles did not contain any S16, but all the functional activities tested were similar to control 30S subunits. Therefore S16 appears to be important for assembly, but not for function. In early experiments, several other proteins (S4, S7, S8 and S9) were also found to be important for physical assembly (Nomura et al. 1969). Omission of any one of these proteins produced particles sedimenting at 20–25S rather than at 30S under the reconstitution conditions employed (incubation at 40°C for 20 min). As expected, these particles showed greatly reduced activity in all the functions tested. However, the question of whether these proteins are required only for assembly or for both assembly and ribosomal function has not been answered. It is possible that some of these proteins, like S16, also play a role only in the assembly process. Of course, it is still possible that a postulated assembly-specific protein, such as S16, has some unknown function in the finished ribosome structure which could not be detected in vitro by the assays used.

Specific alteration of functional activity in single component omission experiments has been observed in some cases. For example, the omission of S12, the protein coded for by the *str* gene, produced a particle that sedimented at 30S and retained activity in chain elongation to a great extent, but had very low activity in chain initiation. Addition of S12 to such S12-deficient particles restored normal activity in chain initiation (Ozaki, Mizushima and Nomura 1969). Thus we can conclude that S12 is important for chain initiation, but not chain elongation. S12-deficient particles also show a pronounced reduction in translational error frequency. In contrast, the omission of protein S11 results in particles that show a pronounced increase in translational error frequency (Nomura et al. 1969). Although the mechanism responsible for translational errors is not known, each of these two proteins may have some unique role, since none of the other proteins that could be tested in a similar way affected the frequency of translation errors. (Concerning the role of ribosomes in translational fidelity, see Gorini, this volume.)

As already discussed, the results of single component omission experiments are sometimes difficult to interpret, since the omission of certain proteins important for assembly produces particles having a drastically altered conformation or deficiency in several other proteins. Thus defects in the function of reconstituted particles in those cases cannot be taken to mean involvement of the omitted protein in the function assayed. However, by controlling reconstitution conditions and analyzing the protein composition of the particles assembled, one can obtain various protein-deficient particles of known protein compositions. In several cases addition of the omitted protein to the protein-deficient particles has been shown to restore the impaired function, demonstrating participation (directly or indirectly) of the omitted protein in the function impaired. For example, it has been shown that S21 is required for full activity of the ribosomes in the initiation of polypeptide synthesis and in several other ribosome functions tested (Held, Nomura and Hershey 1973).

Recognizing the difficulty involved in the reconstitution technique mentioned above, some workers have offered another approach to the identification of the functional roles of individual proteins (Randall-Hazelbauer and Kurland 1972; Van Duin et al. 1972; Kurland 1972). The usual 30S ribosomal preparations are partially deficient in several proteins ("fractional proteins"; Kurland et al. 1969; Voynow and Kurland 1971). Addition of these proteins to such 30S ribosome preparations may stimulate ribosomal functions. Positive stimulation has been taken to imply the direct involvement of those proteins in the function tested. By the use of this approach, it has been proposed that proteins S2, S3 and S14 are part of the 30S A site (Randall-Hazelbauer and Kurland 1972) and that S21 has a unique function in inhibiting fMet-tRNA binding (Van Duin et al. 1972).

It should be noted that this approach is also subject to the same problem as the reconstitution approach. That is, because of cooperative interactions between ribosomal components, stimulation in functional activity obtained by the addition of certain components might be due to changes in the conformation of inactive ribosomal species and exposure of preexisting functional sites. A more serious disadvantage of this alternative approach is the heterogeneity of the initial protein-deficient 30S preparations. Since these preparations are a mixture of several different ribosome species, each of which is different in protein composition (Kurland et al. 1969), the interpretation of the data is more difficult. An absence of stimulation after the addition of certain proteins cannot be taken as proof of the lack of participation of the proteins in the functions being tested, even though the pertinent proteins are present in fractional amounts in the preparation. From these considerations it is clear that the published experiments (Randall-Hazelbauer and Kurland 1972) show neither direct nor unique involvement of the proteins S2, S3 and S14 in the tRNA binding. Thus their conclusion that S2, S3 and S14 are part of the 30S A site is premature.

In earlier reconstitution experiments, an almost absolute requirement for S3 and S14 and a partial requirement for S2 for tRNA binding were demonstrated (Traub et al. 1967; Nomura et al. 1969). However several other proteins, such as S10, were also absolutely required for tRNA binding. In addition to reduced tRNA binding activity, omission of these proteins, S3, S14 or S10, resulted in reduced activity in almost all the ribosomal functions tested (Nomura et al. 1969). In these experiments, addition of the omitted proteins to respective protein-deficient particles completely restored the tRNA binding activity.

The problem just discussed may also explain, in part, the discrepancy between the results obtained by Van Duin et al. (1972) and those obtained by the reconstitution technique (Held, Nomura and Hershey 1973) regarding the role of the 30S protein S21 in the initiation of protein synthesis. In the reconstitution experiments, S21-deficient particles were prepared in the presence of excess amounts of all the 30S proteins except that S21 was entirely omitted. Although reconstituted particles might still be somewhat heterogeneous, the limiting protein is clearly S21, and demonstration of the stimulatory effect of S21 should certainly be easier.

The reconstitution approach has been used in several other studies of ribosome function. For example, Parmeggiani and his coworkers (Marsh and Parmeggiani 1973; Sander, Marsh and Parmeggiani 1973; see also Cohlberg 1974) used the partial reconstitution system to examine the role of each of the split proteins in G factor- or T factor-dependent GTPase activity assayed in the presence of 50S subunits (see Möller, this volume).

Chemical Modification

Chemical modification can be used in combination with the reconstitution technique to obtain information on the functional role of ribosomal components. There are two obvious approaches. (1) Intact 30S subunits can be modified by various reagents and the altered function determined and correlated with the altered component, which can be identified by the reconstitution technique. (2) Individual purified 30S components can be modified and "30S subunits" reconstituted using the modified component and all other unmodified components. The reconstituted "30S subunits" containing the chemically modified component can then be examined to determine whether any ribosomal function has been altered. Using the first approach, Craven and his coworkers inactivated 30S subunits with tetranitromethane, which reacts with tyrosine residues of proteins. By reconstitution techniques, they found that the loss of tRNA binding activity is due to modification of some of the proteins and not the 16S RNA (Craven, Gavin and Fanning 1969). Another example of this approach is the study by Noller and Chaires (1972) of the inactivation of 30S subunits by kethoxal. Since the ribosomes were protected from inactivation by bound tRNA, it was concluded that the crucial modification took place at a tRNA binding site. By reconstitution experiments combined with the use of radioactive kethoxal, it was concluded that modification of the 16S RNA is responsible for the loss of activity.

With the above approach, it is desirable to find suitable reagents or conditions that inactivate only certain functions, but not others, to demonstrate a unique involvement of the altered component in certain functions. Protection by bound tRNA in the kethoxal experiments is certainly interesting and supports the conclusion drawn by these workers. It should be noted, however, that tRNA binding may cause a substantial change in the conformation of ribosomes (Schreier and Noll 1971), and such conformational changes might alter the reactivity of ribosomal components to chemical reagents (Ginzburg, Miskin and Zamir 1973).

The second approach has been used in the modification of 16S RNA (Nomura, Traub and Bechmann 1968; Held, Mansour and Nomura, unpublished experiments cited in Nomura 1970) and each of the purified proteins. An example of the latter is a recent study on the role of SH groups in the 30S subunit (Kahan, Held and Nomura, unpublished experiments). Several 30S proteins have been shown to contain SH groups (Craven et al. 1969; Acharya and Moore 1973), and several reports have suggested that some SH groups in the 30S subunit are functionally important (Tamaoki and Miyazawa 1967; Traut and Haenni, 1967; Moore 1971, 1973; Ginzburg, Miskin and Zamir 1973). In order to identify the proteins that contain important SH groups, purified 30S proteins were treated individually with ethyleneimine and the SH groups converted to *S*-aminoethyl groups. Aminoethylated proteins were then individually tested for their activity by the reconstitution technique. It was found that aminoethylation of many of the cysteine-containing proteins does not impair their ability to participate in the assembly and function of reconstituted particles, as judged by activity in poly(U)-directed polyphenylalanine synthesis. On the other hand, several proteins (S2, S11, S12, S14 and S21) were inactivated by aminoethylation. Therefore some of the SH groups in these proteins are probably important for ribosome assembly and/or function (in poly(U)-directed polyphenylalanine synthesis).

Mutationally or Physiologically Altered Ribosomes

The basic strategy here is to isolate ribosomes with altered properties, either by mutations or by certain physiological events, and to identify the altered component responsible by the reconstitution technique. For example, S12 was identified as the protein coded for by the *str* locus by the reconstitution technique (Ozaki, Mizushima and Nomura 1969). In addition, 30S subunits from certain Str-R mutants show altered activity in vitro in initiation of protein synthesis but not in chain elongation (Zengel and Nomura, unpublished experiments). The finding supports the conclusion obtained from single component omission experiments mentioned above, namely, that S12 is important for chain initiation but not for chain elongation.

Analysis of mutational alterations of ribosomal components by reconstitution techniques may also be useful in defining the structural and functional role of the RNA components of the ribosome. For example, Helser, Davies and Dahlberg (1971) have shown that a mutation resulting in kasugamycin resistance in *E. coli* is due to altered methylation of 16S RNA from the 30S subunit.

Similarly, alteration of ribosome properties by physiological events can be studied by the reconstitution technique. An example of this technique is the study of the inactivation of ribosomes in *E. coli* cells by colicin E3. By the reconstitution technique, it has been established that it is 16S RNA which is altered, and not the proteins, in colicin E3-inactivated 30S ribosomes (Bowman et al. 1971; see also Nomura et al., this volume). Another example, although related to 50S subunits, is the use of the reconstitution technique to demonstrate that physiologically induced lincomycin resistance in *Staphlococcus aureus* is the result of altered methylation of 23S RNA (Lai et al. 1973).

It should be noted that the finding of an alteration of a protein in mutationally altered ribosomes by some other method (such as an alteration in electrophoretic mobility or a direct demonstration of a change in amino acid sequence) does not necessarily prove that the protein containing the *detected* alteration is causally related to the alteration of ribosome properties under study. An undetected alteration in a different protein might be the one responsible for the alteration in function. One should be especially cautious in this respect when the mutants under study were induced by strong mutagens, such as *N*-methyl-*N'*-nitro-*N*-nitrosoguanidine.

Heterologous Reconstitution

Functionally active hybrid 30S ribosomes can be assembled in vitro from 16S RNA of one bacterial species and the ribosomal protein components of another distantly related species (Nomura, Traub and Bechmann 1968). Thus it is possible to assemble functionally active 30S ribosomes using 16S RNA from *B. stearothermophilus* and the protein components of *E. coli* (or the reverse combination), even though hybridization experiments indicated very limited sequence homology in the 16S RNAs from these two bacterial species (Nomura, Traub and Bechmann 1968).

More recent experiments (Higo et al. 1973) have shown that there appears to be a one-to-one correspondence between individual proteins of *E. coli* and *B. stearothermophilus*. It is possible to assemble functionally active 30S ribosomes, judged by activity in poly(U)-directed polyphenylalanine synthesis, in which a single *B. stearothermophilus* protein is substituted for its *E. coli* counterpart, even though individual *B. stearothermophilus* proteins differ greatly in their chemical properties from the corresponding *E. coli* proteins. Similar experiments have been used to show that S7 from *E. coli* strain MRE600, while differing in chemical and electrophoretic properties, is functionally equivalent to S7 from *E. coli* K12 (Sun and Traut 1973; Held and Nomura 1973a). These heterologous reconstitution experiments suggest strongly that the fundamental organization of ribosomes in prokaryotes is very similar and has been highly conserved.

It should be noted that functional substitution may not be completely unambiguous in identifying the corresponding proteins of two bacterial species. Heterologous assembly conditions have not been examined in detail, and possible differences in protein–RNA or protein–protein interactions during assembly may affect the final structure and activity of various hybrid particles. However, both immunochemical and protein sequencing data can be used to confirm and unambiguously identify homologous ribosomal proteins from two different bacterial species. For example, several *B. stearothermophilus* 30S proteins which cross-react with antisera prepared against purified *E. coli* proteins can functionally substitute for their *E. coli* counterparts (Higo et al. 1973; and other unpublished experiments by Held, Kahan and Nomura). Also, *B. stearothermophilus* proteins, which had been identified as homologous to certain *E. coli* proteins on the basis of functional substitution, have been shown to have amino acid sequences homologous to their *E. coli* counterparts (Higo and Loertscher 1974).

In some respects, replacement of a single component to make hybrid ribosomes is analogous to both the chemical modification and mutational approaches discussed above. It is known that ribosomes of different species are somewhat different from each other in their functional properties. The heterologous reconstitution technique offers a method for determining the ribosomal components responsible for these differences. For example, 30S subunits from *E. coli* and those from *B. stearothermophilus* are different with respect to their ability to translate certain mRNAs. Lodish (1970) first showed that *E. coli* 30S subunits can initiate translation at the coat cistron of RNA from RNA phage f2 (or R17) with a high efficiency in a cell-free protein-synthesizing system, whereas *B. stearothermophilus* 30S subunits cannot. By the use of the heterologous reconstitution technique, it has recently been shown that 16S RNA and the protein S12 are the main components responsible for the difference in the efficiency of coat cistron translation between the two types of 30S ribosomes (Held, Gette and Nomura 1974; see also Goldberg and Steitz 1974). Again the important role of S12 in the initiation of protein synthesis has been confirmed. In addition, the observed specific effect on the coat cistron translation caused by 16S RNA replacement suggests that direct or indirect interaction of some parts of 16S RNA with mRNA is involved in the recognition of initiation signals on natural mRNA.

SOME REMARKS ON RIBOSOME STRUCTURE AND FUNCTION

Ribosomes possess several unique structural features, most notably the presence of ribosomal RNA and a complex protein composition. Most of the biochemical reactions in living cells are catalyzed by enzymes or enzyme aggregates which do not contain RNA. Furthermore, all twenty-one of the 30S proteins and most of the approximately thirty 50S proteins exist in one copy (or less) per ribosome (see Wittmann, this volume). Such a structure is very different from the structure of other nucleoprotein particles, such as virus particles, which contain only a few different types of protein molecules, most of them present in many copies per virus particle. Thus one may ask why the ribosome contains RNA and why it has so many different proteins present in single copies.

The functions of the ribosome involve the binding of mRNA and tRNA in specific positions. This binding may occur through direct interactions between the mRNA or tRNA and the ribosomal RNA by mechanisms such as base pairing or base-phosphate interactions, and this may be the basic reason for the presence of RNA in the ribosome. Such a role for ribosomal RNA was first considered by Watson (1964). Subsequently Moore (1966) suggested that 16S RNA might be involved in the binding of poly(U). However the experimental support for this suggestion was indirect, and, in addition, the choice of poly(U) as mRNA was unfortunate because poly(U) binds nonspecifically to basic proteins (Nomura, unpublished experiments), and such nonspecific binding could not be distinguished from specific binding. Yet several reconstitution experiments mentioned above strongly suggest that 16S RNA is important not only for assembly but also for the function of the assembled ribosome. For example, the role of 16S RNA in translation of the R17 coat cistron discussed earlier is consistent with the hypothesis that direct interactions between ribosomal RNA and mRNA are involved in ribosomal functions (Held, Gette and Nomura 1974). The suggestion that tRNA binding to the ribosome involves an interaction between the $Tp\varphi pCpG$ sequence, which is found in most of the tRNAs, and a complementary sequence of 5S RNA has been made repeatedly, and there are some experiments to support this suggestion (see Monier, this volume). In addition, it is conceivable that RNA–RNA interaction is involved in the formation of 30S–50S couples.

The available information even suggests that there are functionally specific regions in the 16S RNA molecule. One such example is the region near the 3' end of the 16S RNA. This region is rich in methylated bases (see chapters by Fellner and Zimmermann, this volume) and one (or a very few) nucleolytic cut by colicin E3 in this region inactivates ribosome functions, especially the ability of the ribosome to initiate on natural mRNA (Bowman et al. 1971; Tai and Davis 1974; Nomura et al., this volume). The same region may also be involved in the binding of kasugamycin (Helser, Davies and Dahlberg 1971; see also above), which inhibits the initiation of protein synthesis (see Davis, Tai and Wallace, this volume). Thus the notion that rRNA has only a structural role and all the functions in the finished ribosome are carried out by proteins is probably incorrect.

The opposite extreme view, which emphasizes the role of ribosomal RNA, is to consider ribosomal proteins as the components whose role is solely to maintain the proper structure of ribosomal RNA. However if ribosomal proteins do not have any direct functional role, it would be difficult to explain why ribosomes contain so many different proteins rather than many copies of a few proteins. In addition, although nucleic acids are involved in binding reactions in various systems, there are no known examples of the actual catalysis of chemical reactions by nucleic acids alone. In fact, experimental studies, for example, those demonstrating the role of S12 in chain initiation mentioned above, strongly suggest that at least some proteins participate directly in ribosome function. The known catalytic ribosomal functions, such as GTPase or peptidyl transferase, almost certainly involve a direct participation of some ribosomal proteins. It is also very likely that some functional sites on the ribosome are composite structures consisting of one or more ribosomal proteins and ribosomal RNA.

The reason for the presence of so many different proteins remains unknown. It is probable that some proteins, such as S16, are involved only in the assembly reaction and may play various roles at several different stages in the assembly. Other proteins (e.g., S12) may be directly involved in one or several ribosome functions. (Of course it is also possible that some proteins have roles in both assembly and function, as noted in the previous section.)

Both the presence of large RNA molecules and the complexity of ribosome protein composition may be related to the size of ribosomes as well as the complexity of the translational process. During protein synthesis the ribosome must interact with several very large structures: aminoacyl tRNA (see Rich, this volume), various protein factors with high molecular weights, such as initiation factors and elongation factors, and, possibly, large regions of mRNA involved in the signals for initiation and termination. The ribosome must also hold a considerable length of growing polypeptides during chain elongation (Malkin and Rich 1967), perhaps in order to prevent possible steric interference of these growing polypeptides with ribosome functions (e.g., the binding of aminoacyl-tRNA).

These two unique structural features which we have been discussing—the presence of large RNA and the complex protein composition—may both be advantageous in building a structure sufficiently large for performing these functions. The large ribosomal RNA serves as a structural backbone that spreads the proteins and functional RNA regions over a large space and orients them in the proper positions for interaction with non-ribosomal components involved in the translational process. The use of many small proteins, rather than a few large proteins, also spreads the functionally active centers over a larger space and, in addition, provides more individual protein molecules to aid the folding of the large rRNA. However in order to consider these questions further, we must await the elucidation of the three-dimensional structure of the ribosome as well as more precise information on the mechanisms involved in ribosome function.

The complexity of the structure in terms of the protein composition may also be related to possible multifunctional properties of the ribosome. It was recently discovered that the ribosome plays a key role in the synthesis of guanosine tetra- and pentaphosphates (Block and Haseltine, this volume). As pointed out previously (Nomura 1973), it is possible that the ribosome has yet some other undiscovered functions not directly related to protein synthesis. The participation of one and the same organelle, such as the ribosome, in several different cellular activities may be of advantage for their coordination and regulation.

RECONSTITUTION OF 50S SUBUNITS

Many unsuccessful attempts have been made to reconstitute functionally active *E. coli* 50S subunits from completely dissociated molecular components. Although there is one report of success in such an attempt (Maruta, Tsuchiya and Mizuno 1971), these experiments have been difficult to reproduce (for example, Chu and Maeba 1973; Mizushima, Erdmann and Nomura, published experiments). On the other hand, the 50S subunit from a thermophilic organism, *Bacillus stearothermophilus,* can be reconstituted using essentially the same techniques that are used for the reconstitution of 30S subunits from *E. coli* or *B. stearothermophilus,* except that incubation is usually done at 60°C for 1–2 hr (Nomura and Erdmann 1970).

It is probable that the difficulty with the reconstitution of *E. coli* 50S subunits reflects a greater complexity of the assembly reaction, as well as higher kinetic energy barriers than 30S assembly, which might be overcome only by longer incubation at even higher temperatures. In fact, the reconstitution of *B. stearothermophilus* 50S subunits, even at the optimal temperature of 60°C, proceeds much more slowly than reconstitution of 30S subunits at 45°C (Nomura and Erdmann 1970). The ribosomal components of *E. coli* 50S subunits, the partially assembled intermediate particles, or the reconstituted 50S subunits themselves might be unstable at such high temperatures, and therefore the method used for *B. stearothermophilus* cannot be applied to the *E. coli* system.

Another problem related to 50S reconstitution may be the difficulty of obtaining and incorporating completely intact 23S RNA. Ceri and Maeba (1973) found that 23S RNA derived from *E. coli* 50S subunits is extensively degraded, whereas that obtained from 70S ribosomes is almost completely intact. They present evidence suggesting that one of the 50S proteins has ribonuclease activity. This may be related to a phenomenon observed with the *B. stearothermophilus* system. Here the activity of the reconstituted 50S subunits, although variable, is always somewhat less than the activity of the subunits from which the RNA and proteins were derived. It has been found consistently that when RNA and proteins derived from 70S ribosomes are used, the reconstituted 50S particles are more active than when 50S subunits prepared from these ribosomes are used as a source of RNA and proteins (Fahnestock, Held and Nomura 1972). However, the addition of 16S RNA and 30S proteins to the 50S components during reconstitution causes no increase in the activity of the reconstituted 50S subunits (see below). Thus it is possible that the better reconstitution with the components derived from 70S ribosomes is due to a greater intactness of the 23S RNA.

It has also been shown (Hosokawa, Kiho and Migita 1973) that physical assembly of *E. coli* 50S subunits from RNA and proteins into particles sedimenting at 48S can be achieved by incubation at 37°C only if polyamines are added to the reconstitution mixture. These 48S particles showed no functional activity. However another study (Chu and Maeba 1973) demonstrated that such physical assembly could also be achieved in the absence of polyamines when the incubation was conducted for brief periods at 60°C.

Finally, it has been suggested that the lack of functional activity of reassembled *E. coli* "50S" subunits is due to an improper conformation of the 5S RNA (Yu and Wittmann 1973).

In the *B. stearothermophilus* 50S reconstitution system, the "RNA" fraction used in the initial reconstitution experiments contained one protein, designated B-L3, still tightly bound to 23S rRNA. The protein B-L3 can be removed from 23S RNA, and complete reconstitution from a mixture of protein-free RNA with the 50S protein fraction plus the purified protein B-L3 has been demonstrated (Fahnestock, Erdmann and Nomura 1973). Although protein B-L3 does not appear to be required for physical assembly of the 50S subunit, it is strongly required for the reconstitution of functionally active particles.

There is some genetic evidence which suggests that 30S subunits (or their precursors or components) play some crucial role in the assembly of 50S subunits in *E. coli* (Nashimoto and Nomura 1970; Kreider and Brownstein 1971; see also Jaskunas, Davies and Nomura, this volume). However in the *B. Stearothermophilus* 50S subunit reconstitution, it was found that the kinetics of reconstitution, using RNA and protein fractions prepared from a mixture of purified 50S and 30S subunits, is not significantly different from the kinetics using components prepared from 50S subunits alone (Fahnestock, Held and Nomura 1972). Thus the assembly of 50S subunits in vitro does not require the presence of 30S subunits (or their components). The role of 30S subunits in 50S assembly in vivo remains unknown.

The mechanism of 50S subunit assembly in the *B. stearothermophilus* system has been studied to some extent. As mentioned earlier, the reconstitution takes place only at temperatures above 50°C (Nomura and Erdmann 1970; Fahnestock, Held and Nomura 1972). At lower temperatures some incomplete particles accumulate. Particles (called "RI_{50}"), which accumulate at 30°C, have been isolated and studied (Fahnestock, Held and Nomura 1972). While active reconstituted "50S" particles sediment only slightly slower (about 48S) than native 50S subunits, RI_{50} particles sediment considerably slower (35–37S). RI_{50} particles show no activity in any of several known partial reactions of 50S subunits, as well as poly(U)-dependent polyphenylalanine synthesis. However despite the large difference in sedimentation coefficients, RI_{50} particles contain 5S RNA and are deficient in only three or four proteins relative to 50S subunits. In addition, heating of the isolated RI_{50} particles at 60°C, without addition of the missing proteins, converts the inactive RI_{50} particles to particles (RI_{50}^*) having considerable activity in polypeptide synthesis and sedimenting at about 43S. Heating the RI_{50} particles in the presence of the missing proteins converts them to particles which have higher activity than the RI_{50}^* particles and sediment at the same rate as the usual reconstituted particles. The increase in sedimentation coefficient observed upon heating the RI_{50} particles indicates the occurrence of a large conformational change, which presumably is related to the appearance of functional activity.

From these and other experimental results, the following scheme has been proposed for the in vitro assembly reaction in this system (Fahnestock, Held and Nomura 1972):

$$23S \text{ RNA} + 5S \text{ RNA} + \text{proteins} \longrightarrow RI_{50}$$

$$\xrightarrow{\text{heat}} RI_{50}^* \xrightarrow{\text{proteins}} RI_{50}^{**} \xrightarrow{\text{heat}} \text{"50S"} \qquad (6)$$

As mentioned above, RI_{50}^* has some activity, but "50S" is more active, and $RI_{50} \rightarrow RI_{50}^*$ represents the major rate-limiting step. However because not all the 50S proteins have been purified in the *B. stearothermophilus* system, detailed analysis of the mechanism of reconstitution has not been possible.

Extensive functional anlyses of the roles of the 50S proteins will probably have to await the availability of these poteins in purified form. However this system has provided a tool for studying the role of 5S RNA. It has been shown that 5S RNA is required for the reconstitution of active 50S subunits (Erdmann et al. 1971). Furthermore, the demonstration that reconstituted 50S subunits containing 5S RNA oxidized at the 3′ terminus are fully active has ruled out the formation during protein synthesis of a covalent complex involving the 3′-OH group of 5S RNA and a growing polypeptide chain (Erdmann, Doberer and Sprinzl 1972; Fahnestock and Nomura 1972). Finally, when 5S RNA is incubated with a mixture of 50S proteins, it forms a complex with proteins B-L5 and B-L22 (Horne and Erdmann 1972), and this complex catalyzes the hydrolysis of both GTP and ATP (Horne and Erdmann 1973). Both activities are inhibited by fusidic acid and thiostrepton. The evidence suggests that one or more of the proteins in the complex forms the catalytic site for the EF-G- and/or EF-Tu-dependent GTPase activities of the ribosome. The complex also shows a tenfold higher affinity than 5S RNA for the oligonucleotide

TpφpCpG, which contains a sequence found in one of the loops of all tRNAs (Erdmann, Sprinzl and Pongs 1973). It has been proposed that this binding interaction is involved in the binding of tRNA to ribosomes.

As mentioned earlier, functional analyses of *E. coli* 50S components are still being performed with various partial reconstitution systems. Alcohol-treated L7, L12-deficient particles have been used to study the role of L7 and L12 in various partial reactions involved in elongation (see Möller, this volume). Various CsCl core particles were used several years ago as a tool for studying the roles of 50S proteins in the peptidyl transferase reaction (Staehelin, Maglott and Monro 1969), but the specific split proteins involved in the restoration of activity were not identified. Recently Nierhaus and his coworkers have extended these early studies using core particles produced by extracting 50S subunits with 0.4 M LiCl ("0.4 c cores") or 0.8 M LiCl ("0.8 c cores") (Nierhaus and Montejo 1973; Nierhaus and Nierhaus 1973). The 0.4 c cores were active in both peptidyl transferase and chloramphenicol binding, whereas the 0.8 c cores were inactive in both these functions. These workers have obtained data suggesting that L16 is involved in chloramphenicol binding and L11 in peptidyl transferase activity. The assignment for L16 agrees with the results of an affinity labeling experiment with a chloramphenicol analog (Pongs et al. 1973). Again, it is difficult to determine whether L11 and L16 are directly involved in these reactions or whether they act indirectly by affecting the conformation of other components present in the 0.8 c core.

IN VITRO ASSEMBLY AND
THE IN VIVO ASSEMBLY REACTION

It is highly probable that information obtained in the in vitro system is pertinent to the in vivo assembly mechanism. Protein compositions of some precursor particles related to 30S subunits (Homann and Nierhaus 1971; Nierhaus, Bordasch and Homann 1973) were found to be approximately consistent with the "sequence" indicated in the assembly map. The 21S particles accumulated by some cold-sensitive mutants have protein compositions very similar to that of RI particles isolated from reconstitution mixtures in vitro (Nashimoto et al. 1971; see Table 1). The latter observation suggests that a rate-determining step in the in vivo assembly under certain conditions is related to the rate-determining step studied in vitro. (For further discussion see Schlessinger, this volume.)

Recent analysis of the 30S proteins required for the methylation of 16S RNA in vitro is also pertinent to the in vivo assembly mechanism (Thammana and Held, manuscript in preparation). Methyl-deficient 16S RNA or 30S ribosomes obtained from a kasugamycin-resistant mutant cannot be methylated in vitro by a partially purified methylase enzyme from wild-type cells. CsCl core particles, however, can be methylated in vitro (Helser, Davies and Dahlberg 1972). The proteins required for methylation were determined by adding purified 30S proteins, singly or in combination, to the methyl-deficient 16S RNA under reconstitution conditions, and then testing whether the resulting particles could be methylated. Eight proteins (S4, S8, S16, S17, S15, S18, S6 and S11) were found to be essential (and together, sufficient) for methylation of the 16S RNA. Other proteins (S9, S14, S10, and S3, present together) were found to inhibit methylation of 16S RNA in the CsCl core particles. Thus the proteins required for methylation are strongly interrelated and occupy early parts of the assembly map, whereas the proteins required for inhibition of methylation, also strongly interrelated, occupy a distal portion of the map (cf. Figure 1). It is highly likely that the methylating enzyme also acts in vivo on intermediate particles that contain the essential eight proteins, and perhaps some other proteins, but lack some or all of the "inhibitory" proteins.

Thus the results of several experiments suggest that assembly of 30S ribosomes in vitro is similar in many respects to that observed in vivo. However as we have already stated, there are clear differences between in vitro assembly and in vivo assmbly. In vivo, ribosomal proteins may interact with rRNA which is still being transcribed. Such rRNA may be very different in its physical and chemical properties from that found in mature ribosomes. In fact, various ribonucleoprotein "precursor" particles that have been isolated in vivo are found to contain ribosomal RNA which is chemically different from mature ribosomal RNA. The fact that the rate of 30S subunit assembly is faster in vivo than in vitro may be related to the use of such precursor RNA (and possibly precursor proteins) in the in vivo process. However substitution of precursor 16S RNA ("17S RNA") from chloramphenicol-treated cells or cold-sensitive mutants for normal 16S RNA in the standard reconstitution mixture did not produce an increase in the rate of assembly. The assembled particles sedimented at 30S but were inactive (Nomura and Lowry, unpublished experiments, cited in Lowry and Dahlberg 1971; P. Sypherd, personal communication). Nierhaus, Bordasch and Homann (1973) have reported that functionally active 30S subunits can be assembled in vitro from isolated 21S precursor particles containing "precursor 16S RNA." Formation of functionally active particles occurred in the presence of 21S precursor particles, total 30S ribosomal proteins, and methylating conditions which included a ribosome-free cell extract, ATP and S-adenosylmethionine. However under identical conditions, the above workers could not obtain reconstitution of active 30S ribosomes starting from "precursor 16S RNA." The problem of the chemical and physical structure of the RNA may be especially pertinent in the case of 50S assembly. It has been observed that in the *B. stearothermophilus* 50S reconstitution system, 23S RNA prepared by the urea-LiCl method is more active in reconstitution than 23S RNA prepared by phenol extraction (Fahnestock, Erdmann and Nomura, unpublished experiments). The difference is not due to some residual bound proteins, such as B-L3, and probably reflects a difference in the conformation of the 23S RNA.

Such observations support the speculation that the assembly in vivo is more efficient because of the use of precursor rRNA which may have a different conformation. In addition, the occurrence of 16S RNA and 23S RNA on a single precursor RNA molecule, as shown recently (Nikolaev, Silengo and Schlessinger 1973; Dunn and Studier 1973), could explain the apparent "coupling" between 30S assembly and 50S assembly, which has been suggested from in vivo genetic experiments but not observed in in vitro experiments (see above).

It is also possible that in vivo assembly is facilitated by nonribosomal elements that are absent from the in vitro system. The role of S16 in 30S assembly described above is pertinent in this speculation. This protein facilitates an energy-requiring structural rearrangement of an intermediate particle, which is formed during 30S assembly but apparently is not required for the function of assembled particles. Because S16 remains bound to the rearranged intermediate and is incorporated into the final assembled ribosome, it is classified as a ribosomal protein. If a protein exists which performs a function similar to that of S16 in facilitating some step in the assembly process but which is released as a result of some structural change during assembly, this protein would not be found among the ribosomal proteins but would be classified as a ribosome assembly "factor."

Since rRNA can be synthesized in vitro using *E. coli* DNA as a template (see Travers, this volume), in vitro reconstitution could possibly be done using "nascent" rRNA synthesized in such a system. Synthesis of ribosomal proteins in vitro using *E. coli* DNA in a DNA-dependent protein-synthesizing system has also been demonstrated recently (Kaltschmidt, Kahan and Nomura 1974). One would hope that complete reproduction in vitro of all the assembly events that occur in vivo would become possible in the future. It would then be possible to study in vitro many unsolved problems related to the biosynthesis of the ribosome.

Acknowledgments

The work from the author's laboratory was supported in part by the College of Agriculture and Life Sciences, University of Wisconsin, and by grants from the National Institute of General Medical Sciences (GM-20427-01) and the National Science Foundation (GB-31086X2). This is paper No. 1723 of the Laboratory of Genetics, University of Wisconsin.

References

Acharya, A. S. and P. B. Moore. 1973. Reaction of ribosomal sulfydryl groups with 5,5'-dithiobis (2-nitrobenzoic acid). *J. Mol. Biol.* **76**:207.
Atsmon, A., P. Spitnik-Elson and D. Elson. 1969. Detachment of ribosomal proteins by salt. II. Some properties of protein-deficient particles formed by the detachment of ribosomal proteins. *J. Mol. Biol.* **45**:125.

Bollen, A., A. Herzog, A. Favre, J. Thibault and F. Gros. 1970. Fluorescence studies on the 30S ribosome assembly process. *FEBS Letters* **11**:49.

Bowman, C. M., J. E. Dahlberg, T. Ikemura, J. Konisky and M. Nomura. 1971. Specific inactivation of 16S ribosomal RNA induced by colicin E3 *in vivo*. *Proc. Nat. Acad. Sci.* **68**:964.

Ceri, H. and P. Y. Maeba. 1973. Association of a ribonuclease with the 50S ribosomal subunit of *Escherichia coli* MRE600. *Biochim. Biophys. Acta* **312**:337.

Chu, F. K. and P. Y. Maeba. 1973. Physical reconstitution of 23S RNA–50S protein complexes from *Escherichia coli*. *Can. J. Biochem.* **51**:129.

Cohlberg, J. A. 1974. Activity of protein-deficient 30S ribosomal subunits in elongation factor G-dependent GTPase. *Biochem. Biophys. Res. Comm.* **57**:225.

Craven, G. R., R. Gavin and T. Fanning. 1969. The t-RNA binding site of the 30S ribosome and the site of tetracycline inhibition. *Cold Spring Harbor Symp. Quant. Biol.* **34**:129.

Craven, G. R., P. Voynow, S. J. S. Hardy and C. G. Kurland. 1969. The ribosomal proteins of *Escherichia coli*. II. Chemical and physical characterization of the 30S ribosomal proteins. *Biochemistry* **8**:2906.

Dunn, J. J. and F. W. Studier. 1973. T7 early RNAs and *Escherichia coli* ribosomal RNAs are cut from large precursor RNAs *in vivo* by ribonuclease III. *Proc. Nat. Acad. Sci.* **70**:3296.

Egberts, E., P. Traub, P. Herrlich and M. Schweiger. 1972. Functional integrity of *Escherichia coli* 30S ribosomes reconstituted from RNA and protein *in vitro* synthesis of S-adenosylmethionine cleaving enzyme. *Biochim. Biophys. Acta* **277**:681.

Erdmann, V. A., H. G. Doberer and M. Sprinzl. 1972. Structure and function of 5S RNA: The role of the 3′ terminus in 5S RNA function. *Mol. Gen. Genet.* **114**:89.

Erdmann, V. A., M. Sprinzl and O. Pongs. 1973. The involvement of 5S RNA in the binding of tRNA to ribosomes. *Biochem. Biophys. Res. Comm.* **54**:942.

Erdmann, V. A., S. Fahnestock, K. Higo and M. Nomura. 1971. Role of 5S RNA in the functions of 50S ribosomal subunits. *Proc. Nat. Acad. Sci.* **68**:2932.

Fahnestock, S. R. and M. Nomura. 1972. Activity of ribosomes containing 5S RNA with a chemically modified 3′ terminus. *Proc. Nat. Acad. Sci.* **69**:363.

Fahnestock, S., V. Erdmann and M. Nomura. 1973. Reconstitution of 50S ribosomal subunits from protein-free ribonucleic acid. *Biochemistry* **12**:220.

Fahnestock, S., W. Held and M. Nomura. 1972. The assembly of bacterial ribosomes. In *The first John Innes symposium on generation of subcellular structures*, ed. R. Markham et al. pp. 179–217. North-Holland, Amsterdam.

Garrett, R. A., K. H. Rak, L. Daya and G. Stöffler. 1971. Ribosomal proteins. XXIX. Specific protein binding sites on 16S rRNA of *Escherichia coli*. *Mol. Gen. Genet.* **114**:112.

Ginzburg, I., R. Miskin and A. Zamir. 1973. *N*-ethyl maleimide as a probe for the study of functional sites and conformations of 30S ribosomal subunits. *J. Mol. Biol.* **79**:481.

Goldberg, M. L. and J. A. Steitz. 1974. Cistron specificity of 30S ribosomes heterologously reconstituted with components from *E. coli* and *B. stearothermophilis. Biochemistry* **13**:2123.

Green, M. and C. G. Kurland. 1973. Molecular interactions of ribosomal components. IV. Cooperative interactions during assembly *in vitro. Mol. Biol. Reports* **1**:105.

Groner, Y., R. Scheps, R. Kamen, D. Kolakofsky and M. Revel. 1972. Host subunit of Qβ replicase is translation control factor i. *Nature New Biol.* **239**:19.

Held, W. and M. Nomura. 1973a. Functional homology between the 30S ribosomal protein S7 from *E. coli* K12 and S7 from *E. coli* MRE600. *Mol. Gen. Genet.* **122**:11.

————. 1973b. Rate-determining step in the reconstitution of *Escherichia coli* 30S ribosomal subunits. *Biochemistry* **12**:3273.

Held, W. A., W. R. Gette and M. Nomura. 1974. Role of 16S ribosomal RNA and the 30S ribosomal protein S12 in the initiation of natural messenger RNA translation. *Biochemistry* **13**:2115.

Held, W. A., S. Mizushima and M. Nomura. 1973. Reconstitution of *Escherichia coli* 30S ribosomal subunits from purified molecular components. *J. Biol. Chem.* **248**:5720.

Held, W. A., M. Nomura and J. W. B. Hershey. 1973. Ribosomal protein S21 is required for full activity in the initiation of protein synthesis. *Mol. Gen. Genet.* **128**:11.

Held, W. A., B. Ballou, S. Mizushima and M. Nomura. 1974. Assembly mapping of 30S ribosomal proteins from *E. coli:* Further studies. *J. Biol. Chem.* **249**:3103.

Helser, T. L., J. E. Davies and J. E. Dahlberg. 1971. Change in methylation of 16S ribosomal RNA associated with mutation to kasugamycin resistance in *Escherichia coli. Nature New Biol.* **233**:12.

————. 1972. Mechanism of kasugamycin resistance in *Escherichia coli. Nature New Biol.* **235**:6.

Higo, K. and K. Loertscher. 1974. Amino terminal sequences of some *Escherichia coli* 30S ribosomal proteins and functionally corresponding *Bacillus stearothermophilus* ribosomal proteins. *J. Bact.* **118**:180.

Higo, K., W. Held, L. Kahan and M. Nomura. 1973. Functional correspondence between 30S ribosomal proteins of *Escherichia coli* and *Bacillus stearothermophilus. Proc. Nat. Acad. Sci.* **70**:944.

Homann, H. H. and K. H. Nierhaus. 1971. Protein composition of biosynthetic precursors and artificial subparticles from ribosomal subunits in *Escherichia coli* K12. *Eur. J. Biochem.* **20**:249.

Horne, J. R. and V. A. Erdmann. 1972. Isolation and characterization of 5S RNA–protein complexes from *Bacillus stearothermophilus* and *Escherichia coli. Mol. Gen. Genet.* **119**:337.

————. 1973. ATPase and GTPase activities associated with a specific 5S RNA–protein complex. *Proc. Nat. Acad. Sci.* **70**:2870.

Hosokawa, K., R. K. Fujimura and M. Nomura. 1966. Reconstitution of functionally active ribosomes from inactive subparticles and proteins. *Proc. Nat. Acad. Sci.* **55**:198.

Hosokawa, K., Y. Kiho and L. K. Migita. 1973. Assembly of *Escherichia coli*

50S ribosomes from ribonucleic acid and protein components. I. Chemical and physical properties affecting the conformation of assembled particles. *J. Biol. Chem.* **248**:4135.

Kaltschmidt, E., L. Kahan and M. Nomura. 1974. *In vitro* synthesis of ribosomal proteins directed by *Escherichia coli* DNA. *Proc. Nat. Acad. Sci.* **71**:446.

Kreider, G. and B. L. Brownstein. 1971. A mutation suppressing streptomycin dependence. II. An altered protein on the 30S ribosomal subunit. *J. Mol. Biol.* **61**:135.

Kurland, C. G. 1972. The structure and function of the bacterial ribosome. *Ann. Rev. Biochem.* **41**:377.

Kurland, C. G., P. Voynow, S. J. S. Hardy, L. Randall and L. Lutter. 1969. Physical and functional heterogeneity of *E. coli* ribosomes. *Cold Spring Harbor Symp. Quant. Biol.* **34**:17.

Lai, C. J., B. Weisblum, S. R. Fahnestock and M. Nomura. 1973. Alteration of 23S ribosomal RNA and erythromycin-induced resistance to lincomycin and spiramycin in *Staphylococcus aureus*. *J. Mol. Biol.* **74**:67.

Lodish, H. F. 1970. Specificity in bacterial protein synthesis: Role of initiation factors and ribosomal subunits. *Nature* **226**:705.

Lowry, C. V. and J. E. Dahlberg. 1971. Structural differences between the 16S ribosomal RNA of *E. coli* and its precursor. *Nature New Biol.* **232**:52.

Malkin, L. I. and A. Rich. 1967. Partial resistance of nascent polypeptide chains to proteolytic digestion due to ribosomal shielding. *J. Mol. Biol.* **26**:329.

Marsh, R. C. and A. Parmeggiani. 1973. Requirements of proteins S5 and S9 from 30S subunits for the ribosome-dependent GTPase activity of elongation factor G. *Proc. Nat. Acad. Sci.* **70**:151.

Maruta, H., T. Tsuchiya and D. Mizuno. 1971. *In vitro* reassembly of functionally active 50S ribosomal particles from ribosomal proteins and RNA's of *Escherichia coli*. *J. Mol. Biol.* **61**:123.

Miall, S. H. and F. O. Walker. 1969. Structural studies on ribosomes. II. Denaturation and sedimentation of ribosomal subunits unfolded in EDTA. *Biochim. Biophys. Acta* **174**:551.

Mizushima, S. and M. Nomura. 1970. Assembly mapping of 30S ribosomal proteins from *E. coli*. *Nature* **266**:1214.

Moore, P. B. 1966. Studies on the mechanism of messenger ribonucleic acid attachment to ribosomes. *J. Mol. Biol.* **22**:145.

―――. 1971. Reaction of *N*-ethyl maleimide with the ribosomes of *Escherichia coli*. *J. Mol. Biol.* **60**:169.

―――. 1973. Protein synthesis catalyzed by thiol-blocked ribosome preparations. *J. Mol. Biol.* **79**:615.

Nashimoto, H. and M. Nomura. 1970. Structure and function of bacterial ribosomes. XI. Dependence of 50S ribosomal assembly on simultaneous assembly of 30S subunits. *Proc. Nat. Acad. Sci.* **67**:1440.

Nashimoto, H., W. Held, E. Kaltschmidt and M. Nomura. 1971. Structure and function of bacterial ribosomes. XII. Accumulation of 21S particles by some cold-sensitive mutants of *Escherichia coli*. *J. Mol. Biol.* **62**:121.

Nierhaus, D. and K. Nierhaus. 1973. Identification of the chloramphenicol-binding protein in *Escherichia coli* ribosomes by partial reconstitution. *Proc. Nat. Acad. Sci.* **70**:2224.

Nierhaus, K. and V. Montejo. 1973. A protein involved in the peptidyl-transferase activity of *Escherichia coli* ribosomes. *Proc. Nat. Acad. Sci.* **70**:1931.

Nierhaus, K. H., K. Bordasch and H. E. Homann. 1973. Ribosomal proteins. XLIII. *In vivo* assembly of *Escherichia coli* ribosomal proteins. *J. Mol. Biol.* **74**:587.

Nikolaev, N., L. Silengo and D. Schlessinger. 1973. Synthesis of a large precursor to ribosomal RNA in a mutant of *Escherichia coli*. *Proc. Nat. Acad. Sci.* **70**:3361.

Noller, H. F. and J. B. Chaires. 1972. Functional modification of 16S ribosomal RNA by kethoxal. *Proc. Nat. Acad. Sci.* **69**:3115.

Nomura, M. 1970. Bacterial ribosome. *Bact. Rev.* **34**:228.

———. 1973. Assembly of bacterial ribosomes. *Science* **179**:864.

Nomura, M. and V. A. Erdmann. 1970. Reconstitution of 50S ribosomal subunits from dissociated molecular components. *Nature* **228**:144.

Nomura, M. and P. Traub. 1968. Structure and function of *E. coli* ribosomes. III. Stoichiometry and rate of the reconstitution of ribosomes from subribosomal particles and split proteins. *J. Mol. Biol.* **34**:609.

Nomura, M., P. Traub and H. Bechmann. 1968. Hybrid 30S ribosomal particles reconstituted from components of different bacterial origins. *Nature* **219**:793.

Nomura, M., S. Mizushima, M. Ozaki, P. Traub and C. V. Lowry. 1969. Structure and function of ribosomes and their molecular components. *Cold Spring Harbor Symp. Quant. Biol.* **34**:49.

Ozaki, M., S. Mizushima and M. Nomura. 1969. Identification and functional characterization of the protein controlled by the streptomycin-resistant locus in *E. coli*. *Nature* **222**:333.

Pongs, O., R. Bald and V. A. Erdmann. 1973. Identification of chloramphenicol-binding protein in *Escherichia coli* ribosomes by affinity labeling. *Proc. Nat. Acad. Sci.* **70**:2229.

Randall-Hazelbauer, L. L. and C. G. Kurland. 1972. Identification of three 30S proteins contributing to the ribosomal A site. *Mol. Gen. Genet.* **115**:234.

Sander, G., R. C. Marsh and A. Parmeggiani. 1973. Role of split proteins from 30S subunits in the EF-T GTPase reaction. *FEBS Letters* **33**:132.

Schaup, H. W., M. Green and C. G. Kurland. 1970. Molecular interactions of ribosomal components. I. Identification of RNA binding sites for individual 30S ribosomal proteins. *Mol. Gen. Genet.* **109**:193.

———. 1971. Molecular interactions of ribosomal components. II. Site-specific complex formation between 30S proteins and ribosomal RNA. *Mol. Gen. Genet.* **112**:1.

Schreier, M. H. and H. Noll. 1971. Conformational changes in ribosomes during protein synthesis. *Proc. Nat. Acad. Sci.* **68**:805.

Schulte, C. and R. A. Garrett. 1972. Optimal conditions for the interaction of ribosomal protein S8 and 16S RNA and studies on the reaction mechanism. *Mol. Gen. Genet.* **119**:345.

Staehelin, T. and M. Meselson. 1966. *In vitro* recovery of ribosomes and of synthetic activity from synthetically inactive ribosomal subunits. *J. Mol. Biol.* **15**:245.

Staehelin, T., D. Maglott and R. E. Monro. 1969. On the catalytic center of peptidyl transfer: A part of the 50S ribosome structure. *Cold Spring Harbor Symp. Quant. Biol.* **34**:39.

Sun, T-T. and R. R. Traut. 1973. The functional and structural homology of ribosomal protein S7 of *E. coli* strains K and MRE600. *Mol. Gen. Genet.* **122**:1.

Tai, P-C. and B. D. Davis. 1974. Activity of colicin E3-treated ribosomes in initiation and in chain elongation. *Proc. Nat. Acad. Sci.* **71**:1021.

Tamaoki, T. and F. Miyazawa. 1967. Dissociation of *Escherichia coli* ribosomes by sulfhydryl reagents. *J. Mol. Biol.* **23**:35.

Traub, P. and M. Nomura. 1968. Structure and function of *E. coli* ribosomes. V. Reconstitution of functionally active 30S ribosomal particles from RNA and proteins. *Proc. Nat. Acad. Sci.* **59**:777.

————. 1969. Structure and function of *E. coli ribosomes*. VI. Mechanism of assembly of 30S ribosomes studied *in vitro*. *J. Mol. Biol.* **40**:391.

Traub, P., K. Hosokawa, G. R. Craven and M. Nomura. 1967. Structure and function of *E. coli* ribosomes. IV. Isolation and characterization of functionally active ribosomal proteins. *Proc. Nat. Acad. Sci.* **58**:2430.

Traut, R. R. and A. L. Haenni. 1967. The effect of sulfhydryl reagents on ribosome activity. *Eur. J. Biochem.* **2**:64.

Van Duin, J. and C. G. Kurland. 1970. Functional heterogeneity of the 30S ribosomal subunit of *E. coli*. *Mol. Gen. Genet.* **109**:169.

Van Duin, J., P. H. van Knippenberg, M. Dieben and C. G. Kurland. 1972. Functional heterogeneity of the 30S ribosomal subunit of *Escherichia coli*. II. Effect of S21 on initiation. *Mol. Gen. Genet.* **116**:181.

Voynow, P. and C. G. Kurland. 1971. Stoichiometry of the 30S ribosomal proteins of *E. coli. Biochemistry* **10**:517.

Watson, J. D. 1964. The synthesis of proteins from ribosomes. *Bull. Soc. Chim. Biol.* **46**:1399.

Weber, H. J. 1972. Stoichiometric measurements of 30S and 50S ribosomal proteins from *Escherichia coli. Mol. Gen. Genet.* **119**:233.

Yu, R. S. T. and H. G. Wittmann. 1973. The structural basis for functional inactivity of reconstituted 50S ribosomal subunits of *Escherichia coli. Biochim. Biophys. Acta* **319**:388.

RNA—Protein Interactions in the Ribosome

Robert A. Zimmermann*

Département de Biologie Moléculaire
Université de Genève
1211 Genève 4, Switzerland

INTRODUCTION

Specific interactions between ribosomal proteins and ribosomal RNAs are essential to the assembly and stability of ribosomal particles. In *Escherichia coli,* approximately 20 of the 54 proteins found in the 30S and 50S subunits are able to bind individually to one or another of the three rRNAs. Many of these interactions are believed to occur early in subunit assembly

* Present address: Department of Biochemistry, University of Massachusetts, Amherst, Massachusetts 01002

and to underlie the ability of the ribosome to attain its active structure. For convenience, the class of ribosomal proteins capable of associating independently with rRNA will be referred to as RNA-binding proteins. This qualification should not be interpreted to exclude the possibility that the other proteins bind to rRNA, however, for it is likely that many of the proteins incorporated later in the assembly sequence are guided and fixed to their proper locations in the subunit structure by specific RNA–protein interactions as well. Although such interactions may therefore be direct, they are not independent in the sense that they cannot be detected in the absence of other ribosomal proteins.

In the following sections, the conditions that promote the formation and maintenance of specific ribosomal RNA–ribosomal protein complexes will be considered in relation to the structure and conformation of both RNA and protein. The discussion will focus upon components of the *E. coli* ribosome, which have been extensively characterized. Six or seven 30S subunit proteins from that organism bind specifically to 16S RNA; ten or eleven 50S subunit proteins bind specifically to 23S RNA, and three others independently interact with 5S RNA. In addition, at least one protein of the 30S subunit binds to the 23S RNA and may thus play a role in the association of the smaller and larger subunits. Furthermore, the ribosomal components of *E. coli* have been found to associate with those from several other bacteria in a specific fashion. The study of heterologous interactions yields valuable information on binding specificity and raises the intriguing possibility that the structural features which define binding sites in both RNA and protein have been conserved throughout evolution. Finally, the location and structure of individual protein binding sites within the rRNA molecules will be discussed. Characterization of the binding sites was made possible both by the isolation of rRNA fragments capable of specific interaction with one or more ribosomal proteins and by the availability of a detailed description of rRNA primary structure (see Fellner, this volume). The relative arrangement of RNA and protein in the ribosome has also been inferred from the investigation of ribonucleoprotein fragments obtained by RNase hydrolysis of 30S and 50S subunits. Together these various avenues of study have revealed certain general characteristics of RNA–protein interactions in the ribosome which may prove helpful in understanding the properties of many other biological systems where specific interactions between nucleic acids and proteins play a major part.

CONDITIONS FOR THE STUDY OF RNA–PROTEIN INTERACTIONS
Optimum Ionic Environment and Temperature

The interaction of individual ribosomal proteins with rRNA has generally been studied in a buffer containing 0.01–0.05 M Tris-HCl pH 7.6–0.02 M $MgCl_2$–0.35 M KCl–0.006 M β-mercaptoethanol (TMK buffer) at 40–

42°C (Mizushima and Nomura 1970; Schaup, Green and Kurland 1970; Garrett et al. 1971; Zimmermann et al. 1972b). These ionic conditions were found to be optimal for the reconstitution of 30S subunits (Traub and Nomura 1968), and it was reasonable to believe that they would also favor interaction of the RNA-binding proteins with their proper binding sites. This assumption has been verified only recently, however, by a systematic investigation of the temperature and solution conditions which promote the specific attachment of proteins S4, S8 and L24 to their respective RNAs (Schulte and Garrett 1972; Schulte, Morrison and Garrett 1974). Criteria of specificity will be discussed later in this section.

K^+.   Concentrations of KCl between 0.25 and 0.35 M are optimal for the specific binding of ribosomal proteins to rRNA. Variation of the KCl concentration from 0–0.4 M does not significantly affect the amount of protein S4 that binds to 16S RNA (Schulte, Morrison and Garrett 1974), but since it has been found that S4 associates with 23S RNA and tRNA as well as 16S RNA at KCl concentrations below 0.01 M (Zimmermann, unpublished results), nonspecific interactions probably play a role at the lower ionic strengths. Moreover, appreciable nonspecific binding of protein S8 has been shown to occur at reduced KCl concentrations (Schaup, Green and Kurland 1970; Schulte and Garrett 1972), and S5, a protein that does not independently interact with ribosomal RNA at high KCl concentrations, binds to the 16S RNA as well under these conditions (Schaup, Green and Kurland 1970; Zimmermann, unpublished results). The lack of specificity at low KCl concentration is presumed to result from increased electrostatic interaction between protein and RNA and may explain why the KCl dependence for 30S subunit reconstitution exhibits a relatively sharp maximum at about 0.30 M (Traub and Nomura 1969).

Mg^{++}.   The interaction of proteins S4 and S8 with 16S RNA and of protein L24 with 23S RNA requires a minimum Mg^{++} concentration of 0.01 M and is optimal at 0.02 M or higher (Schulte and Garrett 1972; Schulte, Morrison and Garrett 1974). Lowering the Mg^{++} concentration from 0.01 to 0.001 M not only produces a striking reduction in the amount of RNA–protein complex formed, but is accompanied both by a significant decrease in the sedimentation coefficient of the 16S RNA (Schulte, Morrison and Garrett 1974) and by a sharp rise in its sensitivity to RNase (Muto et al. 1974). There is no detectable disruption of RNA base pairs during this transition, however (Schulte, Morrison and Garrett 1974). Together these data strongly suggest that high Mg^{++} ion concentrations contribute to binding site stability by inducing the RNA to assume a more compact configuration.

pH.   The optimum pH for the binding of S4 and S8 to 16S RNA and of L24 to 23S RNA has been determined to lie between 7.4 and 7.9 (Schulte and Garrett 1972; Schulte, Morrison and Garrett 1974). Both RNAs were found to aggregate at low pH values and to be hydrolyzed at pH values above 9.0.

Temperature. The temperature dependence of rRNA–protein interaction indicates that maximum binding levels are attained by heating the components to between 35 and 45°C for 30 min (Schulte and Garrett 1972; Schulte, Morrison and Garrett 1974). The complexes themselves, however, were shown to be more stable at lower temperatures once the heating step had been carried out. Although the amount of protein S4 that remains bound to 16S at 40°C is only 20% less than at 5°C, the protein S8–16S RNA complex is four to five times less stable at the higher temperature. These differences imply that the stability, and hence the structure, of the binding sites for S4 and S8 are quite different. Prior heating of the components to 40°C may be necessary only to induce either the protein or the RNA molecules to assume a conformation more favorable for binding (Schulte, Morrison and Garrett 1974; Muto and Zimmermann 1974).

Analysis of RNA–Protein Complexes

A wide variety of methods have been used to fractionate RNA–protein complexes and to determine the quantity of RNA and protein which they contain. The complexes can be conveniently separated from unbound protein and, in the case of RNase-digested material, from one another by sucrose gradient centrifugation (Mizushima and Nomura 1970; Schaup, Green and Kurland 1970; Zimmermann et al. 1972b), gel filtration (Schaup, Green and Kurland 1970; Garrett et al. 1971), or by polyacrylamide gel electrophoresis (Schaup, Green and Kurland 1970; Gray et al. 1973; Branlant et al. 1973). The use of radioactively labeled components permits accurate calculation of the amounts of RNA and protein in a complex from their specific activities and facilitates the study of very small quantities of material. Two procedures for the quantitative analysis of nonradioactive complexes have also been elaborated (Stöffler et al. 1971a,b; Garrett et al. 1971). In both cases, RNA was measured by absorbance at 260 nm. Protein was then estimated either by quantitative immunoprecipitation with specific antisera or by quantitative staining with Coomassie brilliant blue after the complex had been electrophoresed into a polyacrylamide gel.

The detection of nucleic acid–protein complexes by entrapment on nitrocellulose filters has not proved generally suitable for use with intact rRNAs, despite its successful application to many other systems. Thus, although individual RNA-binding proteins of the 30S subunit are retained by the filters, 16S RNA–protein complexes are washed through into the filtrate under a variety of conditions (A. Muto, G. A. Mackie and R. A. Zimmermann, unpublished results). The complex containing 5S RNA, 23S

RNA and certain proteins of the 50S subunit appears to be an exception, however, and can be recovered quite efficiently by the membrane filter technique (Gray et al. 1972). In addition, this method has been successfully employed for the detection of interaction between ribosomal proteins and small specific rRNA fragments (R. A. Garrett, personal communication).

The use of affinity chromatography in the study of ribosomal RNA–protein interactions has been reported by Gyenge, Spiridonova and Bogdanov (1972). When protein S20 was reacted with cyanogen bromide-activated Sepharose, a high coupling yield was obtained. The Sepharose-bound S20 was active in binding and relatively fastidious, exhibiting a tenfold greater affinity for 16S RNA than for 23S RNA. This technique should be particularly well suited to the isolation of RNA fragments containing binding sites for the immobilized protein.

Finally, electron microscopy has permitted the direct visualization of RNA–protein complexes. (Nanninga et al. 1972a,b). Denaturation of the ribosomal RNA with 80% dimethylsulfoxide allowed it to be spread on carbon-coated grids in the absence of basic proteins (Nanninga et al. 1972a). Under these conditions, the RNA molecules appear as long fine threads. When the same technique was applied to complexes between 16S RNA and protein S4, condensed blobs with threadlike tails were observed (Nanninga et al. 1972b). The overall length of these structures was only 50–60% that of free 16S RNA, suggesting that a portion of the RNA chain had been organized around the protein molecules. Despite the strongly denaturing conditions used, control experiments demonstrated the ribonucleoprotein complex to be stable throughout sample preparation.

Criteria of Specificity

The following criteria have been used to establish that RNA-binding proteins interact with specific sites on the rRNA:

(1) A given protein should interact with only one kind of RNA in the presence of others (Garrett et al. 1971). This demonstrates that the ability to bind the protein is embodied in the structure of a particular RNA.

(2) If the RNA contains specific protein binding sites, the molar ratio of protein and RNA in the complex should reach a plateau value in the presence of excess protein (Schaup, Green and Kurland 1970; Garrett et al. 1971). Furthermore, if there is only one such site, the saturation value should not surpass 1:1 (see Figure 1). When a large molar excess of protein must be added to the incubation in order to attain a 1:1 complex, heterogeneity or contamination of the protein fraction may be indicated. By contrast, binding ratios significantly less than 1:1 can be taken as evidence that not all RNA molecules in the population can interact with the protein. Such results might also suggest a low affinity of the protein for

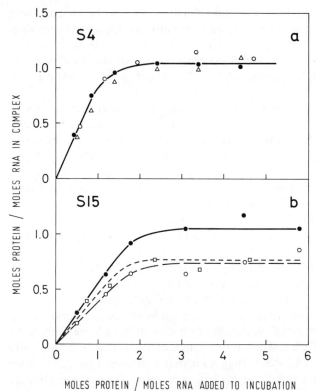

Figure 1   Saturation curves for 30S subunit proteins S4 and S15. (*a*) Inter-action of protein S4 with 16S RNA (•) and with 12S (○) and 9S (△) frag-ments prepared from the 16S RNA by limited T_1 and pancreatic RNase diges-tion, respectively. (*b*) Interaction of protein S15 with 16S RNA (•) and with 12S (○) and 5S (□) fragments isolated from T_1 and pancreatic RNase digests of the 16S RNA, respectively. Procedures for the preparation and analysis of protein–RNA complexes and for the calculation of molar protein:RNA ratios have been described by Muto et al. (1974).

the RNA, but binding experiments are usually performed at component concentrations of 10^{-6} M or above where the equilibrium should strongly favor complex formation if both the RNA and the proteins are active.

(3) The interaction of one protein with the RNA should not be hindered by the presence of another (Schaup, Green and Kurland 1970; Garrett et al. 1971). This condition implies that the two proteins bind to different regions of the RNA. It should be noted, however, that even when two proteins do not directly compete for a given binding site, one may still inhibit or stimulate the binding of a second by other mechanisms.

(4) The protein component of a specific RNA–protein complex should

be directly incorporated into the ribosomal subunit structure without prior dissociation upon addition of an unfractionated mixture of subunit proteins. This requirement has been invoked in connection with studies on the reconstitution of 30S subunits (Mizushima and Nomura 1970; Held et al. 1974).

The first three criteria can be applied as well to the interaction of individual proteins with isolated rRNA fragments (Zimmermann, Muto and Mackie 1974). In this case, it is necessary to modify the first requirement to state that a given protein should interact only with fragments arising from a limited portion of the RNA and not with several fragments from widely separated portions of the RNA chain, assuming that each binding site is localized within a contiguous segment of the molecule.

INTERACTION OF RIBOSOMAL PROTEINS WITH RIBOSOMAL RNA

Interaction of 30S Subunit Proteins with 16S RNA

Independent investigations in four different laboratories have demonstrated that proteins S4, S7, S8, S13, S15, S17 and S20 of the *E. coli* 30S subunit are individually able to bind to 16S RNA (Mizushima and Nomura 1970; Schaup, Green and Kurland 1970, 1971; Nashimoto et al. 1971; Garrett et al. 1971; Zimmermann et al. 1972a; Muto et al. 1974; Held et al. 1974). All of these studies concur in the conclusion that proteins S4, S7, S8, S15 and S20 interact with specific binding sites in the 16S RNA molecule. Stoichiometric measurements, performed under a variety of conditions, showed that complexes containing proteins S4, S8, S15 and S20 become saturated at molar protein:RNA ratios of 0.8:1 to 1:1; the interaction of protein S7 with 16S RNA appears to be less stable, with molar binding ratios falling between 0.6:1 and 0.7:1 (Schaup, Green and Kurland 1970, 1971; Garrett et al. 1971; Muto et al. 1974). Saturation curves for proteins S4 and S15 are illustrated in Figure 1.

The binding properties of proteins S13 and S17 are subject to a greater diversity of interpretation. The interaction of protein S13 with 16S RNA was described by Mizushima and Nomura (1970) and confirmed by Zimmermann et al. (1972a) but has not been detected in two other laboratories (Schaup, Green and Kurland 1971; Garrett et al. 1971). Although results with this protein are variable, an analysis of binding stoichiometry suggests that its interaction with the RNA is a specific one (Muto et al. 1974). The association of protein S17 with 16S RNA was originally judged to be nonspecific because the molar binding ratio surpassed 2:1 in the presence of excess protein without attaining a saturation plateau (Schaup, Green and Kurland 1970) and because S17 was reported to bind to 23S RNA (Garrett et al. 1971). Recent studies have led to different conclusions, however, and there is now substantial evidence that S17 binds only to 16S RNA and that the complex becomes saturated in a normal fashion (Muto et al. 1974;

Held et al. 1974). Furthermore, S17–16S RNA complexes can be incorporated directly into 30S particles without dissociation of the protein when they are incubated with a complete mixture of 30S subunit proteins (Held et al. 1974).

Protein binding sites on the 16S RNA are independent in the sense that none of the RNA-binding proteins directly competes for the site occupied by another (Schaup, Green and Kurland 1970; Garrett et al. 1971). Despite the lack of direct competition, however, one protein may still be able to indirectly influence the attachment of a second to the RNA. Several examples of cooperative stimulation among the RNA-binding proteins have in fact been reported (Mizushima and Nomura 1970; Schaup, Green and Kurland 1970; Garrett et al. 1971; Green and Kurland 1973). By contrast, it has been found that protein S4 strongly inhibits the binding of protein S20 to the 16S RNA under certain conditions (Mackie and Zimmermann 1974a). Proteins S4 and S20 clearly do not compete for a common binding site since the addition of proteins S16 and S17 to the complex completely reverses the inhibition. In addition, all four proteins have been shown to bind simultaneously to a 500-nucleotide segment of the RNA (Zimmermann 1974). Thus although these proteins must all occupy independent sites in the complex, the potential for inhibitory or stimulatory interaction among them is presumed to arise from their proximity in the complex.

Cooperative interactions can be accounted for by assuming that the attachment of a protein imposes specific conformational constraints on portions of the RNA chain adjacent to, as well as within, the actual binding site. If some of the adjacent sequences contribute to the binding site of a second protein, their ability to interact with that protein could be either enhanced or reduced by the particular constraints imposed. The addition of other proteins to the complex might further modify the affinity of one component for another. Although cooperativity may also depend on other factors, such as protein–protein interactions, changes in RNA structure have been found to accompany protein–RNA interaction in a few cases. The attachment of protein S4 to 16S RNA, for instance, leads to alterations in the RNase susceptibility of specific bonds within its binding site (Mackie and Zimmerman 1974a). This interaction also leads to the dissociation of the fluorescent dye ethidium bromide from 16S RNA, a phenomenon that generally reflects changes in nucleic acid secondary structure (Bollen et al. 1970).

The existence of specific protein binding sites in the rRNA can undoubtedly be ascribed to the particular secondary structure and conformation assumed by the RNA molecule as well as to its base sequence. The importance of secondary structure is evident from the fact that 16S RNA loses its ability to bind proteins S4 and S8 when heated to a temperature at which base pairs are disrupted (Schulte and Garrett 1972; Schulte, Morrison and Garrett 1974). The sensitivity of 16S RNA to conformational alterations is illustrated by the following observations: Dialysis of 16S RNA against distilled water eliminates its capacity to interact with proteins S4, S7, S8, S15 and S20. Restoration of Mg^{++} ions to the solution permits the RNA to associate with S8 and S15, but not with S4, S7 and S20, at $0°C$. However if the partially "denatured" RNA is first heated to $40°C$ in the presence of Mg^{++} and then chilled, it binds normal amounts of S4, S7 and S20 at $0°C$ (Muto and Zimmermann 1974). The reversible transition from the "denatured" to the active form produces no detectable changes in the secondary structure of the RNA; however it is accompanied by a decrease in RNase sensitivity and an increase in sedimentation coefficient, both of which indicate that the RNA undergoes a shift in conformation. The concomitant changes in the ability to accept proteins S4 and S20 are consistent with the proximity of their binding sites on the 16S RNA molecule. Schulte, Morrison and Garrett (1974) also noted that the amount of S4 bound to 16S RNA increases as a function of incubation temperature and concluded independently that this change was correlated with changes in RNA conformation.

Interaction of 50S Subunit Proteins with 23S RNA

Proteins L1, L2, L3, L4, L6, L13, L16, L20, L23 and L24 of the *E. coli* 50S subunit have each been found to interact directly and specifically with 23S ribosomal RNA (Stöffler et al. 1971a,b; Garrett et al. 1974; Spierer, Zimmermann and Mackie 1974). Binding stoichiometry was measured in each case by the immunological, electrophoretic and sucrose gradient methods (see earlier discussion), and a detailed analysis of the solution conditions optimal for the attachment of L24 has been made (Schulte, Morrison and Garrett 1974). Very weak binding of protein L18 was also detected by the electrophoretic technique, but the stoichiometry of the complex has not been determined (Gray et al. 1973). Interactions were judged to be specific when a given protein bound exclusively to 23S RNA in the presence of 16S RNA and when saturation of the complex occurred at a molar protein:RNA ratio of 1:1.5 or less. The influence of one protein upon the binding of others has not been systematically investigated for 50S subunit components.

RNA–protein interactions among 50S subunit components can be classified in two groups according to binding stoichiometry at saturation (Garrett et al. 1974). Individual complexes of proteins L1, L2, L3, L13, L20, L23 and L24 with 23S RNA exhibit molar protein:RNA ratios ranging from 0.8:1 to 1.4:1; that is, within normal experimental error, they contain about one molecule of protein per molecule of RNA. By contrast, proteins L4, L6, L16 and, in some circumstances, L2 saturate their binding sites at molar ratios between 0.2:1 and 0.3:1, even in the presence of a

10- to 15-fold excess of protein. In certain cases, the low molar binding ratios were attributed to dissociation of the complex during fractionation. Thus the binding ratio for protein L2 was 1.5:1 when measured by the electrophoretic method but only 0.3:1 when measured by sedimentation. Alternatively, low stoichiometric ratios may reflect the unavailability of binding sites for the corresponding protein in a substantial fraction of the RNA population. This appears to be so for the L4–23S RNA complex, since the molar binding ratio was 0.3:1 under normal conditions but could be as high as 0.7:1 when freshly prepared RNA was used. The limited interaction of both L6 and L16 with the 23S RNA may result from binding site heterogeneity as well.

Considerable variability in the binding properties of 50S subunit proteins has also been observed (Garrett et al. 1974). Successive preparations of the same proteins frequently differed from one another and in some instances, these changes appeared to be related to the purification procedures used. In addition, the high protein:RNA input ratios required to saturate several of the complexes suggested that not all protein molecules in a given preparation were homogeneous in their capacity to interact with 23S RNA.

The association between protein L24 and 23S RNA in the native subunit is of an unusual nature. Although free L24 is readily digested by trypsin, prolonged exposure of 50S particles to the enzyme leaves only L24 of all the subunit proteins almost entirely intact (Crichton and Wittmann 1971). It was suggested that bound L24 is protected against trypsin digestion by virtue of changes in its structural organization when in the complex or by the RNA chain. Protein S20, a 30S subunit component whose molecular characteristics are similar in many ways to those of L24, is also relatively resistant to trypsin digestion when associated with the 16S RNA (R.A. Garrett, personal communication).

An exceptionally stable interaction between *Bacillus stearothermophilus* protein B-L3 and its homologous 23S RNA has also been found in the native subunit structure (Fahnestock, Erdmann and Nomura 1973). Protein B-L3 remained attached to the 23S RNA upon dissociation of the 50S particles with 4 M urea–2 M LiCl, although all other proteins were quantitatively removed. The complex could be disrupted only by the use of extreme conditions: 4 M urea and 0.5 M Mg^{++} at pH 2.0. The equivalent protein in *E. coli,* identified immunologically as L2 (Garrett and Wittmann 1973; Tischendorf, Geisser and Stöffler 1973), can be removed from *E. coli* 23S RNA by the normal urea–LiCl treatment. Unlike most of the RNA-binding proteins of the 30S subunit, protein B-L3 is not required for the attachment of any other protein to the 50S particle but plays an important role in its biological activity (Fahnestock, Erdmann and Nomura 1973). Thus certain ribosomal proteins appear to interact strongly with RNA for purposes other than contributing to physical assembly.

Interaction of 50S Subunit Proteins with 5S RNA

Proteins L5, L18 and L25 of the *E. coli* 50S subunit individually bind to homologous 5S RNA (Gray et al. 1973; Yu and Wittmann 1973). Similar interactions occur among the corresponding components of the *B. stearothermophilus* 50S subunit (Horne and Erdmann 1972). In the presence of L18 and either L5, L6 or L25, complexes containing both 5S and 23S RNAs are formed (Gray et al. 1972, 1973; J. Feunteun, R. Monier and R. A. Garrett, personal communication). The molar ratio of L18 to L25 in such particles was found to be 2:1 (Gray et al. 1973). It is known that L25 binds only to 5S RNA and that L6 binds only to 23S RNA, whereas L18 has been reported to interact with both (Gray et al. 1973). Protein L18 may therefore play a critical role in promoting association of the two RNAs. However since L18 binds only weakly to 23S RNA (Gray et al. 1973; Garrett et al. 1974), additional stability may be conferred by protein–protein interactions or by changes in RNA or protein conformation which occur upon their mutual attachment. Together these interactions appear to mediate the integration of 5S RNA into the 50S subunit structure (Reynier and Monier 1968; Erdmann et al. 1971; Yu and Wittmann 1973). The properties of 5S RNA are discussed in greater detail elsewhere in this volume (see chapter by Monier).

Interaction of 30S Subunit Proteins with 23S RNA

Proteins S11 and S12 have each been shown to bind to 23S RNA alone when incubated with a mixture of 16S and 23S RNAs (Stöffler et al. 1971a; Morrison et al. 1973). At present, the association of S12 with 23S RNA must be regarded as nonspecific by stoichiometric criteria. By contrast, the interaction between S11 and the 23S RNA appears to be specific since the molar protein:RNA ratio of the complex attained a plateau value of 0.6–0.7:1 at saturation (Morrison et al. 1973). This result must be interpreted with caution, however, because saturation required a molar protein:RNA input ratio of more than 20:1, implying either that less than 5% of the protein in the S11 preparation was active in binding or that the complex was very unstable under the conditions of analysis. These data are nonetheless of considerable interest, both because small amounts of S11 and S12 have been identified in dissociated 50S subunits by immunological techniques, and because monovalent Fab antibody fragments directed against S11 completely block, and those against S12 partially block, the reassociation of 30S and 50S subunits (Morrison et al. 1973). Together these findings suggest that at least protein S11 is located at the 30S:50S subunit interface and that it may help stabilize the 70S ribosome via direct interaction with a specific binding site in the 23S RNA.

Heterologous Interactions

Hybrid 30S subunits have been reconstituted with ribosomal components derived from *E. coli, Azotobacter vinelandii* and *B. stearothermophilus* (Nomura, Traub and Bechmann 1968). Particles containing *E. coli* proteins and 16S RNA from either of the two other bacteria were shown to be as active in polypeptide synthesis as those containing *E. coli* RNA. Functional particles were also formed from *E. coli* RNA and heterologous proteins, although their activity was only 40–60% that of subunits reconstituted from homologous components. These results suggested that there was substantial functional homology among the ribosomal constituents of both closely and distantly related bacteria and in particular, that structures mediating interaction between RNA and protein in the corresponding ribosomal particles were similar. However, since the 16S RNAs differed in nucleotide composition and base sequence and since the proteins displayed significant electrophoretic differences, it was concluded that only the particular regions of the 16S RNA and of the ribosomal proteins required for mutual interaction need be conserved (Nomura, Traub and Bechmann 1968). Support for this view was recently provided by the demonstration that functionally equivalent counterparts for most proteins of the *E. coli* 30S subunit, including all those that bind directly to 16S RNA, could be identified among the 30S proteins of *B. stearothermophilus* (Higo et al. 1973). For several of the homologous pairs, no immunological cross-reaction could be detected.

Although the structural basis for the remarkable compatibility among heterologous ribosomal components may reside in short "conserved" nucleotide and amino acid sequences, it could as well be attributed to the presence of common features of secondary structure or, indeed, to similarities in overall conformation, neither of which would require strict sequence homology. In any case, corresponding components from different bacteria must have some structural properties in common, and the study of heterologous RNA–protein interactions should be of use in the analysis of the specific portions of both RNA and proteins that mutually interact. The potential value of this approach can be inferred from the work of Garrett et al. (1971), in which it was shown that *E. coli* proteins S4, S7, S8, S15 and S20 can bind simultaneously to *B. stearothermophilus* 16S RNA. The binding sites for S4, S7 and S15 must be highly conserved since the interaction of these proteins with the heterologous RNA is nearly stoichiometric. Proteins S8 and S20 appear to have less affinity for the heterologous sites, however.

A more comprehensive study has recently been carried out on the interaction of *E. coli* S4 and L24 with 16S and 23S RNAs, respectively, derived from several members each of the Enterobacteriaceae, Bacillaceae and Pseudomonadaceae (Daya-Grosjean et al. 1973). Binding efficiency and specificity were evaluated by determining the molar protein:RNA ratios of the heterologous complexes at saturation. As expected, interaction of the *E. coli* proteins was strongest with RNAs from closely related bacteria of the Enterobacteriaceae, although both also bound well to the RNAs of two Bacilli, suggesting a high degree of binding site homology in these cases as well. The RNA of other Bacilli, and of the Pseudomonads, bound the *E. coli* proteins less efficiently or in several instances, not at all. For a given bacterium, the extent of binding of S4 to 16S RNA was generally well correlated with binding of L24 to 23S RNA, indicating a similar degree of evolutionary change in both kinds of rRNA molecules.

Binding site specificity has also been conserved in the 5S RNA of *E. coli* and *B. stearothermophilus* (Horne and Erdmann 1972). When tested with homologous proteins, *E. coli* 5S RNA interacts mainly with L18 and L25, and *B. stearothermophilus* 5S RNA exclusively with B-L5 and B-L22. Analogous results were obtained when each 5S RNA was incubated with an unfractionated mixture of 50S proteins from the other species: *E. coli* 5S RNA bound only B-L5 and B-L22, whereas *B. stearothermophilus* 5S RNA bound L18, L25 and some L5 as well. These findings are consistent with the observation that in the reconstitution of *B. stearothermophilus* 50S subunits, *E. coli* 5S RNA can efficiently substitute for homologous 5S RNA (Wrede and Erdmann 1973).

As indicated above, one would expect that equivalent protein binding sites in different bacterial RNAs should possess a high degree of similarity either in primary structure or in topography. Accordingly, it should be possible to isolate fragments containing such sites by binding a given protein to different RNAs and digesting the heterologous complexes with RNase. Nucleotide sequence analysis of the fragments could help to define which features of the binding site are conserved across species lines. Elucidation of functionally homologous regions in the RNAs of several strains should also provide information on the evolutionary constraints to which protein binding sites have been subject.

LOCATION OF RIBOSOMAL PROTEIN BINDING SITES

Isolation and Characterization of Protein Binding Sites

RNA fragments containing binding sites for individual proteins of the *E. coli* ribosome have been isolated and characterized by two general methods. In the first, partial RNase digestion is used to fragment either free rRNA or rRNA–protein complexes and the products are fractionated by sucrose gradient centrifugation. Each discrete RNA component is then extracted and tested for its ability to interact with RNA-binding proteins. Finally, its nucleotide sequence is analyzed by fingerprinting (Schaup et al. 1971, 1973; Zimmermann et al. 1972a,b; Muto et al. 1974). In the second method, a two-step electrophoretic procedure is used for the isolation of specific RNA fragments. Complexes between ribosomal RNA and single proteins are hydrolyzed with RNase and the products are separated on polyacrylamide gels under conditions in which protein–RNA fragment complexes remain stable. The ribonucleoproteins are next eluted and subjected to gel electrophoresis in 6–8 M urea, which dissociates the protein and further fractionates the RNA fragments according to size. The resulting bands, which can be repurified under similar conditions if necessary, are then characterized by fingerprint analysis (see Branlant et al. 1973). Because extensive information on the primary sequence of the rRNAs is now available (Fellner et al. 1972; Fellner, Ehresmann and Ebel 1972; Ehresmann et al. 1972; Branlant et al. 1973), it has been possible to define the structural features of a variety of different protein-specific fragments and to determine their positions within the rRNA molecules.

It should be noted that in the electrophoretic technique, only those fragments which are larger than any produced in parallel digests of free 16S RNA are selected for further analysis. This procedure ensures the isolation of fragments that are protected from RNase digestion, at least to some extent, by the presence of bound protein. By contrast, both protected and unprotected fragments are recovered by the sedimentation method. The basis of protection by bound protein, which in some cases extends to sequences of several hundred nucleotide residues, is not completely understood, although it may arise from masking by the protein of a small number of normally labile single-stranded segments that link a series of relatively resistant base-paired loops (Branlant et al. 1973).

In order to establish that an isolated RNA fragment contains the binding site for a particular protein, it is important to know that specific interaction occurs between them. The specificity criteria used are similar to those applied to intact rRNA. (1) Assuming that binding sites are located within limited portions of the RNA molecule, a given protein should not bind to fragments from widely separated regions of the RNA, although it may of course interact with a series of overlapping fragments from the same general region. (2) When increasing amounts of protein are incubated with a fixed quantity of an RNA fragment, the molar ratio of protein and RNA fragment in the complex should reach a plateau value at saturation not in excess of 1:1 (see Figure 1). A lower molar binding ratio may indicate that a fraction of the binding sites have been inactivated during RNase digestion.

RNA fragments that retain the ability to specifically interact with individual ribosomal proteins are not necessarily limited to the sequences actually in contact with the protein molecule. Although these sequences must of course be present, a given fragment may also include portions of the nucleic acid chain required to maintain them in their proper configuration. In this sense, the meaning of "binding site" must be broadened to encompass both the sequences that directly attach to the protein and those elements of secondary and tertiary structure that contribute to their overall conformation. Nonetheless, since some of the fragments may contain non-essential segments in addition, they could still be larger than the binding sites as envisaged in the present interpretation.

The nucleases that have proved most useful for the digestion of ribosomal RNA are RNase T_1 and pancreatic RNase A. Both enzymes are functional at the high K^+ and Mg^{++} concentrations necessary for the formation and stability of specific RNA–protein complexes, yet the activity of each is somewhat curtailed in this ionic environment. Several other RNases, particularly those inhibited by Mg^{++} ions, have been found unsuitable, presumably because the removal of Mg^{++} ions during digestion also disorganizes the RNA structure (Muto et al. 1974; Muto and Zimmermann 1974; R. A. Zimmermann, unpublished results).

Location of Protein Binding Sites on 16S RNA

The unique distribution of 30S subunit binding sites on the 16S RNA was initially inferred from the following observations: Hydrolysis of the 16S RNA with T_1 RNase produced two large fragments of 12S and 8S, which proved to be contiguous sequences representing about 95% of the RNA molecule (Zimmermann et al. 1972b; Muto et al. 1974). The 12S RNA contains a sequence of about 900 nucleotides that begins a few residues from the 5' terminus of the 16S RNA in section L and extends to section O; the 8S RNA encompasses 600 nucleotides and runs from section O' through section A (see Figure 2). Proteins S4, S8, S15 and S20 all bound individually to purified 12S RNA and the specificity criteria were met in each case (Zimmermann et al. 1972b; Zimmermann, Muto and Mackie 1974). The saturation of the 12S fragment with proteins S4 and S15 is depicted in Figure 1. One or both proteins of the S16 + S17 mixture associated with the 12S RNA as well when the other four proteins were present. When complexes between the 16S RNA and either protein S13 or S7 were digested under these conditions, S13 was retained by the 12S RNA and S7 by the 8S RNA. However, neither protein was able to independently reassociate with their respective fragment. Efforts to further delimit protein binding sites within the 16S RNA are described below and a summary scheme which shows the positions of several protein-specific fragments is presented in Figure 2.

The location of the binding site for protein S4 was more precisely delineated by several different experiments (Schaup et al. 1971; Schaup and Kurland 1972; Zimmermann et al. 1972b, 1974). By means of the sedimentation technique, a fragment of 9S was recovered following pancreatic RNase digestion of free 16S RNA or of the S4–16S RNA complex (Zim-

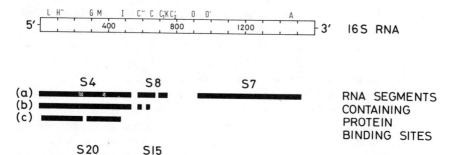

Figure 2   Location of protein binding sites on the 16S RNA. The scheme shows the relative positions within the 16S RNA molecule of RNA fragments capable of specific interaction with five RNA-binding proteins of the 30S subunit. Numbers within the 16S RNA denote distances in nucleotide residues from the 5' terminus of the molecule. Upper case letters refer to sections of the 16S RNA that define the termini and extent of the various fragments (see Figure 1 in Fellner, this volume, for complete nucleotide sequence). Shaded areas within the fragments indicate oligonucleotides which are either present in reduced yield or absent.

S4. (a) 9S fragment produced by pancreatic RNase digestion of 16S RNA or of protein–RNA complexes. (b) Protected fragment isolated from T_1 RNase digest of S4–16S RNA complex. (c) Two fragments isolated by polyacrylamide-urea gel fractionation of the 12S RNA which simultaneously interact with the protein.

S20. (a) Same as for S4. (b) Protected fragment recovered from T_1 RNase digest of S20–16S RNA complex.

S8. (a) 4S–5S RNA resulting from T_1 or pancreatic RNase digestion of 16S RNA or of protein–RNA complexes. (b) Protected fragment recovered after pancreatic RNase digestion of S8–16S RNA complex.

S15. (a) Same as for S8. (b) Protected fragment isolated from pancreatic RNase digest of S15–16S RNA complex.

S7. (a) 8S fragment produced from 16S RNA by T_1 RNase digestion.

mermann et al. 1972b). The 9S RNA encompasses 500 nucleotides and extends from section L to section C" in the 5'-terminal region of the 16S RNA (Figure 2). This fragment interacts specifically with protein S4 (Figure 1) to form a complex containing one mole protein per mole RNA (Muto et al. 1974; Zimmermann, Muto and Mackie 1974). Virtually identical sequences have been isolated from ribonucleoprotein fragments produced by T_1 or pancreatic RNase digestion and prepared by the electrophoretic technique (R. A. Garrett, personal communication; Zimmermann et al. 1974; Mackie and Zimmermann 1974b). Although in all of these cases the RNAs contained a number of "hidden breaks" at specific locations within the nucleic acid chain, the ability of the fragments to reassociate with protein S4 was not impaired (Figure 3). Consistent with these studies, electron micrographs of the S4–16S RNA complex have shown the protein to be located at one end of the RNA molecule, enveloped by an RNA segment of roughly 700 nucleotides (Nanninga et al. 1972b). It was impossible to determine the polarity of the RNA chain in the electron microscopy experiments, but the terminal location of the protein and the amount of RNA associated with it are in good agreement with conclusions about the S4 binding site drawn from the characterization of S4-specific RNA fragments.

It is not known how much of the 500-residue 5′-terminal sequence is required for interaction with protein S4, but it is possible that a substantial amount of it is necessary for maintaining the protein binding site in its proper configuration. The RNA chain is believed to be folded into 15 or more hairpin loops in this part of the molecule (see Figure 2 in Fellner, this volume), and it has been proposed on the basis of indirect evidence that the protein associates with multiple binding sites within this region

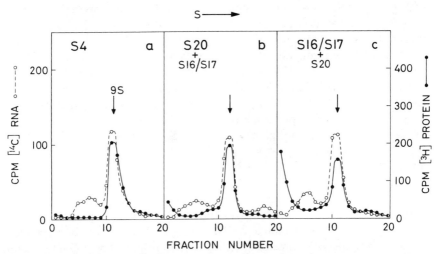

Figure 3   Interaction of proteins S4, S20 and S16 + S17 with isolated 9S RNA. The 9S fragment was produced by pancreatic RNase digestion of a complex containing ^{14}C-labeled 16S RNA and proteins S4, S20, S16 and S17, isolated by sucrose gradient sedimentation and deproteinized by phenol extraction. Purified [^{14}C]9S RNA was then incubated in TMK buffer (0.05 M Tris · HCl, pH 7.6–0.02 M $MgCl_2$–0.35 M KCl-0.006 M β-mercaptoethanol) with a slight molar excess of ^{3}H-labeled (*a*) S4, (*b*) S20 and (*c*) S16 + S17; samples (*b*) and (*c*) contained in addition a threefold molar excess of unlabeled S16 + S17 and S20, respectively. The reaction mixtures were incubated 30 min at 0°C and then separated by sucrose gradient centrifugation. After fractionation of the gradients, the ^{14}C and ^{3}H radioactivity in each tube was assayed by scintillation counting as described by Zimmermann et al. (1972b).

(Schaup and Kurland 1972; Zimmermann et al. 1972b). This suggestion has recently been confirmed by the demonstration that protein S4 can interact simultaneously with two RNA fragments, representing sections L through G and M through I (Mackie and Zimmermann 1974a; Zimmermann et al. 1974), that is, with two separate subsequences derived from the 500 5'-terminal residues (Figure 2). This result clearly shows that protein S4 interacts with at least two distinct sites in the 16S RNA molecule.

The binding site for protein S20 is located within the same region of the 16S RNA as that for protein S4, since S20 too has been found to interact with the 500-nucleotide 9S fragment (Figure 3). The stability of the protein S20–9S RNA complex during sucrose gradient centrifugation was found to depend upon the presence of one or both proteins of the S16 + S17 mixture, however, implying that the site of attachment for either or both of these proteins lies within the 9S fragment as well (Mackie and Zimmermann 1974a). Nonetheless, the interaction of S20 with isolated 9S RNA is a specific one under these conditions. A more precise localization of the S20 binding site has been made by the electrophoretic method (Figure 4). Separation of RNase T_1 digests of the S20–16S RNA complex on polyacrylamide gels yielded a series of protected RNA fragments which together comprise roughly 300 residues and extend from section H'' to section M of the 16S RNA (R. A. Garrett, personal communication; Zimmermann et al. 1974). After deproteinization, these fragments were shown to specifically reassociate with protein S20 by means of the membrane filter assay. Thus the region protected by protein S20 corresponds to the middle portion of the segment that binds protein S4 (Figure 2), suggesting an intimate association of proteins S4, S20 and, possibly, S16 or S17, both among themselves and with the RNA chain.

The binding sites for proteins S15 and S8 appear to be even more closely related to one another in the RNA sequence than those for S4 and S20 (see Figure 2). When the S15–16S RNA complex is partially hydrolyzed with either T_1 or pancreatic RNase (Figure 5), the protein is retained by RNA fragments sedimenting at 4S or 5S, respectively (Zimmermann et al. 1972b; Muto et al. 1974). If protein S8 is present together with protein S15 in the original complex, it too remains bound to the 4S–5S RNA (Figure 5). The 5S pancreatic RNase product has been more extensively characterized. After deproteinization, purified 5S RNA can specifically reassociate with S15 (Figure 1) and can selectively bind S8 and S15 when incubated with an unfractionated mixture of 30S subunit proteins (Zimmermann and Bergmann 1974). The 4S–5S fragments produced by T_1 and pancreatic RNases are very similar to each other, encompassing the 140 residues of sections C, C'_1 and C'_2 but it is possible that they differ slightly in conformation. Section C consists of a long hairpin loop with an extensively hydrogen-bonded stem; sections C'_1 and C'_2 which are highly

(a)   (b)   (c)

S20+16S RNA   16S RNA   PF

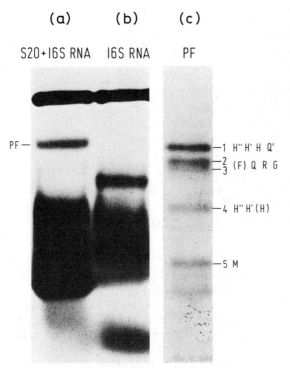

Figure 4 Isolation of RNA fragments protected by protein S20. (*a*) The complex of S20 with ^{32}P-labeled 16S RNA was digested with T_1 RNase at an enzyme:substrate ratio of 1:20 at 0°C in a buffer containing 0.03 M Tris and 0.005 M Mg^{++}. The products were separated by electrophoresis on an 8% polyacrylamide gel in the same buffer. RNA bands were identified by auto-radiography. *PF* marks the band containing the protein and associated RNA fragments. (*b*) Same as (*a*) except that the protein was omitted. Note that there is no band at the position of PF when 16S RNA alone is digested under these conditions. (*c*) Band PF was excised from the gel in (*a*), treated with SDS to dissociate protein-bound RNA, and polymerized into a second gel containing 12% polyacrylamide and 7 M urea. The RNA fragments associated with protein S20 were then fractionated electrophoretically into five bands, and the RNA sequences in each were analyzed by fingerprinting. Letters to the right of each band indicate the sections of the 16S RNA that were found to be present. (Photograph generously supplied by Dr. R. A. Garrett.)

complementary, probably arise from the base-paired stem of an adjoining hairpin loop (see Figure 2 in Fellner, this volume). A fingerprint of the 5S RNA is presented in Figure 6.

The locations of the binding sites for proteins S15 and S8 have been confirmed and further defined by means of the electrophoretic technique (R. A. Garrett, personal communication; Zimmermann et al. 1974). In-

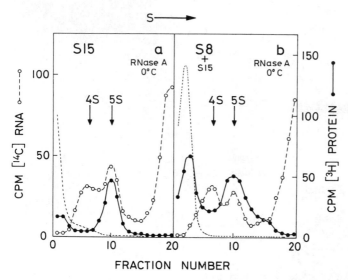

Figure 5 Identification of an RNA fragment containing the binding site for proteins S15 and S8. ^{14}C-labeled 16S RNA was added to 1.5 molar equivalents of (*a*) ^{3}H-labeled S15 and (*b*) ^{3}H-labeled S8 + unlabeled S15 in TMK buffer. Reaction mixtures were incubated 30 min at 40°C, chilled on ice, and treated with pancreatic RNase A at an enzyme:substrate ratio of 1:5 at 0°C. The incubations were then analyzed by sucrose gradient centrifugation as described in the legend to Figure 3. The dotted lines indicate the sedimentation profiles of proteins (*a*) S15 and (*b*) S8 alone under the same conditions.

dividual complexes of each protein with 16S RNA were digested with pancreatic RNase and the products were fractionated on polyacrylamide gels. The RNA associated with protein S15 consisted of two principal subfragments, which were found to arise from sections C, C'_1 and C'_2 (Figure 2). The protected sequences comprise a subset of those contained in the 4S–5S fragments described above and are located within the lower portions of the stems of the two putative hairpin loops (see Figure 2 in Fellner, this volume). The RNA that remained bound to protein S8 after electrophoresis was shown to encompass about 40 nucleotide residues from section C (Figure 2). These sequences derive from two complementary fragments of roughly equal length, which comprise the lower portion of the long base-paired loop believed to occur in that section. Evidence for the protection of a similar region of section C by protein S8 has also been obtained by an independent method (Schaup et al. 1973).

Location of Protein Binding Sites on 23S RNA

Specific fragmentation of the 23S RNA has also been used to position the binding sites for 50S subunit proteins along the nucleic acid chain, and

5S

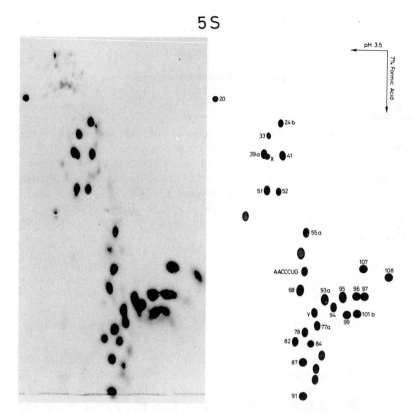

Figure 6 Fingerprint of the 5S RNA fragment. The 5S fragment was produced by pancreatic RNase digestion of a complex between protein S15 and ^{32}P-labeled 16S RNA and isolated by sucrose gradient centrifugation, as described in the legend to Figure 5. After deproteinization with phenol and repurification on a second sucrose gradient, the fragment was completely hydrolyzed with RNase T_1 in the presence of bacterial alkaline phosphatase and fingerprinted by standard procedures (Sanger, Brownlee and Barrell 1965; Brownlee and Sanger 1967). Oligonucleotide spots are identified by number on the accompanying plan according to the system of Fellner, Ehresmann and Ebel (1972).

the results so far obtained are summarized in Figure 7. Partial digestion of *E. coli* 50S subunits with solid-phase pancreatic RNase yields two large fragments of the 23S RNA, which are characterized by sedimentation coefficients of 13S and 18S (Allet and Spahr 1971). Fingerprint analysis revealed that the two products encompass apparently contiguous sequences of 1200 and 2000 nucleotides from the 5′-terminal and 3′-terminal segments of the 23S RNA, respectively, and that they contain all unique sequences present in the original molecule (Spierer, Zimmermann and Mackie 1974). Furthermore, polyacrylamide gel electrophoresis under

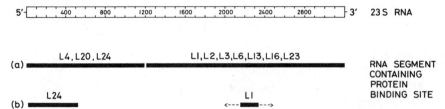

Figure 7 Preliminary map of the binding sites for 50S subunit proteins on the 23S RNA. The sites for ten RNA-binding proteins have been localized as follows: (*a*) Each of the proteins was tested for its ability to specifically interact with the 5′-terminal 13S and 3′-terminal 18S RNA fragments isolated from pancreatic RNase digests of intact 50S subunits. (*b*) Individual complexes of L1 and L24 with the 23S RNA were digested with T_1 RNase. Protein-bound RNA fragments were recovered and purified by the two-step electrophoretic procedure illustrated in Figure 4 and then characterized by fingerprinting (see Branlant et al. 1973). The dashed lines to the right and left of the L1-specific fragment denote the limits of uncertainty in its placement within the molecule. Numbers within the RNA indicate approximate distances in nucleotide residues from the 5′ terminus of the molecule.

denaturing conditions indicated that both fragments contain intact polynucleotide chains uninterrupted by hidden breaks.

Each of the 50S subunit proteins that interact directly with the 23S RNA (see above) was tested individually for its ability to bind to the 13S and 18S RNAs, and the specificity of binding was determined by measurement of stoichiometric ratios at saturation (Spierer, Zimmermann and Mackie 1974). This procedure demonstrated that the binding sites for proteins L4, L20 and L24 occur within the 5′-terminal 13S fragment and that binding sites for proteins L1, L2, L3, L6, L13, L16 and L23 lie within the 3′-terminal 18S fragment (Figure 7). The 18S RNA has also been shown to participate in the formation of a stoichiometric complex with 5S RNA in the presence of proteins L2, L6, L18 and L25 of the *E. coli* 50S subunit (Gray and Monier 1972). Neither the 13S fragment nor the 16S RNA associated to a significant extent with homologous 5S RNA under the same conditions.

The positions of the binding sites for proteins L1 and L24 have been further localized by the isolation of protected RNA fragments following RNase hydrolysis of protein–23S RNA complexes (Figure 7). Protein L1 was found to be associated with a sequence of about 150 nucleotides located between 2000 and 2500 residues from the 5′ terminus of the 23S RNA (C. Branlant and P. Fellner, personal communication). The binding site for protein L24 occurs within the 500 5′-terminal residues of the nucleotide chain (Branlant et al. 1973). Both of these findings are consistent with the results obtained using isolated 13S and 18S fragments (Spierer, Zimmermann and Mackie 1974).

RNA fragments associated with L24 were first prepared from a native ribonucleoprotein complex produced by successive trypsin and RNase digestion of 50S subunits (Crichton and Wittmann 1973). RNA fragments specifically protected by L24 were subsequently isolated by the two-step electrophoretic procedure described in the previous section (Branlant et al. 1973). RNase T$_1$ digests of the L24–23S RNA complex yielded four unique fragments that were not present in digests of free 23S RNA. All of these fragments have been characterized by fingerprinting. A number of other fragments were also recovered, but these were shown to be subsequences of the four main fragments. Together the principal fragments encompass about 500 nucleotides, beginning 15 residues from the 5′ terminus of the 23S RNA. It has not been determined whether the fragments, either individually or in combination, are able to reassociate with protein L24, and it is therefore impossible to know whether the entire sequence is essential for L24 binding. These experiments do, however, define the limits of the binding site for this protein. Since sequence analysis of the L24-specific RNA has not been completed, it is not yet possible to make concrete predictions concerning its secondary structure. Nonetheless, 23S RNA is known to contain about the same percentage of secondary structure as the 16S RNA, and it is probable that the pattern of hairpin loops connected by short single-stranded segments which is thought to prevail in the smaller molecule will also obtain in the larger.

Location of Protein Binding Sites on 5S RNA

Partial pancreatic RNase hydrolysis of complexes containing 5S RNA, 23S RNA and proteins L6, L18 and L25, all from *E. coli,* yields ribonucleoprotein fragments in which sequences 1–11 at the 5′ terminus and 69–120 at the 3′ terminus of the 5S RNA are strongly protected (Gray et al. 1973). Sequences 1–10 and 110–119 are complementary to one another and probably exist as a hydrogen-bonded stem in the native 5S RNA structure, while sequence 69–109 appears to contain three small hairpin loops. When either L18 or L25 is bound individually to 5S RNA, the main products remaining attached to both proteins after digestion arise from sequences 1–11 and 98–120. In addition, protein L18 affords limited protection to sequences between residues 69 and 97, whereas L25 shields very little material from this region. The evidence suggests that both L18 and L25 interact with the double-helical stem formed by the complementary 5′ and 3′ termini of the molecule and with sequence 98–109, while L18 appears to bind to certain sequences between residues 69 and 97 as well. Since nucleotide sequences between positions 12 and 68 in the 5S RNA were never recovered in ribonucleoprotein fragments following RNase treatment, it was concluded that they do not play a significant role in protein–5S RNA interaction.

DISTRIBUTION OF RIBOSOMAL PROTEINS IN RIBONUCLEOPROTEIN PARTICLES

Reconstituted Particles

As we have seen, only about one-third of the ribosomal proteins of *E. coli* possess the ability to bind independently to rRNA. However, proteins that do not *independently* interact with rRNA may nonetheless *directly* associate with specific portions of the nucleic acid chain in the presence of one or more RNA-binding proteins. The delineation of attachment sites for such proteins should consequently contribute to the understanding of overall ribosome structure. In one approach to this question, several different ribonucleoprotein particles were reconstituted in vitro from components of the *E. coli* 30S subunit and then digested with RNase in order to establish the nature of the RNA segments with which particular groups of proteins are associated (Zimmermann, Muto and Mackie 1974; Zimmermann 1974).

Initial studies focused on the distribution of proteins within the reconstitution intermediate (RI) particle. The RI particle as originally defined is the ribonucleoprotein complex that results from the interaction of roughly twelve 30S subunit proteins with 16S RNA in vitro at 0°C (Traub and Nomura 1969); the complex formed under these conditions is referred to in more recent work as the "in vitro 21S particle" (Held and Nomura 1973a). This particle closely resembles naturally occurring intermediates in ribosome assembly (Nashimoto et al. 1971). In addition to the RNA-binding proteins S4, S7, S8, S15 and S20, the RI complex contains proteins S6, S9, S13, S18, S19 and one or both proteins of the S16 + S17 mixture, which were not discriminated by the analytical techniques employed (Nashimoto et al. 1971; Held and Nomura 1973a; Zimmermann, Muto and Mackie 1974). As indicated in Table 1, digestion of RI particles with T_1 RNase produces 14S and 8S subparticles with well-defined RNA and protein compositions (Zimmermann, Muto and Mackie 1974). The 14S subparticle contains proteins S4, S6, S8, S15, S16, S17, S18 and S20 and an RNA fragment consisting of 900 residues from the 5'-terminal portion of the 16S RNA. This fragment was indistinguishable from the 12S RNA isolated from digests of free 16S RNA (Muto et al. 1974). Proteins S7, S9, S13 and S19 were identified in the 8S subparticle, whose RNA component was identical to the 8S fragment found previously to arise from a sequence of 600 nucleotides at the 3' terminus of the 16S RNA (Muto et al. 1974).

A systematic investigation of the binding requirements and the distribution of the RI proteins revealed that they are organized into three independent groups, each of which occupies a specific region of the 16S RNA (Zimmermann 1974). Group I, consisting of proteins S4, S16, S17 and S20, is associated with a 500-residue fragment from the 5' terminus of the 16S RNA. This fragment covers the same sequence as the 9S RNA fragment shown earlier to contain binding sites for S4 and S20 (Zimmer-

Table 1 Protein composition of ribonucleoprotein fragments

Reconstituted RNP fragment	Protein composition[a,b,e]	RNA component[a,b,c]	Native RNP fragment	Protein composition[d,e]	RNA component[f]	Native RNP fragment	Protein composition[g]
14S RI	S4, S6, S8, S15, S16, S17, S18, S20	L–O	13–14	S4, S5, S6, S8, (S11), S15, S16, S17, (S18), S20	—	RNP II	S4, S5, S8, (S13), S15, S16, S17, S20
12S	S4, S8, S15, S16, S17, S20	L–O	—	—	—	—	—
9S	S4, S16, S17, S20	L–C″	—	—	—	—	—
7S	S6, S8, S15, S18	C–O	—	—	—	—	—
5S	S8, S15	C, C$'_1$, C$'_2$	7	S8, S15	—	—	—
8S RI	S7, S9, S13, S19	O′–A	Band III	S7, S9, (S10), S13, S19	D,E′,K,P,E,(A)	RNP III	(S7), S9, (S10), S13, S19
8S	S7, S9, S13, S19	O′–A	—	—	—	—	—

Symbols and abbreviations: RNP, ribonucleoprotein; ( ), protein present in reduced quantities; upper case letters indicate approximate extent of RNA component according to the nomenclature of Fellner and collaborators (see Figure 1 in Fellner, this volume). [a] Zimmermann, Muto and Mackie (1974); [b] Zimmermann (1974); [c] Zimmermann and Bergmann (1974); [d] Morgan and Brimacombe (1972); [e] Morgan and Brimacombe (1973); [f] Szekely, Brimacombe and Morgan (1973); [g] Roth and Nierhaus (1973).

mann et al. 1972b, 1974; Muto et al. 1974). Group II, encompassing proteins S6, S8, S15 and S18, is associated with an RNA fragment of 7S which contains a sequence of about 270 nucleotides lying towards the center of the molecule (Zimmermann et al. 1974; Zimmermann 1974). Group III, comprised of proteins S7, S9, S13 and S19, binds to a fragment of 8S which is derived from the 600 3′-terminal residues of the rRNA and appears to be the same as the 8S RNA isolated both from free 16S RNA (Muto et al. 1974) and from the 8S RI subparticle (Zimmermann, Muto and Mackie 1974). These relationships are summarized in Table 1 and Figure 8.

The RNA segment specific to each protein group was analyzed in two ways. First, complexes between 16S RNA and the members of each group were digested with either T_1 or pancreatic RNase (Figure 9) at temperatures and enzyme:substrate ratios chosen to optimize the recovery of specific ribonucleoprotein subparticles (Zimmermann 1974). Once it was determined that all members of the group were present in a given subparticle, its RNA component was characterized by fingerprinting. Second, it was important to establish that each RNA fragment was competent to bind all the proteins of its related group. Accordingly, the fragments were purified, deproteinized, and incubated at 0°C with an unfractionated mixture of radioactively labeled 30S subunit proteins. Ribonucleoprotein complexes were fractionated on sucrose gradients and the protein composition of each was determined by chromatography and electrophoresis (Figure

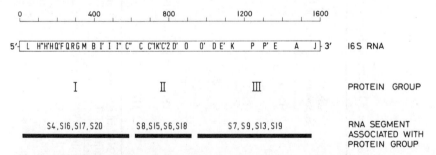

Figure 8 Distribution of 30S subunit proteins along the 16S RNA chain. Proteins that associate with the 16S RNA in the early stages of 30S subunit assembly have been found to comprise three independent groups, each of which can interact with a specific portion of the RNA molecule. All three groups include proteins that bind independently to the 16S RNA as well as those that do not. Evidence for this conclusion is discussed in the text. The scale at the top of the figure measures linear distance from the 5′ terminus of the 16S RNA in nucleotide residues. Sections within the RNA molecule are designated by upper case letters (see Figure 1 in Fellner, this volume). The bars at the bottom of the scheme show the maximum extent of the RNA sequences with which each group associates.

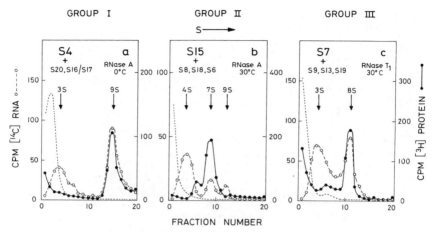

Figure 9 Isolation of the RNA fragments associated with the three protein groups. [^{14}C]16S RNA was incubated with (*a*) [^{3}H]S4, S16 + S17 and S20; (*b*) [^{3}H]S15, S8, S18 and S6; and (*c*) [^{3}H]S7, S9, S13 and S19 for 30 min at 40°C in TMK buffer and chilled on ice. The complexes were then treated with pancreatic RNase at an enzyme:substrate ratio of 1:5 for (*a*) 15 min at 0°C or (*b*) 5 min at 30°C or (*c*) with T$_1$ RNase at an enzyme:substrate ratio of 1:20 for 5 min at 30°C. Partial digestion products were separated by sucrose gradient centrifugation, and the distribution of radioactivity in each gradient was analyzed according to Zimmermann et al. (1972b). The dotted lines, which denote the sedimentation profiles of the corresponding protein mixtures in the absence of RNA, demonstrate that the proteins alone did not form rapidly sedimenting aggregates.

10). The results showed that each of the three fragments could select and bind precisely those proteins originally associated with it, but no others.

In addition to being physically discrete, the three protein groups are also independent of one another in the following sense: Singly or collectively the proteins of one group (1) either stimulate the interaction of other group members with the 16S RNA complex or depend upon them for their own attachment, but (2) do not significantly influence the binding of members of any other group. These results suggest that the early stages of 30S subunit assembly entail formation of three separate protein "nuclei" at different points along the RNA chain. After nucleation, which involves all the RI proteins, more complex interactions may be required to complete assembly since the process advances no further at 0°C, but requires the input of thermal energy, presumably to promote a conformational rearrangement (Traub and Nomura 1969; Held and Nomura 1973a). The closed nature of the groups also indicates that there is less cooperativity in the subunit assembly process, and in particular, among the RNA-binding proteins, than previously reported (Mizushima and Nomura 1970; Schaup,

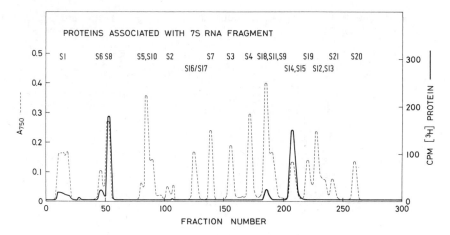

Figure 10 Chromatographic analysis of 30S subunit proteins associated with the 7S RNA fragment. The 7S fragment was prepared as described in the legend to Figure 9b and extracted with phenol. It was next incubated with a slight molar excess of unfractionated ^{3}H-labeled 30S subunit proteins for 60 min at 0°C in TMK buffer. The ribonucleoprotein fragment which resulted was separated from unbound protein by sucrose gradient centrifugation. The bound proteins were extracted, mixed with unlabeled 30S proteins, and chromatographed on phosphocellulose at pH 6.5 by the procedure of Zimmermann, Muto and Mackie (1974). Radioactive protein was located by scintillation counting (——), and unlabeled carrier protein was detected colorimetrically (– – – –).

Green and Kurland 1970, 1971; Garrett et al. 1971; Green and Kurland 1973). With this reservation, however, the physical grouping of proteins described here is generally consistent with the topographical relationships implied by the 30S subunit assembly map (Mizushima and Nomura 1970; Green and Kurland 1973; Held et al. 1974). Moreover, it has been shown that of the proteins which constitute group III, S7 and S9 as well as S13 and S19 can be cross-linked with bifunctional reagents and can thus be considered neighbors in the mature subunit (Lutter et al. 1974; Sun et al. 1974). This result is entirely consistent with the fact that the four proteins are related by cooperative interactions in the assembly sequence and can be isolated together in association with a specific segment of the RNA.

It is not yet known by what mechanism proteins unable to bind independently to 16S RNA, such as S6, S9, S16, S18 and S19, are associated with the ribonucleoprotein complex, although protein–RNA and protein–protein interactions may both play a role. These secondary interactions must be highly specific, however, since the proteins in question are distributed in a unique fashion among the ribonucleoprotein subparticles described above. The importance of RNA–protein interaction in this process is suggested by the following observation. When proteins S6, S8, S15 and S18 (group II) are bound simultaneously to 16S RNA, they protect twice as much RNA from pancreatic RNase digestion as do S8 and S15 (Zimmermann 1974; Zimmermann et al. 1974; Zimmermann and Bergmann 1974). Although this result might be attributed to steric factors, it is more likely that proteins S6 and S18 are themselves able to interact with and protect from RNase attack specific nucleotide sequences once S8 and S15, upon whose presence the attachment of S6 and S18 depends, are properly positioned in their own binding sites. Thus S8 and S15 may create sites of attachment for S6 and S18 by imposing particular conformational constraints on the RNA chain. This mechanism may indeed prove to be of general validity in accounting for interaction between the growing subunit and proteins which have no affinity for the free RNA molecule (see also Green and Kurland 1973).

Native 30S Subunits

The arrangement of ribosomal proteins relative to 16S RNA has also been approached by examining the protein compositions of ribonucleoprotein particles produced by mild RNase digestion of native and unfolded *E. coli* 30S subunits (Möller et al. 1969; Brimacombe et al. 1971; Brimacombe, Morgan and Cox 1971; Schendel, Maeba and Craven 1972; Morgan and Brimacombe 1972, 1973; Székely, Brimacombe and Morgan 1973; Roth and Nierhaus 1973). This methodology was introduced prior to the demonstration that complexes between 16S RNA and select groups of ribosomal proteins could be reconstituted in vitro for similar purposes (Zimmermann, Muto and Mackie 1974; Zimmermann 1974).

The technique used by Brimacombe and his colleagues entailed T_1 or pancreatic RNase digestion of 30S subunits in the absence or presence of 2 M urea, 2 mM EDTA or 10% ethanol (Morgan and Brimacombe 1972, 1973). Ribonucleoprotein fragments were separated by electrophoresis in composite polyacrylamide-agarose slab gels. Strips from this gel were then excised and sliced, and each slice was polymerized into a second gel containing polyacrylamide and sarkosyl for electrophoretic fractionation of the proteins. Fragments included in the study exhibited the following characteristics: (1) Proteins assigned to them were present in equimolar amounts. (2) Their protein:RNA ratios were similar to that of the intact 30S subunit. (3) Their mobilities in the first gel were consistent with the total protein mass they were believed to contain. The ribonucleoprotein subparticles described by Roth and Nierhaus (1973) were prepared by zonal centrifugation following digestion of 30S subunits with RNase T_1. Throughout the purification, the particles were maintained at relatively high ionic strengths. The proteins in each fragment were extracted and identified by two-dimensional gel electrophoresis. The fragments of Schendel, Maeba and Craven (1972) were produced by partial hydrolysis of 30S subunits with solid-phase pancreatic RNase and fractionated on sucrose gradients after the digests were first desalted by gel filtration and heated to 55°C. The identity of fragment-bound ribosomal proteins was established by electrophoresis on polyacrylamide disc gels.

There is good correlation among the proteins identified in fragments 13–14 of Morgan and Brimacombe (1973) and in RNP II of Roth and Nierhaus (1973), and the compositions of both are similar to that of the 14S RI subparticle (Zimmermann, Muto and Mackie 1974) previously discussed (see Table 1). Although the structure of the RNA fragments in the first two ribonucleoproteins has not been determined, they must be similar to the 12S (RI) RNA found in the 14S subparticle. In addition, proteins S8 and S15, which are recovered together in fragment 7 of Morgan and Brimacombe (1972, 1973), have been shown by other means to occupy adjacent or overlapping binding sites within a small segment of the 16S RNA (Zimmermann et al. 1974; Zimmermann and Bergmann 1974). Finally, a close correspondence also exists among proteins of band III described by Morgan and Brimacombe (1972), RNP III isolated by Roth and Nierhaus (1973), and the 8S RI subparticle (Zimmermann, Muto and Mackie 1974). Partial characterization of the RNA component of band III (Székely, Brimacombe and Morgan 1973) indicates that it contains substantially the same nucleotide sequences as the 8S (RI) RNA. The protein groups defined by Schendel, Maeba and Craven (1972) do not correlate particularly well with those discussed above, although some similarities are evident. It is possible that the low ionic strengths and high temperatures utilized to dissociate the partially digested ribonucleoprotein subparticles in the latter experiments promoted nonspecific protein interchange among the various fragments.

Native ribonucleoprotein complexes have been prepared from the 30S subunits of the thermophile *B. stearothermophilus* and the halophile *Halobacterium cutirubrum,* as well as *E. coli,* by stripping off most of the structural proteins with 3.5 M LiCl in EDTA (Chow et al. 1972). These complexes, which contained intact rRNA, also retained four basic proteins (*B. stearothermophilus*), two acidic proteins (*H. cutirubrum*), or two basic proteins (*E. coli*). The last were identified as S8 and S15 by two-dimensional gel electrophoresis. In each case, a series of ribonucleoprotein fragments were isolated from RNase digests of the complexes by gel filtration. The unique partition of proteins among the fragments from *B. stearothermophilus* confirmed the potential utility of this technique for obtaining specific protein–RNA fragment complexes from bacterial ribosomes. Although RNA sequences were not characterized in any of these cases, the fragments to which proteins remained fixed exhibited a G + C content significantly higher than that of the total 16S RNA.

Native 50S Subunits

Attempts have been made to determine the distribution of proteins in the 50S subunit by RNase digestion of native ribonucleoprotein particles. In one experiment, 50S subunits were cleaved into two subparticles by solid-phase pancreatic RNase (Allet and Spahr 1971). After removal of "split" proteins by treatment with 2 M LiCl, the RNA and protein components of the resulting cores were characterized. The RNA fragments, which sedimented at 13S and 18S, were distinct and found to consist of the 40% 5′-terminal and 60% 3′-terminal residues of the 23S RNA, respectively. Chromatographic analysis of the proteins revealed that only four were uniquely situated in one fragment or the other, an additional eight showed a fragment preference, and seven proteins of the ribonucleoprotein cores were distributed more or less equally between the two particles. Reconstitution assays with isolated RNA fragments and unfractionated 50S proteins gave similar results. Although the lack of strict fragment specificity among the core proteins is somewhat puzzling and may have resulted from protein exchange during preparation of the subparticles, these experiments gave an early indication that the distribution of proteins in the 50S subunit is not random.

The relationship of some twenty proteins of the *E. coli* 50S subunit to the 23S RNA has been studied by partial RNase digestion of unfolded subunits (Kagawa, Jishuken and Tokimatsu 1972a; Kagawa, Tokimatsu and Jishuken 1972b). This procedure yielded a series of ribonucleoprotein fragments which were fractionated by polyacrylamide gel electrophoresis at pH 8.5. The complexes were then completely digested with RNase in situ, and the proteins liberated from each were separated by gel electrophoresis in a second dimension at pH 4.7. The proteins in any one complex were assumed to be neighbors in the native subunit, and from a comparison of the protein content of a large number of such ribonucleoprotein fragments, a linear arrangement of the proteins relative to the 23S RNA was deduced. The sense of the array could not be specified, however, since the RNA moiety of the fragments was not analyzed. Here, as in the first experiment, the use of low ionic strength buffers containing EDTA raises the possibility of nonspecific protein exchange during production and isolation of the ribonucleoprotein subparticles.

Because the standard nomenclature for 50S subunit proteins (see Wittmann, this volume) was not used in either of the investigations described above, it is not possible to compare the results with conclusions on protein distribution drawn from the study of individual RNA–protein complexes (see above).

INFLUENCE OF PROTEIN STRUCTURE ON RNA–PROTEIN INTERACTIONS

There is little information available on the structural features of the RNA-binding proteins that permit them to specifically interact with the nucleic acid chain. These proteins are relatively small, with molecular weights ranging from 12,000 to 30,000 daltons (Dzionara, Kaltschmidt and Wittmann 1970), strongly basic, with pK values of 9–12 or higher (Kaltschmidt 1971), and possess considerable secondary structure (Cotter and Gratzer 1969; Dzionara 1970). Few other common characteristics are apparent, although sequence data demonstrate that there is a nonrandom distribution of hydrophilic and hydrophobic amino acids in at least some of the RNA-binding proteins (see Wittmann, this volume). Moreover, as a result of recent studies on enzymatic and chemical modification, certain generalizations can be made about protein groups necessary for interaction with ribosomal RNA (Daya-Grosjean et al. 1974). In particular, the lysine and methionine residues, as well as the C-terminal portions, of S4, S8, S15 and S20 appear to play an essential role in specific protein–RNA complex formation, whereas cysteine and tryptophan, if present, do not. Attempts to define the RNA-binding regions of the proteins by analysis of the interaction between protein fragments and RNA have not been successful so far (Garrett and Wittmann 1973). More detailed inferences about the binding properties of normal and mutationally altered S4, S7 and S8 have been made on the basis of a variety of chemical and physical investigations.

Protein S4

Reversion from streptomycin dependence to independence in *E. coli* is frequently accompanied by an alteration in protein S4 (Birge and Kurland 1970; Deusser et al. 1970; Kreider and Brownstein 1971). It has been established by reconstitution experiments that one such mutant protein, S4-su6, can modify the expression of the S12-mediated, streptomycin-dependent phenotype in vitro (Birge and Kurland 1970). Moreover, the altered S4 has substantially less affinity for its binding site in the 16S RNA than wild-type S4 (Green and Kurland 1971). Subsequent studies on the primary structure of S4-su6 demonstrated that it differed from the normal protein in at least seven amino acid residues, which apparently accounted for its pleiotropic effects (Donner and Kurland 1972). However it is not yet possible to say whether the amino acid replacements occur at positions that are directly involved in interaction with the RNA or whether they affect the secondary and tertiary structures of the protein in a more general way.

The binding properties and conformation of six more altered S4 proteins from streptomycin-independent revertants have recently been described

(Daya-Grosjean et al. 1972). Three of the mutant proteins, which were of normal chain length, were indistinguishable from wild-type RNA in their interaction with 16S RNA. The other three, whose C termini differed in length from that of the wild-type protein by as much as ± 20 amnio acid residues (Funatsu et al. 1972), exhibited reduced affinity for 16S RNA. This evidence, together with the demonstration that the binding of wild-type S4 is eliminated by carboxypeptidase treatment (Daya-Grosjean et al. 1972, 1974), suggests that the C-terminal residues of S4 play a role in its interaction with the RNA. Although fluorescence studies of the mutant proteins with altered chain lengths initially appeared to indicate a correlation between weak binding and changes in the environment of tryptophan residues (Daya-Grosjean et al. 1972), this amino acid is not believed to be actually involved in the binding reaction since its modification in normal S4 does not reduce the affinity of the protein for 16S RNA (Daya-Grosjean et al. 1974).

Altered S4 proteins from two other *E. coli* mutants (Zimmermann, Garvin and Gorini 1971; Zimmermann, Ikeya and Sparling 1973) were unchanged in molecular weight and displayed no differences in either binding affinity or saturation level relative to wild-type S4 (P. Spierer and R. A. Zimmermann, unpublished results). Thus certain mutations causing changes in S4 structure and function apparently occur outside the region of the protein which mediates interaction with the 16S RNA.

The technique of reductive methylation has recently been applied to the identification of RNA-binding regions in protein S4 (Amons et al. 1974). The reactivity of S4 lysine residues was severely curtailed in the native subunit; whereas almost all the lysines in free S4 are available for modification, only three out of the twenty lysine residues studied could be modified to a significant extent in the intact particle. When S4 alone was complexed with 16S RNA, two specific lysine residues within the C-terminal half of the protein chain were masked, although both were available for methylation in free S4.

Hydrodynamic measurements indicate that protein S4 is highly asymmetric, with dimensions of 24 × 170 Å (Paradies, quoted in Garrett and Wittmann 1973), although it is by no means certain that this represents the native configuration of the protein. However, since a prolate molecule would have a substantially larger surface area than a spherical one of the same mass, a pronounced asymmetry in S4 is consistent with the large size and apparent complexity of its binding site in the 16S RNA (Zimmermann et al. 1974).

It has recently been demonstrated by fluorescence polarization studies that protein S4 undergoes a conformational shift when heated in the ribosomal reconstitution buffer to temperatures between 32 and 42°C (Lemieux and Gerard 1973), although the nature of the change is not known.

Protein S7

Although most of the RNA-binding proteins of the 30S subunit appear to be identical in *E. coli* strains B, MRE600 and K12 (Kaltschmidt et al. 1970), protein S7 provides a notable exception. Protein S7 of strain K12 differs from that of strains MRE600 and B in amino acid sequence, molecular weight, and electrophoretic mobility (Leboy, Cox and Flaks 1964; Birge et al. 1969; Sypherd 1969; Kaltschmidt et al. 1970; Sun and Traut 1973). Nonetheless, the S7 proteins from K12 and MRE600 contain common tryptic peptides and immunological determinants, and they are functionally equivalent in 30S subunit assembly (Sun and Traut 1973; Held and Nomura 1973b). Moreover, the structural disparities cited above do not appear to qualitatively or quantitatively alter the interaction of the different S7 proteins with 16S RNA (Garrett et al. 1971; P. Spierer and R. A. Zimmermann, unpublished results). This implies that those portions of the protein molecule that determine its ability to bind to 16S RNA have been conserved in all three strains. Protein S7 has also been reported to undergo an alteration of conformation at temperatures above 23°C (Lemieux and Gerard 1973), but whether or not this is related to its RNA-binding capacities remains to be established.

Protein S8

It has been reported that protein S8 can exist in two conformations, only one of which can bind to 16S RNA at 0°C (Schulte and Garrett 1972). The remainder of the protein population could be converted to the binding form by heating to 42°C. In this case, the reversible loss of binding activity was attributed to storage of S8 under denaturing conditions, and reactivation was interpreted as a unimolecular rearrangement of protein conformation. In a more general sense, however, these observations indicate that at least some ribosomal proteins can acquire a certain degree of structural flexibility at the temperatures used to study RNA–protein interactions and, furthermore, raise the possibility that they may be able to undergo conformational transitions upon interaction with the RNA.

SOME GENERAL CHARACTERISTICS OF RNA–PROTEIN INTERACTIONS IN THE RIBOSOME

Features of Protein Binding Sites in rRNA

As a result of extensive studies on the structure of specific ribosomal RNA fragments, ribosomal protein binding sites are presently believed to differ greatly in size, ranging from 40 nucleotide residues in the case of protein S8 (Schaup et al. 1973; Zimmermann et al. 1974) to several hundred residues in the case of S4 and L24 (Schaup et al. 1971; Schaup and Kurland 1972; Zimmermann et al. 1972b, 1974; Branlant et al. 1973). Moreover, there is strong evidence that RNA secondary and tertiary structure, as well as primary sequence, are critically involved in the interaction with specific ribosomal proteins. The 40-nucleotide fragment that contains the binding site for protein S8, for instance, comprises two highly complementary sequences of 20 residues that arise from the hydrogen-bonded stem of a long hairpin loop (Zimmermann et al. 1974). The RNA segment with which proteins S4 and S20 interact is an order of magnitude larger than that for S8 and contains over a dozen hairpin loops of various sizes (Zimmermann et al. 1972b, 1974). It is difficult to know precisely how much of this segment directly participates in protein binding, but since it has been shown that protein S4 alone interacts with more than one sequence, much of the RNA may be required to maintain the actual sites of attachment in their proper three-dimensional configuration. In this respect, it is worthy of note that fragments containing a number of "hidden breaks" or even small excisions can interact specifically and stoichiometrically with a number of ribosomal proteins (Zimmermann, Muto and Mackie 1974). Binding site integrity does not therefore depend absolutely upon the presence of a continuous sequence, as long as the overall structure remains intact.

A second unexpected characteristic of ribosomal protein binding sites is their apparent clustering along the RNA chain. In particular, the binding sites for proteins S4 and S20 have been found to occupy a similar region of the 16S RNA, while those for proteins S8 and S15 may even overlap to some extent. Proteins S4 and S20 both interact within a 500-nucleotide sequence located at the 5′ terminus of the 16S RNA. Although protein S4 appears to protect the entire fragment from RNase digestion, protein S20 protects only a 300-nucleotide sequence from within it (Zimmermann et al. 1974). Furthermore, one or both proteins of the S16 + S17 mixture can associate with this fragment when S4 and S20 are also present (Zimmermann 1974). Although the disposition of the proteins relative to the RNA molecule is not known, their spatial relationship must be an intimate one. Proteins S8 and S15 bind within a smaller, but very stable, portion of the 16S RNA (Zimmermann et al. 1972b; Schaup et al. 1973; Zimmermann and Bergmann 1974). A double-helical segment in the base-paired stem of one long hairpin loop is specifically protected from RNase attack by protein S8. Protein S15 protects one strand of the same stem as well as the entire base-paired stem of an adjoining loop (Zimmermann et al. 1974). Here again the three-dimensional arrangement of the components is unknown, but the two proteins must be in close contact with each other and with the RNA chain.

The binding site for protein L24 in the 23S RNA is similar in a number of respects to that for protein S4 in the 16S RNA (Branlant et al. 1973). The sequences to which these proteins bind both extend over several hundred nucleotides and both are located at the 5' terminus of their respective RNAs. Moreover, the solution conditions optimal for specific protein–RNA interaction are nearly identical in each case (Schulte, Morrison and Garrett 1974). As mentioned above, S4 associates with at least two distinct sites in the 16S RNA. Although L24 is only half the size of S4, it may bind to multiple sites as well, since it is capable of protecting RNA sequences encompassing approximately 500 residues from RNase hydrolysis (Branlant et al. 1973). A detailed structural similarity between the two binding sites, however, remains to be established. It is of interest, nonetheless, that measurements on the interaction of *E. coli* proteins S4 and L24 with heterologous bacterial 16S and 23S RNAs indicate that, within a given organism, binding sites for the two proteins appear to have been subject to a similar degree of evolutionary change (Daya-Grosjean et al. 1973).

Although binding sites in the 16S RNA have proved to be primarily "local," in the sense that they are situated within limited portions of the molecule, it is possible that certain proteins interact with sequences "dispersed" throughout the primary structure. Protein S13, for example, remains attached to the 5' half of the 16S RNA when the S13–16S RNA complex is digested with T_1 RNase (Muto et al. 1974). If proteins S7, S9 and S19 are also present in the complex, however, S13 is retained by a fragment from the 3' half of the RNA (Zimmermann 1974). A similar partition of protein S13 has been observed when ribonucleoprotein fragments are produced by RNase digestion of intact 30S subunits (Roth and Nierhaus 1973). Protein S13 may therefore interact with widely separated segments of the RNA molecule, and these interactions might be influenced by the protein composition of the initial ribonucleoprotein complex.

The present discussion has been concerned exclusively with the location and properties of binding sites for the small group of proteins that can independently interact with the ribosomal RNA. The majority of the ribosomal proteins cannot associate independently with the nucleic acid chain, however, and the manner in which they are bound in the course of ribosome assembly is not yet understood. There is nonetheless little doubt that these proteins occupy very specific positions in the subunit structure, and while their placement may depend in part upon protein–protein interactions, it is likely that at least some of them associate directly with the rRNA once certain other proteins have attached. Evidence favoring the latter mechanism has been obtained in the case of proteins S6 and S18. S6 and S18 will associate with the 16S RNA only if S8 and S15 are also present, and when the complex is treated with RNase, the four proteins protect twice as much RNA from digestion as do S8 and S15 alone (Zimmermann 1974). It is thus likely that the additional protected sequences, which lie adjacent to those containing the binding sites for proteins S8 and S15, are shielded from RNase attack as a consequence of direct interactions among S6, S18 and the nucleic acid chain. One way of explaining these results would be to assume that the independent attachment of proteins S8 and S15 alters the conformation of the 16S RNA in such a fashion that binding sites become accessible to proteins S6 and S18 in nearby segments of the molecule.

Influence of Molecular Configuration on RNA–Protein Interaction

There can be little doubt that both rRNA and ribosomal proteins must possess suitable secondary and tertiary structures before specific RNA–protein interactions can occur. It is also apparent that the overall configuration of these molecules can vary with temperature, ionic environment and solution conditions. Such flexibility in molecular structure is relevant to the investigation of RNA–protein interactions in at least two ways: (1) If a given component readily undergoes reversible transitions between conformations that are active in binding and those that are not, small changes in environmental conditions could shift the equilibrium in either direction. Variations of this sort could help to explain why different laboratories have reached divergent conclusions about the binding properties of certain ribosomal proteins. (2) If either the RNA or protein molecule undergoes a structural rearrangement upon interaction, a better fit between the components might result. Moreover, a conformational shift under these circumstances could lead to the organization of new binding sites for the attachment of additional components during subunit assembly.

Certain modifications in the conformation of isolated ribosomal components have been directly detected by physicochemical measurements, whereas others have been inferred from reversible changes in binding properties that result from variation of solution conditions. A few examples will suffice to illustrate the kinds of alterations observed. The 16S RNA assumes a less compact configuration when the Mg^{++} ion concentration is reduced from 0.01 M to 0.001 M, although no major shifts in secondary structure occur (Schulte, Morrison and Garrett 1974). Under these circumstances, the RNA molecule undergoes a number of reversible changes: its sedimentation rate decreases, its susceptibility to RNase increases, and its ability to interact with proteins S4, S8 and S20 is drastically curtailed (Muto et al. 1974; Muto and Zimmermann 1974; Schulte, Morrison and Garrett 1974). If the 16S RNA is extensively dialyzed against distilled water, addition of Mg^{++} ions is not sufficient to restore its ability to interact with proteins S4 and S20 at 0°C. Reactivation of the binding sites does occur, however, when the RNA is heated at temperatures above 20°C in the presence of 0.02 M $MgCl_2$; the renaturation reaction is characterized by a high energy of activation and a decrease in sedimentation rate (Muto and Zimmermann 1974). The properties of ribosomal proteins can also be modified by environmental conditions. After storage in high concentrations of urea, a large fraction of the protein S8 population is unable to interact with 16S RNA at 0°C (Schulte and Garrett 1972). Upon heating to 42°C, however, the binding capacity of the inactive S8 molecules can be restored by a conformational rearrangement which entails an activation energy of roughly 12 kcal/mole. Fluorescence polarization studies suggest that proteins S4 and S7 also undergo conformational changes upon heating to 42°C, although the functional significance of these transitions is not yet known (Lemieux and Gerard 1973). It should be noted that while the examples of structural alterations cited above support the notion of molecular flexibility, many of these may be artificial in nature, stemming from the exposure of ribosomal components to nonphysiological conditions during extraction and purification.

There is both direct and indirect evidence suggesting that conformational changes occur during RNA–protein interactions. For instance, the interaction of the RNA-binding proteins of the 30S subunit with 16S RNA provokes dissociation of the fluorescent dye ethidium bromide, indicating a decrease of base-stacking within the nucleic acid molecule (Bollen et al. 1970). Such rearrangements might promote further RNA–protein interactions which would otherwise not take place. The association of 23S RNA with 5S RNA in the presence of 50S subunit proteins L18 and L25 may illustrate the role played by conformational shifts in ribosome assembly (Gray et al. 1972, 1973). Although both L18 and L25 can bind independently to 5S RNA, L18 interacts only weakly, and L25 not at all, with the 23S RNA. Consequently the association of L18 and L25 with the 5S RNA must alter the structure of one or more of the components in such a way that stable attachment to the 23S RNA becomes possible. Evidence that at least one ribosomal protein undergoes a conformational alteration upon interaction with rRNA has recently been provided by the demonstration that *B. stearothermophilus* protein B-L3 possesses different antigenic properties when bound to homologous 23S RNA than when free in solution (Tischendorf, Geisser and Stöffler 1973).

Susceptibility of rRNA to RNase Digestion

The RNA fragments that result from enzymatic digestion of 16S RNA–protein complexes are frequently very similar or identical in structure to those derived from free 16S RNA under the same conditions (Muto et al. 1974). The particular bonds that are hydrolyzed in such cases thus appear to be determined mainly by the configuration of the nucleic acid chain. This inference is clearly supported by the following example. A fragment which sediments at 12S and encompasses the 5′ half of the *E. coli* 16S RNA can be isolated from T_1 RNase digests both of free 16S RNA (Zimmermann et al. 1972b) and of the RI particle (Traub and Nomura 1969), which contains 11–12 proteins bound to the 16S RNA (Nashimoto et al. 1971; Held and Nomura 1973a; Zimmermann, Muto and Mackie 1974). The two fragments yielded identical fingerprints and when they were electrophoresed in polyacrylamide gels under denaturing conditions, they were resolved into several subfragments, indicating that a discrete number of "hidden breaks" had been introduced into each during RNase treatment (Zimmermann, Muto and Mackie 1974). Surprisingly the pattern of subfragments was found to be essentially the same in both cases. These results unambiguously demonstrate that the limited number of sites available for RNase attack in the free 16S RNA are also cleaved in the ribonucleoprotein particle and suggest that the proteins of the RI particle afford little added protection. The relative resistance of the RNA molecule in these experiments can no doubt be attributed to the presence of extensive secondary structure, since endonucleases such as T_1 and pancreatic RNases exhibit a decided preference for single-stranded segments of the RNA under conditions of partial hydrolysis (Ehresmann et al. 1972).

There are nonetheless many instances in which bound proteins clearly protect specific regions within the rRNA. In the most extreme case, otherwise labile tracts within both the 16S and 23S RNAs are resistant to RNase attack when present in the 30S or 50S subunit, respectively (Fellner, Ehresmann and Ebel 1970; Allet and Spahr 1971; Santer and Santer 1973). Moreover, specific sequences of the rRNAs have been found to resist RNase digestion when associated with individual ribosomal proteins (Schaup et al. 1971, 1973; Branlant et al. 1973; Zimmermann et al. 1974; Zimmermann and Bergmann 1974) or small groups of proteins (Zimmermann 1974). The manner in which bound protein protects the RNA is not fully understood, since many of the protected fragments range up to several hundred nucleotides in length and are thus much larger than could be shielded by direct contact with a single protein molecule. Two possible explanations have been offered by Branlant et al. (1973): (1) The protein may induce the RNA to assume a compact configuration, which renders fairly large sequences in the vicinity of contact less accessible to RNase. (2) Roughly 70% of the bases in both 16S and 23S RNAs participate in secondary structure, which is believed to be distributed into a series of hairpin loops connected by short single-stranded segments. Since the base-paired regions are intrinsically quite resistant to RNase (Ehresmann et al. 1972), bound protein might protect substantial RNA sequences by directly blocking access to a small number of labile bonds between one or more hairpins which would be hydrolyzed in the absence of proteins.

Acknowledgments

The author wishes to express his appreciation to his associates and collaborators whose germane suggestions during the preparation of this review and whose contributions to many portions of the work described herein were indispensable. In particular, he would like to cite Drs. A. Tissières,

A. Muto, G. A. Mackie, P. Fellner and C. Ehresmann; Miss K. Bergmann; Mr. P. Spierer. Special thanks are due Dr. R. A. Garrett for thoughtful criticism of this manuscript and for permission to quote numerous experimental observations prior to publication.

References

Allet, B. and P. F. Spahr. 1971. Binding sites of ribosomal proteins on two specific fragments derived from *Escherichia coli* 50S ribosomes. *Eur. J. Biochem.* **19**:250.

Amons, R., W. Möller, E. Schiltz and J. Reinbolt. 1974. Studies on the binding sites of protein S4 to 16S RNA in *Escherichia coli* ribosomes. *FEBS Letters* **41**:135.

Birge, E. A. and C. G. Kurland. 1970. Reversion of a streptomycin-dependent strain of *Escherichia coli*. *Mol. Gen. Genet.* **109**:356.

Birge, E. A., G. R. Craven, S. J. S. Hardy, C. G. Kurland and P. Voynow. 1969. Structural determinant of a ribosomal protein: K locus. *Science* **164**:1285.

Bollen, A., A. Herzog, A. Favre, J. Thibault and F. Gros. 1970. Fluorescence studies on the 30S ribosome assembly process. *FEBS Letters* **11**:49.

Branlant, C., A. Krol, J. Sriwidada, P. Fellner and R. Crichton. 1973. The identification of the RNA binding site for a 50S ribosomal protein by a new technique. *FEBS Letters* **35**:265.

Brimacombe, R., J. M. Morgan and R. A. Cox. 1971. An improved technique for the analysis of ribonucleoprotein fragments from *Escherichia coli* 30S ribosomes. *Eur. J. Biochem.* **23**:52.

Brimacombe, R., J. Morgan, D. G. Oakley and R. A. Cox. 1971. Specific ribonucleoprotein fragment from the 30S subunit of *E. coli* ribosomes. *Nature New Biol.* **231**:209.

Brownlee, G. G. and F. Sanger. 1967. Nucleotide sequences from the low molecular weight ribosomal RNA of *Escherichia coli*. *J. Mol. Biol.* **23**:337.

Chow, C. T., L. P. Visentin, A. T. Matheson and M. Yaguchi. 1972. Specific ribonucleoprotein fragments from the 30S ribosomal subunits of *Halobacterium cutirubrum*, *Escherichia coli* and *Bacillus stearothermophilus*. *Biochim. Biophys. Acta* **287**:270.

Cotter, R. I. and W. B. Gratzer. 1969. An infrared study of the conformation of RNA and protein in the ribosome. *Eur. J. Biochem.* **8**:352.

Crichton, R. R. and H. G. Wittmann. 1971. Ribosomal proteins. XXIV. Trypsin digestion as a possible probe of the conformation of *Escherichia coli* ribosomes. *Mol. Gen. Genet.* **114**:95.

———. 1973. A native ribonucleoprotein complex from *Escherichia coli* ribosomes. *Proc. Nat. Acad. Sci.* **70**:665.

Daya-Grosjean, L., M. Geisser, G. Stöffler and R. A. Garrett. 1973. Heterologous protein-RNA interactions in bacterial ribosomes. *FEBS Letters* **37**:17.

Daya-Grosjean, L., J. Reinbolt, O. Pongs and R. A. Garrett. 1974. A study of the regions of ribosomal proteins S4, S8, S15 and S20 that interact with 16S RNA of *Escherichia coli*. *Biochim. Biophys. Acta* (in press).

Daya-Grosjean, L., R. A. Garrett, O. Pongs, G. Stöffler and H. G. Wittmann.

1972. Properties of the interaction of ribosomal protein S4 and 16S RNA in *Escherichia coli*. Revertants from streptomycin dependence to independence. *Mol. Gen. Genet.* **119**:277.

Deusser, E., G. Stöffler, H. G. Wittmann and D. Apirion. 1970. Ribosomal proteins. XVI. Altered S4 proteins in *Escherichia coli* revertants from streptomycin dependence to independence. *Mol. Gen. Genet.* **109**:298.

Donner, D. and C. G. Kurland. 1972. Changes in the primary structure of a mutationally altered ribosomal protein S4 of *Escherichia coli*. *Mol. Gen. Genet.* **115**:49.

Dzionara, M. 1970. Ribosomal proteins. Secondary structure of individual ribosomal proteins of *E. coli* studied by circular dichroism. *FEBS Letters* **8**:197.

Dzionara, M., E. Kaltschmidt and H. G. Wittmann. 1970. Ribosomal proteins. XIII. Molecular weights of isolated ribosomal proteins of *Escherichia coli*. *Proc. Nat. Acad. Sci.* **67**:1909.

Ehresmann, C., P. Stiegler, P. Fellner and J.-P. Ebel. 1972. The determination of the primary structure of the 16S ribosomal RNA of *Escherichia coli*. (2). Nucleotide sequences of products from partial enzymatic hydrolysis. *Biochimie* **54**:901.

Erdmann, V. A., S. Fahnestock, K. Higo and M. Nomura. 1971. Role of 5S RNA in the functions of 50S ribosomal subunits. *Proc. Nat. Acad. Sci.* **68**:2932.

Fahnestock, S., V. Erdmann and M. Nomura. 1973. Reconstitution of 50S ribosomal subunits from protein-free ribonucleic acid. *Biochemistry* **12**:220.

Fellner, P., C. Ehresmann and J.-P. Ebel. 1970. Nucleotide sequences in a protected area of the 16S RNA within 30S ribosomal subunits from *Escherichia coli*. *Eur. J. Biochem.* **13**:583.

―――. 1972. The determination of the primary structure of the 16S ribosomal RNA of *Escherichia coli*. (1). Nucleotide sequence analysis of T_1 and pancreatic ribonuclease digestion products. *Biochimie* **54**:853.

Fellner, P., C. Ehresmann, P. Stiegler and J.-P. Ebel. 1972. Partial nucleotide sequence of 16S ribosomal RNA from *E. coli*. *Nature New Biol.* **239**:1.

Funatsu, G., W. Puls, E. Schiltz, J. Reinbolt and H. G. Wittmann. 1972. Ribosomal proteins. XXXI. Comparative studies on altered proteins S4 of six *Escherichia coli* revertants from streptomycin dependence. *Mol. Gen. Genet.* **115**:131.

Garrett, R. A. and H. G. Wittmann. 1973. Protein-RNA interactions in bacterial ribosomes. *Sixth Karolinska Symp. Res. Meth. Reproduct. Endocrinol.* **6**:75.

Garrett, R. A., S. Müller, P. Spierer and R. A. Zimmermann. 1974. Binding of 50S ribosomal subunit proteins to 23S RNA of *Escherichia coli*. *J. Mol. Biol.* (in press).

Garrett, R. A., K. H. Rak, L. Daya and G. Stöffler. 1971. Ribosomal proteins. XXIX. Specific protein binding sites on 16S RNA of *Escherichia coli*. *Mol. Gen. Genet.* **114**:112.

Gray, P. N. and R. Monier. 1972. Partial localization of the 5S RNA binding site on 23S RNA. *Biochimie* **54**:41.

Gray, P. N., R. A. Garrett, G. Stöffler and R. Monier. 1972. An attempt at the identification of the proteins involved in the incorporation of 5S RNA during 50S ribosomal subunit assembly. *Eur. J. Biochem.* **28**:412.

Gray, P. N., G. Bellemare, R. Monier, R. A. Garrett and G. Stöffler. 1973. Identification of the nucleotide sequences involved in the interaction between *Escherichia coli* 5S RNA and specific 50S subunit proteins. *J. Mol. Biol.* **77**:133.

Green, M. and C. G. Kurland. 1971. Mutant ribosomal protein with defective RNA binding site. *Nature New Biol.* **234**:273.

————. 1973. Molecular interactions of ribosomal components. IV. Cooperative interactions during assembly *in vitro*. *Mol. Biol. Rep.* **1**:105.

Gyenge, L., V. A. Spiridonova and A. A. Bogdanov. 1972. Association of ribosomal RNA with ribosomal proteins covalently bound to Sepharose. *FEBS Letters* **20**:209.

Held, W. A. and M. Nomura. 1973a. Rate-determining step in the reconstitution of *Escherichia coli* 30S ribosomal subunits. *Biochemistry* **12**:3273.

————. 1973b. Functional homology between the 30S ribosomal protein S7 from *E. coli* K12 and the S7 from *E. coli* MRE600. *Mol. Gen. Genet.* **122**:11.

Held, W. A., B. Ballou, S. Mizushima and M. Nomura. 1974. Assembly mapping of 30S ribosomal proteins from *Escherichia coli:* Further studies. *J. Biol. Chem.* **249**:3103.

Higo, K., W. Held, L. Kahan and M. Nomura. 1973. Functional correspondence between 30S ribosomal proteins of *Escherichia coli* and *Bacillus stearothermophilus*. *Proc. Nat. Acad. Sci.* **70**:944.

Horne, J. R. and V. A. Erdmann. 1972. Isolation and characterization of 5S RNA-protein complexes from *Bacillus stearothermophilus* and *Escherichia coli* ribosomes. *Mol. Gen. Genet.* **119**:337.

Kagawa, H., L. Jishuken and H. Tokimatsu. 1972a. Linear sequence of proteins of *Escherichia coli* 50S ribosomal particles. *Nature New Biol.* **237**:74.

Kagawa, H., H. Tokimatsu and L. Jishuken. 1972b. Partial digestion of 50S ribosomes of *E. coli*. II. Digestion of the CsCl treated ribosomes with RNase T_1. *J. Biochem.* **72**:827.

Kaltschmidt, E. 1971. Ribosomal proteins. XIV. Isoelectric points of ribosomal proteins of *E. coli* as determined by two-dimensional polyacrylamide gel electrophoresis. *Anal. Biochem.* **43**:25.

Kaltschmidt, E., G. Stöffler, M. Dzionara and H. G. Wittmann. 1970. Ribosomal proteins. XVII. Comparative studies on ribosomal proteins of four strains of *Escherichia coli*. *Mol. Gen. Genet.* **109**:303.

Kreider, G. and B. L. Brownstein. 1971. A mutation suppressing streptomycin dependence. II. An altered protein on the 30S ribosomal subunit. *J. Mol. Biol.* **61**:135.

Leboy, P. S., E. C. Cox and J. G. Flaks. 1964. The chromosomal site specifying a ribosomal protein. *Proc. Nat. Acad. Sci.* **52**:1367.

Lemieux, G. and D. Gerard. 1973. Fluorescence studies of *E. coli* ribosomal proteins S-4 and S-7 in regard to reconstitution conditions. *FEBS Letters* **37**:234.

Lutter, L. C., U. Bode. C. G. Kurland and G. Stöffler. 1974. Ribosomal protein neighborhoods. III. Cooperativity of ribosome assembly. *Mol. Gen. Genet.* **129**:167.

Mackie, G. A. and R. A. Zimmermann. 1974a. RNA-protein interactions in the ribosome. IV. Structure and properties of binding sites for proteins S4 and S20 on the 16S RNA. *J. Mol. Biol.* (in press).

———. 1974b. RNA-protein interactions in the ribosome. VI. Characterization of fragments of 16S RNA protected against pancreatic ribonuclease digestion by ribosomal protein S4. *J. Biol. Chem.* (in press).

Mizushima, S. and M. Nomura. 1970. Assembly mapping of 30S ribosomal proteins from *E. coli. Nature* **226**:1214.

Möller, W., R. Amons, J. C. L. Groeng, R. A. Garrett and C. P. Terhorst. 1969. Protein-ribonucleic acid interactions in ribosomes. *Biochim. Biophys. Acta* **190**:381.

Morgan, J. and R. Brimacombe. 1972. A series of specific ribonucleoprotein fragments from the 30S subparticle of *Escherichia coli* ribosomes. *Eur. J. Biochem.* **29**:542.

———. 1973. A preliminary three-dimensional arrangement of the proteins in the *Escherichia coli* 30S ribosomal subparticle. *Eur. J. Biochem.* **37**:472.

Morrison, C. A., R. A. Garrett, H. Zeichhardt and G. Stöffler. 1973. Proteins occurring at, or near, the subunit interface of *E. coli* ribosomes. *Mol. Gen. Genet.* **127**:359.

Muto, A. and R. A. Zimmermann. 1974. RNA-protein interactions in the ribosome. III. Differences in the stability of ribosomal protein binding sites in the 16S RNA. *J. Mol. Biol.* (in press).

Muto, A., C. Ehresmann, P. Fellner and R. A. Zimmermann. 1974. RNA-protein interactions in the ribosome. I. Characterization and RNase digestion of 16S RNA-ribosomal protein complexes. *J. Mol. Biol.* (in press).

Nanninga, N., M. Meyer, P. Sloof and L. Reijnders. 1972a. Electron microscopy of *Escherichia coli* ribosomal RNA: Spreading without a basic protein film. *J. Mol. Biol.* **72**:807.

Nanninga, N., R. A. Garrett, G. Stöffler and G. Klotz. 1972b. Ribosomal proteins. XXXVIII. Electron microscopy of ribosomal protein S4-16S RNA complexes of *Escherichia coli. Mol. Gen. Genet.* **119**:175.

Nashimoto, H., W. Held, E. Kaltschmidt and M. Nomura. 1971. Structure and function of bacterial ribosomes. XII. Accumulation of 21S particles by some cold-sensitive mutants of *Escherichia coli. J. Mol. Biol.* **62**:121.

Nomura, M., P. Traub and H. Bechmann. 1968. Hybrid 30S ribosomal particles reconstituted from components of different bacterial origins. *Nature* **219**:793.

Reynier, M. and R. Monier. 1968. Les ribosomes d'*Escherichia coli*. III. Etude de la dissociation réversible du RNA 5S par le chlorure de lithium. *Bull. Soc. Chim. Biol.* **50**:1583.

Roth, H. E. and K. H. Nierhaus. 1973. Isolation of four ribonucleoprotein fragments from the 30S subunit of *E. coli* ribosomes. *FEBS Letters* **31**:35.

Sanger, F., G. G. Brownlee and B. G. Barrell. 1965. A two-dimensional fractionation procedure for radioactive nucleotides. *J. Mol. Biol.* **13**:373.

Santer, M. and U. Santer. 1973. Action of ribonuclease T_1 on 30S ribosomes of *Escherichia coli* and its role in sequence studies on 16S ribonucleic acid. *J. Bact.* **116**:1304.

Schaup, H. W. and C. G. Kurland. 1972. Molecular interactions of ribosomal components. III. Isolation of the RNA binding site for a ribosomal protein. *Mol. Gen. Genet.* **114**:350.

Schaup, H. W., M. Green and C. G. Kurland. 1970. Molecular interactions of ribosomal components. I. Identification of RNA binding sites for individual 30S ribosomal proteins. *Mol. Gen. Genet.* **109**:193.

————. 1971. Molecular interactions of ribosomal components. II. Site-specific complex formation between 30S proteins and ribosomal RNA. *Mol. Gen. Genet.* **112**:1.

Schaup, H. W., M. L. Sogin, C. G. Kurland and C. R. Woese. 1973. Localization of a binding site for ribosomal protein S8 within the 16S ribosomal ribonucleic acid of *Escherichia coli. J. Bact.* **115**:82.

Schaup, H. W., M. Sogin, C. Woese and C. G. Kurland. 1971. Characterization of an RNA "binding site" for a specific ribosomal protein of *Escherichia coli. Mol. Gen Genet.* **114**:1.

Schendel, P., P. Maeba and G. R. Craven. 1972. Identification of the proteins associated with subparticles produced by mild ribonuclease digestion of 30S ribosomal particles from *Escherichia coli. Proc. Nat. Acad. Sci.* **69**:544.

Schulte, C. and R. A. Garrett. 1972. Optimal conditions for the interaction of ribosomal protein S8 and 16S RNA and studies on the reaction mechanism. *Mol. Gen. Genet.* **119**:345.

Schulte, C., C. A. Morrison and R. A. Garrett. 1974. Protein-ribonucleic acid interactions in *Escherichia coli* ribosomes. Solution studies on S4–16S ribonucleic acid and L24–23S ribonucleic acid binding. *Biochemistry* **13**:1032.

Spierer, P., R. A. Zimmermann and G. A. Mackie. 1974. RNA-protein interactions in the ribosome. VII. Binding of 50S subunit proteins to 5′- and 3′-terminal fragments of the 23S RNA. *Eur. J. Biochem.* (in press).

Stöffler, G., L. Daya, K. H. Rak and R. A. Garrett. 1971a. Ribosomal proteins. XXVI. The number of specific protein binding sites on 16S and 23S RNA of *Escherichia coli. J. Mol. Biol.* **62**:411.

————. 1971b. Ribosomal proteins. XXX. Specific protein binding sites on 23S RNA of *Escherichia coli. Mol. Gen. Genet.* **114**:125.

Sun, T-T. and R. R. Traut. 1973. The functional and structural homology of ribosomal protein S7 of *E. coli* strains K and MRE600. *Mol. Gen. Genet.* **122**:1.

Sun, T-T., A. Bollen, L. Kahan and R. R. Traut. 1974. Topography of ribosomal proteins of the *Escherichia coli* 30S subunit as studied with the reversible crosslinking reagent methyl-4-mercaptobutyrimidate. *Biochemistry* **13**:2334.

Sypherd, P. 1969. Amino acid differences in a 30S ribosomal protein from two strains of *Escherichia coli. J. Bact.* **99**:379.

Székely, M., R. Brimacombe and J. Morgan. 1973. A specific ribonucleoprotein fragment from *Escherichia coli* 30S ribosomes: Location of the RNA component in 16S RNA. *Eur. J. Biochem.* **35**:574.

Tischendorf, G. W., M. Geisser and G. Stöffler. 1973. Comparison of ribsomal RNA-binding protein "L2" isolated from different bacterial species. *Mol. Gen. Genet.* **127**:147.

Traub, P. and M. Nomura. 1968. Structure and function of *E. coli* ribosomes. V. Reconstitution of functionally active 30S ribosomal particles from RNA and proteins. *Proc. Nat. Acad. Sci.* **59**:777.

————. 1969. Structure and function of *Escherichia coli* ribosomes. VI. Mechanism of assembly of 30S ribosomes studied *in vitro*. *J. Mol. Biol.* **40**:391.

Wrede, P. and V. A. Erdmann. 1973. Activities of *B. stearothermophilus* 50S ribosomes reconstituted with prokaryotic and eukaryotic 5S RNA. *FEBS Letters* **33**:315.

Yu, R. S. T. and H. G. Wittmann. 1973. The sequence of steps in the attachment of 5S RNA to cores of *Escherichia coli* ribosomes. *Biochim. Biophys. Acta* **324**:375.

Zimmermann, R. A. 1974. The role of three ribonucleoprotein nuclei in the reconstitution of 30S ribosomal subunits of *Escherichia coli*. *Proc. Nat. Acad. Sci.* (in press).

Zimmermann, R. A. and K. Bergmann. 1974. RNA-protein interactions in the ribosome. V. Location of binding sites for proteins S8 and S15 on the 16S RNA. *J. Mol. Biol.* (in press).

Zimmermann, R. A., R. T. Garvin and L. Gorini. 1971. Alteration of a 30S ribosomal protein accompanying the *ram* mutation in *Escherichia coli*. *Proc. Nat. Acad. Sci.* **68**:2263.

Zimmermann, R. A., Y. Ikeya and P. F. Sparling. 1973. Alteration of ribosomal protein S4 by mutation linked to kasugamycin-resistance in *Escherichia coli*. *Proc. Nat. Acad. Sci.* **70**:71.

Zimmermann, R. A., A. Muto and G. A. Mackie. 1974. RNA-protein interactions in the ribosome. II. Binding of ribosomal proteins to isolated fragments of the 16S RNA. *J. Mol. Biol.* (in press).

Zimmermann, R. A., A. Muto, P. Fellner and C. Branlant. 1972a. The interaction of 30S ribosomal proteins with 16S RNA and RNA fragments. In *Functional units of protein biosynthesis,* Symp. 7th FEBS Meeting (Varna 1971) vol. 23, pp. 53–73. Academic Press, London.

Zimmermann, R. A., A. Muto, P. Fellner, C. Ehresmann and C. Branlant. 1972b. Location of ribosomal protein binding sites on 16S ribosomal RNA. *Proc. Nat. Acad. Sci.* **69**:1282.

Zimmermann, R. A., G. Mackie, A. Muto, R. A. Garrett, E. Ungewickell, C. Ehresmann, P. Stiegler, J.-P. Ebel and P. Fellner. 1974. Location and characteristics of ribosomal protein binding sites in the 16S RNA of *Escherichia coli*. *Nature* (in press).

Protein Topography of Ribosomal Subunit from *Escherichia coli*

R. R. Traut, R. L. Heimark, T-T. Sun,*
J. W. B. Hershey and A. Bollen†
Department of Biological Chemistry
School of Medicine
University of California
Davis, California 95616

1. **The 30S Ribosomal Subunit**
   A. General Principles of Design of the 30S Model
   B. The Three-dimensional Model
2. **The 50S Ribosomal Subunit**
   A. Two-dimensional Projection
3. **Concluding Remarks and Perspectives**

INTRODUCTION

The 70S ribosome of *Escherichia coli* contains 54 different proteins (see Wittmann, this volume). The ribosome may be considered as an organelle whose functional properties are determined by cooperative interactions among its RNA and protein components. There is evidence showing that many of the functional properties of ribosomes, i.e., the capacity to catalyze the specific reactions of peptide bond formation and GTP hydrolysis and to bind a number of protein and RNA ligands, are determined by ribosomal proteins. This does not mean that the role of ribosomal RNA should be ignored. It is nevertheless the aim of this article to summarize available evidence bearing primarily on the spatial arrangement of ribosomal proteins with respect to each other. To achieve this end, models have been constructed which depict the position of all the 30S proteins and half of the 50S ribosomal proteins. It is hoped that the models will prove to be a reasonable approximation of the subunit structures. However at the very least, the models provide a concise and graphic means of summarizing a vast amount of experimental evidence; moreover, they should provide a framework within which to design and correlate the results of future experiments.

In order to simplify construction of models of both the 30S and 50S

Present addresses: * Department of Biology, Massachusetts Institute of Technology, Cambridge, Massachusetts 02139. † Laboratory of Genetics, Free University of Brussels, 67 Rue des Chevaux, 1640 Rhode-St.-Genese, Belgium.

subunits, it has been assumed that all proteins are spheres of roughly equivalent size. This is patently untrue, since it is well established that the proteins range in size from approximately 9000 to 35,000 daltons (S1 excepted; the nomenclature of Wittmann et al. 1971 will be used throughout this article). Moreover, experiments performed with single, pure proteins in solution have indicated axial ratios up to 10:1 for certain proteins (H. G. Wittmann, personal communication), although the shape of a protein in solution need not be the same when incorporated into the ribosome structure. Apparent ambiguities in the arrangement of proteins in the models may well be due to the fact that certain proteins exist in extended rather than globular conformations.

The models are intended primarily to depict neighborhood relationships between proteins, although the 50S model incorporates certain information on the binding of proteins to 5S and 23S RNA. No attempt has been made to indicate actual distances, although studies with bifunctional reagents of defined length do indeed imply maximum distances between proteins. In the case of the 30S subunit, the model conforms grossly to the physical shape of the particle, but no attempt has been made to accommodate ribosomal RNA or the hydration of the particle. On the other hand, placement of each protein in the proximity of one or more other proteins is in itself a reasonable assumption. Regardless of the possible existence of RNA or water of hydration situated between proteins, it has been shown in this laboratory (A. Sommer and R. R. Traut, unpublished results) that most, if not all, of the 30S proteins can undergo protein:protein cross-linking reactions with methyl 4-mercaptobutyrimidate, a cross-linking reagent producing bridges of length 14.6 Å.

The model proposed for the 50S subunit is more hypothetical than that for the 30S subunit. In both cases subjective judgments have been made in evaluating the relevance of various types of experimental data used to construct models. However in the case of the 50S subunit, adequate data from total reconstitution, analysis of ribonucleoprotein subparticles and cross-linking have not been available.

Finally, the question of fractional or repeated proteins has not been considered as a criterion in the construction of either model. The present evidence that certain proteins (fractional) cannot be present as fixed components in all ribosomal subunits, while others (repeated) are present in greater than one copy per subunit, seems indisputable. The models proposed here are based upon qualitative evidence for proximity relationships which apply to these proteins when they *are* incorporated into the ribosome structure. The basic question of the significance of possible ribosome heterogeneity, which is a corollary of the evidence indicative of fractional, marginal and repeated proteins, will not be discussed in detail in this article. However certain apparent correlations of functional properties of the 30S ribosome with the distribution of fractional proteins will become evident later.

PROTEIN TOPOGRAPHY OF THE 30S RIBOSOMAL SUBUNIT

General Principles of Design of the 30S Model

The three-dimensional model (Figure 1), representing twenty of the twenty-one 30S proteins, was constructed from styrofoam balls, the connectors being distinguished by shading or stippling as described in the

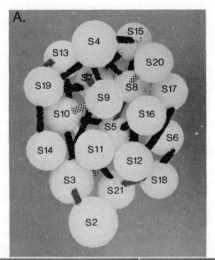

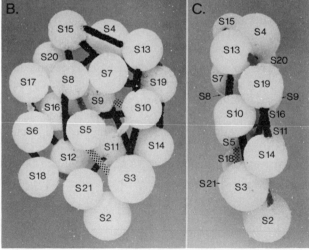

Figure 1   A model of the *E. coli* 30S ribosomal subunit. The model was photographed from (*A*) the top, (*B*) the bottom, and (*C*) the side. Ribosomal proteins are represented by appropriately labeled spheres. Evidence for protein pair relationships is depicted by connectors: grey—cross-linking; stippled—in vitro assembly, binding interdependencies *and* protection from chemical modification; black—in vitro assembly, interdependencies *or* protection.

legend and explained in detail later. In (A) it is shown photographed from the top, in (B) from the bottom, and in (C) from one side. Although all of the evidence shown later in Figures 2 through 11 was considered in constructing the model, the evidence directly depicted in the model by means of the shaded connectors was subjectively evaluated to be of greater importance in arranging pairs of proteins than some of the evidence which is summarized in the two-dimensional projections.

Other models of the arrangement of 30S ribosomal proteins have been constructed: namely, by Morgan and Brimacombe (1973), based largely on their experimental results on the composition of ribonucleoprotein subparticles and supplemented with evidence from cross-linking studies; by C. Cantor and by R. Cox (personal communications), based upon subjective synthesis of all available data, much as has been done here; and finally by computer analysis of data subjectively deemed relevant (A. Bollen, R. Cedergren, D. Sankoff and G. Lapalme, personal communication).

Figure 1C shows that the model is relatively flat; i.e., there is little stacking of proteins and their arrangement is approximately a two-dimensional array. This design principle arose from the physical studies carried out by Van Holde, Hill and others, which are described in detail in the chapter by these authors. Their conclusion, based mainly upon low-angle X-ray scattering studies (Hill, Thompson and Anderegg 1969), was that the 30S particle has dimensions of roughly 220 x 220 x 55 Å. A ribosomal protein of average molecular weight would have a diameter of 30–35 Å, if it exists in a globular (spherical) conformation with hydration, and dimensions similar to other proteins whose structures have been determined crystallographically. Thus extensive stacking of proteins could not occur in a 30S ribosomal structure of thickness 55 Å. Some overlapping or stacking of proteins is evident in the present model, but an attempt has been made to arrange the proteins in the form of a disk approximating the physical data.

Immediately following publication of the assembly map by Nomura and his coworkers (Mizushima and Nomura 1970), several studies on the chemical and enzymatic modification of the proteins in the 30S subunit purported to show that proteins added last in assembly were more reactive, or more exposed, than those proteins added earlier in assembly (Craven and Gupta 1970; Chang and Flaks 1970). Although the experimental results seem indisputable, a variety of investigations of the same general design have not shown a clear distinction between "interior" and "exterior" proteins. Comparison of the modification of ribosomal proteins in the intact particle and free in solution by trypsin (Spitnik-Elson and Breiman 1971; Crichton and Wittmann 1971), by acetic anhydride (Visentin, Yaguchi and Kaplan 1973), by formaldehyde (Moore and Crichton 1973), and by chemical or enzymatic iodination (Miller and Sypherd 1973a,b; C. Cantor, personal communication) have failed to confirm the earlier distinction between interior and exterior proteins. The work of Huang and Cantor

(1972) showed that fluorescein isothiocyanate bound to celite reacted with proteins both early and late in the assembly map. Similarly, both classes of proteins are available for phosphorylation by a protein kinase of molecular weight approximately 40,000 daltons (Traugh, Mumby and Traut 1973; O. Issinger and R. R. Traut, unpublished results). Finally, it has been shown that at least some antigenic sites on all 30S ribosomal proteins in the intact particle react with their cognate antibodies (Stöffler et al. 1973; Hawley and Slobin 1973). It was therefore concluded that the chemical and enzymatic modification studies are not inconsistent with the physical parameters proposed by Van Holde and Hill, and the model was constructed on this premise, i.e., that the arrangement is primarily a two-dimensional array in which at least portions of each protein are exposed on the surface.

The Three-dimensional Model of the 30S Ribosomal Subunit

The model shown in Figure 1 shows the positions of all the 30S proteins of *E. coli* with the exception of S1. The reasons for omitting S1 are as follows: (1) A clear dependence upon specific proteins for the binding of S1 during in vitro assembly has not been demonstrated; (2) protein S1 has not as yet been identified by cross-linking studies in dimers containing other ribosomal proteins; and (3) recent evidence strongly suggests that it may be more appropriate to consider S1 as a soluble factor that is only transiently bound to the 30S particle; indeed it is claimed that S1 is identical both to i factor (H. Inouye, Y. Pollack and J. Petre, personal communications), and to one of the host subunits of Qβ replicase (A. Wahba, personal communication).

In addition to showing in three dimensions the positions of each protein, the model illustrates by means of variously shaded connectors three types of experimental evidence for proximity relationships between the pairs of proteins. These will be discussed individually below.

Cross-linking of 30S Ribosomal Proteins with Bifunctional Reagents. The isolation of dimers consisting of two different ribosomal proteins following reaction of the intact 30S particle with protein-specific bifunctional reagents represents one of the most unambiguous experimental approaches for the determination of proximity relationships between proteins in the ribosome or any other multi-protein assembly. Such reagents are defined chemical compounds whose chemical specificity is well established; the maximum distance bridged by the two functional groups of any specific reagent can be calculated. The use of bifunctional reagents as the sole criterion to construct a model of the 30S subunit would necessitate that each protein be cross-linked to at least one other protein, and that most proteins be linked to more than one protein, thus providing an unambiguous interconnecting network of such relationships. However, only a limited number of protein pairs (or trimers) have been isolated and their components identified at this time.

After the demonstration by Davies and Stark (1970) that bis-imidoesters could be used in the characterization of oligomeric enzyme complexes, several laboratories turned to the application of similar reagents for the study of the ribosome. Among the first were Slobin (1972), Clegg and Hayes (1972), and Bickle, Hershey and Traut (1972). Slobin showed that treatment of both ribosomal subunits with dimethyl suberimidate resulted in the formation of numerous products of high molecular weight. However no results on the composition of any single product were reported. Bickle, Hershey and Traut (1972), in their study of the 30S ribosomal subunit, employed dimethyl suberimidate at concentrations sufficiently low that discrete products of increased molecular weight were apparent on polyacrylamide gels containing dodecyl sulfate (SDS gels). In order to identify the components in the higher molecular weight products, the amidine cross-links were cleaved by ammonolysis, and the monomeric proteins thus generated were identified by gel electrophoresis. Since the ammonolysis reaction proceeded in low yield, the identifications proposed, S5–S9, S11–S19, S6–S7–S9, were only considered as tentative. Nevertheless, the experiments clearly established the usefulness of bifunctional reagents in the study of the protein topography of the ribosome. These data were not used in the construction of the model, since several investigations utilizing reagents both similar and different in chemical specificity and coupled with more unambiguous techniques for the determination of the composition of cross-linked products appeared subsequently.

The data indicating the proximity of specific pairs of ribosomal proteins (or in one case a triplet) which were utilized in construction of the model are briefly summarized in Table 1. A more detailed critical analysis of each experiment appears in the chapter by Kurland in this volume. The protein pairs listed in the table have been placed adjacent to each other in the model and are joined by *grey* connectors. The evidence from cross-linking experiments has been evaluated as representing the strongest evidence for protein proximity relationships. Recent experiments from this laboratory have shown that most, if not all, of the 30S ribosomal proteins can become cross-linked to one or more neighboring proteins with the reagent methyl 4-mercaptobutyrimidate (A. Sommer and R. R. Traut, unpublished results). The pair S5–S9, previously identified by Bickle, Hershey and Traut (1972), has been confirmed. In addition, since the model was constructed seven new protein dimers formed with this reagent have been characterized in this laboratory (A. Sommer and R. R. Traut, unpublished results). With only one exception all the new pairs are consistent with the model. These new results emphasize the utility of the cross-linking approach for the eventual complete solution of the protein topography of the ribosomal particle.

Table 1 Protein:protein interactions identified by cross-linking

Cross-linked proteins	Reagent	Method of identification	Reference
S18–S21	N,N'-p-phenylene-dimaleimide	Coordinate disappearance of proteins analyzed by two-dimensional gel electrophoresis	Chang and Flaks (1972)
	N,N'-p-phenylene-dimaleimide	Immunochemical analysis of purified dimer with antibodies against single pure 30S ribosomal proteins; isotopic labeling of cross-linked proteins	Lutter et al. (1972)
S11–S18–S21	tetranitromethane	Peptide analysis of purified cross-linked product (pair relationship not determined)	Shih and Craven (1973)
S5–S8	dimethyl adipimidate	Immunochemical analysis of purified dimer; isotopic labeling of cross-linked proteins	Lutter et al. (1972)
	methyl 4-mercapto-butyrimidate	Immunochemical analysis of purified dimer; molecular weight analysis of dimer and cleaved products	Sun et al. (1974)
	dimethyl suberimidate	Immunochemical analysis of purified dimer	Sun et al. (1974)
S7–S9	dimethyl suberimidate		Kurland (this volume)
S13–S19	dimethyl suberimidate		Kurland (this volume)
	methyl 4-mercapto-butyrimidate	Immunochemical analysis of purified dimer; molecular weight analysis of dimer and cleaved products	Sun et al. (1974)
S2–S3	methyl 4-mercapto-butyrimidate	Molecular weight analysis and identification by SDS gel electrophoresis of dimer and cleaved products	Sun et al. (1974)

Protein:Protein Interactions Inferred from Ribosome Reconstitution. A detailed description of the *E. coli* 30S subunit assembly map, the methodology used in its construction, and its significance and implications are presented in the article by Nomura and Held in this volume (also see Mizushima and Nomura 1970). The major conclusion drawn from the assembly map relevant to the question of protein topography considered here is the following: In vitro, the binding of a newly added protein to an incomplete particle, assembled by the addition of one or more proteins to 16S RNA, may be strongly dependent on certain specific proteins already present in that particle. This implies the proximity of the proteins interrelated experimentally in this way. These strong interactions are listed in Table 2. The assembly map of Nomura resembles in some respects that presented here and is itself consistent with much other topographical data.

If indeed the above conclusion is correct, then the addition of a later protein to the incomplete particle might protect the protein or proteins upon which its binding is dependent. Craven and his collaborators (see Craven, Rigby and Changchien, this volume) have utilized such an approach in order to determine protein proximity relationships during in vitro assembly by direct chemical analysis of the degree of iodination of specific proteins before and after the binding of additional proteins. These results are also summarized in Table 2.

At the time of writing of this review, there did not appear to be a complete coincidence of the results from the two approaches, dependence *for*

Table 2   Protein:protein interactions during assembly

Strong assembly interactions	Interactions inferred by protection during assembly
S3–S5	S3–S5
S7–S9*	S7–S9*
S8–S20	S8–S20
S9–S10	S9–S10
S3–S10	S3–S9
S5–S8*	S4–S8
S5–S16, S17	S4–S9
S6–S18	S4–S15
S7–S19	S5–S10
S11–S21	S7–S15
S14–S19	S8–S15
	S10–S13
	S13–S15

* Pairs also identified by cross-linking and thus linked by grey connectors in the model (Figure 1).

binding and protection *by* binding. Nevertheless, both approaches were considered as strong evidence for protein:protein proximity relationships, despite the fact that in both cases gross conformational changes induced by the addition of a protein could be invoked to explain the results observed. The protein interactions suggested by both techniques are indicated in the top section of Table 2 and are shown by the *stippled* connectors in Figure 1. Those interactions found in one but not both experimental approaches are listed in the bottom section of Table 2 and are shown by the *black* connectors.

The photograph of the model in Figure 1 shows directly three of the major types of evidence utilized in its construction: cross-linking, assembly and protection during assembly. In addition, other evidence to be summarized below was also taken into account in building the model. In order to facilitate the presentation of the results from a variety of experimental techniques, a series of two-dimensional projections of the top view of the model (Figure 1A) will be utilized. Those results considered most crucial in the construction of the model will be reviewed briefly, and the reader may consult the references given in each section for further details.

Protein Composition of Ribonucleoprotein Subfragments. An additional type of experiment, the protein composition of ribonucleoprotein subribosomal fragments formed by mild ribonuclease digestion of the intact 30S subunit, constituted a significant parameter in defining the proximity of proteins in *groups*. Several reports on the proteins associated with RNA fragments formed by treatment of the 30S ribosome with ribonucleases have appeared (Schendel, Maeba and Craven 1972; Roth and Nierhaus 1973; Morgan and Brimacombe 1973). No attempt will be made here to compare and evaluate these various studies, for there is significant agreement between them. The recent work of Morgan and Brimacombe (1973) represents the most complete study, insofar as first, the digestion conditions employed seemed most likely to preserve the native structure of the subunit; second, the stoichiometric amounts of the proteins found in the ribonucleoprotein subparticles were determined; and third, all subparticles had the same RNA to protein ratio as the intact 30S subunit.

Figure 2 shows the composition of the two largest ribonucleoprotein fragments characterized by Morgan and Brimacombe (1973). They also characterized a number of small fragments; all of these are contained as a subset of one or the other of the major larger fragments, and the reader is referred to Table 2 of the original article (Morgan and Brimacombe 1973) for these details. It is readily apparent that the proteins of the two large ribonucleoprotein fragments are clustered together on opposite sides of the model. Since certain proteins bind directly to 16S RNA and their approximate binding locations in the primary sequence of the RNA are known, some conclusions concerning the distribution of RNA in the model might

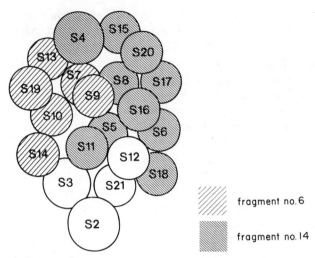

Figure 2 Protein composition of ribosomal subparticles (Brimacombe fragments) produced by ribonuclease.

be drawn. However it is not the purpose of this review to discuss the topography of ribosomal RNA, and the interested reader is referred to the articles by Zimmermann and Fellner in this volume for a complete account of studies bearing on this subject.

Protein Nucleation Sites during Initial Stages of Ribosome Assembly. The work of Zimmermann and his collaborators has shown that the 12 proteins present in the reconstitution intermediate (RI) particles are organized into three independent groups, each of which occupies a specific region of the 16S RNA. Each region must first bind one or more of the proteins (S4, S7, S8, S15 and S20) that had been shown by studies on assembly to bind independently, noncooperatively and in a 1:1 stoichiometry to the 16S RNA (Mizushima and Nomura 1970; Garrett et al. 1971; Zimmermann et al. 1972; Schaup, Green and Kurland 1970, 1971). These proteins are indicated by the heavy circles in Figure 3. The protein composition of the "nucleation sites," each one of which is independent of the other two, is also shown in Figure 3. The model represents these groups as proximal proteins. In addition, the composition of the nucleation sites is consistent with the data of Morgan and Brimacombe (1973); i.e., one (S19, S13, S7, S9) is part of the left-hand large ribonucleoprotein fragment, and the other two (S4, S16, S17, S20) and (S15, S8, S6, S18) are part of the large right-hand fragment.

It is of interest to consider cross-linking data with respect to the evidence presented in Figures 2 and 3. Two cross-linked pairs of proteins, S13–S19 and S7–S9, occur in both the large left-hand fragment of Brimacombe (Figure 2) as well as in one of the nucleation sites of Zimmermann (Fig-

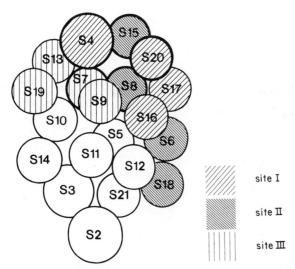

Figure 3   Proteins present in reconstitution intermediate (RI) particles and the Zimmermann nucleation sites. Heavily outlined circles indicate independent binding to 16S RNA.

ure 3). Another cross-linked pair, S5–S8, is found in the right-hand fragment of Brimacombe, but not in any of the nucleation sites. Only one cross-linked protein pair, S5–S9 (Bickle, Hershey and Traut 1972; A. Sommer and R. R. Traut, unpublished results), bridges the two large fragments. The extension of the studies on cross-linking, taken with the protein:RNA binding data, should lead to a more detailed picture of how the RNA is arrayed with respect to the proteins depicted in the present model.

Finally, it should be pointed out that the total composition of the three nucleation sites reported by Zimmermann represents the composition of the reconstitution intermediate (RI) particle previously characterized by Nomura and his coworkers (Held and Nomura 1973). This is not a coincidence, since the experiments of Zimmermann were designed to analyze the attachment of RI proteins; however, the model shown in Figure 3 shows that the 12 proteins of the RI particle were localized at the top and right of the array. This clustering of RI proteins is considered to represent evidence supporting the present model.

Protein Composition of an In Vivo 30S Precursor Particle.   Figure 4 illustrates the protein composition of the in vivo 21S precursor particle isolated by Nierhaus, Bordasch and Homann (1973) and characterized by two-dimensional polyacrylamide gel electrophoresis. The particle contains one of the established cross-linked pairs, S5–S8; the proteins are clustered on the model; proteins from both large ribonucleoprotein subparticles are present; and finally, its composition is similar but not identical

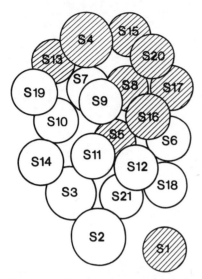

Figure 4   Proteins present in the in vivo "21S" precursor particle.

to that of the RI particle or that of the three Zimmermann nucleation sites. Other precursor particles have been analyzed; e.g., those isolated from cold-sensitive (*sad*) mutants of *E. coli*. The composition of one such particle is as follows: S4, S6, S7, S8, S13, S15, S16, S17, S19, S20 (Nashimoto et al. 1971; Held and Nomura 1973).

Heterogeneity of Ribosomal Proteins.   The model presented here attempts to arrange all the proteins without respect to the question of heterogeneity, i.e., whether the proteins are present in amounts equimolar with 16S RNA ("unit"), less than equimolar (0.5–0.7 copies per copy of 16S RNA) ("marginal"), or less than 0.5 copies per copy of 16S RNA ("fractional"). It is of interest to examine the distribution of "unit," "marginal" and "fractional" proteins in the model presented here. Figure 5 shows the data on protein stoichiometry taken from studies by Traut et al. (1969), Voynow and Kurland (1971), and Weber (1972). Where differences exist between these sets of data, that of the last worker has been represented in the Figure. The unit proteins include S4, S7, S8, S15 and S20, those which have been shown to bind noncooperatively and singly to the 16S RNA. Moreover, they also include other proteins found early in the assembly map, i.e., proteins found in both RI and precursor particles. The unit proteins form a cluster in the model. Fractional and marginal proteins tend to correlate with those late in the assembly map and are also clustered in the model. As will become more evident later, there is a rough correlation between *non*-unit proteins and those implicated in the functions of the subunit.

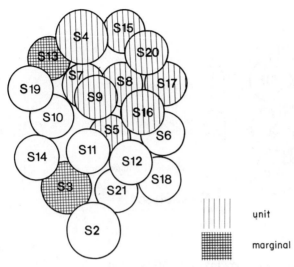

Figure 5 Stoichiometry of 30S ribosomal proteins. Unmarked circles, fractional.

The 30S Proteins Involved in the Interaction of 30S and 50S Subunits.
A property common to all ribosomes is that they are comprised of two
subunits. The functional significance of this fact is still a topic of investigation (see review by Kaempfer in this volume). Since ribosomal subunits
are identifiable as separate entities in cell extracts or lysates, and yet the
activity of the ribosome in protein synthesis resides in the association of the
subunits to form a 70S ribosome (50S:30S couple), it is important to determine which proteins are involved in the interaction between the 30S and
50S subunits. Five different experimental approaches have yielded evidence
relevant to this important structural question; they will be briefly summarized and the proteins implicated by each are represented in Figure 6.

(1) Cross-linking of the 50S subunit to the 30S subunit with formaldehyde
and glutaraldehyde: 50S ribosomal subunits labeled with ^{35}S were
combined with nonradioactive 30S subunits under conditions in which
70S couples were formed. The hybrid 70S ribosomes were incubated
with the reagents mentioned above, and those 70S couples which did
not dissociate at low concentrations of Mg^{++} were isolated. The total
protein was extracted and then tested by Ouchterlony double-diffusion analysis with antibodies against 18 purified ribosomal proteins.
Only the antibody against protein S16 precipitated ^{35}S protein. It
was concluded, therefore, that a protein:protein cross-link was formed
between S16 and one or more 50S proteins (T-T. Sun, L. Kahan and
R. R. Traut, unpublished results).

(2) Protein-deficient core particles were formed by treatment of 30S

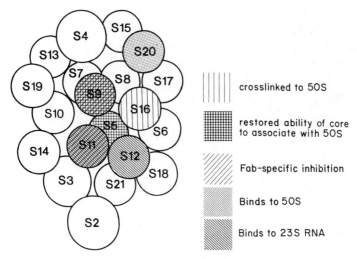

Figure 6   Proteins involved in the binding of the 50S subunit.

subunits with CsCl. They did not recombine with 50S subunits. The specific proteins removed by the high salt concentration ("split" proteins) were tested for their ability to restore the capacity of the 30S core particles to associate with 50S subunits. Proteins S5 and S9 were implicated in the restoration of this property of the 30S subunit (Marsh and Parmeggiani 1973). It should be recalled that a crosslinked dimer of S5 and S9 has been identified (Bickle, Hershey and Traut 1972; A. Sommer and R. R. Traut, unpublished results).

(3) Antibodies specific for each 30S ribosomal protein were allowed to react with 30S particles, and the complexes were then tested for their capacity to reassociate with 50S particles to form 70S couples. Antibodies against proteins S9 and S11 had a significant inhibitory effect on subunit association (Morrison et al. 1974).

(4) Protein S20, although first considered to be a fractional protein when analyzed in 30S particle preparations (Voynow and Kurland 1971), was subsequently shown to be identical to protein L26 found in the 50S particle. When the total amount of protein (S20 + L26) was determined, it was present in amounts indicating one copy per 70S couple (Tischendorf, Stöffler and Wittmann 1974). The fact that S20 is found on both subunits suggests that it is located near the site of interaction of the two subunits.

(5) Proteins S11 and S12 have been shown, uniquely among all 30S proteins tested, to bind to 23S RNA (Stöffler et al. 1971a,b). This result can be interpreted to suggest that these proteins also are involved in the interaction of 30S and 50S subunits. In addition, Ginzburg, Miskin and Zamir (1973) have suggested a role for S18 in subunit association. It is adjacent to the cluster of proteins in Figure 6.

All of the evidence cited above is summarized in Figure 6. The proteins form a cluster or neighborhood in the model. The role of protein:RNA and RNA:RNA interactions in the association of subunits is an important question for future investigation.

30S Proteins Involved in the Binding of Aminoacyl-tRNA to the Ribosome.
Nomura and his collaborators studied the functional roles of 30S ribosomal proteins by constructing particles deficient in single proteins. When protein S21 was omitted from the reconstitution mixture, a 30S particle with markedly decreased capacity to bind aminoacyl-tRNA in the presence of 50S subunits and poly(U) was obtained (Nomura et al. 1969). The function analyzed was therefore the binding of aminoacyl-tRNA to 70S ribosomes, since the lack of S21 has no known effect on subunit association.

Kurland employed a different approach to study the proteins involved in the binding of aminoacyl-tRNA (Randall-Hazelbauer and Kurland 1972). Specific pure 30S proteins were added in excess to intact 70S ribosomes, and the binding of phenylalanyl-tRNA was assayed in the presence of poly(U). Addition of proteins S2, S3 and S14 stimulated the binding of the tRNA.

Rummel and Noller (1973) tested the relative modification of 30S ribosomal proteins by trypsin in the presence and absence of phenylalanyl-tRNA. Proteins S3, S6, S14, S18, S19 and S21 were protected against tryptic attack by the prior binding of the tRNA.

The proteins inferred from these three completely different experimental approaches are shown in Figure 7. In addition, early experiments by Traub et al. (1967) showed that omission of proteins S3, S10 or S14 from protein-deficient particles prepared in CsCl inhibited, and readdition of the

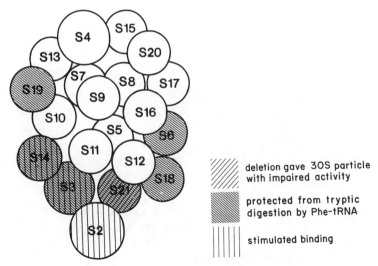

deletion gave 30S particle with impaired activity

protected from tryptic digestion by Phe-tRNA

stimulated binding

Figure 7   Proteins involved in aminoacyl-tRNA binding.

same proteins stimulated, binding of phenylalanyl-tRNA. Of these proteins, S3 and S14 are already represented as implicated from evidence cited above. Protein S10, not indicated in the figure, is proximal to protein S14. By comparison with Figure 5, it is evident that all the proteins implicated as important for tRNA binding are fractional or marginal proteins. They constitute a characteristic neighborhood at the bottom perimeter of the model. However the assays in the first two experiments above were performed not with 30S particles, but with 70S couples. The conclusions are therefore subject to reservations concerning possible conformational changes in the 30S particle taking place as a result of interaction with the 50S subunit.

Studies on the inhibition of binding of aminoacyl-tRNA by antibodies or Fab fragments specific for each 30S protein have been carried out (J. C. Lelong, personal communication). Eleven different proteins (S1, S3, S8, S9, S10, S11, S14, S18, S19, S20 and S21) were implicated in these studies. The combined experimental approaches would thus implicate thirteen of the twenty-one 30S proteins in the binding of phenylalanyl-tRNA in the presence of poly(U).

Ribosomal Proteins Involved in the Binding of fMet-tRNA to the 30S Subunit.   The major experimental evidence implying a set of 30S proteins involved in the binding of fMet-tRNA to the 30S ribosome has been the work of J. C. Lelong et al. (personal communication). Monospecific Fab fragments derived from antibodies against each of the 30S ribosomal proteins were tested for their effect on the binding of the initiator tRNA in the presence of poly(AUG) and the three initiation factors. Fab fragments

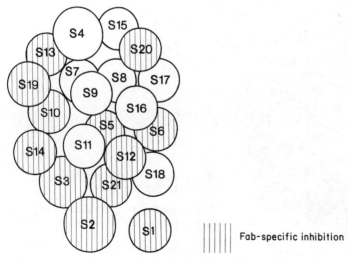

Fab-specific inhibition

Figure 8   Proteins involved in fMet-tRNA binding.

against proteins S1, S2, S3, S5, S6, S10, S12, S13, S14, S19, S20 and S21 appreciably inhibited the binding of fMet-tRNA to 30S subunits (Figure 8).

The results of this investigation show no distinct correlation with the model, primarily because such a large number of proteins are implicated. The results with Fab fragments cited here and in the preceding section raise doubts concerning their applicability to the discrimination of specific functional sites on the ribosome. On the other hand the experiments, although *measuring* the binding of fMet-tRNA to the ribosome, involve the binding of four other ligands. Conformational changes in the formation of the initiation complex are a possibility which cannot be excluded at this time.

30S Proteins Involved in the Binding of Initiation Factors IF-2 and IF-3. Pure initiation factors IF-2 and IF-3 were prepared as described by Hershey and his collaborators (Fakunding and Hershey 1973). The former was radioactively labeled with ^{32}P (Fakunding et al. 1972) and the latter by reductive alkylation with [^{14}C]formaldehyde (Rice and Means 1971; Pon, Friedman and Gualerzi 1972). A complex containing 30S subunits, ^{32}P-IF-2, IF-1 and IF-3 was treated with dimethyl suberimidate; another complex containing 30S subunits and [^{14}C]CH$_3$-IF-3 without the other initiation factors was treated with *N, N'-p-*phenylenedimaleimide. Total ribosomal protein fractions containing the cross-linked radioactive factors were extracted and fractionated and then tested with antisera against single purified 30S proteins in order to determine which antisera would precipitate radioactive initiation factors.

The results of the experiments with ^{32}P-IF-2 (A. Bollen, R. Heimark,

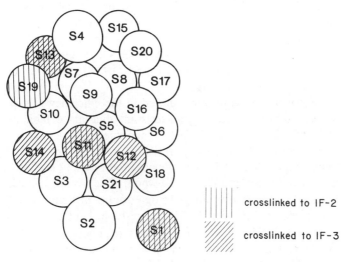

Figure 9 Proteins involved in the binding of initiation factors.

L. Kahan, A. Cozzone, R. Traut and J. Hershey, unpublished results) and [^{14}C]CH$_3$-IF-3 (R. Heimark, L. Kahan, K. Johnston, J. Hershey and R. Traut, unpublished results) are shown in Figure 9; for IF-2: S13, S19, S11 and S1; for IF-3: S13, S14, S11, S12 and S1. The results do not differentiate whether a specific ribosomal protein was cross-linked directly to the radioactive factor or indirectly through a multiple protein complex. A similar observation that S12 is cross-linked to IF-3 has been made by D. A. Hawley, A. Niveleau, L. I. Slobin and A. J. Wahba (personal communication). The proteins associated with either factor do not form closely defined neighborhoods in the model. Proteins S13, S11 and S1 were found associated with both factors. Two of the proteins associated with IF-2, S13 and S19, can form a cross-linked dimer with each other. The two neighborhoods overlap with, but are not coincident with, the fMet-tRNA binding site. This region implicated in factor binding consists primarily of marginal and fractional proteins, not the cluster of unit proteins as found in the RI and precursor particles.

Proximity Relationships among 30S Ribosomal Proteins Derived from Studies with Antibiotics.   Genetic mutations resulting in a single amino acid substitution in protein S12 produce a change from streptomycin sensitivity to resistance (Ozaki, Mizushima and Nomura 1969; Funatsu and Wittmann 1972; for a complete review of the genetics of ribosomal proteins, see chapters by Davis, Tai and Wallace; Gorini; and Jaskunas, Davis and Nomura, this volume). The ribosomes from resistant mutants no longer bind streptomycin (Kaji and Tanaka 1968). This result implies that S12 constitutes part of the binding site for the antibiotic. Mutant strains *dependent* on streptomycin for growth also have a modified S12 protein (Birge and Kurland 1969). Mutations in the gene for protein S4 can suppress the dependence upon streptomycin (Birge and Kurland 1970; Zimmermann et al. 1971). Furthermore, protein S4 is labeled by an affinity analogue of streptomycin (Pongs and Erdmann 1973; O. Pongs, personal communication). As has been suggested by Kurland (1972), these results imply the proximity of proteins S4 and S12.

Streptomycin-resistant mutants are often altered in the efficiency of suppression of amber mutations. Spectinomycin-resistant mutants exhibit an alteration in protein S5 (Bollen et al. 1969). When spectinomycin-resistance mutations are introduced into strains resistant to streptomycin and which contain amber suppressor suII, they alleviate the restriction of suII by the streptomycin-resistance mutation (Kuwano, Endo and Ohnishi 1969). The fact that the spectinomycin gene determines protein S5 and the streptomycin gene determines protein S12 suggests, in light of the evidence presented above, that S5 and S12 may also be neighbors in the ribosome. Furthermore, revertants from streptomycin dependence may arise by mutations in protein S5 as well as in S4 (Kreider and Brownstein 1971;

Deusser, Stöffler and Wittmann 1970; Stöffler et al. 1971c). Therefore there is genetic evidence suggestive of a neighborhood containing S4, S5 and S12.

Chemical evidence, in addition to the affinity labeling of S4, also contributes to a description of the streptomycin binding site. (1) Protein-deficient core particles were unable to bind streptomycin. When proteins S5 and S3 were added back to the core particles, binding activity was restored (Schreiner and Nierhaus 1973). (2) The effect of tryptic digestion of the 30S subunit on the binding of streptomycin was studied, and the loss of binding activity was correlated with the disappearance of proteins S9 and S14 (Chang and Flaks 1970).

All of the information presented above is summarized in Figure 10. While proteins S9, S5 and S12 are neighbors in the model, the other proteins implicated, S3, S4 and S14, are not in this neighborhood. Consistent with the model, proteins S5 and S9 can form a cross-linked dimer. However the distance between the protein labeled by the affinity analogue of streptomycin, S4, and the genetic determinant of the streptomycin phenotype, S12, represents a problem to be solved in future refinements of the model presented here.

Evidence has been reported which indicates two binding sites on the 30S particle for streptomycin (Biswas and Gorini 1972). This result could help to rationalize the multiplicity and distribution of the proteins shown in Figure 10. In any case, the complexity apparent from the evidence on the binding and functional effect of streptomycin may be due to the extensive cooperative interactions between proteins in the ribosome.

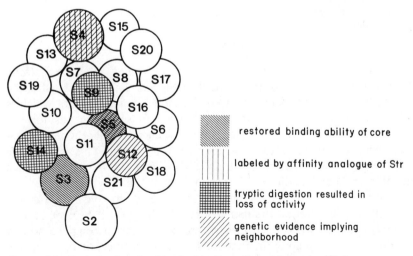

Figure 10   Proteins involved in the binding of streptomycin (Str).

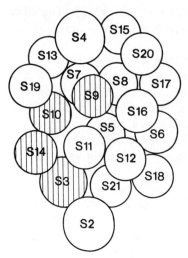

Figure 11   Proteins protecting 3′ end of the 16S RNA from methylation.

Mutations in the *ksg*A locus in *E. coli* lead to kasugamycin resistance by eliminating the activity of a methylase that specifically methylates the two adenine residues in the sequence (AACCUG) close to the 3′ end of 16S RNA (Helser, Davies and Dahlberg 1972). CsCl core particles, but not free undermethylated 16S RNA nor intact 30S subunits (either reconstituted or naturally occurring), were the active substrate for this methylase. By deletion of single proteins from the split protein mixture used in the reconstitution of 30S particles, it was possible to identify S3, S9, S10 and S14 as proteins which protected the 16S RNA in the core particles from methylation (P. Thammana and J. Davies, personal communication). These proteins, shown in Figure 11 as neighbors, may therefore be located at or near the 3′ end of the 16S RNA.

Consistency of the 30S Model with Structural and Functional Data. Brief inspection of the data represented in Figures 2 through 11 shows a rough progression from data which fits the model closely to data which is less consistent with it; i.e., from data which represents the positions of the proteins as clustered in discrete, closely defined neighborhoods to that which represents them as more scattered in the model. This is certainly not by chance; it follows, first of all, from the subjectively evaluated design principles explained earlier. It is useful here to amplify certain aspects of this subjective evaluation. As attempts were made to construct a model consistent with the largest number of data of all types, it became readily apparent that there was much greater self-consistency among the strictly structural data than among the functional data. In essence, then, the model represents a structural arrangement which is subsequently *tested* for consistency with functional data. The fit is not always good; attempts to improve it and retain the structural consistency have been only partially successful. The following points are relevant in attempting to reconcile these discrepancies:

(1) The studies on the binding of aminoacyl-tRNA utilized *70S* ribosomes. Physical data of Van Holde and Hill (this volume) indicate that the 30S subunit when joined to the 50S subunit has a different overall shape than the free subunit. Thus protein proximity relationships in the 70S ribosome may not be identical to those in the isolated 30S particle.

(2) A number of studies on chemical modification of the free 30S subunit versus the 70S couple have purported to identify 30S proteins directly shielded by the 50S particle. An alternative explanation for the change in chemical reactivity of the ribosomal proteins is suggested by studies of Chang (1973), which indicate that the 30S subunit undergoes a slow conformational change after its dissociation from the 50S subunit.

(3) The 30S particle can exist in several conformational states, and the relative availability of certain proteins to specific reagents may differ in these states (Ginzburg, Miskin and Zamir 1973). The group of Elson demonstrated previously that the 30S subunit was inactivated by exposure to 1 mM Mg^{++} and reactivated by heating under appropriate ionic conditions (Zamir, Miskin and Elson 1971).

(4) It is an unanswered question whether the binding of ligands like tRNAs or initiation factors alone without 50S subunits induces conformational changes in the 30S particle like those implied above.

(5) As established long ago by the studies of Nomura on the effect of deleting single proteins from reconstituted particles (Nomura et al. 1969), cooperativity between 30S proteins seems to play an important role in the function of the particle; i.e., deletion of a single protein had effects on several functions; a single function was affected by the deletion of several proteins.

(6) Antibodies or even Fab fragments are large compared to the size of ribosomal proteins. Furthermore, if the ribosome does exist in an equilibrium between different conformational states, it is not difficult to imagine that fixation with an antibody could perturb this equilibrium (Sachs et al. 1972). Thus data from antibody studies have not been weighted heavily in construction of the model.

In conclusion, certain difficulties are inherent in any attempt to represent the 30S ribosomal proteins in a single model, for it seems clear that its conformation is dependent upon a number of factors, including ionic conditions, temperature (Johnson and Walker 1973), and the binding of various ligands. Despite these qualifications and some apparent inconsistencies, the present model is a likely approximation to the protein topography of the free 30S subunit. The mapping of functional information on this model simplifies examination of such data and may provide clues to the nature of possible rearrangements or conformational changes resulting from ligand binding.

PROTEIN TOPOGRAPHY OF THE 50S RIBOSOMAL SUBUNIT

A Two-dimensional Projection

A two-dimensional model showing 17 50S ribosomal proteins has been made (Figure 12). There is at present insufficient evidence to construct a three-dimensional model placing *all* 34 50S ribosomal proteins. The three types of experimental evidence that played a major part in the construction of the 30S subunit model are all lacking for the 50S particle: i.e., total reconstitution of active subunits from purified protein and RNA, extensive data on the cross-linking of 50S ribosomal proteins, and the characterization of ribonucleoprotein subparticles. Therefore a *two*-dimensional arrangement of only those proteins about which experimental evidence is available has been designed. Figure 12 shows the preliminary model used

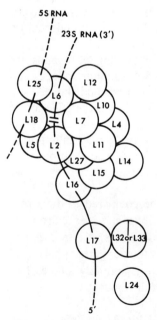

Figure 12   A model of the *E. coli* 50S ribosomal subunit. Ribosomal proteins are represented by the appropriately labeled circles and ribosomal RNAs by the labeled solid lines. The break in the 23S RNA between L6 and L2 represents the observation that L6 binds to the 3′ half, whereas L2, L16 and L17 bind to the 5′ half of the RNA molecule.

here to summarize the present evidence concerning the topography of 17 of the 34 50S ribosomal proteins. It is admittedly more conjectural than that for the 30S, for which more structural evidence was available. The model incorporates the evidence described by Cantor (this volume) with affinity analogues of aminoacyl-tRNA and peptidyl-tRNA and has features in common with the figure in that article. The model resembles even more one presented by M. Pellegrini (1973), whom we thank for aid in the compilation of data summarized here.

The 50S particle contains the active sites for both of the reactions involving covalent bonds in protein synthesis: peptide bond formation, "peptidyl transferase" (Traut and Monro 1964); and GTP hydrolysis (Conway and Lipmann 1964). By virtue of these activities characteristic of the 50S subunit, labeling of functionally important proteins with affinity analogues is a technique that has provided a major part of the information for the 50S model. These results are discussed in detail in the article by Cantor on affinity labeling in this volume and are briefly summarized here in the description of various aspects of the 50S model. Data on the binding of 50S proteins to 23S RNA and 5S RNAs and on the proteins required to bind 5S RNA to the 50S particle is also considered in the model. This subject is reviewed in greater detail in the articles by Zimmermann and Fellner in this volume.

The 50S particle differs from the 30S in several respects relevant to protein topography. There are experiments, more consistent than those for the 30S subunit, on chemical and enzymatic modification which have demonstrated that there may indeed be a set of relatively "protected" or "interior" proteins and another composed of "reactive" or "exposed" proteins (Michalski, Sells and Morrison 1973; Hsiung and Cantor 1973; Litman and Cantor 1973). On the other hand, as with the studies on the 30S subunit, results with different reagents show different patterns of accessibility or protection: no specific protein is completely inaccessible to all of the different reagents tested (see Table III in Litman and Cantor 1973 for a summary of literature bearing on this subject).

The physical evidence concerning the 50S subunit differs in important respects from that for the 30S particle. The results of Van Holde and Hill (see this volume) from low-angle X-ray scattering show the 50S particle to be more spherical than the 30S. It is not a disc; thus it is probable that the protein components are arranged in a *three*-dimensional array. Chemical modification experiments, as mentioned before, are consistent with this conclusion. Furthermore, the results of Moore and Engelman (see this volume) from neutron diffraction experiments indicate that, unlike the 30S particle, the 50S subunit does not have a coincident distribution of RNA and protein. Moore (personal communication and this volume) interprets these results to suggest that there is a protein-enriched region of the 50S subunit which contains many of the proteins involved in translocation, peptidyl transferase, and interaction with the 30S particle. The results summarized here tend to show that this is the case: that a group of 50S proteins, representing not more than half of the total 34 polypeptide species, are localized in the active centers for the three above-mentioned functional properties of the 50S ribosomal subunit.

Evidence from Cross-linking of 50S Ribosomal Proteins.   Acharya, Moore and Richards (1973) treated complexes of 50S subunits and elongation factor EF-G with a variety of cross-linking reagents and found that the factor became cross-linked to a group of 50S proteins which included L7 and L12 (Figure 13). These two proteins have been shown to be essential for all GTP hydrolysis reactions catalyzed by the ribosome (see below).

As will be described later, protein L2 has been shown by affinity labeling experiments to be near the peptidyl transferase site. Since during protein synthesis there is coupling between peptide transfer and GTP hydrolysis and both require a common substrate, peptidyl-tRNA, it is a reasonable hypothesis that the two sites are near each other. Experiments in this laboratory with the cross-linking reagent methyl 4-mercaptobutyrimidate have shown the formation of a dimer between protein L2 and both L7 and L12 (A. Sommer and R. R. Traut, unpublished results). P. Moore (personal communication) found preliminary evidence that L2 was also a

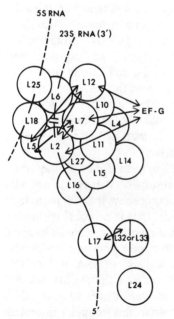

Figure 13   Cross-linking of 50S ribosomal proteins. The double-headed arrow connects proteins identified as cross-linked pairs or complexes.

component of the cross-linked EF-G complex previously described. The new cross-linking results on the isolated 50S particle show directly the proximity of proteins in the peptidyl transferase and translocation sites. Dimers between L5 and both L7 and L12 have also been identified in this laboratory (A. Sommer and R. R. Traut, unpublished results). Protein L5 is also implicated in GTPase activity (see below). Finally, a dimer containing L17 and L32 or L33 has been identified (A. Sommer, unpublished results). Protein L17 has been shown to bind directly to 23S RNA (Stöffler et al. 1971a,b), whereas protein L32 or L33 is one of those labeled with affinity analogues of peptidyl-tRNA of longer chain length (see Cantor, this volume; Pellegrini 1973).

Proteins in the Site Related to GTP Hydrolysis. Proteins L7 and L12 can be removed selectively from the 50S subunit. The resulting protein-deficient particles are inhibited or inactive with respect to the GTP hydrolysis reactions associated with factors EF-G, EF-Tu (Hamel and Nakamoto 1972; Hamel, Koda and Nakamoto 1972; Sander, Marsh and Parmeggiani 1972; Brot et al. 1972) and IF-2 (Fakunding, Traut and Hershey 1973). Furthermore, GTP hydrolysis associated with EF-G is blocked when ribosomes are treated with antibodies specific for L7 and L12 (Kischa, Möller and Stöffler 1971). These antibodies prevent the interaction of EF-G with the 50S subunit (Highland et al. 1973). It is inferred from all this evidence that proteins L7 and L12 are an integral part of the site(s) for GTP hydrolysis on the 50S subunit, for all three factors mentioned above.

More recent evidence indicates additional proteins in the site of GTP hydrolysis. Proteins L2, L6, L18 and L25 are required for the integration of 5S RNA into the 50S ribosome (Gray et al. 1972). Of these proteins, L2 and L6 bind directly to the 23S RNA (Stöffler et al. 1971a,b). Proteins L5, L18 and L25 bind directly to 5S RNA (Yu and Wittmann 1973a; Horne and Erdmann 1972). Recent results of O. Pongs (personal communication) have demonstrated that a complex consisting of 5S RNA plus proteins L18 and L25 has GTPase activity which is inhibited by thiostrepton and fusidic acid. Similar complexes of 5S RNA and 50S proteins which have GTPase activity have been characterized in *Bacillus stearothermophilus* (Horne and Erdmann 1973).

An affinity analogue of GDP, 5'-guanylyl-*p*-azidophenylphosphate, has been synthesized by W. Möller and his collaborators (personal communication; see review by Möller, this volume). When this reagent was photoactivated in a complex containing EF-G, fusidic acid and 70S ribosomes, the reagent reacted with proteins L5, L11, L18 and L16. The proteins L7 and L12, already implicated in functions related to GTP, were not labeled with the affinity analogue. Of those proteins which reacted, L5 binds to 5S RNA and a 5S RNA:protein complex has GTPase activity; L11 is implicated by experiments with SNAP-tRNA and reconstitution to be part of the peptidyl transferase site (see below); and L16 was shown to become labeled by experiments using BAP-tRNA bound in the A site and by monoiodoamphenicol (see below). The multiplicity of proteins labeled with the GTP affinity analogue is somewhat surprising. However there is evidence to suggest the existence of at least two GTPase sites in ribosomes: an IF-2-specific site and an EF-G site (Lockwood and Maitra 1974). The large number of proteins labeled with the GTP analogue may be related to the possible existence of more than one site for GTP hydrolysis.

Proteins L6 and L10 have been shown to stimulate the binding of L7 and L12 to protein-deficient 50S core particles (Schrier, Maassen and Möller 1973). This finding, plus the fact that L6 is also involved in the integration of the 5S RNA, led to the arrangement of these proteins shown in the model. All of the data described above are presented in Figures 14 and 15.

Specific details from the two preceding sections and from studies on the proteins that bind to 5S and to 23S RNA (see reviews by Monier and by Zimmermann, this volume) were used in constructing the 50S model. (1) The proteins that bind to 5S RNA (L5, L25 and L18) and to 23S RNA (L2, L6, L16 and L17) are shown as such. (2) Proteins L2 and L6, which stabilize the interaction of 5S RNA with 23S RNA or with core particles, are located such as to be near both RNA molecules. (3) Proteins L6 and

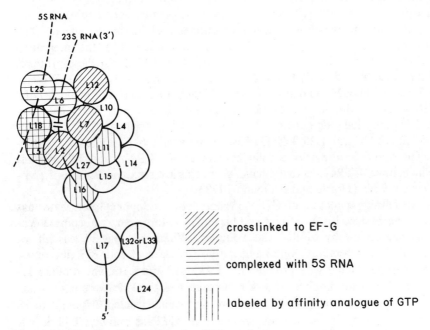

Figure 14   Proteins involved in the GTPase site.

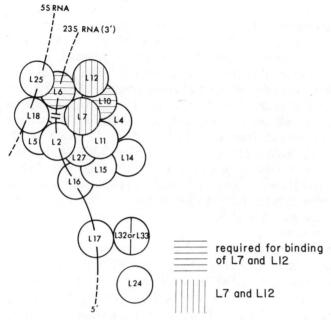

Figure 15   Neighborhood of L7 and L12.

L10, which are required to bind L7 and L12, are shown as neighbors of L7 and L12. (4) The model incorporates unpublished cross-linking data showing the dimers L2–(L7/L12), L2–L5, and L17–(L32/33).

Proteins Implicated near Peptidyl Transferase.   The distinction between A site (acceptor site) and P site (donor site) rests on the operational definition that substrates in the P site, fMet-tRNA or peptidyl-tRNA, may transfer their aminoacyl derivatives to substrates in the A site, i.e., aminoacyl-tRNA or puromycin. However the structural elements of these two sites are inferred to be at least partially overlapping, since in order for peptide transfer to occur, both substrates must interact with the same peptidyl transferase center.

P Site.   Cantor and his collaborators have employed two affinity analogues of peptidyl-tRNA in order to identify proteins in the P site of the 50S ribosome. These experiments are described in detail in the article by Cantor in this volume. An important point in these studies was the finding that the affinity analogues could participate in peptide bond synthesis (Oen et al. 1973). Therefore the proteins identified are the result of specific binding. Furthermore, since peptide transfer still took place, the reagents evidently did not react with and inhibit the catalytic sites of peptidyl transferase.

Studies with bromoacetyl-phenylalanyl-tRNA (BAP-tRNA) and bromoacetyl-methionyl-tRNA led to identification of proteins L2 and L27 linked to the affinity analogue (Pellegrini, Oen and Cantor 1972; Oen et al. 1973; C. R. Cantor, personal communication). Appropriate controls demonstrated conclusively that the reaction depended upon the specific binding of the affinity analogue to an intact active ribosome. Since reaction of *these* affinity analogues required the presence of appropriate reactive amino acid residues, additional experiments were carried out with a photoactivatable affinity analogue of peptidyl-tRNA, SNAP-tRNA (O-[2-nitro-4-azido-phenyl]p-hydroxyphenylacetyl-phenylalanyl-tRNA). The major products with this reagent were L11 and L18 (N. Hsiung, J. Reines and C. Cantor, personal communication). Thus the proteins L2, L27, L11 and L18 can all be inferred to be in the neighborhood of the P site. Since the photoaffinity analogue (SNAP-tRNA) does not require the presence of reactive amino acid residues, the results with it may more accurately implicate proteins in the immediate vicinity of the active center.

Using different affinity analogues of peptidyl-tRNA, p-nitrophenyl-carbamyl-methionyl-tRNA$_f^{met}$ (PNPC-Met-tRNA) and p-nitrophenyl-carbamyl-phenylalanyl-tRNA (PNPC-Phe-tRNA) (Czernilofsky et al. 1974), Kuechler and his collaborators concluded that protein L27 was the major product with the methionine analogue, and L27 plus the additional proteins L2, L16 and L14 were the major products labeled with the phenylalanine analogue (E. Kuechler, personal communication). These results expand the list of proteins in the peptidyl transferase or P site neighborhood to include L2, L27, L11, L18, L16 and L14.

Affinity analogues of antibiotics known to inhibit the peptidyl transferase reaction have also been employed to identify 50S proteins in this neighborhood. Two affinity analogues of chloramphenicol (see also A site below) have been tested. Monoiodoamphenicol reacts specifically with protein L16 (Pongs, Bald and Erdmann 1973), one of the proteins implicated in the experiments of Kuechler. Bromamphenicol, on the other hand, reacted with proteins L2 and L27 (Sonenberg, Wilchek and Zamir 1973), proteins also implicated with BAP-tRNA employed by Cantor and his collaborators.

Partial reconstitution of protein-deficient 50S core particles by addition of individual split proteins has been used by Nierhaus and Nierhaus (1973) to identify the chloramphenicol-binding protein. This was identified as L16. Similar techniques (Nierhaus and Montejo 1973) were used to demonstrate that protein L11 restored peptidyl transferase activity to protein-deficient core particles.

All of these results are summarized in Figure 16. The results as depicted here clearly imply that a number of proteins cooperatively comprise the ribosomal P site. The combined evidence from the experiments with SNAP-tRNA and with reconstitution make L11 one of the most likely candidates for the actual peptidyl transferase protein. Comparison of Figures 14 and

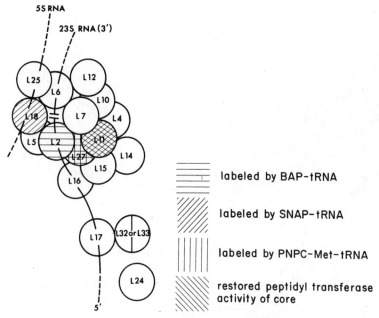

Figure 16   Proteins involved in the binding of peptidyl-tRNA ("P site").

16 clearly indicates the overlap of GTPase and peptidyl transferase sites. Finally it should be mentioned that certain erythromycin-resistant mutants have impaired peptidyl transferase activity correlated with an alteration in protein L4 (Wittmann et al. 1973). For this reason, L4 has been depicted in the model in the position shown.

A Site.   Under appropriate ionic conditions, bromoacetyl-Phe-tRNA will bind to the A site. When binding is performed under such conditions, protein L16 reacted with the affinity analogue (as cited in Pellegrini 1973).

Since the antibiotics puromycin and chloramphenicol compete for the same binding site as the terminal fragment of phenylalanyl-tRNA (CCA-phe) (Pestka 1969; Fernandez-Munoz et al. 1971; Lessard and Pestka 1972), their affinity analogues have been used to identify the proteins contributing to the A site on the 50S subunit (Figure 17) (Nierhaus and Nierhaus 1973). The results obtained with bromamphenicol, an analogue of chloramphenicol, have already been cited; the proteins labeled by it, L2 and L27, were identical to those identified in the P site by affinity analogues of peptidyl-tRNA. On the other hand, another analogue of chloramphenicol, monoiodoamphenicol, reacted with protein L16 (Pongs, Bald and Erdmann 1973), which was required in order to bind unmodified chloramphenicol. Thus L16 is labeled both with an affinity analogue of aminoacyl-tRNA directed to the A site and chloramphenicol. The different proteins labeled

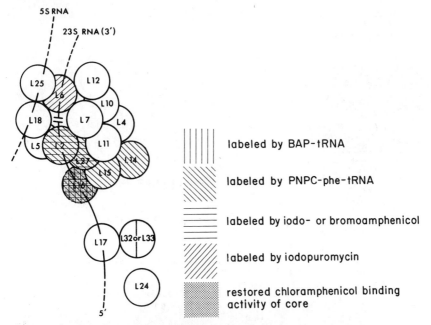

5S RNA

23S RNA (3')

L25   L12

L6   L10

L18   L7   L4

L5   L2   L11

L27   L14

L15

L17   L32 or L33

5'

L24

|||||| labeled by BAP-tRNA

labeled by PNPC-phe-tRNA

labeled by iodo- or bromoamphenicol

labeled by iodopuromycin

restored chloramphenicol binding activity of core

Figure 17   Proteins involved in aminoacyl-tRNA binding ("A site").

by similar affinity analogues of chloramphenicol may be explained by the results of Pestka indicating that there are two modes of binding of chloramphenicol to 50S ribosomes. Both of these inhibit peptidyl transferase activity (Lessard and Pestka 1972).

Pongs (Pongs et al. 1973; personal communication) has found that the affinity analogue of puromycin, N-iodoacetyl puromycin, reacted with protein L6. This protein stabilizes the binding of 5S RNA to the ribosome and supports the binding of L7 and L12 to the 50S particle. To the extent that the modified puromycin is an affinity analogue of aminoacyl-tRNA, this result implies that protein L6 is also in the A site.

50S Proteins Implicated in the Interaction of 50S with 30S Subunits. Although it is beyond the scope of this article to discuss in detail the role of ribosomal RNA, the suggestion was made in the previous discussion of the 30S:50S interaction that RNA might play an important role. This is especially true in the case of the 50S subunit. First, 5S RNA may be involved in subunit association, since alteration of its native conformation in the 50S particle leads to loss of capacity to reassociate with the 30S subunit (Yu and Wittmann 1973b). Second, several 30S ribosomal proteins bind to 23S RNA, suggesting that 23S RNA also plays a role in subunit association.

Three types of evidence depicted in Figure 18 implicate certain proteins

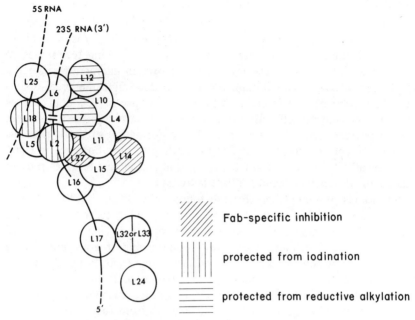

Figure 18   Proteins involved in the binding of the 30S subunit.

that may be involved in the interaction between the subunits. First, W. Möller (personal communication) has shown that some lysine residues of proteins L7 and L12 are protected against chemical modification by the binding of the 30S to the 50S subunit. On the other hand, antibodies against L7 and L12 do not inhibit the reassociation of subunits (Highland et al. 1974), and 50S particles deficient in L7 and L12 can reassociate (Fakund- ing, Traut and Hershey 1973). Second, antibodies against 50S proteins L14, L19, L23 and L27 inhibit the capacity of the 50S particle to associate with the 30S. And third, it has been shown that proteins L2, L26, L28 and L18 are protected against modification by iodination by the association of the 30S to the 50S particle. There appears to be some ambiguity in the identification of protein L26, and it is possible that the protein is indeed L27 (M. Pellegrini, personal communication). The protein L27 and not L26 has been depicted on the 50S model for this reason.

Since all of the approaches used in the experiments cited in this section are open to the qualification that the results might be due to conformational changes induced in the 50S subunit by subunit interaction or by Fab frag- ments, it is clear that chemical cross-linking of the subunits as performed in this laboratory can contribute substantially toward the identification of 50S components involved in subunit association.

CONCLUDING REMARKS AND PERSPECTIVES

The gaps in our knowledge of the protein topography of the two subunits are likely to be filled soon by a variety of experimental approaches. These will likely include: (1) the increased use of cross-linking techniques to determine an interlocking network of protein:protein interactions; (2) the use of protein:RNA cross-linking reagents to determine the precise sites with which each protein interacts with RNA; the use of (3) fluorescence and (4) slow neutron diffraction techniques to determine precise distances between protein pairs; (5) studies on the conformation of single proteins and possible changes therein induced by binding to the ribosome; and (6) the use of affinity analogues to label the actual amino acid residues present at the peptidyl transferase and GTPase active sites.

The models presented here for the topography of all the 30S proteins and 17 of the 50S proteins fulfill several purposes. They serve to summarize a variety of experimental data on ribosomal proteins. They can be tested against new data and may predict new relationships or point out critical experiments by which they can be tested. They represent, in the opinion of the authors, a close approximation to the best models that can be constructed at this time. They may represent, therefore, reasonable approximations of the actual structures which will emerge from future experiments.

Acknowledgments

The authors' work has been supported by grants from the National Institute of General Medical Sciences and the American Cancer Society.

References

Acharya, A. S., P. B. Moore and F. M. Richards. 1973. Crosslinking of elongation factor EF-G to the 50S ribosomal subunit of *E. coli. Biochemistry* **12**:3108.

Bickle, T. A., J. W. B. Hershey and R. R. Traut. 1972. Spatial arrangement of ribosomal proteins: Reaction of the *Escherichia coli* 30S subunit with *bis*-imidoesters. *Proc. Nat. Acad. Sci.* **69**:1327.

Birge, E. A. and C. G. Kurland. 1969. Altered ribosomal protein in streptomycin-dependent *Escherichia coli. Science* **166**:1282.

———. 1970. Reversion of a streptomycin-dependent strain of *Escherichia coli. Mol. Gen. Genet.* **109**:356.

Biswas, D. K. and L. Gorini. 1972. The attachment site of streptomycin to the 30S ribosomal subunit. *Proc. Nat. Acad. Sci.* **69**:2141.

Bollen, A., J. Davies, M. Ozaki and S. Mizushima. 1969. Ribosomal protein conferring sensitivity to the antibiotic spectinomycin in *Escherichia coli. Science* **165**:85.

Brot, N., E. Yamasaki, B. Redfield and H. Weissbach. 1972. The properties of an *E. coli* ribosomal protein required for the function of factor G. *Arch. Biochem. Biophys.* **148**:148.

Chang, F. N. 1973. Conformational changes in ribosomal subunits following dissociation of the *Escherichia coli* 70S ribosome. *J. Mol. Biol.* **78**:563.

Chang, F. N. and J. G. Flaks. 1970. Topography of the *Escherichia coli* 30S ribosomal subunit and streptomycin binding. *Proc. Nat. Acad. Sci.* **67**:1321.

————. 1972. The specific crosslinking of two proteins from the *Escherichia coli* 30S ribosomal subunit. *J. Mol. Biol.* **68**:177.

Clegg, C. and D. Hayes. 1972. Introduction de ponts covalents entre proteins voisines du ribosome 50S d'*E. coli*. *C. R. Acad. Sci.* **275**:1819.

Conway, T. W. and F. Lipmann. 1964. Characterization of a ribosome-linked guanosine triphosphate in *Escherichia coli* extracts. *Proc. Nat. Acad. Sci.* **52**:1462.

Craven, G. R. and V. Gupta. 1970. Three-dimensional organization of the 30S ribosomal proteins from *Escherichia coli*. I. Preliminary classification of the proteins. *Proc. Nat. Acad. Sci.* **67**:1329.

Crichton, R. R. and H. G. Wittmann. 1971. Ribosomal proteins. XXIV. Trypsin digestion as a possible probe of the conformation of *Escherichia coli* ribosomes. *Mol. Gen. Genet.* **114**:95.

Czernilofsky, A. P., E. E. Collatz, G. Stöffler and E. Kuechler. 1974. Proteins at the tRNA binding sites of *Escherichia coli* ribosomes. *Proc. Nat. Acad. Sci.* **71**:230.

Davies, G. E. and G. R. Stark. 1970. Use of dimethyl suberimidate, a cross-linking reagent, in studying the subunit structure of oligomeric proteins. *Proc. Nat. Acad. Sci.* **66**:651.

Deusser, E., G. Stöffler and H. G. Wittmann. 1970. Ribosomal proteins. XVI. Altered S4 proteins in *Escherichia coli* revertants from streptomycin dependence to independence. *Mol. Gen. Genet.* **109**:298.

Fakunding, J. L. and J. W. B. Hershey. 1973. The interaction of radioactive initiation factor IF-2 with ribosomes during initiation of protein synthesis. *J. Biol. Chem.* **248**:4206.

Fakunding, J., R. R. Traut and J. W. B. Hershey. 1973. Dependence of initiation factor IF-2 activity on proteins L7 and L12 from *Escherichia coli* 50S ribosomes. *J. Biol. Chem.* **248**:8555.

Fakunding, J. L., J. A. Traugh, R. R. Traut and J. W. B. Hershey. 1972. Phosphorylation of initiation factor IF-2 from *Escherichia coli* with skeletal muscle kinase. *J. Biol. Chem.* **247**:6365.

Fernandez-Munoz, R., R. E. Monro, R. Torres-Pinedo and D. Vazquez. 1971. Substrate and antibiotic-binding of *Escherichia coli* ribosomes. Studies on the chloramphenicol, lincomycin, and erythromycin site. *Eur. J. Biochem.* **23**:185.

Funatsu, G. and H. G. Wittmann. 1972. Ribosomal proteins. XXXIII. Location of amino-acid replacements in protein S12 isolated from *Escherichia coli* mutants resistant to streptomycin. *J. Mol. Biol.* **68**:547.

Garrett, R. A., K. H. Rak, L. Daya and G. Stöffler. 1971. Ribosomal proteins. XXIX. Specific protein binding sites on 16S RNA of *Escherichia coli*. *Mol. Gen. Genet.* **114**:112.

Ginzburg, I., R. Miskin and A. Zamir. 1973. N-ethyl maleimide as a probe for the study of functional sites and conformations of 30S ribosomal subunits. *J. Mol. Biol.* **79**:481.

Gray, P. N., R. A. Garrett, G. Stöffler and R. Monier. 1972. An attempt at the identification of the proteins involved in the incorporation of 5S RNA during 50S ribosomal subunit assembly. *Eur. J. Biochem.* **28**:412.

Hamel, E. and T. Nakamoto. 1972. Studies on the role of an *Escherichia coli* 50S ribosomal component in polypeptide chain elongation. *J. Biol. Chem.* **247**:6810.

Hamel, E., M. Koba and T. Nakamoto. 1972. Requirement of an *Escherichia coli* 50S ribosomal protein component for effective interaction of the ribosome with T and G factors and with guanosine triphosphate. *J. Biol. Chem.* **247**:805.

Hawley, D. A. and L. I. Slobin. 1973. The immunological reactivity of 30S ribosomal proteins in 70S ribosomes from *Escherichia coli. Biochem. Biophys. Res. Comm.* **55**:162.

Held, W. A. and M. Nomura. 1973. Rate-determining step in reconstitution of *Escherichia coli* 30S ribosomal subunit. *Biochemistry* **12**:3273.

Helser, T. L., J. E. Davies and J. E. Dahlberg. 1972. Mechanism of kasugamycin resistance in *Escherichia coli. Nature New Biol.* **235**:6.

Highland, J. H., J. W. Bodley, J. Gordan, R. Hasenbank and G. Stöffler. 1973. Identification of the ribosomal proteins involved in the interaction with elongation factor G. *Proc. Nat. Acad. Sci.* **70**:142.

Highland, J. H., E. Ochsner, J. Gordon, J. Bodley, R. Hasenbank and G. Stöffler. 1974. Inhibition of elongation factor G function by antibodies specific for several ribosomal proteins. *Proc. Nat. Acad. Sci.* **71**:627.

Hill, W. E., J. D. Thompson and J. W. Anderegg. 1969. X-ray scattering study of ribosomes from *Escherichia coli. J. Mol. Biol.* **44**:89.

Horne, J. R. and V. A. Erdmann. 1972. Isolation and characterization of 5S RNA-protein complexes from B. *stearothermophilus* and *E. coli* ribosomes. *Mol. Gen. Genet.* **119**:337.

————. 1973. ATPase and GTPase activities associated with a specific 5S RNA protein complex. *Proc. Nat. Acad. Sci.* **70**:2870.

Hsiung, N. and C. R. Cantor. 1973. Reaction of celite-bound fluorescein isothiocyanate with the 50S *E. coli* ribosomal subunit. *Arch. Biochem. Biophys.* **157**:125.

Huang, K. H. and C. R. Cantor. 1972. Surface topography of the 30S *Escherichia coli* ribosomal subunit: Reactivity toward fluorescein isothiocyanate. *J. Mol. Biol.* **67**:265.

Johnson, P. M. and I. O. Walker. 1973. The reactivity of *Escherichia coli* ribosomal sulfhydryl groups with 5,5′-dithiobis(2-nitrobenzoic acid). *Eur. J. Biochem.* **38**:459.

Kaji, H. and Y. Tanaka. 1968. Binding of dihydrostreptomycin to ribosomal subunits. *J. Mol. Biol.* **32**:221.

Kischa, K., W. Möller and G. Stöffler. 1971. Reconstitution of a GTPase activity by a 50S ribosomal protein. *Nature* **233**:62.

Kreider, G. and B. L. Brownstein. 1971. A mutation suppressing streptomycin dependence. II. An altered protein on the 30S ribosomal subunit. *J. Mol. Biol.* **61**:135.

Kurland, C. G. 1972. Structure and function of the bacterial ribosome. *Annu. Rev. Biochem.* **41**:793.

Kuwano, M., H. Endo and Y. Ohnishi. 1969. Mutations to spectinomycin resistance which alleviate the restriction of an amber suppressor by streptomycin resistence. *J. Bact.* **97**:940.

Lessard, J. L. and S. Pestka. 1972. Studies on the formation of transfer ribonucleic acid-ribosome complexes. XXIII. Chloramphenicol, amino-acyl-oligonucleotides and *Escherichia coli* ribosomes. *J. Biol. Chem.* **247**:6909.

Litman, D. J. and C. R. Cantor. 1973. Surface topography of the *E. coli* ribosome: Enzymatic iodination of the 50S subunit. *Biochemistry* **13**:512.

Lockwood, A. H. and U. Maitra. 1974. Relation between the ribosomal sites involved in initiation and elongation of polypeptide chains. *J. Biol. Chem.* **249**:346.

Lutter, L. C., H. Zeichhardt, C. G. Kurland and G. Stöffler. 1972. Ribosomal protein neighborhoods. I. S18 and S21 as well as S5 and S8 are neighbors. *Mol. Gen. Genet.* **119**:357.

Marsh, R. C. and A. Parmeggiani. 1973. Requirement of proteins S5 and S9 from 30S subunit for the ribosome-dependent GTPase activity of elongation factor G. *Proc. Nat. Acad. Sci.* **70**:151.

Michalski, C. J., B. H. Sells and M. Morrison. 1973. Molecular morphology of ribosomes. Localization of ribosomal proteins in 50S subunits. *Eur. J. Biochem.* **33**:481.

Miller, R. V. and P. S. Sypherd. 1973a. Chemical and enzymatic modification of proteins in the 30S ribosome of *Escherichia coli*. *J. Mol. Biol.* **78**:527.

———. 1973b. Topography of the *Escherichia coli* 30S ribosome revealed by the modification of ribosomal proteins. *J. Mol. Biol.* **78**:539.

Mizushima, S. and M. Nomura. 1970. Assembly mapping of 30S ribosomal proteins from *E. coli*. *Nature* **226**:1214.

Moore, G. and R. R. Crichton. 1973. Reductive methylation: A method for preparing functionally active radioactive ribosomes. *FEBS Letters* **37**:74.

Morgan, J. and R. Brimacombe. 1973. A preliminary three-dimensional arrangement of the proteins in the *E. coli* 30S ribosomal sub-particle. *Eur. J. Biochem.* **37**:472.

Morrison, C. A., R. A. Garrett, H. Zeichhardt and G. Stöffler. 1974. Proteins occurring at, or near, the subunit interface of *E. coli* ribosome. *Mol. Gen. Genet.* **127**:359.

Nashimoto, H., W. Held, E. Kaltschmidt and M. Nomura. 1971. Structure and function of bacterial ribosomes. XII. Accumulation of 21S particles by some cold-sensitive mutants of *Escherichia coli*. *J. Mol. Biol.* **62**:121.

Nierhaus, D. and K. H. Nierhaus. 1973. Identification of the chloramphenicol-binding protein in *Escherichia coli* ribosomes by partial reconstitution. *Proc. Nat. Acad. Sci.* **70**:2224.

Nierhaus, K. H. and V. Montejo. 1973. A protein involved in the peptidyl tranferase activity of *Escherichia coli* ribosomes. *Proc. Nat. Acad. Sci.* **70**:1931.

Nierhaus, K. H., K. Bordash and H. E. Homann. 1973. Ribosomal proteins. XLIII. *In vivo* assembly of *Escherichia coli* ribosomal proteins. *J. Mol. Biol.* **74**:587.

Nomura, M., S. Mizushima, M. Ozaki, P. Traub and C. V. Lowry. 1969. Structure and function of ribosomes and their molecular components. *Cold Spring Harbor Symp. Quant. Biol.* **34**:49.

Oen, H., M. Pellegrini, D. Eilat and C. Cantor. 1973. Identification of 50S proteins at the peptidyl-tRNA binding site of *Escherichia coli* ribosomes. *Proc. Nat. Acad. Sci.* **70**:2799.

Ozaki, M., S. Mizushima and M. Nomura. 1969. Identification and functional characterization of the protein controlled by the streptomycin-resistent locus in *E. coli. Nature* **222**:333.

Pellegrini, M. 1973. Affinity label studies with bacterial ribosomes. *Ph.D. dissertation,* Columbia University, New York.

Pellegrini, M., H. Oen and C. R. Cantor. 1972. Covalent attachment of a peptidyl-transfer RNA analog to the 50S subunit of *Escherichia coli* ribosomes. *Proc. Nat. Acad. Sci.* **69**:837.

Pestka, S. 1969. Studies on the formation of transfer ribonucleic acid-ribosome complexes. XI. Antibiotic effects on phenylalanyl-oligonucleotide binding to ribosomes. *Proc. Nat. Acad. Sci.* **64**:709.

Pon, C. L., S. M. Friedman and C. Gualerzi. 1972. Studies on the interaction between ribosomes and $^{14}CH_3$-F_3 initiation factor. *Mol. Gen. Genet.* **116**:192.

Pongs, O. and V. A. Erdmann. 1973. Affinity labelling of *E. coli* ribosomes with a streptomycin-analogue. *FEBS Letters* **37**:47.

Pongs, O., R. Bald and V. A. Erdmann. 1973. Identification of chloramphenicol-binding protein in *Escherichia coli* ribosomes by affinity labeling. *Proc. Nat. Acad. Sci.* **70**:2229.

Pongs, O., R. Bald, T. Wagner and V. A. Erdmann. 1973. Irreversible binding of *N*-iodoacetyl-puromycin to *E. coli* ribosomes. *FEBS Letters* **35**:137.

Randall-Hazelbauer, L. L. and C. G. Kurland. 1972. Identification of three 30S proteins contributing to the ribosomal A site. *Mol. Gen. Genet.* **115**:234.

Rice, R. H. and G. E. Means. 1971. Radioactive labelling of proteins *in vitro. J. Biol. Chem.* **246**:831.

Roth, H. E. and K. H. Nierhaus. 1973. Isolation of four ribonucleoprotein fragments from the 30S subunit of *E. coli* ribosomes. *FEBS Letters* **31**:35.

Rummel, D. P. and H. F. Noller. 1973. Functional mapping of the *E. coli* ribosome: Protection of the 30S ribosomal proteins by transfer RNA. *Nature New Biol.* **245**:72.

Sachs, D. H., A. N. Schechter, A. Eastlake and C. B. Anfinsen. 1972. An immunologic approach to the conformational equilibria of polypeptide. *Proc. Nat. Acad. Sci.* **69**:3790.

Sander, G., R. C. Marsh and A. Parmeggiani. 1972. Isolation and characterization of two acidic proteins from the 50S subunit required for GTPase activities of both EFG and EFT. *Biochem. Biophys. Res. Comm.* **47**:866.

Schaup, H. W., M. Green and C. G. Kurland. 1970. Molecular interactions of ribosomal components. I. Identification of RNA binding sites for individual 30S ribosomal proteins. *Mol. Gen. Genet.* **109**:193.

―――. 1971. Molecular interactions of ribosomal components. II. Site-specific complex formation between 30S proteins and ribosomal RNA. *Mol. Gen. Genet.* **112**:1.

Schendel, P., P. Maeba and G. R. Craven. 1972. Identification of the proteins associated with subparticles produced by mild ribonuclease digestion of 30S ribosomal particles from *Escherichia coli*. *Proc. Nat. Acad. Sci.* **69**:544.

Schreiner, G. and K. H. Nierhaus. 1973. Protein involved in the binding of dihydrostreptomycin to ribosomes of *Escherichia coli*. *J. Mol. Biol.* **81**:71.

Schrier, P. I., J. A. Maassen and W. Möller. 1973. Involvement of 50S ribosomal proteins L6 and L10 in the ribosome dependent GTPase activity of elongation factor G. *Biochem. Biophys. Res. Comm.* **53**:90.

Shih, G. T. and G. R. Craven. 1973. Identification of neighbor relationships in the 30S ribosome: Intermolecular cross-linkage of three proteins induced by tetranitromethane. *J. Mol. Biol.* **78**:651.

Slobin, L. I. 1972. Use of bifunctional imidoesters in the study of ribosome topography. *J. Mol. Biol.* **64**:297.

Sonenberg, N., M. Wilchek and A. Zamir. 1973. Mapping of *Escherichia coli* ribosomal components involved in peptidyl transferase activity. *Proc. Nat. Acad. Sci.* **70**:1423.

Spitnik-Elson, P. and A. Breiman. 1971. The effect of trypsin on 30S and 50S ribosomal subunits of *Escherichia coli*. *Biochem. Biophys. Acta* **254**:457.

Stöffler, G., L. Daya, K. H. Rak and R. A. Garrett. 1971a. Ribosomal proteins. XXVI. The number of specific protein binding sites on 16S and 23S RNA of *Escherichia coli*. *J. Mol. Biol.* **62**:411.

————. 1971b. Ribosomal proteins. XXX. Specific protein binding sites on 23S RNA of *Escherichia coli*. *Mol. Gen. Genet.* **114**:125.

Stöffler, G., E. Deusser, H. G. Wittmann and D. Apirion. 1971c. Ribosomal Proteins. XIX. Altered S5 ribosomal protein in an *Escherichia coli* revertant from streptomycin dependence to independence. *Mol. Gen. Genet.* **111**:334.

Stöffler, G., R. Hasenbank, M. Lütgehaus, R. Maschler, C. A. Morrison, H. Zeichhardt and R. A. Garrett. 1973. The accessibility of proteins of the *Escherichia coli* 30S ribosomal subunit to antibody binding. *Mol. Gen. Genet.* **127**:89.

Sun, T-T., A. Bollen, L. Kahan and R. R. Traut. 1974. Topography of ribosomal proteins of the *E. coli* 30S subunit as studied with the reversible cross-linking reagent methyl-4-mercaptobutyrimidate. *Biochemistry* **13**:2334.

Tischendorf, G. W., G. Stöffler and H. G. Wittmann. 1974. Structural homology between two proteins of 30S and 50S ribosomal subunits from *E. coli* as determined by immunochemical methods. *Mol. Gen. Genet.* (in press).

Traub, P., K. Hosokawa, G. R. Craven and M. Nomura. 1967. Structure and function of *E. coli* ribosomes. IV. Isolation and characterization of functionally active ribosomal proteins. *Proc. Nat. Acad. Sci.* **58**:2430.

Traugh, J. A., M. Mumby and R. R. Traut. 1973. Phosphorylation of ribosomal proteins by substrate-specific protein kinases from rabbit reticulocytes. *Proc. Nat. Acad. Sci.* **70**:373.

Traut, R. R. and R. Monro. 1964. The puromycin reaction and its relation to protein synthesis. *J. Mol. Biol.* **10**:63.

Traut, R. R., C. Ahmad-Zadeh, T. A. Bickle, P. Pearson and A. Tissières. 1969. Ribosomal proteins of *E. coli:* Stoichiometry and implications for ribosomal structure. *Cold Spring Harbor Symp. Quant. Biol.* **34**:25.

Visentin, L. P., M. Yaguchi and H. Kaplan. 1973. Structure of the ribosome: Exposure of ribosomal proteins determined by competitive labelling. *Can. J. Biochem.* **51**:1487.

Voynow, P. and C. G. Kurland. 1971. Stoichiometry of the 30S ribosomal proteins of *Escherichia coli. Biochemistry* **10**:517.

Weber, H. J. 1972. Stoichiometric measurements of 30S and 50S ribosomal proteins from *Escherichia coli. Mol. Gen. Genet.* **119**:233.

Wittmann, H. G., G. Stöffler, D. Apirion, L. Rosen, K. Tanaka, M. Tamaki, R. Takata, S. Dekio, E. Otaka and S. Osawa. 1973. Biochemical and genetic studies on two different types of erythromycin-resistant mutants of *Escherichia coli* with altered ribosomal proteins. *Mol. Gen. Genet.* **127**:175.

Wittmann, H. G., G. Stöffler, I. Hindennach, C. G. Kurland, L. Randall-Hazelbauer, E. A. Birge, M. Nomura, E. Kaltschmidt, S. Mizushima, R. R. Traut and T. A. Bickle. 1971. Correlation of 30S ribosomal proteins of *Escherichia coli* isolated in different laboratories. *Mol. Gen. Genet.* **111**:327.

Yu, R. S. T. and H. G. Wittmann. 1973a. The sequence of steps in the attachment of 5S RNA to cores of *Escherichia coli* ribosomes. *Biochem. Biophys. Acta* **324**:375.

―――. 1973b. The structural basis for functional inactivity of reconstituted 50S ribosomal subunits of *Escherichia coli. Biochem. Biophys. Acta* **319**:388.

Zamir, A., R. Miskin and D. Elson. 1971. Inactivation and reactivation of ribosomal subunits: Amino acyl-transfer RNA binding activity of the 30S subunit of *Escherichia coli. J. Mol. Biol.* **60**:347.

Zimmermann, R. A., R. T. Garwin and L. Gorini. 1971. Alteration of a 30S ribosomal protein accompanying the *ram* mutation in *Escherichia coli. Proc. Nat. Acad. Sci.* **68**:2263.

Zimmermann, R. A., A. Muto, P. Fellner, C. Ehresmann and C. Branlant. 1972. Location of ribosomal protein binding sites on 16S ribosomal RNA. *Proc. Nat. Acad. Sci.* **69**:1282.

Functional Organization of the 30S Ribosomal Subunit

C. G. Kurland
The Wallenberg Laboratory
University of Uppsala
Uppsala, Sweden

INTRODUCTION

Studies of ribosome structure have as their principal goal an understanding of the mechanism of protein synthesis. An ancillary, but by no means uninteresting, aspect of the ribosome is its assembly. Indeed, that aspect of ribosome structure which so thoroughly complicates the analysis of ribosome function is precisely the property which makes its assembly so fascinating, namely, its complexity. One objective of the present chapter is to suggest that what we have learned about the mechanism of ribosome assembly provides important clues to the functional organization of the ribosome.

Although most attempts to attribute specific functional contributions to individual ribosomal components have centered around the proteins, I will suggest that it is the RNA which may provide the more interesting active sites of the ribosome. Of course, enzymatic functions such as peptidyl transferase and the nucleoside triphosphate esterase activities that are expressed in the course of protein synthesis must be protein activities. However, the characteristic functions of the ribosome are not these enzymatic ones. It is the binding of specific aminoacyl-tRNA molecules, complexed with factors, and the movement of these macromolecules to different sites, all guided by coded signals in the messenger RNA, that constitute the unique functions of ribosomes. Here the contributions of ribosomal RNA may have been obscured by a number of accidents.

Once each of the ribosomal proteins has been purified and thoroughly characterized, it is easy enough to reconstitute ribosomes with each of the proteins missing in turn or to study the effects of chemical modification and mutations on the functions of individual proteins. But how do you omit from a reconstitution mixture a 100-nucleotide sequence in the middle of the 1500-nucleotide sequence of 16S RNA? How do you select mutants with ribosomal RNA alterations if the organism carries six or more copies of the structural determinants for ribosomal RNA? These or equivalent manipulations of RNA will no doubt be feasible in the future. Until they are available, however, there will be an unrealistic preoccupation with the minor fraction of the ribosome made up of proteins. In effect, the distortion in our present view of the ribosome comes from the fact that the proteins are found in many small pieces, but the RNA comes in a few gigantic pieces.

I will try in this chapter to develop the following ideas. First, the proteins of the small ribosomal subunit are probably dispersed in a matrix of RNA and not, in general, solely in intimate contact with each other. Second, it follows from this that most, if not all, of the proteins will have their own RNA binding sites. Third, it would also follow that cooperative interactions between proteins either during assembly or protein synthesis are not necessarily a consequence of protein–protein interactions. Instead, cooperativity could also be a consequence of more subtle protein–RNA–protein interactions. Finally, it would seem that the large complement of small proteins found in the ribosome has as a principal function the folding of RNA. Accordingly, it may turn out to be more profitable to study the functional contributions of the ribosomal RNA to the mechanism of protein synthesis than was previously suspected.

SPATIAL ARRANGEMENTS

Gross Features

The gross structural feature of the bacterial ribosome most relevant to the present discussion is simple to state: The major fraction of the ribosome mass is made up of RNA (Tissières et al. 1959). Thus the 30S subunit contains a single 16S RNA molecule with a mass close to 560,000 daltons (Kurland 1960; Stanley and Bock 1965). The protein compliment of the 30S subunit contains at most one copy each of 21 different proteins, whose average mass is close to 18,000 daltons (Kurland et al. 1969; Sypherd, O'Neill and Taylor 1969; Traut et al. 1969; Kaltschmidt and Wittmann 1970; Wittmann et al. 1971). The aggregate mass of 30S ribosomal proteins is difficult to estimate with precision because of the apparent heterogeneity of the purified subunits.

If one copy of each 30S protein were present in the subunit, the aggregate mass of protein would be greater than 400,000 daltons. However, purified subunits can contain as little as 260,000 daltons of protein (Kurland 1966). The obvious inference that some of the proteins must be missing from some classes of the purified 30S subnuits (Hardy et al. 1969; Craven et al. 1969) has received extensive experimental support (Kurland et al. 1969; Traut et al. 1969; Voynow and Kurland 1971; Weber 1972). It is not clear, however, whether this heterogeneity is an artifact or a reflection of a functional specialization. In order to avoid unnecessary distractions, we will assume for the calculations to be presented in this section that all 30S subunits are the same: each contains one copy of 21 proteins.

The property of the ribosomal proteins that we are most concerned with here is the volume occupied by an average protein. The anhydrous volume can be obtained from the average molecular weight (17,600) and from the average $\bar{v}$ (0.74), calculated from the amino acid compositions of the proteins (Craven et al. 1969). Such a protein would occupy a volume close to 20,000 Å³. If it were a compact spherical protein, it would have a diameter of 34 Å. It turns out that there are not so many different ways to arrange 21 such proteins in a 30S subunit.

The principal limitation on the ways of distributing the 30S proteins is the geometry of the subunit. Physical studies by Hill et al. (1969a, b, 1970) and by Smith (1971) show that the 30S subunit is quite asymmetric. The X-ray scattering data are best accommodated by assuming that the 30S subunit in solution is an oblate ellipsoid of revolution with dimensions 56 Å × 224 Å × 224 Å. The average thickness of such an ellipsoid would be approximately 43 Å. This means that the 34 Å proteins can be accommodated as a monolayer, but not in two layers. In general, the 30S proteins are too large to be arranged so that some are buried in an inner core and others in a surrounding shell (Kurland 1972).

The data concerning the distribution of the proteins in the functional subunit support this interpretation. Antibodies specific for each of the ribosomal proteins have been raised in rabbits and sheep (see Stöffler, this volume) as well as in mice (Hawley and Slobin 1973). These antibodies, in the form of IgG molecules or the smaller Fab fragments, have been used to study the accessibility of the proteins in the ribosomes (see Stöffler, this volume). All the 30S proteins and most of the 50S proteins can be shown by several independent assays to have at least some of their antigenic sites accessible to the corresponding antibodies in the functional ribosome. We will discuss this data in more detail below; for the moment, its principal value is to eliminate from discussion models of ribosome structure that postulate the existence of "buried" proteins.

Having said this, I am obliged to account for data suggesting that there are "buried" proteins in the 30S subunit. Chang and Flaks (1970) and

Craven and Gupta (1970) have identified a partially overlapping set of proteins that are masked from the effects of trypsin and chemical derivatizing agents in the intact 30S subunit. But do such data tell us anything about the location of the masked proteins?

If one or more side chains in a protein are masked, this does not mean that the rest of the protein is also masked. Furthermore, intimate interaction with RNA could account for some masking effects, as has already been demonstrated with a 50S protein by Crichton and Wittmann (1973). Finally, Miller and Sypherd (1973) have shown that the masking effects they observe with proteins in the 30S subunit are not dependent on the retention of the intact structure of the subunit. The proteins in an unfolded particle are as well masked as those in the functional subunit. Clearly the positive results obtained with antibodies should take precedence in our thinking about protein arrangements over the negative results obtained in "masking" experiments.

If we assume that the proteins are globular, arranged in a monolayer, and all partially exposed on the surface of the 30S subunit, we can arrange them as shown in Figure 1. Here we have represented the proteins as slightly elongated prolate ellipsoids of revolution, oriented with their long axis parrallel to the minor axis of the subunit. The average length of each protein is taken as 43 Å, which corresponds to the average thickness of the subunit along this axis. Accordingly, the average diameter of the proteins at their midsection would be 32 Å.

Protein Separations

The next problem we consider concerns the separation of the 30S proteins from each other. If we take the model in Figure 1 as the basis of our calculations, we can reduce the problem to a two-dimensional one. Thus we calculate the average minimum distance between proteins with diameters of 32 Å in a circle with a diameter of 224 Å. Such a circle has an area 3.9×10^4 Å^2. If the 21 proteins are distributed uniformly within this circle, each will be associated with a domain close to 1800 Å^2. This would be equivalent to a square with sides roughly 42 Å. Hence, the smallest distance between two average proteins centered in adjoining squares would be 10 Å.

This calculation is based on two unsupported assumptions: one, that the proteins are uniformly distributed in the 30S subunit and two, that the proteins are globular. For the purposes of the present analysis, however, both assumptions are dispensable. For example, an alternative arrangement of proteins is seen in the chapter by Van Holde and Hill (this volume). There the proteins are aligned in rows which are separated by RNA. These and many other kinds of arrangements of either globular or fibrous proteins with or without protein–protein contacts would permit extensive interactions between proteins and RNA. Another class of models can also be imagined with the distinctive feature that most of the 30S proteins would be bound to each other and not in contact with RNA. Thus without evidence to the contrary, we are obliged to consider models in which only a small subset of the 30S proteins is in contact with RNA and the majority are densely packed in RNA-free domains. There are, however, two lines of evidence which are difficult to reconcile with this latter class of models.

One way to study the separation of the proteins in the ribosome is to employ an homologous series of cross-linking reagents which have the same functional groups but differ only in length. By comparing the lengths of reagents that do and do not cross-link proteins, some rough idea of the relative distances of the proteins from each other can be obtained. For example, a series of diimidoesters have been used by two groups (Bickle, Hershey and Traut 1972; Lutter et al. 1972; Lutter and Kurland 1973; Lutter et al. 1974). The reagents DMM (dimethylmalonimidate), DMA (dimethyladipimidate) and DMS (dimethylsuberimidate) can bridge distances between the proteins corresponding to approximately 5 Å, 9 Å and

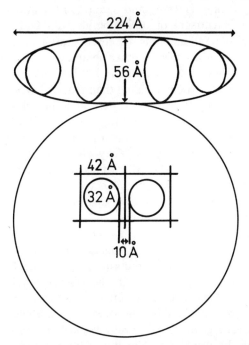

Figure 1   One possible way of distributing the proteins of the 30S ribosomal subunit of *E. coli*. The subunit is represented as an oblate ellipsoid with dimensions 56 × 224 × 224 Å. The top view is a cross section in a plane containing the minor axis, and the bottom view is a cross section in the plane defined by the major axes. The proteins are represented as prolate ellipsoids with average dimensions 43 × 32 × 32Å. This corresponds to the average volume calculated from the molecular weights and partial specific volumes of the 30S proteins, excluding the aberrant S1. The spacings between the proteins were calculated as described in the text.

12 Å, respectively. A comparison of the effects of these reagents on the proteins of the 30S subunit, along with some simple control experiments, has turned out to be quite provocative.

The imidoesters form covalent bonds with the ε-amino group of lysine, which is one of the most common amino acids in ribosomal proteins. Nearly one out of every ten amino acids in this system is a lysine; this corresponds to approximately sixteen lysines for every protein. Fully two-thirds of these lysines are reactive with imidoesters in the intact 30S subunit (Slobin 1972), and all of the 30S proteins will be labeled by a radioactive monofunctional imidoester which is an analog of DMA (Lutter 1973). Therefore an average 30S protein should have approximately eleven sites for reactivity with a diimidoester. Accordingly, we would expect reagents such as DMM, DMA and DMS to create extensive intermolecular cross-links if the 30S proteins are concentrated in one or a few RNA-free domains and are in intimate contact with each other.

Nothing like this is observed (Bickle, Hershey and Traut 1972; Lutter et al. 1972; Lutter et al. 1974). Treatment of 30S subunits with DMS results in the production of substantial amounts of three cross-linked pairs of proteins along with traces of a number of other complexes. (The identification and interpretation of these pairs of proteins will be taken up in the next section.) The effect of DMA is less striking than that of DMS, and DMM creates no cross-linked complexes that are detectable in poly-acrylamide gels. Thus bifunctional reagents which are known to form covalent bonds with more than 200 lysines per 30S subunit are not particularly effective in forming intermolecular cross-links. They do form intra-molecular cross-links, however, and even DMM, which fails to form any cross-linked complexes, seems to form extensive intramolecular cross-links (Slobin, personal communication). Clearly factors other than the chemical reactivity of the proteins limit the yield of the complexes.

The observation that those complexes formed by this series of reagents are produced in greatest yield by the largest reagent suggests that one of the limiting factors is the distance between the neighboring proteins. Since we know nothing about the shapes of the proteins in the ribosome nor the three-dimensional distribution of the lysines within the proteins, we cannot conclude that the proteins are nowhere in contact with each other. However, the data suggest that arrangements of densely packed proteins which are held in position solely by protein–protein interactions are unlikely.

The Mixed Neighborhood Model

One consequence of models in which the proteins are dispersed in some manner through a matrix of RNA is that each of the proteins would be available for direct interaction with the RNA.

In contrast, a model which concentrates the proteins in domains deficient in RNA requires most of the proteins to have binding sites on other proteins. The two models predict rather different results in an experiment involving the destruction of 30S subunits with ribonuclease. The present model would predict that ribonuclease might release fragments of RNA protected from nuclease digestion by associated proteins. The alternative model predicts that most of the RNA should be hydrolized and that protein-rich fragments or free proteins should be released by nuclease digestion.

The results of the relevant experiments are rather clear (Morgan and Brimacombe 1972; Roth and Nierhaus 1973; Schendel, Maeba and Craven 1972). Nuclease digestion of 30S subunits, even under conditions which disrupt the normal structure of the subunits, leads to the release of a large population of ribonucleoprotein fragments. Careful analysis of the compositions of these fragments shows that they contain very nearly the same relative amounts of protein and RNA as the native 30S subunit. This is true for both large and small fragments. These results suggest that a majority of the 30S proteins have RNA binding sites. However, it could be argued that the ribonucleoprotein fragments recovered in these experiments are random aggregates created after the destruction of the ribosome. Considerable evidence indicates that the fragments are not artifacts of this sort. To justify this remark the data on ribosome assembly in vitro as well as the correlated protein neighborhoods must be discussed. This will be done in the next section.

The data and calculations discussed so far can be summarized as follows: The proteins of the *E. coli* 30S subunit make up a minor fraction of its volume. Accordingly, the proteins could be arranged so that contacts with each other are limited. This might account for the low yields of cross-linked complexes produced by bifunctional reagents of the sort studied so far. Although it is not yet possible to determine to what extent the 30S proteins interact directly with each other, it does seem likely that most of them are available for interaction with sites on the 16S RNA. For convenience I will refer to all arrangements of RNA and protein with these properties as the mixed neighborhood model. The implications and support for this model will be explored in succeeding sections.

MOLECULAR INTERACTIONS AND ASSEMBLY

The arrangements of the proteins and RNA in the functional ribosome should be reflected in the kinds of molecular interactions observed during assembly. Indeed, the particular model suggested in the preceding section makes quite specific predictions about the kinds of interactions and intermediates expected during the assembly of 30S subunits. In general, we expect protein–protein interactions to be supplemented, if not dominated, by protein–RNA interactions in this system.

The most obvious requirement of the present model is that all multicomponent intermediates in the assembly of 30S subunits should contain RNA. Although no RNA-free intermediates in ribosome assembly have been observed to date (see Kjeldgaard and Gausing, Nomura and Held, this volume), this is less than convincing. The failure to detect multicomponent precursors containing only proteins could be a consequence of the bias of current experimental techniques, which depend on the large size of the ribosomal RNA for identification of the objects of interest. Clearly one way of destroying the present model would be to detect or construct such protein-rich precursors to 30S subunits. It should also be noted, however, that the detection of only a few proteins that aggregate in a specific way before being incorporated into the subunit would not constitute a significant challenge to the present model.

A variety of different experimental results can be used to relate the structure of the ribosome to its assembly mechanism. Here we will first take up the distinction between proteins that independently bind to RNA and those which are dependent on cooperative interactions for their incorporation into the ribosome. Then the evidence indicating that even this latter class of proteins may be bound to RNA will be considered. Next the identification of protein neighborhoods and their correlation with the assembly data will be discussed. Finally, the connection between the protein neighborhoods and the RNA binding sites will be made by considering the structural effects of binding proteins to the 16S RNA.

Assembly Clusters

Several groups have studied the interactions of individual 30S proteins with 16S RNA under conditions that facilitate reconstitution of 30S subunits in vitro (Traub and Nomura 1968). There is now general agreement that five of the proteins (S4, S7, S8, S15 and S20) can bind in a site-specific manner to 16S RNA in the absence of other proteins (Mizushima and Nomura 1970; Schaup, Green and Kurland 1970; Schaup, Green and Kurland 1971; Garrett et al. 1972; Zimmermann et al. 1972). Two others are more problematic.

Although very weak binding of S13 was observed by Mizushima and Nomura (1970) and Zimmermann et al. (1972), no evidence for the site specificity of this complex was described. More recently Held, Ballou, Mizushima and Nomura (personal communication) reported that the binding of S13 was variable in their hands. Binding of S13 at a very low level in the absence of other proteins was also observed by Green and Kurland (1973); however, the presence of additional proteins, particularly S7, enhanced the site-specific binding of S13. Finally, binding of S17 to 16S RNA in the absence of additional proteins has been observed by several groups. Schaup, Green and Kurland (1970) and Garrett et al. (1972) could not find evidence that this binding was site-specific. More recently Held et al. (personal communication) reported experiments indicating that S17 can bind in a site-specific manner to the 16S RNA in the absence of additional proteins.

Accordingly, there is evidence that at least five and at most seven proteins will bind in a site-specific manner to 16S RNA without cooperative effects of other proteins. This group of proteins has been called HIP (helper independent positioning) proteins in order to distinguish them from proteins which bind to RNA with the assistance of other proteins (Green and Kurland 1973). However, the functional distinction between HIP and non-HIP proteins may not be as important as was previously believed; we will return to this point below.

The cooperative interactions required for the binding of the non-HIP proteins are, to say the least, complicated. These are described in detail in the chapter by Nomura and Held (this volume), and I have chosen a subset for use in the present discussion. There are at least three groups of proteins within which there is evidence of cooperative interactions during assembly in vitro. One such group contains S4 and S20, which bind at the 5′-terminal quarter of the 16S RNA (Schaup et al. 1972; Zimmermann et al. 1972) and together facilitate the binding of S16 (Green and Kurland 1973; Held et al., personal communication). The binding sites for S8 and S15 have been located in the middle of the 16S RNA molecule (Zimmermann et al. 1972; Schaup et al. 1973), and S5 will bind to RNA in the presence of S8, as well as S16 (Mizushima and Nomura 1970; Held et al., personal communication). Finally, the binding site for S7 has been located on the 3′ half of the 16S RNA molecule (Zimmermann et al. 1972). The non-HIP proteins which directly or indirectly respond to the presence of S7 are S9, S13, S14 and S19 (Mizushima and Nomura 1970; Green and Kurland 1973; Held et al., personal communication).

The observation that nuclease degradation of 30S subunits releases a complex series of ribonucleoprotein particles was introduced in the previous section as an indication that both HIP and non-HIP proteins have RNA binding sites. We are now in a position to justify the conclusion that these fragments are not random aggregates created after the destruction of the ribosome.

Morgan and Brimacombe (1972, 1973) have analyzed the compositions of a large number of ribonucleoprotein fragments, ranging in size from those containing only two proteins to others with as many as nine proteins. The member proteins of these fragments can be related to the pattern of cooperativity manifest between the same proteins during assembly in vitro. For example, fragments containing various combinations of S9, S10, S13, S14 and S19 are recovered, and all of these contain S7. Similarly, S5, S8 and S15 are found in different combinations and still larger fragments contain these three proteins as well as S4, S16 (or S17) and S20. In other words, the three clusters of cooperative proteins deduced from assembly data are represented in related clusters as the RNP fragments obtained by nuclease treatment of 30S subunits.

These data, as well as the neighborhood assignments that will be discussed below, make it quite unlikely that the fragments are random aggregates. Similarly, an analysis of the nucleotide sequences recovered from one fragment containing the cluster of proteins organized around S7 (Szekely, Brimacombe and Morgan 1973) shows that this RNA comes from that part of the 16S molecule where S7 is known to bind (Zimmermann et al. 1972).

It has been suggested that the groups of proteins recovered on these fragments represent near neighbors in the ribosome. This is of course possible, particularly for those few proteins recovered on the smallest fragments. However, the larger fragments present a problem. The RNA in these fragments, if extended, could be as long or longer than any dimension of the 30S subunit. As a consequence, proteins distributed along such RNA segments could be positioned at quite some distance from each other in the ribosome. Therefore we require some sort of independent evidence that these clusters of proteins are in fact near neighbors in the intact ribosome. Such evidence can be obtained from cross-linking experiments.

Protein Neighborhoods

The basic approach employing bifunctional reagents to study neighbor relations between proteins in the functional ribosome is described above. Although it is simple enough on paper, the identification of the proteins which are coupled to each other in a cross-linked complex is not trivial. Lutter and coworkers (Kurland et al. 1972a,b; Lutter et al. 1972, 1974; Lutter and Kurland 1973; Lutter 1973) have provided evidence for the cross-linking of the following pairs of proteins: S5–S8, S7–S9, S13–S19 and S18–S21. The first three are obtained with DMA and DMS, while the last is obtained with a sulfhydryl reagent PDM (phenylenedimaleimide).

The criteria employed by Lutter and coworkers to identify the crosslinked pairs were as follows: immunological reactivity of purified complexes, isotopic labeling of specific complexes, fractionation of complexes based on their immunological reactivity, and comparison of molecular weights of complexes with those of member proteins as well as measurements of the molar ratios of members in purified complexes. The application of independent criteria, all of which agree, lends considerable weight to the identification of the member proteins in these complexes.

It does not follow, however, that such well-identified complexes represent neighborhoods in the functional ribosome. Only a fraction of the purified ribosomes are active and the cross-linking reagents, after all, modify the structure of the ribosomes. Therefore additional evidence is required to determine whether or not these neighborhoods exist in the functional ribosome. One way to obtain such evidence has been described by Lutter and Kurland (1973). If it can be shown that a cross-linked complex can replace its member proteins in the reconstitution of active ribosomes, it is likely that this complex represents a meaningful neighborhood. It is possible to reconstitute 30S subunits containing the complex of S5 and S8; such ribosomes are very nearly as active as their counterparts reconstituted with the individual proteins S5 and S8 (Lutter and Kurland 1973). Similarly, the complex of S13 and S19 can be incorporated into 30S subunits that are approximately 60% as active as subunits containing the corresponding individual proteins (Lutter 1973).

Such data is still wanting for the other two cross-linked pairs described above. However in the case of the S7–S9 complex, it will be shown below that this neighborhood, as well as others, can be correlated with the cooperativity expressed between proteins during 30S assembly in vitro. Correlations such as these provide an additional way of controlling the character of the protein neighborhoods determined in cross-linking experiments, since it is likely that proteins which interact with each other during assembly will be near neighbors in the ribosome.

Unfortunately, these data are not consistent with the preliminary identification of complexes purified by Bickle, Hershey and Traut (1972). The principal problem seems to reside in the procedure which they used to study the complexes created by DMS. Their purified complexes were cleaved by ammonolysis at high pH, and the member proteins so released were analyzed on one-dimensional polyacrylamide gels. In all cases there was partial agreement with the results of Lutter and coworkers, but in no case was the agreement complete. Thus the complex which corresponds to S5–S8 was identified by them as S5–S9, that which corresponds to S7–S9 was thought to contain S6 or S7 and S9, and that which corresponds to S13–S19 was thought to contain S11 or S12 and S19. In the hands of Lutter et al. (1972), ammonolysis of the cross-linked complex appears to have two serious limitations. First, the recovery of member proteins from the cross-linked complex is less than 10%. This means that contaminants of a putative complex may be mistaken for member proteins. Second, the exposure of the proteins to ammonolysis effects the electrophoretic mobility of the proteins, which subsequently migrate in gels in a diffuse band. This makes the identification of such proteins on one-dimensional gels very difficult. It is significant that recent experiments by Traut and his coworkers (personal communication) have detected cross-linked complexes of S5 and S8 as well as S13 and S19.

Still less satisfying was the procedure used by Chang and Flaks (1972) to characterize the complex which they identified as S18–S21. Two facts were used here. First, they found that S18 and S21 are strongly diminished in amounts after treatment of ribosomes with PDM. Second, they discovered a new electrophoretic component with a molecular weight close to the sum of that of S18 and S21. That S18 and S21 have indeed been identified in a cross-linked complex (Kurland et al. 1972a; Lutter et al. 1972) does not mean that the particular complex identified by Chang and Flaks is S18 and S21. Thus several other components are cross-linked by PDM in addition to S18 and S21 (Kurland et al. 1972a,b; Lutter 1973). Furthermore, the sum of molecular weights of many pairs of proteins will be approximately the same as that of S18 and S21.

For the reasons described above, I take as established the following cross-linked pairs of proteins: S5–S8, S7–S9, S13–S19 and S18–S21. Although these represent only a minor fraction compared to the neighborhoods remaining to be identified, a pattern is already emerging. Three of these cross-linked pairs can be correlated with the assembly data and the fragment analysis discussed earlier. Thus S5 and S8 cooperate for assembly and are found on the same small RNP fragments released by nuclease digestion. Similarly, S7, S9, S13 and S19 are members of a common assembly cluster and these too can be found in various combinations on a large series of RNP fragments. Furthermore, in this particular case we have direct evidence that these proteins share a common region of the 16S RNA located at the 3' end of the molecule.

Such results provide strong support for the mixed neighborhood model of the 30S ribosomal subunit, at least for these six proteins. The fact that these clusters of proteins are cross-linkable in the intact ribosome also suggests that the rather large stretches of RNA with which they are associated are not extended in the ribosome. Thus pairs of the proteins must have lysines which are positioned at least within 10 Å of each other; this could be accomplished by folding the RNA during the assembly process.

RNA Folding and Cooperativity

If we assume for the moment that the mixed neighborhood model is correct, then it is far from obvious that all of the cooperative interactions seen between different proteins during ribosome assembly are a consequence of direct protein–protein interactions. It is, for that matter, not clear whether any of the interactions are of this type. Therefore it would seem appropriate to consider briefly the kinds of interactions between proteins that could be mediated indirectly by RNA. It turns out that the little that is known about the RNA binding sites for some of the HIP proteins suggests at least one way that one protein could provide a binding site for another without a direct contact between the two proteins.

The availability of a simple procedure for the formation of one-to-one complexes of individual HIP proteins and 16S RNA could be exploited to identify protein-specific RNA binding sites (Schaup et al. 1972). Here, S4 was bound to 16S RNA and the complex digested with pancreatic ribonuclease. A core of RNA resistant to the nuclease and still bound to S4 can be recovered after this treatment: this RNA is called S4aR. It turns out that S4aR is not a short sequence of nucleotides. Instead, it is made up of at least six fragments, each with a unique nucleotide sequence, and the aggregate has a mass well over 100,000 daltons. The remarkable property of the S4aR fragments is that following purification they can be incubated again with S4 and they will reform a site-specific complex (Schaup and Kurland 1972). No other HIP protein will interact with S4aR. Thus several large fragments carry the discontinuous nucleotide sequences specific for the binding of S4.

It is unthinkable that a small protein such as S4, with a molecular weight of 27,000, can physically shield more than 100,000 daltons of RNA from nuclease. On the other hand, such a protein could bind to short, discontinuous nucleotide sequences on the RNA in such a way that helical regions of the RNA between the bound sequences were not disrupted. Since helical RNA would be relatively resistant to ribonuclease, exposed single-stranded RNA could be preferentially cleaved by nuclease digestion of the complex. As a consequence, the actual binding sites of the protein would be recovered along with a much larger mass of helical RNA. These would appear on separate fragments because extraneous single-stranded regions between the binding sites would have been cleaved (Schaup et al. 1972; Schaup and Kurland 1972).

The binding of a small protein to multiple sites distributed along one-fourth of the 16S RNA should have the effect of collapsing that part of the RNA into a more compact structure. This prediction is directly supported by the structural studies of Nanninga et al. (1972), which showed that S4 can organize approximately one-fourth of the 16S RNA into a compact structure clearly visible in the electron microscope. Accordingly, S4 may provide an extreme example of what may be an important function of many of the ribosomal proteins; i.e., to organize the tertiary structure of associated RNA sequences.

If one protein can organize the folding of the RNA to which it binds, it might create binding sites for other proteins without a direct interaction with the dependent proteins. For example, S16 requires the presence of S4 and S20 before it can associate with the 16S RNA. The binding of S4 and S20 to the RNA could bring into close and stable apposition sequences of nucleotides that can associate specifically with S16. In the absence of S4 and S20 these same sequences may be unable to form the binding site for S16 simply because the requisite configuration is not energetically favored.

In summary, the RNP fragments released by nuclease digestion are concrete expressions of the mixed neighborhood model. Here, groups of proteins are thought to cooperate in their attachment to common regions of the 16S RNA. The cooperative interactions between such clusters of proteins are thought to reflect their mutual dependence on a correct folding of the RNA. Surprisingly, it remains to be seen whether or not extensive protein–protein interactions are also at play within these mixed neighborhoods.

FUNCTIONAL INTERACTIONS

The models developed in the preceding sections oblige us to consider the possible contributions of the 16S RNA to the functions of the 30S subunit in protein synthesis. Indeed, it is possible to develop the viewpoint that the functional sites of this subunit are to some extent made up of RNA sequences that are held in effective configuration by the proteins. Since the functions of the proteins are considered at length elsewhere in this volume, the present stress on RNA function may have some heuristic value.

Three operationally distinguishable functions of 30S subunits are obvious at the outset: the binding of mRNA, the binding of tRNA, and the interaction with the 50S subunit. Each of these functions is likely to be a dynamic one, since movements of the tRNA, mRNA and the subunits relative to each other are required during protein synthesis. Much effort has gone into the assignment of one or more of these functions to individual components of the 30S subunit.

The first serious attempt to make functional assignments for the individual 30S components (Nomura et al. 1969) exploited the effects of omitting one component at a time from reconstitution mixtures and assaying the incomplete assembly products so obtained. The assays were designed to uncover defects in polypeptide synthesis, tRNA binding, and the assembly process per se. As a consequence, certain selective effects of omitting one component at a time were masked. For example, a protein which contributes to the binding of mRNA but not to the binding of tRNA would be scored in the same way as proteins with the reverse contributions.

The results of subsequent experiments illustrate the need for appropriate assays to distinguish the functions of different components. Four 30S proteins, either separately or in combinations, stimulate the binding of tRNA by the ribosome: S1, S2, S3 and S14. However, these proteins act in different ways. S1 stimulates the binding of mRNA to 30S subunits and, as a consequence, stimulates the binding of the appropriate tRNA to a system programmed either by poly(U) or phage mRNA (Van Duin and Kurland 1970; Tal et al. 1972; Van Duin and Van Knippenberg 1974). In contrast, S2, S3 and S14 have no effect on mRNA binding but do enhance the binding of the appropriate tRNA in the absence of factors, with an EF-Tu-dependent system and with an IF_2-dependent system (Randall-Hazelbauer and Kurland 1972; Van Duin et al. 1972).

The formal assignment of functional contributions for these four proteins only brings us to a much more serious ambiguity: We cannot decide on the basis of these sorts of experiments whether any of the proteins are directly participating in the binding of tRNA or mRNA. We can simplify this problem by assuming that the sole function of 16S RNA is to provide binding sites for the 30S proteins. Then it would be likely that at least some of the 30S proteins are providing binding sites for tRNA, mRNA and the 50S subunit. However, this assumption begs the question that most concerns us here.

Messenger RNA Binding

We can do a little better than this in the case of mRNA binding. Unlike the situation with tRNA, the binding of mRNA would require a stable interaction between the ribosome and any permutation of nucleotide sequences. Since we would expect nucleic acids to interact via base pairs, it is difficult to imagine how a fixed sequence of nucleotides in 16S RNA could provide a binding site for random sequences of nucleotides found in mRNA. Similarly, the binding of a triplet nucleotide codon by the ribosome would probably require ligands to the backbone of the codon so that the bases are available for the selection of tRNA. For these reasons, it seems reasonable to guess that one or more proteins bind mRNA to the 30S subunit. Although it is not excluded that backbone interactions between mRNA and 16S RNA contribute as well, the available data support the guess that proteins are involved.

So far, only one protein (S1) has been implicated in mRNA binding. In addition to the data described above showing that S1 stimulates the binding of translatable mRNA, there is evidence that a bound mRNA and S1 are rather intimately associated on the ribosome. First, there appears to be a one-to-one relationship between the ribosomes which will form polysomes with poly(U) and the amount of S1 on the ribosomes (Van Duin and Van Knippenberg 1974). Second, S1 is protected from photooxidation by rose Bengal in the presence of poly(U) (Noller et al. 1971). Third, the only protein of the 30S subunit which is protected by poly(U) from the proteolytic action of trypsin is S1 (Rummel and Noller 1973). Finally, there are indications that S1 can directly bind poly(U), even when separate from the 30S subunit (see Figure 3 in Van Duin and Kurland 1970).

It might be added for the sake of completeness that the anomolous properties of S1 have prompted the suggestion that this protein might be more conveniently viewed as a supernatant factor rather than ribosomal protein (see Wittmann, this volume). An earlier discussion of the pros and cons of this view may be found in Van Duin and Kurland (1970). However in the present context, it seems to make little difference which category S1 occupies, since the problem that most concerns us here is the nature of the functional particle involved in protein synthesis.

Transfer RNA Binding

The data concerning the tRNA binding site are a good deal more complicated. There is little evidence indicating a direct interaction between one or more 30S proteins and tRNA. However, more than half of the proteins of the 30S subunit are either contributing to this function or, at the very least, are in close physical proximity to the bound tRNA-factor complexes (see Stöffler, this volume). Finally, there is some evidence that the 16S RNA may be involved in this function.

Noller and Chaires (1972) have found that treatment of 30S subunits with kethoxal leads to a loss of tRNA binding capacity, attended by the modification of several guanine residues in the 16S RNA. Significantly there is no concomitant effect on mRNA binding, and preincubation of the ribosomes with tRNA protects the 16S RNA from the kethoxal-mediated inactivation. These data suggest that one or more nucleotide sequences of the 16S RNA may be involved in the binding of tRNA.

A similar, although weaker, conclusion can be drawn from the data concerning the effects of streptomycin on ribosome function. It has been known for quite some time that streptomycin can influence the codon-specific binding of tRNA. Recently Biswas and Gorini (1973) have suggested on the basis of binding experiments that the binding site for streptomycin is the 16S RNA, and not a 30S protein. They interpret the effects of mutations in the ribosomal proteins that influence the expression of streptomycin-induced functional alterations as indirect effects dependent on protein–RNA interactions. Such an interpretation is consistent with the observation that several proteins seem to influence the accessibility of the streptomycin binding site on the ribosome (see Stöffler, this volume).

Finally, mutations expressed as alterations of S4 are responsible for changes in the fidelity of tRNA binding (Zimmermann, Garvin and Gorini 1971). Thus the Ram phenotype, which corresponds to an enhanced error frequency in protein synthesis, is associated with an alteration in the protein that seems to have the most extensive interactions with the 16S RNA, namely, S4.

It need hardly be stressed that the foregoing observations do not establish an unambiguous function for the 16S RNA in the binding of tRNA by the ribosome. On the other hand, there appears to be a pattern in the recent experiments: whenever it is possible to experimentally manipulate the 16S RNA, there is a strong hint of an involvement in tRNA binding.

Although this may be interpreted as an indirect effect following from the interaction of the RNA with proteins, the alternative interpretation should not be ignored.

Subunit Interactions

The pronounced dependence of the 30S–50S couple on divalent cation concentration suggests that its formation leads to the close apposition of RNA sequences from each subunit (Tissières et al. 1959). On the other hand, it is known that specific antibodies block couple formation (see Stöffler, this volume). Unfortunately, the large size of the antibodies precludes a decision on whether the proteins to which the antibodies bind provide the attachment sites for subunit interaction. It is quite possible that the effect of the antibodies is merely a form of steric hindrance. Indeed, there is one property of the 16S RNA which recommends it as a potential binding site for the 50S subunit. This property is the presence of repeated sequences in the RNA (see Fellner; Zimmermann, this volume). A brief side excursion will be needed to justify this comment.

Two models for the function of the ribosome have been proposed recently. One of these, the AC-DC model, employs a single tRNA binding site on the 30S subunit and two tRNA binding sites on the 50S subunit to effect a cycle of protein synthesis (Kurland et al. 1972a,b). During initiation, the 30S site is thought to function in conjunction with the puromycin-sensitive 50S site, but during chain elongation it is thought to function in conjunction with the puromycin-resistant 50S site. The two functionally distinguishable states are expected to correspond to two alternative ways of forming a 30S–50S couple.

A similar requirement is seen in the "rotating ribosome" model (Woese 1973), in which a single 30S tRNA binding site is thought to operate alternatively with each of two 50S tRNA binding sites. According to this model, successive peptide bonds are formed on a 30S–50S couple in which the 30S subunit is in effect rotated $180°$ after each step.

It is not my intention to explore these models in any more detail. However, they do illustrate what is likely to be an important property of the 30S–50S interaction: the couples would be expected to function in more than one configuration. One way to facilitate the apposition of subunits in multiple configurations would be to use repeated structures in one subunit to provide the binding sites for the other subunit.

A number of independent estimates of the stoichiometry of each of the 30S proteins indicate that there is at most one copy of each protein per ribosome (Voynow and Kurland 1971; Weber 1972). Therefore it seems unlikely that the 30S proteins themselves could provide the kinds of repeated attachment sites for which we are looking. In contrast, there are two rather long nucleotide sequences in 16S RNA which are iterated several times each (see Fellner; Zimmermann, this volume). Therefore, it does not seem too far-fetched to suggest that these repeated sequences could provide a structural basis for the formation of alternative configurations of the 30S–50S couple.

Functional Cooperativity

Finally, intimate interaction of proteins and RNA would permit the functional interactions between different 30S proteins to be mediated by the RNA as, for example, was considered in the section on assembly interaction. The analysis of the interactions between mutations effecting the response of the ribosome to streptomycin is consistent with this inference. Mutations to streptomycin dependence or streptomycin resistance are expressed in the structure and function of 30S protein S12 (Birge and Kurland 1969; Ozaki, Mizushima and Nomura 1969). Reversion of these phenotypes is a consequence of nonallelic suppressor mutations (Hashimoto 1960), which are also expressed as modifications of ribosomal components (Brownstein and Lewandowski 1967). One such suppressed mutant could be shown to have an alteration in S4, as well as the original alteration in S12 (Birge and Kurland 1970; Donner and Kurland 1971). More significant was the finding that the altered protein, S4su6, can modify the expression of the streptomycin-dependence phenotype of ribosomes in complementation tests performed in vitro (Birge and Kurland 1970). A number of subsequent studies uncovered electrophoretic alterations in S4 and S5 that are correlated in vivo with the suppression of S12-mediated streptomycin dependence (Deusser et al. 1970; Funatsu et al. 1972; Krieder and Brownstein 1971).

Here, then, is an appropriate system to study the functional interactions between two proteins. Is their interaction a direct one, or is it mediated by RNA? An important clue came from an analysis of the growth properties of the mutant containing S4su6 which seemed to be pleiotropic. Not only does S4su6 suppress the streptomycin phenotype, but it appears to create a defect in ribosome assembly (Birge and Kurland 1970). This observation led to the demonstration that S4su6 has a much lower affinity for its RNA binding site than does the wild-type S4 (Green and Kurland 1971).

Thus a mutation selected by suppression of a functional response to an antibiotic produces a seemingly independent but marked change in the suppressor protein's interaction with its RNA binding site. An accident? Perhaps, but several similar pleiotropic forms of S4 have been found subsequently (Daya-Grosjean et al. 1972). That some S4 suppressors of streptomycin dependence fail to exhibit a marked change in their affinity for RNA does not indicate that their interaction with RNA is normal. It only means that a rather crude assay has failed to detect any alteration in this interaction. While not eliminating the possibility of direct interactions, the apparent correlation between a defect in RNA binding and suppression of streptomycin dependence suggests that the functional interaction between S4 and S12 may be mediated at least in part by RNA.

Some Ambiguities

In conclusion, the data discussed in this section consistently reflect an ambiguity inherent in experiments performed to date to assign functions to the 30S ribosomal components. These experiments can uncover the influence of a component on some function, but they say little about the mechanism through which this influence is effected. In particular, such experiments do not permit the identification of the contacts on the ribosome to which are directed the macromolecules that help mediate the separate steps in protein synthesis. Therefore while it may be inferred that the structure of the 16S RNA influences different ribosome functions, it remains to be determined whether this RNA provides contacts for the functional apposition of factors, tRNA and the 50S subunit during protein synthesis. It must be added that precisely the same ambiguity pertains to each of the 30S proteins.

Acknowledgments

I am indebted to Drs. C. I. Brändén, L. C. Lutter and Ms. I. Winkler for helpful criticism.

References

Bickle, T. A., J. W. B. Hershey and R. R. Traut. 1972. Spatial arrangement of ribosomal proteins: Reaction of the *E. coli* 30S subunit with bis-imidoesters. *Proc. Nat. Acad. Sci.* **69**:1327.

Birge, E. A. and C. G. Kurland. 1969. Altered ribosomal protein in streptomycin-dependent *Escherichia coli. Science* **166**:1282.

―――. 1970. Reversion of a streptomycin-dependent strain of *Escherichia coli. Mol. Gen. Genet.* **109**:356.

Biswas, D. K. and L. Gorini. 1973. The attachment site of streptomycin to the 30S ribosomal subunit. *Proc. Nat. Acad. Sci.* **69**:2141.

Brownstein, B. and L. J. Lewandowski. 1967. A mutation suppressing streptomycin dependence. I. An effect on ribosome function. *J. Mol. Biol.* **25**:99.

Chang, F. N. and J. G. Flaks. 1970. Topography of the *Escherichia coli* 30S ribosomal subunit and streptomycin binding. *Proc. Nat. Acad. Sci.* **67**:1321.

―――. 1972. The specific cross-linking of two proteins from the *E. coli* 30S ribosomal subunit. *J. Mol. Biol.* **68**:177.

Craven, G. R. and V. Gupta. 1970. Three-dimensional organization of the 30S ribosomal proteins from *Escherichia coli*. I. Preliminary classification of the proteins. *Proc. Nat. Acad. Sci.* **67**:1329.

Craven, G. R., P. Voynow, S. J. S. Hardy and C. G. Kurland. 1969. The ribosomal proteins of *Escherichia coli*. II. Chemical and physical characterization of 30S ribosomal proteins. *Biochemistry* **8**:2906.

Crichton, R. and H. G. Wittmann. 1973. A native ribonucleoprotein complex from *Escherichia coli* ribosomes. *Proc. Nat. Acad. Sci.* **70**:665.

Daya-Grosjean, L., R. A. Garrett, O. Pongs, G. Stöffler and H. G. Wittmann. 1972. Properties of the interaction of ribosomal protein S4 and 16S RNA in *Escherichia coli* mutants from streptomycin dependence to independence. *Mol. Gen. Genet.* **119**:277.

Deusser, E., G. Stöffler, H. G. Wittmann and D. Apirion. 1970. Ribosomal proteins. XVI. Altered S4 proteins in *Escherichia coli* revertants from streptomycin dependence to independence. *Mol. Gen. Genet.* **109**:298.

Donner, D. and C. G. Kurland. 1971. Changes in the primary structure of a mutationally altered ribosomal protein S4 of *Escherichia coli*. *Mol. Gen. Genet.* **115**:49.

Funatsu, G., W. Puls, E. Schiltz, J. Reinbolt and H. G. Wittmann. 1972. Ribosomal proteins. XXXI. Comparative studies on altered proteins S4 of six *Escherichia coli* revertants from streptomycin dependence. *Mol. Gen. Genet.* **115**:131.

Garrett, R. A., K. H. Rak, L. Daya and G. Stöffler. 1972. Ribosomal proteins. XXIX. Specific protein binding sites on 16S RNA of *Escherichia coli*. *Mol. Gen. Genet.* **114**:112.

Green, M. and C. G. Kurland. 1971. Mutant ribosomal protein with defective RNA binding site. *Nature New Biol.* **234**:273.

――――. 1973. Molecular interactions of ribosomal components. IV. Cooperative interactions during assembly *in vitro*. *Mol. Biol. Reports* **1**:105.

Hardy, S. J. S., C. G. Kurland, P. Voynow and G. Mora. 1969. The ribosomal proteins of *Escherichia coli*. I. Purification of the 30S ribosomal proteins. *Biochemistry* **8**:2897.

Hashimoto, K. 1960. Streptomycin resistance in *Escherichia coli* analyzed by transduction. *Genetics* **45**:49.

Hawley, D. A. and L. I. Slobin. 1973. The immunological reactivity of 30S ribosomal proteins in 70S ribosomes from *Escherichia coli*. *Biochem. Biophys. Res. Comm.* **55**:162.

Hill, W. E., J. W. Anderegg and K. E. Van Holde. 1970.. Effects of solvent environment and mode of preparation on the physical properties of ribosomes from *Escherichia coli*. *J. Mol. Biol.* **53**:107.

Hill, W. E., G. P. Rossetti and K. E. Van Holde. 1969a. Physical studies of ribosomes from *Escherichia coli*. *J. Mol. Biol.* **44**:263.

Hill, W. E., J. D. Thompson and J. W. Anderegg. 1969b. X-ray scattering study of ribosomes from *Escherichia coli*. *J. Mol. Biol.* **44**:89.

Kaltschmidt, E. and H. G. Wittmann. 1970. Ribosomal proteins. XII. Number of proteins in small and large ribosomal subunits of *Escherichia coli* as determined by two-dimensional gel electrophoresis. *Proc. Nat. Acad. Sci.* **67**:1276.

Krieder, G. and B. L. Brownstein. 1971. A mutation suppressing streptomycin dependence. II. An altered protein on the 30S ribosomal subunit. *J. Mol. Biol.* **61**:135.

Kurland, C. G. 1960. Molecular characterization of ribonucleic acid from *Escherichia coli* ribosomes. *J. Mol. Biol.* **2**:83.

————. 1966. The requirements for specific sRNA binding by ribosomes. *J. Mol. Biol.* **18**:90.

————. 1972. Structure and function of the bacterial ribosome. *Amer. Rev. Biochem.* **41**:377.

Kurland, C. G., P. Voynow, S. J. S. Hardy, L. Randall and L. Lutter. 1969. Structural and functional heterogeneity *of E. coli* ribosomes. *Cold Spring Harbor Symp. Quant. Biol.* **34**:17.

Kurland, C. G., M. Green, H. W. Schaup, D. Donner, L. Lutter and E. A. Birge. 1972a. Molecular interaction between ribosomal components. *FEBS Symp.* **23**:75.

Kurland, C. G., D. Donner, J. Van Duin, M. Green, L. Lutter, L. Randall-Hazelbauer, H. W. Schaup and H. Zeichhardt. 1972b. Structure and function of the ribosome. *FEBS Symp.* **27**:225.

Lutter, L. C. 1973. Structural studies of the ribosome of *Escherichia coli. Ph.D. dissertation,* University of Wisconsin, Madison.

Lutter, L. C. and C. G. Kurland. 1973. Reconstitution of active ribosomes with cross-linked proteins. *Nature New Biol.* **243**:15.

Lutter, L. C., U. Bode, C. G. Kurland, and G. Stöffler. 1974. Ribosomal neighborhoods. III. Cooperativity of ribosome assembly. *Mol. Gen. Genet.* **129**:167.

Lutter, L. C., H. Zeichhardt, C. G. Kurland and G. Stöffler. 1972. Ribosomal protein neighborhoods. I. S18 and S21 as well as S5 and S8 are neighbors. *Mol. Gen. Genet.* **119**:357.

Miller, R. V. and P. S. Sypherd. 1973. Topography of the *Escherichia coli* 30S ribosome revealed by the modification of ribosomal proteins. *J. Mol. Biol.* **78**:539.

Mizushima, S. and M. Nomura. 1970. Assembly mapping of 30S ribosomal proteins from *Escherichia coli. Nature* **226**:1214.

Morgan, J. and R. Brimacombe. 1972. A series of specific ribonucleoprotein fragments from the 30S subparticle of *Escherichia coli* ribosomes. *Eur. J. Biochem.* **29**:542.

————. 1973. A preliminary three-dimensional arrangement of the proteins in the *Escherichia coli* 30S ribosomal subunit. *Eur. J. Biochem.* **37**:472.

Nanninga, N., R. A. Garrett, G. Stöffler and G. Klotz. 1972. Ribosomal proteins. XXXVIII. Electron microscopy of ribosomal proteins S4-16S RNA complexes of *Escherichia coli. Mol. Gen. Genet.* **119**:174.

Noller, H. F. and J. B. Chaires. 1972. Functional modification of 16S RNA by kethoxal. *Proc. Nat. Acad. Sci.* **69**:3115.

Noller, H. F., C. Chang, G. Thomas and J. Aldridge. 1971. Chemical modification of the transfer RNA and polyuridylic acid binding site of *Escherichia coli* S30 ribosomal subunits. *J. Mol. Biol.* **61**:669.

Nomura, M., S. Mizushima, M. Ozaki, P. Traub and C. V. Lowry. 1969. Structure and function of ribosomes and their molecular components. *Cold Spring Harbor Symp. Quant. Biol.* **34**:49.

Ozaki, M., S. Mizushima and M. Nomura. 1969. Identification and functional characterization of the protein controlled by the streptomycin-resistant locus in *Escherichia coli. Nature* **222**:333.

Randall-Hazelbauer, L. L. and C. G. Kurland. 1972. Identification of three 30S proteins contributing to the ribosomal A site. *Mol. Gen. Genet.* **115**:234.

Roth, H. and K. Nierhaus. 1973. Isolation of four ribonucleoprotein fragments from the 30S subunit of *E. coli* ribosomes. *FEBS Letters* **31**:35.

Rummel, D. P. and H. F. Noller. 1973. Use of protection of 30S ribosomal proteins by tRNA for functional mapping of the *E. coli* ribosome. *Nature New Biol.* **245**:72.

Schaup, H. W. and C. G. Kurland. 1972. Molecular interactions of ribosomal components. III. Isolation of the RNA binding site for a ribosomal protein. *Mol. Gen. Genet.* **114**:350.

Schaup, H. W., M. Green and C. G. Kurland. 1970. Molecular interactions of ribosomal components. I. Identification of RNA binding sites for individual 30S ribosomal proteins. *Mol. Gen. Genet.* **109**:193.

————. 1971. Molecular interactions of ribosomal components. II. Site-specific complex formation between 30S proteins and ribosomal RNA. *Mol. Gen. Genet.* **112**:1.

Schaup, H. W., M. L. Sogin, C. G. Kurland and C. R. Woese. 1973. Localization of a binding site for ribosomal protein S8 within the 16S ribosomal ribonucleic acid of *Escherichia coli. J. Bact.* **115**:82.

Schaup, H. W., M. Sogin, C. Woese and C. G. Kurland. 1972. Characterization of an RNA "binding site" for a specific ribosomal protein of *Escherichia coli. Mol. Gen. Genet.* **114**:1.

Schendel, P., P. Maeba and G. R. Craven. 1972. Identification of the proteins associated with subparticles produced by mild ribonuclease digestion of 30S ribosomal particles from *Escherichia coli. Proc. Nat. Acad. Sci.* **69**:544.

Slobin, L. 1972. Use of bifunctional imidoesters in the study of ribosome topography. *J. Mol. Biol.* **64**:297.

Smith, W. S. 1971. An X-ray scattering study of the smaller ribosomal subunit of *Escherichia coli. Ph.D. dissertation,* University of Wisconsin, Madison.

Stanley, W. and R. Bock. 1965. Isolation and physical properties of the ribosomal ribonucleic acid of *E. coli. Biochemistry* **4**:1302.

Sypherd, P. S., D. M. O'Neill and M. M. Taylor. 1969. The chemical and genetic structure of bacterial ribosomes. *Cold Spring Harbor Symp. Quant. Biol.* **34**:77.

Szekely, M., R. Brimacombe, and J. Morgan. 1973. A specific ribonucleoprotein fragment from *Escherichia coli* 30S ribosomes. *Eur. J. Biochem.* **35**:574.

Tal, M., M. Aviram, A. Kanorek and A. Weiss. 1972. Polyuridylic acid binding and translating by *Escherichia coli* ribosome: Stimulation by protein 1, inhibition by aurintricarboxylic acid. *Biochim. Biophys. Acta* **281**:381.

Tissières, A., J. D. Watson, D. Schlessinger and B. R. Hollingworth. 1959. Ribonucleoprotein particles from *E. coli. J. Mol. Biol.* **1**:221.

Traub, P. and M. Nomura. 1968. Structure and function of *Escherichia coli* ribosomes. V. Reconstitution of functionally active 30S ribosomal particles from RNA and protein. *Proc. Nat. Acad. Sci.* **59**:777.

Traut, R. R., H. Delius, C. Ahmad-Zadeh, T. A. Bickle, P. Pearson and A. Tissières. 1969. Ribosomal proteins of *E. coli*. Stoichiometry and implications for ribosome structure. *Cold Spring Harbor Symp. Quant. Biol.* **34**:25.

Van Duin, J. and C. G. Kurland. 1970. Functional heterogeneity of the 30S ribosomal subunit of *Escherichia coli. Mol. Gen. Genet.* **109**:169.

Van Duin, J. and P. Van Knippenberg. 1974. Functional heterogeneity of the 30S ribosomal subunit of *Escherichia coli.* III. Requirement of protein S1 for translation. *J. Mol. Biol.* **84**:185.

Van Duin, J., P. H. Van Knippenberg, M. Dieben and C. G. Kurland. 1972. Functional heterogeneity of the 30S ribosomal subunit of *Escherichia coli.* II. Effect of S21 on initiation. *Mol. Gen. Genet.* **116**:181.

Voynow, P. and C. G. Kurland. 1971. Stoichiometry of the 30S ribosomal proteins of *Escherichia coli. Biochemistry* **10**:517.

Weber, J. 1972. Stoichiometric measurements of 30S and 50S ribosomal proteins from *Escherichia coli. Mol. Gen. Genet.* **119**:233.

Wittmann, H. G., G. Stöffler, I. Hindennach, C. G. Kurland, L. Randall-Hazelbauer, E. A. Birge, M. Nomura, E. Kaltschmidt, S. Mizushima, R. R. Traut and T. A. Bickle. 1971. Correlation of 30S ribosomal proteins of *Escherichia coli* isolated in different laboratories. *Mol. Gen. Genet.* **111**:327.

Woese, C. 1973. The rotating ribosome: A gross mechanical model for translation. *J. Theoret. Biol.* **38**:203.

Zimmermann, R. A., R. Garvin and L. Gorini. 1971. Alteration of a 30S ribosomal protein accompanying the *ram* mutation in *Escherichia coli. Proc. Nat. Acad. Sci.* **68**:2263.

Zimmermann, R. A., A. Muto, P. Fellner, C. Ehresmann and C. Branlant. 1972. Location of ribosomal protein binding sites on 16S ribosomal RNA. *Proc. Nat. Acad. Sci.* **69**:1282.

Genetics of Bacterial Ribosomes

S. R. Jaskunas, M. Nomura and J. Davies
Institute for Enzyme Research and
Departments of Biochemistry and Genetics
University of Wisconsin
Madison, Wisconsin 53704

INTRODUCTION

For an organelle as complex as the ribosome, genetic studies can play an important role in providing information about the components, how they are put together, and how synthesis and assembly may be controlled. Bacterial mutants with altered ribosomes, in spite of their scarcity, have proved to be of considerable utility in the study of ribosome structure and functions. Since the state of our knowledge in this field has been reviewed recently (Davies and Nomura 1972), this chapter will summarize recent work in the context of earlier findings. In addition, information on many mutations related to ribosomes is compiled in Table 1.

ORGANIZATION OF rRNA STRUCTURAL GENES

Evidence for 16S + 23S rRNA Transcriptional Units

Analysis of the residual rRNA synthesis after inhibition with rifampicin (Pato and von Meyenberg 1970; Doolittle and Pace 1970, 1971; Bremer and Berry 1971), the reinitiation of rRNA synthesis following amino acid starvation of stringent cells (Kossman, Stamato and Pettijohn 1971), and hybridization of rRNA with DNA fragments from *B. subtilis* (Colli, Smith and Oishi 1971; Colli and Oishi 1969) have indicated that rRNA genes in bacteria are organized into transcriptional units containing one gene for each type of rRNA, with the gene order 16S–23S–5S. Electron microscopic

Table I (a) Mutations affecting properties or synthesis of ribosomes

Gene	Phenotype	Map position E. coli	Map position B. subtilis	Altered component	References
bry	resistance to bryamycin (Bry-R) (thiostrepton)		cysA-str region	70S ribosome resistant	Smith and Smith (1973)
cam	resistance to chloramphenicol (Cam-R) and in some cases cross-resistance with erythromycin (Ery-R)		I not known; II, III, IV, V, VI in cysA-str region	unknown-ribosomes sensitive; 50a, 50b, 50c, 50e, 50f altered 50S ribosomal proteins	Osawa et al. (1973)
ery	resistance to erythromycin (Ery-R); some cross-resistance with chloramphenicol and lincomycin; impaired peptidyl transferase activity	linked to spcA		eryA:L4, eryB:L22 } 50S proteins	Wittmann et al. (1973)
			cysA-str region	50d-50S protein	Tanaka et al. (1973)
"K"	K12 strains of E. coli distinct from B strains	linked to strA		S7-30S protein	Mayuga et al. (1968) Birge et al. (1969)
kan	resistance to kanamycin (Kan-R) (sensitivity to neomycin is not known)		cysA-str region	unknown	Goldthwaite and Smith (1972a,b)

ksgA	resistance to kasugamycin (Ksg-R)	linked to *thr-leu* (minute 1)	methylating enzyme altered, fails to methylate 16S RNA in resistant strain	Helser et al. (1972)
mac	dependence for growth on presence of erythromycin and related antibiotics (Mac-D)	near minute 25, close to *trp*	unknown	Sparling and Blackman (1973)
mic	resistance to antibiotic micrococcin (Mic-R)	*cysA-str* region	70S ribosome resistant	Goldthwaite and Smith (1972b) Goldthwaite et al. (1970)
nea	resistance to antibiotic neamine (Nea-R)	*neaA* linked to *spcA* *neaB* linked to *strA*	unknown	Cabezon, Cannon and Bollen (personal communication) Goldthwaite and Smith (1972b)
nek	co-resistance to neomycin and kanamycin; masks expression of Spc-R	linked to *strA*	unknown	Apirion and Schlessinger (1968); Tyler and Ingraham (1973)
neo	resistance to neomycin (Neo-R) (sensitivity to kanamycin not known)	*cysA-str* region	unknown	Goldthwaite et al. (1970)
ole	resistance to oleandomycin	*cysA-str* region	70S ribosome resistant	Goldthwaite and Smith (1972b)

335

Table 1 (a) Mutations affecting properties or synthesis of ribosomes (*continued*)

Gene	Phenotype	Map position		Altered component	References
		E. coli	B. subtilis		
ramA ("S4")	(a) ribosomal ambiguity suppression of missense and nonsense mutations, potentiates genetic suppressors	linked to *spcA*		S4-30S protein	Rosset and Gorini (1969) Gorini (1971) Zimmermann et al. (1971)
	(b) suppression of Str-D [whether the same mutation causes both (a) and (b) phenotype is unknown]				Birge and Kurland (1970) Funatsu et al. (1972b) Hasenbank et al. (1973)
ramB	not known	linked to *ksgA*		S4-30S protein	Zimmermann et al. (1973)
relC	partially relaxed	linked to *rif*		possibly 50S component	Friesen, Haseltine and Fiil (personal communication); see Block and Haseltine (this volume)
spcA ("S5")	(a) resistance to antibiotic spectinomycin (Spc-R) (b) suppression of Str-D (c) cold-sensitivity of growth (not all mutants show these three phenotypes)	linked to *aroE*	*cysA-str* region	altered S5-30S protein	Davies et al. (1965) Sparling et al. (1968) Funatsu et al. (1972a) Hasenbank et al. (1973) Nashimoto and Nomura (1970)
				altered 30a protein	Kimura et al. (1973)

strA	resistance to antibiotic streptomycin (Str-R); restriction of genetic suppression;	minute 64	S12-30S protein	Ozaki et al. (1969) Breckenridge and Gorini (1970) Momose and Gorini (1971) Birge and Kurland (1969)
	dependence on streptomycin (Str-D)	cysA-str region	altered 30S protein	Smith et al. (1969)
"S12"	temperature sensitivity for growth	linked to strA	altered S12(?)	Kang (1970a,b)
"S18"	altered electrophoretic mobility of S18	between minutes 76–88	altered S18-30S protein	Bollen et al. (1973) Kahan et al. (1973)
"S20"	suppression of temperature-sensitive alanyl-tRNA synthetase	not known	altered S20	Buckel et al. (1972); see also Wittmann and Wittmann-Liebold (this volume)
ts7 ts9	temperature sensitivity for growth	between minutes 76–78	altered 50S protein(?)	Flaks et al. (1966)
sad	subunit assembly defective: precursors of ribosome subunits accumulate in cells grown under nonpermissive conditions (cold-sensitive)	sad 19: linked to spcA sad 38: linked to aroE	unknown except for certain Spc-R mutants with this phenotype	Guthrie et al. (1969a,b)

Table 1 (a) Mutations affecting properties or synthesis of ribosomes (*continued*)

Gene	Phenotype	Map position		Altered component	References
		E. coli	B. subtilis		
rim	defective maturation of 50S subunit	rimA linked to ilvD (minute 75) rimB linked to aroD (minute 32) rimC linked to purB(?) (minute 25) rimD, between ilvD and malB (minutes 75–81)		unknown	Bryant and Sypherd (1974) Bryant et al. (1974)
	Cold-sensitive				
csr	cold-sensitive mutations in S. typhimurium; some accumulate ribosomal precursors sedimenting between 4S and 30S at nonpermissive temperatures csrA,B,C,D	map between cysG and aroC; linked to strA and spcA loci (in S. typhimurium)		unknown	Tai et al. (1969) Tyler and Ingraham (1973)

defect in ribosome assembly	accumulates 43S ribonucleoprotein particles during growth	close to *xyl* (minute 70)	Turnock (1969)
cqs (5S RNA)	different oligonucleotide sequences in different 5S RNA alleles	*cqs*A, near *metE*; *cqs*B, near *ilv*; *cqs*C, near *aroB*; *cqs*D, near *aroE*	Jarry and Rosset (1973a,b)
5S RNA	unknown	between *purB* and *ery*	Smith et al. (1968)
16S and 23S RNA	unknown	(a) near *metB* (see Fig. 1); (b) 54–59 minutes; between *purB* and *ery*	Birnbaum and Kaplan (1971); Unger et al. (1972); Deonier et al. (1973); Matsubara, Takata and Osawa (1972); Smith et al. (1968)

339

Table 1 (b) Mutants of elongation factors EF-G and EF-T

Gene	Phenotype	Map position		Altered component	References
		E. coli	*B. subtilis*		
fus	resistance to fusidic acid or thermolability for growth	near *strA*	*cysA-str* region	elongation factor EF-G	Tocchini-Valentini and Mattoccia (1968) Bernadi and Leder (1970) Kuwano et al. (1971) Tanaka et al. (1971) Osawa (personal communication) Aharonowitz and Ron (1972)
eft	thermolability for growth	not known		elongation factor EF-T	Kuwano et al. (1973)

observations of DNA in the process of being transcribed are consistent with these conclusions (Miller and Hamkalo 1972). Several recent observations have provided more direct evidence for 16S + 23S rRNA transcriptional units. On the other hand, direct evidence of the inclusion of 5S rRNA genes in these transcriptional units remains elusive.

Several 16S + 23S rRNA gene sets in *B. subtilis* have been seen with the electron microscope by Chow and Davidson (1973). Double-strand regions containing 4830 ± 250 base pairs formed when bacterial DNA renatured with itself. The sum of the lengths of 16S and 23S rRNA is approximately 4800 bases. Thus these double-strand duplexes have the size expected for a segment of DNA that contains one gene for 16S plus one gene for 23S rRNA. No duplexes of the size expected for a single 16S or single 23S rRNA gene were observed. Therefore these workers concluded that all 16S and 23S rRNA genes in *B. subtilis* are organized into 16S + 23S rRNA gene sets and the order of the two genes is the same for all gene sets. The double-stranded segments are apparently formed by out-of-register renaturation between different gene sets.

Using similar techniques, Davidson's group has also observed 16S + 23S gene sets in *E. coli* (Deonier et al. 1973). Two gene sets were found on the *E. coli* episome F14 by observing double-strand duplexes between rRNA and episomal DNA in the electron microscope. This episome covers the region between *ilv* and *arg*H, approximately 75–78.5 min, where rRNA genes have been previously mapped (Yu, Vermeulen and Atwood 1970; Birnbaum and Kaplan 1971). Because rRNA:DNA hybrids were observed directly, it was possible to measure the spacer region between the 16S and 23S cistrons. Both gene sets contain one gene for 16S and one gene for 23S rRNA with lengths of 1800 (±200) and 3400 (±300) base pairs, respectively. In both gene sets, the 16S and 23S sequences were separated by a spacer of length 300 (±50) base pairs. However in DNA:DNA hybrids, between the gene sets there were two double-strand regions separated by a loop of single-strand DNA. Thus the spacers for these two gene sets are nonhomologous. This is different from the situation in *B. subtilis* where any spacer is essentially homologous (Chow and Davidson 1973). The total size of the gene sets in *E. coli* as measured in this way is 5500 (±550) base pairs. Both in the case of *B. subtilis* and *E. coli,* these experiments do not provide any direct confirmation of the existence of a 5S rRNA gene in each of these gene sets. However the experimental error is such that there is room for a 5S rRNA cistron. As mentioned above, the physiological experiments have suggested a 5S gene should follow the 23S gene.

Biochemical evidence that these gene sets constitute a single transcriptional unit comes from the discovery of a 30S precursor rRNA in a mutant strain of *E. coli* that competes with 16S and 23S rRNA in RNA-DNA hybridization experiments (Nikolaev, Silengo and Schlessinger 1973; Dunn and Studier 1973). The experiments suggest that this precursor rRNA contains the sequences of one 16S rRNA and one 23S rRNA. It has an apparent molecular weight of 1.8–2.3 × 10^6 as measured by electrophoretic mobility in a gel. This corresponds to approximately 5200–6600 bases, which is close to the size of the two gene sets in *E. coli* measured by Deonier et al. (1973). The best evidence for inclusion of 5S genes in 16S + 23S transcriptional units would be to find 5S-specific sequences on this precursor rRNA. Other aspects of the finding of a large precursor rRNA are discussed in the section on mechanism of assembly in this chapter and by Schlessinger (this volume).

Location of rRNA Genes

It has been known for some time that rRNA genes in bacteria are redundant. Bacterial rRNA hybridizes with more DNA than one would expect if there were only one structural gene per genome for each of the three rRNA molecules (Yankofsky and Spiegelman 1962). There are approximately 4–10 genes for each rRNA as calculated from hybridization data (Smith et al. 1968; Pace and Pace 1971; Spadari and Ritossa 1970; Jarry, Vola and Rosset, personal communication). The electron microscopic investigation of self-annealed duplexes in *B. subtilis* (Chow and Davidson 1973) and the molar yields of minor 5S rRNA sequences in *E. coli* (Jarry and Rosset 1971) are consistent with these estimates of the extent of redundancy.

The hybridization experiments have indicated that there are an equal number of genes for 16S and 23S rRNA (Pace and Pace 1971; Spadari and Ritossa 1970). This would be the expected observation if all 16S and 23S rRNA genes are organized into transcriptional units containing one gene of each. Similar experiments have given conflicting results as to whether there are the same number of 5S rRNA genes (Pace and Pace 1971; Attardi, Huang and Kabat 1965; Smith et al. 1968; Morell et al. 1967). Since there is one of each type of rRNA in the 70S ribosome, and they appear to be synthesized at approximately the same rate (Galibert et al. 1967), the simplest situation would be to have the same number of genes for each type of rRNA.

Despite the redundancy of rRNA genes, the sequences of the mature rRNAs that are produced from these genes differ from one another at only a few positions. So far 4 residues in the 5S rRNA population (Jarry and Rosset 1971; Brownlee, Sanger and Barrell 1967) and approximately 16 residues in the 16S rRNA population of *E. coli* K12 have been found to be heterogeneous. A similar level of heterogeneity has been found for 23S rRNA. (See the chapters by Fellner and Zimmermann for more details on the location of the heterogeneous residues.)

Microheterogeneity has been detected by the appearance of minor oligomers in the ribonuclease digest of the rRNA population that have a yield much less than one oligomer per RNA molecule. One of these minor oligomers from the 5S rRNA in *E. coli* K12 is not present in the ribonuclease digest of 5S rRNA from MRE600. However this digest contains another minor oligomer that is not present in the digest of 5S rRNA from K12. Thus there are strain-specific minor oligomers for 5S rRNA from K12 and MRE600 (Jarry and Rosset 1971).

Jarry and Rosset (1973a,b) have mapped genes for a strain-specific oligomer from the 5S rRNA of MRE600 at three loci by transferring genes for the oligomer from MRE600 to K12 by P1 transduction. These genes map at the following positions (see Figure 1): *cqs*B between *rbs* and *ilv* (74 min), *cqs*C between *aro*B and *mal*A (65 min), and *cqs*D between *arg*R and *aro*E (63 min). Similarly, they mapped a 5S rRNA gene in K12, *cqs*A, between *met*E and *rha* (76 min).

These experiments can unambiguously map only the genes in the strain of *E. coli* from which the strain-specific oligomer originates. They do not prove, for example, that the *cqs*B, *cqs*C and *cqs*D genes in MRE600 have alleles in K12. Recently Jarry, Vola and Rosset (personal communication) have shown that there is probably a 5S rRNA gene in K12 that is an allele to the *cqs*B gene in MRE600. They isolated an *E. coli* episome that covered the region where the *cqs*B gene is located. By direct hybridization, they found the episome does indeed carry one gene for 5S rRNA and one or two genes for 16S and 23S rRNA.

Using a strain-specific oligomer, Matsubara, Takata and Osawa (1972) mapped by conjugation two 16S rRNA genes in *E. coli* K12 that seem to be linked to *met*B. Sypherd, O'Neill and Taylor (1969) had previously used the same technique to limit a 23S rRNA gene in *E. coli* to the region between *xyl* and *pro*B (70–88 min). For further discussion on this subject, see Sypherd and Osawa (this volume).

The two rRNA gene sets found by Deonier et al. (1973) on the *E. coli* episome F14 were also mapped by electron microscopic techniques. The first set, *rib*1, is located 80×10^3 base pairs clockwise from *ilv*A, approximately midway between *ilv*A and *met*BJF. The mapping of the second set, *rib*2, has not been unambiguously determined. It could be between *arg*CBH and *gly*T or between *trk*D and *ilv,* or there could be 16S + 23S gene sets at both loci. These alternatives are indicated in Figure 1.

The question we can ask of all the mapping data on rRNA genes is whether it supports the hypothesis that all rRNA genes are organized into transcriptional units containing one gene of each type. If that is the case, we would expect to find a 5S RNA gene at each of the 16S + 23S loci and vice versa. In support of the hypothesis, there is some correlation between the mapped positions of two 5S rRNA loci and 16S + 23S gene sets (Figure 1). The *cqs*A locus may correspond to the *rib*1 gene set and one of the 16S rRNA genes found near *met*B (Matsubara, Takata and Osawa 1972). The second gene set, *rib*2, may correspond to *cqs*B and the 16S and

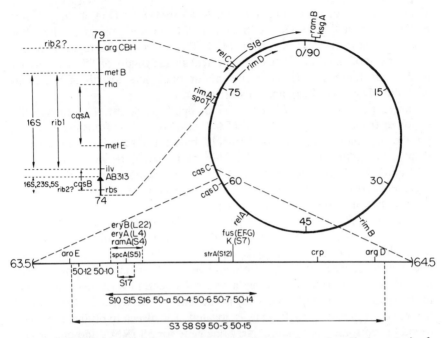

Figure 1   Genetic map of *E. coli* showing the location of well-characterized genes for rRNA and ribosomal proteins. Several 50S protein genes are identified by the nomenclature of Osawa et al. (1972), e.g., 50-a.

23S rRNA genes that Jarry, Vola and Rosset (personal communication) found anticlockwise from the origin of Hfr AB313. These results are consistent with earlier hybridization experiments indicating that several rRNA genes were located in this region (Yu, Vermeulen and Atwood 1970; Birnbaum and Kaplan 1971). If the hypothesis is correct, we would also expect to find 16S and 23S rRNA genes at the *cqs*C and *cqs*D loci. Unfortunately, hybridization experiments have given conflicting results as to whether there are any 16S and 23S genes in this region. Gorelic (1970) concluded there were 16S or 23S rRNA genes between 60 and 66 min, which includes the *cqs*C and *cqs*D loci, and none in the region of 55 to 61 min. Just the opposite conclusion was reached by Unger et al. (1972). They found 30% of the 16S and 23S rRNA cistrons between 54 and 59 min and none in the region of *cqs*C and *cqs*D. Both conclusions were based on the increased amount of rRNA that could be hybridized to DNA from diploids compared to DNA from haploids. Thus the evidence for or against the inclusion of 5S genes in 16S + 23S gene sets is not entirely consistent.

Assuming that 16S and 23S genes are found at each of the four known 5S rRNA loci in *E. coli,* it is reasonable to ask whether we will have found all the rRNA genes in *E. coli*. The best data at present (Pace and Pace

1971; Spadari and Ritossa 1970; Jarry and Rosset 1971) indicate there are six genes for each of the rRNAs in *E. coli*. Thus we will be missing two gene sets. One of the missing sets might be *rib2*, if it does not correspond to the *cqs*B locus. The other gene set(s) may be between 54 and 59 min, as indicated by the hybridization experiments of Unger et al. (1972).

Genes for rRNA have been mapped in *B. subtilis* by measuring the hybridization ability of newly synthesized DNA taken from synchronized cultures at various times after the start of replication (Oishi and Sueoka 1965; Dubnau, Smith and Marmur 1965, Smith et al. 1968). These experiments indicated 60–80% of the genes for 5S, 16S and 23S rRNA were located close to the origin of replication near the *str–spc* region where genes for some ribosomal proteins are located (see Figure 2). The remainder appeared to be in the last quarter to be replicated. These conclusions are consistent with the observations of Chow and Davidson (1973), who noted that at least two of the 16S + 23S rRNA gene sets in *B. subtilis* were close to the attachment site for the phage SPO2, which is linked to the erythromycin locus, *ery*-1 (Inselburg et al. 1969), and close to the origin of replication. Furthermore, they were able to measure the distance separating the various gene sets. These regions are nonhomologous and of

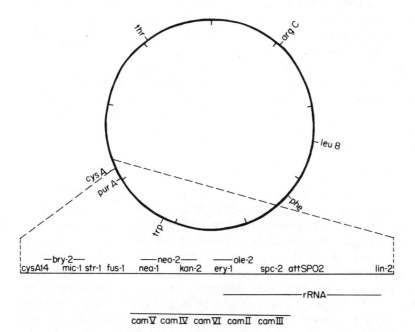

Figure 2 Genetic map of *B. subtilis* showing the location of well-characterized genes for rRNA and ribosomal proteins. Cistrons for rRNA are linked to *att*SPO2 (Chow and Davidson 1973). However, their locations relative to other genes in this region are not known.

varying lengths, including 100, 300, 600, 1000, 5000, 15,000 and 55,000 base pairs. Although the linkage of several of the gene sets with respect to each other could be deduced from the observations, the data were not extensive enough to conclusively establish the linkage of all gene sets.

The simplest model to account for the available data would be if all bacterial rRNA genes are collected into transcriptional units containing one gene of each type with the order 16S–23S–5S. One aspect of this model, for which we have no direct evidence, is the inclusion of 5S rRNA genes in the transcriptional unit. It should be noted that in eukaryotic cells, the 18S, 28S and 7S rRNA are transcribed as a single precursor rRNA (Birnstiel, Chipchase and Speirs 1971). However, the mammalian 5S rRNA is made from a separate transcriptional unit (Attardi and Amaldi 1970). Thus we might conceivably find bacterial 5S rRNA genes in transcriptional units separate from the 16S–23S transcriptional units.

STRUCTURAL GENES FOR RIBOSOMAL PROTEINS

In bacteria the ribosome contains some 54 discrete protein components (see Wittmann, this volume): studies on sequence and immunological cross-reactions show that the ribosomal proteins in *E. coli,* for example, have few similarities with one another. As mentioned earlier, mutations in ribosomal protein genes are not easy to obtain, and selection for antibiotic resistance has provided the major source of such mutants (Davies and Nomura 1972). However it is unlikely that an antibiotic-resistant mutant will be found for every ribosomal protein gene, and analyses of temperature-sensitive mutants can be predicted to play an important role in future studies, particularly since biochemical techniques for the recognition of altered ribosomal components are now well established and sensitive.

Genes Clustered Near *str*A and *spc*A

The region near minute 64 on the *E. coli* genetic map has received the greatest attention in ribosome studies since at least four antibiotic-resistance markers are clustered here. In addition, interstrain or interspecies crosses have indicated that a large number of the ribosomal protein genes may be clustered within this region of the chromosomes of various bacteria (see Sypherd and Osawa, this volume, and Figure 1).

The *str*A locus has been examined most closely, and many mutations in this locus have been mapped, notably by Gorini and his coworkers, who have studied a number of different *str*A alleles with either the drug-resistent or dependent phenotype (Breckenridge and Gorini 1970; Momose and Gorini 1971). These alleles are spaced at various places throughout the locus, and there appears to be no division of the locus into "resistant" or "dependent" regions. The pleiotropy of mutations at the *str*A locus has long been recognized but remains little understood. For example, streptomycin-resistant (Str-R) mutations can restrict the action of tRNA suppressors (Gorini 1971; Strigini and Brickman 1973; see also Gorini, this volume). In addition, the regulation of expression of alkaline phosphatase is anomalous in the presence of some *str*A alleles (Rosenkranz 1963; Rosenkranz et al. 1964). These results could be explained in terms of effects occurring at the ribosome during initiation of translation or in fidelity of translation. It might be expected that alterations in protein S12, the product of the *str*A locus (Ozaki, Mizushima and Nomura 1969), may cause perturbations during initiation since S12 plays an important role in this process (Ozaki, Mizushima and Nomura 1969; see also Nomura and Held, this volume).

The protein S12 has been sequenced and a number of mutant proteins responsible for resistant or dependent (Str-D) phenotypes have been found to have amino acid replacements at different places; the replacement data is consistent with the genetic mapping of Gorini and his coworkers (see Wittmann and Wittmann-Liebold, this volume). No ribosomal Str-R mutation in *E. coli* has been shown to be altered in other than the *str*A locus. On the other hand, reversions from Str-D to independence have been shown to occur by several methods. In addition to true back-revertants, second-site mutations that cause alterations in proteins S4 or S5 lead to independence (Deusser et al. 1970; Birge and Kurland 1970; Funatsu et al. 1972b; Kreider and Brownstein 1972). Among 100 revertants of Str-D, 24 were shown to have altered S4 protein and 16 an altered S5 protein (Hasenbank et al. 1973). The alterations in S5 were apparently single amino acid replacements, whereas alterations in S4 cause shortening or lengthening of the polypeptide chain. There are several possible mechanisms for these large alterations, as discussed by Wittmann and Wittmann-Liebold (this volume).

Str-R mutants are probably the most easily obtained and best-characterized ribosome mutants and have been isolated and studied in other bacterial species. The *str*A locus in *S. typhimurium* has been characterized by in vitro studies and is strictly analogous to *str*A in *E. coli,* although the altered ribosomal protein has not been characterized. Thus far in *S. typhimurium,* pleiotropy of Str-R mutants has not been found with *str*A mutations, but it does occur with mutations at a nonribosomal locus *str*B (Yamada and Davies 1971). Osawa, Smith and Sueoka and their coworkers have isolated in *B. subtilis* a number of antibiotic-resistant mutants that have altered ribosomes and are clustered near the origin of chromosome replication (see Davies and Nomura 1972). Mutations at the *str*-1 locus were demonstrated by polyacrylamide gel electrophoresis to have an altered 30S ribosomal protein (Smith et al. 1969). However, no pure, altered ribosomal proteins have been identified and isolated. Str-R mutants have also been isolated in Pneumococcus and Proteus, but little or no biochemical detail is available. In *B. stearothermophilus,* spontaneous Str-R mutants have been shown by the reconstitution technique to have an alteration in one of the 30S proteins (Gette and Nomura, unpublished experiments). This protein (*B.* S12) is functionally equivalent to protein S12 from *E. coli,* as shown by heterologous reconstitution (Higo et al. 1973). Thus we expect that Str-R mutations in other bacterial species involve 30S ribosomal proteins functionally equivalent to *E. coli* S12.

The mechanism of dominance/recessivity relationships of Str-S, Str-R and Str-D mutations in merodiploids has been a source of some speculation. In general, Str-S is dominant to Str-R, although it has been claimed that the inhibitory action of streptomycin is different in such diploids (Breckenridge and Gorini 1969). In addition, Str-R/Str-D diploids require streptomycin for growth. Several explanations for these dominance relationships have been suggested. Recently it was found that in Str-S/Str-R diploids, unequal synthesis (or assembly) of ribosomes occurred with the bias strongly towards Str-S ribosomes (Chang et al. 1974). This finding neatly explains the dominance of Str-S, but it is not clear if this situation occurs in all such diploids. It also raises questions as to how ribosome assembly and the synthesis of ribosomal proteins are regulated.

The *spc*A locus is the gene for another structural protein, S5 of the 30S ribosome subunit (Bollen et al. 1969). As with mutations at the *str*A locus, mutations to spectinomycin resistance (Spc-R) are often pleiotropic. Some Spc-R mutants are also cold sensitive and defective in ribosome assembly (Nashimoto and Nomura 1970; Nashimoto et al. 1971). In another case, a Spc-R mutation has been reported to abolish the restriction effect (see above) of the Str-R mutation (Kuwano et al. 1969). There is also a form of incompatibility between Spc-R and neomycin-kanamycin resistance (Nek-R); if Spc-R and Nek-R mutations are present together, the Spc-R phenotype is masked. This is considered as evidence for functional interactions between ribosomal proteins (Apirion and Schlessinger 1969; Tyler and Ingraham 1973). Since the product of the *nek* locus is not known and no biochemical evidence is available for ribosomal alterations in Nek-R mutants, this conclusion is somewhat premature. Although some suppressors of Str-D have been shown to be alterations in S5 (the "spectinomycin protein"), none of them are Spc-R, and no Spc-R mutations have been shown to suppress Str-D (Kreider and Brownstein 1972). Thus these represent two independent sets of mutations within the same gene. A number of amino acid replacements have been reported in S5 proteins from Spc-R strains. The Spc-R alterations are located in one tryptic peptide (T10) and the Str-D reversions in another (T1) (see Wittmann and Wittmann-Liebold, this volume). It will be of interest to study fine-structure mapping of mutations in the *spc*A locus; Spc-R mutations and mutations to suppress Str-D are known to be capable of recombination with each other (Kreider and Brownstein 1972).

Spc-R mutations have been analyzed in some detail in *S. typhimurium* and *B. subtilis*. In the latter, altered ribosomes have been shown to be the result of mutations at *spc*-2 (Goldthwaite and Smith 1972b). More recently, Kimura et al. (1973) have shown that three independent Spc-R mutants in *B. subtilis* were altered in a single ribosomal protein (30A) of the 30S subunit. In *S. typhimurium,* mutations in either of two loci, *spc*A and *spc*B, are known to lead to a Spc-R phenotype (Yamada and Davies 1971; Tyler and Ingraham 1973), and although both loci are in the "ribosome" region of the chromosome, only *spc*A mutations have been shown to lead to altered 30S ribosomes (Yamada and Davies 1971). Transductional analyses have shown that both *spc*A and *spc*B are linked to *aro*C in *S. typhimurium*. This region has been shown to contain a number of other putative ribosomal genes. The nature of the *spc*B locus is not understood. No ribosomal protein or reconstitution studies have been carried out with *S. typhimurium*.

Closely linked to the *spc*A locus in *E. coli* is the *ram*A locus, which is the structural gene for protein S4 (Zimmermann et al. 1971). The *ram* mutation was originally discovered as a mutation which overcame the restriction effect of Str-R mutations on streptomycin-induced phenotypic suppression of an amber mutation (Rosset and Gorini 1969). In the absence of the Str-R mutation, it was shown that mutations in the *ram*A locus suppress all three types of nonsense as well as missense mutations. Mutations in this locus are pleiotropic, probably because they decrease the fidelity of ribosomal translation. The efficiency of misreading by tRNA suppressors, and probably misreading by some wild-type tRNAs (in su⁻ strains), is increased in a strain carrying a mutation in the *ram*A locus (Strigini and Brickman 1973; for further discussion, see Gorini, this volume). In addition, mutations at the *ram* locus have an effect on the expression of frameshift mutations in β-galactosidase. It was shown that many *lac* frameshift mutants are leaky. This leakiness is restricted by the presence of *str*A, and this restriction by the *str*A allele is counteracted by the *ram* mutation (Atkins, Elseviers and Gorini 1972).

As has been noted earlier, some revertants of Str-D are mutations in S4; these lead to an S4 protein which is drastically altered. It is not known whether *ram*A phenotype and Str-D reversion in S4 can be associated with the same mutation. Protein sequence studies of altered S4 isolated from *ram* mutants should be interesting.

Protein S4 binds directly to 16S RNA during ribosome assembly, so it is reasonable to suppose that additions or deletions to its sequence may have effects on 30S assembly. Several revertants of Str-D strains are known to be defective in ribosome subunit assembly (Lewandowski and Brownstein 1969; Nashimoto and Nomura 1970).

Erythromycin resistance (Ery-R) is the only mutation known to affect proteins of the 50S subunit in *E. coli*. These mutations all map near minute 64 and have been shown to have alterations in L4 or L22. Ery-R mutations leading to altered L4, but not L22, appear to have a defective peptidyl transferase activity on the ribosome and have a reduced affinity for erythromycin binding (Wittmann et al. 1973). There is currently some conflict in the ordering of the *ery* loci relative to *str*A and *spc*A: some workers have shown that the order is *ery–spc*A–*str*A and others have indicated an order *spc*A–*ery–str*A (Wittmann et al. 1973). It has also been reported that different gene orders are obtained depending on which strains are used as recipients in P1-transduction experiments (Wittmann et al. 1973). Because of the expected interactions among ribosomal proteins, certain gene combinations might be difficult to recover in transduction experiments. For example, the phenotypic expression of Ery-R mutations is masked in several bacterial strains. It will not be possible to resolve this conflict until different techniques, such as deletion mapping using diploid strains similar to the ones used for bacteriophage Mu insertion experiments (see below), are used and all the results compared.

Whatever the outcome, there is no compelling evidence at the moment to support the notion that the structural genes for 30S and 50S proteins in *E. coli* are separately clustered (Osawa et al. 1972). In *B. subtilis,* the genes for 30S and 50S proteins appear to be scattered among each other (Goldthwaite and Smith 1972a).

In *B. subtilis,* Ery-R mutations have been isolated and shown to be alterations of the 50S subunit. Tanaka and his collaborators (Tanaka et al. 1973) have found that 13 independent Ery-R mutants all had alterations in one 50S ribosomal protein (50d in Osawa's nomenclature). None of the mutants showed reduced binding of erythromycin to ribosomes. As noted above, the same result was obtained with those Ery-R mutants of *E. coli* known to have an altered L22 protein. In addition, ribosomes of *B. subtilis* Ery-R mutants were sensitive to the drug in vitro. Tanaka et al. have suggested that the mechanism of resistance to erythromycin could involve either (1) a supernatant factor which binds to ribosomes and prevents erythromycin binding as a result of the ribosomal protein change, or (2) a change in membrane permeability as a secondary effect of an alteration in a ribosomal protein. The *ery* locus in *B. subtilis* maps between *spc*-2 and *kan*-2 and close to *ole*-2 in the "ribosome" gene cluster. It is possible that the *ole* locus is the same as the *ery* locus, since erythromycin and oleandomycin are closely related macrolide antibiotics.

The K-protein (30S protein S7) is different in its electrophoretic mobility between *E. coli* K12 and B strains. The genetic locus determining this protein has been mapped by crossing K12 and B strains and is placed close to the *str*A locus.

In *E. coli,* several other putative ribosomal mutants have been reported to map in the *str–spc* ribosome cluster. However no ribosomal protein alteration has been determined, and resistant phenotypes have not been demonstrated in cell-free extracts of the resistant strains. Mutants in this category include Nek-R (Apirion and Schlessinger 1968) and Nea-R (Cabezon, Cannon and Bollen, personal communication). The latter have been classified into two groups, *neaA* and *neaB*. The *neaA* maps close to *spcA* and the *neaB* close to *strA*. Mutations at the *nek, neaA* and *neaB* genes, like those at *str, spc* and *ery,* are recessive to wild-type in *E. coli* merodiploid strains. Nek-R mutants have been isolated in *S. typhimurium,* and although not shown to possess altered ribosomes, they are closely linked to *strA* and show a strong interaction (phenotype masking effect) when combined with *spcA* mutants, such that the strains behave as Spc-S; the Spc-R mutation can be recovered in appropriate genetic crosses (Tyler and Ingraham 1973).

Probably because of its increased susceptibility to antibiotics (compared to *E. coli*), more types of resistant mutants have been isolated in *B. subtilis*. These include bryamycin (thiostrepton), chloramphenicol, and separate mutants for neomycin and kanamycin resistance. The work of Osawa, Smith and Sueoka and their coworkers has led to extensive genetic mapping of the ribosome gene cluster in this organism. Not all of the mutants have been clearly demonstrated to give rise to ribosomes with altered properties or altered components. Since the antibiotics used are known to affect steps in protein synthesis by binding to ribosomes, it is reasonable to assume that strains resistant to these antibiotics, having mutations linked (clustered) to known ribosome mutations, have alterations in ribosomal structural genes. In *E. coli* there have been no reports of mutants resistant to chloramphenicol, which is a well-studied protein synthesis inhibitor. However in *B. subtilis* a variety of interesting mutants have been isolated, and Osawa et al. (1973) have classified chloramphenicol-resistant mutants (CM^R) of *B. subtilis* into six groups. Five of the groups had altered 50S proteins (50a, b, c, e and f) as determined by chromatography on carboxymethyl-cellulose; the remaining group had no demonstrable protein change. Several of the mutant ribosomes had impaired ability to bind both chloramphenicol and erythromycin. Ery-R mutants isolated by the same group of workers were found to have an altered 50d protein and were not affected in erythromycin binding capacity.

Ribosomal Genes Mapping Outside the *str–spc* Region

It had been thought originally that all ribosomal protein genes might map in the *str–spc* cluster, but it has now been shown that the structural gene for at least one ribosomal protein is located at another place on the *E. coli* chromosome. In addition, the gene(s) for modifying enzyme(s) for ribosomal RNA maps elsewhere. Factors involved in ribosome maturation may exist and map outside of this region, e.g., the *rim* loci, which might be structural genes for ribosome maturation factors (Bryant and Sypherd 1974).

The structural gene for protein S18 of the 30S subunit is located between 76 and 88 min. on the *E. coli* genetic map. A mutant has been isolated having an S18 protein with electrophoretic properties different from those of S18 from a wild-type strain. Amino acid and tryptic peptide analyses have shown a cysteine for arginine substitution at an internal position in the polypeptide chain. This alteration accounts for the altered electrophoretic properties of the mutant protein. The *E. coli* strain with altered S18 has no recognizable phenotypic alteration (Bollen et al. 1973; Kahan et al. 1973).

Mutations in the *ram*B locus map close to minute 1 (near *ksg*A) on the *E. coli* chromosome and produce an altered protein S4 (Zimmermann et al. 1973). Since the structural gene for S4 is known to map in the *str–spc* cluster (*ram*A), it would appear that the *ram*B locus might determine an enzyme which modifies S4. Since the altered S4 obtained from *ram*B mutants is distinguishable from wild-type S4 on chromatography, it should soon be possible to isolate this altered protein and determine the nature of the change. Since the *ram*B mutants have no recognizable phenotype, it will be interesting to determine the function of this gene.

Mutations at the *ksg*A locus co-transduce with *ram*B, and it is possible that a group of ribosome-modifying enzymes map in this region. The *ksg*A locus determines an enzyme that dimethylates two adenine residues in a specific sequence (—AACCUG—) near the 3′ end of 16S ribosomal RNA. Failure to methylate this sequence in *ksg*A mutants leads to resistance to the antibiotic kasugamycin (Helser et al. 1971, 1972). Studies on the methylation of various ribonucleoprotein particles indicate that methylation of 16S RNA with the *ksg*A enzyme occurs at a specific stage of ribosome assembly (Thammana and Held, personal communication; see Nomura and Held, this volume).

Certain conditionally lethal mutants of *E. coli,* which map in various regions of the chromosome outside the *str–spc* region, are thought to be ribosome mutants. These include Mac-D (dependence on erythromycin) (Sparling and Blackman 1973) and several temperature-sensitive mutants that fail to synthesize protein at 42°C (Flaks et al. 1966; Phillips et al. 1969). However in none of these cases has the ribosome been directly implicated. If these do prove to be ribosome mutations, the genetics and biochemistry of ribosomes may turn out to be more complicated than presently realized.

At present, the structural genes for about 20 ribosomal proteins in *E. coli* have been located in the "*spc–str*" region, mainly using interspecies and intergeneric crosses (see Figure 1 and also Sypherd and Osawa, this volume). However the resolving power of these methods is not sufficient to determine with accuracy how many of the ribosomal protein genes of *E. coli* are within the *str–spc* region.

REGULATION OF THE BIOSYNTHESIS OF RIBOSOMES

Several observations indicate that the biosynthesis of ribosomes in bacteria is regulated. First, the number of ribosomes per genome is proportional to growth rate (Maaløe and Kjeldgaard 1966; Kjeldgaard and Gausing, this volume). Second, mature ribosomes contain stoichiometric amounts of most components (Wittmann, this volume). Thus the biosynthesis of the components might be coordinated in some fashion. Finally, stringent bacteria do not accumulate rRNA during amino acid starvation (Edlin and Broda 1968; Cashel and Gallant, this volume).

Ribosomal RNA

The only observation for which we have a satisfactory explanation is coordinate biosynthesis of the three types of rRNA. As discussed previously, it is becoming increasingly clear that rRNA genes exist in gene sets that contain one gene for 16S and one gene for 23S rRNA. A gene for 5S rRNA may also be included. Thus the stoichiometric relationship of the production (Galibert et al. 1967) and incorporation of rRNA into ribosomes can be easily understood.

Given the redundancy of the rRNA gene sets, it would be interesting to know whether they are all expressed equally at all growth rates. The sequence studies of Jarry and Rosset (1971) are applicable to this question. They found that residues 11 to 13 in 5S rRNA from MRE600 have three different sequences. Thus at least three different structural genes are expressed in this organism. Furthermore, the molar yields of the three sequences are not the same. The data is consistent with the view that there are six or seven 5S rRNA genes equally expressed in MRE600. However the data could also be consistent with more complicated models.

During amino acid starvation the accumulation of rRNA is inhibited in "stringent" bacteria but not in "relaxed" mutants, that is, mutants at the classical *rel*A locus near 53 min in *E. coli* (Stent and Brenner 1961; Edlin and Broada 1968; Cashel and Gallant, this volume). This locus is also called RC or *rel*-1. Amino acid starvation also results in an increase in the concentration of ppGpp (magic spot 1 or MS-1) and pppGpp (magic spot 2 or MS-2) in "stringent" bacteria but not in "relaxed" mutants (Cashel and Kalbacher 1970; Cashel and Gallant 1969; Lazzarini and Cashel 1971; Cashel 1969). It has not been proven that these nucleotides are directly or indirectly responsible for the stringent phenomenon rather than an effect of this phenomenon. However it seems likely that they are a causative factor. Travers has attempted to demonstrate a role for these nucleotides in the regulation of rRNA synthesis using an in vitro system. These experiments are discussed in the chapter by Travers (this volume). The biochemistry of their synthesis and degradation is reviewed in this volume by Cashel and Gallant.

In vitro experiments have indicated that ppGpp and pppGpp are made on ribosomes in the presence of the "stringent factor," which appears to be the product of the *rel*A gene (Haseltine et al. 1972). This is consistent with the observation that the *rel*A mutation is recessive to wild type. The in vitro synthesis of ppGpp by ribosomes is described in this volume by Block and Haseltine.

If ribosomes are responsible for the synthesis of ppGpp, then one may expect to find mutations in ribosomal components that confer a "relaxed" phenotype upon the bacteria. Friesen, Haseltine and Fiil (personal communication) have isolated such a mutant. It has a partially "relaxed" phenotype but does not map at *rel*A. The locus for this phenotype, called *rel*C, cotransduces with *rif*. The accumulation of rRNA and ppGpp during amino acid starvation of the *rel*C mutant is intermediate between the response of "stringent" and *rel*A "relaxed" bacteria. The evidence that *rel*C may be a locus for a ribosomal protein is that the 70S ribosomes from the mutant are unable to convert GTP to ppGpp and pppGpp in an in vitro reaction. The lesion appears to lie in the 50S subunits. However the mutant component has not been identified as yet (see Block and Haseltine, this volume).

Mutations at two other loci are known to effect the metabolism of ppGpp: *rel*B and *spo*T. Mutants at the *rel*B locus were originally isolated by Lavalle (1965) and have not been extensively studied. Laffler and Gallant (personal communication) have recently characterized a mutant at the *spo*T locus that maps close to 72 min in *E. coli*. Its gene product appears to play a role in the conversion of ppGpp to pppGpp. The *spo*T mutant is discussed by Cashel and Gallant in this volume.

Chaney and Schlessinger (personal communication) have isolated temperature-sensitive mutants that appear to have lesions in genes whose products are specifically involved in the synthesis or accumulation of rRNA and not other RNA species. Further investigation of these mutants should be interesting.

A major development relevant to the investigation of the regulation of rRNA biosynthesis has been the isolation of an rRNA transducing phage by Soll (personal communication). This ϕ80 defective phage carries the *rib*2 16S + 23S gene set mapped on the *E. coli* episome F14 as well as the su7+ and *ilv*A cistrons (Deonier et al. 1973). We expect this phage will be useful for in vivo and in vitro investigations on the regulation and mechanisms of rRNA biosynthesis.

Ribosomal Protein

Most of the attention in the investigation of the regulation of the biosynthesis of ribosomes has been centered on rRNA. Recently several experiments have been performed relating to the genetic organization and physiological regulation of ribosomal protein genes.

The clustering of many ribosomal protein genes in the *str–spc* region of both *E. coli* and *B. subtilis* suggests the genes may be organized into a transcriptional unit. The first direct indication of this came from the mu phage experiments of Nomura and Engbaek (1972). Mu phage can integrate into many sites on the *E. coli* chromosome, perhaps randomly (Taylor 1963; Boram and Abelson 1971; Bukhari and Zipser 1972). It inactivates genes into which it is inserted and has a strong polar effect when integrated into a transcriptional unit such as the *lac* operon (Daniell and Abelson 1973). Nomura and Engbaek (1972) used a merodiploid that contained resistant genes for *ery*B, *spc*A, *str*A and *fus* on the episome and sensitive genes on the chromosome, i.e., F' *ery*^{r} *spc*^{r} *str*^{r} *fus*^{r}/*ery*^{s} *spc*^{s} *str*^{s} *fus*^{s}. This "4R" diploid is sensitive to all four antibiotics. Selection was made for insertion of mu phage into the region of the sensitive genes on the chromosome by selecting for drug resistance. The observed pattern of drug resistance could be explained if the drug-resistant genes that were used are part of a transcription unit that is read in the order promoter–*ery*B–*spc*A–*str*A–*fus*⟶. The *ery*B, *spc*A and *str*A loci code for ribosomal proteins (see above), and the *fus* locus is the gene for a nonribosomal chain elongation factor, EF-G. Other ribosomal protein genes that are known to map in this region could also be part of the transcriptional unit, either interspersed between the known markers or at either end. However there is no direct information on the size of this transcriptional unit. More recently, Jørgensen (personal communication) obtained essentially the same results using a different diploid. In addition, in one case he was able to rescue an inactivated gene that was distal from the insertion of mu. This result would appear to eliminate any possibility that these observations are due to large deletions or recombination between episome and chromosome mediated by mu (van de Putte and Gruijthuijsen 1972; Faelen, Toussaint and Couturier 1971).

The model of a ribosomal protein transcriptional unit proposed by Nomura and Engbaek (1972) makes several predictions. Polar mutations arising from nonsense, frameshift or insertion mutations should predominately affect genes distal from the mutation. Thus it should be possible to obtain polar mutants that inactivate the *str* and *fus* genes without affecting the *ery* and *spc* genes. However, polar mutations inactivating *ery* and *spc* would probably also inactivate *str* and *fus*. Furthermore, it should be possible to delete the *str–fus* end, together with neighboring genes outside the transcriptional unit, without affecting the expression of the promoter proximal genes *ery* and *spc*. On the other hand, it would not be possible to delete the *ery–spc* end, together with outside neighboring genes, without affecting the expression of the *str* and *fus* genes. Some of these predictions have been confirmed.

The pattern of spontaneous drug-resistant mutants that occur in "4R" diploids, F' *ery*^{r} *spc*^{r} *str*^{r} *fus*^{r}/*ery*^{s} *spc*^{s} *str*^{s} *fus*^{s}, has been analyzed (Nomura and Engbaek 1972; Jaskunas, Engbaek and Nomura, unpublished observations). Pleiotropic mutants have been obtained with drug phenotypes RRRR, SRRR and SSRR. These mutants may have resulted from deletions of the inactivated genes or polar mutations. No mutants, such as RRSS, have been observed that would suggest other organizations of these genes. The effect of deletions covering these genes has been specifically analyzed. Using a merodiploid strain with an episome covering the *str–spc* region, many spontaneous mutants with deletions covering the *fus* and *str* genes on the episome were located and analyzed (Zengel and Nomura, unpublished observations). In some mutants, the deletion also included the *spc* gene or genes such as *aro*B, which are outside the ribosomal protein gene cluster. In all cases the expression of the remaining *ery* or *spc* genes on the episome was not detectably reduced. All these observations are consistent with the model of Nomura and Engbaek (1972). A better test of the model, however, would be to isolate polar mutants resulting from known causes, such as nonsense mutations or deletions of the promoter, and demonstrate that expression of the remaining distal genes was eliminated or reduced.

Attempts have been made to isolate polar amber mutants of this transcriptional unit using a "4R" diploid (Jaskunas, Engbaek and Nomura, unpublished observations). One suppressible mutant having an amber mutation in or near *str*A (the structural gene for S12) was studied in detail. After introduction of the suppressor gene, the episome was eliminated and the rate of synthesis of individual ribosomal proteins was compared to the control haploid strain constructed from the parent strain in the same way. The results of such experiments indicated a polar effect on *str*A and also the structural gene for 30S protein S7 (the K locus). Thus at least these two proteins are translated from a polycistronic mRNA. It was also found that such haploid cells show an increase in the rate of synthesis of many other ribosomal proteins examined. This observation may reflect the coordinate regulation of ribosomal proteins. Perhaps the mutant cells "respond" to decreased synthesis of the ribosomal proteins such as S12 and S7 and compensate by stimulating the expression of all the ribosomal genes.

Other physiological experiments have been performed that concern the organization and regulation of ribosomal protein genes. These experiments are relevant to two principle questions: How large are the transcriptional units for ribosomal proteins? What is the order of "induction" of ribosomal proteins following a shift-up (from poor media to rich) or following release of rifampicin inhibition of transcription? Molin et al. (1974) investigated the kinetics of the decay of ribosomal protein mRNA in *E. coli* treated with rifampicin (rifampicin "run-out" experiments) and the kinetics of the formation of individual ribosomal proteins following removal of rifampicin and the resumption of RNA synthesis. Similarly, Dennis (personal communication) examined the kinetics of the increased synthesis of individual ribosomal proteins following a shift-up from a poor media to a rich media.

In the rifampicin "run-out" experiments, Molin et al. (1974) did not find any evidence for an exceptionally long transcriptional unit that might code for nearly all the ribosomal proteins. Furthermore, Molin et al. (1974) and Dennis (personal communication) observed that all ribosomal proteins could be "induced" in approximately 1–4 min. Assuming that RNA polymerase transcribes ribosomal protein transcriptional units at the same rate at which it transcribes other mRNAs (50 nucleotides/sec/-polymerase), Molin et al. concluded that ribosomal protein genes are organized into transcriptional units containing up to 10 genes each and that there may be one larger unit containing as many as 20 genes. It is not clear whether these observations are inconsistent with conclusions of the mu phage experiments of Nomura and Engbaek (1972). Although at least 22 ribosomal genes appear to map in the *str–spc* region of *E. coli* (see above and Sypherd and Osawa, this volume), it is not known with any certainty how many of these (and other genes) are part of the transcriptional unit defined by the mu phage experiments. There could be more than one transcriptional unit in this region of the chromosome. For another perspective on these experiments, see the chapter by Kjeldgaard and Gausing (this volume).

In both investigations, the order of appearance of the individual proteins following "induction" was not what might be expected from the mu phage experiments. Those experiments suggested, for example, that the gene for S5 should be closer to the promoter than the genes for S12 and S7. Thus S5 should appear before S12 and S7, if the observed "induction" results from increased transcription and the order of "induction" reflects the linear arrangement of the genes in the transcriptional unit. In fact, just the opposite was observed. On the other hand, S12 and S7 were "induced" at about the same time, which is consistent with the physiological experiments on the ribosomal protein amber mutant (see above), indicating that the genes for S12 and S7 were close and on the same polycistronic mRNA. One way to reconcile these experiments with the mu phage experiments would be to postulate internal promoters that are operative during a shift-up or other times of increased demand for ribosomal proteins. A different order of induction than the linear arrangement of the genes could also occur if the ribosomal protein mRNA is cut into smaller mRNAs during or following transcription. In any case, it does not seem possible to reconcile these physiological experiments with the mu phage experiments without complicating the picture of the regulation of the biosynthesis of ribosomal proteins.

MECHANISM OF ASSEMBLY

Information on the mechanism of ribosomal assembly has been obtained from the investigation of the reconstitution of ribosomes in vitro (see this volume, Nomura and Held; Schlessinger), the identification of in vivo ribosomal intermediates with pulse-chase kinetic experiments (see Schlessinger, this volume), and the isolation of mutants that are defective in the assembly of ribosomes.

The isolation of mutants can help us define the various steps and factors involved in ribosome assembly. Altered assembly could conceivably result from mutations in genes for three classes of molecules involved in the assembly of ribosomes: (1) ribosomal protein and rRNA; (2) enzymes required for post-translational or post-transcriptional modification of ribosomal protein or rRNA; (3) other factors involved in the assembly of ribosomes that are not normally constituents of mature ribosomes.

The high energy of activation for the reconstitution of 30S subunits in vitro suggested that conditional mutants defective in ribosomal maturation might be found among cold-sensitive mutants (Traub and Nomura 1969). Subsequently many ribosomal assembly mutants of *E. coli* (Nashimoto and Nomura 1970; Guthrie, Nashimoto and Nomura 1969a,b; Bryant and Sypherd 1974; Berman, Budzilowicz and Chang 1973) and *S. typhimurium* (Tai, Kessler and Ingraham 1969; Tyler and Ingraham 1973) have been found among this type of conditional mutant. Some of these have been designated *sad* for subunit assembly defective (Guthrie, Nashimoto and Nomura, 1969a,b; Nashimoto and Nomura 1970) or *rim* for ribosome maturation defective (Bryant and Sypherd 1974) or *csr* for cold-sensitive ribosome assembly (Tyler and Ingraham 1973). The phenotype of ribosomal assembly mutants has been taken to be the inability to form 30S and/or 50S subunits at the nonpermissive temperature (usually 20°C in the case of cold-sensitive mutants) and the accumulation of ribonucleoprotein particles of intermediate size.

Initially three classes of defective ribosome assembly were observed in cold-sensitive mutants of *E. coli* (Guthrie, Nashimoto and Nomura 1969a): (1) mutants that accumulated 32S particles instead of 50S particles at 20°C; (2) mutants that accumulated 43S particles instead of a 50S particle at 20°C; and (3) mutants that accumulated 32S and 21S particles instead of 50S and 30S particles at 20°C. For the third class, a single mutation has apparently resulted in defective 30S and 50S assembly. These mutants were selected only for cold sensitivity, and the altered components have not been identified. Several mutants have been shown to map outside the *str–spc* region (Guthrie, Nashimoto and Nomura 1969b).

Four cold-sensitive mutants of *S. typhimurium* have been mapped at separate loci in the *str–spc* region (Tyler and Ingraham 1973). Thus they may be mutants of four different ribosomal proteins that result in cold sensitivity. However, the altered components are unknown.

Several cold-sensitive ribosomal assembly mutants of *E. coli* have been obtained in which the altered component is clearly a ribosomal protein. In these cases the initial selection was for an altered ribosomal protein. These mutants were then screened to find those that had also become cold sensitive and defective in ribosome assembly. Thus many spontaneous spectinomycin-resistant mutants, which have an altered S5, are also cold sensitive and defective in ribosome assembly (Nashimoto and Nomura 1970; Nashimoto et al. 1971). Some revertants from streptomycin dependence, that presumably have an altered S4 or S5 in addition to an altered S12, are also defective in ribosome assembly and cold sensitive in some cases (Lewandowski and Brownstein 1966, 1969; Nashimoto and Nomura 1970). It has been demonstrated that one of these mutants does have an altered S4 (Kreider and Brownstein 1971). As in the case for other cold-sensitive mutants, defective 50S assembly resulted in the accumulation of 43S or 32S particles; defective 30S assembly resulted in the accumulation of 26S or 21S particles.

The RNA in the 21S particle that accumulates in one of the cold-sensitive, spectinomycin-resistant mutants is a submethylated precursor to 16S rRNA (Lowry and Dahlberg 1971; Nashimoto and Nomura 1970). The precursor is larger than mature 16S rRNA by at least 94 residues (Lowry and Dahlberg 1971). These observations indicate that the final cleavage steps and methylations of 16S rRNA probably occur during or after assembly. The protein composition of this 21S particle is similar to that of an intermediate (the "RI" particle) in the reconstitution of 30S subunits (Traub and Nomura 1969). Thus the order of addition of proteins appears to be similar or identical for the assembly of 30S subunits in vitro and in vivo (see this volume, Nomura and Held; Schlessinger for further discussions of this subject).

Several mutants that have altered 30S ribosomal proteins are nevertheless defective in 50S assembly as well as 30S assembly (Nashimoto and Nomura 1970; Nashimoto et al. 1971; Kreider and Brownstein 1971). This suggests that 30S ribosomes or precursors have a role in the assembly of 50S particles. However, there are mutants defective in 30S assembly but not 50S assembly (Rosset et al. 1971) and mutants defective in 50S assembly but not 30S assembly (Guthrie, Nashimoto and Nomura 1969a). Furthermore, it is possible to assemble 30S and 50S ribosomes in vitro in the absence of the other subunit (see Nomura and Held, this volume). Thus the significance and mechanism of this apparent coupling between 30S and 50S assembly is not clear.

Perhaps the best way to identify nonribosomal factors involved in ribosome maturation is through genetic techniques. Bryant and Sypherd (1974) have isolated four cold-sensitive mutants of *E. coli* that map outside the *str–spc* region. All of these mutants are defective in 50S assembly at 20°C. The location of three of these mutants, designated *rim* for ribosome maturation, is given in Figure 1. The altered components in these mutants have not been identified. However from three observations, the authors suggest they are not ribosomal proteins. First, they map far from the *str–spc* region, where structural genes for many ribosomal proteins are located. Second, the mobilities of the ribosomal proteins of the mutants are the same as those from the parent in a two-dimensional gel (Bryant, Fujisawa and Sypherd 1974). Finally, all of the mutant alleles are recessive to wild type. Since none of the above observations exclude the possibility of these mutations being in ribosomal protein genes, the suggestion that the mutant genes may be cistrons for enzymes or factors involved in ribosome assembly remains to be proven.

A new development in our understanding of the mechanism of assembly has been the isolation of a 30S precursor rRNA containing the sequences of one 16S and one 23S molecule (Nikolaev, Silengo and Schlessinger 1973; Dunn and Studier 1973). As discussed previously, this demonstrates the existence of transcriptional units containing 16S and 23S cistrons. This precursor rRNA could be isolated from an *E. coli* mutant with a defective RNase III (Kindler, Keil and Hofschneider 1973) but not from its parent. Furthermore, incubation of the precursor with purified RNase III in vitro resulted in the formation of two RNAs of the size of precursor 16S rRNA and 23S rRNA. Other smaller RNAs (5S rRNA?) were also released. These experiments demonstrate a role for RNase III in the post-transcriptional cleavage of precursor rRNA. Kinetic studies suggested that the precursor is cleaved in vivo into two fragments of the size 25S and 17.5S, which then leads to the formation of 23S and 16S rRNA. In the mutant, the cleavage of the 30S precursor rRNA is either done by the mutationally altered RNase III (perhaps at a reduced rate) or by another (substitute?) endonuclease. If assembly with ribosomal protein were to begin before the precursor was cut, it could explain the indicated role of 30S subunits in the assembly of 50S ribosomes (see above and Nashimoto and Nomura 1970).

Another enzyme involved in the maturation of ribosomal components is the specific 16S–rRNA methylase that is altered in kasugamycin-resistant strain, i.e., *ksg*A mutants (Sparling 1968; Helser, Davies and Dahlberg 1971, 1972). The product of the *ram*B locus, which maps near *ksg*A, may also fit in this category (Zimmermann, Ikeya and Sparling 1973). It appears to be an enzyme that modifies ribosomal protein S4. See the section in this chapter on ribosomal proteins for a discussion of the *ksg*A and *ram*B genes.

In summary, several conditional mutants have been isolated that are defective in ribosomal maturation at the nonpermissive temperature. In all cases where the altered component has been identified, it has been a ribosomal protein. There are also mutants of enzymes that modify ribosomal components. No mutants of other nonribosomal factors that do not modify ribosomal components but are involved in the assembly of ribosomes have been identified. Some *rim* or *sad* mutants, which map outside the *str–spc* region, may fall into this classification. However there is no direct evidence for the existence of such factors.

CONCLUDING REMARKS

Genetic approaches to ribosome research were limited in the past. Because ribosomes are essential for cellular growth, isolation of useful mutants was difficult. In addition, because of the complexity of the ribosome structure, biochemical characterization of isolated mutants was also not an easy task. In the past few years, we have witnessed much progress in several aspects in the studies of bacterial ribosomes. Various biochemical techniques to separate and characterize ribosomal components are now available. More importantly, several pertinent biological questions have been clearly formulated. Thus we expect extensive genetic studies in the near future and progress in solving several problems discussed in this article.

Acknowledgments

We would like to acknowledge the assistance of Janice Zengel and Dr. I. Smith in the preparation of the two figures and thank the numerous investigators who made unpublished data available to us. Work done in the laboratory of M. Nomura has been supported in part by grant No. GB-31086X2 from the National Science Foundation and grant No. GM-20427-01 from the National Institutes of Health. Work done in the laboratory of J. Davies was supported by grant No. AI-10076 from the National Institutes of Health. S. R. Jaskunas is a recipient of a National Institutes of Health Special Postdoctoral Fellowship. This is paper No. 1728 of the Laboratory of Genetics.

References

Aharonowitz, Y. and E. Z. Ron. 1972. A temperature-sensitive mutant in *Bacillus subtilis* with an altered elongation factor G. *Mol. Gen. Genet.* **119**:131.
Apirion, D. and D. Schlessinger. 1968. Co-resistance to neomycin and kanamycin by mutations in an *Escherichia coli* locus that affects ribosomes. *J. Bact.* **96**:768.
———. 1969. Functional interdependence of ribosomal components of *Escherichia coli*. *Proc. Nat. Acad. Sci.* **63**:794.
Atkins, J. F., D. Elseviers and L. Gorini. 1972. Low activity of β-*galactosidase* in frameshift mutants of *Escherichia coli*. *Proc. Nat. Acad. Sci.* **69**:1192.

Attardi, G. and F. Amaldi. 1970. Structure and synthesis of ribosomal RNA. *Ann. Rev. Biochem.* **39**:183.

Attardi, G. A., P. C. Huang and S. Kabat. 1965. Recognition of ribosomal RNA sites in DNA. I. Analysis of the *E. coli* system. *Proc. Nat. Acad. Sci.* **53**:1490.

Berman, D., C. Budzilowicz and F. N. Chang. 1973. Accumulation of different precursor ribonucleoprotein particles by various cold-sensitive, antibiotic-resistant mutants. *Biochem. Biophys. Res. Comm.* **54**:991.

Bernadi, A. and P. Leder. 1970. Protein biosynthesis in *Escherichia coli*. Purification and characteristics of a mutant G factor. *J. Biol. Chem.* **245**:4263.

Birge, E. A. and C. G. Kurland. 1969. Altered ribosomal protein in strepto-mycin-dependent *Escherichia coli*. *Science* **166**:1282.

———. 1970. Reversion of a streptomycin-dependent strain of *Escherichia coli*. *Mol. Gen. Genet.* **109**:356.

Birge, E. A., C. R. Craven, S. J. S. Hardy, C. G. Kurland and P. Voynow. 1969. Structure determinant of a ribosomal protein: K locus. *Science* **164**:1285.

Birnbaum, L. S. and S. Kaplan. 1971. Localization of a portion of the ribosomal RNA genes in *Escherichia coli*. *Proc. Nat. Acad. Sci.* **68**:925.

Birnstiel, M. L., M. Chipcase and J. Speirs. 1971. In *Progress in nucleic acid research and molecular biology* (ed. J. N. Davidson and W. E. Cohn) vol. 11, The ribosomal RNA cistrons, pp. 351–390. Academic Press, New York.

Bollen, A., J. Davies, M. Ozaki and M. Mizushima. 1969. Ribosomal protein confirming sensitivity to the antibiotic spectinomycin in *Escherichia coli*. *Science* **165**:85.

Bollen, A., M. Faelen, J. P. Lecocq, A. Herzog, J. Zengel, L. Kahan and M. Nomura. 1973. The structural gene for the ribosomal protein S18 in *Escherichia coli*. I. Genetic studies on a mutant having an alteration in the protein S18. *J. Mol. Biol.* **76**:463.

Boram, W. and J. Abelson. 1971. Bacteriophage Mu integration: On the mechanism of Mu-induced mutation. *J. Mol. Biol.* **62**:171.

Breckenridge, L. and L. Gorini. 1969. The dominance of streptomycin sensitivity re-examined. *Proc. Nat. Acad. Sci.* **62**:979.

———. 1970. Genetic analysis of streptomycin resistance in *Escherichia coli*. *Genetics* **65**:9.

Bremer, H. and L. Berry. 1971. Co-transcription of 16S and 23S ribosomal RNA in *Escherichia coli*. *Nature New Biol.* **234**:81.

Brownlee, G. G., F. Sanger and B. G. Barrell. 1967. Nucleotide sequence of 5S-ribosomal RNA from *Escherichia coli*. *Nature* **215**:735.

Bryant, R. E. and P. S. Sypherd. 1974. Genetic analysis of cold-sensitive, ribosome maturation mutants of *Escherichia coli*. *J. Bact.* **117**:1082.

Bryant, R. E., T. Fujisawa and P. S. Sypherd. 1974. Ribosomal proteins and RNA of ribosome maturation mutants of *Escherichia coli*. *Biochemistry* **13**:2110.

Buckel, P., D. Ruffler, W. Piepesberg and A. Böck. 1972. RNA overproducing revertants of an alanyl-tRNA synthetase mutant of *Escherichia coli*. *Mol. Gen. Genet.* **119**:323.

Bukhari, A. and D. Zipser. 1972. Random insertion of Mu-1 DNA within a single gene. *Nature New Biol.* **236**:240.

Cashel, M. 1969. The control of ribonucleic acid synthesis in *Escherichia coli*. IV. Relevance of unusual phosphorylated compounds from amino acid-starved stringent strains. *J. Biol. Chem.* **244**:3133.

Cashel, M. and J. Gallant. 1969. Two compounds implicated in the function of the RC gene of *Escherichia coli*. *Nature* **221**:838.

Cashel, M. and B. Kalbacher. 1970. The control of ribonucleic acid synthesis in *Escherichia coli*. V. Characterization of a nucleotide associated with the stringent response. *J. Biol. Chem.* **245**:2309.

Chang, F. N., Y. J. Wang, C. J. Fetterolf and J. G. Flaks. 1974. Unequal contribution to ribosomal assembly of both *str* alleles in *Escherichia coli* merodiploids and its relationship to the dominance phenomenon. *J. Mol. Biol.* **82**:273.

Chow, L. T. and N. Davidson. 1973. Electron microscope mapping of the distribution of ribosomal genes of the *Bacillus subtilis* chromosome. *J. Mol. Biol.* **75**:265.

Colli, W. and M. Oishi. 1969. Ribosomal RNA genes in bacteria: Evidence for the nature of the physical linkage between 16S and 23S RNA genes in *Bacillus subtilis*. *Proc. Nat. Acad. Sci.* **64**:642.

Colli, W., I. Smith and M. Oishi. 1971. Physical linkage between 5S, 16S and 23S ribosomal RNA genes in *Bacillus subtilis*. *J. Mol. Biol.* **56**:117.

Daniell, E. and J. Abelson. 1973. *Lac* messenger RNA in *lac Z* gene mutants of *Escherichia coli* caused by insertion of bacteriophage Mu. *J. Mol. Biol.* **76**:319.

Davies, J. and M. Nomura. 1972. The genetics of bacterial ribosomes. *Ann. Rev. Genet.* **6**:203.

Davies, J., P. Anderson and B. D. Davis. 1965. Inhibition of protein synthesis by spectinomycin. *Science* **149**:1096.

Deonier, R. C., L. Soll, E. Ohtsubo, H. J. Lee and N. Davidson. 1973. Mapping ribosomal RNA and other genes on defined segments of the *E. coli* chromosome. *Fed. Proc.* **32**:663 Abs.

Deusser, E., G. Stöffler, H. G. Wittmann and D. Apirion. 1970. Ribosomal proteins. XVI. Altered S4 proteins in *Escherichia coli* revertants from streptomycin dependence to independence. *Mol. Gen. Genet.* **109**:298.

Doolittle, W. F. and N. R. Pace. 1970. Synthesis of 5S ribosomal RNA in *Escherichia coli* after rifampicin treatment. *Nature* **228**:125.

———. 1971. Transcriptional organization of the ribosomal RNA cistron in *Escherichia coli*. *Proc. Nat. Acad. Sci.* **68**:1786.

Dubnau, D., I. Smith and J. Marmur. 1965. Gene conservation in Bacillus species. II. The location of genes concerned with the synthesis of ribosomal components and soluble RNA. *Proc. Nat. Acad. Sci.* **54**:724.

Dunn, J. J. and F. W. Studier. 1973. T7 early RNAs and *Escherichia coli* ribosomal RNAs are cut from large precursor RNAs *in vivo* by ribonuclease III. *Proc. Nat. Acad. Sci.* **70**:3296.

Edlin, G. and P. Broda. 1968. Physiology and genetics of the ribonucleic acid control locus in *Escherichia coli*. *Bact. Rev.* **32**:206.

Faelen, M., A. Toussaint and M. Couturier. 1971. Mu-1 promoted integration of a λ-gal phage in the chromosome of *E. coli*. *Mol. Gen. Genet.* **113**:367.

Flaks, J. G., P. S. Leboy, E. A. Birge and C. G. Kurland. 1966. Mutations and genetics concerned with the ribosome. *Cold Spring Harbor Symp. Quant. Biol.* **31**:623.

Funatsu, G., E. Schlitz and H. G. Wittmann. 1972a. Ribosomal proteins. XXVII. Localization of the amino acid exchanges in protein S5 from two *Escherichia coli* mutants resistant to spectinomycin. *Mol. Gen. Genet.* **114**:106.

Funatsu, G., W. Puls, E. Schiltz, J. Reinbolt and H. G. Wittmann. 1972b. Ribosomal proteins. XXXI. Comparative studies on altered proteins S4 of six *Escherichia coli* revertants from streptomycin dependence. *Mol. Gen. Genet.* **115**:131.

Galibert, F., J. C. Lelong, C. J. Larsen and M. Boiron. 1967. Position of 5-S RNA among cellular ribonucleic acid. *Biochim. Biophys. Acta* **142**:89.

Goldthwaite, C. and I. Smith. 1972a. Genetic mapping of aminoglycoside and fusidic acid resistant mutations in *Bacillus subtilis*. *Mol. Gen. Genet.* **114**:181.

————. 1972b. Physiological characterization of antibiotic resistant mutants of *Bacillus subtilis*. *Mol. Gen. Genet.* **114**:190.

Goldthwaite, C., D. Dubnau and I. Smith. 1970. Genetic mapping of antibiotic resistance markers in *Bacillus subtilis*. *Proc. Nat. Acad. Sci.* **65**:96.

Gorelic, L. 1970. Chromosomal location of ribosomal RNA cistrons in *Escherichia coli*. *Mol. Gen. Genet.* **106**:323.

Gorini, L. 1971. Ribosomal discrimination of tRNAs. *Nature New Biol.* **234**:261.

Guthrie, C., H. Nashimoto and M. Nomura. 1969a. Structure and function of *E. coli* ribosomes. VIII. Cold-sensitive mutants defective in ribosome assembly. *Proc. Nat. Acad. Sci.* **63**:384.

————. 1969b. Studies on the assembly of ribosomes in vivo. *Cold Spring Harbor Symp. Quant. Biol.* **34**:69.

Haseltine, W. A., R. Block, W. Gilbert and K. Weber. 1972. MSI and MSII made on ribosome in idling step of protein synthesis. *Nature* **238**:381.

Hasenbank, R., C. Guthrie, G. Stöffler, H. G. Wittmann, L. Rosen and D. Apirion. 1973. Electrophoretic and immunological studies on ribosomal proteins of 100 *Escherichia coli* revertants from streptomycin dependence. *Mol. Gen. Genet.* **127**:1.

Helser, T. L., J. E. Davies and J. E. Dahlberg. 1971. Change in methylation of 16S ribosomal RNA associated with mutation to kasugamycin resistance in *Escherichia coli*. *Nature New Biol.* **233**:12.

————. 1972. Mechanism of kasugamycin resistance in *Escherichia coli*. *Nature New Biol.* **235**:6.

Higo, K., W. Held, L. Kahan and M. Nomura. 1973. Functional correspondence between 30S ribosomal proteins of *Escherichia coli* and *Bacillus stearothermophilus*. *Proc. Nat. Acad. Sci.* **70**:944.

Inselburg, J. W., T. Eremenko-Volpe, L. Greenwald, W. L. Meadow and J. Marmur. 1969. Physical and genetic mapping of the SPO2 prophage on the chromosome of *Bacillus subtilis* 168. *J. Virol.* **3**:627.

Jarry, B. and R. Rosset. 1971. Heterogeneity of 5S RNA in *Escherichia coli*. *Mol. Gen. Genet.* **113**:43.

————. 1973a. Localization of some 5S RNA cistrons on *Escherichia coli* chromosome. *Mol. Gen. Genet.* **121**:151.

————. 1973b. Further mapping of 5S RNA cistrons in *Escherichia coli. Mol. Gen. Genet.* **126**:29.

Kahan, L., J. Zengel, M. Nomura, A. Bollen and A. Herzog. 1973. The structural gene for the ribosomal protein S18 in *Escherichia coli*. II. Chemical studies on the protein S18 have an altered electrophoretic mobility. *J. Mol. Biol.* **76**:473.

Kang, S-S. 1970a. A mutant of *Escherichia coli* with temperature-sensitive streptomycin protein. *Proc. Nat. Acad. Sci.* **65**:544.

————. 1970b. Temperature-sensitive alteration of 30S subunits demonstrated by *in vitro* reassociation of functional ribosomes. *Nature* **225**:1132.

Kimura, A., K. Kobata, R. Takata and S. Osawa. 1973. Genetic and chemical studies of ribosomes from spectinomycin resistant mutants of *Bacillus subtilis. Mol. Gen. Genet.* **124**:107.

Kindler, P., T. U. Keil and P. H. Hofschneider. 1973. Isolation and characterization of a ribonuclease III deficient mutant of *Escherichia coli. Mol. Gen. Genet.* **126**:53.

Kossman, C. R., T. D. Stamato and D. E. Pettijohn. 1971. Tandem synthesis of the 16S and 23S ribosomal RNA sequences of *Escherichia coli. Nature New Biol.* **234**:102.

Kreider, G. and B. L. Brownstein. 1971. A mutation suppressing streptomycin dependence. II. An altered protein on the 30S ribosomal subunit. *J. Mol. Biol.* **61**:135.

————. 1972. Ribosomal proteins involved in the suppression of streptomycin dependence in *Escherichia coli. J. Bact.* **109**:780.

Kuwano, M., H. Endo and Y. Ohnishi. 1969. Mutations to spectinomycin resistance which alleviate the restriction of an amber suppressor by streptomycin resistance. *J. Bact.* **97**:940.

Kuwano, M., D. Schlessinger, G. Rinaldi, L. Felicetti and G. P. Tocchini-Valentini. 1971. G factor mutants of *Escherichia coli:* Map location and properties. *Biochem. Biophys. Res. Comm.* **42**:441.

Kuwano, M., M. Ono, M. Yamamoto, H. Endo, T. Kamiya and K. Hori. 1973. Elongation factor T altered in a temperature-sensitive *Escherichia coli* mutant. *Nature New Biol.* **244**:107.

Lavallé, R. 1965. Nouveaux mutants de regulation de la synthèse de l'ARN. *Bull. Soc. Chim. Biol.* **47**:1567.

Lazzarini, R. A. and M. Cashel. 1971. On the regulation of guanosine tetraphosphate levels in stringent and relaxed strains of *Escherichia coli. J. Biol. Chem.* **246**:4381.

Lewandowski, L. J. and B. L. Brownstein. 1966. An altered pattern of ribosome synthesis in a mutant of *E. coli. Biochem. Biophys. Res. Comm.* **25**:554.

————. 1969. Characterization of a 43S ribonucleoprotein component of a mutant of *Escherichia coli. J. Mol. Biol.* **41**:277.

Lowry, C. V. and J. E. Dahlberg. 1971. Structural differences between the 16S ribosomal RNA of *E. coli* and its precursor. *Nature New Biol.* **232**:52.

Maaløe, O. and N. O. Kjeldgaard. 1966. *Control of macromolecular synthesis.* W. A. Benjamin, New York.

Matsubara, M., R. Takata and S. Osawa. 1972. Chromosomal loci for 16S ribosomal RNA in *Escherichia coli. Mol. Gen. Genet.* **117**:311.

Mayuga, C., D. Meier and T. Wang. 1968. *Escherichia coli:* The K12 ribosomal protein and the streptomycin region of the chromosome. *Biochem. Biophys. Res. Comm.* **33**:203.

Miller, O. L., Jr. and B. A. Hamkalo. 1972. Visualization of RNA synthesis on chromosomes. *Int. Rev. Cytol.* **32**:1.

Molin, S., K. von Meyenburg, K. Gulløv and O. Maaløe. 1974. The size of transcriptional units for ribosomal proteins in *Escherichia coli. Mol. Gen. Genet.* **129**:11.

Momose, H. and L. Gorini. 1971. Genetic analysis of streptomycin dependence in *Escherichia coli. Genetics* **67**:19.

Morell, P., I. Smith, D. Dubnau and J. Marmur. 1967. Isolation and characterization of low molecular weight ribonucleic acid species from *Bacillus subtilis. Biochemistry* **6**:258.

Nashimoto, H. and M. Nomura. 1970. Structure and function of bacterial ribosomes. XI. Dependence of 50S ribosomal assembly on simultaneous assembly of 30S subunits. *Proc. Nat. Acad. Sci.* **67**:1440.

Nashimoto, H., W. Held, E. Kaltschmidt and M. Nomura. 1971. Structure and function of bacterial ribosomes. XII. Accumulation of 21S particles by some cold-sensitive mutants of *Escherichia coli. J. Mol. Biol.* **62**:121.

Nikolaev, N., L. Silengo and D. Schlessinger. 1973. Synthesis of a large precursor to ribosomal RNA in a mutant of *Escherichia coli. Proc. Nat. Acad. Sci.* **70**:3361.

Nomura, M. and F. Engbaek. 1972. Expression of ribosomal protein genes as analyzed by bacteriophage Mu-induced mutations. *Proc. Nat. Acad. Sci.* **69**:1526.

Oishi, M. and N. Sueoka. 1965. Location of genetic loci of ribosomal RNA on *Bacillus subtilis* chromosome. *Proc. Nat. Acad. Sci.* **54**:483.

Osawa, S., R. Takata, K. Tanaka and M. Tamaki. 1973. Chloramphenicol resistant mutants of *Bacillus subtilis. Mol. Gen. Genet.* **127**:163.

Osawa, S., E. Otaka, R. Takata, S. Dekio, K. Matsubara, T. Itoh, A. Muto, K. Tanaka, H. Teraoka and M. Tamaki. 1972. Ribosomal protein genes in bacteria. *FEBS Symp.* **23**:313.

Ozaki, M., S. Mizushima and M. Nomura. 1969. Identification and functional characterization of the protein controlled by the streptomycin-resistant locus in *E. coli. Nature* **222**:333.

Pace, B. and N. R. Pace. 1971. Gene dosage for 5S ribosomal ribonucleic acid in *Escherichia coli* and *Bacillus megaterium. J. Bact.* **105**:142.

Pato, M. L. and K. von Meyenberg. 1970. Residual RNA synthesis in *Escherichia coli* after inhibition of initiation of transcription by rifampicin. *Cold Spring Harbor Symp. Quant. Biol.* **35**:497.

Phillips, S. L., D. Schlessinger and D. Apirion. 1969. Mutants in *Escherichia coli* ribosomes: A new selection. *Proc. Nat. Acad. Sci.* **62**:772.

Rosenkranz, H. S. 1963. Unusual alkaline phosphatase levels in streptomycin-dependent strains of *E. coli. Biochemistry* **2**:122.

Rosenkranz, H. S., A. J. Bendich and H. S. Carr. 1964. The isolation of two streptomycin-resistant mutants of *Escherichia coli* ML35 differing in constitutive enzymes. *Biochim. Biophys. Acta* **82**:110.

Rosset, R. and L. Gorini. 1969. A ribosomal ambiguity mutation. *J. Mol. Biol.* **39**:95.

Rosset, R., C. Vola, J. Feunteun and R. Monier. 1971. A thermosensitive mutant defective in ribosomal 30S subunit assembly. *FEBS Letters* **18**:127.

Smith, I. and H. Smith. 1973. Location of the SPO2 attachment site and the bryamycin resistance marker on the *Bacillus subtilis* chromosome. *J. Bact.* **114**:1138.

Smith, I., C. Goldthwaite and D. Dubnau. 1969. The genetics of ribosomes in *Bacillus subtilis. Cold Spring Harbor Symp. Quant. Biol.* **34**:85.

Smith, I., D. Dubnau, P. Morell and J. Marmur. 1968. Chromosomal location of DNA base sequence complementary to transfer RNA and to 5S, 16S, and 23S ribosomal RNA in *Bacillus subtilis. J. Mol. Biol.* **33**:123.

Spadari, S. and F. Ritossa. 1970. Clustered genes for ribosomal ribonucleic acids in *Escherichia coli. J. Mol. Biol.* **53**:357.

Sparling, P. F. 1968. Kasugamycin resistance: A 30S ribosomal mutation with an unusual location on the *Escherichia coli* chromosome. *Science* **167**:56.

Sparling, P. F. and E. Blackman. 1973. Mutation to erythromycin dependence in *Escherichia coli* K-12. *J. Bact.* **116**:74.

Sparling, P. F., J. Modolell, Y. Takeda and B. D. Davis. 1968. Ribosomes from *Escherichia coli* merodiploids heterozygous for resistance to streptomycin and to spectinomycin. *J. Mol. Biol.* **37**:407.

Stent, G. S. and S. Brenner. 1961. A genetic locus for the regulation of ribonucleic acid synthesis. *Proc. Nat. Acad. Sci.* **47**:2005.

Strigini, P. and E. Brickman. 1973. Analysis of specific misreading in *Escherichia coli. J. Mol. Biol.* **75**:659.

Sypherd, P. S., D. M. O'Neill and M. M. Taylor. 1969. The chemical and genetic structure of bacterial ribosomes. *Cold Spring Harbor Symp. Quant. Biol.* **34**:77.

Tai, P., D. P. Kessler and J. Ingraham. 1969. Cold-sensitive mutations in *Salmonella typhimurium* which offset ribosome synthesis. *J. Bact.* **97**:1298.

Tanaka, N., G. Kawano and T. Kinoshita. 1971. Chromosomal location of a fusidic acid resistant marker in *Escherichia coli. Biochem. Biophys. Res. Comm.* **42**:664.

Tanaka, K., M. Tamaki, S. Osawa, A. Kimwa and R. Takata. 1973. Erythromycin resistant mutants of *Bacillus subtilis. Mol. Gen. Genet.* **127**:157.

Taylor, A. L. 1963. Bacteriophage-induced mutation in *Escherichia coli. Proc. Nat. Acad. Sci.* **50**:1043.

Tocchini-Valentini, G. P. and E. Mattoccia. 1968. A mutant of *E. coli* with an altered supernatant factor. *Proc. Nat. Acad. Sci.* **61**:146.

Traub, P. and M. Nomura. 1969. Structure and function of *E. coli* ribosomes. VI. Mechanism of assembly of 30S ribosomes studied *in vitro. J. Mol. Biol.* **40**:391.

Turnock, G. 1969. A genetic analysis of a mutant of *Escherichia coli* with a defect in the assembly of ribosomes. *Mol. Gen. Genet.* **104**:295.

Tyler, B. and J. L. Ingraham. 1973. Studies on ribosomal mutants of *Salmonella typhimurium* LT-2. *Mol. Gen. Genet.* **122**:197.

Unger, M., L. S. Birnbaum, S. Kaplan and A. Pfister. 1972. Location of the ribosomal RNA cistron of *Escherichia coli:* A second site. *Mol. Gen. Genet.* **119**:377.

van de Putte, P. and M. Gruijthuijsen. 1972. Chromosome mobilization and integration of F-factors in the chromosome of Rec A strains of *E. coli* under the influence of bacteriophage Mu-1. *Mol. Gen. Genet.* **118**:173.

Wittmann, H. G., G. Stöffler, D. Apirion, L. Rosen, K. Tanaka, M. Tamaki, R. Takata, S. Dekio, E. Otaka and S. Osawa. 1973. Biochemical and genetic studies on two different types of erythromycin resistant mutants of *Escherichia coli* with altered ribosomal proteins. *Mol. Gen. Genet.* **127**:175.

Yamada, T. and J. Davies. 1971. A genetic and biochemical study of streptomycin resistance in *Salmonella typhimurium. Mol. Gen. Genet.* **110**:197.

Yankofsky, S. A. and S. Spiegelman. 1962. The identification of the ribosomal RNA cistron by sequence complementarity. II. Saturation of and competitive interaction at the RNA cistron. *Proc. Nat. Acad. Sci.* **48**:1466.

Yu, M. T., C. W. Vermeulen and K. C. Atwood. 1970. Location of the genes ribosomal protein accompanying the *ram* mutation in *Escherichia coli. Proc. Nat. Acad. Sci.* **67**:26.

Zimmermann, R. A., R. T. Garvin and L. Gorini. 1971. Alteration of a 30S ribosomal protein accompanying the *ram* mutation in *Escherichia coli. Proc. Nat. Acad. Sci.* **68**:2263.

Zimmermann, R. A., Y. Ikeya and P. F. Sparling. 1973. Alteration of ribosomal protein S4 by mutation linked to kasugamycin resistance in *Escherichia coli. Proc. Nat. Acad. Sci.* **70**:71.

Regulation of Biosynthesis of Ribosomes

Niels Ole Kjeldgaard
Department of Molecular Biology
Aarhus University
Aarhus, Denmark

Kirsten Gausing
Institute of Microbiology
University of Copenhagen
Copenhagen, Denmark

INTRODUCTION

The ribosomes account for 40–50% of the cell mass of rapidly growing bacteria. It is therefore not surprising that the biosynthesis of ribosomes is a precisely regulated process requiring a strict coordination of the production of the 50–60 different molecules forming part of the ribosomal structures.

At our present stage of ignorance, we have limited information about the regulatory processes governing ribosome biosynthesis in prokaryotes:

Are all structural molecules of the ribosomes, RNA as well as proteins, regulated through one and the same basic mechanism?

Is the biosynthesis of rRNA regulated at the level of initiation of transcription, at the level of the substrates, or does a breakdown mechanism of the nascent rRNA play a role?

Is the biosynthesis of the ribosomal proteins regulated according to the Jacob-Monod model, or is there only an indirect control of the transcription frequency of the ribosomal protein operons through a common competition among all open promoters for the RNA polymerase molecules, as suggested by Maaløe (1969)?

Other questions might also be asked:

Does the regulatory mechanism primarily hit the ribosomal RNAs which in turn trigger the production of the ribosomal proteins? Inversely, are the ribosomal protein operons the prime target for the regulation, and do these proteins, or at least one of them, turn on the production of rRNAs?

Are the ribosomal proteins protecting the nascent rRNAs against degradation, leaving excess rRNA chains at the prey of nucleases?

The coordination between the rRNAs and the ribosomal proteins is a focal point in our understanding of the biosynthesis of ribosomes. How is this coordination assured with ribosomal protein messengers giving rise to multiple copies of the proteins coded by the messengers? What mechanism assures the exact stoichiometry between 30S and 50S ribosomal subunits?

A few of these questions, which of course are more or less overlapping, can be answered with confidence, but much, very much indeed, remains to be learned.

RIBOSOME FORMATION

During balanced growth of bacterial cultures, the number of ribosomes per cell is observed to vary in a regular fashion with the growth rate of the culture. This was originally observed with *Salmonella typhimurium* (Schaechter, Maaløe and Kjeldgaard 1958), but closely similar variations have been found for a number of other bacterial strains (Rosset, Monier and Julien 1964; Rosset, Julien and Monier 1966; Forchhammer and Lindahl 1971; Skjold, Juarez and Hedgecoth 1973).

Table 1 exemplifies these variations, giving the number of 70S ribosomes, tRNA molecules and amino acids in peptide linkage as calculated per genome equivalent of RNA for *E. coli* 15 TAU grown in different media.

Based on the early measurements, Maaløe and Kjeldgaard (1966) suggested that the number of ribosomes is controlled in such a way that the very same efficiency of the ribosomes in protein synthesis is maintained at all growth rates. An average efficiency of ribosomes in polypeptide formation at 37°C of about 17 amino acids per second was calculated under the assumption that all ribosomes were active. A constant polypeptide chain growth rate at all growth rates defines a relationship between number of ribosomes per genome (R) and the growth rate (μ) as $R = k\mu$ and, as emphasized by Koch (1970) and Maaløe (1969), a rate of ribosomal synthesis proportional to μ^2. There is no indication, as seen from the values in Table 1, that the number of ribosomes approaches 0 at $\mu \to 0$. Furthermore, it can be calculated that the average chain growth rate actually decreases with decreasing growth rate. This decrease in chain growth rate might be a true reflection of a more sluggish behavior of all ribosomes or

Table 1   Cell composition of *Escherichia coli*

Medium	μ	Genomes per cell	70S ribosomes	Molecules tRNA	Amino acid molecules in proteins	α_r	Q value
				Per genome			
Broth	2.7	5.0	32,800	190,000	63.3×10^7	0.26	0.43
Casamino acids	2.1	3.6	21,100	170,000	60.6×10^7	0.20	0.46
Glucose	1.5	2.5	14,100	130,000	60.7×10^7	0.15	0.55
Succinate	0.9	1.8	9,000	94,000	58.2×10^7	0.10	0.72
Acetate	0.5	1.4	7,100	82,000	59.5×10^7	0.07	0.86

The figures given are idealized values based on the measurements on *E. coli* 15 TAU by Skjold, Juarez and Hedgecoth (1973). The number of genomes per cell was calculated using a genome molecular weight of 3×10^9 daltons. The number of ribosomes and tRNA molecules was calculated from the content of rRNA and tRNA using a molecular weight of 1.5×10^6 for 16S + 23S rRNA and 25×10^3 for tRNA.

The number of amino acids in peptide linkage was calculated from the protein content using an average molecular weight for the amino acids in *E. coli* proteins of 110 daltons.

The α_r values are based on unpublished data by Dennis and Bremer, Gausing, and Bennett and Maaløe. The "Q values" are from Pato and von Meyenburg (1970), with a later re-evaluation of the data for casamino acid-grown cultures (von Meyenburg, unpublished data).

might be caused by a decrease in the fraction of active ribosomes, which by themselves maintain a high chain growth rate. A third possibility to account for the decrease in the average chain growth rate involves a combination of maintenance of the true chain growth rate and an increase in the turnover rate of cellular proteins at slow growth rates.

Direct measurements of the chain growth rate have been made in the case of β-galactosidase, giving values of 16–20 amino acids per second at 37°C over the range of growth rates between $\mu = 0.5$ and $\mu = 1.9$ (Lacroute and Stent 1968; Engbæk, Kjeldgaard and Maaløe 1973). Using SDS gel electrophoresis of total cell proteins from *E. coli* grown in minimal glucose medium, Gausing (1972) determined the chain growth rate for a number of molecular weight size classes. For all classes a value of 12 amino acids per second at 30°C was measured, corresponding to 17 amino acids per second at 37°C. By measurements of the time-delay from the induction of β-galactosidase until the first enzyme molecule was completed, Coffman, Norris and Koch (1971) found only small variations for chemostat culture with doubling times varying from 50 min to 24 hr. This indicates that neither the chain growth rates for translation nor transcription vary with the growth rate.

The number of active ribosomes has been measured by following the percentage of ribosomes in polysomes in cultures growing at different rates (Forchhammer and Lindahl 1971; Varrichio and Monier 1971). At all growth rates tested, close to 80% of the ribosomes were found in polysome structures. Recent experiments by Harvey (1973) show a decrease in the number of polysomal ribosomes with decreasing growth rate. Furthermore, measurements of the amount of nascent peptides which can be released from the ribosomes by puromycin treatment also indicate that the number of active ribosomes decreases with the growth rate (Harvey 1973).

A corresponding decrease in growth rate and the average chain growth rate, together with a nearly conserved actual chain growth rate, favors variations in the number of active ribosomes. However the problem is still open for further experimentation.

Measurements of the amount of total mRNA in the cells (Forchhammer and Kjeldgaard 1968; Nierlich 1972b) suggest that the amount of mRNA per ribosome is constant at growth rates above $\mu = 1.5$ and decreases at lower growth rates. This also supports the notion of a decrease in number of active ribosomes with the growth rate.

The turnover of cellular proteins has been measured by Nath and Koch (1970) under conditions where a rapid equilibrium between the internal amino acid pool and the growth medium was achieved. Over a range of growth rates they found a turnover of a few percent of the cellular proteins with a half-life at 37°C of about 1 hr. At very low growth rates the half-life decreased to about 30 min. It is therefore doubtful that turnover of finished protein chains can contribute to a decrease of the average polypeptide chain growth rate. It should be emphasized, however, that prematurely terminated polypeptide chains can be degraded very rapidly in the cells (Goldberg 1972).

In summary, it is most likely that the number of active ribosomes decreases with decreasing growth rate, which again indicates that there is no simple correlation between the control of ribosome synthesis and the function of the ribosomes in polypeptide formation. In fact, it may be argued that the low growth rates result in an overproduction of ribosomes (Koch 1971).

In a few cases an overproduction of ribosomes has been observed in the sense that cells grown at a certain growth rate contain more ribosomes per cell than a normal culture. MacDonald, Turnock and Forchhammer (1967) described the isolation of a mutant of *E. coli* 15 that was a slow grower and accumulated a precursor to the 50S ribosomal subunit. The mutation seems to be a defect in the ribosome assembly, mapping at about 73 min on the *E. coli* genome (Turnock 1969). The mutant contains 30S subunits in amounts corresponding to the sum of 50S and precursor particles (MacDonald, Turnock and Forchhammer 1967) and an amount of mRNA proportional to the total number of particles (Forchhammer and Kjeldgaard 1968).

Departing from an *E. coli* strain carrying a temperature-sensitive alanyl-tRNA ligase with a strongly increased K_m for tRNA, Buckel et al. (1972) isolated slow-growing revertants with a highly increased RNA content. A few of these revertants have been analyzed, and also in these cases an accumulation of immature 50S particles was found. The genetic defect is located in the *str* region and involves a mutation in the L26, S20 ribosomal protein (see Wittmann, this volume).

By addition of low concentrations of fusidic acid (0.25 μg/ml) to a growing culture of *E. coli,* a new steady state of growth is established with a lower growth rate (Bennett and Maaløe, unpublished data). The fusidic acid causes protein synthesis to occur with a lower average chain growth rate, which again is reflected in a lowered basal level of ppGpp and in an overproduction of ribosomes. In this case only normal ribosomal particles are found.

RIBOSOMAL RNA SYNTHESIS

Steady State Conditions

The nucleotide sequences of ribosomal RNAs are coded by six identical but widely scattered operons accounting for less than 1% of the genome. The labeling pattern of the rRNAs after rifampicin treatment of sensitive cultures shows that the rRNA operons are transcribed in the order 16S–23S rRNA. Recent experiments by Nikolaev, Silengo and Schlessinger (1973a, b) suggest that the active operons are transcribed as one large polynucleotide chain, which is subsequently processed by the RNase III and RNase II (Corte et al. 1971; Yuki 1971) to 16S and 23S rRNA (see Schlessinger, this volume).

Ratio of Stable to Labile RNA. In *E. coli* growing exponentially in glucose minimal medium, labeling kinetics show that the rate of stable RNA synthesis is about 50% of the total rate of RNA synthesis (Kennell 1968; Mueller and Bremer 1968; Winslow and Lazzarini 1969; Norris and Koch 1972; Nierlich 1972a,b).

A much less complicated method for relative measurements of the rate of synthesis of labile and stable RNA species involves the simultaneous addition of rifampicin and radioactive uridine to a sensitive culture of *E. coli* (Figure 1) (Pato and von Meyenburg 1970; Bremer, Berry and Dennis 1973). Measurements on cultures grown at different rates give a simple linear correlation between the growth rate and the fraction of total RNA synthesis taken up by the labile RNA species. These "Q values" range from 0.43 at $\mu = 2.5$ to 0.80 at $\mu = 0.5$, extrapolating towards 1 at $\mu \rightarrow 0$ (Table 1).

RNA Chain Growth Rate. This parameter has been measured in a number of cases to about 50 nucleotides per second at 37°C and is found to

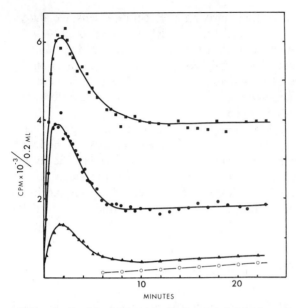

Figure 1   Residual RNA synthesis after inhibition of the initiation of transcription by rifampicin. *E. coli* B AS19 was grown in casamino acid medium (■), $\mu = 2.4$, glucose minimal medium (●), $\mu = 1.4$, and acetate minimal medium (▲), $\mu = 0.5$. Rifampicin (20 μg/ml) and [³H]uridine and nalidixic acid (10 μg/ml) were added to the culture at time 0. Samples of 0.2 ml were lysed with SDS at 95°C, precipitated with trichloroacetic acid, and the radioactivity measured. All values are normalized to a culture density of A_{450}mμ = 0.4. The "Q value" is determined as the ratio $a/a + b$, where $a + b$ is the maximum amount of radioactivity incorporated and b is the amount of radioactivity at the plateau. To correct for residual incorporation, [³H]uridine was added to a parallel culture 5 min after rifampicin addition (○). Later observations (von Meyenburg, unpublished) have shown that cells grown in casamino acid medium require a higher concentration of rifampicin (100 μg/ml) to completely stop the initiation of RNA synthesis (Q = 0.46 at $\mu = 2.1$). Figure redrawn after Pato and von Meyenburg (1970).

show only slight variations with the growth rate of the cultures (Bremer and Yuan 1968; Manor, Goodman and Stent 1969; Bremer and Berry 1971; Rose, Mosteller and Yanofsky 1970; Jacquet and Kepes 1971). Because of the techniques used, these determinations are mostly a measure of the value of mRNA chain growth rate. The chain growth rate for rRNA was measured directly by following the decrease in the amount of radioactivity incorporated into rRNA when the label was added to cultures of *E. coli* B/r at different times after rifampicin inhibition of the initiation of new RNA chains (Dennis and Bremer 1973). A chain growth rate almost twice that of mRNA was found, varying with the growth rate from about

70 nucleotides per second at $\mu = 0.7$ to about 100 nucleotides per second at $\mu = 2.1$.

Confirmation of these findings is of great importance for the evaluation of many experimental results, but this must await further independent experimental approaches.

RNA Polymerase in Transcription.   The translation of the "Q values" into number of RNA polymerase molecules involved in the synthesis of the various RNA species requires knowledge of both the size of the molecule transcribed and of the chain growth rate of polynucleotide synthesis.

Using the high chain growth rate to estimate the number of polymerase molecules per ribosomal RNA operon, Dennis and Bremer (1973) found values of 4 at $\mu = 0.7$, 14 at $\mu = 1.4$, and 24 at $\mu = 2.2$. The latter value corresponds to about 150 RNA polymerase molecules per genome engaged in rRNA transcription. Other measurements give higher numbers of RNA polymerase molecules involved in rRNA synthesis. During in vitro synthesis of rRNA by a crude DNA–protein complex from *E. coli* grown in casamino acid medium, Pettijohn et al. (1970) estimated the number of growing rRNA chains to about 480 chains per genome. Due to the handling of the system, which appears only to allow for the termination of RNA chains initiated before harvesting and lysing the cells, this estimate must be a minimum value. In the electron micrographs of Miller, Hamkalo and Thomas (1970) and Hamkalo and Miller (1973), 60–70 RNA fibers are seen attached to regions that presumably are identical to the rRNA cistrons.

The total number of RNA polymerase molecules involved in RNA transcription was estimated from the "Q curves" (Figure 1) to about 2300 for *E. coli* B cells grown in glucose medium (Pato and von Meyenburg 1970). Measurements of the ratio of 3' UMP to internal uridine in RNA from uridine pulse-labeled cells have been used to evaluate the number of nascent RNA chains (Winslow and Lazzarini 1969). For cells grown in glucose minimal medium at 27°C, this method gave a number of 5100 for *E. coli* 15 TAU and 2500 for *E. coli* K12.

The total number of RNA polymerase molecules in *E. coli* B and K12 has been estimated by measurements of the β and β' subunits to 7000 per cell grown in glucose medium, corresponding to 4500 molecules per genome (Matzura, Hansen and Zeuthen 1973). The ratio of β, β' subunits to total protein (α_p) shows a slight decrease with decreasing growth rate from 0.015 at $\mu = 2.1$ to 0.010 at $\mu = 0.5$ (Matzura, Hansen and Zeuthen 1973; Dalbow 1973). It thus seems that only a minor fraction of the RNA polymerase molecules present in the cells is actually participating in transcription. As the relative amounts of the α and σ subunits seem to follow the amount of β, β' (F. Engbæk, unpublished data), other factors might be important for the in vivo function of the RNA polymerase.

Nucleotide Pool.   Measurements of the intracellular nucleotide pool in *E. coli* K12 show that all four ribonucleoside triphosphates decrease strongly in concentration with decreasing growth rate, extrapolating to zero at $\mu \to 0$ (Bagnara and Finch 1973). Other measurements of nucleotide pools confirm these results for *E. coli* K12 but indicate that not all bacterial strains show the same strong growth rate relationship (J. Neuhard, unpublished data). The concentrations of ppGpp and pppGpp have been found to increase considerably with decreasing growth rate (Cashel and Gallant, this volume).

Transition States

Shift-up. In many bacterial species, the transition from balanced growth at a low growth rate to fast growth conditions is characterized by a rapid increase in the overall rate of RNA synthesis, occurring within a few seconds after the shift (Maaløe and Kjeldgaard 1966; Koch 1965; Nierlich 1972a). In some strains the shift-up to a rich medium results in an initial exponential rate of RNA synthesis higher than the eventual rate in the new medium, and in others the new steady state rate is immediately established; in some strains the transition occurs rather slowly, and in a few cases only after a prolonged delay. These properties seem to be governed by at least two different genetic loci (R. Lavallé, unpublished data).

In the fast responders the fraction of RNA polymerase molecules involved in stable RNA formation, as measured by the "Q values," increases rapidly after the upshift and reaches or even overshoots the steady state value for the new growth medium within 4–5 min after the upshift (Pato and von Meyenburg 1970; Dennis and Bremer, unpublished data). Similar results were obtained by Nierlich (1972b) using hybridization techniques and base ratio measurements of pulse-labeled RNA for the characterization of mRNA and rRNA.

The transition therefore involves a repartitioning of the RNA molecules over the messenger RNA genes and the rRNA operons. If the chain growth rate for rRNA is independent of the growth rate, the rapid increase in net RNA synthesis requires the mobilization of new RNA polymerase molecules. The variations in rRNA chain growth rate observed by Dennis and Bremer (1973) overcome the need for such a mobilization (Bremer, Berry and Dennis 1973).

During an upshift from minimal medium to rich medium, the nucleotide pool shows a remarkable and rapid drop in the concentration of the four ribonucleoside triphosphates, which is only reconstituted to the steady state value after about 60 min (Beck et al. 1973).

As the upshift is accompanied by a rapid synthesis of RNA, this finding seems to exclude that the actual concentration of the triphosphates has a controlling function in RNA synthesis.

Amino Acid Starvation. Treatments, such as amino acid starvation of an auxotrophic strain, leading to an increased concentration of one or several species of uncharged tRNA, result in *rel*$^+$ strains in a strong reduction of the net rate of RNA synthesis to about 5% of the unstarved rate. This behavior has always been assumed to represent an extreme reflection of the regulatory functions working under balanced growth conditions.

Due to pool problems during labeling experiments, there existed a long-standing argument as to whether the reduction in net RNA synthesis did reflect a coordinate decrease in the rate of synthesis of stable as well as labile RNA species. It has now been firmly established by a number of different techniques that amino acid starvation does not dramatically affect the rate of mRNA synthesis (Forchhammer and Kjeldgaard 1968; Morris and Kjeldgaard 1968; Lavellé and DeHauwer 1968; Edlin et al. 1968; Lazzarini and Dahlberg 1971; Nierlich 1972b).

On the other hand, amino acid starvation results in a coordinate reduction in the rate of synthesis of all stable species of RNA (Raué and Gruber 1971). In *rel*$^-$ strains accumulation of uncharged tRNA within the cells has little effect on RNA synthesis, and rRNA accumulates in the cells for about 60 min at 37°C in the form of aberrant particles (relaxed particles). If strong antibiotic inhibitors of protein synthesis, such as chloramphenicol, tetracycline, fusidic acid, erythromycin or puromycin, are added to amino acid starved, sensitive *rel*$^+$ *E. coli* strains, stable RNA synthesis resumes and aberrant particles resembling and most likely identical to the relaxed particles appear (see Schlessinger, this volume). It thus seems that the accumulation of rRNA is not dependent on the simultaneous formation of ribosomal proteins.

Under all conditions so far tested, the block of stable RNA synthesis in *rel*$^+$ strains is accompanied by a remarkable accumulation of guanosine polyphosphates (Cashel and Gallant, this volume). In *rel*$^-$ strains no accumulation is observed under the same circumstances, and during antibiotic treatment of the *rel*$^+$ strains, guanosine polyphosphates disappear. The accumulation of the guanosine polyphosphates is always accompanied by a strong decrease in the GTP levels.

Regulatory Mechanisms

Excluding a regulatory function of nucleoside triphosphate substrates on RNA synthesis, two opposite mechanisms of control can be visualized to explain the variation in the rate of rRNA synthesis described above; regulation at the level of initiation of new rRNA chains, or regulation at the level of breakdown of nascent rRNA.

Ribosomal RNA Breakdown. The rRNA found in ribosomes is under normal conditions metabolically stable. It is known, however, that under certain conditions of stress, such as Mg^{++} starvation, even this rRNA is broken down, although at a rather slow rate (Kennell and Magasanik 1962;

McCarthy 1962). That the rRNA molecules, like the messenger RNA, might be labile before the rRNA chains are protected by the ribosomal proteins has been suggested (Rosset, Julien and Monier 1966).

Several attempts to observe a breakdown have been undertaken using hybridization techniques to characterize the nascent rRNA. Norris and Koch (1972) measured the amounts of and the rate of synthesis of rRNA in glucose-limited chemostat cultures of *E. coli* B. The labeled RNA was hybridized to *E. coli* DNA at different RNA/DNA ratios (Kennell 1968) and in the presence of competing nonradioactive rRNA. At the very low growth rates ($\mu \simeq 0.1$), their results suggest that the rRNA is synthesized 5- to 6-fold faster than it is accumulating.

During amino acid starvation of a rel^+ strain, using hybridization to *E. coli* DNA at different RNA/DNA ratios to characterize the rRNA, Donini (1972) found that the rate of rRNA synthesis was only slightly affected by the starvation. Therefore he concluded that the restricted accumulation of rRNA during amino acid starvation was caused by a breakdown of the major part of the nascent rRNA.

Determination of rRNA by hybridization to *E. coli* DNA is always hampered by the presence of mRNA, which in pulse-labeled RNA has a specific radioactivity about 50-fold higher than that of rRNA. To avoid this complication, a specific method for rRNA hybridization using DNA from *Vibrio metschnikovii* was developed (Pedersen and Kjeldgaard 1972). In this assay system *E. coli* mRNA shows only weak cross-hybridization, whereas there exists a large degree of homology between Vibrio and *E. coli* rRNA permitting an efficient hybridization of the latter to the DNA.

To test the metabolic stability of nascent rRNA during amino acid starvation, rifampicin and radioactive uridine were added simultaneously to a starved culture of *E. coli* B AS19, and RNA was isolated at intervals and hybridized to Vibrio DNA. If nascent rRNA is labile, the hybridization values would be expected to follow the shape of a normal "Q curve" (Figure 1). In this case the initial slope corresponds to the total rate of rRNA synthesis, and an elevated maximum level of radioactivity would be expected, followed by a decay to the plateau value of the stable rRNA molecules. This course of events is barely seen (Figure 2). The initial slope of the curve is 8% of the total RNA synthesis, and at most 20% of the nascent rRNA is broken down (Pedersen and Kjeldgaard, unpublished data). Similar results were obtained by Stamato and Pettijohn (1971) in experiments involving hybridization of rRNA pulse labeled during amino acid starvation to *E. coli* DNA enriched for the ribosomal genes.

Experiments performed on rifampicin-treated cultures grown at different growth rates show that almost all nascent rRNA molecules are stable in rapidly growing cells, whereas a significant fraction (up to 40%) is undergoing a decay with a half-life like that of mRNA in cells grown in minimal media (μ from 0.5 to 1.5) (Figure 3) (Pedersen, unpublished data). It is possible that the decay might function as a controlling element in rRNA

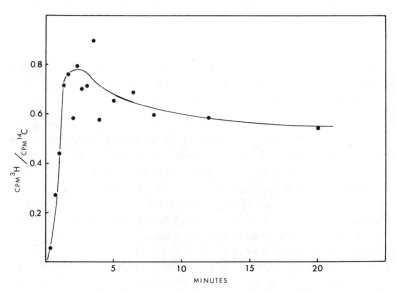

Figure 2   Stability of nascent rRNA during amino acid starvation. A culture of *E. coli* B AS19 *leu⁻* was starved of the required amino acid. After 5 min starvation, rifampicin (40 μg/ml), [³H]uridine and nalidixic acid (10 μg/ml) were added (time 0). At intervals samples were lysed in SDS at 95°C and mixed with a given amount of cells labeled with [¹⁴C]uracil in rRNA; RNA was prepared and hybridized to Vibrio DNA-containing filters. The figure shows the ³H/¹⁴C ratio of the RNA hybridized and competed by cold rRNA (Pedersen and Kjeldgaard, unpublished data).

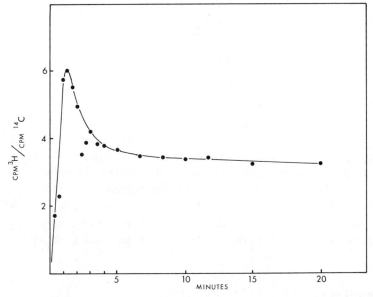

Figure 3   Decay of nascent rRNA during growth in glucose minimal medium. A culture of *E. coli* B AS19 grown exponentially in glucose minimal medium was treated with rifampicin and labeled as described in Figure 2 (Pedersen, unpublished data).

synthesis during slow exponential growth. However, it cannot be excluded that the decay observed is induced by the presence of rifampicin, e.g., through its inhibition of ribosomal protein synthesis.

Transcriptional Control.   A decay of nascent rRNA to the extent mentioned above is not sufficient to account for the observed variations of the rate of net rRNA synthesis. These variations, which are clearly expressed in the "Q values," therefore have to reflect true changes in the frequency of transcription of the rRNA operons. Similarly, the restart of rRNA synthesis after, e.g., chloramphenicol addition to amino acid starved *rel*[+] cells must be a reflection of a strong increase in frequency of rRNA transcription. This in fact can be directly observed on electron micrographs of the rRNA operons (Hamkalo and Miller 1973).

As emphasized above, there is a suggestive correlation between synthesis/nonsynthesis of rRNA and absence/presence of the guanosine polyphosphates (see Cashel and Gallant, this volume). It was therefore suggested by Cashel and Gallant (1968) that ppGpp might function as the direct effector of the control of rRNA transcription. However, no definite evidence in support of this model has as yet been obtained. In cells made permeable to nucleotides by a cold shock, addition of ppGpp resulted in a decrease in the rate of RNA synthesis, but no preferential inhibition of rRNA synthesis was observed (Lazzarini and Johnson 1973). Suggestion of a specific involvement of the elongation factor Tu and the Tu-ppGpp complex in the initiation of rRNA transcription is not yet firmly established (see Travers, this volume).

SYNTHESIS OF RIBOSOMAL PROTEINS

Studies of biosynthesis of ribosomal protein and ribosomal RNA have progressed along different lines simply because the difficulties encountered are of a different nature for the two kinds of components. In general, much less is known about the protein part. The protein content of crude ribosomes is $37 \pm 3\%$ (Tissières et al. 1959), and more extensively purified particles contain 30–33% protein (Hardy et al. 1969; Mora et al. 1971). Calculations based on the cell content of rRNA (Forchhammer and Lindahl 1971) give estimates for the ribosomal protein as a fraction of total protein of about 0.09 at $\mu = 0.6$, 0.14 at $\mu = 1.5$, and 0.22 at $\mu = 2.5$. Thus the fraction of total protein which is ribosomal is expected to increase with the growth rate, and this observation is borne out by direct measurements.

Measurements of α_r

Whereas determination of the rate of synthesis of rRNA has been complicated by the turnover of mRNA, the ribosomal proteins (r proteins) belong to a class of comparatively stable molecules, making it possible to measure rates of synthesis with relative ease. Thus the first method developed for determinations of ribosomal proteins as a fraction of total protein was designed to measure the parameter α_r = rate of synthesis of ribosomal protein/rate of synthesis of total protein (Schleif 1967). Only in steady states of growth, where all cellular components increase proportionally, is α_r equal to the ratio of amount of ribosomal protein to amount of total protein.

The protocol for α_r determination involves incorporation of radioactive amino acid followed by a chase with excess cold amino acid for a period sufficient to incorporate all radioactive ribosomal proteins into mature ribosomes, separation of the ribosomes from nonribosomal protein, and calculation of the fraction of radioactivity incorporated into ribosomes. Schleif (1967) used centrifugation of complete cell lysates through low Mg^{++}, D_2O-sucrose gradients, where the separation of ribosomal particles from nonribosomal protein is much better than in H_2O gradients. Further improvement of this separation is obtained when the D_2O gradients are formed at high Mg^{++} concentration since the 70S region is practically free of nonribosomal protein; the protein to RNA ratio in the 70S region is then used to correct for the contamination of nonribosomal protein in the subunit region (Bennett and Maaløe, unpublished data).

Other gradient techniques have been developed recently. In one, 70S particles are partially purified on H_2O-sucrose gradient and α_r is calculated from the ratio of protein to RNA in these particles and the ratio of protein to rRNA in whole cells; the contamination of the 70S particles with soluble protein was determined by centrifugation after dialysis against low Mg^{++} (Dennis and Bremer, unpublished data). The other technique consists of centrifugation of complete lysates on preformed CsCl gradients (ρ = 1.37–1.71) in 40 mM Mg^{++} run for a relatively short time; the ribosomal particles are well separated from nonribosomal protein without breakdown of the particles (S. Reeh, unpublished data).

Electrophoresis of complete cell lysates in pure agarose gels at pH 6.5 has been found very convenient for separation of ribosomal particles and nonribosomal protein (Figure 4). Because of their high content of negative phosphate groups, the ribosomes electrophorese well ahead of all nonribosomal protein, and since most nonribosomal proteins from *E. coli* are acidic (Waller 1963), very little protein escapes the gel during electrophoresis. The recovery of RNA and protein is virtually 100%, and α_r is calculated as the fraction of total protein in the gel found in the ribosomal bands (Gausing 1974).

The renewed interest in α_r determinations warrants a few remarks about the definition of ribosomal protein and assessment of absolute values of α_r. At the present stage of ignorance of the precise composition of ribosomes

Figure 4 Agarose gel electrophoresis of a complete cell lysate. Autoradiogram of a dried gel showing the separation of ribosomes from the bulk of cellular proteins. Cells grown in glucose minimal medium were pulse-chase labeled with (*a*) [^{14}C]uracil and (*b*) [^{14}C]leucine. The cells were sonicated and the lysate was mixed with warm (40°C) agarose and applied to an 0.85% agarose slab gel. Electrophoresis was for 1 hr at 9 V/cm in a Tris succinate buffer pH 6.5. The gel was dried and an autoradiogram prepared. *S* marks the slot and *B* the position of the tracking dye (Gausing, 1974).

inside the cells, the absolute values of α_r obtained with different methods can only reflect the composition of particles isolated in each case.

The parameter α_r is useful in physiological studies in attempts either to evaluate the fraction of total protein production that is reserved for the translation machinery and/or to study the regulation of a class of proteins that are synthesized coordinately. In particular, translation factors EF-T and EF-G could be included, as they are found in close to one copy per ribosome over a range of different growth rates (Gordon 1970). The gene coding for EF-G is closely linked to *str*A (Kuwano et al. 1971; Tanaka, Kuwano and Kinoshita 1971) and possibly regulated by the same mechanism as ribosomal protein (Nomura and Engbæk 1972). Aminoacyl-tRNA synthetases, on the other hand, should probably be treated as a separate group of proteins; although their level increases with the growth rate, they are not found in a constant amount per ribosome at different growth rates (Parker and Neidhardt 1972).

α_r has been determined in different strains of *E. coli* B and in *S. typhimurium* LT-2 at growth rates between $\mu = 0.4$ and 2.4 (Bennett and Maaløe; Dennis and Bremer; Reeh: all unpublished data; Gausing 1974). All these determinations are in fairly good agreement, and the values of α_r are 0.08–0.09 at $\mu = 0.6$, 0.14–0.16 at $\mu = 1.5$, and 0.20–0.23 at $\mu = 2.2$ (Table 1). The values obtained by Schleif (1967, 1968) are about 25% higher at medium and fast growth rates and identical at $\mu = 0.6$. Comparison of α_r and cell content of rRNA shows that the two kinds of ribosome components are indeed maintained in well-balanced amounts at different growth rates. This balance is preserved during a shift of the cells from slow to fast growth. When the accumulation of ribosomes adjusts to a higher rate, both components are produced at an increased rate very rapidly after the upshift (Schleif 1967; Gausing, unpublished data).

Pool Size of Ribosomal Protein

Estimates of the size of the pool of free ribosomal protein have been obtained from two types of experiments: direct measurement of particle-free supernatant fractions and kinetics of appearance of radioactive amino acid in ribosomal particles (including precursor particles) in pulse-chase experiments. Voynow and Kurland (1971) screened the nonribosomal cell fraction for ribosomal protein; they concluded that less than 10% of the cell content of ribosomal protein is present as free protein and cited an immunochemical test by Stöffler indicating that approximately 1% of ribosomal protein is free in the cell. With antisera against split and core fractions of the 30S and 50S protein, Gupta and Singh (1972) detected 7–9% of the ribosomal protein in the soluble protein fraction, whereas their results from kinetic studies of disappearance of ribosomal protein from the soluble proteins and appearance in the ribosomal fraction gave estimates of 1–2%. Kinetic experiments where the appearance of radioactive amino acid, incorporated during a short pulse into the ribosomal particles, is followed during the chase period consistently yield evidence of a very small pool. The rapid entry of ribosomal proteins into ribosomal particles results in the incorporation of a relatively large fraction of radioactive ribosomal protein into particles during pulses of 0.5–1 min. When growing peptide chains are not removed from the ribosomes, the kinetics early after the chase are inaccessible to analysis. Thus Schleif (1967, 1968) and Gierer and Gierer (1968) could not assign minimum values for the pool of free ribosomal protein and arrived at maximum estimates of 4% of total ribosomal protein. With very short pulse length (10 sec at 30°C), the kinetics have been partly resolved (Gausing 1974): a minimum of two subfractions of the free ribosomal protein was detected, 60% of the ribosomal protein appeared in ribosomal particles with a half-life of less than 30 sec, and 40% of the ribosomal protein appeared with a half-life of 1.3–2.4 min (at 30°C). This corresponds to a maximum pool size of 2.1% for cells growing in glucose minimal medium. The observation that the size of the free pool is not the same for all ribosomal proteins could explain why different ribosome maturation pathways are deduced from analysis of protein composition of isolated precursor particles (Nierhaus, Bordasch and Homann 1973) and the single time point analysis of the labeling patterns of proteins on mature particles (Marvaldi, Pichon and Marchis-Mouren 1972; Pichon, Marvaldi and Marchis-Mouren 1972).

In the experiments of Gausing, the time course of appearance of ribosomal protein in ribosomal particles was measured at three different growth rates (between $\mu = 0.45$ and 1.5 at 30°C). The results indicate that the fraction of ribosomal protein in the free pool increases linearly with the growth rate, and since α_r also increases linearly with growth rate, the absolute concentration of free ribosomal protein is roughly proportional to the square of the growth rate.

Aside from the free pool and mature particles, ribosomal proteins exist on polysomes during their synthesis and in different ribosomal precursors. At a growth rate of $\mu = 1.5$, the rate of ribosomal protein synthesis is equivalent to about 240 ribosomes/min and since an average ribosomal protein is made in 10 sec (at 37°C), it follows that ribosomal protein corresponding to 0.15% of the total ribosomal protein is being synthesized at any one time. From the kinetics of labeling of mature ribosomal particles (70S), Gierer and Gierer (1968) found about 5% of the ribosomal protein in the free pool + precursor particles. The 2% in the free pool leaves 3% for the ribosome precursor particles, a reasonable figure compared to other estimates of the size of the pool of precursor particles (Lindahl, unpublished data; Schlessinger, this volume).

Synthesis of Individual Ribosomal Proteins

Approximately half of the 30S proteins and one-third of the 50S proteins are found in significantly less than one copy per isolated particle (fractional proteins), even when the particles are isolated by relatively mild procedures (Kurland et al. 1969; Traut et al. 1969; Voynow and Kurland 1971; Weber 1972). Since the size of the free pools is of the order of 1–3%, classification of fractional proteins as unit proteins presupposes that 15–20% of these ribosomal proteins is lost from the ribosomes when cells are opened. A partial loss cannot be excluded, but Voynow and Kurland (1971) did not find sufficient 30S protein in the nonribosomal cell fractions to account for a massive loss of these proteins. The observation that some 50S proteins are found in more than one copy per ribosome shows that the cells contain different numbers of molecules of the various ribosomal proteins.

Thus it appears that mechanisms exist for maintaining coordinate but unequal numbers of the ribosomal proteins in vivo. With respect to the synthesis of individual ribosomal proteins and the composition of the ribosomes as a function of the growth rate, Deusser (1972) detected differences in the degree of heterogeneity for some proteins in isolated particles, whereas Dennis (unpublished results) finds that the ribosomal proteins are produced and incorporated into ribosomes in the same proportions at all growth rates.

Differences in the stoichiometry between ribosomal proteins could be created either on the level of synthesis or by breakdown. With the demonstration of repeat proteins (Weber 1972), the heterogeneity of ribosomal proteins now ranges from 0.2–3 molecules per ribosome. This heterogeneity cannot be explained by a breakdown mechanism alone, as it requires a protein breakdown much higher than the limited turnover observed in growing cells (Nath and Koch 1970).

With the scant knowledge of any of the facets of the regulatory mechanisms governing ribosomal protein synthesis, a great number of models can be proposed for the subproblem of unequal synthesis of different ribosomal proteins. Information about the arrangement of ribosomal protein genes in transcriptional units is necessary for our understanding of these processes. Attempts to answer this question have started along two different lines, genetic and physiological, with apparently conflicting results.

At least 20, and probably more, ribosomal protein genes are located in a region near 64 minutes on the *E. coli* genetic map (see Jaskunas, Nomura and Davies, this volume), but at least one protein (S18) maps outside this region (Bollen et al. 1973). The clustering of the ribosomal protein genes has been tied in with the coordinate expression of these genes. However, analysis of other systems has shown that a close genetic linkage is not a prerequisite of a coordinate expression. The close linkage of the ribosomal protein genes might be required for other reasons, e.g., to maintain nearly the same gene dosage during the cell cycle.

By analysis of mutants with the bacteriophage Mu-1 inserted in the *ery, spc, str, fus* gene cluster, evidence was obtained that these four markers, and probably several ribosomal protein genes interspaced between them, are located on one transcriptional unit with an anticlockwise direction of transcription (Nomura and Engbæk 1972). Initial studies with amber mutants isolated in a strain diploid for the *str* region also support the notion of one transcriptional unit covering the four antibiotics markers (Jaskunas, Engbæk and Nomura, unpublished data). Starting with the antibiotic-sensitive markers on the episome and the corresponding resistant markers on the chromosome, analysis of five markers in an Mu-1-induced resistant mutant again places the markers on one transcriptional unit (P. Jørgensen, unpublished data).

The kinetics of appearance of individual ribosomal proteins has also been used to study the organization of ribosomal protein genes in transcriptional units. Molin et al. (1974) measured the appearance of individual ribosomal proteins during relief from rifampicin treatment which had depleted the cells of mRNA before the experiments. The time difference in appearance between the earliest and latest ribosomal protein was 2–2.5 min (at 37°C), and the overall pattern places the majority of the ribosomal protein in operons containing 5–10 ribosomal cistrons. If all ribosomal proteins were organized in one single operon, this region would contain 22,000 base pairs and the messenger transcribed from the operon would account for about 15% of all mRNA. After the simultaneous addition of rifampicin and radioactive uridine to an *E. coli* culture, an mRNA of this size should become clearly apparent, as label is expected to be incorporated into this size class of labile RNA for up to 8 min. This is not observed; less than 1% of the mRNA is transcribed from operons longer than 12,000 base pairs (Molin et al. 1974).

The number of cistrons between *ery* and *fus* is not known, but it is undoubtedly larger than the numbers compatible with the results of Molin et al. (1974). Additional experiments, both genetic and physiological, are needed to solve the apparent paradox of long versus short transcriptional units for the ribosomal protein genes and may also provide information on how the inequality of ribosomal protein arises.

It is well known that functional ribosomes can be assembled in vitro from RNA and protein isolated from mature particles (Nomura and Held, this volume). But just as rRNA is processed during ribosome maturation in vivo, it is possible that ribosomal proteins undergo post-translational modifications. One case of modification has been described, namely, the acetylation of the N-terminal serine of L12 (Brot and Weissbach 1972). Other modifications may be detected in the DNA-directed cell-free system which can produce ribosomal proteins (Kaltschmidt, Kahan and Nomura 1974).

Regulation of Ribosomal Protein Synthesis

It is evident from the results discussed above that the plurality of the ribosomal proteins and the complexity of their genetic organization have created difficulties in obtaining relevant data and in formulating directly testable models for the control of ribosomal protein synthesis.

The rapid increase in both α_r and rRNA production after an upshift from poor to rich medium suggests that the control of ribosomal protein synthesis acts swiftly and in coordination with the control of rRNA synthesis. Conditions can be established where rRNA synthesis occurs in the absence of ribosomal protein production. The reverse case, where ribosomal proteins are produced in the absence of rRNA synthesis, has not yet been seen since most conditions blocking the RNA synthesis also result in a block of protein synthesis. There is, however, at least one example where this is not the case. Heating of a wild-type strain of *E. coli* K12 to 47–48°C results in a block of net RNA synthesis, whereas protein synthesis continues at a decreased but still appreciable rate (Lund and Kjeldgaard 1972). Measurements of α_r under these conditions might give additional information about the coordination problem. Similarly, measurements of the α_r values in strains diploid for the *str* region are still missing. Attempts have been made to measure α_r during the cell cycle in synchronized cultures of *E. coli* B/r. Schleif (1967) found no variations in α_r, whereas Zaritsky and von Meyenburg (1974) have seen a pattern of variation which agrees in time with the replication of the *str* region.

An assay of ribosomal protein mRNA has not yet been developed, and the transcription of the ribosomal protein genes has still not been approached experimentally. Thus it is not known whether ribosomal protein mRNA, like many other mRNA species, is formed during amino acid starvation or other conditions where protein synthesis is inhibited. Therefore we observe that rRNA can be produced in the absence of ribosomal protein production, but we cannot decide whether the synthesis of rRNA and ribosomal protein mRNA are coordinated.

Two general models exist for the control of ribosomal protein synthesis at the level of transcription: (1) a specific regulation of the Jacob-Monod type governing the initiation of the transcription of the ribosomal protein genes, and (2) a constitutive formation of ribosomal protein mRNA, where the frequency of transcription is determined by a general competition for the available RNA polymerase molecules among all open promoters (Maaløe 1969).

Much work needs to be done before we can hope to distinguish between these and possibly other models. With the development of a cell-free system for DNA-directed synthesis of ribosomal proteins (Kaltschmidt, Kahan and Nomura 1974) and with the renewed interest in α_r determinations, new aspects of the control of ribosomal protein synthesis certainly will emerge in the near future.

REDUNDANT REMARKS

In many cases we can predict how ribosome biosynthesis will respond to changes in the environmental conditions of the bacteria. Nevertheless, the studies of ribosome synthesis are at a point where more questions are raised than are answered by every new paper appearing in the literature.

Regulatory mutants have been the key to our understanding of the control of many metabolic circuits. In the present case, the isolation of such mutants is severely hampered or even impossible due to the central role of the ribosomes in cell metabolism. Furthermore, the isolation of promoter and operator type mutants must be particularly difficult since the rRNA genes are redundant and more than one operon exists for the ribosomal proteins.

Even with the odds against and in spite of the magnitude of the labor involved in the isolation and screening of regulatory mutants in ribosome biosynthesis, it is of prime importance to further development that an intensive search is conducted.

The coordination between the synthesis of all ribosomal components is another major problem open for new experimental approaches. The incorporation of episomes into minicells gives us a means to dissociate and separately test the function of different genomic regions. Thus Hori et al. (1974) have found a production of rRNA in minicells containing an episome carrying rRNA genes. This method might be generally useful.

The role of guanosine tetraphosphate in the control of rRNA synthesis is still an unsolved and intriguing problem. Circumstantial evidence is overwhelming; accumulation of MS (magic spots: see Cashel and Gallant, this volume) and arrest of rRNA synthesis always go together. Direct evidence, however, is equally meager, and intensive effort is needed to clarify this question both with crude and with purified systems for the in vitro synthesis of rRNA.

References

Bagnara, A. S. and L. R. Finch. 1973. Relationship between intracellular contents of nucleotides and 5-phosphoribosyl-1-pyrophosphate in *Escherichia coli. Eur. J. Biochem.* **36**:422.

Beck, C., J. Ingraham, O. Maaløe and J. Neuhard. 1973. Relationship between the concentration of nucleoside triphosphates and the rate of synthesis of RNA. *J. Mol. Biol.* **78**:117.

Bollen, A., M. Faelen, J. D. Lecocq, A. Herzog, J. Zengel, L. Kahan and M. Nomura. 1973. The structural gene for the ribosomal protein S18 in *Escherichia coli.* I. Genetic studies on a mutant having an alteration in the protein S18. *J. Mol. Biol.* **76**:463.

Bremer, H. and L. Berry. 1971. Co-transcription of 16S and 23S ribosomal RNA in *Escherichia coli. Nature New Biol.* **234**:81.

Bremer, H. and D. Yuan. 1968. RNA chain growth-rate in *Escherichia coli. J. Mol. Biol.* **38**:163.

Bremer, H., L. Berry and P. P. Dennis. 1973. Regulation of ribonucleic acid synthesis in *Escherichia coli* B/r: An analysis of a shift-up. II. Fraction of RNA polymerase engaged in the synthesis of stable RNA at different steady-state growth rates. *J. Mol. Biol.* **75**:161.

Brot, N. and H. Weissbach. 1972. The enzymatic acetylation of *E. coli* ribosomal protein L12. *Biochem. Biophys. Res. Comm.* **49**:673.

Buckel, P., D. Ruffler, W. Piepersberg and A. Böck. 1972. RNA overproducing revertants of an alanyl-tRNA synthetase mutant of *Escherichia coli. Mol. Gen. Genet.* **119**:323.

Cashel, M. and J. Gallant. 1968. Control of RNA synthesis in *Escherichia coli.* I. Amino acid dependence of the synthesis of the substrates of RNA polymerase. *J. Mol. Biol.* **34**:317.

Coffman, R. L., T. E. Norris and A. L. Koch. 1971. Chain elongation rate of messenger and polypeptides in slowly growing *Escherichia coli. J. Mol. Biol.* **60**:1.

Corte, G., D. Schlessinger, D. Longo and P. Venkov. 1971. Transformation of 17S to 16S ribosomal RNA using ribonuclease II of *Escherichia coli. J. Mol. Biol.* **60**:325.

Dalbow, D. G. 1973. Synthesis of RNA polymerase in *Escherichia coli* B/r growing at different rates. *J. Mol. Biol.* **75**:181.

Dennis, P. P. and H. Bremer. 1973. Regulation of ribonucleic acid synthesis in *Escherichia coli* B/r: An analysis of a shift-up. I. Ribosomal RNA chain growth rates. *J. Mol. Biol.* **75**:145.

Deusser, E. 1972. Heterogeneity of ribosomal populations in *Escherichia coli* cells grown in different media. *Mol. Gen. Genet.* **119**:249.

Donini, P. 1972. Turnover of ribosomal RNA during the stringent response in *Escherichia coli. J. Mol. Biol.* **72**:553.

Edlin, G., G. S. Stent, R. F. Baker and C. Yanofsky. 1968. Synthesis of a specific messenger RNA during amino acid starvation of *Escherichia coli*. *J. Mol. Biol.* **37**:257.

Engbæk, F., N. O. Kjeldgaard and O. Maaløe. 1973. Chain growth rate of β-galactosidase during exponential growth and amino acid starvation. *J. Mol. Biol.* **75**:109.

Forchhammer, J. and N. O. Kjeldgaard. 1968. Regulation of messenger RNA synthesis in *Escherichia coli. J. Mol. Biol.* **37**:245.

Forchhammer, J. and L. Lindahl. 1971. Growth rate of polypeptide chains as a function of the cell growth rate in a mutant of *Escherichia coli. J. Mol. Biol.* **55**:563.

Gausing, K. 1972. Efficiency of protein and messenger RNA synthesis in bacteriophage T4-infected cells of *Escherichia coli. J. Mol. Biol.* **71**:529.

————. 1974. Ribosomal protein synthesis and pool size at different growth rates. *Mol. Gen. Genet.* **129**:61.

Gierer, L. and A. Gierer. 1968. Synthesis of ribosomal proteins and formation of ribosomes in *Escherichia coli. J. Mol. Biol.* **34**:293.

Goldberg, A. L. 1972. Degradation of abnormal proteins in *Escherichia coli*. *Proc. Nat. Acad. Sci.* **69**:422.

Gordon, J. 1970. Regulation of the *in vivo* synthesis of the polypeptide chain elongation factors in *Escherichia coli. Biochemistry* **9**:912.

Gupta, R. S. and U. N. Singh. 1972. Biogenesis of ribosomes: Free ribosomal protein pools in *Escherichia coli. J. Mol. Biol.* **69**:279.

Hamkalo, B. A. and O. L. Miller. 1973. Visualization of genetic transcription. In *Gene expression and its regulation* (ed. F. T. Kenney et al.) pp. 63–74. Plenum Press, New York.

Hardy, S. J. S., C. G. Kurland, P. Voynow and G. Mora. 1969. The ribosomal proteins of *Escherichia coli*. I. Purification of the 30S ribosomal proteins. *Biochemistry* **8**:2897.

Harvey, R. J. 1973. Fraction of ribosomes synthesizing protein as a function of specific growth rate. *J. Bact.* **114**:287.

Hori, H., R. Takata, A. Muto and S. Osawa. 1974. Ribosomal RNA synthesis in the F'14 episome containing minicells of *Escherichia coli. Mol. Gen. Genet.* **128**:341.

Jacquet, M. and A. Kepes. 1971. Initiation, elongation and inactivation of *lac* messenger RNA in *Escherichia coli* studied by measurements of its β-galactosidase synthesizing capacity *in vivo. J. Mol. Biol.* **60**:453.

Kaltschmidt, E., L. Kahan and M. Nomura. 1974. *In vitro* synthesis of ribosomal proteins directed by *Escherichia coli* DNA. *Proc. Nat. Acad. Sci.* **71**:446.

Kennell, D. 1968. Titration of the gene sites on DNA by DNA–RNA hybridization. II. The *E. coli* chromosome. *J. Mol. Biol.* **34**:85.

Kennell, D. and B. Magasanik. 1962. The relation of ribosome content to the rate of enzyme synthesis in *Aerobacter aerogenes. Biochim. Biophys. Acta* **55**:139.

Koch, A. L. 1965. Kinetic evidence for a nucleic acid which regulates RNA biosynthesis. *Nature* **205**:800.

————. 1970. Overall controls on the biosynthesis of ribosomes in growing bacteria. *J. Theoret. Biol.* **28**:203.

————. 1971. The adaptive responses of *Escherichia coli* to a feast and famine existence. In *Advances in microbial physiology* (ed. A. H. Rose and J. F. Wilkinson) vol. 6, pp. 147–217. Academic Press, New York.

Kurland, C. G., P. Voynow, S. J. S. Hardy, L. Randall and L. Lutter. 1969. Physical and functional heterogeneity of *E. coli* ribosomes. *Cold Spring Harbor Symp. Quant. Biol.* **34**:17.

Kuwano, M., D. Schlessinger, G. Rinaldi, L. Felicetti and G. P. Tocchini-Valentini. 1971. G factor mutants of *Escherichia coli:* Map location and properties. *Biochem. Biophys. Res. Comm.* **42**:441.

Lacroute, F. and G. S. Stent. 1968. Peptide chain growth of β-galactosidase in *Escherichia coli. J. Mol. Biol.* **35**:165.

Lavallé, R. and G. DeHauwer. 1968. Messenger RNA synthesis during amino acid starvation in *Escherichia coli. J. Mol. Biol.* **37**:269.

Lazzarini, R. A. and A. E. Dahlberg. 1971. The control of ribonucleic acid synthesis during amino acid deprivation in *Escherichia coli. J. Biol. Chem.* **246**:420.

Lazzarini, R. A. and L. D. Johnson. 1973. Regulation of ribosomal RNA synthesis in cold-shocked *E. coli. Nature New Biol.* **243**:17.

Lund, E. and N. O. Kjeldgaard. 1972. Metabolism of guanosine tetraphosphate in *Escherichia coli. Eur. J. Biochem.* **28**:316.

Maaløe, O. 1969. An analysis of bacterial growth. *Develop. Biol.* (suppl.) **3**:33.

Maaløe, O. and N. O. Kjeldgaard. 1966. *Control of macromolecular synthesis.* W. A. Benjamin, New York.

MacDonald, R. E., G. Turnock and J. Forchhammer. 1967. The synthesis and function of ribosomes in a new mutant of *Escherichia coli. Proc. Nat. Acad. Sci.* **57**:141.

Manor, H., D. Goodman and G. S. Stent. 1969. RNA chain growth rates in *Escherichia coli. J. Mol. Biol.* **39**:1.

Marvaldi, J., J. Pichon and G. Marchis-Mouren. 1972. The *in vivo* order of addition of ribosomal proteins in the course of *Escherichia coli* 30S subunit biogenesis. *Biochim. Biophys. Acta* **269**:173.

Matzura, H., B. S. Hansen and J. Zeuthen. 1973. Biosynthesis of the β and β′ subunits of RNA polymerase in *Escherichia coli. J. Mol. Biol.* **74**:9.

McCarthy, B. J. 1962. The effect of magnesium starvation on the ribosome content of *Escherichia coli. Biochim. Biophys. Acta* **55**:880.

Miller, O. L., B. A. Hamkalo and C. A. Thomas. 1970. Visualization of bacterial genes in action. *Science* **169**:392.

Molin, S., K. von Meyenburg, K. Gulløv and O. Maaløe. 1974. The size of transcriptional units for ribosomal proteins in *Escherichia coli. Mol. Gen. Genet.* **129**:11.

Mora, G., D. Donner, P. Thammana, L. Lutter and C. G. Kurland. 1971. Purification and characterization of 50S ribosomal proteins of *Escherichia coli. Mol. Gen. Genet.* **112**:229.

Morris, D. W. and N. O. Kjeldgaard. 1968. Evidence for the non-coordinate regulation of ribonucleic acid synthesis in stringent strains of *Escherichia coli. J. Mol. Biol.* **31**:145.

Mueller, K. and H. Bremer. 1968. Rate of synthesis of messenger ribonucleic acid in *Escherichia coli*. *J. Mol. Biol.* **35**:329.

Nath, K. and A. L. Koch. 1970. Protein degradation in *Escherichia coli*. I. Measurement of rapidly and slowly decaying components. *J. Biol. Chem.* **245**:2889.

Nierhaus, K. H., K. Bordasch and H. E. Homann. 1973. Ribosomal proteins. XLIII. *In vivo* assembly of *Escherichia coli* ribosomal proteins. *J. Mol. Biol.* **74**:587.

Nierlich, D. P. 1972a. Regulation of ribonucleic acid synthesis in growing bacterial cells. I. Control over the total rate of RNA synthesis. *J. Mol. Biol.* **72**:751.

―――. 1972b. Regulation of ribonucleic acid synthesis in growing bacterial cells. II. Control over the composition of the newly made RNA. *J. Mol. Biol.* **72**:765.

Nikolaev, N., L. Silengo and D. Schlessinger. 1973a. Synthesis of a large precursor to ribosomal RNA in a mutant of *Escherichia coli*. *Proc. Nat. Acad. Sci.* **70**:3361.

―――. 1973b. A role for RNase III in processing of rRNA and mRNA precursors in *Escherichia coli*. *J. Biol. Chem.* **248**:7967.

Nomura, M. and F. Engbæk. 1972. Expression of ribosomal protein genes as analyzed by bacteriophage μ-induced mutations. *Proc. Nat. Acad. Sci.* **69**:1526.

Norris, T. E. and A. L. Koch. 1972. Effect of growth rate on the relative rates of synthesis of messenger, ribosomal and transfer RNA in *Escherichia coli*. *J. Mol. Biol.* **64**:633.

Parker, J. and F. C. Neidhardt. 1972. Metabolic regulation of aminoacyl-tRNA synthetase formation in bacteria. *Biochem. Biophys. Res. Comm.* **49**:495.

Pato, M. L. and K. von Meyenburg. 1970. Residual RNA synthesis in *Escherichia coli* after inhibition of initiation of transcription by rifampicin. *Cold Spring Harbor Symp. Quant. Biol.* **35**:497.

Pedersen, S. and N. O. Kjeldgaard. 1972. A hybridization assay specific for ribosomal RNA from *Escherichia coli*. *Mol. Gen. Genet.* **118**:85.

Pettijohn, D. E., K. Clarkson, C. R. Kossman and O. G. Stonington. 1970. Synthesis of ribosomal RNA on a protein–DNA complex isolated from bacteria: A comparison of ribosomal RNA synthesis *in vitro* and *in vivo*. *J. Mol. Biol.* **52**:281.

Pichon, J., J. Marvaldi and G. Marchis-Mouren. 1972. The *in vivo* order of addition of ribosomal proteins in the course of *E. coli* 50S subunit biogenesis. *Biochem. Biophys. Res. Comm.* **47**:531.

Raué, H. A. and M. Gruber. 1971. Control of stable RNA synthesis and maturation of 5S RNA in *Salmonella typhimurium* RC[str] and RC[rel]. *Biochim. Biophys. Acta* **232**:314.

Rose, J. K., R. D. Mosteller and C. Yanofsky. 1970. Tryptophan messenger RNA elongation rates and steady-state levels of tryptophan operon enzymes under various growth conditions. *J. Mol. Biol.* **51**:541.

Rosset, R., J. Julien and R. Monier. 1966. Ribonucleic acid composition of bacteria as a function of growth rate. *J. Mol. Biol.* **18**:308.

Rosset, R., R. Monier and J. Julien. 1964. RNA composition of *Escherichia coli* as a function of growth rate. *Biochem. Biophys. Res. Comm.* **15**:329.

Schaechter, M., O. Maaløe and N. O. Kjeldgaard. 1958. Dependency on medium and temperature of cell size and chemical composition during balanced growth of *Salmonella typhimurium. J. Gen. Microbiol.* **19**:592.

Schleif, R. 1967. Control of production of ribosomal protein. *J. Mol. Biol.* **27**:41.

————. 1968. Origin of chloramphenicol particles protein. *J. Mol. Biol.* **37**:119.

Skjold, A. C., H. Juarez and C. Hedgecoth. 1973. Relationship among deoxy-ribonucleic acid, ribonucleic acid and specific transfer ribonucleic acid in *Escherichia coli* 15T⁻ at various growth rates. *J. Bact.* **115**:177.

Stamato, T. D. and D. E. Pettijohn. 1971. Regulation of ribosomal RNA synthesis in stringent bacteria. *Nature New Biol.* **234**:99.

Tanaka, N., G. Kuwano and T. Kinoshita. 1971. Chromosomal location of a fusidic acid resistant marker in *Escherichia coli. Biochem. Biophys. Res. Comm.* **42**:664.

Tissières, A., J. D. Watson, D. Schlessinger and B. R. Hollingworth. 1959. Ribonucleoprotein particles from *Escherichia coli. J. Mol. Biol.* **1**:221.

Traut, R. R., H. Delius, C. Ahmad-Zadeh, T. A. Bickle, P. Pearson and A. Tissières. 1969. Ribosomal proteins of *E. coli*: Stoichiometry and implication for ribosome structure. *Cold Spring Harbor Symp. Quant. Biol.* **34**:25.

Turnock, G. 1969. A genetic analysis of a mutant of *Escherichia coli* with a defect in the assembly of ribosomes. *Mol. Gen. Genet.* **104**:295.

Varrichio, F. and R. Monier. 1971. Ribosome pattern in *Escherichia coli* growing at various rates. *J. Bact.* **108**:105.

Voynow, P. and C. G. Kurland. 1971. Stoichiometry of the 30S ribosomal proteins of *Escherichia coli. Biochemistry* **10**:517.

Waller, J.-P. 1963. The NH$_2$-terminal residues of the proteins from cell-free extracts of *E. coli. J. Mol. Biol.* **7**:483.

Weber, H. J. 1972. Stoichiometric measurements of 30S and 50S ribosomal proteins from *Escherichia coli. Mol. Gen. Genet.* **119**:233.

Winslow, R. M. and R. A. Lazzarini. 1969. The rates of synthesis and chain elongation of ribonucleic acid in *Escherichia coli. J. Biol. Chem.* **244**:1128.

Yuki, A. 1971. Tentative identification of a "maturation enzyme" for precursor 16S ribosomal RNA in *Escherichia coli. J. Mol. Biol.* **62**:321.

Zaritsky, A. and K. von Meyenburg. 1974. Synthesis of ribosomal protein during the cell cycle of *Escherichia coli* B/r. *Mol. Gen. Genet.* **129**:217.

Ribosome Formation in *Escherichia coli*

David Schlessinger
Department of Microbiology
Washington University School of Medicine
St. Louis, Missouri 63110

INTRODUCTION

During the formation of an organelle that involves more than 70 interacting elements, some steps might be expected to occur indifferently with respect to others, while others are rate-limiting. In theory, any step can become rate-limiting if a component is in short supply, or if a jam-up of a series of complex steps occurs. The purpose of this chapter is to try to analyze which steps are rate-limiting for ribosome formation during exponential growth and how various other events are timed in relation to the rate-limiting ones.

In an attempt to order the facts, opinions and preferences are stated below. However, they are clearly labeled and segregated in order to make it easier to ignore or spurn them as further evidence accumulates. Also, there are only oblique references to comparisons of *E. coli* and eukaryotic paradigms. The attentive reader may notice the appearance in the discussion of *E. coli* of a large RNA that contains both 16S- and 23S-specific sequences. In further analogy to eukaryotic cells, the large RNA precursor can be split in absence of protein synthesis, and final maturation steps in the formation of 30S ribosomes may occur "out in the cytoplasm," etc. Nevertheless, the eukaryotic rRNAs are still of different size, have different methylation patterns, and cannot substitute for bacterial rRNA in reconstitution reactions: oversimplifiers, beware!

Much of the material and preliminary discussion is covered in recent review articles by Pace (1973) and Nomura (1970, 1973). The nomenclatures current there for precursors of rRNA species or ribosomes are used here or are redefined. In addition, we refer to the "30S pre-rRNA"—the primary transcription product that contains both 16S and 23S rRNA sequences (Kossman, Stamato and Pettijohn 1971; Dunn and Studier 1973; Nikolaev, Silengo and Schlessinger 1973a,b)—and to "p_o 30S-50S" as defined below.

THE COMPONENTS

The final structural components of the ribosome are discussed in detail elsewhere in this volume.

Initial RNAs

In confirmation of various studies by transcriptional mapping or by completion in vitro of rRNA chains begun in vivo (see review by Pace 1973), 16S and 23S rRNAs now appear to arise from a single RNA transcript, 30S pre-rRNA, of molecular weight 2.1×10^6 (Nikolaev, Silengo and Schlessinger 1973a,b). The RNA probably contains the 16S sequence nearer the 5' end than the 23S sequence (Kossman, Stamato and Pettijohn 1971; Pettijohn and Kossman, unpublished data) and may also contain 5S rRNA, as suggested in other cases (Pace 1973). It is also very undermethylated compared to mature rRNA (Lowry and Dahlberg 1971).

Studies of the flow of ^{3}H-labeled precursors into 16S and 23S rRNA of growing cells indicated that the two received label at equal rates, in contrast to the expectation for independent synthesis of the two (Mangiarotti et al. 1968; Adesnik and Levinthal 1969). Earlier explanations for this discrepancy were based on suggestions that RNA polymerase might move faster on 23S sequences, or that 23S rRNA might be assembled from smaller-sized units. These suggestions are now discarded since pool effects and cleavage of the initial 30S pre-rRNA before it is completed can explain the kinetics of labeling in vivo.

From the molecular weight data, the 30S pre-rRNA would contain about 20% more nucleotide sequences than are present in mature 16S and 23S rRNA. The base composition has been determined for some of these extra sequences from the 5' and 3' termini of p16S, a precursor of 16S RNA that seems to be one of the intermediates formed after cleavage of 30S pre-rRNA. p16S RNA can be obtained as a pulse-labeled species from normally growing cells (Sogin et al. 1971; Brownlee and Cartwright 1971; Hayes et al. 1971) or as an accumulated RNA species from cells treated with chloramphenicol (Lowry and Dahlberg 1971). 5S RNA also exhibits more initial sequences than found in its final form. Under some conditions, precursors can be detected that have 1, 2, or 3 additional 5'-nucleotides (see Monier, this volume).

Initial Proteins

As yet, there is no substantive indication that larger precursor proteins are split to yield the various ribosomal proteins. The initial protein products therefore probably contain formyl methionine groups and often additional amino acids at the presumed N-terminal end that are cleaved to give the mature protein chains (see Yaguchi et al. 1973). Some nascent chains lack methyl groups present in several 50S proteins; 85% of the methylation seems to be in protein L11, with subsidiary amounts detected in L3 and L5 of a K12 strain (J. H. Alix and D. Hayes, unpublished data). Also some of the mature L7–L12 proteins appear with an *N*-acetylated N-terminus (Terhorst, Wittmann-Liebold and Möller 1972) that is presumably added after the synthesis of the polypeptide chain.

THE FIRST AND LAST STEPS IN FORMATION OF ACTIVE RIBOSOMES

For convenience, the synthesis of ribosomes will be discussed as if it were divided into three phases:

(1) The initial steps—formation of the rRNA and binding of the first protein chains, up to the cleavage of the p_0 precursor.

(2) The formation of particles that can sediment at 30S and 50S. By this criterion, the particles formed during this phase are structurally and conformationally complete ribosomes, though they are not yet active.

(3) The formation of functional ribosomes. Some modifications of primary structure seem to be required to bring nascent 30S and 50S *particles* to *active* ribosomes. Some authors have even suggested that final cosmetic steps which make new ribosomes indistinguishable from the bulk of pre-existing ones may occur even after the ribosomes are already functioning in protein synthesis.

These steps are considered in the next section.

The major kinetic holdups in ribosome biosynthesis occur en route to tightly organized 30S and 50S *particles* during phase 2. Those rate-limiting steps are the primary subject here and are treated extensively below. Phases 1 and 3 are discussed first, because the first events (phase 1) occur before rate-limiting steps in growing cells, whereas the last steps (phase 3) occur after ribosomal particles have been finished.

The First Steps

In the detailed studies of Mangiarotti et al. (1968) and Lindahl (1973a), precursor rRNA—even in the form of incomplete, heterogeneous fragments —was always found bound to proteins. This suggested early on that ribosome formation begins even while rRNA is being transcribed. Further support for this suggestion came from the direct observations by Miller and Hamkalo (1972) of what may be rRNA cistrons in active transcription, in which the transcripts are covered with protein. Such results are not surprising now, since a number of ribosomal proteins bind directly to isolated segments of rRNA (Mizushima and Nomura 1970; Scheup, Green and Kurland 1971; Stöffler et al. 1971; Zimmermann, Ehresmann and Branlant 1972; Fellner, Zimmermann, this volume). Presumably this results in the formation of a very short-lived putative precursor, "p_o 30S-50S," containing all of the 30S pre-rRNA that is synthesized before the time at which its 5′ portion, containing the 16S rRNA sequence (Pettijohn and Kossman, unpublished data), is cleaved from the partially finished nascent 23S rRNA segment. From several types of experiment (Pettijohn and Kossman, unpublished data; Nikolaev, Silengo and Schlessinger 1973b), cleavage is inferred to occur most probably when RNA polymerase is part way through the 23S-specific region.

Very likely because of its early cleavage, 30S pre-rRNA has not been detected in wild-type cells (D. Pettijohn, personal communication). However, the cleavage of 30S pre-rRNA is specifically slowed in strain AB105, a mutant originally isolated by Keil and Hofschneider (1973) as deficient in RNase III activity against double-stranded replicative forms of RNA phage. Three reports specify that the 30S pre-rRNA indeed contains the rRNA sequences and that the isolated RNA contains sequences specifically recognized by the endonuclease RNase III (Nikolaev, Silengo and Schlessinger 1973a,b; Dunn and Studier 1973). In these studies, RNase III cleaves isolated 30S pre-rRNA to produce electrophoretic peaks with the mobility (25S and 17.5S) of initial precursors of 23S and 16S rRNA. Since RNase III has a known preference for double-stranded RNA structures (Robertson, Webster and Zinder 1968), the suggestion has been made that the signal for cleavage involves a double-stranded recognition site on the RNA.

Starting from AB105, Nikolaev and coworkers have purified bulk quantities of 30S pre-rRNA and the products of cleavage by RNase III. The molecular weights by formamide gel electrophoresis are 2.1, 1.2 and 0.65 × 10⁶ for the 30S, 25S and 17.5S species, respectively; direct measurement in the electron microscope gives confirmatory estimates of molecular weight of 2.07, 1.2 and 0.65 × 10⁶ for the three species (Nikolaev, Wellauer and Schlessinger, unpublished results).

Relevant to the present discussion are trials in which the availability of bulk 30S pre-rRNA has permitted the determination of the capacity of proteins to bind to the 16S and 23S rRNA sequences while the two are covalently linked. Under appropriate conditions, homogeneous particles are formed that sediment in sucrose gradients at about 36S for the complex of 30S pre-rRNA with 30S proteins and at 45S for the complex with 50S proteins (Figure 1). The peaks are consistently relatively symmetrical and show a constant ratio of protein to RNA across their width, indicative of the relative amounts of protein they contain. Compared to ribosomes, the

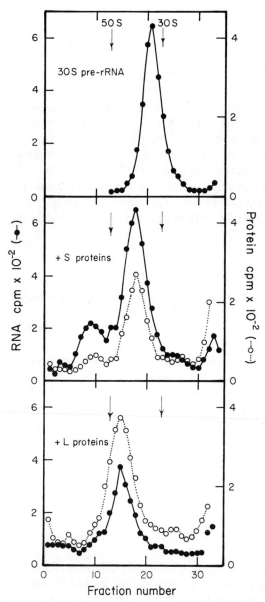

Figure 1 Zonal sedimentation in sucrose gradients of purified 30S pre-rRNA (*top*) and its complexes with purified 30S proteins (*middle*) or with purified 50S proteins (*bottom*). Reaction mixtures contained ³H-30S pre-rRNA (16 μg containing 2100 ³H (U) cpm/μg) and ¹⁴C-protein (8000 cpm/μg, employing 4.2 μg S proteins or 8.6 μg L proteins). Incubation for 35 min at 37°C in appropriate conditions; sedimentation in 17 ml 10–30% sucrose gradients for 13 hr in the SW27 Spinco rotor at 25,000 rpm. The indicated positions of authentic 30S and 50S ribosomes were determined from gradients run in parallel (N. Nikolaev, unpublished data).

peaks contain about 0.8 equivalents of 30S or 50S ribosomal proteins per mole of RNA.

As discussed above, cleavage of 30S pre-rRNA ordinarily occurs before a complete complex with both 30S and 50S proteins can form. It is important to note, however, that complete ribosomes are made in strain AB105 starting from 30S pre-rRNA (Nikolaev, Silengo and Schlessinger 1973b), so that the specific complexes of 30S pre-rRNA and proteins may more closely resemble complexes in vivo than do those initial complexes formed with partially or fully matured 16S or 23S rRNA. The binding sites of individual proteins are presumably the same in the various cases and should be those observed in reconstitution experiments (Nomura 1973), but the kinetic route may be different, with certain sites selectively exposed or masked.

Modifications and the Last Steps

The next two sections below deal with the requirements for the formation of particles sedimenting at 30S and 50S. Here additional steps are mentioned that are apparently required to make such 30S and 50S particles *functional*.

The steps required certainly include methylation. Beaud and Hayes (1971) showed that inactive 30S and 50S particles formed during growth of *E. coli met*[-] in the presence of ethionine contain normal amounts of 5S, 16S and 23S RNA and complete sets of 30S and 50S proteins in which all methionines are replaced by ethionines, and that these particles cannot function until their 16S and 23S RNAs and a small number of their proteins are methylated by provision of *S*-adenosyl methionine and appropriate enzyme factors in cells or extracts. A more direct demonstration of the requirement for methylation comes from experiments with the major precursors isolated from in vivo p30S and $p_2$50S. Nierhaus, Bordasch and Homann (1973) have reported that these particles can form 30S and 50S particles in vitro in reconstitution conditions, but the particles are active only when final methylation occurs.

Specific methylases for the various methylation steps have not been extensively assigned. One exception is the activity studied by Davies and his coworkers which catalyzes a methylation otherwise absent in mutants resistant to kasugamycin (Helser, Davies and Dahlberg 1972). They find that a core particle containing 16S RNA from kasugamycin-resistant cells can be methylated, but 16S RNA cannot. These results are again consistent with the observation that much methylation occurs late during ribosome formation. However, this methylation is obviously not itself one of those obligate for formation of active ribosomes since the *ksg*[r] mutants are viable.

Kinetic studies show that final cleavages of some extra sequences of RNA also occur very late. Hayes and coworkers have shown that a high rate of maturation of p30S and cleavage of p16S RNA requires protein-synthetic conditions (Hayes and Vasseur 1973), and Lindahl (1973a,b) and Mangiarotti et al. (1974) have found 30S particles containing p16S RNA in cell lysates. In the most extensive study to date, Mangiarotti et al. show that such particles can be found in polyribosomes extracted from cells; that they can also form polyribosomes with T4 mRNA in vitro; and that the entry of such particles into polyribosomes on T4 mRNA in vitro accelerates the conversion of p16S to 16S RNA. p16S RNA can also form inactive 30S particles in reconstitution trials with 30S proteins (Wireman and Sypherd, unpublished data); thus formation of *active* ribosomes probably requires at least some cleavage of precursor RNA sequences, and efficient final maturation in vivo may even require the prior functioning, or at least incorporation, of immature particles in polyribosomes.

One qualification of the need for complete maturation of precursor RNA comes from the work of Feunten in studies of a temperature-sensitive mutant 219 isolated by Vola and Rosset (Rosset et al. 1971). At 42°C, incomplete 30S particles are formed containing a precursor form of 16S RNA with extra sequences at both 5′ and 3′ termini (J. Feunten, personal communication). However even at the permissive temperature (30°C), the 5′-oligonucleotides of 16S RNA are still somewhat abnormal. Many chains contain one or two extra residues at the 5′ terminus (pUUAAAUUG and pUAAAUUG). The extra 5′-residues are reminiscent of the extra nucleotides found in some precursors of 5S RNA.

In the case of the pre-16S RNA described by Feunten, in analogy to results for some methylases, trimming steps are not all totally obligate for ribosome function. But the assignment of enzymes to individual trimming steps, like the assignment of methylases, is only beginning. Apart from the tentative assignment of RNase III to an initial cleavage step (see above), at least one further loss of extra sequences can occur from 25S RNA, even in cells treated with chloramphenicol (Pace 1973). Therefore that cleavage probably also occurs rather early (with no proteins required?). One other activity of possible relevance to rRNA processing has been reported; it gives an endo cleavage of p16S and is temperature sensitive in a mutant (Yuki 1971; Corte et al. 1971). It tends to purify with RNase II but can be separated from it by fractionation (Venkov, Waltschewa and Schlessinger 1972) or by genetic analysis (Weatherford et al. 1972). However, this activity acts on isolated p16S, whereas the cleavage of p16S RNA occurs late in 30S ribosome formation in vivo. Thus a different enzyme may prove to be involved in the intact cell.

RATE-LIMITING STEPS AND COMPONENTS IN FORMATION OF 30S AND 50S PARTICLES: SOURCES OF INFORMATION

In principle, *many* steps in ribosome formation *could* be rate-limiting: one could imagine limiting levels of RNase III or another modifying enzyme, rate-limiting formation of particular proteins (like the 30S protein S7 or

50S L19, see below), etc. Furthermore, there may be alternate pathways for part or all of the process in different cells or under different growth conditions. For example, the numbers of minor ribosome precursors differ somewhat in kinetic analyses by different authors (see below). However, every one of the techniques discussed below tends to indicate the presence of strong, *major* kinetic holdup points that probably are universal.

Reconstitution

The "assembly map" worked out by Mizushima and Nomura (1970; see also Nomura and Held, this volume) for 30S reconstitution from isolated 16S RNA and ribosomal proteins almost certainly reflects the relative order of binding of certain components, based on the capacity of some proteins to bind directly to RNA and of some to bind only when preceded by specified others. Particularly when one considers the branches of the assembly map, each beginning from a protein that binds to a domain of isolated RNA, it seems inescapable that, in general, *whenever* a component binds, it *must* bind in the way specified by reconstitution trials.

However, the relation of the assembly map to the case in vivo must be qualified in two ways: (1) Particularly as the kinetic path gets increasingly far from the initial liaisons of proteins that bind directly to isolated RNA, the map may not include all the arrows, or may add some, or may miss some dictated by the action of auxiliary components (for further discussion, see Nomura 1973). (2) There is no way to determine from the assembly map what the order of availability of components in cells will be. In other words, the problem of biosynthesis in these terms is: Which branches of the assembly path are picked in which order, and why?

In reconstitution studies, a rate-limiting step has been observed at which some but not all proteins are bound. The possible relation of the resultant particle, "RI," to ribosome formation in vivo is discussed in the next section.

Genetic

The most useful class of mutants (see Jaskunas, Nomura and Davies, this volume) has been those with assembly defects, especially the cold-sensitive *sad* mutants of the type isolated by Guthrie, Nashimoto and Nomura (1969) and by Tai, Kessler and Ingraham (1969). The technique was exploited by Guthrie and coworkers (1969) as a way to obtain mutants in many ribosomal proteins and to observe partially assembled ribosomes that accumulate in some of the strains at the nonpermissive temperature (Nashimoto et al. 1971).

The original intuition of Guthrie, Nashimoto and Nomura (1969) was that since ribosome assembly has a "temperature-sensitive" step, cold-sensitive mutants would be a likely source of those with defects in ribosome formation. Alternatively or in addition, the richness of the cold-sensitive class may be related to the fact that cold sensitivity often affects the organization of structures (Tyler and Ingraham 1973). Perhaps as a result, a successful extension of the use of cold-sensitive mutants has been developed by Bryant and Sypherd (1973) to isolate ribosome maturation mutants (*rim*) in other *loci* of *E. coli*. A 43S precursor from one such mutant can be made to sediment at 50S by the addition of a soluble factor from wild-type cells. Some such mutations might be in genes coding for methylases, for example; others could represent lesions in "morphopoietic factors."

Especially since such components are generally not found in finished ribosomes, the genetic approach is the surest one to try to saturate the catalog of elements involved in ribosome formation.

Some of the results of genetic studies are not easy to foresee—for example, the revelation from some mutants in 30S function that 30S formation interacts somehow with 50S formation, or the example of mutant AB105, which permits the isolation of the large 30S pre-rRNA; the latter also demonstrates that the whole cell can form 30S and 50S ribosomes starting from intact 30S pre-rRNA, even though the 30S rRNA ordinarily does not exist in growing cells.

Much more tenuous logically is the attempt to use mutants to define the exact sequence of events during ribosome formation in vivo. While a series of such mutants might give a cumulative, self-consistent picture of successive points of blockage, mutated elements might also tend to induce aberrant steps—perhaps even leading to common aberrant subribosomal particles. As a result, the jump from observations of incomplete ribosomes in mutants to discussions of rate-limiting steps in growing cells is difficult. For example, as discussed below, a particle accumulated at low temperature in vivo in a *sad* mutant contains protein components very similar to those of the RI particle formed as an intermediate in reconstitution reactions at low temperature in vitro (Nomura and Held, this volume). But the consistency of the compositions cannot logically be used to show that the cell has a rate-limiting step similar to that observed in reconstitution systems in vitro.

The same objection applies to the attempts to use mutants in auxiliary factors (e.g., Corte et al. 1971) to define rate-limiting steps in vivo. Presumably a mutation in a rate-limiting component would lead to accumulation of a "true" intermediate in ribosome formation. But how is it possible to tell which intermediate in which mutant is the true one without an auxiliary criterion?

Physiological Studies

Physiological experiments have provided the major means to supplement the genetic and reconstitution studies. A variety of techniques have been

employed to try to see when various components are added during biosynthesis, when various modifications or trimming of RNA or protein occurs,
and especially to find out the composition of the partially finished subparticles that have arrived at kinetic holdup points.

Several groups have looked at the flow of labeled amino acids into
individual proteins of finished ribosomes (Marvaldi, Pichon and Marchis-
Mouren 1972; Sells and Davis 1970; Pichon, Marvaldi and Marchis-
Mouren 1972). The proteins labeled latest in finished ribosomes are
presumably the first to add to precursors. Kinetically, the determination of
the first proteins to add would be expected to be relatively the most accurate, since specific labeling would tend to blur at later times. Furthermore, at least two complications may confuse the results: (1) many
fractional proteins may cycle from one ribosome to another or from a ribosome to a precursor particle; and (2) even if ribosomal proteins are made
from only one or a few polycistronic mRNA transcriptional units
("operons"), the relative production of different proteins may yield steady-
state pools of somewhat variable size for different proteins. Perhaps because of these problems, the results of labeling experiments deviate
considerably from those by other techniques (see Table 1).

A more extensive source of information about rate-limiting steps in ribosome formation in vivo has come from the analysis of the flow of labeled
precursors into rRNA and thence into ribosomal particles during exponential growth. Kinetic holdup points are invariably seen in such analyses.
Building on the preliminary studies and mathematical formulation of Mc-
Carthy, Britten and Roberts (1962), Mangiarotti et al. (1968) were able
to specify one major precursor state for the 30S ribosome and two for the
50S, a minor one and a major one. "Major" means that the nascent ribosome stops for a considerable time in a pool that fills and empties with exponential kinetics.

The average stopping time of a particle in the major precursor pool is
directly related to the fraction of 30S-specific or 50S-specific ribosomal
mass in that particle. The major precursors contain about 5% of the total
mass of ribosomal material in the cell, with equivalent fractions in the
precursor pools for 30S and 50S ribosomes: the time to make a ribosome
is the same for 30S and 50S ribosomes. In the very slow-growing cultures
used by Mangiarotti et al., the average formation time was about 10 to 20
min.

All subsequent analyses have confirmed the existence of similar major
precursors, though a second, and even a third, minor 30S precursor have
also been sometimes reported (Osawa 1968; Osawa et al. 1969; Lindahl
1973a). The precursor of 30S, originally named "26S," and those of 50S,
originally called "32S" and "43S," have been renamed p30S, $p_1$50S and
$p_2$50S by Hayes and Hayes (1971) to avoid problems caused by variability
in sedimentation behavior. Most important, Nierhaus, Bordasch and Ho-

Table 1   Proteins in some complexes relevant to 30S ribosomes

	From in vitro experiments		From in vivo experiments			
	CsCl core[a]	Binding to 16S rRNA[b]	RI[b]	p30S[c]	*sad* particle[d]	Appearance time[e]
S1	(−)		(−)	+	(−)	(−)
S2	−		−	−	−	−
S3	−		−	−	−	−
S4	+	+	+	+	+	+
S5	+		+	+	(−)	+
S6	−		(+)	−	(+)	(+)(+)
S7	(+)	+	(+)	−	(+)	−
S8	+	+	+	+	+	+
S9	−		(+)	−	(±)	(+)
S10	−		−	−	−	(+)
S11	(+)		(+)	−	−	−
S12	(+)		−	−	−	−
S13	+		+	+	+	(−)
S14	−		−	−	−	−
S15	+	+	+	+	+	+ +
S16	+		+	+	+	+ +
S17	+	+?	+	+	+	+ +
S18	(+)		(+)	−	(±)	(+)(+)
S19	(+)		(+)	−	(+)	(+)
S20	+	+	+	+	+	+ +
S21	+		(−)	+	(−)	(−)

The protein list is taken from the standard catalog of Wittmann et al. (see this volume). Entries in parentheses indicate differences from the composition of p30S.

[a] Traub et al. 1967.   [b] From Nomura 1973 as modified from Mizushima and Nomura 1970.   [c] Nierhaus, Bordasch and Homann 1973.   [d] Nashimoto et al. 1971 and Nomura 1973.   [e] Marvaldi, Pichon and Marchis-Mouren 1972. The entries on the left indicate those proteins inferred to add first; the entries on the right include additional proteins inferred to add next.

mann (1973) have isolated each of these particles and determined their protein and RNA compositions. These are discussed below.

In these analyses, the positions and characteristics of precursors were judged by their content of newly labeled RNA. The analysis has been refined in recent studies by Lindahl (1973b) and by Mangiarotti et al. (1974) by tracing the location of precursors of rRNA (particularly the p23S and p16S RNAs) as markers for nascent ribosomes. All these analyses have indicated that final maturation of 16S and 23S RNA is a late step in biosynthesis. As a specific example, Lindahl found that in his conditions the major precursors were either more tightly folded or more nearly complete than in other analyses and sedimented nearly at 30S and 50S. Nevertheless they could be recognized by their content of precursor RNA and their higher electrophoretic mobility. (They may well correspond to the major p30S and p$_2$50S of other published reports, according to comparative trials of Lindahl and Nierhaus, personal communication.)

The analyses of Lindahl (1973b) also extend the estimates of amounts and synthetic times for precursors and ribosomes. Precursors of each particle again contain about the same fraction of total 16S or 23S RNA (here 2%), both from direct measurements and from kinetic experiments. Most of the pRNA is of course in the major precursors. The levels correspond to an average biosynthetic time of 1–2 min at a doubling time of 30 min at 37°C.

In the recent analyses by Michaels, by Lindahl and by Mangiarotti, the time of formation of ribosomes varied directly with growth rate. The available information on the relative time of maturation of RNA is summarized above. The late steps in trimming and methylation should continue to provide an additional handle to follow precursors.

RATE-LIMITING STEPS: COMPARISON OF PRESENT INFORMATION FROM VARIOUS SOURCES

The Problem of Fractional Proteins in 30S Ribosomes

Discussions of ribosome formation have generally not faced up to the problem posed by fractional proteins. Table 1 lists the reported content of 30S proteins (S proteins) in various subribosomal particles and in the isolated p30S. If one compares lists of fractional proteins (Voynow and Kurland 1970; Wittmann, this volume) to the data of Table 1, one finds, for example, that S20 is one of the proteins with a binding site on pure 16S RNA and is present in RI particles, in CsCl cores, in a *sad* particle, and in the p30S precursor purified from in vivo; yet it is fractional. Nomura (1973) has pointed out that S14 is required for activity for every ribosome function that has been tested; yet it is fractional.

It is especially troubling that of the ten proteins isolated with the p30S precursor, at least one is "marginal," three are "fractional," and no information is available about the stoichiometry of two others (S13 and S15).

So far as I can tell, no one has suggested that the cell has a way to form a fractional ribosome—containing 0.3 chain of S5, for example. There are at least three alternatives:

1. Only certain unit proteins are truly involved in ribosome formation, and the others add to ribosomes later. Using a line of argument from any column of Table 1, this seems unlikely.

2. Heterogeneity is introduced during biosynthesis: The requirements of the pathway for ribosome formation and stable structure may be rather flexible, and different particles might take different routes; or possibly,

some proteins might play a role during synthesis and then immediately begin to cycle off. However, at least for those marginal or fractional proteins found in the core after CsCl treatment (including at least S6, S11, S18, S19, S21), this type of notion seems unlikely.

3. One might wonder whether some of the difference is due to differential losses in preparation; the finding of a protein in the p30S, at the rate-limiting step, would suggest that it is a unit protein in the cell. Consistent with this suggestion are (a) the genetic hint that ribosomal proteins may be produced from a single (or a very few) transcriptional unit, which would tend to result in comparable translation yields; and (b) the studies of several groups which have shown the requirement for stoichiometric (equimolar) amounts of a variety of proteins for maximal activity of reconstituted 30S ribosomes. For example, Held, Mizushima and Nomura (1973) have demonstrated the requirements for an equimolar amount of S19 (a "fractional" protein) in order to achieve maximal activity in polyphe synthesis, even though repurified RI particles, like repurified 30S ribosomes, contain less than stoichiometric levels of this protein. In other words, one might suggest that *every nascent ribosome has a binding site for every ribosomal protein.*

This puzzle of the "true" levels of fractional proteins in the cell might be clarified by immunological or other verification of the total cellular content of each ribosomal protein.

A Model for the Rate-limiting Step in 30S Ribosome Formation

In the meantime, perhaps the simplest operational inference can be drawn from the studies of the p30S: Any component present in the precursor cannot *itself* be the rate-limiting factor during biosynthesis. And possibly many more proteins are unit, with some final dynamic ones (like S1) added later (Nomura 1973).

The use of the p30S as a guide for our understanding of the biosynthetic route, *beyond the initial steps,* is deliberate and is forced by the comparison of the columns in Table 1. Figure 2 shows the relation of these analyses to the assembly map derived from in vitro reconstitution experiments. The proteins in the natural precursor show some similarity to the early phase of proteins added as analyzed by labeling kinetics (Marvaldi et al. 1972; last column of Table 1). Four of the first six proteins added according to labeling studies in vivo are present in p30S, and even this weak correlation breaks down rapidly thereafter. There is some trend of agreement with the CsCl core particle, though it contains rather more protein chains. However, there is no way to avoid the discordance between the composition of p30S and that of the RI obtained from reconstitution experiments. The two disagree with respect to eight of 21 proteins: two are present in p30S, but not in the RI; six more are present in RI and absent in p30S. The *"sad* particle" shows similar discrepancies.

Nevertheless, there are two strong hints from the comparison: The whole family of proteins (S7, S9, S10, S14, S19) that assemble in vitro led by S7 is missing in the p30S and partially present (S7, S9, S19) in the *sad* particle and the RI. Similarly, S6 and S18 are missing in p30S and present in the others.

There are at least two interpretations of this comparison that seem tenable:

1. The p30S in vivo is very similar to the RI, but some proteins, especially those of the assembly branch based on S7, have been lost during the repeated centrifugations required to purify the precursor. In fact, an earlier analysis by Homann and Nierhaus (1971) gave results for RI that lacked the S7-dependent assembly branch. Certainly everyone who has worked with ribosomes and precursors would agree that it is easy to lose some proteins from them during washing procedures, and when the measured numbers of different proteins disagree (cf. e.g., the analyses of $p_2$50S in Osawa et al. [1969] with that in Table 2), the higher number is very likely closer to that in vivo.

2. The holdup point in biosynthesis is such that the S7-dependent "assembly branch" (or the S18, S6 addition) cannot occur in vivo. According to this view, since S7 has a direct binding site to 16S RNA that survives both CsCl treatment and comparable purification of RI or *sad* particles (Table 1), its absence in purified p30S suggests that it is also missing in the cell.

Perhaps a fair statement of the merits of these alternatives would be that the composition of p30S could be *either* of those suggested; but in any case, the rate-limiting step in ribosome formation seems to be different in vivo from the models in vitro, so that agreement would be fortuitous.

In order to discuss the mechanism of limitation of ribosome formation, one can make the minimal hypothesis, following the approach of Traub and Nomura (1969), that the cell should be able to do anything the investigator can, when given the same components at the same temperature. It is then very suggestive that the reconstitution systems tend to go further than the system in vivo before they arrive at a rate-limiting step. Consider the case of the p30S precursor: (1) The 17S RNA is very likely as methylated as need be for 30S particle formation (though not for *active* 30S ribosome formation; Beaud and Hayes 1971); and (2) the 30S proteins are not methylated and the proteins in p30S show the positions of normal 30S proteins in two-dimensional electrophoresis. As a result, it is not surprising that in reconstitution conditions, p30S can be brought to active 30S particles (Nierhaus, Bordasch and Homann 1973) and 17S RNA will form 30S particles with ribosomal proteins (Wireman and Sypherd, unpublished data). Why then does the cell accumulate the p30S particle?

Table 2 Proteins in rRNA–protein complexes relevant to 50S ribosomes

	Direct binding to rRNA[a]	p$_1$50S[b]	p$_2$50S[b]	2 M LiCl core[c]	Appearance time[d]
L1		+	+		
L2	+	−	−		(+)
L3		+*	+	+	+
L4		+	+	+	+
L5		+	+*	+	
L6	+	−	−		
L7		±*	+		
L8 } L9		+	+*		
L10		+*	+		
L11		±*	+		
L12		−	−		
L13		+	+	+	+
L14		−	+		
L15		−	+		(+)
L16	+	−	−		(+)
L17	+	+	+	+	+
L18		+	+	+	
L19	+	−	+*	(+)	(+)
L20	+	+	+		
L21		+	+	+	+
L22		+	+	+	+
L23	+	+*	+	+	+
L24	+	+	+	+	+
L25		+	+		
L27		+	+*		
L28		−	−		
L29		+	+	+	+
L30		+	+		+
L31		−	−		
L32		−	−	(+)	(+)
L33		−	+		

The protein list is taken from the standard catalog of Kaltschmidt and Wittmann. Entries in parentheses indicate some differences from the reported composition of p$_1$50S. Components reported as present in low or variable amounts are shown with an asterisk.

[a] Stöffler et al. 1971. [b] Nierhaus, Bordasch and Homann 1973. [c] Homann and Nierhaus 1971. [d] Pichon, Marvaldi and Marchis-Mouren 1972.

One can restate this question by saying that the rate-limiting step in vivo is apparently different from that in vitro. Furthermore, if the concentrations of ions and ribosomes and the temperature are indeed comparable in cells and in reconstitution reactions, then unless the cell contains a negative (repressive) control on ribosome formation, only two possibilities remain: (1) The process is limited by a conformational change, requiring a high activation energy. This suggestion, from the model of reconstitution, has been discussed by Nomura (1973; see comparison of RI and *sad* particles in Table 1 references). (2) The process is limited by provision of a rate-limiting component, such as synthesis of a particular protein.

An argument for the second alternative can be inferred from the experiments of Michaels (1971), Lindahl (1973) and Mangiarotti (personal communication). In their trials, in accord with an earlier suggestion (Schlessinger and Apirion 1969), the rate of appearance of new 30S and 50S ribosomes at a given temperature varies with growth rate. As Mangiarotti first pointed out, a process with first-order kinetics, like the activation-rearrangement required for the RI to RI* transition, would not show such a dependence on growth rate. Therefore the rate-limiting step in ribosome formation is more likely to be a limitation for a required component.

Continuing this line of thought, based again on the supposed capacity of the cell to carry out reconstitutions at similar temperatures to those used in vitro, the capacity of p30S (and $p_2$50S; see next section) to be reconstituted in vitro suggests that the limitation in the cell is not for a morphopoietic factor or modifying enzyme but for the provision of one or more ribosomal proteins.

A likely missing component for the formation of a 30S particle can be judged from the list of those proteins most required in "single omission" reconstitution experiments: S4, S8, S7, S16 and S9 (Nomura et al. 1969). Three of these (S4, S8 and S16) are present in the reported analysis of p30S. S9 should be dependent on S7 for its function. Thus S7 is a possible candidate for the rate-limiting component. (This would necessarily require that S7, or a group of proteins, has a lower pool content than many others. The required regulation of the levels of limiting proteins would remain a mystery.)

Whether or not such a hypothetical construct should prove accurate, Figure 2 schematizes the relation of ribosome formation in vivo to the "assembly map" inferred from in vitro reconstitution trials.

The Problem of 50S Ribosome Formation

More of its constituent proteins are stoichiometric ("unit"), and the formation of the 50S ribosome is in that sense more straightforward. But it is otherwise more complex.

The verification by Nierhaus, Bordasch and Homann (1973) of the earlier detection of two strong kinetic holdup points at 32S ($p_1$50S) and 43S ($p_2$50S) has been extended by their careful analysis of protein composition. As Table 2 shows, the first precursor stage contains a total of

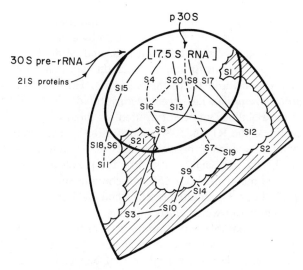

Figure 2 Biosynthesis of 30S ribosome compared to the assembly map and to subparticles made by various means. This schematic, whimsically outlined as a "30S particle," is based on the data summary of Table 1 and the assembly map primarily from Nomura 1973 (a refined assembly map is given in Nomura and Held, this volume). Strong relations are shown by solid lines. The heavy line encircles the protein complement of p30S (see text). RI: all but the shaded area. CsCl core: like RI, but including S21 and excluding S5 and S9.

about 17 proteins. Following the reasoning presented for the 30S particle, a hypothetical suggestion can be made that the construction of $p_1 50S$ is based on branches resulting from the binding of L17, L20, L23 and L24, four of the eight proteins that bind directly to 23S rRNA, and of L18 and L25 (Table 2), two proteins that complex with 5S RNA and are necessary and sufficient to link it to 23S rRNA (Gray et al. 1972). This process is presented schematically in Figure 3.

In a similar way, the binding of seven more proteins in the transition of $p_1 50S$ to $p_2 50S$ may be based in part on a branch that begins with the direct binding of one of them, L19, to the RNA, and the rate-limiting step in the 32S to 43S transition could be the provision of protein(s), perhaps L19. Concerning the content of 5S RNA, the finding that 5S RNA is already present in $p_1 50S$ (Hayes and Hayes 1971) could be a result of its possible presence in 30S pre-rRNA or alternatively could be due to early availability of L18 and L25.

However, the final stage in formation of a 50S particle ($p_2 50S$ to 50S) is the major holdup in vivo, and it is unclear whether the failure of the last seven proteins to bind is related merely to the absence of certain proteins— perhaps of three of the seven that can bind directly to 23S rRNA.

The RI particle formed during reconstitution trials with 50S (L) proteins and 23S and 5S RNAs seems likely to be similar in composition to the p₂50S precursor. No detailed composition of the RI formed by a mixture of 23S RNA and 50S-specific proteins has been published, and as discussed elsewhere in this volume, the 50S reconstitution has been difficult to achieve with *E. coli* components (Nomura and Held, this volume). However, the 50S particle of *B. stearothermophilus* has been reconstituted. Especially since the 30S proteins of the Bacillus can be substituted for those of *E. coli* in 30S reconstitution reactions, the RI for 50S reconstitution is also likely to be comparable with components from the two organisms. Therefore it is extremely suggestive that the RI for Bacillus 50S, like the p₂50S precursor obtained from whole cells, sediments at about 45S and contains most of the 50S proteins; of a total of about 37 tentatively classified by Fahnestock, Held and Nomura (1973), only four were clearly missing.

Of course the absence of comparable small numbers of proteins in both RI and p₂50S remains only a hint. There may be a closer relationship between the 50S RI and 50S ribosome formation than that between 30S RI and 30S particle. But on the other hand, the time of formation of 50S ribosomes in vivo is again related to growth rate, so that the rate-limiting step in vivo is not likely to be a unimolecular RI to RI* rearrangement (Traub and Nomura 1969). Also, as in the case of 30S ribosome formation, some proteins that can bind directly to 23S RNA appear to bind relatively late during the in vivo process (Figure 3). Therefore once again the provision of a component in short supply seems more likely to be rate-limiting in vivo.

An apparent difference between p₂50S and "RI" from *E. coli* is underlined by the report, as yet unconfirmed, that p₂50S, unlike RI, can be brought to complete 50S ribosomes in reconstitution conditions (Nierhaus, Bordasch and Homann 1973). Thus, assuming once again that the cell can carry out reconstitution reactions, a limiting provision of the last com-

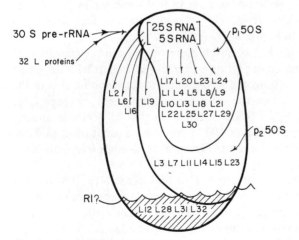

Figure 3 Biosynthesis of 50S ribosome, indicating probable entry times of various likely assembly branches. Based on data summary in Table 2 and text.

ponents is enough to explain the existence of a major kinetic holdup point, while RI seems to follow a different structural or kinetic path. In vivo, a simple limitation, for example, for production of L2–L6–L16 would of course be sufficient to limit the rate of 50S ribosome formation.

As far as the comparison with other sources of information goes, as Table 2 shows, there is even less correlation for the 50S than for the 30S ribosome between the protein composition of the two major precursors isolated from in vivo and constructs based on release of protein from complete ribosomes by high salt or times of labeling of individual ribosomal proteins. Nevertheless, comparable trends and strengths of these approaches are noticeable. For example, the proteins most difficult to release by salt (e.g., those remaining after 2 M or 4 M LiCl treatment) include almost exclusively proteins found in $p_1$50S (Homann and Nierhaus 1971); but the correlation breaks down when low or intermediate salt levels are used. Similarly, if the kinetic (Pichon, Marvaldi and Marchis-Mouren 1972) rather than the structural analysis is examined in detail, the group of nine proteins that are very slowly labeled (that is, the ones added first to 23S rRNA sequences) are all among the 17 proteins of $p_1$50S; they include three of the four that bind directly to 23S rRNA. The next group to bind, as inferred from labeling rates, include three proteins found only in complete 50S and two found in $p_2$50S.

Thus in the intermediate ground between the joining of the first 10 to 12 proteins to 23S-specific rRNA and the formation of the final ribosome, the analyses of precursor protein composition remain the only detailed information available at present from whole cells.

The Kinetic Path in the Cell

If it is true that the cell limits 30S and 50S particle formation by a slow supply of certain components, constituent or morphopoietic, then how does it avoid the high activation energies that seem to limit 30S and 50S reconstitution reactions in vitro? Traub and Nomura (1969) of course realized the discrepancies in activation energy in vivo and in vitro. From recent data the discrepancy can be better quantitated in various ways. For example, the amounts of precursor particles observed at 37°C (Lindahl 1973a) are an order of magnitude lower than those predicted by the in vitro reconstitution reactions.

The solution to this puzzle is not yet known, but it is the role of many enzymes and structural features of cells precisely to lower activation energies. Three specific possibilities are:

1. Structural modification. It has been suggested that binding and action of a methylase or nuclease could be arranged to provide a favorable kinetic route (Corte et al. 1971).

2. "Favored environment." A possibility would be a special hydrophobic membrane site. As detailed information about a process increases in any field, the invocation of "membrane effects" usually disappears. Nevertheless, several groups have reported that membrane fractions from lysed cells tend to be enriched for ribosomal precursor RNAs (Di Girolamo, Hinckley and Busiello 1968; Haywood 1971).

3. Morphopoietic factors. Possible candidates include small molecules, like the polyamines that aid in 50S particle formation (Hosokawa, personal communication), as well as protein factors, like those suggested by the genetic analysis of Bryant and Sypherd (1973) and others (see above).

One interesting speculative possibility that combines features of all three of these lines of thought comes from the data previously mentioned. Earlier genetic studies had shown that mutants in 30S protein components frequently led to a partial failure of 50S ribosome formation as well. The reverse effect has been noted rarely or not at all (see Davies and Nomura 1972; Rosset et al. 1971). Most of the mutations studied thus far are specific for 30S components, so that this disparity may be invalid on statistical grounds. Nevertheless, several groups have pointed out, on the basis of these results, that somehow the correct formation of 30S ribosomes is linked to proper 50S ribosome formation (Nashimoto and Nomura 1970; Kreider and Brownstein 1971). Such a possibility could be achieved, for example, if the two subunits were formed near one another at a membrane site.

However the establishment of the 30S pre-rRNA as an entity containing 16S and 23S rRNA and binding sites for various ribosomal proteins suggests a different possibility: the nascent p30S particle, *while still joined* to a growing 23S rRNA sequence, could help to ensure, or could itself be modified by, an appropriate conformation of $p_1$50S. It is at this level that such an action would have to take place, for $p_1$50S then exists free of p30S in the cell. Compared to the alternatives stated above, this one would employ the nascent 30S pre-rRNA as a morphopoietic factor (3) and favored environment (2) for a kinetic route to products based on nuclease action (1).

Note Added in Proof

Several recent studies are consistent with the suggestion that provision of a rate-limiting component—a ribosomal protein or morphopoietic factor—sets the rate of ribosome formation. Marvaldi et al. (1974) have specified that r-protein pools are generally low and undetectable for some protein species; and Mangiarotti (personal communication) has found that p30S precursors isolated from whole cells in vivo will form ribosomes when incubated with preparations of r-proteins even at $0°C$.

References

Adesnik, M. and C. Levinthal. 1969. Synthesis and maturation of ribosomal RNA in *Escherichia coli*. *J. Mol. Biol.* **46**:281.

Beaud, G. and D. Hayes. 1971. Proprietés des ribosomes et du RNA synthetisés par *Escherichia coli* cultivé en presence d'ethionine. *Eur. J. Biochem.* **19**:323.

Brownlee, G. G. and E. Cartwright. 1971. Sequence studies on precursor 16S ribosomal RNA of *Escherichia coli*. *Nature New Biol.* **232**:50.

Bryant, R. E. and P. S. Sypherd. 1973. Genetic analysis of cold-sensitive, ribosome maturation mutants of *Escherichia coli*. *J. Bact.* **117**:1082.

Corte, G., D. Schlessinger, D. Longo and P. Venkov. 1971. Transformation of 17S to 16S ribosomal RNA using ribonuclease II of *Escherichia coli*. *J. Mol. Biol.* **60**:325.

Davies, J. and M. Nomura. 1972. The genetics of bacterial ribosomes. *Ann. Rev. Genet.* **6**:203.

DiGirolamo, M., E. Hinckley and E. Busiello. 1968. Localization of ribosome precursors in *Escherichia coli*. *Biochim. Biophys. Acta* **196**:387.

Dunn, J. J. and F. W. Studier. 1973. T_7 early RNAs and *E. coli* ribosomal RNAs are cut from large precursor RNAs *in vivo* by RNase III. *Proc. Nat. Acad. Sci.* **70**:3296.

Fahnestock, S., W. Held and M. Nomura. 1973. The assembly of bacterial ribosomes. In *Generation of subcellular structures*, First John Innes Symposium (ed. R. Markham et al.) p. 179. North-Holland/American Elsevier, New York.

Gray, P. N., R. A. Garrett, G. Stöffler and R. Monier. 1972. An attempt at the identification of the proteins involved in the incorporation of 5S RNA during 50S ribosomal subunit assembly. *Eur. J. Biochem.* **28**:412.

Guthrie, C., H. Nashimoto and M. Nomura. 1969. Studies on the assembly of ribosomes *in vivo*. *Cold Spring Harbor Symp. Quant. Biol.* **34**:69.

Hayes, F. and D. Hayes. 1971. Biosynthesis of ribosomes in *E. coli*. I. Properties of ribosomal precursor particles and their RNA components. *Biochimie* **53**:369.

Hayes, F. and M. Vasseur. 1973. Etude in vitro de la maturation du précurseur 17S du RNA ribosomique 16S d'*E. coli*. *Compt. Rend. Acad. Sci.* **277D**:881.

Hayes, F., D. Hayes, P. Fellner and C. Ehresmann. 1971. Additional nucleotide sequences in precursor 16S ribosomal RNA from *Escherichia coli*. *Nature New Biol.* **232**:55.

Haywood, A. M. 1971. Cellular site of *Escherichia coli* ribosomal RNA synthesis. *Proc. Nat. Acad. Sci.* **68**:435.

Held, W. A., S. Mizushima and M. Nomura. 1973. Reconstitution of *Escherichia coli* 30S ribosomal subunit from purified molecular components. *J. Biol. Chem.* **248**:5720.

Helser, T. L., J. E. Davies and J. E. Dahlberg. 1972. Mechanism of kasugamycin resistance in *Escherichia coli*. *Nature New Biol.* **235**:6.

Homann, H. and K. Nierhaus. 1971. Protein compositions of biosynthetic precursors and artificial subparticles from ribosomal subunits in *Escherichia coli* K12. *Eur. J. Biochem.* **20**:249.

Keil, T. U. and P. H. Hofschneider. 1973. Secondary structure of RNA phage M12 replicative intermediates *in vivo*. *Biochim. Biophys. Acta* **312**:297.

Kossman, C. R., T. D. Stamato and D. E. Pettijohn. 1971. Tandem synthesis of the 16S and 23S ribosomal RNA sequences of *Escherichia coli*. *Nature New Biol.* **234**:102.

Kreider, G. and B. L. Brownstein. 1971. A mutation suppressing streptomycin dependence. II. An altered protein on the 30S ribosomal subunit. *J. Mol. Biol.* **61**:135.

Lindahl, L. 1973a. Two new ribosomal precursor particles in *Escherichia coli*. *Nature New Biol.* **243**:170.

————. 1973b. Studies on intermediates and time kinetics of the assembly of ribosomes in *Escherichia coli*. *Ph.D. thesis,* University of Copenhagen, Denmark.

Lowry, C. V. and J. E. Dahlberg. 1971. Structural differences between the 16S ribosomal RNA of *E. coli* and its precursor. *Nature New Biol.* **232**:52.

Mangiarotti, G., D. Apirion, D. Schlessinger and L. Silengo. 1968. Biosynthetic precursors of 30S and 50S ribosomal particles in *Escherichia coli*. *Biochemistry* **7**:456.

Mangiarotti, G., E. Turco, A. Ponzetto and F. Altruda. 1974. Precursor 16S RNA in active 30S ribosomes. *Nature* **247**:147.

Marvaldi, J., J. Pichon and G. Marchis-Mouren. 1972. The *in vivo* order of addition of ribosomal proteins in the course of *Escherichia coli* 30S subunit biogenesis. *Biochim. Biophys. Acta* **269**:173.

Marvaldi, J., J. Pichon, M. Delaage and G. Marchis-Mouren. 1974. Individual ribosomal protein pool size and turnover rate in *Escherichia coli*. *J. Mol. Biol.* **84**:83.

McCarthy, B. J., R. J. Britten and R. B. Roberts. 1962. The synthesis of ribosomes in *E. coli*. III. Synthesis of ribosomal RNA. *Biophys. J.* **2**:57.

Michaels, G. A. 1971. Ribosome maturation of *Escherichia coli* growing at different growth rates. *J. Bact.* **110**:889.

Miller, O. L., Jr. and B. A. Hamkalo. 1972. Visualization of RNA synthesis on chromosomes. *Int. Rev. Cytol.* **32**:1.

Mizushima, S. and M. Nomura. 1970. Assembly mapping of 30S ribosomal proteins from *E. coli*. *Nature* **226**:1214.

Nashimoto, H. and M. Nomura. 1970. Structure and function of bacterial ribosomes. XI. Dependence of 50S ribosomal assembly on simultaneous assembly of 30S subunits. *Proc. Nat. Acad. Sci.* **67**:1440.

Nashimoto, H., W. Held, E. Kaltschmidt and M. Nomura. 1971. Structure and function of bacterial ribosomes. XII. Accumulation of 21S particles by some cold-sensitive mutants of *Escherichia coli*. *J. Mol. Biol.* **62**:121.

Nierhaus, K. H., K. Bordasch and H. E. Homann. 1973. Ribosomal proteins. XLIII. *In vivo* assembly of *Escherichia coli* ribosomal proteins. *J. Mol. Biol.* **74**:587.

Nikolaev, N., L. Silengo and D. Schlessinger. 1973a. A role for ribonuclease III in processing of ribosomal ribonucleic acid and messenger ribonucleic acid precursors in *Escherichia coli*. *J. Biol. Chem.* **248**:7907.

————. 1973b. Synthesis of a large precursor to ribosomal RNA in a mutant of *Escherichia coli*. *Proc. Nat. Acad. Sci.* **70**:3361.

Nomura, M. 1970. Bacterial ribosome. *Bact. Rev.* **34**:228.

———. 1973. Assembly of bacterial ribosomes. *Science* **179**:864.

Nomura, M., S. Mizushima, M. Ozaki, P. Traub and C. V. Lowry. 1969. Structure and function of ribosomes and their molecular components. *Cold Spring Harbor Symp. Quant. Biol.* **34**:49.

Osawa, S. 1968. Ribosome formation and function. *Ann. Rev. Biochem.* **37**:109.

Osawa, S., E. Otaka, T. Itoh and T. Fukui. 1969. Biosynthesis of 50S ribosomal subunit in *Escherichia coli*. *J. Mol. Biol.* **40**:321.

Pace, N. R. 1973. The structure and synthesis of the ribosomal RNA of prokaryotes. *Bact. Rev.* **37**:562.

Pichon, J., J. Marvaldi and G. Marchis-Mouren. 1972. The *in vivo* order of addition of ribosomal proteins in the course of *E. coli* 50S subunit biogenesis. *Biochem. Biophys. Res. Comm.* **47**:531.

Robertson, H. D., R. Webster and N. Zinder. 1968. Purification and properties of ribonuclease III from *Escherichia coli*. *J. Biol. Chem.* **243**:82.

Rosset, R., C. Vola, J. Feunten and R. Monier. 1971. A thermosensitive mutant defective in ribosomal 30S subunit assembly. *FEBS Letters* **18**:127.

Schaup, H. W., M. Green and C. G. Kurland. 1971. Molecular interactions of ribosomal components. II. Site-specific complex formation between 30S proteins and ribosomal RNA. *Mol. Gen. Genet.* **112**:1.

Schlessinger, D. and D. Apirion. 1969. *Escherichia coli* ribosomes: Recent developments. *Ann. Rev. Microbiol.* **23**:287.

Sells, B. H. and F. C. Davis. 1970. Biogenesis of 50S particles in exponentially growing *Escherichia coli*. *J. Mol. Biol.* **47**:155.

Sogin, M., B. Pace, N. R. Pace and C. R. Woese. 1971. Primary structural relationship of p16 to m16 ribosomal RNA. *Nature New Biol.* **232**:48.

Stöffler, G., L. Daya, K. H. Rak and R. A. Garrett. 1971. Ribosomal proteins. XXVI. The number of specific protein binding sites on 16S and 23S RNA of *Escherichia coli*. *J. Mol. Biol.* **62**:411.

Tai, P., D. P. Kessler and J. Ingraham. 1969. Cold-sensitive mutation in *Salmonella typhimurium* which affects ribosome synthesis. *J. Bact.* **97**:1298.

Terhorst, C., B. Wittmann–Liebold and W. Möller. 1972. 50S ribosomal proteins. Peptide studies on two acidic proteins, A_1 and A_2, isolated from 50S ribosomes of *E. coli*. *Eur. J. Biochem.* **25**:13.

Traub, P. and M. Nomura. 1969. Structure and function of *Escherichia coli* ribosomes. VI. Mechanism of assembly of 30S ribosomes studied *in vitro*. *J. Mol. Biol.* **40**:391.

Traub, P., K. Hosokawa, G. R. Craven and M. Nomura. 1967. Structure and function of *E. coli* ribosomes. IV. Isolation and characterization of functionally active ribosomal proteins. *Proc. Nat. Acad. Sci.* **58**:2430.

Tyler, B. and J. L. Ingraham. 1973. Studies of ribosomal mutants of *Salmonella typhimurium* LT-2. *Mol. Gen. Genet.* **122**:197.

Venkov, P., L. Waltschewa and D. Schlessinger. 1972. Polyribosome metabolism in *Escherichia coli*. In *Functional units in protein biosynthesis*, FEBS Symp., vol. 23, p. 379. Academic Press, New York.

Voynow, P. and C. G. Kurland. 1970. Stoichiometry of the 30S ribosomal proteins of *Escherichia coli*. *Biochemistry* **10**:517.

Weatherford, S. C., L. Rosen, L. Gorelic and D. Apirion. 1972. *Escherichia coli* strains with thermolabile ribonuclease II activity. *J. Biol. Chem.* **247**:5404.

Yaguchi, M., C. Roy, A. T. Matheson and L. P. Visentin. 1973. The amino acid sequence of the N-terminal region of some 30S ribosomal proteins from *Escherichia coli* and *Bacillus stearothermophilus*. *Can. J. Biochem.* **51**:1215.

Yuki, A. 1971. Tentative identification of a "maturation enzyme" for precursor 16S ribosomal RNA in *Escherichia coli*. *J. Mol. Biol.* **62**:321.

Zimmermann, R. A., C. Ehresmann and C. Branlant. 1972. Location of ribosomal protein binding sites on 16S ribosomal RNA. *Proc. Nat. Acad. Sci.* **69**:1282.

Structure and Function
of Eukaryotic Ribosomes

Ira G. Wool
Department of Biochemistry
University of Chicago
Chicago, Illinois 60637

Georg Stöffler
Max-Planck-Institut für Molekulare Genetik
Berlin (Dahlem), Germany

INTRODUCTION

The unraveling of the structure and insight into the details of the function of prokaryotic ribosomes is derived from three technological advances, each an extraordinary achievement: the isolation, purification and characterization of the 55 proteins of *E. coli* ribosomes (see Wittmann, this volume); the reconstitution of active ribosomal subunits from their molecular components, i.e., from RNA and protein (see Nomura and Held, this volume); and finally the application of immunochemical techniques to their analysis (see Stöffler, this volume). The critical achievement was the separation of the ribosomal proteins; it made everything else (including reconstitution and the immunochemistry) possible.

The structure of ribosomal 5S RNA was determined by Sanger and his associates (Brownlee, Sanger and Barrell 1968) some time ago; the sequence of nucleotides in 16S RNA has been all but completed (Fellner

Dedicated to our colleagues Dieter Rot and Oswald Wiener.

et al. 1972) and that of 23S RNA is well under way (see Fellner, this volume). Thus there is a great likelihood that the primary structure of all the components of prokaryotic ribosomes will soon be known. The situation with respect to eukaryotic ribosomes is not nearly so far advanced. The individual proteins of eukaryotic ribosomes have not been isolated (in significant amounts), nor have they been purified or characterized, and that is a major impediment to further progress in analysis of the structure and function of the particles. There is no practical means of reconstituting functional eukaryotic ribosomes from their components, and only very limited use has been made of antibodies to study animal ribosomes.

All that we know of eukaryotic ribosomes supports the contention that they do not differ from prokaryotic ribosomes in a fundamental way. Prokaryotic and eukaryotic ribosomes perform the same function and, as far as one knows, by the same biochemical means. Why then are eukaryotic ribosomes larger, why do they contain a greater number of proteins, and why are the protein and RNA molecules bigger? No one knows. It has been suggested that in eukaryotic cells protein synthesis is more frequently controlled during translation and that some of that control might be mediated by changes in the structure of the ribosome. A larger ribosome with more molecular elements might facilitate the regulation of translation. However the evidence for such control, no less that it is modulated by affecting ribosome structure, is at best far from conclusive.

GENERAL CHARACTERISTICS OF EUKARYOTIC RIBOSOMES

Eukaryotic ribosomes are composed of two subunits, which together contain three molecules of RNA[1] and some seventy proteins. The sedimentation coefficient of the monomer and the subparticles probably varies somewhat from species to species (as do other of the characteristics) but are close to 80 for the ribosome and 60 and 40 for the subunits (Petermann 1964); those numbers are frequently used to designate the particles. The small ribosomal subunit contains one molecule of 18S RNA of molecular weight 0.7×10^6 (Weinberg and Penman 1970) and about 30 proteins whose combined mass is 0.78×10^6 daltons (see below). Thus the calculated molecular weight is about 1.5×10^6, which is just what Hamilton, Pavlovec and Petermann (1971) measured (Table 1). The protein and RNA content of the 40S subunit was determined (Table 2) and the percentage of RNA in the particle reckoned to be 45.5. The buoyant density of the small subparticle was found to be 1.515 (Table 2), from which the percent RNA can be calculated to be 44.9. The values obtained by Hamilton, Pavlovec and Petermann (1971) with subunits prepared in a slightly different way (with urea rather than a high concentration of potassium chloride) were similar (see also Bielka et al. 1968).

[1] The large subunit of some eukaryotic ribosomes may have a fourth molecule of RNA with a sedimentation coefficient of 7 (cf. Maden et al., this volume).

Table 1 Physical properties of eukaryotic ribosomes

	Ribosomal particle[a]			
	80S	60S	40S	References
RNA				
sedimentation coefficient		28S 5S	18S	Perry (1967)
molecular weight		1.7×10^6 3.2×10^4	0.7×10^6	Weinberg and Penman (1970)
Proteins				
number	70	40	30	Sherton and Wool (1972)
molecular weight		1.37×10^6	0.78×10^6	Lin and Wool (1974)
Particle				
molecular weight	4.5×10^6	3×10^6	1.5×10^6	Hamilton, Pavlovec and Petermann (1971)
sedimentation coefficient[b]	80	60	40	Petermann (1964)

[a] The data are for rat liver or HeLa cell ribosomes.

[b] The sedimentation coefficient of eukaryotic ribosomes and ribosomal subunits are actually heterogeneous (cf. Cammarano et al. 1972a,b,c) but are usually referred to by the values given (which are for mammalian ribosomes).

Table 2 Protein and RNA content and buoyant density of rat liver 80S ribosomes and ribosomal subunits

Ribosome particle	Protein[a] (μg per A_{260})	RNA[a] (μg per A_{260})	RNA[b] (%)	Buoyant density[c] (gm·cm^{-3})	RNA[c] (%)
40S	57.3 ± 1.2 (6)	45.5 ± 0.5 (3)	44.3	1.515	44.9
60S	36.2 ± 0.6 (6)	52.1 ± 0.4 (3)	59.0	1.600	59.4
80S	51.7 ± 0.9 (6)	52.3 ± 2.0 (3)	50.3	1.577	55.6

[a] The values are the mean ± S.E.; the number of experiments is given in parentheses.

[b] Calculated from the chemical determinations.

[c] Calculated from the buoyant density in cesium chloride (E. Harris and I. G. Wool, unpublished data) using the formula of Perry and Kelly (1966).

The large subunit of eukaryotic ribosomes contains one molecule each of 28S and 5S RNA; the molecular weights of the RNAs (Table 1) are 1.7×10^6 and about 32,000, respectively (Weinberg and Penman 1970). The particle has about 40 proteins whose mass is approximately 1.37×10^6

daltons. Thus the total mass of the 60S subunit is calculated to be 3.1 ×
10^6: the actual value obtained by Hamilton, Pavlovec and Petermann
(1971) for the particle was 3.0 × 10^6. The 60S ribosome has a buoyant
density of 1.600 (Table 2). The RNA content from the buoyant density
was 59.4%, and the chemical determination gave the same value (Table
2). Thus the large subparticle has an appreciably greater proportion of
RNA than the 40S subunit. Our values are somewhat higher than those
obtained by Hamilton, Pavlovec and Petermann (1971).

Eukaryotic ribosomes are in fact a heterogeneous group; "80S ribosomes" actually range in aggregate mass from about 3.9 (for plants) to
4.55 × 10^6 daltons (for mammals) (Cammarano et al. 1972a, b, c). The
change in ribosome mass is the result of an increase in the size of the large
subunit from 2.4 to 3.05 × 10^6 daltons: the increase is in the mass of RNA
(0.4 × 10^6) and protein (0.2 × 10^6) (Cammarano et al. 1972b; Loening
1968). The 40S particle, on the other hand, has not changed appreciably
in size or conformation during eukaryotic evolution.

THE PROTEINS OF EUKARYOTIC RIBOSOMES

The isolation and purification of ribosomal proteins is a prerequisite for
analysis of the structure and function of the organelle. A start has been
made on their characterization (Cohn and Simpson 1963; Low and Wool
1967; Cohn 1967; Hamilton and Ruth 1967; Bielka and Welfle 1968;
Low, Wool and Martin 1969; Kanai et al. 1969; Westermann, Bielka and
Böttger 1969, 1971; Welfle, Bielka and Böttger 1969; Welfle, Stahl and
Bielka 1971; Gould 1970; King, Gould and Shearman 1971; Huynh-Van-
Tan, Delaunay and Schapira 1971; Martini and Gould 1971; Warner 1971;
Westermann 1971; Welfle 1971a, b; Bickle and Traut 1971; Terao and
Ogata 1972; Westermann and Bielka 1973). The results would indicate
that ribosomal proteins from eukaryotic cells have the same general properties as those from prokaryotic organisms. The number of proteins contained in the subunits of eukaryotic ribosomes and their molecular weight
has been determined by two-dimensional polyacrylamide gel electrophoresis. It is certain that the proteins are heterogeneous—that each ribosome
has approximately 70 unique proteins (see Table 3 for references).

A practical procedure for the isolation of pure eukaryotic ribosomal
proteins in significant amounts has not yet been devised. There is a claim
(Westermann and Bielka 1973) for the isolation and characterization of
24 of the proteins of the small subunit of rat liver ribosomes; however, the
report must be discounted since it lacks data on the chromatography of
the proteins and evidence of their purity. Amino acid compositions and
tryptic peptides (which are given) hardly suffice in the absence of rigorous
evidence of purity. The purification carried out by Terao and Ogata (1972)
is far better documented, but only 12 of the small subunits proteins were
isolated.

Table 3 Number of proteins in eukaryotic ribosomes and ribosomal subunits

| Source of ribosomes | Number of proteins | | | Method | Reference |
	40S	60S	80S		
Rat liver	30	40	70	1	Sherton and Wool (1972, 1974)
Rat liver	27	34	61	1	Welfle, Stahl and Bielka (1971)
Rat liver	31	39	70	1	Welfle, Stahl and Bielka (1972)
Rat liver	—	38	—	2	Martini and Gould[a]
Rat muscle	31	38	69	1	Sherton and Wool (1974)
Rabbit liver	—	—	75	1	Huynh-Van-Tan, Delaunay and Schapira (1971)
Rabbit liver	30	38	68	2	Martini and Gould[a]
Rabbit reticulocytes	32	38	70	2	Martini and Gould (1971)[a]
Rabbit reticulocyte	32	39	71	1	Howard, Traut and Traugh
Rabbit reticulocyte	33	40	73	1	Chatterjee, Kazemie and Matthaei (1973)
Yeast	30	50	80	3	Warner (1971)
Mouse liver	29	38	67	2	Martini and Gould[a]
Chicken liver	30	36	66	2	Martini and Gould[a]
Xenopus liver	29	—	—	2	Martini and Gould[a]

Methods: (1) Two-dimensional polyacrylamide gel electrophoresis in urea; (2) two-dimensional polyacrylamide gel electrophoresis in sodium dodecyl sulfate; (3) one-dimensional polyacrylamide gel electrophoresis.

[a] Reported at the Cold Spring Harbor meeting, September 1973.

Number of Proteins in Eukaryotic Ribosomes and Ribosomal Subunits

The 40S subunits of rat liver ribosomes contain 30 proteins and the 60S subunit 40 (Figure 1 and Sherton and Wool 1972, 1974). Since the proteins of the two subunits are very likely unique (see below), the 80S ribosome monomer contains 70 different proteins.[2] Rat liver ribosome monomers prepared in a manner likely to remove initiation and elongation factors, nascent peptide and protein bound to the ribosome factitiously during isolation of the particle actually contained the 70 proteins of the sub-

[2] For reasons which we give later, we now include L20 with the small subunit proteins designating it S31; however we no longer consider S8 a small subunit protein since it is likely to be the same as L13.

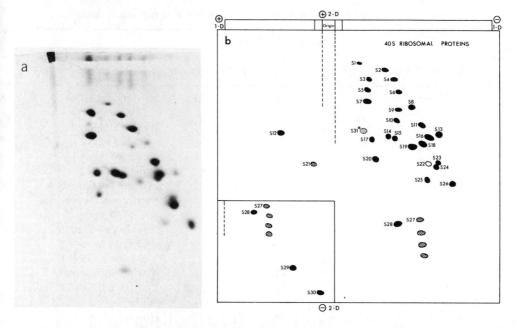

Figure 1   Two-dimensional electropherograms and schematics of rat liver ribosomal proteins. The electropherograms are of proteins from (*a*) the 40S subunit, (*c*) the 60S subunit, and (*e*) 80S ribosomes. The schematics are of (*b*) 40S subunit proteins, (*d*) 60S proteins, and (*f*) 80S proteins. The *solid* spots on the schematics were always seen; the *cross-hatched* spots either varied in location or intensity of staining; the *open* spots were only seen when the conditions of electrophoresis were changed. The diagrams include spots that are difficult to see in the photographs. The intensity of the staining is not represented with fidelity. The conditions of electrophoresis are described in detail in Sherton and Wool (1972).

units and three additional spots (albeit the three stained lightly). The identity of those three proteins is not known, but there is no reason to believe they are ribosomal proteins since they are not present in either subunit and the subunits function normally, at least in the translation of polyuridylic acid (Martin and Wool 1968; Martin et al. 1969) and encephalomyocarditis RNA (Leader et al. 1972).

Others have estimated the number of proteins in eukaryotic ribosomes and in ribosomal subunits (Table 3), and the agreement is very good. For example, Martini and Gould (1971) reported that rabbit reticulocyte ribosomes have 70 proteins: 32 in the small subunit and 38 in the large. Traut, Howard and Traugh (1974) had very similar results: 32 proteins in the 40S subunit and 39 in the large, for a total of 71. Welfle, Stahl and Bielka (1971) had originally estimated that the 40S subunit had 27 proteins and the 60S 34. Moreover, they had found that 15 proteins were actually con-

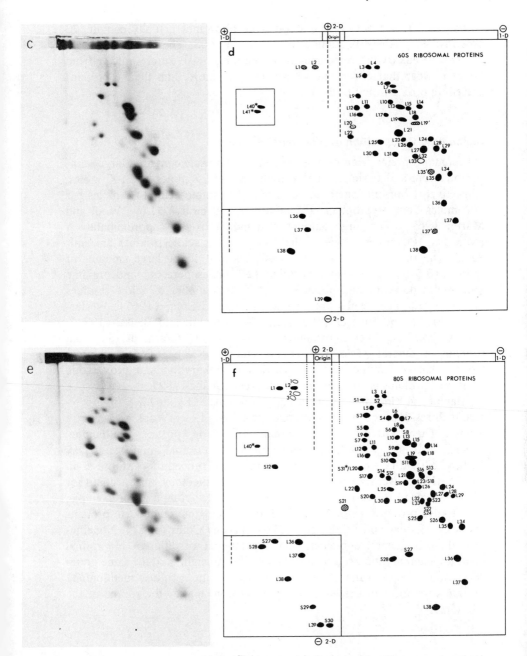

tained in, or shared by, both subunits. Their conclusions were clearly in error, the result of analyzing impure subunit fractions prepared with EDTA (an unsatisfactory procedure). After being appraised of Sherton's finding (Sherton and Wool 1972), they amended their estimate (Welfle, Stahl and

Bielka 1972) and now find that rat liver ribosomes actually contain 70 proteins: 31 in the 40S subunit and 39 in the 60S. The exact number of proteins in eukaryotic ribosomes is not certain, but is surely not far from 70. It is clear that eukaryotic ribosomes have more than the 54 proteins present in prokaryotic ribosomes.

Amino Acid Composition of Ribosomal Proteins

Ribosomal proteins from rat liver and skeletal muscle contain a large number of acidic and basic residues (about 18 and 27 moles per 100 moles, respectively) but only small amounts of sulfur amino acids (1.7 moles per 100 moles) and tryptophan (0.5 mole per 100 moles) (Low, Wool and Martin 1969). The large amount of amide nitrogen (approximately 5 moles per 100 moles) suggests that many of the acidic residues are amidated. The ribosomal proteins are basic because of the large amounts of lysine (10.6 moles percent), histidine (2.5 moles percent) and arginine (9.6 moles percent) and because a significant number of acidic residues are amidated. The results are similar to those of Cohn (1967), Westermann, Bielka and Böttger (1971) and Westermann and Bielka (1973), who analyzed rat liver ribosomal proteins, and are typical of ribosomal proteins in general (Ts'o, Bonner and Dintzis 1958). It is indeed noteworthy that the amino acid composition of eukaryotic and prokaryotic ribosomal proteins is not remarkably different (Spahr 1962).

Alanine, glycine and serine account for most (e.g., 67%) of the N-terminal residues of liver ribosomal proteins (Low, Wool and Martin 1969). There are also lesser amounts of valine, proline, methionine, aspartic acid and glutamic acid. Some of the N-terminal amino acids may be blocked and hence not detectable by the usual procedures (Moore et al. 1968) which, together with the limitations of the methods, may account for discrepancies in the results that have been reported; compare, for example, the relative proportions of N-terminal amino acids of reticulocyte proteins given by Mathias and Williamson (1964) and by Cohn (1967). Nonetheless, there is general agreement that the number of different N-terminal residues in eukaryotic ribosomal proteins is five to eight, and the same group of six amino acids (alanine, glycine, serine, valine, proline, methionine) accounts for most of the N-terminal residues no matter the source of the ribosomes.

Molecular Weights of Rat Liver Ribosomal Proteins

The proteins of rat liver ribosomal subunits have been separated by two-dimensional polyacrylamide gel electrophoresis in urea; the protein spots were cut out and compared with molecular weight standards by one-dimensional electrophoresis in polyacrylamide gels containing sodium

Table 4   Molecular weights of liver ribosomal proteins

<table>
<tr><td colspan="7" align="center">Molecular weight ($\times 10^{-3}$)</td></tr>
<tr><td colspan="3" align="center">40S subunit</td><td colspan="4" align="center">60S subunit</td></tr>
<tr><td>S1</td><td>44.0</td><td>S22</td><td>—</td><td>L1</td><td>38.6</td><td>L22</td><td>—</td></tr>
<tr><td>S2</td><td>41.0</td><td>S23</td><td>23.9</td><td>L2</td><td>32.4</td><td>L23</td><td>23.3</td></tr>
<tr><td>S3</td><td>38.1</td><td>S24</td><td>22.6</td><td>L3</td><td>53.0</td><td>L24</td><td>24.0</td></tr>
<tr><td>S4</td><td>35.3</td><td>S25</td><td>22.1</td><td>L4</td><td>53.7</td><td>L25</td><td>23.7</td></tr>
<tr><td>S5</td><td>29.8</td><td>S26</td><td>19.1</td><td>L5</td><td>45.8</td><td>L26</td><td>25.6</td></tr>
<tr><td>S6</td><td>38.5</td><td>S27</td><td>16.9</td><td>L6</td><td>48.5</td><td>L27</td><td>21.9</td></tr>
<tr><td>S7</td><td>31.3</td><td>S28</td><td>11.5</td><td>L7</td><td>38.3</td><td>L28</td><td>22.7</td></tr>
<tr><td>S8[a]</td><td>32.5</td><td>S29</td><td>10.0</td><td>L8</td><td>35.8</td><td>L29</td><td>24.0</td></tr>
<tr><td>S9</td><td>27.2</td><td>S30</td><td>12.4</td><td>L9</td><td>31.8</td><td>L30</td><td>21.7</td></tr>
<tr><td>S10</td><td>27.3</td><td>S31[a]</td><td>25.3</td><td>L10</td><td>34.8</td><td>L31</td><td>20.8</td></tr>
<tr><td>S11</td><td>26.3</td><td></td><td></td><td>L11</td><td>26.5</td><td>L32</td><td>20.7</td></tr>
<tr><td>S12</td><td>38.5</td><td></td><td></td><td>L12</td><td>23.0</td><td>L33</td><td>22.0</td></tr>
<tr><td>S13</td><td>21.2</td><td></td><td></td><td>L13[a]</td><td>33.9</td><td>L34</td><td>14.5</td></tr>
<tr><td>S14</td><td>24.9</td><td></td><td></td><td>L14</td><td>32.3</td><td>L35</td><td>19.7</td></tr>
<tr><td>S15</td><td>25.3</td><td></td><td></td><td>L15</td><td>30.0</td><td>L36</td><td>18.1</td></tr>
<tr><td>S16</td><td>20.5</td><td></td><td></td><td>L16</td><td>24.8</td><td>L37</td><td>16.8</td></tr>
<tr><td>S17</td><td>22.5</td><td></td><td></td><td>L17</td><td>29.5</td><td>L38</td><td>10.5</td></tr>
<tr><td>S18</td><td>21.5</td><td></td><td></td><td>L18</td><td>29.0</td><td>L39</td><td>10.0</td></tr>
<tr><td>S19</td><td>20.5</td><td></td><td></td><td>L19</td><td>32.0</td><td>L40</td><td>25.5</td></tr>
<tr><td>S20</td><td>10.1</td><td></td><td></td><td>L20[a]</td><td>27.0</td><td>L41</td><td>25.5</td></tr>
<tr><td>S21</td><td>18.8</td><td></td><td></td><td>L21</td><td>28.0</td><td></td><td></td></tr>
</table>

40S: $\overline{M}_n$ = 25,400   60S: $\overline{M}_n$ = 28,000

$\overline{M}_w$ = 28,400 $\overline{M}_w$ = 31,500

ΣMW = ~0.78 $\times$ 10^6 ΣMW = ~1.37 $\times$ 10^6

[a] S8 and L13, and S31 and L20 are likely to be the same proteins.

dodecyl sulfate (Lin and Wool 1974). From the results, we were able to estimate the molecular weights of the ribosomal proteins (Table 4). The number average molecular weight for 29 of the 30 proteins[3] of the 40S subunit was 25,400 and for 40 of the 41 proteins[2] of the 60S subunit 28,000. The weight average molecular weight for the proteins of the 40S particle was 28,400 and for the 60S particle 31,500. The total daltons of protein in the 40S subunit is approximately 0.78 $\times$ 10^6 and for the 60S subunit about 1.37 $\times$ 10^6. The amounts are slightly in excess of the estimates from physicochemical data (King, Gould and Shearman 1971): 0.64 $\times$ 10^6 and 1.09 $\times$ 10^6 for the 40S and 60S particles, respec-

[3] We do not yet have an estimate of the molecular weight of S22 or L22. In computing the mass of protein in the subunits, we assumed a molecular weight of 20,000 for S22 and 25,000 for L22.

tively. However the discrepancy is not so great, given the uncertainty of the determinations, as to necessitate postulating that the proteins are present in other than molar amounts. Of course the data are far from constituting proof of that proposition.

Two other groups reported data on the molecular weights of eukaryotic ribosomal proteins at the Cold Spring Harbor meeting: Howard et al. from the University of California, Davis, and Martini and Gould from London. It is unfortunate that the numbers used to designate ribosomal proteins by the several groups are not uniform and hence the results are not easy to compare directly. In this review the nomenclature is from Sherton and Wool (1972, 1974; and Figure 1). However the general impression is of relatively good agreement. Howard et al. made their determination on proteins from reticulocyte ribosomal subunits by the same method as Lin and Wool (1974). The number average molecular weight for the 40S subunit was 22,000 and for the 60S, 27,000 (slightly lower than our calculations). Martini and Gould have determined the molecular weights of the proteins of liver ribosomes from several species (mouse, chicken, rabbit and Xenopus) and from two rabbit tissues (reticulocytes and liver). The estimations were made directly since the second dimension of their gel electrophoresis was in SDS. It is hard to epitomize the results, but the molecular weights were probably in general somewhat lower than Lin and Wool (1974) or Howard et al. For example, the number average molecular weight for reticulocyte ribosomal proteins was 20,300 for the 40S subunit, and for the 60S subunit, 24,600. Westermann and Bielka (1973) have reported the molecular weight for 24 of the proteins of the 40S subunit and Terao and Ogata (1972) for 12 of them.

It is obvious that number and weight average molecular weights of eukaryotic ribosomal proteins are greater than for prokaryotes, which are 14,000 and 21,000 for the 30S subunit and 16,000 and 21,000 for the 50S (Bickle and Traut 1971). Thus eukaryotic ribosomes not only have more proteins, but they are larger.

Reconstitution of Eukaryotic Ribosomes

A practical, reproducible means of reconstituting functional eukaryotic ribosomes from their components has not yet been devised, although there have been some reports of success (Lerman 1968; Terao and Ogata 1971; Reboud et al. 1972). The failure, which has not been for lack of effort, has hindered progress in analyzing the function of individual molecular components, especially of the proteins. Of course, no one knows why reconstitution has failed. However, it is possible that the failure reflects the fact that eukaryotic ribosomes are not assembled on mature 18 and 28S RNA, but rather that assembly is initiated on 45S precursor ribosomal RNA (Perry 1967; Darnell 1968; Weinberg and Penman 1970). It may be that 45S RNA has information required for proper assembly, information that is lacking in 18 and 28S RNA. Unfortunately, the possibility is hard to test since appreciable amounts of 45S RNA are difficult to come by, and if the RNA were available, one still would have to face the formidable task of fashioning the precursor ribonucleoprotein particle into functional ribosomal subunits.

Ribosomal Proteins of Different Species and Tissues

Some of the proteins of eukaryotic ribosomes are peculiar to a species. There are, for example, marked differences in the ribosomal proteins of so distantly related species as rat and protozoa (Martin and Wool 1969). There are also distinct, definite differences in the number and size of ribosomal proteins from animals and plants (Cammarano et al. 1972a, b, c; Jones, Nagabhushan and Zalik 1972; Gualerzi et al. 1974). It seems unlikely that ribosomal proteins have been so assiduously conserved as, say, histones. Whether more closely related species differ is still moot: Delaunay, Cruesot and Schapira (1973) found no definite difference in the pattern of ribosomal proteins on two-dimensional electropherograms between man, rat, birds and lizards; however, when Noll and Bielka (1970b) compared ribosomal proteins from even more similar species (rat, mouse, bovine, chicken) by immunochemical techniques, they found definite differences. It may be that sensitive techniques will be required for an exact comparison of species differences in ribosomal proteins. The problem will be discussed later in this chapter.

It is not certain whether the proteins of ribosomes in separate organs of a single animal differ. In general, one-dimensional polyacrylamide gel electrophoresis of eukaryotic ribosomal proteins from divers tissues or organs (Low and Wool 1967; Mutolo et al. 1967; Bielka and Welfle 1968; Di Girolamo and Cammarano 1968; Low, Wool and Martin 1969; Houssais 1971; MacInnes 1972) and from various stages of development (Mutolo et al. 1967; Welfle and Bielka 1968; MacInnes 1973) has failed to show any qualitative differences, although quantitative alterations have been observed (Low and Wool 1967; Low, Wool and Martin 1969; Houssais 1971; MacInnes 1972, 1973). An exception is the report of an increase in the number of ribosomal proteins in membrane-bound as opposed to free liver ribosomes (Burka and Bulova 1971)—a difference which could have been due to contamination of the bound ribosomes with membrane protein.

Two-dimensional polyacrylamide gel electrophoresis is extremely useful in comparing eukaryotic ribosomes since each protein is displayed as a single zone (Martini and Gould 1971; Huynh-Van-Tan, Delaunay and Schapira 1971; Bielka, Stahl and Welfle 1971; Welfle, Stahl and Bielka 1971, 1972; Sherton and Wool 1972, 1974; Hultin 1972; Hultin and Sjöqvist 1972; Delaunay and Schapira 1972; Delaunay, Mathieu and Schapira 1972; Lambertsson 1972; Jones, Nagabhushan and Zalik 1972; Rodgers 1973; Chatterjee, Kazemie and Matthaei 1973; Gualerzi et al.

1974). The technique was employed by Huynh-Van-Tan, Delaunay and Schapira (1971) and Delaunay, Mathieu and Schapira (1972) to compare the proteins of 80S ribosomes from rabbit reticulocytes, liver, kidney and caecal appendix; a number of differences (although not precisely defined) were found in the proteins of the ribosomes from the four tissues, particularly between the proteins of reticulocyte ribosomes and those of ribosomes from other tissues. Rodgers (1973) has compared the proteins of 80S ribosomes from several mouse tumors and from liver, kidney, intestine and brain by two-dimensional gel electrophoresis. He found differences in the apparent quantity of ribosomal proteins of the normal tissues as well as an additional protein in brain ribosomes. The mouse tumor ribosomes had a number of unique proteins and lacked some present in the ribosomes of normal tissues. On the other hand, no differences were found in the proteins of ribosomes from normal liver and hepatomas of rats (Bielka, Stahl and Welfle 1971; Delaunay and Schapira 1972). The possibility of differences in the proteins of ribosomes from normal and malignant cells is discussed again later (see section on antibodies).

The proteins of rat skeletal muscle and liver ribosomes were compared by two-dimensional polyacrylamide gel electrophoresis (Sherton and Wool 1974). Some differences were found, but most of the apparent variance could be accounted for either by differential partition of selected proteins to one or the other of the two subunits (see below) or by differences in the methods required to prepare pure muscle ribosomes in high yield and those used to prepare liver ribosomes. If liver and muscle ribosomes and ribosomal subunits were prepared by the same method, there were few, if any, differences in the proteins. It would be surprising if one were to discover significant differences in the two-dimensional electrophoretic pattern of ribosomal proteins of different tissues of the same species since no differences are found in the ribosomal proteins of man and lizard (Delaunay et al. 1973). We cannot be certain that the proteins of liver and muscle ribosomes are identical, but it is unlikely that there are more than minor variations, hardly sufficient to account for the marked differences in the character of protein synthesis in the tissues.

Partition of Proteins between Subunits

The comparison of the proteins of liver and muscle ribosomes led to a consideration of the possibility that certain ones might be partitioned to one or the other or both subparticles when ribosome couples are disassociated (Sherton and Wool 1974). The observations are of import not only for the bearing they have on the presumptive localization of the proteins, e.g., at the ribosome subunit interface, but also because they indicate that some proteins may be shared by the subparticles. The latter has implications for the role of the proteins in ribosome function. Four proteins, S8, L13, L20 and S31, partition between subunits.

The position on two-dimensional electropherograms of the muscle small subunit protein S31 corresponded exactly with that occupied by the liver large subunit protein L20. Whether S31 occurred with the small subunit or L20 with the large subparticle was conditioned by the method used to prepare the particles, especially the ratio of potassium to magnesium. Thus it seems reasonable to conclude, although the conclusion cannot be supported with chemical evidence, that the proteins in zones S31 and L20 are the same. It would follow that the partition of the protein between subunits is conditioned by the concentration of ions in which the subparticles are prepared. The behavior of the protein may be a reflection of its topographical location: S31/L20 may occupy the interface between the two subunits, and hence the protein might actually be shared by the two subparticles. It has been shown (Stöffler, this volume) that the spots S20 and L26 on two-dimensional electropherograms of *Escherichia coli* ribosomal proteins are actually a single, identical protein, although the former occurs with the small subunit and the latter with the large. Moreover it has been established that S20 is an interface protein (Stöffler, this volume). The analogy with the results for the eukaryotic proteins is striking.

Similar reasoning and arguments can be applied to the zones S_m1 and L31. S_m1, a component of the 40S subunit of muscle ribosomes, falls in a position on electropherograms occupied by L31. Although L31 is also found with the large subunit proteins in muscle, it still may be the same protein as S_m1; the single protein may be apportioned between the two subunits of muscle ribosomes, whereas it goes exclusively with the large subunit of liver ribosomes.

In the analysis of the proteins of liver ribosomes and ribosomal subunits (Sherton and Wool 1972), only a single pair of proteins, S8 and L13, was not separated in any of the conditions used for electrophoresis. It was assumed that S8 and L13 might have the same size and charge, but unique sequences of amino acids. It is now likely that S8 and L13 are in fact the same protein and that a portion of the protein partitions with the small subunit while the remainder is found with the large subparticle (like *E. coli* S20 and L26). It needs to be noted that the molecular weights of S8 and L13, and of L20 and S31, are very similar (Table 4). If we reassign the proteins—S8/L13 to the large subunit and L20/S31 to the small subunit—then the number of proteins in the 40S subparticle is 30 and the number in the 60S particle is 40. Differential apportionment of specific proteins between subunits during their preparation could account for minor variations in the number of proteins assigned to each subunit.

CHEMICAL MODIFICATION OF EUKARYOTIC RIBOSOMAL PROTEINS

There are a number of ways in which the structure, and hence the function, of ribosomes might be altered: by addition or deletion of proteins which amplify ribosome activity (Kurland 1970; Van Duin and Kurland 1970;

Randall-Hazelbauer and Kurland 1972), by a change in the conformation
of a ribosomal protein or of the particle itself (Zamir, Miskin and Elson
1971; Chuang and Simpson 1971); or by chemical modification, i.e.,
phosphorylation (Kabat 1970; Eil and Wool 1971; Blat and Loeb 1971),
acetylation (Deusser 1972; Liew and Gornall 1973) or methylation
(Terhorst et al. 1973), of ribosomal proteins. There is evidence that each
of those changes can occur. It has been suggested that modification of its
proteins may be the one characteristic that distinguishes eukaryotic ribosomes; indeed, that phosphorylation of ribosomal proteins may be the
means by which control of translation is achieved in animal cells. (For
phosphorylation of ribosomal proteins, see chapter by Krystosek et al., this
volume.)

Phosphorylation of Eukaryotic Ribosomal Proteins

The proteins of rat liver ribosomes and ribosomal subunits can be phosphorylated in vitro by cyclic AMP-activated protein kinases isolated from
the cytosol (Loeb and Blat 1970; Kabat 1971; Walton et al. 1971; Li and
Amos 1971; Yamamura et al. 1972; Eil and Wool 1971, 1973a, b; Stahl,
Welfle and Bielka 1972; Delaunay et al. 1973; Martini and Gould 1973;
Traugh, Mumby and Traut 1973; Traugh and Traut 1974). The enzymes
catalyze the transfer of the terminal phosphate of ATP to serine and
threonine residues in a number of proteins in the 40S and 60S subunits
(Eil and Wool 1971, 1973a). The enzymes are quite likely to be the same
ones found by Chen and Walsh (1971), by Chambaut, Leray and Hanoune
(1971) and by Yamamura et al. (1972) to phosphorylate a number of
substrates, among which are histones and ribosomal proteins (as well as
glycogen phosphorylase b kinase, glycogen synthetase and lipase). In no
case has the enzyme that phosphorylated ribosomal proteins in vitro been
shown to be specific for that substrate; quite to the contrary, a lack of substrate preference is a prominent characteristic of the enzyme. It is possible,
however, that the enzyme has specificity in the cell because of some unique
feature of the substrate or the enzyme in vivo.

One might ask whether there is a unique ribosomal protein kinase bound
to the particle; whether, in fact, protein kinase might not be a ribosomal
protein. There are a number of reports (Kabat 1971; Eil and Wool 1971;
Li and Amos 1971; Jergil 1972; Traugh and Traut 1974) of protein kinase
activity associated with ribosomes. However in each case, the enzyme can
be separated from the ribosome by treatment with high ionic strength
buffers (0.5–1.0 M KCl, for example). A number of intracellular enzymes
are adventitiously bound to ribosomes (Elson 1958; Nue and Heppel 1964;
Hardy and Kurland 1966), and there is no certain evidence that the ribosome-associated protein kinase in eukaryotic cells is not a cytosol enzyme
casually bound to the particle.

It is important that criteria for phosphorylation of ribosomal proteins be established. We suggest the following three as a minimum: (1) that the phosphate bound to the ribosome survive repeated centrifugal washing of the particles in buffers containing high concentrations of monovalent cation and low concentrations of magnesium, procedures which remove proteins fortuitously bound to ribosomes; (2) that the phosphate should be shown to be covalently bound to serine or threonine (or both) in ribosomal proteins; and (3) that the covalently bound phosphate should be shown to migrate with ribosomal proteins during electrophoresis in polyacrylamide gels. The satisfaction of these criteria is particularly important for experiments undertaken in vivo since the chance of contamination of ribosomes with ^{32}P-labeled compounds is especially great. In only a small number of the reports (Kabat 1970, 1971; Eil and Wool 1971, 1973a) of the phosphorylation of ribosomal proteins have all three criteria been observed.

The Function of Phosphorylated Ribosomes

It is implicit in a large number of experiments that the function of eukaryotic ribosomes might be regulated by reversible phosphorylation of their proteins. The evidence for that impression, however, is at best circumstantial. A direct test of the proposition was undertaken by Eil and Wool (1973b). Although all the functions of the particles were not analyzed, a number were examined, and no appreciable and consistent difference in the activity of phosphorylated and nonphosphorylated ribosomes was found. The ability of phosphorylated ribosomes to use the elongation factors (EF-1 and EF-2) was assessed by measuring the synthesis of polyphenylalanine at high concentrations of magnesium. The assay provides a gauge of the ability of ribosomes to bind aminoacyl-tRNA, to catalyze peptide bond formation, and to carry out the translocation of peptidyl-tRNA. No difference in the competence of phosphorylated and nonphosphorylated ribosomes for the elongation reactions in polyphenylalanine synthesis was detected. It is important that the test was made in two ways, i.e., where ribosomes restricted synthesis and where elongation factors were limiting. The latter tested the ability of phosphorylated ribosomes to use reduced amounts of factors and is probably a more discriminative trial of the effect of phosphorylation on ribosome function.

The ability of phosphorylated ribosomes to use initiation factors was next surveyed. Modulation of ribosome function as a means to regulate protein synthesis, if it occurs, is more likely to affect initiation than the other partial reactions, because the most efficient means to control translation is to influence initiation. Moreover, it is a post-transcriptional means of controlling the synthesis of specific proteins. It is perhaps worthy of note that those small effects of phosphorylation that were observed were in initiation, however, they were neither appreciable nor consistent (Eil and Wool 1973b).

As with all negative results, the failure to observe a meaningful effect of phosphorylation on ribosomal function may well have been due to unsuitable experimental conditions.

There has been only the one test of the effect of phosphorylation on the function of eukaryotic ribosomes in vitro (Eil and Wool 1973b). There are, on the other hand, several reports of an association of phosphorylation with an alteration of ribosome function in vivo (Kabat 1970, 1972; Blat and Loeb 1971; Li and Amos 1971; Bitte and Kabat 1972; Correze, Pinell and Nunez 1972). Correlation is not synonymous with causation, and in no instance has the function of the ribosome actually been shown to be altered by phosphorylation of a protein.

If it is assumed for the moment that eukaryotic ribosomes are phosphorylated in vivo and that phosphorylation alters their activity, how then do we rationalize the failure to detect the change in vitro? It is possible that liver ribosomes are nearly maximally phosphorylated when they are isolated and that the phosphorylation that occurs in vitro is, therefore, only a small fraction of that which can occur. If liver ribosomal subunits are treated with bacterial alkaline phosphatase or with ascites cell cytosol (which is rich in ribosomal phosphoprotein phosphatase), the treatment changes neither the ability of the particles to serve as a substrate for protein kinase nor the activity of the particles. It is conceivable that the enzymes used in the in vitro experiments were not the physiologically relevant ribosomal protein kinases, and thus the proper serine and threonine residues were not phosphorylated. Moreover, it may be necessary to phosphorylate and dephosphorylate ribosomal proteins cyclically to alter function, and that may require the coupling of ribosomal phosphoprotein phosphatase to the kinase; one or both enzymes may have been lacking from the in vitro assay.

It is a distinct possibility that the failure to detect an effect of phosphorylation on ribosome activity was the result of omitting to test the proper function. Among the many functions that were not analyzed are the translation of specific cellular mRNAs, the binding of ribosomes to the endoplasmic reticulum, and the termination of protein synthesis. The suggestion that phosphorylation of ribosomal proteins might affect those functions is not frivolous. It should be kept in mind that while negative results are only rarely decisive, they are not infrequently correct; if phosphorylation does not affect ribosome function (a conclusion supported by all the available evidence), then negative results are all that can be expected.

The Phosphorylation of Rat Liver Ribosomes In Vivo

The failure to find an effect of phosphorylation in vitro with protein kinase and ATP on the function of liver ribosomes raised a serious question as to whether liver ribosomal proteins were, in fact, phosphorylated in vivo. It is perhaps apropos that bacterial ribosomes seem not to be phosphorylated.

E. coli ribosomes and isolated proteins can be phosphorylated in vitro with rabbit muscle protein kinase (Traugh and Traut 1972); however, Gordon (1971) detected no incorporation of [^{32}P]orthophosphate into ribosomal proteins during growth of bacteria.

An attempt was made to determine if phosphorylation of liver ribosomal proteins occurred in vivo and if so, the identity of the proteins (Gressner and Wool 1974). Since phosphorylation might require the synthesis of new ribosomes, experiments were also carried out with hepatectomized rats. At about 24 hours after removal of 70% of the liver, there is an order of magnitude increase in the synthesis of hepatic ribosomes (Chaudhuri and Lieberman 1968). Hepatectomized rats were injected with 2 mCi of [^{32}P]orthophosphoric acid and the liver ribosomes were isolated and subunits prepared: the specific activity of the proteins from regenerating liver was eleven times greater than from sham-operated animals.

The ribosomal pellets contained a considerable amount of radioactivity. However, when the particles were rigorously purified—by treatment with high concentrations of potassium chloride to remove contaminating non-ribosomal phosphoproteins, with hot and cold trichloroacetic to remove nucleic acids and various small molecules that contain ^{32}P, and with organic solvents to remove phospholipids—very little of the radioactivity remained, indeed not enough to say whether phosphorylation of ribosomal proteins had actually occurred. The results emphasize the care that must be taken in in vivo experiments in which the labeling of macromolecules with ^{32}P is studied.

The importance of the experiments with hepatectomized rats was that they enabled us to prepare ribosomal proteins containing sufficient radioactivity so as to make analysis by two-dimensional polyacrylamide gel electrophoresis possible. Although the total number of counts was small, we reckoned that if they were in a few proteins it might be possible to identify them. [^{32}P]Orthophosphoric acid was injected into hepatectomized rats and proteins were extracted from liver ribosomal subunits; the proteins were separated by two-dimensional polyacrylamide gel electrophoresis and radioautographs were made of the gels. A single radioactive zone was found on the radioautographs of 40S ribosomal proteins (Figure 2a). When the radioautograph was compared with a stained gel (Figure 2b), it was obvious the zone included S6 but also had a tail extending toward the origin or the anode. The intensity of the zone on the radioautograph (and hence the radioactivity) described a gradient which was greatest in the anodal portion and least where one would expect the normal S6. Protein S6 on the stained gel resembled the radioactive zone, except that the intensity of staining was just the reverse of the radioactivity. No radioactive zones were found on radioautographs of 60S subunit proteins prepared in the same way. Incidentally, it was shown in a separate experiment that S6 is phosphorylated in normal rat liver ribosomes.

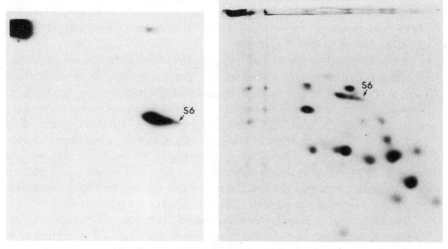

Figure 2 Radioautograph and electropherogram of ^{32}P-labeled 40S ribosomal proteins from the liver of partially hepatectomized rats separated by two-dimensional polyacrylamide gel electrophoresis. [^{32}P]orthophosphoric acid (2 mCi) was injected into rats 24 hr after they had been hepatectomized, and the animals were killed 30 min later. Liver ribosomes were prepared, subunits made from the ribosomes, and the proteins extracted from the 40S particles. The proteins were separated by two-dimensional polyacrylamide gel electrophoresis (*b*) and (*a*) radioautographs were made of the gel (Gressner and Wool 1974). The arrow indicates the approximate normal position of S6.

The results to this point supported the proposition that there were a number of phosphorylated derivatives of S6, the derivatives perhaps containing increasing numbers of radioactive phosphoserines. The radioactivity was likely to be in protein since it was resistant to RNase and DNase but sensitive to trypsin and alkaline phosphatase. Furthermore, when radioactive 40S ribosomal subunit protein (labeled in vitro with ^{32}P) was hydrolyzed with 6 N HCl for 7 hours at 110°C and the amino acids separated by paper electrophoresis and radioautographs made, [^{32}P]-*o*-phosphoserine was found.

The specific activity of S6 and its derivatives was determined. Proteins of the 40S subunit were labeled in vivo with ^{32}P and then in vitro by reductive alkylation with [^{14}C]formaldehyde. The area containing S6 and its derivatives was cut from an electropherogram; 1-mm slices were made and the ratio of ^{32}P to ^{14}C determined. The ^{32}P radioactivity should give the amount of phosphoserine and the ^{14}C the amount of S6 protein. The results clearly indicated a gradient of [^{32}P]phosphate extending from the anode to the normal position of S6—the most anodal fraction had the greatest amount of ^{32}P and the least ^{14}C radioactivity. There were at least three distinct species of S6. Finally, it was shown that if ^{32}P-labeled 40S ribosomal protein was treated with alkaline phosphatase and then separated by two-dimensional polyacrylamide gel electrophoresis, the derivatives disappeared and normal S6 was regenerated.

Only a single liver ribosomal protein (S6) is phosphorylated in vivo,[4] whereas a large number are phosphorylated in vitro by protein kinase (Eil and Wool 1971, 1973a). The disparity suggests that either another enzyme catalyzes phosphorylation in vivo or the conditions in vitro do not approximate the physiological circumstances. The findings clearly are different than those of Kabat (1970, 1971, 1972); he found that five reticulocyte ribosomal proteins were phosphorylated in vivo and that the same proteins were phosphorylated in vitro. Moreover, the pattern of phosphorylation of ribosomes from sarcoma 180 tumor cells resembled that for reticulocyte ribosomes (Bitte and Kabat 1972). The reason for the difference in the phosphorylation of reticulocyte and liver ribosomal proteins in vivo is not known.

Kabat (1972) had first reported that cyclic AMP did not affect the phosphorylation of reticulocyte ribosomal proteins; however, he and his associates now find (Cawthon et al. 1974) that the cyclic nucleotide actually stimulates the phosphorylation of one among the five phosphoprotein bands (designated protein II)—there is, of course, no certainty that the bands on a one-dimensional sodium lauryl sulfate polyacrylamide gel are single proteins. The phosphoprotein (band) is associated with the small subunit of reticulocyte ribosomes, thus raising a suspicion it might be S6 and its derivatives. The derivatives and S6 would form a single band if the analysis was by electrophoresis in sodium lauryl sulfate. However, protein II from reticulocyte ribosomes has a molecular weight of 27,500, which is considerably less than the 36,000 of S6 from liver ribosomes. For the moment, then, there is no reason to believe the phosphoproteins are the same.

A single liver ribosomal protein band was also found by Loeb and Blat (1970) after administration of [^{32}P]orthophosphoric acid to rats. The report was the first to suggest that eukaryotic ribosomal proteins were phosphorylated; however, no precaution was taken to ensure that the phosphoprotein was in fact a ribosomal protein. Nonetheless the band Loeb and Blat found (1970) may well be S6. Certainly our results accord with their finding.

We do not know the significance of the phosphorylation of S6 and of the phosphorylated derivatives of the protein, although it is obviously important to determine their function. One possibility is that S6 is phosphorylated prior to, or during, assembly of ribosomes, that phosphorylation

[4] We recognize the possibility, although we think it less likely, that additional ribosomal proteins are phosphorylated, but that the phosphate residues are hydrolyzed by phosphatases during the preparation of ribosomes.

in some way facilitates that process, and that the phosphate groups are removed after transport to the cytoplasm. It is conceivable that the phosphorylation of S6 serves a function in the transport of ribosomes from nucleus to the cytoplasm. It may be that during regeneration the process of ribosome synthesis is sufficiently accelerated so that intermediates in the removal of phosphates from S6 can be detected. It is important that in normal rat liver S6 is phosphorylated and trace amounts of the derivatives are present, although they can only be detected on the radioautographs of ^{32}P-labeled proteins. In normal rat liver S6 is more radioactive than the derivatives, even though the latter presumably contain a greater number of phosphoserine residues, because there is so much more S6 than the derivatives. The effect of hepatectomy, it would seem, is to decrease the relative amount of S6 and to increase the amount of the derivatives; we cannot tell whether hepatectomy increases phosphorylation or decreases dephosphorylation of S6.

One can imagine that the phosphorylation of S6 (e.g., the increase in the number of phosphorylated derivatives) is causally, or fortuitously, related to the acceleration of protein synthesis that occurs during hepatic regeneration. Phosphorylation of S6 may have significance for the synthesis of specific proteins, perhaps even for the binding of initiator-tRNA or of mRNA. That it is a small subunit protein that is phosphorylated encourages one in the belief that the process may be related to initiation of peptide synthesis. Blat and Loeb (1971) discovered that glucagon increased the phosphorylation of a single liver ribosomal protein band (designated M). Glucagon administration is known to induce several hepatic enzymes, among which are tyrosine aminotransferase and serine dehydratase. The phosphorylation of the M-band protein might condition the binding and the translation of the mRNA for the specific enzymes.

The finding that there are as many as six forms of S6, which differ in the number of phosphoserine residues, raises the possibility that the small subunit of eukaryotic ribosomes are heterogeneous, at least with respect to the degree of chemical modification of S6. There is no reason to believe that eukaryotic ribosomes contain more than one copy of S6; hence the several forms of the protein are likely to be components in different 40S subparticles. We do not know if S6 is phosphorylated before or after it is incorporated into a ribonucleoprotein particle, although we recognize that the distinction is important since it might help to decide whether the forms of S6 are transient intermediates of maturation or whether ribosomes containing different forms of S6 serve different functions.

Phosphoryl Transfer from GTP to 40S Ribosomal Proteins

Traugh (Traugh, Mumby and Traut 1973; Traugh and Traut 1974) discovered an enzyme (protein kinase III) in the supernatant of reticulocytes that catalyzes the transfer of the γ-phosphoryl of GTP to ribosomal protein.

The enzyme, which has dual nucleotide specificities (it uses both ATP and GTP), phosphorylates one protein band in the 40S subunit and four in the large subparticle (the latter may be more apparent than real—see below). There is a similar enzyme activity in rat liver cytosol (Ventimiglia and Wool 1974).

The rat liver enzyme (protein kinase-GTP) catalyzes the transfer of the γ-phosphoryl group of GTP to a single protein band in the 40S subunit of rat liver ribosomes. In addition, a number of proteins in the enzyme preparation are phosphorylated (the latter migrate a shorter distance from the anode than even the least basic of ribosomal proteins). Phosphorylation of the 40S subunit protein band does not require the 60S subparticle. The proteins in the enzyme preparation that are phosphorylated associate preferentially with the 60S subunit; they are not removed by centrifugation through 20% sucrose containing 0.5 M KCl.

Although radioautographs of one-dimensional polyacrylamide gels had revealed only a single phosphorylated ribosomal protein band in the 40S subunit, we knew that the exact number and identity of the phosphorylated proteins were best determined by two-dimensional polyacrylamide gel electrophoresis. 40S ribosomal proteins, phosphorylated in the reaction with [^{32}P]GTP and the rat liver enzyme, were separated by two-dimensional polyacrylamide gel electrophoresis and radioautographs were made: there were four radioactive zones (Figure 3). When we compared the radioactive spots with the ribosomal proteins on the stained, dried gel slabs, it was apparent that the phosphoproteins were shifted toward the origin or anode of

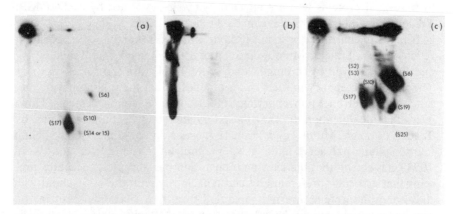

Figure 3 Radioautographs of phosphorylated 40S and 60S ribosomal proteins separated by two-dimensional polyacrylamide gel electrophoresis. (*a*) 40S or (*b*) 60S ribosomal subunits were incubated in 1 ml of medium for 20 min at 37°C with protein kinase-GTP and 50 μM [γ-^{32}P]GTP. In (*c*) 40S ribosomal subunits were incubated with 50 μM [γ-^{32}P]ATP, 10^{-5} M cyclic AMP, and protein kinase-ATP. The ribosomes were extracted, separated by two-dimensional polyacrylamide gel electrophoresis, and radioautographs made (Ventimiglia and Wool 1974).

the gel. That is the change in electrophoretic mobility to be expected of proteins that have gained negative charge(s) as a result of phosphorylation. Similar changes in the electrophoresis of ribosomal proteins have been demonstrated before (Stahl, Welfle and Bielka 1972).

Because the radioactive zones and the stained proteins are not congruent, their identification can only be tentative. What is required is a method of identifying ribosomal proteins apart from their net charge (Czernilofsky 1974). Keeping this reservation in mind, the 40S ribosomal proteins which were phosphorylated by protein kinase-GTP were S6, S10, S14 or 15, and S17 (Figure 3a). Several additional proteins, of which S19 was the most prominent, were phosphorylated by protein kinase-ATP (Figure 3c).

60S subunits were incubated with [^{32}P]GTP and protein kinase-GTP; the ribosomal proteins were separated once again by two-dimensional polyacrylamide gel electrophoresis and radioautographs of the gels made (Figure 3b). The radioactive phosphoproteins hardly migrated in the first dimension, indicating they are far less basic than the bulk of 60S ribosomal proteins. No radioactive zone corresponded to an authentic ribosomal protein, even if allowance was made for change in charge due to phosphorylation. The finding distinguishes protein kinase-GTP from protein kinase-ATP, since the latter catalyzes the phosphorylation of a number of 60S ribosomal proteins.

The radioactive phosphorylated proteins formed in the reaction with the protein kinase-GTP preparation were shown to contain ^{32}P-labeled o-phosphoserine and o-phosphothreonine. The transfer of the γ-phosphate of GTP to ribosomal proteins was not affected by cyclic AMP nor by cyclic GMP; Traugh and Traut (1974) have made the same finding.

It is likely that the ribosomal protein kinase-GTP activities from rabbit reticulocytes (Traugh, Mumby and Traut 1973; Traugh and Traut 1974) and from rat liver (Ventimiglia and Wool 1974) are the same enzymes. The major discrepancy in their characteristics is that the rabbit enzyme, but not the rat liver one, phosphorylates 60S ribosomal subunit proteins. However, the identification of the phosphorylated large subunit proteins by Traugh (Traugh, Mumby and Traut 1973) was on one-dimensional sodium lauryl sulfate polyacrylamide gels. Ventimiglia (Ventimiglia and Wool 1974) has shown that the enzyme preparation has phosphate-acceptor proteins that associate preferentially and tenaciously with the 60S subparticle; those phosphoproteins are not separated from 60S ribosomal proteins on one-dimensional sodium lauryl sulfate polyacrylamide gels (Ventimiglia and Wool, unpublished data). Ventimiglia and Wool (1974) found that the enzyme did not transfer phosphate from GTP to 60S proteins if the analysis was by electrophoresis on one-dimensional or two-dimensional urea gels.

A large number of ribosomal proteins are phosphorylated when the enzyme is protein kinase-ATP; a far smaller number are phosphorylated in the reaction catalyzed by protein kinase-GTP. The results suggest protein kinase-GTP has some specificity and, perhaps, that the reaction is important for ribosome function. An additional consequence of the observations is to raise a question as to the phosphoryl donor for phosphorylation of ribosomal proteins in the cell. One no longer can be certain it is ATP, as has been assumed (Kabat 1972); it is possible that it is GTP. If GTP is in fact the donor, that would tend to marry the reaction to protein synthesis.

Acetylation of Eukaryotic Ribosomal Proteins

It is possible that other postsynthetic chemical modifications of ribosomal proteins occur and that the occurrence is used to regulate protein synthesis. The *E. coli* proteins L7 and L12 differ only in that the former has an *N*-acetyl, rather than a free seryl, residue at the amino-terminus (Terhorst, Wittmann-Liebold and Möller 1972). The ribosomes from more rapidly growing bacteria have relatively less L7 (the acetylated form) and more L12 than particles from slowly growing bacteria, suggesting that the acetylation of L7 is a physiologically controlled modification of L12 (Deusser 1972; Thammana et al. 1973).

Rat liver ribosomes appear to be acetylated in vivo and in vitro (Liew and Gornall 1973); acetylation is of both the α and ε amino groups of lysine. Although the evidence is good that the acetylation was of ribosomal proteins, the identity of the proteins has not been established. It would, of course, be important to determine if L40 and L41, the putative eukaryotic homologs, are acetylated, or better, if the difference in their observed electrophoretic behavior is to be attributed to acetylation of the N-terminal amino acid of L40. Acetylation of a single large subunit protein cannot account for the results of Liew and Gornall (1973), since proteins of both subunits were acetylated; indeed the specific activity (after labeling with [³H]acetate) for the 40S subunit was greater than that of the large subparticle. The functional significance of acetylation of ribosomal proteins, if indeed there is any, remains to be determined.

IMMUNOCHEMICAL STUDIES OF EUKARYOTIC RIBOSOMES

The use of antibodies to ribosomal proteins for analysis of the structure and function of the particles has extraordinary potential. New applications are being found all the time. Antibodies and immunochemical techniques have been helpful in confirming that *E. coli* ribosomal proteins do not contain extensive homologous sequences (L7/L12 excepted), they have been useful as reagents to probe the topography of prokaryotic ribosomes, and they have been valuable in defining the function of specific proteins (Stöffler and Wittmann 1971a,b; Wittmann and Stöffler 1972; Stöffler, this volume).

Production of Antibodies to Eukaryotic Ribosomes

Antibodies to eukaryotic ribosomes have been prepared before (Dodd et al. 1962; Bigley, Dodd and Geyer 1963; Alberghina and Suskind 1967; Meyer-Bertenrath 1967; Meyer-Bertenrath and Würz 1967; Lamon and Bennett 1970; von der Decken and Hultin 1971; Wikman-Coffelt 1972; Wikman-Coffelt, Howard and Traut 1972; Busch et al. 1971, 1972; Noll and Bielka 1969, 1970a, 1970b; Noll, Heumann and Bielka 1973), and the antisera have been employed to compare the ribosome and the ribosomal proteins from different species (Alberghina and Suskind 1967; Lamon and Bennett 1970; Noll and Bielka 1969, 1970b) and from different tissues of the same species (Noll and Bielka 1970b). For example, Noll and Bielka (1970b) made antibodies to rat and bovine liver ribosomes in rabbits and used the antisera to demonstrate a difference in the ribosomes from several species (rat, rabbit, mouse, bovine, chicken);[5] on the other hand, they found the ribosomes from different tissues (liver and kidney) of the same species to have immunological identity.

There has been much interest in the possibility that ribosomes in malignant cells contain a different set of proteins than the particles in normal tissues, and, of course, in the corollary proposition that the extra or missing proteins account for the abnormal rate and lack of regulation of protein synthesis that characterizes cancer cells. Immunological procedures have been used to test the proposition. Noll and Bielka (1970a) found no immunological differences in the ribosomes or ribosomal proteins from normal rat liver and those from hepatoma, a finding at variance with that of Meyer-Bertenrath (1967), who observed an alteration in the antigenic properties of ribosomes from the liver of rats fed a carcinogen (*N*-nitrosomorpholine). H. and R. K. Busch and their colleagues (1971, 1972) prepared antibodies to 40S and 60S ribosomal subunits from Novikoff hepatoma ascites cells by immunization of goats. The antisera did not discriminate between subunits from normal rat liver and ascites cells; however, the antisera markedly inhibited the growth of hepatoma cells. These findings are very exciting, not only for the potential therapeutic importance of the application of the observation, but also because they suggest antibodies might be used to localize ribosomes in intact cells. Wikman-Coffelt, Howard and Traut (1972), in an elaboration of the same studies, reported two antigenic differences in the proteins of the 60S ribosomal subunits of normal and hepatoma ribosomes. The proteins responsible for the immunological differences were isolated by absorbing tumor 60S antiserum with normal 60S ribosomal proteins and then react-

---

[5] The dissimilarities in the ribosomal proteins of different species are difficult to demonstrate by two-dimensional polyacrylamide gel electrophoresis (Delaunay, Creusot and Schapira 1973). The results would indicate again the great advantage of immunochemical techniques for comparative studies of species differences in ribosomal proteins.

ing the absorbed antiserum with tumor 60S ribosomal proteins. The two tumor-specific ribosomal proteins responsible for the difference had molecular weights of 30,000 and 65,000.

Finally, it should be noted that antibodies to ribosomes, primarily but not exclusively to ribosomal RNA, have been found to occur in certain diseases, especially systematic lupus erythematosus (Sturgill and Carpenter 1965; Schur, Moroz and Kunkel 1967; Sturgill and Preble 1967). Autoimmune disease can follow immunization of rabbits with rat or rabbit liver ribosomes (Dodd et al. 1962) but, obviously, this is not inevitable. There is a convenient, and potentially most useful, radioimmunoassay for ribosomes (Wikman-Coffelt 1972).

Antibodies were prepared in rabbits and in sheep to highly purified preparations of rat liver ribosomes, to ribosomal subunits, and to mixtures of proteins from the particles. The immunization was successful: a number of individual preparations of antisera of relatively high titer to 80S ribosomes, to 60S and 40S ribosomal subunits, and to mixtures of proteins (termed TP80, TP60, TP40; TP, for total proteins) were raised in rabbits and sheep. The antisera were characterized by immunodiffusion on Ouchterlony plates, by quantitative immunoprecipitation, and by immunoelectrophoresis. The antibodies in the antisera were against ribosomal proteins, not ribosomal RNA. At least half of the proteins in the immunogen (ribosomal particles or mixtures of ribosomal proteins) elicited antibodies. There was no appreciable cross-reaction between antisera against 40S and 60S subunits. Rat liver ribosomes and ribosomal proteins were far more antigenic in sheep than rabbits. It is noteworthy that mixtures of eukaryotic ribosomal proteins are more antigenic than are ribosomal particles, whereas the reverse is the case for *E. coli* mixtures (e.g., they are less antigenic than particles). It is also the case that 60S particles and 60S proteins are more antigenic than the small subunit or its proteins.

The ribosomal proteins of rats, rabbits and sheep appear to be remarkably similar: the presumption of similarity comes from comparison of the proteins by two-dimensional polyacrylamide gel electrophoresis (Delaunay, Creusot and Schapira 1973). Unfortunately, no immunological test of the relatedness of rat, sheep and rabbit ribosomal proteins has yet been made. If their proteins are similar, why then should rat liver ribosomes be more antigenic in sheep than in rabbits? The finding suggests that there will prove to be immunologic differences.

A number of purposes underlaid the preparation of the antibodies to eukaryotic ribosomal proteins. It was anticipated that they could be employed to determine whether the proteins of eukaryotic and prokaryotic ribosomes are homologous—structurally or functionally homologous or both—and to ascertain the nature of the pool of free ribosomal proteins in the cell cytosol, and if there were free ribosomal proteins, their identity and the amount of each. Finally, the antibodies might be used to explore the topography of eukaryotic ribosomes and to define the function of specific proteins, just as has been done with prokaryotic ribosomes.

Nature of the Pool of Free Ribosomal Proteins

Dice and Schimke (1972) had reported that as much as 17% of the protein in rat liver cytosol was ribosomal; that there was far more (nine times as much) ribosomal protein free in the cytosol than in particles; that the free proteins exchanged rapidly with the proteins of the ribosome—indeed, that every few minutes 90% of the proteins of polysomes exchanged with free ribosomal proteins while the synthesis of nascent chains was proceeding; and finally, that ribosomal proteins turn over while free rather than as ribosome constituents. The observation raised the possibility that regulation of protein synthesis in animal cells might occur by addition and deletion of ribosomal proteins. We wished to determine whether the observation was correct and, if so, the identity and the amount of individual proteins. The tactics were to carry out quantitative immunoprecipitation of the putative ribosomal proteins in the cytoplasm using the antibodies that had been prepared.

The procedure selected was to absorb the antibodies in the antiserum with liver cytosol containing 2, 5 or 10 mg of protein. The absorbed antisera were then used to carry out immunoprecipitation with 10 A_{260} units (450 μg rRNA) of 80S ribosomes. The decision as to the quantity of cytosol protein and ribosomes to use was based on the following: if it is assumed that 5% of the soluble protein in rat liver cytoplasm is ribosomal protein [recall that Dice and Schimke (1972) estimated that it was as great as 17%], then in 10 mg of the cytoplasmic protein approximately 450 μg will be ribosomal. Since 10 A_{260} units of 80S ribosomes also has 450 μg of protein, absorption of the antiserum with 10 mg of cytoplasmic protein should completely inhibit the subsequent immunoprecipitation. It should be noted that the amount of specific antibody in each antiserum was determined by quantitative immunoprecipitation and the amount used in the experiments was carefully selected so that an increase in antibody gave an increase in immunoprecipitate, e.g., the amount of antiserum (and not of antigen) was actually limiting in the final reaction. Thus one could be confident that if ribosomal proteins were present in the cytosol and antibodies were absorbed from the antiserum, there would be a reduction in the amount of immunoprecipitate. For the moment two assumptions are required: that the ribosomal proteins are present in the cytosol in the same proportion as in ribosomes and that the antisera have specific antibodies for a reasonable proportion of the ribosomal proteins. The absorption experiment had an advantage: nonprecipitating, as well as precipitating, antibodies are absorbed; the former will not then coprecipitate with the latter in the second reaction.

In the first experiment (Table 5), antiserum to rat liver 60S ribosomal subunits (A-60S) was absorbed with 2, 5 and 10 mg of cytoplasmic protein, prepared in the same way as Dice and Schimke (1972), had, and with the same amounts of cytosol prepared according to Leader, Wool and Castles (1970). The two procedures were used so as to be sure the method of preparation of the cytosol did not prejudice the results. The absorbed antisera were next incubated with 80S ribosomes and the amount of RNA and protein in the immunoprecipitate was determined. It is very important that the antisera used in this and subsequent experiments contained no precipitating antibodies against RNA; however, if ribosomes were used as antigen, then the antisera caused RNA to coprecipitate with

Table 5 Determination of ribosomal proteins in rat liver cytosol by immunoprecipitation with antisera to 60S ribosomal subunits

| | Immunoprecipitate | |
| | Protein | RNA |
Antisera	(μg)	
Nonabsorbed	830	186
Absorbed		
Cytosol preparation I		
2 mg protein	974	197
5 mg protein	735	226
10 mg protein	725	335
Cytosol preparation II		
2 mg protein	786	197
5 mg protein	735	206
10 mg protein	730	208
Nonimmune		
Cytosol preparation I		
10 mg protein	230	14
Cytosol preparation II		
10 mg protein	370	22

Antiserum (0.5 ml) to rat liver 60S ribosomes prepared in sheep was absorbed with cytosol containing 2, 5 or 10 mg of protein prepared according to Dice and Schimke (1972) (referred to as Preparation I) or Leader, Wool and Castles (1970) (referred to as Preparation II). The samples were allowed to stand for 48 hr at 0°C, and the absorbed antisera were separated by centrifugation and used in an immunotitration with 10 A_{260} units (450 μg of rRNA) of 80S ribosomes; again the samples were allowed to stand at 0°C for 48 hr. The amount of protein and RNA in the immunoprecipitate was determined, and the former was corrected for nonspecific precipitation. The values are the mean of two separate determinations, each done in duplicate.

ribosomal proteins. If the amount of RNA is examined first, it is apparent that the absorbed antisera are at least as effective in precipitating 80S ribosomes as is the nonabsorbed antiserum.

The protein in the immunoprecipitate includes both antibody and antigen, probably in a ratio of about five to one. From the data (Table 5), it would appear that absorption of the antiserum caused a decrease in the effectiveness with which it acted to precipitate 80S ribosomes. However the decrease is small and seemingly not related to the amount of cytoplasmic protein used for the absorption (compare the results with 5 and 10 mg of cytoplasmic protein) and, therefore, probably not significant. The small decrease in protein in the precipitate may be the result of absorption of specific antibodies in the antiserum: antibodies were formed against supernatant proteins, proteins presumably tightly bound to the ribosomes which had originally been used as antigens. Nonimmune serum caused no appreciable precipitation of 80S ribosomes. Exactly the same results were obtained when the experiment was repeated with a second antiserum against 60S ribosomal subunits (this one raised in a rabbit rather than in a sheep), and perhaps of even more significance, when an antiserum against a mixture of 60S ribosomal proteins (A-TP60) was used. The purpose in using antisera to ribosomal particles and to mixtures of ribosomal proteins and in using antisera raised in different animals, was to increase the antigenic spectrum and hence the variety and range of antibodies. If the array of antibodies were broadened, the chance of overlooking individual ribosomal proteins in the cytosol would be decreased.

A similar experiment was carried out with antiserum to 40S ribosomal subunits (A-40S). Absorption of the antiserum caused a decrease (of about 25%) in the amount of 80S ribosomal RNA in the immunoprecipitate (Table 6). In this experiment there was no decrease in the amount of protein in the precipitate. Once again the quantity of ribosomal RNA that was precipitated was not related to the amount of protein in the cytoplasm used for the absorption, as would be expected if the cytoplasm contained ribosomal proteins. But even if the decrease in the precipitation of ribosomal RNA, as a result of absorption of the antiserum with cytosol, is attributed entirely to the presence of ribosomal protein, still less than 0.5% of cytoplasmic protein is 40S ribosomal protein. The same results were obtained when other antisera were used, i.e., against 40S ribosomal subunits (A-40S) or against mixtures of 40S ribosomal proteins (A-TP-40).

The cytoplasm contains few, if any, of the ribosomal proteins to which we have made antibodies. The exact number and identity of the ribosomal proteins that elicited antibodies is not known. The antisera contain specific antibodies to a large number of proteins. The conclusion is based on the observation that the curve describing quantitative immunoprecipitation of ribosomes with antiserum is very broad; that immunoelectrophoresis gives

Table 6   Determination of ribosomal proteins in rat
liver cytosol by immunoprecipitation with antisera
to 40S ribosomal subunits

	Immunoprecipitate	
Antisera	Protein	RNA
	(μg)	
Nonabsorbed	478	148
Absorbed		
2 mg cytosol protein	570	116
5 mg cytosol protein	495	105
10 mg cytosol protein	489	114
Nonimmune		
5 mg cytosol protein	107	8

The experiments were carried out as in Table 5, except
that antiserum to rat liver 40S ribosomal subunits prepared
in a rabbit was used. The cytosol was prepared as described
before (Leader, Wool and Castles 1970).

six to ten distinct lines and a large, poorly resolved precipitate that is likely
to derive from a large number of additional proteins. It needs to be emphasized that the same results were obtained with several antisera—antisera
to ribosomal particles and to mixtures of proteins—thereby increasing the
probability of the detection of individual ribosomal proteins in the cytoplasm; the several immunized animals are likely to recognize and make
antibodies to a different subset of proteins. Preliminary analyses by affinity
chromatography (using agarose beads to which antibodies were bound)
would indicate that the antisera contain antibodies against at least half the
ribosomal proteins. Although the cytosol lacks material amounts of most
ribosomal proteins, the possibility that it contains a few cannot be excluded yet.

The immunochemical results would indicate there is no appreciable pool
of free ribosomal proteins and hence no exchange of ribosomal proteins.
The results are clearly different from those of Dice and Schimke (1972)
and Garrison, Bosselman and Kaulenas (1972). What is the explanation
of the discrepancy? It is hard to be sure, but contamination of the ribosomes with cytoplasmic proteins would certainly account for the disparity
in the findings. It is only fair to concede that Dice and Schimke (1972)
took precautions on that score, but nonetheless, the ribosomal proteins were
separated on one-dimensional SDS-polyacrylamide gels, which is not a
satisfactory way of distinguishing them from cytoplasmic proteins that
might have contaminated the particles. Delaunay (personal communication) was unable to detect ribosomal proteins in liver cytoplasm by two-
dimensional polyacrylamide gel electrophoresis even when very large

amounts of protein were used; nor did he find any exchange of proteins when care was taken to purify the ribosomes.

Ribosomal proteins are synthesized in the cytoplasm and then, one presumes, transported to the nucleolus where they are assembled on ribosomal RNA. Unless the ribosomal proteins are released directly from the ribosome to the nucleus, there must be a finite amount in the cytoplasm. The amount is likely to be so small as to be beyond detection by the techniques we have used. Finally, the experiments clearly do not mitigate against there being a significant pool of free ribosomal proteins in the nucleolus.

Homology between the Prokaryotic Ribosomal Proteins L7/L12 and Eukaryotic Proteins

The large subunit of *E. coli* and other prokaryotic ribosomes contains a pair of acidic proteins which have been designated L7 and L12 from their position on two-dimensional electropherograms. The proteins were first recognized and characterized by Möller and Widdowson (1967) and then by Kaltschmidt, Dzionara and Wittmann (1970). Each has a molecular weight of 12,000; their amino acid composition and peptide maps are identical (Möller et al. 1972; Terhorst, Wittmann-Liebold and Möller 1972); and since antibodies to the two proteins did not distinguish between them, it was assumed they had closely related primary structures (Stöffler and Wittmann 1971b). That presumption has now been confirmed. The primary sequence of the two proteins was determined (Terhorst et al. 1973) and L7 found to differ from L12 only in that its N-terminal serine is acetylated. The proteins, which are present in at least two copies (but perhaps as many as four) per ribosome (Thammana et al. 1973), have a number of striking chemical features: a very high content of alanine (28 of 120 residues), a large nonpolar region at the center of the molecule, the presence of monomethyllysine, and a high α-helical content (60%). A good deal is known of the function of proteins L7/L12 (see chapters by Möller and by Stöffler, this volume).

It was decided to determine whether that protein has been conserved during evolution from prokaryotes to eukaryotes and, at the same time, to find out if any other ribosomal proteins had been preserved.

The Identification of the Yeast Proteins Homologous with E. coli *L7/L12.* An acidic protein had been isolated from yeast ribosomes in Professor H.-G. Wittmann's laboratory in Berlin. From the position of the protein on two-dimensional gel electropherograms (a "typical" pair of acidic spots) and from its amino acid composition, it was reasonable to suspect it might be homologous with *E. coli* L7/L12. For the moment, we refer to it as yeast acidic protein. Our surmise was strengthened when it was found that an antiserum to *E. coli* L7/L12 gave a precipitation line when developed against yeast TP80 in the Ouchterlony double-diffusion test. Moreover, when the same antisera (A-L7/L12) was placed in the center well and developed against yeast TP80 and yeast acidic protein in adjacent wells on the Ouchterlony plate, the precipitation line fused—evidence that they are immunologically identical. The antiserum to *E. coli* L7/L12 (in the center well) was next tested against the yeast acidic protein and pure *E. coli* L7/L12, again in adjacent wells: the results revealed partial immunological identity (Figure 4A). A spur was formed between the precipitation lines of the homologous reactants (A-L7/L12 *E. coli* and *E. coli* L7/L12) and the yeast acidic protein, clear evidence that the proteins are similar but not identical. Of nine antisera against *E. coli* L7/L12, four did not react against yeast TP80, four reacted weakly, and one reacted strongly (the one we used in these experiments).

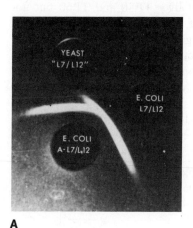

A

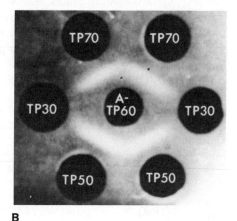

B

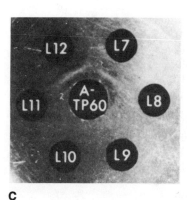

C

Figure 4 Homology between prokaryotic ribosomal proteins determined by Ouchterlony immunodiffusion. In (*A*) the center well had 100 μl of anti-*E. coli* L7/L12; the peripheral wells contained yeast acidic protein or *E. coli* L7/L12. In (*B*) the center well had 100 μl of five-times enriched anti-rat liver TP60 γ-globulin; the peripheral wells contained *E. coli* TP70, TP50 or TP30 as indicated. In (*C*) the center well had 100 μl of ten-times enriched γ-globulin fraction of rat liver anti-TP60; the peripheral wells had pure single *E. coli* 50S ribosomal proteins in amounts (1.5–10 μg) that gave good precipitation with specific *E. coli* antisera.

An antiserum was prepared against the yeast acidic protein and it re-acted (although weakly) with *E. coli* TP70 and TP50, but not at all with TP30. The circle of evidence was completed when it was found that the antiserum to yeast acidic protein gave a precipitation line when developed against pure *E. coli* L7/L12 by Ouchterlony double-diffusion (Table 7). There is then a structural homology between *E. coli* L7/L12 and yeast acidic protein. It is only fair to point out that in a prescient experiment Alberghina and Suskind (1967) had shown that antisera raised against *Neurospora crassa* ribosomes cross-reacted with a mixture of *E. coli* ribo-somal proteins.

The Identification of the Rat Liver Proteins Homologous with E. coli *L7/L12.* There was an additional exciting finding: an antiserum to *E. coli* proteins L7/L12 gave a precipitation line when tested by double-diffusion on Ouchterlony plates against rat liver TP80 and TP60 but not against TP40 (Table 7). It is perhaps of some significance that two of four *E. coli* L7/L12 antisera reacted strongly with rat liver TP60, whereas only one reacted well with yeast acidic protein. The results suggested that the large subunit of liver ribosomes also contained protein(s) homologous to *E. coli* L7/L12. The supposition was strengthened by the results of the

Table 7 Homology between eukaryotic and prokaryotic ribosomal proteins

Antiserum	Antigen	Immunoprecipitation
E. coli A-L7/L12	Rat liver TP80	+
	TP60	+
	TP40	0
Rat liver A-TP60	*E. coli* TP70	+
	TP50	+
	TP30	0
E. coli A-L7/L12	Yeast acidic protein	+
E. coli A-L7/L12	Rat liver TP60	+
Yeast A-acidic protein	*E. coli* L7/L12	+
Yeast A-acidic protein	Rat liver TP60	+
Rat liver A-TP60	*E. coli* L7/L12	+
Rat liver A-TP60	Yeast acidic protein	+
Rat liver A-L40/L41	*E. coli* 70S ribosomes	+
	50S ribosomes	+
	30S ribosomes	0
	L7/L12	+
E. coli A-S1 . . . A-S21[a]	Rat liver TP40	0
Rat liver A-TP40	*E. coli* S1 . . . S21[a]	0
E. coli A-L1 . . . A-L33[b]	Rat liver TP60	0
Rat liver A-TP60	*E. coli* L1 . . . L33[b]	0

[a] A-S17 and S17 and [b] A-L7/L12 and L7/L12 were not tested.

reciprocal experiment (Table 7): antisera to rat liver TP60 reacted in the same test with *E. coli* 70S and 50S ribosomal proteins but not with the proteins of the small subunit (Figure 4B). Moreover, the same antisera (A-TP60) cross-reacted with pure *E. coli* L7 and L12 (Figure 4C) and no other single *E. coli* large subunit protein.

There are two acidic proteins in rat liver 60S ribosomal subunits that have electrophoretic properties similar to *E. coli* L7/L12 (Figure 1, numbered L40 and L41). The proteins had not been detected before on two-dimensional electropherograms. It was only after the immunological results had indicated that they must be there that the gels were reviewed and the proteins located. L40 and L41 are by far the most acidic proteins of eukaryotic ribosomes, which accounts for the failure to find them earlier, since they are generally lost in the anode buffer chamber unless the time of electrophoresis in the first dimension is reduced from 40 to 20 hr (Sherton and Wool 1974). Moreover, the proteins stain lightly and so are easily overlooked (and are hard to see and photograph).

Proteins L40 and L41 have been prepared from rat liver ribosomes. They form separate zones after two-dimensional polyacrylamide gel electrophoresis in urea (Figure 5), but a single zone if electrophoresis in the second dimension is in SDS (Figure 5), behavior that mimics *E. coli* L7 and L12 (Geisser et al. 1973). Three separate antisera were raised against rat liver ribosomal proteins L40/L41: each precipitated *E. coli* 70S ribosomes and 50S subunits, but not 30S particles; moreover, they formed a precipitation line with pure L7 and L12 from *E. coli* (Table 7). There is good evidence then that L40 and L41 are the homologs of *E. coli* L7 and L12.

Comparison of the Yeast and Rat Liver Proteins Homologous with E. coli *L7/L12.* The antiserum to yeast acidic protein that gave precipitation lines when tested against *E. coli* L7/L12 also reacted against rat liver TP60 (Table 7). In a correlative experiment an antiserum to rat liver TP60 cross-reacted with pure *E. coli* L7/L12 and with yeast acidic protein. We conclude there are structural homologies in the primary sequences of *E. coli* L7/L12 and proteins in yeast and mammalian ribosomes (L40/L41). There is, however, an important difference between *E. coli* L7/L12 and rat liver L40/L41: the molecular weight of the prokaryotic proteins is 12,000, whereas the molecular weight of the eukaryotic proteins is approximately 25,500 (Table 4 and Figure 5).

Homologies between Other Prokaryotic and Eukaryotic Ribosomal Proteins. An attempt was next made to determine if there were structural homologies between any other prokaryotic and eukaryotic ribosomal proteins. The test was conducted in two ways: antisera to each of the individual proteins of the 30S subunit of *E. coli* ribosomes were tested by immunodiffusion for reaction with rat liver 40S ribosomal proteins; antisera to

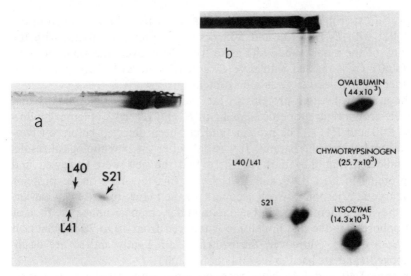

Figure 5 Two-dimensional electropherograms of proteins prepared from rat liver 60S ribosomal subunits with ethanol. A large preparation of rat liver 60S ribosomal subunits was precipitated with 0.2 volumes of ethanol. The ethanol supernatant was collected and evaporated to dryness. The proteins in the extract were separated (*a*) by two-dimensional polyacrylamide gel electrophoresis in urea (Sherton and Wool 1972) or (*b*) in SDS (Martini and Gould 1971). The molecular weight of L40 and L41 was determined by comparison with the standards to be about 25,500. S21 was identified by its molecular weight; it was present because the 60S subunits were not repurified.

separate *E. coli* 50S proteins were tested against rat liver 60S ribosomal proteins (Table 7). In the reciprocal experiment, antiserum to rat liver TP40 was tested for immunoprecipitation against each of the pure *E. coli* small subunit proteins, and antiserum to rat liver TP60 against the proteins of the large subunit of *E. coli* ribosomes (Table 7). All of the tests were negative (L7/L12 excepted). Thus the only prokaryotic and eukaryotic ribosomal proteins for which there is evidence of homology are L7 and L12. There is, of course, a measure of uncertainty in the results. There may be homologies less than sufficient to constitute determinants. Moreover, homologies might exist and not be detected by the immunochemical techniques that were used. More sensitive immunological assays as well as tests of inhibition of function need to be done.

Functional Homology of Prokaryotic and Eukaryotic Ribosomal Proteins. We sought to determine if the homologous prokaryotic and eukaryotic proteins had similar functions. The assay selected was EF-G-dependent binding of [³H]GDP to *E. coli* 50S subunits in the presence of fusidic acid (Highland et al. 1973). The reaction was inhibited by antisera to *E. coli* L7/L12, to yeast acidic protein, to rat liver TP60, and to rat liver 60S ribosomes (Figure 6). Moreover, if the antisera to rat liver TP60 was

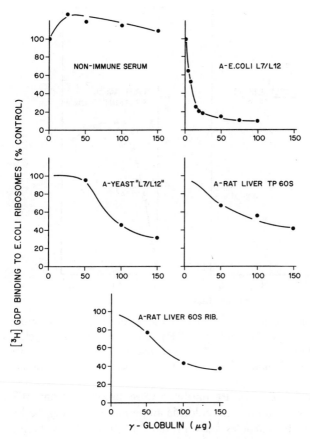

Figure 6   The effect of γ-globulin on the formation of a ribosome · EF-G ·
[³H]GDP complex. The reaction mixture contained in 50 μl of buffer (10 mM
Tris-HCl pH 7.4–10 mM magnesium acetate–60 mM NH₄Cl–1 mM dithiothreitol): 0.43 A₂₆₀ of *E. coli* 50S ribosomal subunits, 11 units EF-G, 3 mM
fusidic acid, 10 pmoles [³H]GTP, and the amount of γ-globulin indicated (1 mg
of γ-globulin protein is the equivalent of 1.2 A₂₈₀ units). The mixture was
incubated for 5 min at 0°C, and the complex (ribosome · EF-G · [³H]GDP)
was collected on Millipore filters (Highland et al. 1973). The binding of
[³H]GDP to the filters was limited by the amount of ribosomes. In the absence
of γ-globulin about 12,000 cpm (100%) of [³H]GDP was bound. *A-E. coli*
L7/L12, an antisera against *E. coli* ribosomal proteins L7/L12; A-Yeast "L7/
L12," an antisera against an acidic protein(s) prepared from yeast ribosomes;
A-rat liver TP60S, as antisera against all of the proteins of the 60S subunit
of rat liver ribosomes; A-rat liver 60S rib., an antisera against rat liver 60S ribosomal subunits.

absorbed with *E. coli* proteins L7/L12, the inhibitory activity was lost.
The prokaryotic (L7/L12) and eukaryotic (L40/L41) proteins are not
only structurally related but also functionally similar.

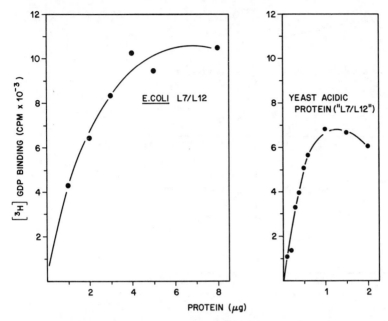

Figure 7 The effect of prokaryotic (L7/L12) and eukaryotic (yeast acidic protein) ribosomal proteins on the binding of [³H]GDP to *E. coli* PI particles. The reaction was carried out as in Figure 6, except that 0.43 A_{260} of *E. coli* PI particles (Hamel, Koka and Nakamoto 1972) were used instead of 50S subunits.

In another experiment, *E. coli* PI particles were prepared from 50S subunits by treatment with ammonium chloride and ethanol (Hamel, Koka and Nakamoto 1972). PI particles lack proteins L7/L12. Fully active ribosomes can be reconstituted if proteins L7/L12 are added to *E. coli* PI particles (Figure 7; also Hamel, Koka and Nakamoto 1972). If yeast acidic proteins were added to *E. coli* PI particles, hybrid ribosomes were formed and they were active in EF-G-dependent binding of GDP. The results reinforce the conclusion that the prokaryotic and eukaryotic proteins are functionally as well as structurally homologous. Moreover, the experiment demonstrates that active hybrid ribosomes can be formed with prokaryotic and eukaryotic proteins.

Acknowledgments

We acknowledge our indebtedness to the many persons with whom we have had the pleasure of working: Ralph Di Camelli, Charles Eil, Axel Gressner, Renata Hasenbank, Alan Lin, Karl-Heinz Rak, Corinne Sherton, Lawrence Swanson and Frank Ventimiglia. We are especially grateful to Axel Gressner, Renata Hasenbank, Alan Lin, Karl-Heinz Rak and Frank Ventimiglia for allowing us to describe the results of unpublished experiments.

The expenses of the research carried out in the Department of Biochemistry, University of Chicago, were met by grants from the National Institutes of Health (AM-04842) and from The John A. Hartford Foundation.

References

Alberghina, F. A. M. and S. R. Suskind. 1967. Ribosomes and ribosomal protein from *Neurospora crassa*. I. Physical, chemical and immunochemical properties. *J. Bact.* **94**:630.

Bickle, T. A. and R. R. Traut. 1971. Differences in size and number of 80S and 70S ribosomal proteins by dodecyl sulfate gel electrophoresis. *J. Biol. Chem.* **246**:6828.

Bielka, H. and H. Welfle. 1968. Characterization of ribosomal proteins from different tissues and species by electrophoresis on polyacrylamide gel. *Mol. Gen. Genet.* **102**:128.

Bielka, H., J. Stahl and H. Welfle. 1971. Studies on proteins of animal ribosomes. IX. Proteins of ribosomal subunits of some tumors characterized by two-dimensional polyacrylamide gel electrophoresis. *Arch. Geschwulstforsch.* **38**:109.

Bielka, H., H. Welfle, M. Böttger and W. Förster. 1968. Structurveränderungen und Dissoziation von Leberribosomen in Abhängigkeit von der Mg⁺⁺-Konzentration. *Eur. J. Biochem.* **5**:183.

Bigley, N. J., M. C. Dodd and V. B. Geyer. 1963. The immunological specificity of antibodies to liver ribosomes and nuclei. *J. Immunol.* **90**:416.

Bitte, L. and D. Kabat. 1972. Phosphorylation of ribosomal proteins in sarcoma 180 tumor cells. *J. Biol. Chem.* **247**:5345.

Blat, C. and J. E. Loeb. 1971. Effect of glucagon on phosphorylation of some rat liver ribosomal proteins *in vivo*. *FEBS Letters* **18**:124.

Brownlee, G. G., F. Sanger and B. G. Barrell. 1968. The sequence of 5S ribosomal ribonucleic acid. *J. Mol. Biol.* **34**:379.

Burka, E. R. and S. I. Bulova. 1971. Heterogeneity of reticulocyte ribosomes. *Biochem. Biophys. Res. Comm.* **42**:801.

Busch, H., R. K. Busch, W. H. Spohn, J. Wikman and Y. Daskal. 1971. Antibodies produced by immunization of goats with 60S ribosomal subunits from Novikoff hepatoma ascites cells. *Proc. Soc. Exp. Biol. Med.* **137**:1470.

Busch, R. K., W. H. Spohn, Y. Daskal and H. Busch. 1972. Antibodies produced by immunization of goats with 40S ribosomal subunits from Novikoff hepatoma ascites cells. *Proc. Soc. Exp. Biol. Med.* **140**:1030.

Cammarano, P., S. Pons, A. Romeo, M. Galdieri and C. Gualerzi. 1972a. Characterization of unfolded and compact ribosomal subunits from plants and their relationship to those of lower and higher animals: Evidence for physicochemical heterogeneity among eukaryotic ribosomes. *Biochim. Biophys. Acta* **281**:571.

Cammarano, P., A. Romeo, M. Gentile, A. Felsani and C. Gualerzi. 1972b. Size heterogeneity of the large ribosomal subunits and conservation of the small subunits in eukaryotic evolution. *Biochim. Biophys. Acta* **281**:597.

Cammarano, P., A. Felsani, M. Gentile, C. Gualerzi, A. Romeo and G. Wolf.

1972c. Formation of active hybrid 80S particles from subunits of pea seedlings and mammalian liver ribosomes. *Biochim. Biophys. Acta* **281**:625.

Cawthon, M. L., L. F. Bitte, A. Krystosek and D. Kabat. 1974. Effect of cyclic adenosine 3′:5′-monophosphate on ribosomal protein phosphorylation in reticulocytes. *J. Biol. Chem.* **249**:275.

Chambaut, A. M., F. Leray and J. Hanoune. 1971. Relation between cyclic AMP dependent protein kinase(s) and cyclic AMP binding protein(s) in rat liver. *FEBS Letters* **15**:328.

Chatterjee, S. K., M. Kazemie and H. Matthaei. 1973. Studies on rabbit reticulocyte ribosomes. II. Separation of the ribosomal proteins by two-dimensional electrophoreses. *Hoppe-Seyler's Z. Physiol. Chem.* **354**:481.

Chaudhuri, S. and I. Lieberman. 1968. Control of ribosome synthesis in normal and regenerating liver. *J. Biol. Chem.* **243**:29.

Chen, L.-J. and D. A. Walsh. 1971. Multiple forms of hepatic cyclic AMP dependent protein kinase. *Biochemistry* **10**:3614.

Chuang, D.-M. and M. V. Simpson. 1971. A translocation-associated ribosomal conformational change detected by hydrogen exchange and sedimentation velocity. *Proc. Nat. Acad. Sci.* **68**:1474.

Cohn, P. 1967. Properties of ribosomal proteins from two mammalian sources. *Biochem. J.* **102**:735.

Cohn, P. and P. Simpson. 1963. Basic and other proteins in microsomes of rat liver. *Biochem. J.* **88**:206.

Correze, C., P. Pinell and J. Nunez. 1972. Effects of thyroid hormones on phosphorylation of liver ribosomal proteins and on protein phosphokinase activity. *FEBS Letters* **23**:87.

Czernilofsky, A. P., E. E. Collatz, G. Stöffler and E. Kuechler. 1974. Proteins at the tRNA binding sites of *Escherichia coli* ribosomes. *Proc. Nat. Acad. Sci.* **71**:230.

Darnell, J. E. 1968. Ribonucleic acids from animal cells. *Bact. Rev.* **32**:262.

von der Decken, A. and T. Hultin. 1971. Immunological properties of isolated protein fractions from *Artemia* ribosomes. *Exp. Cell Res.* **64**:179.

Delaunay, J. and G. Schapira. 1972. Rat liver and hepatoma ribosomal proteins. Two-dimensional polyacrylamide gel electrophoresis. *Biochim. Biophys. Acta* **259**:243.

Delaunay, J., F. Creusot and G. Schapira. 1973. Evolution of ribosomal proteins. *Eur. J. Biochem.* **39**:305.

Delaunay, J., C. Mathieu and G. Schapira. 1972. Eukaryotic ribosomal proteins. Interspecific and intraspecific comparisons by two-dimensional polyacrylamide gel electrophoresis. *Eur. J. Biochem.* **31**:561.

Delaunay, J., J. E. Loeb, M. Pierre and G. Schapira. 1973. Mammalian ribosomal proteins: Studies on the *in vitro* phosphorylation patterns of ribosomal proteins from rabbit liver and reticulocytes. *Biochim. Biophys. Acta* **312**:147.

Deusser, E. 1972. Heterogeneity of ribosomal populations in *Escherichia coli* cells grown in different media. *Mol. Gen. Genet.* **119**:249.

Dice, J. F. and R. T. Schimke. 1972. Turnover and exchange of ribosomal proteins from rat liver. *J. Biol. Chem.* **247**:98.

Di Girolamo, M. and P. Cammarano. 1968. The protein composition of ribosomes and ribosomal subunits from animal tissues. Electrophoretic studies. *Biochim. Biophys. Acta* **168**:181.

Dodd, M. C., N. J. Bigley, V. B. Geyer, F. W. McCoy and H. E. Wilson. 1962. Autoimmune response in rabbits injected with rat and rabbit liver ribosomes. *Science* **137**:688.

Eil, C. and I. G. Wool. 1971. Phosphorylation of rat liver ribosomal subunits: Partial purification of two cyclic AMP activated protein kinases. *Biochem. Biophys. Res. Comm.* **43**:1001.

————. 1973a. Phosphorylation of liver ribosomal proteins. Characteristics of the protein kinase reaction and studies of the structure of phosphorylated ribosomes. *J. Biol. Chem.* **248**:5122.

————. 1973b. Function of phosphorylated ribosomes. The activity of ribosomal subunits phosphorylated *in vitro* by protein kinase. *J. Biol. Chem.* **248**:5130.

Elson, D. 1958. Latent ribonuclease activity in a ribonucleoprotein. *Biochim. Biophys. Acta* **27**:216.

Fellner, P., C. Ehresmann, P. Stiegler and J.-P. Ebel. 1972. Partial nucleotide sequence of 16S ribosomal RNA from *E. coli. Nature New Biol.* **239**:1.

Garrison, N. E., R. A. Bosselman and M. S. Kaulenas. 1972. The effect of ribosomal protein exchange on the activity of *Xenopus laevis* ribosomes. *Biochem. Biophys. Res. Comm.* **49**:171.

Geisser, M., G. W. Tischendorf, G. Stöffler and H.-G. Wittmann. 1973. Immunological and electrophoretical comparison of ribosomal proteins from eight species belonging to *Enterobacteriaceae. Mol. Gen. Genet.* **127**:111.

Gordon, J. 1971. Determination of an upper limit to the phosphorus content of polypeptide chain elongation factor and ribosomal proteins in *Escherichia coli. Biochem. Biophys. Res. Comm.* **44**:579.

Gould, H. J. 1970. Proteins of rabbit reticulocyte ribosomal subunits. *Nature* **227**:1145.

Gressner, A. M. and I. G. Wool. 1974. The phosphorylation of liver ribosomal proteins *in vivo*. Evidence that only a single small subunit protein (S6) is phosphorylated. *J. Biol. Chem.* (in press).

Gualerzi, C., H. G. Janda, H. Pasow and G. Stöffler. 1974. Studies on the protein moiety of plant ribosomes. *J. Biol. Chem.* **249**:3347.

Hamel, E., M. Koka and T. Nakamoto. 1972. Requirement of an *Escherichia coli* 50S ribosomal protein component for effective interaction of the ribosome with T and G factors and with guanosine triphosphate. *J. Biol. Chem.* **247**:805.

Hamilton, M. G. and M. E. Ruth. 1967. Characterization of some of the proteins of the large subunit of rat liver ribosomes. *Biochemistry* **6**:2585.

Hamilton, M. G., A. Pavlovec and M. L. Petermann. 1971. Molecular weight, buoyant density, and composition of active subunits of rat liver ribosomes. *Biochemistry* **10**:3424.

Hardy, S. J. S. and C. G. Kurland. 1966. The relationship between poly A polymerase and the ribosomes. *Biochemistry* **5**:3676.

Highland, J. H., J. Bodley, J. Gordon, R. Hasenbank and G. Stöffler. 1973. Identity of the ribosomal proteins involved in the interaction with elongation factor G. *Proc. Nat. Acad. Sci.* **70**:147.

Houssais, J. F. 1971. Nature de l'hétérogénéite interspécifique des protéines des ribosomes de mammifères. *Eur. J. Biochem.* **24**:232.

Hultin, T. 1972. Evidence for disulfide interaction *in situ* between two adjacent proteins in mammalian 60S ribosomal subunits. *Biochim. Biophys. Acta* **269**:118.

Hultin, T. and A. Sjöqvist. 1972. Two-dimensional polyacrylamide gel electrophoresis of animal ribosomal proteins based on charge inversion. *Anal. Biochem.* **46**:342.

Huynh-Van-Tan, J. Delaunay and G. Schapira. 1971. Eukaryotic ribosomal proteins. Two-dimensional electrophoretic studies. *FEBS Letters* **17**:163.

Jergil, B. 1972. Protein kinase from rainbow trout testis ribosomes. Partial purification and characterization. *Eur. J. Biochem.* **28**:546.

Jones, B. L., N. Nagabhushan and S. Zalik. 1972. Two-dimensional acrylamide gel electrophoresis of wheat leaf cytoplasmic and chloroplast ribosomal proteins. *FEBS Letters* **23**:167.

Kabat, D. 1970. Phosphorylation of ribosomal proteins in rabbit reticulocytes. Characterization and regulatory aspects. *Biochemistry* **9**:4160.

———. 1971. Phosphorylation of ribosomal proteins in rabbit reticulocytes. A cell-free system with ribosomal protein kinase activity. *Biochemistry* **10**:197.

———. 1972. Turnover of phosphoryl groups in reticulocyte ribosomal phospho-proteins. *J. Biol. Chem.* **247**:5338.

Kaltschmidt, E., M. Dzionara and H.-G. Wittmann. 1970. Ribosomal proteins. XV. Amino acid compositions of isolated ribosomal proteins from 30S and 50S subunits of *Escherichia coli. Mol. Gen. Genet.* **109**:292.

Kanai, K., J. J. Castles, I. G. Wool, W. S. Stirewalt and A. Kanai. 1969. The proteins of liver and muscle ribosomal subunits: Partial separation by carboxymethyl-cellulose column chromatography. *FEBS Letters* **5**:68.

King, H. W. S., H. J. Gould and J. J. Shearman. 1971. Molecular weight distribution of proteins in rabbit reticulocyte ribosomal subunits. *J. Mol. Biol.* **61**:143.

Kurland, C. G. 1970. Ribosome structure and function emergent. *Science* **169**:1171.

Lambertsson, A. G. 1972. The ribosomal proteins of *Drosophila melanogaster.* II. Comparison of protein patterns of ribosomes from larvae, pupae and adult flies by two-dimensional polyacrylamide gel electrophoresis. *Mol. Gen. Genet.* **118**:215.

Lamon, E. W. and J. C. Bennett. 1970. Antibodies to homologous RNA in the rabbit following stimulation by exogenous RNA. *Proc. Soc. Exp. Biol. Med.* **134**:968.

Leader, D. P., I. G. Wool and J. J. Castles. 1970. A factor for the binding of aminoacyl-transfer RNA to mammalian 40S ribosomal subunits. *Proc. Nat. Acad. Sci.* **67**:523.

Leader, D. P., H. Klein-Bremharr, I. G. Wool and A. Fox. 1972. Distribution of initiation factors in cell fractions from mammalian tissues. *Biochem. Biophys. Res. Comm.* **46**:215.

Lerman, M. I. 1968. Dissociation of structural proteins from the ribosomes of rat liver and *in vitro* reconstitution of biologically active ribosomes from protein-deficient derivatives of RNP particles and stripped proteins. *Mol. Biol.* (USSR) **2**:171.

Li, C.-C. and H. Amos. 1971. Alteration of phosphorylation of ribosomal proteins as a function of variation of growth conditions of primary cells. *Biochem. Biophys. Res. Comm.* **45**:1398.

Lin, A. and I. G. Wool. 1974. The molecular weights of rat liver ribosomal proteins determined by "three-dimensional" polyacrylamide gel electrophoresis. *Mol. Gen. Genet.* (in press).

Liew, C.-C. and A. G. Gornall. 1973. Acetylation of ribosomal proteins. I. Characterization and properties of rat liver ribosomal proteins. *J. Biol. Chem.* **248**:977.

Loeb, J. E. and C. Blat. 1970. Phosphorylation of some rat liver ribosomal proteins and its activation by cyclic AMP. *FEBS Letters* **10**:105.

Loening, U. E. 1968. Molecular weights of RNA in relation to evolution. *J. Mol. Biol.* **38**:355.

Low, R. B. and I. G. Wool. 1967. Mammalian ribosomal protein: Analysis by electrophoresis on polyacrylamide gel. *Science* **155**:330.

Low, R. B., I. G. Wool and T. E. Martin. 1969. Skeletal muscle ribosomal proteins: General characteristics and effect of diabetes. *Biochim. Biophys. Acta* **194**:190.

MacInnes, J. W. 1972. Difference between ribosomal subunits from brain and those from other tissues. *J. Mol. Biol.* **65**:157.

―――. 1973. Mammalian brain ribosomes are behaviorally and structurally heterogeneous. *Nature New Biol.* **241**:244.

Martin, T. E. and I. G. Wool. 1968. Formation of active hybrids from subunits of muscle ribosomes from normal and diabetic rats. *Proc. Nat. Acad. Sci.* **60**:569.

―――. 1969. Active hybrid 80S particles formed from subunits of rat, rabbit and protozoan (*Tetrahymena pyriformis*) ribosomes. *J. Mol. Biol.* **43**:151.

Martin, T. E., F. S. Rolleston, R. B. Low and I. G. Wool. 1969. Dissociation and reassociation of skeletal muscle ribosomes. *J. Mol. Biol.* **43**:135.

Martini, O. H. W. and H. J. Gould. 1971. Enumeration of rabbit reticulocyte ribosomal proteins. *J. Mol. Biol.* **62**:403.

―――. 1973. Phosphorylation of rabbit reticulocyte ribosomal proteins *in vitro*. *Biochim. Biophys. Acta* **295**:621.

Mathias, A. P. and R. Williamson. 1964. Studies on the protein of rabbit reticulocyte ribosomes. *J. Mol. Biol.* **9**:498.

Meyer-Bertenrath, J. G. 1967. Alterationen der Ribosomenstruktur während der durch Nitrosomorpholin induzierten Carcinogenese. *Hoppe-Seyler's Z. Physiol. Chem.* **348**:645.

Meyer-Bertenrath, J. G. and H. Würz. 1967. Aggregationsstufen und Biosyntheseleistungen von Rattenleber-ribosomen unter dem Einfluss Ribosomen-spezifischer Antikörper. *Z. Naturforschg.* **22b**:1153.

Möller, W. and J. Widdowson. 1967. Fractionation studies of the ribosomal proteins from *Escherichia coli*. *J. Mol. Biol.* **24**:367.

Möller, W., A. Groene, C. Terhorst and R. Amons. 1972. 50S ribosomal proteins. Purification and partial characterization of two acidic proteins, A_1 and A_2, isolated from 50S ribosomes of *Escherichia coli*. *Eur. J. Biochem.* **25**:5.

Moore, P. B., R. R. Traut, H. Noller, P. Pearson and H. Delius. 1968. Ribosomal proteins of *Escherichia coli*. II. Proteins from the 30S subunit. *J. Mol. Biol.* **31**:441.

Mutolo, V., G. Giudice, V. Hopps and G. Donatuti. 1967. Species specificity of embryonic ribosomal proteins. *Biochim. Biophys. Acta* **138**:214.

Noll, F. and H. Bielka. 1969. Antigeneigenschaften ribosomaler Proteine aus tierischen Geweben. *Acta Biol. Med. Germ.* **23**:15.

――――. 1970a. Untersuchungen über Proteine tierischer ribosomen. IV. Antigeneigenschaften von Ribosomen und ribosomalem Protein aus Leber und Hepatom. *Arch. Geschwultsforsch.* **35**:338.

――――. 1970b. Studies of proteins of animal ribosomes. III. Immunochemical analyses of ribosomes from different tissues and species of animals. *Mol. Gen. Genet.* **106**:106.

Noll, F., W. Heumann and H. Bielka. 1973. On the nature of antiribosome antibodies and some properties of their reaction with ribosomes. *Immunochemistry* **10**:9.

Nue, H. C. and L. A. Heppel. 1964. Some observations on the "latent" ribonuclease of *Escherichia coli. Proc. Nat. Acad. Sci.* **51**:1267.

Perry, R. P. 1967. The nucleolus and the synthesis of ribosomes. *Prog. Nucleic Acid Res. Mol. Biol.* **6**:219.

Perry, R. P. and D. E. Kelley. 1966. Buoyant densities of cytoplasmic ribonucleoprotein particles of mammalian cells: Distinctive character of ribosome subunits and the rapidly labeled components. *J. Mol. Biol.* **16**:255.

Petermann, M. L. 1964. *The physical and chemical properties of ribosomes.* Elsevier, New York.

Randall-Hazelbauer, L. L. and C. G. Kurland. 1972. Identification of three 30S proteins contributing to the ribosomal A site. *Mol. Gen. Genet.* **115**:234.

Reboud, A. M., M. Buisson, M. J. Amoros and J. P. Reboud. 1972. Partial *in vitro* reconstitution of active 40S ribosomal subunits from rat liver. *Biochem. Biophys. Res. Comm.* **46**:2012.

Rodgers, A. 1973. Ribosomal proteins in rapidly growing and nonproliferating mouse cells. *Biochim. Biophys. Acta* **294**:292.

Schur, P. H., L. A. Moroz and H. G. Kunkel. 1967. Precipitating antibodies to ribosomes in the serum of patients with systemic lupus erythematosus. *Immunochemistry* **4**:447.

Sherton, C. C. and I. G. Wool. 1972. Determination of the number of proteins in liver ribosomes and ribosomal subunits by two-dimensional polyacrylamide gel electrophoresis. *J. Biol. Chem.* **247**:4460.

――――. 1974. A comparison of the proteins of rat skeletal muscle and liver ribosomes by two-dimensional polyacrylamide gel electrophoresis. *J. Biol. Chem.* **249**:2258.

Spahr, P. F. 1962. Amino acid composition of ribosomes from *Escherichia coli. J. Mol. Biol.* **4**:395.

Stahl, J., H. Welfle and H. Bielka. 1972. Studies on proteins of animal ribosomes. XIV. Analysis of phosphorylated rat liver ribosomal proteins by two-dimensional polyacrylamide gel electrophoresis. *FEBS Letters* **26**:233.

Stöffler, G. and H.-G. Wittmann. 1971a. Sequence differences of *Escherichia coli* 30S ribosomal proteins as determined by immunochemical methods. *Proc. Nat. Acad. Sci.* **68**:2283.

――――. 1971b. Ribosomal proteins. XXV. Immunological studies on *Escherichia coli* ribosomal proteins. *J. Mol. Biol.* **62**:407.

Sturgill, B. C. and R. R. Carpenter. 1965. Antibody to ribosomes in systemic lupus erythematosus. *Arthritis Rheum.* **8**:213.

Sturgill, B. C. and M. R. Preble. 1967. Antibody to ribosomes in systemic lupus erythematosus: Demonstration by immunofluorescence and precipitation in agar. *Arthritis Rheum.* **10**:538.

Terao, K. and K. Ogata. 1971. Studies on the small subunit of rat liver ribosomes: Some biochemical properties with specific reference to the reconstruction of the small subunit. *Biochim. Biophys. Acta* **254**:278.

―――. 1972. Characterization of the proteins of the small subunits of rat liver ribosomes. *Biochim. Biophys. Acta* **285**:473.

Terhorst, C., B. Wittmann-Liebold and W. Möller. 1972. 50S ribosomal proteins. Peptide studies on two acidic proteins, A_1 and A_2, isolated from 50S ribosomes of *Escherichia coli. Eur. J. Biochem.* **25**:13.

Terhorst, C., W. Möller, R. Laursen and B. Wittmann-Liebold. 1973. The primary structure of an acidic protein from 50S ribosomes of *Escherichia coli* which is involved in GTP hydrolysis dependent on elongation factor G and T. *Eur. J. Biochem.* **34**:138.

Thammana, P., C. G. Kurland, E. Deusser, J. Weber, R. Maschler, G. Stöffler and H.-G. Wittmann. 1973. Structural and functional evidence for a repeated 50S subunit ribosomal protein. *Nature New Biol.* **242**:47.

Traugh, J. A. and R. R. Traut. 1972. Phosphorylation of ribosomal proteins of *Escherichia coli* by protein kinase from rabbit skeletal muscle. *Biochemistry* **11**:2503.

―――. 1974. Characterization of protein kinases from rabbit reticulocytes. *J. Biol. Chem.* **249**:1207.

Traugh, J. A., M. Mumby and R. R. Traut. 1973. Phosphorylation of ribosomal proteins by substrate-specific protein kinases from rabbit reticulocytes. *Proc. Nat. Acad. Sci.* **70**:373.

Traut, R. R., G. A. Howard and J. A. Traugh. 1974. Ribosomal protein from rabbit reticulocyte: Numbers, molecular weights, relative amounts and phosphorylation by protein kinase. In *Third international symposium on the metabolic introconversion of enzymes* (ed. E. Fisher et al.) Springer Verlag, Berlin (in press).

Ts'o, P. O., J. Bonner and H. Dintzis. 1958. On the similarity of amino acid composition of microsomal nucleoprotein particles. *Arch. Biochem. Biophys.* **76**:225.

Van Duin, J. and C. G. Kurland. 1970. Functional heterogeneity of the 30S ribosomal subunit of *E. coli. Mol. Gen. Genet.* **109**:169.

Ventimiglia, F. and I. G. Wool. 1974. A kinase that transfers the γ-phosphoryl group of GTP to proteins of eukaryotic 40S ribosomal subunits. *Proc. Nat. Acad. Sci.* **71**:350.

Walton, G. M., G. N. Gill, I. B. Abrass and L. D. Garren. 1971. Phosphorylation of ribosome-associated protein by an adenosine 3' 5' cyclic AMP-dependent protein kinase: Location of the microsomal receptor and protein kinase. *Proc. Nat. Acad. Sci.* **68**:880.

Warner, J. R. 1971. The assembly of ribosomes in yeast. *J. Biol. Chem.* **246**: 447.

Weinberg, R. A. and S. Penman. 1970. Processing of 45S nucleolar RNA. *J. Mol. Biol.* **47**:169.

Welfle, H. 1971a. Studies on proteins of animal ribosomes. XII. Determination of molecular weights of ribosomal proteins by SDS gel electrophoresis. *Acta Biol. Med. Germ.* **27**:211.

―――. 1971b. Studies on proteins of animal ribosomes. XI. A simple method of two-dimensional polyacrylamide gel electrophoresis of ribosomal proteins of rat liver. *Acta Biol. Med. Germ.* **27**:547.

Welfle, H. and H. Bielka. 1968. Characterisierung von Proteinen aus Leber- und Hepatomribosomen durch Polyacrylamid-Gelelektrophorese. *Z. Naturforsch.* **23b**:690.

Welfle, H., H. Bielka and M. Böttger. 1969. Studies on proteins of animal ribosomes. Separation of ribosomal proteins from rat liver by preparative polyacrylamide gel electrophoresis and some properties of the protein fractions. *Mol. Gen. Genet.* **104**:165.

Welfle, H., J. Stahl and H. Bielka. 1971. Studies on proteins of animal ribosomes. VIII. Two-dimensional polyacrylamide gel electrophoresis of ribosomal proteins of rat liver. *Biochim. Biophys. Acta* **243**:416.

―――. 1972. Studies on proteins of animal ribosomes. XIII. Enumeration of ribosomal proteins of rat liver. *FEBS Letters* **26**:228.

Westermann, P. 1971. Studies on proteins of animal ribosomes. VI. Comparison of the proteins of the 30S and 50S subunit of rat liver ribosomes by chromatography and electrophoresis. *Acta Biol. Med. Germ.* **26**:617.

Westermann, P. and H. Bielka. 1973. Studies on proteins of animal ribosomes. XV. Proteins of the small subunit of rat liver ribosomes: Isolation, amino acid composition, tryptic peptides and molecular weights. *Mol. Gen. Genet.* **126**:349.

Westermann, P., H. Bielka and M. Böttger. 1969. Studies on proteins of animal ribosomes. I. Separation of ribosomal proteins from rat liver by chromatography on CM-cellulose and properties of some protein components. *Mol. Gen. Genet.* **104**:157.

―――. 1971. Studies on proteins of animal ribosomes. VII. Isolation of proteins from rat liver ribosomes, their molecular weights, amino acid compositions and secondary structure. *Mol. Gen. Genet.* **111**:224.

Wikman-Coffelt, J. 1972. Radioimmunoassay for ribosomes. *Analyt. Biochem.* **48**:339.

Wikman-Coffelt, J., G. A. Howard and R. R. Traut. 1972. Comparison of antigenic properties of ribosomal proteins from Novikoff hepatoma and normal liver. *Biochim. Biophys. Acta* **277**:671.

Wittmann, H.-G. and G. Stöffler. 1972. Structure and function of bacterial ribosomal proteins. In *The mechanism of protein synthesis and its regulation* (ed. L. Bosch) pp. 285–351. North-Holland, Amsterdam.

Yamamura, H., Y. Inoue, R. Shimomura and Y. Nishizuka. 1972. Similarity and pleotropic actions of cyclic-AMP-dependent protein kinase from mammalian tissues. *Biochem. Biophys. Res. Comm.* **46**:589.

Zamir, A., R. Miskin and D. Elson. 1971. Inactivation and reactivation of ribosomal subunits: Aminoacyl-transfer RNA binding activity of the 30S subunit of *Escherichia coli. J. Mol. Biol.* **60**:347.

The Assembly of Ribosomes in Eukaryotes

Jonathan R. Warner
Departments of Biochemistry and Cell Biology
Albert Einstein College of Medicine
Bronx, New York 10461

1. **Components of the Ribosome**
   A. Ribosomal RNA
   B. 5S RNA
   C. Ribosomal Proteins
2. **Interaction of Ribosomal Precursor RNA and Protein**
   A. Electron Microscopic Observations
   B. Isolation of Nucleolar RNP
   C. Appearance of New Ribosomes in the Cytoplasm
   D. Kinetics of Ribosomal Protein Appearance
3. **Ribosome Synthesis under Various Conditions**
   A. Ribosome Synthesis and the Cell Cycle
   B. Effect of Temperature on Ribosome Synthesis
   C. In Vitro Ribosome Synthesis
4. **Regulation of Ribosome Biosynthesis**
   A. General Considerations on Regulation
   B. Regulation of rRNA Production
   C. Relationship of Ribosome Synthesis to Protein Synthesis in Cultured Cells
   D. Migration to the Cytoplasm
   E. Ribosomal Precursor RNA "Wastage" in Cultured Lymphocyte
   F. Ribosomal Precursor RNA in Normal and Regenerating Liver
   G. Regulation of rRNA Synthesis: Summary
   H. Synthesis and Fate of Ribosomal Proteins
   I. Regulation of Synthesis of Ribosomal Proteins
   J. Turnover of Ribosomal Constituents
   K. New Approaches—Genetics
5. **Speculations**
   A. Variety of Regulatory Sites
   B. Model of Ribosome Assembly

INTRODUCTION

This review is primarily concerned with the assembly of ribosomes, that is, the interaction of RNA and protein leading to the production of a functional ribosomal subunit. Yet this interaction has such a profound influence on the synthesis of ribosomal precursor RNA and its maturation to ribosomal RNA (rRNA), that it will be necessary to consider nearly every aspect of ribosome structure and biosynthesis to some degree. Fortunately a number of topics are discussed in greater detail in this volume. Reeder describes the organization of the genes for rRNA, and Maden, Salim and Robertson discuss the chemical relationship between ribosomal precursor RNA and rRNA.

Recent comprehensive reviews have been published by Maden (1971) and by Craig (1973).

COMPONENTS OF THE RIBOSOME

Ribosomal RNA

It now appears that nearly all RNA in eukaryotic cells is produced by the cleavage of longer transcripts (reviewed by Weinberg 1973). This concept was derived first from the observations on the formation of ribosomal RNA in mammalian cells from large ribosomal precursor RNA. The reader is referred to several recent reviews for the evidence demonstrating the pathways of maturation for various species (Craig 1973; Maden 1971; Attardi and Amaldi 1970). Elsewhere in this volume Maden, Salim and Robertson describe more recent work in this area.

In summary, the pathway is similar in all eukaryotes, from the mammals to the fungi (see Figure 1). A single transcript is formed in the nucleolus by a specific polymerase (reviewed by Jacob 1973). This is methylated, either during (Greenberg and Penman 1966) or after (Udem and Warner 1972) transcription, and is then cleaved sequentially at several sites. Some fragments, representing nearly 50% of the total in mammalian cells, but less than 20% in the cells of lower eukaryotes, must be rapidly degraded, since they have never been observed. These often have a distinct base composition and have no methyl groups. Eventually three stable RNAs are produced from each transcript, the 18S RNA of the small subunit, the 25–

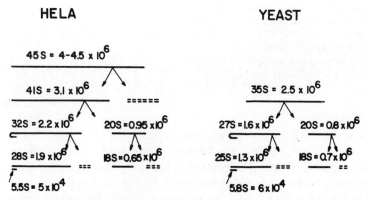

Figure 1   The formation of rRNA in HeLa cells and yeast cells. The results for HeLa cells are taken from Weinberg and Penman (1970) and those for yeast cells are taken from Udem and Warner (1972).

28S RNA of the large subunit, and the 5.5–5.8S RNA, which is non-covalently bonded to the RNA of the large subunit. Most of the processing occurs in the nucleolus, although in yeast the final step in the formation of 18S RNA appears to take place in the cytoplasm (Udem and Warner 1973).

5S RNA

5S RNA in eukaryotes appears to be an exception in that it is not derived from a precursor molecule (Weinberg 1973). In yeast, its genes are interspersed among the ribosomal precursor RNA genes (Rubin and Sulston 1973). In higher eukaryotes, however, the genes for 5S RNA are separated from those of rRNA on the chromosome (Brown and Weber 1968) or are on different chromosomes (Tartof and Perry 1970). Their transcription is not coordinated with that of the ribosomal precursor RNA (Perry and Kelley 1968). On the other hand, there appears to be a large pool of 5S RNA to supply the ribosome assembly process (Leibowitz, Weinberg and Penman 1973). In any case, 5S RNA becomes associated with the ribosome at an early stage in its development in the nucleolus (Warner and Soeiro 1967).

Ribosomal Proteins

The protein complement of eukaryotic ribosomes is discussed in detail elsewhere in this volume by Wool and Stöffler and by Traut et al. Eukaryotic ribosomes have considerably more protein than those of prokaryotes, as they have larger RNA molecules and a higher protein to RNA ratio. There are approximately 70 ribosomal proteins. Little is yet known about individual proteins. As a whole, they are very basic, generally smaller than other proteins in the cell, but rather larger than the ribosomal proteins of *E. coli*.

For the purpose of studying the assembly of ribosomes, it is important to distinguish between those proteins that are assembled with the RNA during the original synthesis of the ribosome and any proteins that later become associated with the ribosome in a transient way. It is well known that many factors involved in protein synthesis, as well as a myriad of other proteins, are loosely bound to ribosomes in a cell extract but can be removed if the ribosomes are washed at high ionic strength. Other proteins are present in less than molar amounts (Kurland et al. 1969). This probably reflects heterogeneity not of the ribosome itself, but of the stage of protein synthesis in which it was involved at the time the cell was broken. These two classes of protein therefore appear to be reversibly associated with the ribosome.

In eukaryotic cells it has been possible to distinguish directly between permanent and transient ribosomal proteins. In HeLa cells ribosome synthesis can be blocked while protein synthesis continues by adding actinomycin D (Warner 1966) or in yeast by having certain mutants at a restrictive temperature (Warner and Udem 1972). When radioactive amino acids are incorporated by such cells, only a limited number of labeled proteins are found on washed ribosomes. In both cells, only three or four newly made proteins are associated with old ribosomes. The rest of the proteins on the ribosomes are not labeled, even though other experiments show that they are being synthesized (Craig 1971; Warner and Udem 1972). The absence of radioactivity in most ribosomal proteins shows that these proteins do not exchange with cytoplasmic pools of proteins but must be assembled with ribosomal precursor RNA during ribosome synthesis. These observations are in direct conflict with the claim of Dice and Schimke (1972) that 90% of the protein of rat liver ribosomes is exchangeable with a pool of ribosomal proteins in the cytoplasm, which constitutes 17% of the total soluble proteins of the cell. In criticizing these experiments for lack of adequate purification of the ribosomes, Terao et al. (1974) also find that while there may be some binding of cytoplasmic proteins to ribosomes, there is no exchange of the major ribosomal proteins. Furthermore, the proteins that are bound are different from the major ribosomal proteins on two-dimensional gel electrophoresis. Wool and Stöffler (see this volume) come to a similar conclusion on immunological grounds. The evidence at present favors the view that the major portions of the ribosome form an integral, stable structure.

INTERACTION OF RIBOSOMAL PRECURSOR RNA AND PROTEIN

Electron Microscopic Observations

Electron microscopists have long distinguished between fibrillar and granular regions of the nucleolus, the former being somewhat sensitive to DNase (Busch and Smetana 1970). High resolution autoradiographs, after short pulses with tritiated nucleosides, indicated that RNA synthesis occurs in the fibrillar region and only later becomes associated with the granular part (Granboulan and Granboulan 1965). Since the granules bore some resemblance to cytoplasmic ribosomes, it was natural to conclude that the ribosomal precursor RNA was transcribed from DNA in the fibrillar region and assembled with protein in the granular region before migration to the cytoplasm. Recent electron micrographs, however, indicate that proteins become associated with ribosomal precursor RNA during transcription (Miller et al. 1970).

Isolation of Nucleolar RNP

The nucleolus is a very prominent organelle in the cell, its high refractive index indicating a high concentration of macromolecules, but it is not enclosed in a membrane. It now seems likely that its structure is maintained in large part by disulfide bonds, because treatment of isolated nucleoli with high concentrations of dithiothreitol causes extensive disaggregation and releases ribonucleoprotein particles (RNP) (Warner and Soeiro 1967) from the granular portion (Narayan and Birnstiel 1969). Polyvinyl sulfate also releases RNP containing ribosomal precursor RNA from the nuclei of FL cells (Yoshikawa-Fukada 1967). These RNPs contain all species of ribosomal precursor RNA as well as many, if not all, of the ribosomal proteins (Warner and Soeiro 1967). Little (if any) free protein or RNA is released. By the use of sulfhydryl reagents with or without EDTA, nucleolar RNPs have been obtained from L cells (Liau and Perry 1969), rat liver (Narayan and Birnstiel 1969; Prestayko, Lewis and Busch 1972; Tsurugi, Morita and Ogata 1973), amphibian oocytes (Rogers 1968) and peas (Takahashi et al. 1972). Mirault and Scherrer (1971) have obtained improved purification of the nucleolar RNP using gel electrophoresis.

When the nucleoli are extracted with dithiothreitol in the presence of EDTA, the only products are two species of RNP which sediment at 80S and 55S. These ribonucleoprotein particles have been extensively analyzed in terms of their RNA and protein content (Warner and Soeiro 1967; Kumar and Warner 1972).

The 80S nucleolar particles (110S in 5×10^{-4} M Mg^{++}) contain 45S RNA, 5S RNA, ribosomal proteins and nucleolar proteins. The detailed protein composition remains unclear because these particles are (a) only one-sixth as abundant as 55S particles in the nucleolus, (b) extracted in less than 50% yield from the nucleoli, and (c) relatively labile to nuclease degradation. The 55S particle, which sediments at 65–75S in the presence of 5×10^{-4} M Mg^{++} (Liau and Perry 1969), contains exclusively 32S RNA and 5S RNA. Its protein components are of two classes, distinguished by their size and by their kinetics of labeling (Kumar and Warner 1972). The bulk of the proteins appears to be newly formed ribosomal proteins when compared by electrophoresis with ribosomal proteins from the 60S subunit. This has been confirmed by fingerprint analysis (Shepherd and Maden 1972). When particles have been isolated from nucleoli of cells briefly labeled with an amino acid, the specific activity of these proteins is far higher than that of the cytoplasmic ribosomes. However after a chase with nonradioactive amino acid, the specific activity of the nucleolar particles falls, while that of the cytoplasmic subunits rises. These considerations led us to conclude that the nucleolar RNPs are immature ribosomes, undergoing processing within the nucleolus before migrating to the cytoplasm (Warner and Soeiro 1967).

On analyzing the proteins remaining on the nucleolar RNP after a long chase, it became evident that these particles have another set of proteins, which do not pass out to the cytoplasm, and which, by electrophoretic analysis, are not ribosomal proteins. They are in general larger than the ribosomal proteins, become labeled more slowly, and do not chase out of the nucleolar particles (Kumar and Warner 1972). At least four such proteins have been identified by two-dimensional gel electrophoresis (Tsurugi, Morita and Ogata 1973). Since the RNPs constantly leave the nucleolus and are replaced with new ones, these proteins must be continuously reutilized as described by the model proposed in Figure 2. Nothing is known about the topology of the binding of either set of proteins to the ribosomal precursor RNA.

Because the 45S and 32S ribosomal precursor RNAs have sequences which are discarded during maturation, it is not unexpected to find nucleolar proteins attached to the newly forming ribosomal subunits. In fact, the buoyant density of the nucleolar RNP is less than that of the cytoplasmic ribosomes (Liau and Perry 1969; Craig and Perry 1970; Kumar and Warner 1972), indicating that the protein to RNA ratio is greater in nucleolar particles. The phenomenon is not understood, although control experiments have shown that the particles as isolated have not been formed by the aggregation of nucleolar protein onto the ribosomal precursor RNA.

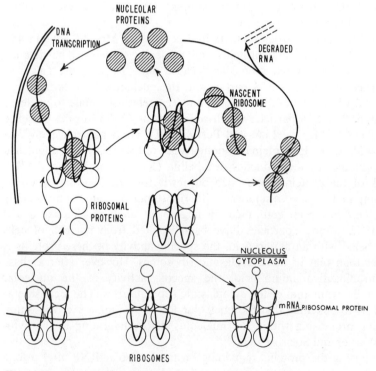

Figure 2   The reutilization of nucleolar proteins during ribosome formation. The diagram has been simplified by omitting the nucleolar 80S ribonucleoprotein particles and the 40S cytoplasmic ribosomal subunit. The topological arrangement of RNA and proteins is arbitrary.

Nucleolar RNP isolated from cells in which protein synthesis is inhibited (Pederson and Kumar 1971) or stopped (Craig and Perry 1970) have a more heterogeneous density distribution.

One possible explanation for these results is that the interaction of ribosomal proteins with the ribosomal RNA is more intimate, such that each protein molecule interacts with the RNA at several sites, leading to a lower protein/RNA ratio than obtains on the nonconserved regions, where the proteins attach to only a single site on the RNA. Such a speculation is supported by electron micrographs of nucleolar RNP (Narayan and Birnstiel 1969; Kumar, unpublished observations) in which the particles appear on the whole more amorphous and less compact than ribosomes. This would explain why the 80S nucleolar particle cosediments with the 80S ribosome, which is half its molecular weight.

Appearance of New Ribosomes in the Cytoplasm

In the earliest studies of the functional distribution of ribosomes in cultured cells, it was found that 80–90% were on polysomes, engaged in protein synthesis, while most of the rest were single, 80S ribosomes, presumably idling. A small proportion, never more than 5% and often less, were present as native 60S and 40S subunits (Girard et al. 1965). It is now known that subunits exist only fleetingly while taking part in the initiation of polypeptide chains (Henshaw, Goinez and Hirsch 1973).

New ribosomal RNP entering the cytoplasm from the nucleus appear first as free subunits (Girard et al. 1965), such that after brief labeling times the subunits have a specific activity that far exceeds that of the polysomes. Newly labeled 40S subunits appear in the cytoplasm before 60S subunits, presumably because the maturation of their RNA requires less time. Yet new subunits are fundamentally different from old ones in that they spend several minutes as subunits, whereas old subunits recycle onto polysomes within seconds. The reason for this is now clear in the case of yeast 40S subunits, for they are not mature when they enter the cytoplasm. They contain 20S precursor RNA (Udem and Warner 1973). When the 20S RNA is converted to 18S RNA, the subunits immediately become part of a polysome. In vitro, mature but not new 40S subunits interact with 60S subunits and poly(U) to form 80S particles (Warner and Skogerson, unpublished observations). A similar situation may obtain for the final methylation step of 18S RNA in HeLa, which apparently occurs in the cytoplasm (Maden, Salim and Summers 1972).

The RNA of the 60S subunit appears to be mature when it enters the cytoplasm, suggesting that some final maturation of the protein moiety is necessary. There is no direct evidence on this matter since new ribosomal subunits cannot be physically separated from their mature counterparts. However there are several proteins which reversibly associate with the ribosome, as discussed above. At least one, and probably all, of these proteins is not involved in the nuclear assembly of the ribosomes (Warner and Soeiro 1967). Perhaps several minutes are required for the newly formed 60S subunit to acquire a full complement of these proteins.

Kinetics of Ribosomal Protein Appearance

In order to understand the dynamics of ribosomal synthesis, I have studied the kinetics of appearance of ribosomal protein in cytoplasmic ribosomes in both HeLa (Warner 1966) and yeast (Warner 1971). When cells are permitted to incorporate radioactive amino acids for a brief period, the ribosomal proteins made at that time appear as part of newly formed ribosomes during a period of approximately 10% of a generation time in each case. This is approximately twice the minimal time necessary to assemble a subunit. Since the passage of ribosomal precursor RNA through the various stages of maturation is thought to be a stochastic process, ribosomal protein appears in the cytoplasm with about the same kinetics as ribosomal RNA made at the same time, except for the original lag period. These considerations, as well as the relatively insignificant assembly of ribosomes in the absence of protein synthesis and the instability of new ribosomal proteins in the absence of ribosomal precursor RNA, make it unlikely that there are significant pools of free ribosomal protein. At the time of a previous suggestion that such pools exist (Warner 1966), I was unaware that most of the ribosomal proteins in the nucleolus were bound to ribosomal precursor RNA.

RIBOSOME SYNTHESIS UNDER VARIOUS CONDITIONS

Ribosome Synthesis and the Cell Cycle

Does the rate of ribosomal precursor RNA transcription increase smoothly during the cell cycle, or is there an abrupt increase when rDNA is synthesized? Although conflicting views exist, there is good evidence now that there is a smooth increase for most types of cells (Mitchison 1971), implying that the rDNA of a cell is not the limiting factor in rRNA synthesis during G_1.

In yeast, as well as other simple eukaryotes, there is evidence that many enzymes are synthesized at a unique, limited stage of the cell cycle (Halvorson, Carter and Tauro 1971). We have recently shown, however, that this is not true of ribosomal proteins, which, individually and collectively, are synthesized throughout the cell cycle (Shulman, Hartwell and Warner 1973). Evidence to be presented below suggests that the supply of ribosomal proteins limits the assembly of ribosomes and the synthesis of ribosomal precursor RNA. Presumably, therefore, the synthesis of these proteins increases continually during the cell cycle as well.

In most eukaryotic cells as the chromosomes condense during mitosis, the nuclear membrane breaks down and the nucleoli become indistinct. At this time no ribosomal precursor RNA is transcribed, although 5S RNA continues to be (Zylber and Penman 1971). Previously synthesized 45S and 32S RNA remain intact, apparently as RNP loosely associated with the chromosomes, and mature into ribosomes when the nucleoli reform in G_1 (Fan and Penman 1971). Thus chromosome condensation and mitosis bring about a suspension of ribosome formation. It is not known if nascent transcripts remain attached to the DNA under these conditions.

Effect of Temperature on Ribosome Synthesis

A number of studies have been concerned with the effect of temperature on ribosomal precursor RNA maturation. When the temperature is increased from 37 to 42°C, ribosomal precursor RNA synthesis continues, but no stable rRNA is formed (Warocquier and Scherrer 1969). There is little change in the level of nucleolar precursors (Amalric, Simard and Zalta 1969), leading to the conclusion that there is effective degradation of excess RNA. Electron micrographs suggest that the nucleolar RNP are reversibly unfolded at the higher temperature (Simard and Bernhard 1967; Amalric, Simard and Zalta 1969).

Studying the effects of lowered temperature, Stevens and Amos (1971) found that the incorporation of both [^{14}C]uridine and [^{3}H]methylmethionine into ribosomal precursor RNA had the same temperature coefficient, the rate being halved at approximately 29°C. The conversion of 45S to 32S RNA was less, and of 32S to 28S RNA was more sensitive to a decrease in temperature.

When yeast are shifted from 23 to 36°C, they undergo a transient inhibition of ribosomal precursor RNA synthesis and maturation (Warner and Udem 1972). By 20 min they have returned to their original level, and by 40 min both processes occur at double the 23°C rate.

In none of these cases, however, is it clear whether the primary influence on maturation involves a temperature-sensitive enzyme or a temperature-sensitive RNA–protein interaction which presents the incorrect site to an enzyme.

In Vitro Ribosome Synthesis

Processing of Ribosomal Precursor RNA. A number of laboratories have investigated the processing of ribosomal precursor RNA in vitro, either in whole nucleoli (Vesco and Penman 1968; Liau, Craig and Perry 1968) or in extracts containing nucleolar RNP (Mirault and Scherrer 1972). All obtain smaller RNA species, and in the latter case, this is dependent on a partially purified enzyme preparation. However the nucleus is a veritable zoo of nucleases. To identify the right one(s), it will eventually be necessary to show that each product has the right molecular weight, base composition, methylation and end groups. This question is discussed elsewhere in this volume by Schlessinger. The results thus far are disappointing.

Reconstitution of Ribosomal Subunits.   Although progress in this field has been excruciatingly slow, Terao and Ogata (1971) have recently succeeded in a partial reconstitution of the 40S subunit from 18S RNA and proteins. Now that improved assays for ribosomal function and improved purification of ribosomal proteins (Wool and Stöffler, this volume) are available, there is more hope for the future.

THE REGULATION OF RIBOSOME BIOSYNTHESIS

General Considerations on Regulation

Most of our ideas about the regulation of macromolecular synthesis have developed from the studies of prokaryotic cells, which led to concepts such as the operon and the stringent control of ribosome synthesis. In transposing such concepts to eukaryotic cells, it is essential to bear in mind two critical features of the latter.

1. In the eukaryotic cell, the genome is within a nucleus, separated from the protein synthetic apparatus by a double nuclear membrane (reviewed by Hamkalo and Miller 1973). Thus the tight coupling of transcription and translation, which plays a major role in bacterial regulation, cannot exist.

2. The evolution of higher organisms has required regulatory mechanisms that will maintain homeostasis within the organism, *preventing* cell growth in most cases. In contrast, the selective pressures for bacteria insist on efficiency of growth at almost any cost. Some eukaryotes, such as yeast, may fall into this category as well. By comparing the regulatory mechanisms of yeast with those of higher eukaryotes, we may be able to distinguish those which pertain to the separation of nucleus and cytoplasm from those adapted for the special requirements of homeostatic organisms.

In assessing the type of regulation to be found in a eukaryotic cell, it is important to know whether the process being regulated places great demands on the cell. Ribosomes themselves are only indirectly responsible for changes in the constituent molecules of the cell. That is the task of the mRNA and its products. Therefore the major reason to regulate the synthesis of ribosomes appears to be the economic one of conserving nucleotides, amino acids and ATP. Yet no data is available to estimate how much of the cell's energy supply is used to sustain RNA and protein synthesis. There is, however, one experimental fact which implies that the cell does not seem to require very efficient regulation of RNA synthesis. In actively growing HeLa cells, ribosomal precursor RNA synthesis accounts for less than 25% of the total RNA synthesis in the cell (Soeiro et al. 1968). The bulk represents heterogeneously sedimenting nuclear RNA (HnRNA), some of which is precursor to mRNA (reviewed by Weinberg 1973). But the synthesis of HnRNA exceeds the production of mRNA by two orders of magnitude. The rest of the HnRNA is degraded within minutes. Superficially at least, the efficient utilization of RNA does not seem to have a high priority in cultured cells. On the other hand, in simple eukaryotes such as the slime mold (Firtel and Lodish 1973) and yeast (Warner, unpublished), HnRNA synthesis exceeds mRNA production by no more than a factor of 2.

Regulation of rRNA Production

The regulation of rRNA production has been extensively studied in many eukaryotic organisms (see reviews by Maden 1971; Craig 1973; Perry 1973). The dramatic switching-on of rRNA synthesis at gastrulation in the sea urchin and toad embryos shows the role that such regulation can play in development and in cell proliferation. This review, however, will be confined to only a few systems, in which the regulation appears to involve maturation of ribosomal precursor RNA or its assembly with proteins.

There are three major ways in which the production of new ribosomal RNA can be regulated:

(1) By changing the number of genes, as best exemplified by the amplification of rRNA genes of the developing oocyte in Xenopus. Such genetic changes are discussed at length in this volume by Reeder.

(2) By influencing transcription of ribosomal precursor RNA through changes in the rate of either initiation or elongation of polynucleotide chains.

(3) By altering the efficiency of the process in which ribosomal precursor RNA matures to rRNA.

It is often difficult to distinguish experimentally between the latter, since there is clearly some mechanism that maintains a balance between the rate of transcription and the rate of processing. Furthermore in many cases ribosomal precursor RNA which is not processed promptly is degraded. Thus if rRNA synthesis decreases, it is necessary to distinguish between an inhibition of ribosomal precursor RNA transcription and an increase in its degradation; if rRNA synthesis increases, it is necessary to distinguish between a stimulation of ribosomal precursor RNA transcription and in increase in the efficiency of its processing.

Relationship of Ribosome Synthesis to Protein Synthesis in Cultured Cells

The obvious drawback of studying the regulation of ribosome synthesis in freely growing animal cells in culture is that such cells, by their nature, have escaped from some normal regulatory process. Yet such cells offer a great versatility of experimental approach and have often been used to develop concepts which can then be applied to more complex systems.

Much work has been carried out in the HeLa cell, where it has been possible to measure accurately many of the parameters involved. Greenberg and Penman (1966) have shown, using pulses of [³H]methionine, that RNA polymerase requires 2.3 min to transcribe a 45S RNA molecule. After an average lifetime of 20 min, it is converted to 32S RNA, which, in turn, has an average lifetime of 45 min before being converted to 28S RNA. A HeLa cell synthesizes ribosomes at a rate tenfold greater than a liver or kidney cell, and since it requires 2–3 times longer to process 45S ribosomal precursor RNA to mature cytoplasmic ribosomes, there must be 20–30 times as many intermediate species (Chaudhuri and Lieberman 1968). This fact has facilitated the study of ribosomal precursor RNA metabolism in HeLa cells since one can readily follow the ribosomal precursor RNA both by UV absorbance and by radioactivity.

Total inhibition of protein synthesis by cycloheximide in either HeLa (Warner et al. 1966) or L cells (Ennis 1966) causes a two- to threefold inhibition of ribosomal precursor RNA synthesis and an even more severe inhibition of ribosomal precursor RNA maturation. In both cell types a small number of complete ribosomes can be synthesized in the complete absence of protein synthesis, many citations to the contrary notwithstanding. If cells are pulsed with [³H]uridine between 5 and 10 min after cycloheximide treatment, some label appears in cytoplasmic rRNA, but less than 10% of the untreated control. Messenger RNA synthesis and transport, however, appear to be unaffected (Warner et al. 1966). The residual ribosome formation is probably due to the small pools of ribosomal protein available in the nucleolus (Warner 1966) and to the limited exchangeability of the proteins on the nucleolar RNP.

Unexpectedly, puromycin causes less inhibition of ribosomal precursor RNA synthesis, but more inhibition of its processing (Soeiro, Vaughan and Darnell 1968). While 45S is converted to 32S RNA, essentially no mature rRNA is formed.

In a more complete study of the effects of cycloheximide in HeLa cells, Willems, Penman and Penman (1969) found that the decreased rate of 45S RNA synthesis was due to a decreased number of polymerase molecules transcribing at the same rate. This reduced rate of synthesis was just balanced by an increased processing time, leading to a constant amount of 45S RNA in the nucleolus. Similarly the amount of 32S RNA was unchanged. These results suggest that newly synthesized proteins are involved in the initiation of transcription and in maturation, but probably not in transcription itself or in the release of 45S RNA from the template, as suggested by Perry (1973).

In yeast, the maturation of ribosomal precursor RNA is more closely tied to protein synthesis. While some 35S ribosomal precursor RNA can be synthesized in the presence of cycloheximide (deKloet 1966), it does not mature and the maturation of previously made ribosomal precursor RNA ceases within one minute (Udem and Warner 1972). Cycloheximide attacks the 60S subunit of the ribosomes (Rao and Grollman 1967). Since conversion of 20S to 18S RNA is inhibited as fully as other maturation steps, cycloheximide must act by causing an immediate limitation of the supply of critical proteins for each subunit.

In a recent publication, Yu and Feigelson (1973) have proposed that cycloheximide inhibits ribosomal precursor RNA synthesis in rat liver because the RNA polymerase or one of its subunits is rapidly turning over. They isolated nucleoli from the livers of rats treated for increasing periods of time with cycloheximide and observed a progressive decline in the incorporation of nucleoside triphosphates into RNA. Nucleoli from rats treated for 80 min had approximately half the potential for incorporation as the controls. While the final answer awaits identification and purification of the polymerase involved, it is important to bear in mind that proteins are intimately associated with the growing ribosomal precursor RNA chains (Miller et al. 1970), and that as the supply of such proteins decreases, there is likely to be a decreased endogenous rate of synthesis in the nucleolus. It will be important to determine if the reaction being studied by Yu and Feigelson is the elongation of previously initiated ribosomal precursor RNA transcripts, the initiation of new ones, or the synthesis of nonspecific RNA.

When cultured cells are starved for an amino acid, net protein synthesis stops, but actual protein synthesis continues at 25–30% of normal due to turnover of intracellular portein. Consequently, formation of both ribosomal subunits continues as well, but at a reduced rate (Maden et al. 1969). Lack of the amino acid slows both the rate of synthesis of 45S RNA, measured directly, and the rate of conversion of 45S to 32S RNA, measured after actinomycin D treatment. The total amount of and ratio between 45S and 32S RNA are barely affected. It was concluded that "the availability of some proteins may regulate the rates of synthesis and of cleavage of ribosomal precursor RNA."

The lack of an amino acid generally affects the synthesis of both 28S and 18S RNA equally, but insufficient valine causes an additional deficit of 28S RNA, and insufficient lysine an additional deficit of 18S RNA (Maden 1972). A lack of methionine, on the other hand, completely prevents the formation of any mature RNA, although 45S RNA is formed and is converted to 32S RNA (Vaughan et al. 1967). Both are severely deficient in methyl groups. One can rescue some of this ribosomal precursor RNA by adding methionine to the cells. Thus while submethylated 45S RNA can give rise to 32S RNA, the lack of methyl groups in some way prevents the formation of a stable mature subunit. The 55S nucleolar RNP containing submethylated 32S RNA appears to contain most of its usual complement of proteins (Maden and Vaughan 1969). Thus it is not yet clear whether the lack of methylation prevents the attachment of certain key proteins or whether it exposes the RNA to nuclease attack at incorrect sites.

A number of drugs which are either incorporated into ribosomal precursor RNA, such as toyocamycin, an adenine analog (Tavitian, Uretsky and Acs 1968), and azacytidine (Reichman, Karlan and Penman 1973), or which intercalate into ribosomal precursor RNA, such as proflavine (Snyder, Kann and Kohn 1971), cause abnormal maturation of RNA. It will be important to determine if the conformational changes induced by these drugs modify the interaction with the cleavage enzymes directly, or whether they cause a modified interaction with nucleolar proteins.

Migration to the Cytoplasm

Several observations indicate that the migration of newly completed subunits to the cytoplasm is more complex than originally envisaged. Weinberg et al. (1967) have shown that even during the earliest stages of poliovirus infection, ribosomal precursor RNA processing becomes aberrant, leading to an accumulation of intermediates and of 18S RNA in the nucleolus. Kumar and Wu (1973) found that the 45S RNA made just prior to inhibition of RNA synthesis with camptothecin is processed normally except that the newly formed 28S RNA remains in the nucleus. On removal of camptothecin, the 28S RNA promptly migrates to the cytoplasm, even in the absence of protein synthesis, but not in the presence of a concentration of actinomycin low enough to block only ribosomal precursor RNA synthesis. Superficially at least, there appears to be a direct connection between the first step of rRNA synthesis, transcription, and the last step, migration of the mature particle to the cytoplasm.

Ribosomal Precursor RNA "Wastage" in Cultured Lymphocytes

The cultured lymphocyte is a relatively dormant cell until stimulated with a mitogen such as phytohemagglutinin. From a careful study of the synthesis of rRNA in resting and dividing lymphocytes, Cooper (1969a,b) has concluded that much of the ribosomal precursor RNA synthesized in resting lymphocytes does not reach maturity and apparently is degraded by a process he calls "wastage." The evidence for this wastage is indirect, based on the finding that after brief labeling times the ratio of newly formed 28 and 32S RNA to newly formed 18S RNA is significantly greater than the steady state value. Yet no excess 28S RNA ever accumulates in the cytoplasm.

Cooper (1969a,b) interprets these results in the following way: 45S ribosomal precursor RNA is transcribed and normal processing ensues, producing 32S RNA and 18S RNA. Some of the latter is destroyed because of a deficiency of some factor, probably ribosomal proteins. Thus the molar ratio of $(32 + 28)/18$ is temporarily greater than one and provides a measure of the portion of the 18S RNA that has been degraded. Subsequently as 28S RNA matures, an equal amount of it is destroyed, so that the cytoplasmic ratio of 28S/18S is one. If this interpretation is correct, the measured wastage will depend critically on the duration of labeling and on the rate of processing, and will be a minimum estimate of the actual wastage.

Emerson (1971) has interpreted his and Cooper's data in a different way, as showing that in slowly growing cells the transit time of the RNA polymerase transcribing ribosomal precursor RNA is greatly increased. While Cooper's rebuttal (1971) is cogent, the most telling argument against this view is that Wellauer and Dawid (1973) have now shown that the 28S portion of ribosomal precursor RNA is near the 5' end of the molecule, rather than at the 3' end as predicted by Emerson.

Cooper and Gibson (1971) have now measured more directly the synthesis of ribosomal precursor RNA and its wastage and conclude that at least half the ribosomal precursor RNA molecules transcribed in resting cells never become cytoplasmic ribosomes. Upon stimulation with phytohemagglutinin, the ribosomal precursor RNA transcription increases threefold, and the efficiency of conversion to mature rRNA becomes nearly 100 percent. By 20 hr, however, when the growth rate is still increasing, the wastage again begins to rise (Cooper 1969b).

Treatment of the cells with cycloheximide causes an increase of wastage, whereas putting the cells at a higher temperature (to increase protein synthesis) or treating them with very low doses of actinomycin D (to decrease ribosomal precursor RNA transcription) causes a decrease of wastage. Therefore Cooper concludes that the lack of a protein(s) in limiting supply brings about the instability of newly formed subunits. Unfortunately in this system there has been no study either of ribosomal protein synthesis itself or of the nucleolar RNP.

Ribosomal Precursor RNA in Normal and Regenerating Liver

In mammals the liver responds to partial hepatectomy by a dramatic increase in macromolecular synthesis, leading to a rapid hyperplasia (for a recent review see Bucher and Malt 1971). Within a few hours, RNA and protein synthesis have increased substantially, and by 24 hr, more than 80% of the cells are undergoing DNA synthesis and mitosis. The rate of new ribosome formation is nearly fivefold greater than in the normal liver. Yet careful measurements by Chaudhuri and Lieberman (1968) and by Busch and Smetana (1970) indicate that there is at most a twofold increase in the synthesis of ribosomal precursor RNA. Chaudhuri and Lieberman (1968) stated, "It is more likely that ribosomal precursor RNA is rapidly destroyed in normal as well as regenerating liver unless it is able to combine with some protective protein." While the precision of the data may be in some doubt due to contaminating HnRNA and due to changes in the uptake of isotope into nucleotide pools (Bucher and Swaffield 1969), it is evident that the "wastage" phenomenon occurs in tissues as well as cultured cells.

Regulation of rRNA Synthesis: Summary

In summary, the following facts about regulation of rRNA synthesis may serve as a framework on which to build any general theory.

(1) The rate of transcription of ribosomal precursor RNA does vary, although usually less than the rate of synthesis of ribosomes. Is this due solely to the rate of initiation, or does transcription play a role?

(2) The relative amounts of the intermediate RNA species are nearly invariant in cultured cells. Change in the absolute amounts is far less than the change in transcription.

(3) The rate of both synthesis and processing of ribosomal precursor RNA is related to the protein complement of the nucleolus.

(4) If ribosomal precursor RNA synthesis exceeds the available ribosomal proteins, some or all of the ribosomal precursor RNA is degraded within or near the nucleolus.

(5) The migration of completed 28S RNA from the nucleolus to the cytoplasm appears somehow dependent on the transcription of ribosomal precursor RNA.

Synthesis and Fate of Ribosomal Proteins

Ribosomal proteins are important not only because of the vital role they play in protein synthesis, but also because they make up a significant portion of the protein complement of the cell, from 5% in growing HeLa cells to 15% or more in yeast. It is therefore essential to understand both the mechanics and the regulation of the synthesis of these proteins.

Ribosomal proteins are found in the nucleolus shortly after they are synthesized, and for a considerable fraction of a generation time after labeling, they make up the bulk of the radioactive proteins in that organelle (Wu, Kumar and Warner 1971; Soeiro and Basile 1973).

It is now evident that neither the nucleus nor the nucleolus contains functioning ribosomes (Hamkalo and Miller 1973). It no longer appears that there is any protein synthesis within the nucleus. Cytoplasmic synthesis of nuclear proteins, such as histones, is well established (Robbins and Borun 1967). Therefore earlier results, interpreted as showing that nucleolar proteins are synthesized within the nucleolus (discussed by Izawa and Kawashima 1969), are probably not valid. The synthesis of both ribosomal and nonribosomal nucleolar proteins, but not the incorporation of amino acids into isolated nucleoli, is fully inhibited by cycloheximide (Izawa and Kawashima 1969). More direct evidence that ribosomal proteins are synthesized in an entirely conventional way comes from two laboratories (Heady and McConkey 1971; Craig and Perry 1971) which have shown independently that purified cytoplasmic polyribosomes can, in vitro, incorporate labeled amino acids into proteins having the chromatographic and electrophoretic behavior of ribosomal proteins. The authors believe that they are observing only the termination of nascent polypeptides. Approximately 6% of the incorporated radioactivity had the properties of ribosomal protein.

Radioactive protein in the cell is the sum of two components, the nascent proteins on the polyribosomes and the completed proteins. Whereas incorporation into total protein is linear during brief labeling periods, the accumulation of label into completed proteins is proportional to the square of the time until the nascent polypeptides are fully labeled, at 80–100 sec in HeLa cells. The kinetics of the accumulation of labeled protein in the nucleolus is identical to that of completed protein in the cytoplasm (Wu and Warner 1971). This result confirms the notion that the synthesis of ribosomal proteins occurs on polyribosomes and indicates that the migration of ribosomal protein to the nucleolus is rapid compared with the synthesis time of the protein. This was substantiated by following the accumulation of protein in the nucleus after a pulse was terminated with cycloheximide. Because migration is prevented at 0°C, we infer that some active process is involved.

The nucleolus contains 1% of the protein of a HeLa cell, the ribosomes 3–5%. Since all ribosomal proteins pass through the nucleolus while being assembled into ribosomes, the nucleolus accumulates new protein at a greater rate, per unit protein, than any other part of the cell (Wu, unpublished results).

Regulation of Synthesis of Ribosomal Proteins

In the most direct measurement of the coupling of rRNA and ribosomal protein synthesis, Craig and Perry (1971) measured the synthesis of ribosomal proteins in vitro on polysomes from cells treated with a low concentration of actinomycin. With either a low concentration of actinomycin, blocking only rRNA synthesis, or a high concentration, blocking all RNA synthesis, the synthesis of ribosomal proteins, as well as other proteins, was unaffected for several hours (Craig, Kelley and Perry 1971).

However Craig (1971) found that few proteins, synthesized while rRNA synthesis in L cells had been inhibited by actinomycin D, were found in ribosomes formed after rRNA synthesis was allowed to resume. These results indicated that ribosomal protein was made, but could not be utilized, when rRNA synthesis resumed several hours later. Maisel and McConkey (1971) supported this notion by showing that ribosomal proteins did not accumulate in the nucleolus in the presence of actinomycin. Final confirmation was made possible by the discovery of a rapidly reversible inhibitor of RNA synthesis in mammalian cells, camptothecin (Horwitz, Chang and Grollman 1971). Using this, we found that when rRNA synthesis is inhibited, the nucleolus accumulates less than 20% as much radioactive protein during a 20-min label as do control cells. Furthermore, most of the protein which does appear is not ribosomal but the structural, recycling proteins of the nucleolus (Wu, Kumar and Warner 1971). When camptothecin is removed, ribosome synthesis resumes within 5 min, yet few ribosomal proteins labeled in the presence of camptothecin are found on the newly made ribosomes. We conclude from these results that ribosomal proteins are "fixed" in the nucleolus by attachment to newly made ribosomal precursor RNA. If there is no ribosomal precursor RNA, they pass out of the nucleus, are unable to return, and must eventually be destroyed. In our hands the functional half-life of an unattached ribosomal protein is less than 5 min. The physical half-life we have not been able to measure. This concept implies that there is some special feature that permits or facilitates the entry only of new ribosomal proteins into the nucleolus. Data on the turnover of ribosomes supports this view. Tsurugi et al. (1973) find that the RNA and all protein parts of the ribosome turn over with the same half-life, suggesting that proteins released from ribosomes undergoing degradation are unable to find their way back to the nucleolus to reassociate with new RNA.

These observations must influence our interpretation of other data on the regulation of ribosomal protein synthesis, although this may not be a universal phenomenon. Hallberg and Brown (1969) have done an extensive study of the accumulation of ribosomal proteins in embryos of *Xenopus laevis* at stages of development before and after rRNA synthesis commences. They find that ribosomal proteins accumulate only when rRNA synthesis is occurring. Are ribosomal proteins synthesized only at that time? It can be argued that the developing embryo is more likely than the L cell to regulate the synthesis of ribosomal proteins, but the data cannot be considered proof that such regulation occurs.

When HeLa cells are starved of an amino acid, ribosome synthesis continues at a reduced rate, due to free amino acids released through protein turnover (Maden et al. 1969). During valine starvation, the proteins in new 60S subunits account for only 1.0% of the total proteins synthesized, compared to 1.6% for growing cells (Pawlowski and Vaughan 1972). The authors suggest that two of the proteins of that subunit are made in limiting amounts. Since fewer 60S subunits than 40S subunits are made in these cells (Maden 1972), it is likely that most of the ribosomal proteins are made in relatively normal amounts, but that a portion of the 60S proteins find themselves on nucleolar RNP which, because they lack one or two of the limiting proteins, are eventually degraded.

On the other hand, in methionine-deprived cells, where no completed ribosomes are formed because the ribosomal precursor RNA which is made is submethylated, ribosomal proteins continue to be made and to assemble with 32S RNA to form nucleolar RNP (Maden and Vaughan 1969). If methionine is restored, the RNA of these particles is methylated, and they mature to cytoplasmic ribosomes (Vaughan et al. 1967).

In yeast one can prevent ribosome synthesis reversibly in certain temperature-sensitive mutants. In cells labeled at the restrictive temperature, only about 25% of the normal proportion of radioactive protein appears on ribosomes when the temperature is lowered and ribosome synthesis resumes (Warner and Udem 1972). Further experiments are needed to distinguish between a specific regulation of ribosomal protein synthesis and general degradation of excess ribosomal proteins.

Most eukaryotic tissues are in a relatively homeostatic condition, where major fluctuations in ribosome synthesis occur rarely (e.g., liver regeneration) if at all. Thus once a balanced synthesis of ribosomal proteins and RNA exists, it may not be necessary to keep the relative rates of synthesis finely tuned. A slight excess of either RNA or protein can be rectified by degradation, as discussed above. Thus when the system is drastically unbalanced, e.g., by actinomycin D, it is perhaps not surprising that the cell does not cope in the economical way expected from studies on *E. coli*.

However in situations of anticipated rapid change, such as the development of an egg or the exposure of yeast to a new environment, one might expect the cell to be able to couple rRNA and ribosomal protein syntheses more directly. It is important to develop the experimental methods to measure directly and easily the synthesis of ribosomal protein.

Turnover of Ribosomal Constituents

Most cells of an adult animal do not proliferate. Therefore for each new ribosome which is assembled, an old one must be degraded. Considering, then, that the turnover of a ribosome plays a role equal to the synthesis of a ribosome, there has been surprisingly little study of this process. Perhaps the major drawback has been the difficulty in carrying out such studies with growing cultured cells.

Hirsch and Hiatt (1966) showed some time ago that in rat liver, both the RNA and the protein portion of the ribosomes had the same half-life of approximately 5 days, suggesting that the whole ribosome is turned over as a unit. There is no difference between free and membrane-bound ribosomes (Mishra et al. 1972). More recently, Tsurugi, Morita and Ogata (1974) have extended these observations. They pooled ribosomes from rat liver labeled with [³H]leucine 6 days previously with those labeled for 4 hr with

[^{14}C]leucine and separated the purified proteins on two-dimensional gel electrophoresis. The ^{14}C to ^{3}H ratio was the same for each protein. Therefore the rate of degradation of each protein is the same and is also the same as that of both major rRNA molecules. In their hands, ribosomal proteins have a half-life of 4.5 days, compared with an average of 2.5 days for proteins from the cell sap.

The decay is logarithmic, suggesting that a single event, occurring on a randomly selected ribosome, brings about the degradation of the whole ribosome or, more likely, of a whole ribosomal subunit. Apparently the ribosomal proteins cannot be utilized in the assembly of new ribosomes.

Weber (1972) has shown that the rate of degradation of ribosomes is under some regulation. In growing monolayer cells, rRNA appears to be stable, but as the cells reach confluence, ribosomes begin to turn over with a half-life of about 40 hr. Such degradation permits the ribosome assembly system to keep operating without accumulating an excess of ribosomes in the cell. It will be interesting to know how the cell initiates this degradative process.

New Approaches—Genetics

Mammalian Cells.   Toniolo, Meiss and Basilico (1973) observed that the nucleoli of one of their temperature-sensitive lines of BHK cells were enlarged at the restrictive temperature. This appears to be due to the inability of these cells to carry out the cleavage of 32S to 28S RNA. The nucleoli thus accumulate enormous amounts of 32S RNA, while 18S RNA migrates normally to the cytoplasm. It will be interesting to learn not only about the protein responsible for the temperature-sensitive lesion itself, but also why the accumulated 32S RNA is stable and does not prevent the synthesis of new 45S RNA.

Yeast.   In *S. cerevisiae* it has been possible to identify a number of temperature-sensitive mutants that are specifically defective in ribosome synthesis (Hartwell, McLaughlin and Warner 1970). Although these mutants fall into ten distinct complementation groups, the phenotypes of all appear similar, at the present stage of investigation. At the restrictive temperature, there is a very severe inhibition of RNA maturation, a less severe inhibition of 35S ribosomal precursor RNA synthesis, and a degradation of that RNA which is made (Warner and Udem 1972). Ribosomal precursor RNA made at the restrictive temperature can, however, be rescued if the cells are returned to permissive conditions. Recent work with these mutants is discussed by McLaughlin elsewhere in this volume.

In spite of the rather unsatisfactory beginnings, it is clear that the use of such mutants in both types of cells will facilitate our understanding of the regulation of ribosome synthesis.

SPECULATIONS

Variety of Regulatory Sites

Each cell has many complex assemblies of macromolecules, the ribosome being one of the simplest. Understanding the control of the synthesis of its constituents may provide insight into general mechanisms of cross-regulation among macromolecules.

The steps which lead to the assembly of a eukaryotic ribosome are outlined in Figure 3. Potential regulatory sites for the appearance of RNA in ribosomes include:

(1) the transcription of ribosomal precursor RNA,
(2) the methylation of ribosomal precursor RNA,
(3) the availability of nucleolar proteins and of new ribosomal proteins with which the ribosomal precursor RNA can assemble,
(4) the cleavage of ribosomal precursor RNA and the fate of the fragments produced,
(5) the passage of new ribosomal subunits from the nucleus to the cytoplasm.

Potential regulatory sites for the appearance of ribosomal protein in ribosomes include:

(1) the transcription of the HnRNA molecules containing information for ribosomal protein,
(2) the processing of these RNAs to form mRNA for ribosomal protein,
(3) the migration of the mRNA from the nucleus to the cytoplasm,
(4) the translation of the mRNA to produce ribosomal protein,
(5) the migration of newly formed ribosomal protein to the nucleolus,
(6) the availability of ribosomal precursor RNA with which the ribosomal protein can assemble,

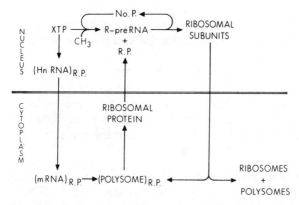

Figure 3   Schematic representation of ribosome synthesis in eukaryotes.

(7) the availability of all ribosomal proteins in order to form a complete, stable ribosomal subunit.

The experimental data described in this review lead to the conclusion (Perry 1973; Warner et al. 1973) that in mammalian cells much of the regulation of rRNA occurs at steps after transcription. In many cases ribosomal precursor RNA is made and then degraded, due to lack of protein. While under certain experimental conditions ribosomal protein is made and then not utilized, this may be infrequent in vivo. Some results, e.g., Cooper's (1970), imply that the control over ribosome production is the supply of one or more of the ribosomal proteins, presumably, but not necessarily, because of a limiting supply of its mRNA. Since polycistronic mRNA is rare in eukaryotic cells, it is likely that there are 70 or more mRNAs for ribosomal protein. If these limit the production of ribosomes, we must ask whether there can be (1) equimolar synthesis of the mRNA for each ribosomal protein; (2) equimolar synthesis of each ribosomal protein; (3) relatively uncoordinated synthesis of the ribosomal proteins, ribosome synthesis being limited by one or a few, while the excess molecules of the others are degraded. Perhaps the production of new ribosomes is ultimately controlled through the level of a single mRNA.

Model of Ribosome Assembly

Present evidence leads to the following proposal for ribosome assembly in the nucleolus:

(1) The transcription of ribosomal precursor RNA occurs freely, subject only to inhibition by the ribosomal precursor RNA already present in the nucleolus.
(2) New ribosomal proteins enter the nucleolus and bind to the appropriate site on the ribosomal precursor RNA.
(3) This binding is not permanent, however, as a single protein may move from one RNA molecule to another (Pederson and Kumar 1971). Thus the ribosomal precursor RNA molecules compete for any proteins which are limiting.
(4) Those molecules of ribosomal precursor RNA with a complete set of proteins are properly cut by a nuclease and sent to the cytoplasm. However an RNA molecule without a complete set of proteins has a finite probability of an incorrect cleavage, leading to its degradation. The proteins on such a molecule could either rejoin the nucleolar pool or be degraded.

The applicability of this proposal is particularly clear in the case where more than one ribosomal protein is limiting. If five proteins are available in half the molar amount of the ribosomal precursor RNA and are distributed randomly, only $\frac{1}{2}^5 = \frac{1}{32}$ of the ribosomal precursor RNA would have a complete set. To optimize the yield of complete ribosomes, the proteins must be able to move from one RNA molecule to another. The reaction forming complete ribosomes could be driven by the increased binding of limiting proteins to more complete RNP and/or by the cleavage of the ribosomal precursor RNA, preventing further protein exchange. The slow processing of RNA seen during amino acid starvation may be due to the increased time necessary for a ribosomal precursor RNA to acquire a full complement of proteins. Incomplete RNP could inhibit the transcription of new ribosomal precursor RNA to avoid more acute competition for the limiting ribosomal protein.

Nucleolar Proteins

It will be of great interest to determine whether the recycling nucleolar proteins are concerned only with determining the conformation of the RNA or whether they play a direct role in assembling the ribosomal proteins, as does the gene 8 protein of phage P 22 (King, Lenk and Botstein 1973).

Acknowledgments

The work from the author's laboratory has been supported by grants from the American Cancer Society #NP-72C and the National Science Foundation #GB 32276X. The author is a Faculty Research Awardee of the American Cancer Society. I am grateful to my many colleagues who read and criticized an earlier draft of this article and to Mrs. Grace Sullivan for her success in overcoming the chaos of my manuscript pages.

References

Amalric, F., R. Simard and J. P. Zalta. 1969. Effet de la temperature supra-optimale sur les ribonucleoproteines et le RNA nucleolaise. *Exp. Cell Res.* **55**:370.

Attardi, G. and J. Amaldi. 1970. Structure and synthesis of ribosomal RNA. *Ann. Rev. Biochem.* **39**:183.

Brown, D. D. and C. S. Weber. 1968. Gene linkage by DNA-RNA hybridization. *J. Mol. Biol.* **34**:661.

Bucher, N. L. R. and R. A. Malt. 1971. *Regeneration of liver and kidney.* Little, Brown and Co., Boston.

Bucher, N. L. R. and M. N. Swaffield. 1969. Ribonucleic acid synthesis in relation to precursor pools in regenerating rat liver. *Biochim. Biophys. Acta* **174**:491.

Busch, H. and K. Smetana. 1970. *The nucleolus,* Chap. 2. Academic Press, New York.

Chaudhuri, S. and I. Lieberman. 1968. Control of ribosome synthesis in normal and regenerating liver. *J. Biol. Chem.* **243**:29.

Cooper, H. L. 1969a. Ribosomal ribonucleic acid production and growth regulation in human lymphocytes. *J. Biol. Chem.* **244**:1946.

———. 1969b. Ribosomal ribonucleic acid wastage in resting and growing lymphocytes. *J. Biol. Chem.* **244**:5590.

———. 1970. Control of synthesis and wastage of ribosomal RNA in lymphocytes. *Nature* **227**:1105.

Cooper, H. L. and E. M. Gibson. 1971. Control of synthesis and wastage of ribosomal ribonucleic acid in lymphocytes. *J. Biol. Chem.* **246**:5059.

Craig, N. C. 1971. On the regulation of the synthesis of ribosomal proteins in L cells. *J. Mol. Biol.* **55**:129.

———. 1973. Ribosomal RNA synthesis in eukaryotes and its regulation. In *Biochemistry of nucleic acids* (ed. K. Burton) vol. 6. University Park Press, Baltimore.

Craig, N. C. and R. P. Perry. 1970. Aberrant intranucleolar maturation of ribosomal precursors in the absence of protein synthesis. *J. Cell Biol.* **45**:554.

———. 1971. Persistent cytoplasmic synthesis of ribosomal proteins during the selective inhibition of ribosomal RNA synthesis. *Nature New Biol.* **229**:75.

Craig, N. C., D. E. Kelley and R. P. Perry. 1971. Lifetime of the messenger RNA which codes for ribosomal protein in L cells. *Biochim. Biophys. Acta* **246**:493.

deKloet, S. R. 1966. Ribonucleic acid synthesis in yeast. *Biochem. J.* **99**:566.

Dice, J. F. and R. T. Schimke. 1972. Turnover and exchange of ribosomal proteins from rat liver. *J. Biol. Chem.* **247**:98.

Emerson, C. P. 1971. Regulation of the synthesis and stability of ribosomal RNA during contact inhibition of growth. *Nature New Biol.* **232**:101.

Ennis, H. L. 1966. Synthesis of ribonucleic acid in L cells during inhibition of protein synthesis by cycloheximide. *Mol. Pharmacol.* **2**:543.

Fan, H. and S. Penman. 1971. Regulation of synthesis and processing of nucleolar components in metaphase-arrested cells. *J. Mol. Biol.* **59**:27.

Firtel, R. A. and H. Lodish. 1973. A small nuclear precursor of messenger RNA in the cellular slime mold *Dictyostelium discoideum*. *J. Mol. Biol.* **79**:295.

Girard, M., H. Latham, S. Penman and J. E. Darnell. 1965. Entrance of newly formed messenger RNA and ribosomes into HeLa cell cytoplasm. *J. Mol. Biol.* **11**:187.

Granboulan, N. and P. Granboulan. 1965. Cytochimie ultrastructurale du nucleole. *Exp. Cell Res.* **38**:604.

Greenberg, H. and S. Penman. 1966. Methylation and processing of ribosomal RNA in HeLa cells. *J. Mol. Biol.* **21**:527.

Hallberg, R. L. and D. D. Brown. 1969. Co-ordinated synthesis of some ribosomal proteins and ribosomal RNA in embryos of *Xenopus laevis*. *J. Mol. Biol.* **46**:393.

Halvorson, H. O., B. L. A. Carter and P. Tauro. 1971. Synthesis of enzymes during the cell cycle. *Adv. Microb. Physiol.* **6**:47.

Hamkalo, B. A. and O. L. Miller. 1973. Electronmicroscopy of genetic activity. *Ann. Rev. Biochem.* **42**:379.

Hartwell, L. H., C. S. McLaughlin and J. R. Warner. 1970. Identification of ten genes that control ribosome formation in yeast. *Mol. Gen. Genet.* **109**:42.

Heady, J. E. and E. H. McConkey. 1971. Completion of nascent HeLa ribosomal proteins in a cell-free system. *Biochem. Biophys. Res. Comm.* **40**:30.

Henshaw, E. C., D. G. Goinez and C. A. Hirsch. 1973. The ribosome cycle in mammalian protein synthesis. *J. Biol. Chem.* **248**:4367.

Hirsch, C. A. and H. H. Hiatt. 1966. Turnover of liver ribosomes in fed and in fasted rats. *J. Biol. Chem.* **241**:5936.

Horwitz, S., C. Chang and A. Grollman. 1971. Studies on camptothecin: Effects on nucleic acid and protein synthesis. *Mol. Pharmacol.* **7**:632.

Izawa, M. and K. Kawashima. 1969. Incorporation of L-^{14}C-leucine into the nucleoli of mouse ascites tumor cells. *Biochim. Biophys. Acta* **190**:139.

Jacob, S. J. 1973. Mammalian RNA polymerases. *Prog. Nucleic Acid Res. Mol. Biol.* **13**:93.

King, J., E. V. Lenk and D. Botstein. 1973. Mechanism of head assembly and DNA encapsulation in Salmonella phage P22. *J. Mol. Biol.* **80**:697.

Kumar, A. and J. R. Warner. 1972. Characterization of ribosomal precursor particles from HeLa cell nucleoli. *J. Mol. Biol.* **63**:233.

Kumar, A. and R. Wu. 1973. Role of ribosomal RNA transcription in ribosome processing in HeLa cells. *J. Mol. Biol.* **80**:265.

Kurland, C. G., P. Voynow, S. J. S. Hardy, L. Randall and L. Lutter. 1969. Physical and functional heterogeneity of *E. coli* ribosomes. *Cold Spring Harbor Symp. Quant. Biol.* **34**:17.

Leibowitz, R. D., R. Weinberg and S. Penman. 1973. Unusual metabolism of 5S RNA in HeLa cells. *J. Mol. Biol.* **73**:139.

Liau, M. C. and R. P. Perry. 1969. Ribosome precursor particles in nucleoli. *J. Cell Biol.* **42**:272.

Liau, M. C., N. C. Craig and R. P. Perry. 1968. The production of ribosomal RNA from high molecular weight precursors. *Biochim. Biophys. Acta* **169**:196.

Maden, B. E. H. 1971. The structure and formation of ribosomes in animal cells. *Prog. Biophys. Mol. Biol.* **22**:127.

————. 1972. Effects of amino acid starvation on ribosome formation in HeLa cells. *Biochim. Biophys. Acta* **281**:396.

Maden, B. E. H. and M. H. Vaughan. 1969. Synthesis of ribosomal proteins in the absence of ribosome maturation in methionine deficient HeLa cells. *J. Mol. Biol.* **38**:431.

Maden, B. E. H., M. Salim and D. F. Summers. 1972. Maturation pathway for ribosomal RNA in the HeLa cell nucleolus. *Nature New Biol.* **237**:5.

Maden, B. E. H., M. H. Vaughan, J. R. Warner and J. E. Darnell. 1969. Effects of valine deprivation on ribosome formation in HeLa cells. *J. Mol. Biol.* **45**:265.

Maisel, J. C. and E. H. McConkey. 1971. Nucleolar protein metabolism in actinomycin D treated HeLa cells. *J. Mol. Biol.* **61**:251.

Miller, O. L., B. R. Beatty, B. A. Hamkalo and C. A. Thomas. 1970. Electron microscopic visualization of transcription. *Cold Spring Harbor Symp. Quant. Biol.* **35**:505.

Mirault, M. E. and K. Scherrer. 1971. Isolation of preribosomes from HeLa cells. *Eur. J. Biochem.* **23**:372.

————. 1972. In vitro processing of HeLa cell preribosomes by a nucleolar endoribonuclease. *FEBS Letters* **20**:233.

Mishra, R. K., J. F. Wheldrake and L. A. W. Feltham. 1972. RNA turnover in endoplasmic reticulum-bound and free ribosomes. *Biochim. Biophys. Acta* **281**:393.

Mitchison, J. M. 1971. *The biology of the cell cycle,* pp. 117–122. Cambridge University Press, Cambridge.

Narayan, K. S. and M. L. Birnstiel. 1969. Biochemical and ultrastructural characteristics of ribonucleoprotein particles isolated from rat liver cell nucleoli. *Biochim. Biophys. Acta* **190**:470.

Pawlowski, P. J. and M. H. Vaughan. 1972. Comparison of the relative synthesis of the proteins of the 50S ribosomal subunit in growing and valine-deprived HeLa cells. *J. Cell Biol.* **52**:409.

Pederson, T. and A. Kumar. 1971. Relationship between protein synthesis and ribosome assembly in HeLa cells. *J. Mol. Biol.* **61**:655.

Perry, R. P. 1973. The regulation of ribosome content in eukaryotes. *Biochem. Soc. Symp.* (in press).

Perry, R. P. and D. E. Kelley. 1968. Persistent synthesis of 5S RNA when production of 28S and 18S ribosomal RNA is inhibited by low doses of actinomycin D. *J. Cell. Physiol.* **72**:235

Prestayko, A. W., B. C. Lewis and H. Busch. 1972. Endoribonuclease activity associated with nucleolar ribonucleoprotein particles from Novikoff hepatoma. *Biochim. Biophys. Acta* **269**:90.

Rao, S. S. and A. P. Grollman. 1967. Cycloheximide resistance in yeast: A property of the 60S ribosomal subunit. *Biochem. Biophys. Res. Comm.* **29**:696.

Reichman, M., D. Karlan and S. Penman. 1973. Destructive processing of the 45S ribosomal precursor in the presence of 5 aza-cytidine. *Biochim. Biophys. Acta* **299**:173.

Robbins, E. and T. W. Borun. 1967. The cytoplasmic synthesis of histones in HeLa cells and its temporal relationship to DNA replication. *Proc. Nat. Acad. Sci.* **57**:409.

Rogers, M. E. 1968. Ribonucleoprotein particles in the amphibian oocyte nucleus. *J. Cell Biol.* **36**:421.

Rubin, G. M. and J. E. Sulston. 1973. Physical linkage of the 5S cistrons to the 18S and 28S RNA cistrons in *Saccharomyces cerevisiae. J. Mol. Biol.* **79**:521.

Shepherd, J. and B. E. H. Maden. 1972. Ribosome assembly in HeLa cells. *Nature* **236**:211.

Shulman, R. W., L. H. Hartwell and J. R. Warner. 1973. Synthesis of ribosomal proteins during the yeast cell cycle. *J. Mol. Biol.* **73**:513.

Simard, R. and W. Bernhard. 1967. A heat sensitive cellular function located in the nucleolus. *J. Cell Biol.* **34**:61.

Snyder, A., H. Kann and K. Kohn. 1971. Effects of proflavin and other intercalating molecules on maturation. *J. Mol. Biol.* **58**:555.

Soeiro, R. and C. Basile. 1973. Non-ribosomal nucleolar proteins in HeLa cells. *J. Mol. Biol.* **79**:507.

Soeiro, R., M. H. Vaughan and J. E. Darnell. 1968. The effect of puromycin on intranuclear steps in ribosome biosynthesis. *J. Cell Biol.* **36**:91.

Soeiro, R., M. Vaughan, J. Warner and J. Darnell. 1968. The turnover of nuclear DNA-like RNA in HeLa cells. *J. Cell Biol.* **39**:112.

Stevens, R. and H. Amos. 1971. RNA metabolism in HeLa cells at reduced temperature. *J. Cell Biol.* **50**:818.

Takahashi, N., T. Shimada, M. Higo and S. Tanifuji. 1972. Ribosomal precursor particles from plant nucleoli (pea). *Biochim. Biophys. Acta* **262**:502.

Tartof, K. D. and R. P. Perry. 1970. The 5S RNA genes of *Drosophila melanogaster*. *J. Mol. Biol.* **51**:171.

Tavitian, A., S. C. Uretsky and G. Acs. 1968. Selective inhibition of ribosomal RNA synthesis in mammalian cells. *Biochim. Biophys. Acta* **157**:33.

Terao, K. and K. Ogata. 1971. Studies on the small subunit of rat liver ribosomes: Some biochemical properties with specific references to the reconstruction of the small subunit. *Biochim. Biophys. Acta* **254**:278.

Terao, K., K. Tsurugi and K. Ogata. 1974. Non-exchangeability of ribosomal structural proteins with cell sap proteins of rat liver *in vitro*. *Japan. J. Biochem.* (in press).

Toniolo, D., H. K. Meiss and C. Basilico. 1973. A temperature-sensitive mutation affecting 28S ribosomal RNA production in mammalian cells. *Proc. Nat. Acad. Sci.* **70**:1273.

Tsurugi, K., T. Morita and K. Ogata. 1973. Identification and metabolic relationship between proteins of nucleolar 60S particles and of ribosomal large subunits of rat liver by means of two-dimensional disc electrophoresis. *Eur. J. Biochem.* **32**:555.

————. 1974. Mode of degradation of ribosomes in regenerating rat liver *in vivo*. *Eur. J. Biochem.* **45**:119.

Udem, S. A. and J. R. Warner. 1972. Synthesis and processing of ribosomal RNA in *Saccharomyces cerevisiae*. *J. Mol. Biol.* **65**:227.

————. 1973. The cytoplasmic maturation of a ribosomal precursor ribonucleic acid in yeast. *J. Biol. Chem.* **248**:1412.

Vaughan, M. H., R. Soeiro, J. R. Warner and J. E. Darnell. 1967. The effects of methionine deprivation on ribosome synthesis in HeLa cells. *Proc. Nat. Acad. Sci.* **58**:1527.

Vesco, C. and S. Penman. 1968. The fractionation of nuclei and the integrity of purified nucleoli in HeLa cells. *Biochim. Biophys. Acta* **169**:188.

Warner, J. R. 1966. The assembly of ribosomes in HeLa cells. *J. Mol. Biol.* **19**:383.

————. 1971. The assembly of ribosomes in yeast. *J. Biol. Chem.* **246**:447.

Warner, J. R. and R. Soeiro. 1967. Nascent ribosomes from HeLa cells. *Proc. Nat. Acad. Sci.* **58**:1984.

Warner, J. R. and S. A. Udem. 1972. Temperature-sensitive mutations affecting ribosome synthesis in *Saccharomyces cerevisiae*. *J. Mol. Biol.* **65**:243.

Warner, J. R., M. Girard, H. Latham and J. E. Darnell. 1966. Ribosome formation in HeLa cells in the absence of protein synthesis. *J. Mol. Biol.* **19**:373.

Warner, J. R., A. Kumar, S. A. Udem and R. S. Wu. 1973. Ribosomal proteins and the assembly of ribosomes in eukaryotes. *Biochem. Soc. Symp.* **37**:3.

Warocquier, R. and K. Scherrer. 1969. RNA metabolism in mammalian cells at elevated temperature. *Eur. J. Biochem.* **10**:362.

Weber, M. J. 1972. Ribosomal RNA turnover in contact inhibited cells. *Nature New Biol.* **235**:58.

Weinberg, R. A. 1973. Nuclear RNA metabolism. *Ann. Rev. Biochem.* **42**:329.

Weinberg, R. A. and S. Penman. 1970. Processing of 45S nucleolar RNA. *J. Mol. Biol.* **47**:169.

Weinberg, R. A., U. Leoning, M. Willems and S. Penman. 1967. Acrylamide gel electrophoresis of HeLa cell nucleolar RNA. *Proc. Nat. Acad. Sci.* **58**:1088.

Wellauer, D. and I. Dawid. 1973. Secondary structure maps of RNA: Processing of HeLa ribosomal RNA. *Proc. Nat. Acad. Sci.* **70**:2827.

Willems, M., M. Penman and S. Penman. 1969. The regulation of RNA synthesis and processing in the nucleolus during inhibition of protein synthesis. *J. Cell Biol.* **41**:177.

Wu, R. S. and J. R. Warner. 1971. Cytoplasmic synthesis of nuclear proteins. *J. Cell Biol.* **51**:643.

Wu, R. S., A. Kumar and J. R. Warner. 1971. Ribosome formation is blocked by camptothecin, a reversible inhibitor of RNA synthesis. *Proc. Nat. Acad. Sci.* **68**:3009.

Yoshikawa-Fukada, T. 1967. The intermediate state of ribosome formation in animal cells in culture. *Biochim. Biophys. Acta* **145**:651.

Yu, F. and P. Feigelson. 1973. The rapid turnover of RNA polymerase of rat liver nucleolus, and of its messenger RNA. *Proc. Nat. Acad. Sci.* **69**:2833.

Zylber, E. and S. Penman. 1971. Synthesis of 5S and 4S RNA in metaphase arrested HeLa cells. *Science* **172**:947.

Ribosomes from Eukaryotes: Genetics

Ronald H. Reeder
Department of Embryology
Carnegie Institution of Washington
Baltimore, Maryland 21210

INTRODUCTION

This review concerns the genes that control the synthesis of eukaryotic ribosomes. Due to the general intractability of animal cells to formal genetic experiments, coupled with the difficulty of obtaining mutants in ribosomal loci, the amount of classical genetic data in this area is rather limited. In compensation, however, the structural genes for the ribosomal RNAs are perhaps the most convenient to manipulate biochemically of any known genes, animal, bacterial or viral. Because of these circumstances, the bulk of this review will be concerned with information gained by techniques other than standard genetics. In this context the word "genetic" has been interpreted broadly to include all information regarding the structure, location, function and inheritance of the genes for eukaryotic ribosomes.

Two exceptions to the above generalizations on animal genetics are Drosophila and yeast. The ribosomal genes of these organisms are the subject of separate reviews in this volume (Spear, Drosophila; McLaughlin, yeast) and will not be covered here. In addition, this review is only concerned with nuclear ribosomal genes and will not include literature concerning the rDNA of mitochondria or chloroplasts.

Throughout this review ribosomal DNA (rDNA) refers to the DNA containing the genes for 18S and 28S RNA of ribosomes plus spacer sequences; ribosomal RNA (rRNA) refers to the RNA product of rDNA. 5S DNA refers to the DNA coding for the 5S RNA of ribosomes plus spacer sequences.

THE GENES FOR 18S AND 28S RIBOSOMAL RNA
Chromosomal Location of rDNA

The genes for ribosomal RNA are unique in that they are associated in interphase cells with one or more distinctive nuclear organelles called nucleoli, which are easily visible in the light microscope and whose size and overall morphology is correlated roughly with the transcriptive activity of the rDNA. The morphology of nucleoli is the subject of a large literature, which has been reviewed by Busch and Smetana (1970). In general, nucleoli as seen in the electron microscope contain a "fibrillar component" with fibers 50–80 Å in width and a "particulate component" with particle diameters of about 150 Å (Swift 1966). In some nucleoli, these regions are segregated from each other, the fibrillar component being on the inside and the particulate component on the outside. There is no limiting membrane around the nucleolus. Nucleoli often contain threadlike elements (about 0.1 μm diameter), composed of either or both fibrillar or particulate components, which are called nucleolonemata. In Triturus and Xenopus the fibrillar component has been shown to contain the rDNA (Miller 1966; Miller and Beatty 1969). The particulate component probably represents ribosome precursor particles in some stage of assembly and processing (see article by Warner, this volume). Miller (1966) has shown that the amplified oocyte nucleoli of Triturus also contain a membraneous component of unknown function.

The ability to form a nucleolus is present at a specific chromosomal site (or sites) called the "nucleolus organizer." During telophase a nucleolus usually forms at each nucleolar organizer site. In many cells, however, these nucleoli rapidly fuse, so that each interphase nucleus contains only one nucleolus regardless of the number of organizers present. At the next mitosis the nucleoli disappear as the chromosomes begin to condense. The presence of an active nucleolus appears to interfere with condensation of the chromosome at the organizer site, resulting in a thin "secondary constriction" at each organizer that previously organized a nucleolus. (Apparently other factors can also cause constrictions at other chromosomal loci; therefore the correspondence between constrictions and organizers is not perfect.) These various terms are defined and discussed by McClintock (1934) in a paper that reviews much of the earlier work on nucleolar morphology and inheritance. She was also able to show that the organizer of *Zea mays* could be split by a translocation and that each piece could organize a separate nucleolus. This work was probably the first evidence (although unrecognized at the time) that the genes for rRNA are present in multiple copies.

Several lines of work have converged to establish that the nucleolus is the site of rRNA synthesis and ribosome assembly and that the rDNA is physically located therein. One line of evidence came from work on the frog *Xenopus laevis,* in which a mutant was found that lacked the ability to organize a nucleolus (Elsdale, Fischberg and Smith 1958) and did not develop a secondary constriction (Kahn 1962; Rafferty and Sherwin 1969). The anucleolate mutation was inherited like a simple Mendelian factor. In the homozygous state these animals synthesize no rRNA and die as tadpoles (Brown and Gurdon 1964). Later it was shown by Wallace and Birnstiel (1966) that the homozygous mutants are completely deleted for rDNA. Ritossa and Spiegelman (1965) were able to obtain Drosophila stocks that contained from one to four doses of the nucleolar organizer region. The maximum amount of rRNA that could be hybridized specifically to the DNA from each stock was shown to be related directly to the number of organizers present.

More recently the technique of in situ hybridization has been used to demonstrate unequivocally that the rDNA is located physically within the nucleolus (Pardue et al. 1970; Gerbi 1971; Amaldi and Buongiorno-Nardelli 1971; Durante, Cionini and D'Amato 1972; Brady and Clutter 1972; Ryser, Fakan and Braun 1973) and also at the secondary constrictions, which are nucleolus associated (Henderson, Warburton and Atwood 1972; Pardue, Brown and Birnstiel 1973). These studies clearly show that all true nucleoli contain rDNA. The most intriguing observation is that in at least two instances significant amounts of rDNA were seen at chromosomal loci that did not organize nucleoli. One case was a plant, *Phaseolus coccineus* (Durante, Cionini and D'Amato 1972), and the other was a dipteran, *Rhynchosciara hollaenderi* (Pardue et al. 1970). Are these "silent" genes ever active in other tissues? Have they diverged in nucleotide sequence from the "active" rDNA? Their presence is reminiscent of the nucleolar dominance phenomenon seen in some interspecific crosses (see below).

In order to explain her observations on the interaction between nucleolar organizer fragments in the same nucleus, McClintock (1934) proposed that the organizers were merely collection sites for nucleolar "matrix," which was synthesized at other chromosomal loci. Like many good theories, this "matrix hypothesis" has turned out to be partially true and partially false. In light of present knowledge, we·know that the rDNA is physically located at the organizer site and directs the synthesis of rRNA, the major component of ribosomes. But the other components of ribosomes, the 5S RNA and the ribosomal proteins, are synthesized elsewhere in the cell and are collected at the nucleolus for assembly into ribosomes.

Structure of rDNA

In all eukaryotes so far examined, the genes for rRNA are present in multiple copies. Birnstiel, Chipchase and Speirs (1971) have tabulated most of the known examples. Since their review, additional studies have been reported for *Dictyostelium discoideum* (Firtel and Bonner 1972),

Physarum (Zellweger, Ryser and Braun 1972), *Ilyanassa obsoleta* (Collier 1971), and for several species of fish (Pedersen 1971). In most organisms the repetitious rDNA has a higher buoyant density than does the main band DNA and is often associated with a distinct satellite band in CsCl density gradients. The buoyant densities of the rDNA from a number of species have been reported (Sinclair and Brown 1971; Birnstiel, Chipchase and Speirs 1971).

The presence of multiple copies plus their high buoyant density has made it possible to isolate the rDNA from several species and to study its primary structure in considerable detail. The most extensive work has been done on the rDNA from the frog, *Xenopus laevis*. In diploid somatic cells of this animal, there are two chromosomes that bear organizers (Kahn 1962), each of which contains about 450 tandemly repeated copies of the genes for 18S and 28S rRNA (Wallace and Birnstiel 1966; Brown and Weber 1968). In somatic cells the rDNA comprises about 0.2% of the total nuclear DNA (Birnstiel et al. 1966; Dawid, Brown and Reeder 1970). In oocytes of *X. laevis,* the rDNA is specifically amplified several thousandfold so that it comprises about two-thirds of the total DNA in the tetraploid oocyte nucleus (Brown and Dawid 1969; Gall 1968) (see next section). Within the limits of present techniques, the somatic rDNA is identical to the oocyte rDNA, except that the somatic rDNA is methylated (5% of all bases are 5-methylcytosine), whereas oocyte rDNA is not (Dawid, Brown and Reeder 1970).

Isolation of *X. laevis* rDNA was first accomplished by repeated density gradient centrifugation in neutral CsCl (Birnstiel et al. 1968). For the isolation of rDNA from somatic DNA, some preliminary fractionation capable of handling large amounts of DNA is useful since the rDNA is present in such low concentration. Selective precipitation of the lower G + C bulk DNA by polylysine has been used for this purpose (Dawid, Brown and Reeder 1970). Final purification was obtained by CsCl gradient centrifugation. The most convenient source of rDNA is from the ovaries of young frogs at one to two months after metamorphosis. At this stage, rDNA comprises 5–10% of the total ovarian DNA (Gall 1968) and can be completely separated from all other DNA by only one or two cycles of CsCl gradient centrifugation.

Another method capable of handling large quantities of DNA is selective heat denaturation of the lower G + C bulk DNA, followed by separation of the native rDNA and denatured bulk DNA with a two-phase polyethylene glycol-dextran system. This technique has been applied successfully to sea urchin rDNA (Patterson and Stafford 1970).

Not all species contain rDNA that separates from the bulk DNA in neutral CsCl. In such cases density gradients of Cs_2SO_4 containing silver or mercury ions or CsCl gradients containing DNA-binding drugs, such as actinomycin D, may prove useful (Brown, Wensink and Jordan 1971;

Brown and Sugimoto 1973). $Hg-Cs_2SO_4$ gradients have been used to increase the resolution of rDNA in yeast (Schweizer, MacKechnie and Halvorson 1969). It is also possible to isolate one strand of any rDNA by denaturing the DNA, hybridizing with rRNA, and isolating the RNA-DNA hybrid (Spadari et al. 1971; Chattopadhyay, Kohne and Dutta 1972).

The structure of *X. laevis* rDNA is shown in Figure 1. Biochemical studies (Birnstiel et al. 1968; Dawid, Brown and Reeder 1970) on the purified rDNA have established that it is a tandemly repeated structure, each repeat containing a sequence that is transcribed as a single 40S precursor molecule (gene region) and a sequence that is not transcribed (nontranscribed spacer). Within each gene region is a sequence for 28S rRNA, a sequence for 18S rRNA, and two small regions which are discarded (transcribed spacer) during processing of the 40S precursor. The gene region has a lower G + C content (about 60%) than the nontranscribed spacer region, about 73%). This heterogeneity is reflected in the hyperchromic melting profile of the DNA, which contains two sharp transitions separated by a small plateau.

The rate at which the rDNA renatures (Birnstiel et al. 1969), the fidelity with which it renatures (Dawid, Brown and Reeder 1970; Forscheit, personal communication), and the uniformity of the repeat length (Wensink and Brown 1971) are all consistent with the hypothesis that all repeats are identical. However, such techniques would probably not be able to detect subtle sequence heterogeneity.

The most reliable measurements of the size, arrangement and function of the various regions of *X. laevis* rDNA have come from electron microscopic observations. The most accurate estimate of repeat length is $8.7 \pm 0.6 \times 10^6$ daltons, as derived from denaturation mapping experiments (Wensink and Brown 1971). This number is in good agreement with earlier estimates arrived at by biochemical methods (Dawid, Brown and Reeder 1970; Birnstiel et al. 1971). The denaturation mapping technique also offers dramatic confirmation of the large base composition difference between the gene region and the spacer region. The alternation of transcribed with nontranscribed regions in *X. laevis* rDNA is shown most clearly in the electron micrographs of dispersed oocyte nucleolar cores (Miller and Beatty 1969). These micrographs show regions of the DNA having about 80–100 nascent RNA chains interspersed with spaces containing no nascent chains.

Recently Wellauer and Dawid (1973) have been able to map the various regions in *X. laevis* rDNA by direct observation in the electron microscope. They discovered that the various ribosomal RNAs, as well as single strands of rDNA, have characteristic regions which reproducibly form double-strand loops when spread under the proper conditions. Thus it is possible to identify the 28S, 18S and various spacer regions clearly and to observe directly their arrangement in the rDNA or in various precursor rRNA molecules. By partial digestion of the RNA molecules with a 3′-specific exonuclease, they were also able to establish the polarity of the sequences. The structure of *X. laevis* rDNA shown in Figure 1 is based on their observations. The 28S sequence is located on the extreme 5' end of the 40S precursor, followed by a short stretch of transcribed spacer, the 18S sequence, and another short transcribed spacer on the 3' end. A region of nontranscribed spacer intervenes before the start of the next 28S sequence and so forth. The assignment of the 28S sequence to the 5' end of the 40S precursor disagrees with two previous assignments that were made by indirect means (Reeder and Brown 1970; Hecht and Birnstiel 1972) but seems to be inarguable.

There is little base asymmetry between the strands of *X. laevis* rDNA, and consequently, they do not separate in alkaline CsCl (Dawid, Brown and Reeder 1970). The strand transcribed in vivo (H strand) can be separated from the nontranscribed strand (L strand) by hybridization with rRNA and separation of the RNA-DNA hybrid in neutral CsCl gradients (Reeder and Brown 1970). Poly(IG) also binds preferentially to the H strand, but the density shift is not great enough for practical separation (Dawid and Reeder, unpublished).

In somatic chromosomes the rDNA is probably covalently inserted as part of the continuous DNA chain present in a chromosome. In yeast, one entire chromosome may be composed of rDNA (Finkelstein, Blamire and Marmur 1972). In the amplified nucleoli the rDNA is present as covalently closed circles of various sizes (Miller and Beatty 1969; Hourcade, Dressler and Wolfson 1973).

The rDNA from *X. mulleri,* a species closely related to *X. laevis,* has also been isolated and its structure analyzed (Brown, Wensink and Jordan 1972). The structures of both rDNAs are very similar. Both have the same repeat length and code for the same size 40S precursor molecule. The 28S and 18S sequences cannot be distinguished between the two species by RNA-DNA hybridization techniques. However both the transcribed and the nontranscribed spacers show only about 10% cross homology. The fact that *X. laevis* can be crossed with *X. mulleri* to yield fertile hybrid frogs

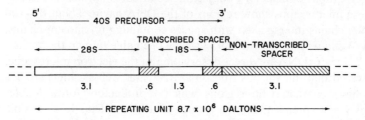

Figure 1 Structure of one repeating unit of *X. laevis* rDNA. Molecular weights are from electron microscope observations (Wensink and Brown 1971; Wellauer and Dawid, personal communication).

makes it possible to do limited genetic experiments in which the inheritance of the primary gene structure is followed directly.

Wellauer and Dawid (1973) have also mapped the secondary structure of the 45S rRNA precursor from HeLa cells, and the results are shown in Figure 2. It is interesting to note that the relative positions of the 28S, 18S and transcribed spacer sequences are the same in both HeLa and Xenopus precursor rRNA. In HeLa, however, each of these sequences, except the 18S, has increased in size. From the observations of Miller and Bakken (1972), we can surmise also that the HeLa rDNA contains large sequences of nontranscribed spacer separating the gene regions. The secondary structure map shown here is in excellent agreement with the map of the 45S molecule derived through study of rRNA processing in nucleoli (Weinberg and Penman 1970; Maden, Salim and Summers 1972).

The secondary structure of the 45S precursor from mouse L cells has also been examined and is indistinguishable from that of HeLa, except in the detailed structure of some of the loops. The size and arrangement of the major sequences is in general agreement with the preferred model of Perry and Kelley (1972).

The rDNA from the sea urchin *Lytechinus variegatus* has been isolated (Patterson and Stafford 1970) and partially characterized (Patterson and Stafford 1971). Measurements of the reannealing kinetics of this rDNA suggest a repeating unit of about 5.2×10^6 daltons. Saturation hybridization experiments suggest that about 75% of this sequence codes for 26S and 18S rRNA.

Fragmentary data on the structure of the gene region of rDNA in a wide range of species is available through studies of their rRNA molecules. This subject has been reviewed elsewhere (Attardi and Amaldi 1970; Birnstiel, Chipchase and Speirs 1971) and will be treated only briefly in this article. The 18S and 28S rRNAs of all eukaryotes are initially synthesized as part of a single large precursor molecule, which is subsequently processed to yield the two mature rRNAs. The processing is nonconservative, and some sequences (transcribed spacer) are discarded during maturation. There

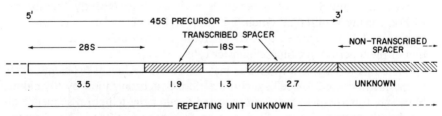

Figure 2 Structure of one repeating unit of HeLa rDNA. Molecular weights are from electron microscope observations of Wellauer and Dawid (1973). The presence of an undetermined length of nontranscribed spacer is deduced from observations of Miller and Bakken (1972).

appears to be an evolutionary pattern in the amount of transcribed spacer. Birds and mammals generally have a large primary precursor, of which roughly 50% is lost during maturation. Other phyla generally have a smaller primary precursor, of which only 10–15% is lost.

The relatedness of the 18S and 28S rRNA sequences has been examined over a wide range of plants and animals (Bicknell and Douglas 1970; Bendich and McCarthy 1970; Sinclair and Brown 1971; Vodkin and Katterman 1971). In general, the 18S and 28S sequences are highly conserved. *X. laevis* rRNA, for example, hybridizes to a significant extent with organisms as far removed as pumpkin and yeast (Sinclair and Brown 1971). No homology was detected between eukaryote and prokaryote rRNA, however. In contrast to the 18S and 28S sequences, the spacer regions show rapid divergence even between two interfertile Xenopus species (Brown, Wensink and Jordan 1972) and show no homology at all between *X. laevis* and HeLa (Reeder and Brown, unpublished). In a study of nine species of the genus Saccharomyces, Bicknell and Douglas (1970) showed that whereas the bulk DNAs showed little, if any, cross homology, the rRNA retained at least 90% homology within the genus.

Replication and Amplification of rDNA

The replication of rDNA has been examined in several cell types to determine when it occurs in the cell cycle. At present no clear picture emerges from these studies. In Physarum the rDNA is replicated throughout the mitotic cycle (Newlon, Sonenshein and Holt 1973), with the possible exception of about one hour at the beginning of S (Zellweger, Ryser and Braun 1972). In yeast, Gimmler and Schweizer (1972) found that rDNA replicated along with the bulk chromosomal DNA. In HeLa cells rDNA was also found to replicate throughout the S period (Balazs and Schildkraut 1971), whereas in a cell line from the rat kangaroo, the rDNA was reported to replicate at the end of S after almost all of the bulk DNA (Giacomoni and Finkel 1972). In Chinese hamster cells, Amaldi, Giacomoni and Zito-Bignami (1969) reported that the rDNA replicates only between 1.5 and 3 hr of a 6.5-hr S period. More recently Stambrook (1974), using a different synchronization technique and a more sensitive assay for rDNA, reported that rDNA replicates throughout the early portion of the S period in Chinese hamster cells.

Some of the variation in results may be due to technical problems, such as the method used for cell synchronization. Synchronization by thymidine blockage, for instance, has been reported to alter the temporal sequence of some cellular events (see discussion in Stambrook 1974). The chromosomal distribution of the rDNA may also be responsible for some of the differences seen between cell types. In HeLa cells, for example, the rDNA is located on at least five different chromosomes (Henderson et al. 1972) and might be expected to replicate over a longer time span than in Chinese hamster cells, which appear to have only one nucleolar organizer per haploid set (Hsu et al. 1967).

The rDNA undergoes a dramatic replication during oogenesis in many animals when it is specifically amplified (Brown and Dawid 1968; Gall 1968). The literature on this subject has been reviewed previously by Gall (1969), Birnstiel et al. (1971) and MacGregor (1972). Once again the best studied case is that of *X. laevis*. Bird and Birnstiel (1971) and Watson-Coggins and Gall (1972) have examined the time sequence of rDNA amplification in this animal. The ultrastructure of amplifying cells has been described recently by Van Gansen and Schram (1972) and by Coggins (1973). In the primordial germ cells no amplification is detectable, and the normal diploid number of 1–2 nucleoli per cell is present. A low level of amplification is first detectable by in situ hybridization and [^{3}H] thymidine labeling when sexual differentiation occurs. About the same amount is found in both oogonia and spermatogonia (Pardue and Gall 1969; Pardue 1969; Kalt 1972 and personal communication). Amplification in the gonial cells continues during a number of mitotic divisions until the cells enter a premeiotic S phase. At this point, both male and female gonial cells lose most of their amplified rDNA, although the loss may not be complete in the female cells. Male cells undergo no further amplification. Female cells replicate their total DNA during premeiotic S phase to become tetraploid oocytes. There follows a 4-day leptotene and a 5-day zygotene stage, at the end of which the condensed chromosomes are bent double with all of their ends clustered together at one side of the nucleus. During the ensuing pachytene stage (about 18 days), a dense rDNA-containing cap forms at the pole of the nucleus opposite the chromosome ends. By the end of pachytene, the cap contains about 30 pg of rDNA (compared to 12 pg of DNA in the tetraploid chromosomes) and amplification ceases. In the succeeding diplotene stage the cap material disperses, and gradually about 1500 extrachromosomal nucleoli appear. The amplified rDNA is present throughout the rest of oogenesis and can still be detected in the unfertilized egg (Brown and Dawid 1968). By the gastrula stage of development the amplified rDNA is no longer detectable. Presumably it is degraded or diluted out during early development. Several studies have shown that the amplified rDNA is in the form of covalently closed circles (Miller and Beatty 1969; Callan 1966; MacGregor and Vlad 1972).

In oocytes of Acheta, a large chromatin body is seen which contains amplified rDNA (Lima-de-Faria, Birnstiel and Jaworska 1969; Cave 1972; Allen and Cave 1972; Hansen-Delkeskamp 1972; Lima-de-Faria, Jaworska and Gustafsson 1973). In this insect, however, it is clear that genes for rRNA are only a small part of the amplified material. Similar amplification of large amounts of non-rRNA gene DNA has been reported in the water beetles Dytiscus and Columbetes (Gall, MacGregor and Kidston 1969). It is not clear whether this DNA represents extremely large spacer sequences between the rRNA genes or amplification of other DNA unlinked to the rDNA.

The number of species in which rDNA amplification is known or suspected to occur is large. Gall (1969) listed the cases known at that time, and since then only one new example, the tailed frog *Ascaphus truei,* has been added (MacGregor and Kezer 1970). In general, rDNA amplification seems to occur in species that synthesize and store large amounts of ribosomes in their oocytes. Other species either do not store excessive amounts of ribosomes or else the ribosomes may be supplied by the surrounding nurse cells (Bier et al. 1967). No evidence has been found for rDNA amplification at any stage of maize or wheat development (Ingle and Sinclair 1972). The degree of amplification varies widely from species to species. In *X. laevis,* amplified rDNA is present in about 2500-fold excess over the diploid amount, while in the marine worm Urechis (Dawid and Brown 1970), only about tenfold amplification occurs.

The mechanism by which rDNA amplification occurs has been the subject of several investigations. Two different laboratories have raised the possibility that some of the amplified rDNA may persist throughout embryonic development and be sequestered in the germ cells as an episome (Wallace et al. 1971; Brown and Blackler 1972). Since the sperm carry no detectable amplified rDNA, the simplest version of this model predicts that amplified rDNA would be inherited maternally. Experiments by Brown and Blackler (1972), using interspecific hybrids between *X. laevis* and *X. mulleri,* show that, in fact, the amplified rDNA is not inherited maternally and suggest that it is copied from the chromosomal rDNA. A peculiar finding of these experiments was that only *X. laevis* rDNA was ever amplified regardless of which species was the female parent. Since *X. laevis* rDNA is also transcriptionally dominant in the somatic cells of hybrid frogs (Honjo and Reeder 1973), this suggests that transcription may play some role in the amplification process.

The mechanism for producing the first extrachromosomal copies of rDNA is at present a subject of some controversy. Crippa, Tocchini-Valentini and their coworkers have published experiments to support the idea that RNA-dependent DNA polymerase (reverse transcriptase) plays a key role in the early stages of amplification (Crippa and Tocchini-Valentini 1971; Mahdavi and Crippa 1972; Brown and Tocchini-Valentini 1972). They envision that DNA-dependent RNA polymerase must first transcribe at least one entire repeat of the chromosomal rDNA. This RNA transcript is copied by reverse transcriptase to yield DNA copies, and these DNA copies are then joined by ligase to form the amplified rDNA. They report finding structures in the ovaries of young frogs corresponding to the various nucleic acid and polymerase complexes predicted by this model.

Bird, Rogers and Birnstiel (1973) have searched for these structures independently with no success and they question the entire model. This disagreement may be due to the fact that one group used animals sex-reversed by estrogen and the other group used untreated animals. There is evidence that estrogen treatment may accelerate the amplification process (Tocchini-Valentini et al. 1974). More recently, rolling circle structures have been found in amplified rDNA (Hourcade, Dressler and Wolfson 1974; Bird, Rochaix and Bakken 1973). This finding suggests that at least some stage of amplification occurs by this mechanism. It does not, however, rule out other mechanisms for the origin of the first extrachromosomal copies.

Any discussion of rDNA replication must consider the problem of multiple copies and how they are kept identical within a species in the face of mutational pressure to diverge. One model has been proposed by Buongiorno-Nardelli, Amaldi and Lava-Sanchez (1972). They have presented some intriguing numerology concerning the number of rDNA repeats present in the diploid somatic genome of a few species of amphibians. These numbers fell into groupings which are very close to various powers of 2. Then they counted the number of extrachromosomal nucleoli present in oocytes of the same species. These numbers also fell into groupings that were powers of 2. To explain these numbers they proposed that during amplification each repeat of the chromosomal rDNA is excised from the chromosome and forms an individual monogenic circle. These circles then go through n rounds of replication by some mechanism which causes them to double in size at each round. The number n is determined genetically, and when amplification is complete, one of the extrachromosomal circles is reinserted back into the chromosome. Thus the chromosomal rDNA is rectified each generation.

The model is ingenious, but data from other laboratories seem to contradict it. For example, Miller and Beatty (1969) and Hourcade, Dressler and Wolfson (1974) report that amplified rDNA in *X. laevis* displays an arithmetical progression of sizes, rather than the geometric progression required by the model. Data from Perkowska et al. (1968) and Miller and Beatty (1969) show that, in *X. laevis* at least, each nucleolus may contain several cores (circles of rDNA), thus casting doubt on the significance of counting intact nucleoli. Furthermore, the amount of somatic rDNA has been reported to vary widely from individual to individual within a species (Miller and Brown 1969). The model as presented can only account for rectification in genes which are amplified. Some other mechanism is required to account for homogeneity in nonamplified sequences, such as the sequence for somatic 5S RNA.

Callan has proposed that tandemly repetitious genes could be kept homogeneous if one copy was the "master" and the rest were "slaves" that were somehow compared to the master copy once each generation for correction (Callan and Loyd 1960; Callan 1967). He proposed that this correction could take place sometime during meiosis and could account for some of the structural peculiarities seen in diplotene lampbrush chromosomes. Callan originally proposed that all genes were members of repetitious master-slave sets. Subsequent work indicates that, on the contrary, much of the genome consists of unique or single-copy DNA, which obviates the need for any general use of a master-slave mechanism for the entire chromosome, although it is still possible for tandemly repetitious genes such as rDNA.

Several authors have proposed that multiple copies may be maintained by the mechanism of unequal crossing-over (Brown, Wensink and Jordan 1972; Smith 1974). Smith has shown how this mechanism could produce exactly repeated sequences (such as rDNA) or a sequence of heterogeneous repeats by varying the rate of crossing-over versus the mutation rate.

Transcription of rDNA

The nuclei of all eukaryotes contain at least two major types of DNA-dependent RNA polymerase, which have been termed I and II (Roeder and Rutter 1969). Polymerase I is localized in nucleoli (Roeder and Rutter 1970), is completely resistant to the fungal toxin α-amanitin (Kedinger et al. 1970), and by several criteria appears to be the polymerase responsible for rRNA synthesis (Blatti et al. 1970; Beermann 1971; Zylber and Penman 1971a; Reeder and Roeder 1972).

Electron micrographs of dispersed oocyte nucleoli from *X. laevis* show that about 100 polymerase molecules, each with its attached nascent rRNA chain, are present on each gene region and not on the intervening spacer regions (Miller and Bakken 1972). The nascent rRNA molecules are arranged in a gradient of lengths, with the shortest ones presumably near the promoter site for RNA polymerase and the longest ones at the termination site. The short to long polarity of these nascent chains is in the same direction over the longest stretches of DNA visible. This argues that transcription does not switch from strand to strand. Occasionally small intact circles are seen with the same polarity of nascent chains all around the circle (Miller and Beatty 1969). Structures similar to those in Xenopus oocyte nucleoli have also been found in HeLa nuclei (Miller and Bakken 1972). The regions of presumed nascent rRNA chains are longer than in Xenopus, but they still seem to be separated from each other by long stretches of nontranscribed spacer.

Photographs by Miller and coworkers offer the most direct evidence that the gene regions in Xenopus rDNA are separated by nontranscribed spacer sequences. They also seem to rule out the possibility that both strands of the rDNA are transcribed with rapid degradation of the wrong strand copy, as has been reported in mitochondria (Aloni and Attardi 1971) and SV40 transcription (Aloni 1972).

Recently Scheer, Trendelenburg and Franke (1973) have published photographs of dispersed nucleoli from Triturus. In their experiments they regularly see an additional, shorter region of nascent chains located just before the main region of transcription. This suggests that additional spacer may be transcribed in Triturus, but not as part of the large rRNA precursor.

During mitosis the nucleolus disappears and is reformed at the beginning of telophase. Synthesis and processing of rRNA also stops during metaphase (Fan and Penman 1971). The various rRNA precursors are stable during this period, and a normal nucleolus can reform at telophase in the absence of rRNA synthesis (Phillips and Phillips 1973).

The transcriptional activity of rDNA varies over a wide range during development. In *X. laevis* oocytes, the amplified rDNA appears to be inactive in rRNA synthesis immediately after amplification and up to the point where active yolk accumulation begins (Mairy and Denis 1971; Thomas 1970; Ford 1971). The oocytes then go through a period of intense rRNA synthesis which persists in the mature oocyte (LaMarca, Smith and Strobel 1973). In unfertilized eggs and early cleavage embryos, rRNA synthesis is indetectable up to the gastrula stage, at which point there is an apparent turn-on of the rDNA. Turn-on of the rDNA at gastrulation has been reported to occur in a number of other organisms. This subject has been reviewed by Brown and Dawid (1969). In organisms such as the mouse which do not store such massive quantities of ribosomes during oogenesis, rRNA synthesis begins much earlier at about the fourth cleavage division.

The hypothesis that rDNA is turned on at gastrulation has been challenged by Emerson and Humphreys (1970, 1971), who have reexamined the question in sea urchins. They detected newly made rRNA considerably earlier than the gastrula stage and proposed that, on a per nucleus basis, the rate of rRNA synthesis may be constant from fertilization onward. They suggest that earlier methods simply were not sensitive enough to detect the rRNA made during cleavage when heterogeneous nuclear RNA synthesis is unusually active.

Other laboratories have reexamined this problem in sea urchins (Sconzo et al. 1970; Sconzo and Giudice 1972) and in amphibians (Knowland 1970; Miller 1972). They have employed different methods than those used by Emerson and Humphreys but still found no rRNA synthesis prior to gastrulation.

Morphological evidence has also been used to support the claim of rDNA activation at gastrulation. It is generally believed that rDNA transcription is correlated with the appearance of a granular component in the fine structure of the nucleolus. Since cleavage-stage nucleoli are small and have almost no granular component, this has been taken as evidence that they are not engaged in rRNA synthesis (for review, see Hay 1968). During cleavage the rate of mitosis is much higher than in later embryos. Since nucleoli disperse during mitosis and rRNA synthesis is arrested, it has been proposed that the abnormal appearance of cleavage nucleoli is due to the fact that they do not have sufficient time in interphase to reform properly. Emerson and Humphreys (1971) have shown that sea urchin blastulae will form much larger nucleoli when mitosis is slowed with low doses of FdU. Brachet et al. (1972) claim, however, that such nucleoli still lack a granular component.

Recently Honjo and Reeder (1973) have examined the turn-on of rRNA synthesis in *X. laevis–X. mulleri* hybrid embryos, which were specially constructed so that they had no *X. laevis* rDNA. Using RNA-DNA hybridization to detect rRNA synthesis, they were able to show that there was no synthesis of *X. mulleri* rRNA until sometime after gastrulation. Presumably this delay in turn-on was due to the presence of the *X. laevis* cytoplasm contributed by the egg. In these embryos it seems certain that the *mulleri* rDNA transcription was turned on at a particular stage.

Two different laboratories have reported the isolation of substances that specifically repress rRNA synthesis. One was isolated from Xenopus oocytes (Crippa 1970) and the other from cleaving Xenopus embryos (Shiokawa and Yamana 1967). Several laboratories have been unable to repeat the results of Shiokawa and Yamana (Landesman and Gross 1968; Van Snick and Brachet 1972; Hill and McConkey 1972). Laskey, Gerhart and Knowland (1973) reported they could detect no inhibitor that diffused from intact cells as originally claimed. However, extraction of blastulae with perchloric acid yielded a substance that adsorbed to charcoal and depressed rRNA synthesis relative to 4S synthesis when applied to neurulae cells. Definitive characterization of molecules controlling rDNA transcription will require the establishment of an accurate in vitro transcription system for these genes.

Mutations of the rDNA

The only mutations reported so far in the rRNA structural genes are deletion mutants of varying degrees of severity. One well-known case is that of the anucleolate mutation in *X. laevis* mentioned earlier in this review (Elsdale, Fischberg and Smith 1958). This mutation has been shown to be a complete deletion of the rDNA (Wallace and Birnstiel 1966), and homozygous embryos carrying this mutation synthesize no rRNA. Due to the large supply of ribosomes which they inherit from the egg, however, they are able to survive until the swimming tadpole stage before dying (Brown and Gurdon 1964; Elsdale, Fischberg and Smith 1958). In the heterozygous state, with only half the normal amount of rDNA, embryos produce a single nucleolus of twice the normal size and accumulate rRNA just as rapidly as do wild-type embryos. Other mutations have been reported in *X. laevis* which are only partial deletions of the rDNA. Miller and Knowland (1970) and Knowland and Miller (1970) have described two partial deletions, each of which makes the normal amount of rRNA when in the heterozygous state. Embryos that carry both of these mutations, however, die at an early stage and make rRNA at a reduced rate even though they have 70% of the normal level of rDNA. This suggests that the nucleolus contains some locus essential for transcription at the full rate. It would be of great interest, of course, to find a mutant which behaved phenotypically as a partial (or complete) deletion but which contained a normal complement of rDNA. Such animals have not yet been found. Miller and Brown (1969) have presented evidence that the amount of rDNA per cell is not rigidly controlled and may fluctuate widely from animal to animal.

A mutant which organizes an abnormally small nucleolus has been reported in the Axolotl by Humphrey (1961). Miller and Brown (1969) have shown that this mutant is a partial deletion of the rDNA. Deletions of the rDNA have also been reported in Chironomus (Pelling and Beermann 1966), Sciara (Gerbi 1971) and Drosophila. In Drosophila, partial deletion of the rDNA gives rise to a phenotype called "bobbed." The bobbed mutants have been the subject of considerable study and are reviewed elsewhere in this volume.

Nucleolar Dominance

One major obstacle to developing a genetics of ribosomes is the difficulty of detecting nonlethal mutations in the structural genes for rRNA. A possible route around this obstacle is through the use of interspecific hybrids between species that have evolved differences in their rDNA. Each species may thus be considered as a naturally occurring, nonlethal mutant.

The initial work in this area was done by Navashin (1934), who was following the inheritance of secondary constrictions in interspecific hybrids of the plant genus Crepis. This and other cytological literature has been reviewed and further results added by Wallace and Langridge (1971). Navashin observed that in Crepis each haploid set of mitotic chromosomes contained one chromosome with a secondary constriction on the short arm. In an interspecific cross, such as C. capillaris (n = 3) × C. neglecta (n = 4), seven chromosomes appeared at metaphase, each morphologically attributable to one of the parents. However only the C. capillaris set contained a chromosome with a secondary constriction. Navashin showed that the C. capillaris chromosome was also dominant in reciprocal crosses, thus excluding any maternal effect. The effect could be produced by only a part of the C. capillaris chromosome set as long as the secondary constriction-bearing chromosome was present. And the effect was reversible since, in appropriate back-crosses, the C. neglecta chromosome would again develop a secondary constriction.

Altogether 30 different hybrids of Crepis have been studied and can be arranged in a self-consistent hierarchy of nucleolar dominance. Thus a given species may be dominant to some species, subordinate to others, and codominant to still others (in the latter case the hybrid cells contain two secondary constrictions at metaphase). In most cells of Crepis, nucleolar fusion occurs during interphase, and only one nucleolus is visible regardless of the number of nucleolar organizers (secondary constrictions) present. In rapidly growing tissue such as root tips, however, a minor proportion of cells display two nucleoli in the parent stocks. In the hybrids there is a perfect correspondence between the number of secondary constrictions visible at metaphase and the maximum number of nucleoli found during the interphase.

Observations on meiosis in Crepis hybrids suggest that the repressed chromosome can resume its ability to form a nucleolus as soon as it is separated from the dominant chromosome by only a nuclear membrane and the intervening cytoplasm; isolation by a cell wall is not necessary for such recovery (Wallace and Langridge 1971).

Similar cytological observations of nucleolar dominance have been made in the plant genera Ribes (Keep 1971) and Triticum (Crosby 1957; Longwell and Svihla 1960).

In Rhynchosciara (Pardue et al. 1970) and in Phaseolus (Durante, Cionini and D'Amato 1972), in situ hybridization has revealed the presence of rDNA which is apparently not transcribed (at least in the cell types examined). These might be cases of dominance by one locus over the other. Alternatively, it is possible that the silent locus is expressed only in certain tissues and may even have different nucleotide sequences. The 5S genes in Xenopus are an example of this type of behavior (see below). It has also been reported that no nucleolus is associated with the Y chromosome rDNA in Drosophila salivary glands (Viinikka, Hannah-Alava and Arajarvi 1971), but this point is in some dispute (Spear, personal communication).

The first clear report of nucleolar dominance in animal cells was made recently by Blackler (Blackler and Gecking 1972; Cassidy and Blackler, personal communication) in hybrids between the two frog species, *Xenopus laevis* and *Xenopus mulleri*. In both of these species there is one nucleolar organizer-bearing chromosome per haploid set. During early development of either parent stock (after gastrulation), two prominent nucleoli are present in diploid somatic cells. In hybrids only one nucleolus appears. Since the rDNA of *X. laevis* has some nucleotide sequences which differ from those of *X. mulleri,* it was possible for Brown and Blackler (1972) to show that rDNA from both parents was present in the hybrids. Honjo and Reeder (1973) were able to show, however, that only the *X. laevis* rDNA was transcribed. As in Crepis, no maternal effect was found; *X. laevis* was dominant as either male or female parent. The repression of *X. mulleri* rDNA was leaky; at about the same stage (swimming tadpole) that Cassidy and Blackler (personal communication) began to see significant numbers of two nucleolate cells in the hybrids, Honjo and Reeder found a low level of *X. mulleri* rRNA synthesis. In adult frog livers, anywhere from zero to 50% *X. mulleri* rRNA synthesis was observed in individual animals.

By use of the rDNA deletion mutant of *X. laevis,* it is possible to construct hybrids that contain a complete set of chromosomes from each species except that *X. laevis* rDNA is absent; only *X. mulleri* rDNA is present. In such animals, cytological and biochemical studies show that the *X. mulleri* rDNA is expressed, but the time of its turn-on is subject to maternal influence. If the female parent is *X. laevis,* turn-on is considerably delayed beyond the gastrula stage. In the reverse cross there appears to be no delay in turn-on at gastrulation.

The occurrence of nucleolar dominance also raises an opportunity to ask what other genes, if any, are controlled coordinately with the transcription of rDNA. This question may be answerable for some of the ribosomal protein genes in Xenopus since Hallberg (personal communication) has shown that 2–3 of the ribosomal proteins are electrophoretically distinguishable between *X. laevis* and *X. mulleri.* Wall and Blackler (1974) have looked at several dehydrogenases that display electrophoretic variants between the two species and so far have found both parental types expressed in the hybrids.

The data from both Crepis and Xenopus are compatible with the notion that it is the rDNA of the dominant species which is itself responsible for the repression of the dominated rDNA. This could be effected through some product of the rDNA (rRNA?). Alternatively, as initially suggested by McClintock (1934), it could be due to competition between the two types of rDNA for some essential but scarce substance within the cell. It does not seem to be due to some locus outside of the nucleolar organizer, although the existence of such loci has been described in *Zea mays* (McClintock 1934).

What is needed is an experimental system in which biochemical analysis can be combined with extensive genetics. A system of great potential is Drosophila. A number of hybrid crosses which yield fertile offspring have been recorded in this genus (Patterson and Stone 1952). Presumably some of these species will have evolved sequence differences in their rDNA, and thus suitable biochemical assays could be developed as for Xenopus. If any of these crosses turn out to exhibit nucleolar dominance, a powerful analysis of the control of rRNA synthesis would seem possible.

An alternative method for maneuvering two different types of rDNA into the same cytoplasmic or nuclear environment is by somatic cell fusion using inactivated virus as the fusing agent. The initial product of such fusion is a heterokaryon, in which two or more nuclei reside in the same cytoplasm. The general result of such experiments so far is that the rDNAs remain codominant; each nucleus retains its nucleolus and seems to continue rRNA synthesis. In a heterokaryon formed between a cell with an active nucleolus (i.e., HeLa) and a cell with an inactive nucleolus (i.e., chick erythrocyte), the inactive nucleolus is turned on and begins rRNA synthesis (Harris et al. 1966). Harris and coworkers further believe that activation of the chick nucleolus is obligatory before the RNA specifying other chick antigens can be transported to the cytoplasm for translation into protein (Deak, Sidebottom and Harris 1972).

In some cases the nuclei in heterokaryons will fuse, yielding a somatic hybrid cell that is capable of multiplying to form a cell line. Analyses of several somatic hybrid lines, necessarily done at least 10 or 20 generations after the initial fusion event, show that synthesis of one species of rRNA is sometimes (Elicieri and Green 1969; Bramwell and Handmaker 1971), but not always (Elicieri 1972), dominant over the other. However none of these cell lines has been analyzed to see which species of rDNA it contains. Therefore we cannot decide at present whether the observed rRNA dominance is due to selective transcription or selective loss of rDNA.

THE GENES
FOR 5S RIBOSOMAL RNA
Chromosomal Location

A ribosome contains one molecule each of the large 18S and 28S RNAs plus one smaller molecule called the 5S RNA. In most eukaryotic cells these three RNAs accumulate in an equimolar ratio, suggesting that they might be closely linked on the chromosome and be synthesized as part of a single polycistronic precursor RNA. This does not appear to be the case, however. The 5S DNA from *X. laevis* bands in CsCl on the opposite side of the bulk DNA from the rDNA. Furthermore, frogs which have completely lost their rDNA still have a normal complement of 5S DNA (Brown and Weber 1968).

The technique of in situ hybridization has allowed more precise localization of the 5S DNA. In *X. laevis* the 5S DNA is localized at the extreme end of the long arm of many (if not all) of the metaphase chromosomes (Pardue, Brown and Birnstiel 1973). This location is in sharp contrast to the rDNA, which is all at one site in the middle of the short arm of only one chromosome of each haploid set. In normal *X. laevis* chromosomes no 5S DNA is detected near the rDNA. But in a cell line derived from *X. laevis* kidney, an apparent translocation has placed some 5S DNA at the distal end of the nucleolar secondary constriction. During the pairing stages of meiosis in *X. laevis* ovaries, the chromosomes align in such a way that their telomeric ends are tightly clustered. In situ hybridization shows that all the 5S DNA is located in this cluster. The dense nuclear cap of amplified rDNA forms on the opposite side of the nucleus from the 5S DNA cluster and contains no detectable 5S DNA.

In HeLa cells (Aloni, Hatlen and Attardi 1971), 5S RNA was shown to hybridize with DNA isolated from chromosomes of all size classes. Therefore 5S DNA is probably located on most of the chromosomes in this cell type also. In contrast, the 5S DNA of *Drosophila melanogaster* is entirely localized in a single band (56F) on the right arm of the second chromosome (Wimber and Steffensen 1970; Quincey 1971).

Structure of 5S DNA

The 5S DNA, like rDNA, consists of many tandemly repeated copies. In *X. laevis,* the number of repeats per haploid complement has been estimated by saturation hybridization as around 24,000 (Brown and Weber 1968) and by analysis of hybridization kinetics as 9000 (Birnstiel, Sells and Purdom 1972). About 2000 copies per haploid genome have been estimated for HeLa cells (Hatlen and Attardi 1971) and about 200 copies for Drosophila (Tartof and Perry 1970; Quincey 1971).

The 5S DNAs from *X. laevis* and *X. mulleri* have been isolated and their structure studied (Brown, Wensink and Jordan 1971; Brown and Sugimoto 1973). The *X. laevis* 5S DNA consists of tandemly repeating units of molecular weight 0.5–0.6 $\times$ 10^6 daltons. About $\frac{1}{7}$ of each repeat (0.08 $\times$ 10^6 daltons) is a structural gene for a 5S RNA molecule; the rest is spacer of unknown function. The 5S DNA repeats are heterogeneous, both in the gene and spacer regions. Although the 5S RNA synthesized by cultured cells is homogeneous (Brownlee et al. 1972; Wegnez, Monier and Denis 1972), oocytes are known to synthesize at least three other species of 5S RNA in addition to the somatic sequence (Wegnez, Monier and Denis 1972; Ford and Southern 1973). The ratio of somatic sequences to oocyte sequences in the 5S DNA is unknown at present, although preliminary evidence suggests that oocyte sequences are greatly predominant (Brownlee and Brown, personal communication). Heterogeneity in the spacer sequences is shown by the fact that reannealing of 5S DNA results in a moderate degree of mismatching. Therefore the spacer regions also consist of a family of similar but not identical sequences (Brown and Sugimoto 1973). A fundamental question not yet answered is whether the heterogeneity exists only between 5S DNA on different chromosomes or whether it exists within a locus on a single chromosome. If the heterogeneity is found to exist within a single locus, this will tend to favor a crossing-over mechanism for maintaining the 5S DNA as a closely related family of sequences (Brown and Sugimoto 1973).

Some structural aspects of *X. mulleri* 5S DNA have been compared with *X. laevis* 5S DNA (Brown and Sugimoto 1973). Preliminary sequence data show that the somatic 5S RNAs of the two species are very similar; however the spacer regions of the two species are so far diverged that no cross-hybridization can be detected between them in RNA-DNA annealing experiments. The average repeat length of *X. mulleri* is also 1.2–1.5 $\times$ 10^6 daltons, about 2½ times longer than that of *X. laevis*.

Transcription of 5S DNA

Several experiments have indicated that there is no precursor to somatic 5S RNA in eukaryotes. Perry and Kelley (1968) could detect no precursor after brief labeling of cultured mouse cells, and several groups have shown that a fraction of 5S RNA can be isolated with a triphosphate group at its 5′ terminus (Hatlen, Amaldi and Attardi 1968; Brownlee et al. 1972; Wegnez, Monier and Denis 1972). Hybridization experiments designed to search for such a precursor among pulse-labeled nuclear RNA have also failed to find one (Brown and Sugimoto 1973). Wegnez (1973), however, has reported finding in oocytes of *X. laevis* a precursor form of 5S RNA which contains 15 extra nucleotides on the 3′ end of the molecule before it is cleaved to the mature length.

Since in most cells the 5S RNA accumulates in amounts equimolar with the 18S and 28S rRNAs, it has often been assumed that there is transcriptional coupling between these genes even though they are not physically linked. In fact, in situ hybridization experiments suggest that during interphase in some cells, the 5S DNA is associated with the nucleolus (Wimber and Steffensen 1970; Amaldi and Buongiorno-Nardelli 1971; Pardue, Brown and Birnstiel 1973). Numerous other experiments have shown, however, that 5S synthesis can occur in the absence of rRNA synthesis. Treatment of cells with low doses of actinomycin D will selectively inhibit rRNA synthesis without blocking 5S RNA synthesis (Perry and Kelley 1968). Synthesis of rRNA is absent in metaphase-arrested HeLa cells, but synthesis of 5S RNA continues at nearly normal rates (Zylber and Penman 1971b). By use of temperature-sensitive mutants of yeast that are defective in ribosome synthesis, Helser and McLaughlin (personal communication) have concluded that 5S synthesis is not directly controlled by the availability of any maturing ribosome assembly products.

During oogenesis in *X. laevis,* very little rRNA is synthesized in the period between rDNA amplification and the beginning of vitellogenesis. During this time, the oocyte accumulates about 25% of its RNA as 5S RNA (and a further 25% as 4S RNA). Both RNAs accumulate in the cytoplasm in a 42S ribonucleoprotein particle (Ford 1971; Mairy and Denis 1971; Denis and Mairy 1972; Wegnez 1973). Thus at this stage, there is a massive uncoupling of 5S and rRNA synthesis. However in the mature oocyte, the normal equimolar situation has been restored (Brown and Littna 1966).

Synthesis of 5S RNA is not detectable during the early cleavage stages of embryogenesis in *X. laevis* and presumably these genes are turned off, as are the genes for rRNA (Brown and Littna 1966; Miller 1973). Synthesis of both 5S and rRNA is first detectable at gastrulation. In the anucleolate mutant of *X. laevis* it was originally believed that 5S synthesis did not turn on at gastrulation (Brown 1967). Recently more sensitive experiments show that 5S synthesis does turn on at gastrulation in these animals, providing one more example of uncoupling between 5S and rRNA synthesis (Miller 1973).

Price and Penman (1972) have suggested that 5S RNA may be synthesized by an RNA polymerase distinct from the nucleolar polymerase I. Leibowitz, Weinberg and Penman (1973) estimate that in exponentially growing HeLa cells, 5S RNA is made in molar amounts about fourfold greater than rRNA. Presumably the excess 5S RNA is degraded. Amplification of 5S DNA has never been detected in any cell type. In *X. laevis* this can be rationalized by assuming that the somatic 5S RNA sequences are sufficient for ribosome synthesis in most cells. During oogenesis, when the rDNA is amplified, 5S synthesis keeps pace by turning on a large new set of genes which are silent in most other cells.

THE GENES FOR RIBOSOMAL PROTEIN

At present very little is known about these genes. Hallberg and Brown (1969) were unable to detect ribosomal protein synthesis during early cleavage in *X. laevis* embryos or in later stage embryos which lacked rDNA and therefore made no rRNA. In normal embryos ribosomal protein synthesis is first detectable at gastrulation. Therefore they proposed that there is tight coordination between rRNA and ribosomal protein synthesis. Although ribosomes are assembled in the nucleolus, ribosomal proteins are synthesized on cytoplasmic ribosomes like most other proteins (Craig and Perry 1971).

By use of interspecies hybrids between *Drosophila melanogaster* and *D. simulans,* Steffensen (1973) was able to show that the genes for at least seven ribosomal proteins map within the proximal segment of the X chromosome not far from the nucleolus organizer.

Some efforts have been made to see if the same ribosomal proteins are synthesized at all stages of development. Gray and Landesman (1972) found no differences in monosome proteins throughout *X. laevis* development but did detect minor differences between monosome and polysome proteins. DeLauney, Mathieu and Schapira (1972) examined the ribosomal proteins of several species and found no variation within an organ as it developed, although there were some minor differences between organs and between species. Lambertsson (1972), working with Drosophila, reported finding some differences between ribosomal proteins from larvae, pupae and adult flies. In all of these experiments the detection of minor differences is impossible to interpret until the number and function of eukaryotic ribosomal proteins is better defined.

There are stages during oogenesis in *X. laevis* when a large fraction of total oocyte protein synthesis is ribosomal protein (~30%; R. Hallberg, personal communication). It may be possible to isolate ribosomal protein messenger RNA from such oocytes and thereby gain a tool for studying the ribosomal protein genes in greater detail.

References

Allen, E. R. and M. D. Cave. 1972. Nucleolar organization in oocytes of gryllid crickets: Subfamilies *Grillinae* and *Nemobiinae. J. Morphol.* **137**:433.

Aloni, Y. 1972. Extensive symmetrical transcription of simian virus 40 DNA in virus-yielding cells. *Proc. Nat. Acad. Sci.* **69**:2404.

Aloni, Y. and G. Attardi. 1971. Expression of the mitochondrial genome in HeLa cells. II. Evidence for complete transcription of mitochondrial DNA. *J. Mol. Biol.* **55**:251.

Aloni, Y., L. E. Hatlen and G. Attardi. 1971. Studies of fractionated HeLa cell metaphase chromosomes. II. Chromosomal distribution of sites for transfer RNA and 5S RNA. *J. Mol. Biol.* **56**:555.

Amaldi, F. and M. Buongiorno-Nardelli. 1971. Molecular hybridization of Chinese hamster 5S, 4S and "pulse-labeled" RNA in cytological preparations. *Exp. Cell Res.* **65**:329.

Amaldi, F., D. Giacomoni and R. Zito-Bignami. 1969. The duplication of ribosomal RNA cistrons in Chinese hamster cells. *Eur. J. Biochem.* **11**:419.

Attardi, G. and F. Amaldi. 1970. Structure and synthesis of ribosomal RNA. *Ann. Rev. Biochem.* **39**:183.

Balazs, I. and C. L. Schildkraut. 1971. DNA replication in synchronized cultured mammalian cells. II. Replication of ribosomal cistrons in thymidine-synchronized HeLa cells. *J. Mol. Biol.* **57**:153.

Beermann. W. 1971. Effect of α-amanitin on puffing and intranuclear RNA synthesis in Chironomus salivary glands. *Chromosoma* **34**:152.

Bendich, A. J. and B. J. McCarthy. 1970. Ribosomal RNA homologies among distantly related organisms. *Proc. Nat. Acad. Sci.* **65**:349.

Bicknell, J. N. and H. C. Douglas. 1970. Nucleic acid homologies among species of *Saccharomyces. J. Bact.* **101**:505.

Bier, K., W. Kunz and D. Ribbert. 1967. Struktur und Funktion der oocyten-chromosomen und nukleolen sowie der extra-DNS wahrend der oogenese panoisticher und meroisticher insekten. *Chromosoma* **23**:214.

Bird, A. P. and M. L. Birnstiel. 1971. A timing study of DNA amplification in *Xenopus laevis* oocytes. *Chromosoma* **35**:300.

Bird, A. P., J.-D. Rochaix and A. H. Bakken. 1973. The mechanism of gene amplification in *Xenopus laevis* oocytes. In *Molecular cytogenetics* (ed. B. H. Hamkalo and J. Papaconstantinou), pp. 49–58. Plenum Press, New York and London.

Bird, A., E. Rogers and M. Birnstiel. 1973. Is gene amplification RNA-directed? *Nature New Biol.* **242**:226.

Birnstiel, M. L., M. Chipchase and J. Speirs. 1971. The ribosomal RNA cistrons. *Prog. Nucleic Acid Res. Mol. Biol.* **11**:351.

Birnstiel, M. L., B. H. Sells and I. F. Purdom. 1972. Kinetic complexity of RNA molecules. *J. Mol. Biol.* **63**:21.

Birnstiel, M., M. Grunstein, J. Speirs and W. Hennig. 1969. Family of ribosomal genes of *Xenopus laevis. Nature* **223**:1265.

Birnstiel, M. L., H. Wallace, J. L. Sirlin and M. Fischberg. 1966. Localization of the ribosomal DNA complements in the nucleolar organizer region of *Xenopus laevis. Nat. Cancer Inst. Monogr.* **23**:431.

Birnstiel, M., J. Speirs, I. Purdom, K. Jones and U. E. Loening. 1968. Properties and composition of the isolated ribosomal DNA satellite of *Xenopus laevis*. *Nature* **219**:454.

Blackler, A. W. and C. A. Gecking. 1972. Transmission of sex cells of one species through the body of a second species in the genus *Xenopus*. II. Interspecific matings. *Develop. Biol.* **27**:385.

Blatti, S. P., C. J. Ingles, T. J. Lindell, P. W. Morris, R. F. Weaver, F. Weinberg and W. J. Rutter. 1970. Structure and regulatory properties of eukaryotic RNA polymerase. *Cold Spring Harbor Symp. Quant. Biol.* **35**:649.

Brachet, J., D. O'Dell, G. Steinert and R. Tencer. 1972. Cleavage nucleoli and ribosomal RNA synthesis in sea urchin eggs. *Exp. Cell Res.* **73**:463.

Brady, T. and M. E. Clutter. 1972. Cytolocalization of ribosomal cistrons in plant polytene chromosomes. *J. Cell Biol.* **53**:827.

Bramwell, M. E., and S. D. Handmaker. 1971. Ribosomal RNA synthesis in human-mouse hybrid cells. *Biochim. Biophys. Acta* **232**:580.

Brown, D. D. 1967. The genes for ribosomal RNA and their transcription during amphibian development. In *Current topics in developmental biology* (ed. A. Monroy and A. Moscona) vol. 2, p. 47. Academic Press, New York.

Brown, D. D. and A. W. Blackler. 1972. Gene amplification proceeds by a chromosome copy mechanism. *J. Mol. Biol.* **63**:75.

Brown, D. D. and I. B. Dawid. 1968. Specific gene amplification in oocytes. Oocyte nuclei contain extrachromosomal replicas of the genes for ribosomal RNA. *Science* **160**:272.

———. 1969. Developmental genetics. *Ann. Rev. Genet.* **3**:127.

Brown, D. D. and J. M. Gurdon. 1964. Absence of ribosomal RNA synthesis in the anucleolate mutant of *Xenopus laevis*. *Proc. Nat. Acad. Sci.* **51**:139.

Brown, D. D. and E. Littna. 1966. Synthesis and accumulation of low molecular weight RNA during embryogenesis of *Xenopus laevis*. *J. Mol. Biol.* **20**:95.

Brown, D. D. and K. Sugimoto. 1973. 5S DNAs of *Xenopus laevis* and *Xenopus mulleri:* Evolution of a gene family. *J. Mol. Biol.* **78**:397.

Brown, D. D. and C. S. Weber. 1968. Gene linkage by RNA-DNA hybridization. I. Unique DNA sequences homologous to 4S RNA, 5S RNA and ribosomal RNA. *J. Mol. Biol.* **34**:661.

Brown, D. D., P. C. Wensink and E. Jordan. 1971. Purification and some characteristics of 5S DNA from *Xenopus laevis*. *Proc. Nat. Acad. Sci.* **68**:3175.

———. 1972. A comparison of the ribosomal DNAs of *Xenopus laevis* and *Xenopus mulleri*—the evolution of tandem genes. *J. Mol. Biol.* **63**:57.

Brown, R. D. and G. P. Tocchini-Valentini. 1972. The role of RNA in gene amplification. *Proc. Nat. Acad. Sci.* **69**:1746.

Brownlee, G. G., E. Cartwright, T. McShane and R. Williamson. 1972. The nucleotide sequence of somatic 5S RNA from *Xenopus laevis*. *FEBS Letters* **25**:8.

Buongiorno-Nardelli, M., F. Amaldi and P. A. Lava-Sanchez. 1972. Amplification as a rectification mechanism for the redundant rRNA genes. *Nature New Biol.* **238**:134.

Busch, H. and K. Smetana. 1970. *The nucleolus*. Academic Press, New York.

Callan, H. G. 1966. Chromosomes and nucleoli of the Axolotl, *Ambystoma mexicanum*. *J. Cell Sci.* **1**:85.

————. 1967. The organization of genetic units in chromosomes. *J. Cell Sci.* **2**:1.

Callan, H. G. and L. Loyd. 1960. Lampbrush chromosomes of crested newts *Triturus cristatus* (Laurenti). *Phil. Trans. Roy. Soc.* **B 243**:135.

Cave, M. 1972. Localization of ribosomal DNA within oocytes of the house cricket, *Acheta domesticus. J. Cell Biol.* **55**:310.

Chattopadhyay, S. K., D. E. Kohne and S. K. Dutta. 1972. Ribosomal RNA genes of *Neurospora:* Isolation and characterization. *Proc. Nat. Acad. Sci.* **69**:3256.

Coggins, L. W. 1973. An ultrastructural and radioautographic study of oogenesis in the toad *Xenopus laevis. J. Cell Sci.* **12**:71.

Collier, J. R. 1971. Number of ribosomal RNA cistrons in the marine mud snail *Ilyanassa obsoleta. Exp. Cell Res.* **69**:181.

Craig, N. and R. P. Perry. 1971. Persistent cytoplasmic synthesis or ribosomal proteins during the selective inhibition of ribosomal RNA synthesis. *Nature New Biol.* **229**:75.

Crippa, M. 1970. Regulatory factor for the transcription of the ribosomal genes in amphibian oocytes. *Nature* **227**:1138.

Crippa, M. and G. P. Tocchini-Valentini. 1971. Synthesis of amplified DNA that codes for ribosomal RNA. *Proc. Nat. Acad. Sci.* **68**:2769.

Crosby, A. R. 1957. Nucleolar activity of lagging chromosomes in wheat. *Amer. J. Bot.* **44**:813.

Dawid, I. B. and D. D. Brown. 1970. The mitochondrial and ribosomal DNA components of oocytes of *Urechis caupo. Devlop. Biol.* **22**:1.

Dawid, I. B., D. D. Brown and R. H. Reeder. 1970. Composition and structure of chromosomal and amplified ribosomal DNA's of *Xenopus laevis. J. Mol. Biol.* **51**:341.

Deak, I., E. Sidebottom and H. Harris. 1972. Further experiments on the role of the nucleolus in the expression of structural genes. *J. Cell Sci.* **11**:379.

Delaunay, J., C. Mathieu and G. Schapira. 1972. Eukaryotic ribosomal proteins. Interspecific and intraspecific comparisons by two-dimensional polyacrylamide gel electrophoresis. *Eur. J. Biochem.* **31**:561.

Denis, H. and M. Mairy. 1972. Recherches biochimiques sur l'oogenese. I. Distribution intracellulaire du RNA dans les petits oocytes de *Xenopus laevis. Eur. J. Biochem.* **25**:524.

Durante, M., P. G. Cionini and F. D'Amato. 1972. Cytological localization of ribosomal cistrons in polytene chromosomes of *Phaseolus coccineus. Chromosoma* **39**:191.

Elicieri, G. L. 1972. The ribosomal RNA of hamster-mouse hybrid cells. *J. Cell Biol.* **53**:177.

Elicieri, G. L. and H. Green. 1969. Ribosomal RNA synthesis in human-mouse hybrid cells. *J. Mol. Biol.* **41**:253.

Elsdale, T. R., M. Fischberg and S. Smith. 1958. A mutation that reduces nucleolar number in *Xenopus laevis. Exp. Cell Res.* **14**:642.

Emerson, C. P., Jr. and T. Humphreys. 1970. Regulation of DNA-like RNA and the apparent activation of ribosomal RNA synthesis in sea urchin embryos— quantitative measurements of newly synthesized RNA. *Develop. Biol.* **23**:86.

————. 1971. Ribosomal RNA synthesis and the multiple atypical nucleoli in cleaving embryos. *Science* **171**:898.

Fan, H. and S. Penman. 1971. Regulation of synthesis and processing of nucleolar components in metaphase-arrested cells. *J. Mol. Biol.* **59**:27.

Finkelstein, D. B., J. Blamire and J. Marmur. 1972. Location of ribosomal RNA cistrons in yeast. *Nature New Biol.* **240**:279.

Firtel, R. A. and J. Bonner. 1972. Characterization of the genome of the cellular slime mold *Dictyostelium discoideum*. *J. Mol. Biol.* **66**:339.

Ford, P. J. 1971. Noncoordinated accumulation and synthesis of 5S ribonucleic acid by ovaries of *Xenopus laevis*. *Nature* **233**:561.

Ford, P. J. and E. M. Southern. 1973. Different sequences for 5S RNA in kidney cells and ovaries of *Xenopus laevis*. *Nature New Biol.* **241**:7.

Gall, J. G. 1968. Differential synthesis of the genes for ribosomal RNA during amphibian oogenesis. *Proc. Nat. Acad. Sci.* **60**:553.

————. 1969. The genes for ribosomal RNA during oogenesis. *Genetics* **61**:121.

Gall, J. G., H. C. MacGregor and M. E. Kidston. 1969. Gene amplification in the oocytes of dytiscid water beetles. *Chromosoma* **26**:169.

Gerbi, S. A. 1971. Localization and characterization of the ribosomal RNA cistrons in *Sciara coprophila*. *J. Mol. Biol.* **58**:499.

Giacomoni, D. and D. Finkel. 1972. Time of duplication of ribosomal RNA cistrons in a cell line of *Potorous tridactylis* (rat kangaroo). *J. Mol. Biol.* **72**:725.

Gimmler, G. M. and E. Schweizer. 1972. rDNA replication in a synchronized culture of *Saccharomyces cerevisiae*. *Biochem. Biophys. Res. Comm.* **46**:143.

Gray, S. and R. Landesman. 1972. An analysis of ribosomal protein during the development of *Xenopus laevis*. *Mol. Gen. Genet.* **115**:324.

Hallberg, R. L. and D. D. Brown. 1969. Coordinated synthesis of some ribosomal proteins and ribosomal RNA in embryos of *Xenopus laevis*. *J. Mol. Biol.* **46**:393.

Hansen-Delkeskamp, E. 1972. RNS-synthese und funktion der extra-DNS waehrend der oogenese von *Acheta domesticus* L. *Wilhelm Roux Arch. Entwicklungsmech. Org.* **170**:344.

Harris, H., J. F. Watkins, C. E. Ford and G. I. Schoefli. 1966. Artificial heterokaryons of animal cells from different species. *J. Cell Sci.* **1**:1.

Hatlen, L. and G. Attardi. 1971. Proportion of the HeLa cell genome complementary to transfer RNA and 5S RNA. *J. Mol. Biol.* **56**:535.

Hatlen, L. E., F. Amaldi and G. Attardi. 1968. Oligonucleotide pattern after pancreatic ribonuclease digestion and the 3′ and 5′ termini of 5S ribonucleic acid from HeLa cells. *Biochemistry* **8**:4989.

Hay, E. D. 1968. The nucleolus in developing cells. In *The nucleus*, vol. 3 of *Ultrastructure in biological systems* (ed. A. J. Dalton and F. Haguenau). Academic Press, New York.

Hecht, R. M. and M. L. Birnstiel. 1972. Integrity of the DNA template, a prerequisite for the faithful transcription of *Xenopus* rDNA *in vitro*. *Eur. J. Biochem.* **29**:489.

Henderson, A. S., D. Warburton and K. C. Atwood. 1972. Location of ribosomal DNA in the human chromosome complement. *Proc. Nat. Acad. Sci.* **69**:3394.

Hill, R. N. and E. H. McConkey. 1972. Coordination of ribosomal RNA synthesis in vertebrate cells. *J. Cell. Physiol.* **79**:15.

Honjo, T. and R. H. Reeder. 1973. Preferential transcription of *Xenopus laevis* ribosomal RNA in interspecies hybrids between *X. laevis* and *X. mulleri. J. Mol. Biol.* **80**:217.

Hourcade, D., D. Dressler and J. Wolfson. 1974. The nucleolus and the rolling circle. *Cold Spring Harbor Symp. Quant. Biol.* **38**:537.

Hsu, T. C., B. R. Brinkley and E. F. Arrighi. 1967. The structure and behavior of the nucleolus organizers in mammalian cells. *Chromosoma* **23**:137.

Humphrey, R. R. 1961. A chromosomal deletion in the Mexican Axolotl (*Siredon mexicanum*) involving the nucleolar organizer and the gene for dark color. *Amer. Zool.* **1**:361.

Ingle, J. and J. Sinclair. 1972. Ribosomal RNA genes and plant development. *Nature* **235**:30.

Kahn, J. 1962. The nucleolar organizer in the mitotic chromosome complement of *Xenopus laevis. Quart. J. Micr. Sci.* **103**:407.

Kalt, M. R. 1972. Germ cell development and premeiotic rDNA amplification in *Xenopus laevis. J. Cell Biol.* **55**:128A.

Kedinger, C., M. Gniazdowski and J. L. Mandel, Jr. 1970. α-Amanitin: A specific inhibitor of one of two DNA-dependent RNA polymerase activities from calf thymus. *Biochem. Biophys. Res. Comm.* **38**:165.

Keep, E. 1971. Nucleolar suppression: Its inheritance and association with taxonomy and sex in the genus *Ribes. Heredity* **26**:443.

Knowland, J. S. 1970. Polyacrylamide gel electrophoresis of nucleic acids synthesized during the early development of *Xenopus laevis* daudin. *Biochim. Biophys. Acta* **204**:416.

Knowland, J. and L. Miller. 1970. Reduction of ribosomal RNA synthesis and ribosomal RNA genes in a mutant of *Xenopus laevis* which organizes only a partial nucleolus. I. Ribosomal RNA synthesis in embryos of different nucleolar types. *J. Mol. Biol.* **53**:321.

La Marca, J. M., L. D. Smith and M. C. Strobel. 1973. Quantitative and qualitative analysis of RNA synthesis in stage 6 and stage 4 oocytes of *Xenopus laevis. Develop. Biol.* **34**:106.

Lambertsson, A. G. 1972. The ribosomal proteins of *Drosophila melanogaster*. II. Comparison of protein patterns of ribosomes from larvae, pupae and adult flies by two-dimensional polyacrylamide gel electrophoresis. *Mol. Gen. Genet.* **118**:215.

Landesman, R. and P. R. Gross. 1968. Patterns of macromolecule synthesis during development of *Xenopus laevis.* I. Incorporation of radioactive precursors into dissociated embryos. *Develop. Biol.* **18**:571.

Laskey, R. A., J. C. Gerhart and J. S. Knowland. 1973. Inhibition of ribosomal RNA synthesis in neurula cells by extracts from blastulae of *Xenopus laevis. Develop. Biol.* **33**:241.

Leibowitz, R. D., R. A. Weinberg and S. Penman. 1973. Unusual metabolism of 5S RNA in HeLa cells. *J. Mol. Biol.* **73**:139.

Lima-de-Faria, A., M. Birnstiel and H. Jaworska. 1969. Amplification of ribosomal cistrons in the heterochromatin of *Acheta. Genetics* **61**: 145.

Lima-de-Faria, A., H. Jaworska and T. G. Gustafsson. 1973. Release of amplified ribosomal DNA from the chromomeres of Acheta. *Proc. Nat. Acad. Sci.* **70**:80.

Longwell, A. C. and G. Svihla. 1960. Specific chromosomal control of the nucleolus and of the cytoplasm in wheat. *Exp. Cell Res.* **20**:294.

MacGregor, H. C. 1972. The nucleolus and its genes in amphibian oogenesis. *Biol. Rev.* **47**:177.

MacGregor, H. C. and J. Kezer. 1970. Gene amplification in oocytes with 8 germinal vesicles from the tailed frog *Ascaphus truei* (Stejneger). *Chromosoma* **29**:189.

MacGregor, H. C. and M. Vlad. 1972. Interlocking and knotting of ring nucleoli in amphibian oocytes. *Chromosoma* **39**:205.

Maden, B. E., M. Salim and D. F. Summers. 1972. Maturation pathway for ribosomal RNA in the HeLa cell nucleolus. *Nature New Biol.* **237**:5.

Mahdavi, V. and M. Crippa. 1972. An RNA-DNA complex intermediate in ribosomal gene amplification. *Proc. Nat. Acad. Sci.* **69**:1749.

Mairy, M. and H. Denis. 1971. Recherches biochimiques sur l'oogenese. I. Synthese et accumulation du RNA pendant l'oogenese du crapaud sudafricain *Xenopus laevis. Develop. Biol.* **24**:143.

McClintock, B. 1934. The relationship of a particular chromosomal element to the development of the nucleoli in *Zea mays. Zeit. Zellforsch. mik. Anat.* **21**:294.

Miller, L. 1972. Initiation of the synthesis of ribosomal ribonucleic acid precursor in different regions of frog (*Rana pipiens*) gastrulae. *Biochem. J.* **127**:733.

———. 1973. Control of 5S RNA synthesis during early development of anucleolate and partial nucleolate mutants of *Xenopus laevis. J. Cell Biol.* **59**:624.

Miller, L. and D. D. Brown. 1969. Variation in the activity of nucleolar organizers and their ribosomal gene content. *Chromosoma* **28**:430.

Miller, L. and J. Knowland. 1970. Reduction of ribosomal RNA synthesis and ribosomal RNA genes in a mutant of *Xenopus laevis* which organizes only a partial nucleolus. II. The number of ribosomal RNA genes in animals of different nucleolar types. *J. Mol. Biol.* **53**:329.

Miller, O. L., Jr. 1966. Structure and composition of peripheral nucleoli of salamander oocytes. *Nat. Cancer Inst. Monogr.* **23**:53.

Miller, O. L., Jr. and A. H. Bakken. 1972. Morphological studies of transcription. *Karolinska Symp.: Res. methods reproductive endocrinol.* **5**:155.

Miller, O. L., Jr. and B. R. Beatty. 1969. Visualization of nucleolar genes. *Science* **164**:955.

Navashin, M. 1934. Chromosome alterations caused by hybridization and their bearing upon certain general genetic problems. *Cytologia* **5**:169.

Newlon, C. S., G. E. Sonenshein and C. E. Holt. 1973. Time of synthesis of genes for ribosomal ribonucleic acid in Physarum. *Biochemistry* **12**:2338.

Pardue, M. L. 1969. Nucleic acid hybridization in cytological preparations. *J. Cell Biol.* **43**:101a.

Pardue, M. L. and J. G. Gall. 1969. Molecular hybridization of radioactive RNA to the DNA of cytological preparations. *Proc. Nat. Acad. Sci.* **64**:600.

Pardue, M. L., D. D. Brown and M. L. Birnstiel. 1973. Location of the genes for 5S ribosomal RNA in *Xenopus laevis. Chromosoma* **42**:191.

Pardue, M. L., S. A. Gerbi, R. A. Eckhardt and J. C. Gall. 1970. Cytological localization of DNA complementary to ribosomal RNA in polytene chromosomes of Diptera. *Chromosoma* **29**:268.

Patterson, J. B. and D. W. Stafford. 1970. Sea urchin satellite deoxyribonucleic acid. Its large-scale isolation and hybridization with homologous ribosomal ribonucleic acid. *Biochemistry* **9**:1278.

————. 1971. Characterization of sea urchin ribosomal satellite deoxyribonucleic acid. *Biochemistry* **10**:2775.

Patterson, J. T. and W. S. Stone. 1952. *Evolution in the genus Drosophila.* Macmillan, New York.

Pedersen, R. A. 1971. DNA content, ribosomal gene multiplicity, and cell size in fish. *J. Exp. Zool.* **177**:65.

Pelling, C. and W. Beermann. 1966. Diversity and variation of the nucleolar organizing regions in *Chironomids. Nat. Cancer Inst. Monogr.* **23**:393.

Perkowska, E., H. C. MacGregor and M. L. Birnstiel. 1968. Gene amplification in the oocyte nucleus of mutant and wild-type *Xenopus laevis. Nature* **217**:649.

Perry, R. P. and D. E. Kelley. 1968. Persistent synthesis of 5S RNA when production of 28S and 18S ribosomal RNA is inhibited by low doses of actinomycin D. *J. Cell. Physiol.* **72**:235.

————. 1972. The production of ribosomal RNA from high molecular weight precursors. III. Hydrolysis of pre-ribosomal and ribosomal RNA by a 3'-OH specific exoribonuclease. *J. Mol. Biol.* **70**:265.

Phillips, D. M. and S. G. Phillips. 1973. Repopulation of post mitotic nucleoli by preformed RNA. II. Ultrastructure. *J. Cell Biol.* **58**:54.

Price, R. and S. Penman. 1972. A distinct RNA polymerase activity synthesizing 5.5S, 5S and 4S RNA in nuclei from adenovirus 2-infected HeLa cells. *J. Mol. Biol.* **70**:435.

Quincey, R. V. 1971. The number and location of genes for 5S ribonucleic acid within the genome of *Drosophila melanogaster. Biochem. J.* **123**:227.

Rafferty, K. A., Jr. and R. W. Sherwin. 1969. The length of secondary chromosomal constrictions in normal individuals and in a nucleolar mutant of *Xenopus laevis. Cytogenetics* (Basel) **8**:427.

Reeder, R. H. and D. D. Brown. 1970. Transcription of the ribosomal RNA genes of an amphibian by the RNA polymerase of a bacterium. *J. Mol. Biol.* **51**:361.

Reeder, R. H. and R. G. Roeder. 1972. Ribosomal RNA synthesis in isolated nuclei. *J. Mol. Biol.* **67**:433.

Ritossa, F. M. and S. Spiegelman. 1965. Localization of DNA complementary to ribosomal RNA in the nucleolus organizer region of *Drosophila melanogaster. Proc. Nat. Acad. Sci.* **53**:737.

Roeder, R. G. and W. J. Rutter. 1969. Multiple forms of DNA-dependent RNA polymerase in eukaryotic organisms. *Nature* **224**:234.

————. 1970. Specific nucleolar and nucleoplasmic RNA polymerases. *Proc. Nat. Acad. Sci.* **65**:675.

Ryser, U., S. Fakan and R. Braun. 1973. Localization of ribosomal RNA genes by high resolution autoradiography. *Exp. Cell Res.* **78**:89.

Scheer, U., M. F. Trendelenburg and W. W. Franke. 1973. Transcription of ribosomal RNA cistrons. *Exp. Cell Res.* **80**:175.

Schweizer, E., C. MacKechnie and H. O. Halvorson. 1969. The redundancy of ribosomal and transfer RNA genes in *Saccharomyces cerevisiae. J. Mol. Biol.* **40**:261.

Sconzo, G. and G. Giudice. 1972. Synthesis of ribosomal RNA in sea urchin embryos. V. Further evidence for an activation following the hatching blastula stage. *Biochim. Biophys. Acta* **254**:447.

Sconzo, G., A. M. Pirrone, V. Mutolo and G. Giudice. 1970. Synthesis of ribosomal RNA during sea urchin development. III. Evidence for an activation of transcription. *Biochim. Biophys. Acta* **199**:435.

Shiokawa, K. and K. Yamana. 1967. Inhibitor of ribosomal RNA synthesis in *Xenopus laevis* embryos. *Develop. Biol.* **16**:389.

Sinclair, J. H. and D. D. Brown. 1971. Retention of common nucleotide sequences in the ribosomal deoxyribonucleic acid of eukaryotes and some of their physical characteristics. *Biochemistry* **10**:2761.

Smith, G. P. 1974. Unequal crossover and the evolution of families of repeated genes. *Cold Spring Harbor Symp. Quant. Biol.* **38**:507.

Spadari, S., V. Sgaramella, G. Mazza and A. Falaschi. 1971. Enzymic purification of a hybrid between ribosomal ribonucleic acid and deoxyribonucleic acid. *Eur. J. Biochem.* **19**:294.

Stambrook, P. 1974. The temporal replication of ribosomal genes in synchronized Chinese hamster cells. *J. Mol. Biol.* **82**:303.

Steffensen, D. M. 1973. Mapping genes for the ribosomal protein of *Drosophila. Nature New Biol.* **244**:231.

Swift, H. 1966. Nomenclature of nucleoli. *Nat. Cancer Inst. Monogr.* **23**:573.

Tartof, K. D. and R. P. Perry. 1970. The 5S RNA genes of *Drosophila melanogaster. J. Mol. Biol.* **51**:171.

Thomas, C. 1970. Ribonucleic acids and ribonucleoproteins from small oocytes of *Xenopus laevis. Biochim. Biophys. Acta* **224**:99.

Tocchini-Valentini, G. P., V. Mahdavi, R. Brown and M. Crippa. 1974. The synthesis of amplified ribosomal DNA. *Cold Spring Harbor Symp. Quant. Biol.* **38**:551.

Van Gansen, P. and A. Schram. 1972. Evolution of the nucleoli during oogenesis in *Xenopus laevis* studied by electron microscopy. *J. Cell Sci.* **10**:339.

Van Snick, P. and J. Brachet. 1972. Attempt to detect an inhibitor of ribosomal RNA synthesis in cleaving amphibian eggs. *Arch. Biol.* **82**:173.

Viinikka, Y., A. Hannah-Alava and P. Arajarvi. 1971. A re-investigation of the nucleolus organizing regions in the salivary gland of *Drosophila melanogaster. Chromosoma* **36**:34.

Vodkin, M. and F. R. H. Katterman. 1971. Divergence of ribosomal RNA sequences within Angiospermae. *Genetics* **69**:435.

Wall, D. A. and A. W. Blackler. 1974. Enzyme patterns in two species of *Xenopus* and their hybrids. *Develop. Biol.* **36**:379.

Wallace, H. and M. L. Birnstiel. 1966. Ribosomal cistrons and the nucleolar organizer. *Biochim. Biophys. Acta* **114**:296.

Wallace, H. and W. H. R. Langridge. 1971. Differential amphiplasty and the control of ribosomal RNA synthesis. *Heredity* **27**:1.

Wallace, H., J. Morray and W. H. Langridge. 1971. Alternative model for gene amplification. *Nature New Biol.* **230**:201.

Watson-Coggins, L. and J. G. Gall. 1972. The timing of meiosis and DNA synthesis during early oogenesis in the toad, *Xenopus laevis. J. Cell Biol.* **52**:569.

Wegnez, M. 1973. Synthése du RNA 5S pendant l'oogenese du crapaud sud-africain *Xenopus laevis. Ph. D. thesis,* University of Liége, Belgium.

Wegnez, M., R. Monier and H. Denis. 1972. Sequence heterogeneity in the 5S RNA in *Xenopus laevis. FEBS Letters* **25**:13.

Weinberg, R. A. and S. Penman. 1970. Processing of 45S nucleolar RNA *J. Mol. Biol.* **47**:169.

Wellauer, P. K. and I. B. Dawid. 1973. Secondary structure maps of RNA: Processing of HeLa ribosomal RNA. *Proc. Nat. Acad. Sci.* **70**:2827.

Wensink, P. C. and D. D. Brown. 1971. Denaturation map of the ribosomal DNA of *Xenopus laevis. J. Mol. Biol.* **60**:235.

Wimber, D. E. and D. M. Steffensen. 1970. Localization of 5S RNA genes on *Drosophila* chromosomes by RNA-DNA hybridization. *Science* **170**:639.

Zellweger, A., U. Ryser, and R. Braun. 1972. Ribosomal genes of Physarum: Their isolation and replication in the mitotic cycle. *J. Mol. Biol.* **64**:681.

Zylber, E. A. and S. Penman. 1971a. Products of RNA polymerases in HeLa cell nuclei. *Proc. Nat. Acad. Sci.* **68**:2861.

―――. 1971b. Synthesis of 5S and 4S RNA in metaphase-arrested HeLa cells. *Science* **172**:947.

Biosynthesis
of Organelle Ribosomes

Nam-Hai Chua and David J. L. Luck
The Rockefeller University
New York, New York 10021

INTRODUCTION

Compared with the status in prokaryotic and eukaryotic ribosomes, knowledge of the biosynthesis of organelle ribosomes is in early development. Information concerning the protein components of these ribosomes is still incomplete. Since animal cells contain two potential sites for structural genes and for protein synthesis, and plant cells contain three, the question of biosynthetic origin and genetic specification of the ribosomal proteins is potentially complex. The RNA components are better characterized, and there is evidence that in each case they are transcription products of the organelle DNA. Yet little is known concerning RNA precursor molecules, and the questions of control of organelle ribosome assembly or coordination of biosynthesis of the structural components have not yet been approached. This review will attempt to summarize current knowledge of these topics and to point out areas that require development.

STRUCTURE AND PROPERTIES OF ORGANELLE RIBOSOMES

It has been known for some time that mitochondria and chloroplasts both contain ribosomes that have properties distinctly different from those of their counterparts in the cell sap. The structures and functions of these ribosomes have been extensively reviewed in several recent articles (Borst and Grivell 1971; Küntzel 1971; Boulter, Ellis and Yarwood 1972; Borst 1972; Kroon, Agsteribbe and De Vries 1972; Mahler 1973).

Chloroplast ribosomes were first isolated from spinach by Lyttleton in 1962, and since then they have been isolated and characterized from several higher plants and from green algae (Boulter, Ellis and Yarwood 1972). In contrast to the 80S ribosomes of the cell sap, these ribosomes have a sedimentation coefficient of approximately 70S. Furthermore, they are sensitive to antibiotics which inhibit protein synthesis on bacterial ribosomes but resistant to those which interfere with 80S ribosomal functions. In these respects they are similar to ribosomes of eubacteria and are therefore often referred to as prokaryotic ribosomes. At low Mg^{++} concentrations (Grivell and Groot 1972) or in the presence of high-salt buffer (Chua, Blobel and Siekevitz 1973), chloroplast ribosomes readily dissociate into two unequal subunits with sedimentation coefficients of 50S and 33S. The subunits obtained by either procedure are active in the synthesis of polyphenylalanine in the presence of polyuridylic acid (Grivell and Groot 1972; Chua, Blobel and Siekevitz 1973). Several groups of workers have reported that chloroplast ribosomes contain two high molecular weight RNAs: 23S and 16S (Hoober and Blobel 1969; Rawson and Stutz 1969; Bourque, Boynton and Gillham 1971; Leaver and Ingle 1971). In addition, there is also a small molecular weight RNA, the 5S RNA, which is localized in the large ribosomal subunits (Payne and Dyer 1971; Galling and Jordon 1972). Another small ribosomal RNA component, 5.8S, which is hydrogen-bonded to the large ribosomal RNA of 80S ribosomes, has not been detected in chloroplast ribosomes, thus further emphasizing the difference between these two types of ribosomes (Payne and Dyer 1972). Chemical analysis of chloroplast ribosomal RNAs reveals a G + C content of about 51–56 mole percent (Rossi and Gualerzi 1970) and a high degree of base methylation relative to ribose methylation (Rijven and Zwar 1973). Both these features are also characteristic of bacterial ribosomes (Attardi and Amaldi 1970).

In contrast to the chloroplast ribosomes, mitochondrial ribosomes show a greater variation in sedimentation coefficients among the various phylogenetic groups (Borst and Grivell 1971; Kroon, Agsteribbe and De Vries 1972; Mahler 1973). Mitochondrial ribosomes of fungi and higher plants are characterized by a sedimentation coefficient of 70–75S, whereas those of animal cells have a smaller sedimentation coefficient of 55–60S. Under appropriate conditions, both types of ribosomes can be dissociated into active subunits (Grivell, Reijnders and Borst 1971; Swanson and Dawid 1970; Ibrahim and Beattie 1973; Greco et al. 1973). Physiochemical studies carried out by several workers demonstrate that all mitochrondrial ribosomal RNAs have a low G + C content, a low degree of secondary structure, and a higher proportion of base methylation compared to ribose methylation (Borst and Grivell 1971; Borst 1972; Kroon, Agsteribbe and De Vries 1972; Mahler 1973; Freeman, Mitra and Bartoov 1973). In addition, mitochondrial ribosomes of both fungi and animal cells do not seem to contain the two low molecular weight RNA components 5S and 5.8S, both of which are present in the ribosomes of the corresponding cell sap (Lizardi and Luck 1971; Borst 1972). However recent evidence by Gray and Attardi (1973) suggests that HeLa cell mitochondrial ribosomes or large subunits contain small molecular weight RNAs that are smaller than mitochondrial 4S RNAs. One of these small molecular weight RNAs may represent an equivalent to 5S RNA. Preliminary evidence indicates that its methyl content is low and that, like the ribosomal RNAs, it appears to be a transcription product of the H strand of mitochondrial DNA (Gray and Attardi, unpublished data).

The 55S ribosomes of animal cells contain ribosomal RNAs considerably smaller than those of mitochondrial ribosomes of fungi (Borst 1972; Kroon, Agsteribbe and De Vries 1972). This finding, together with the observation of a low sedimentation coefficient, has led to the assumption that the 55S ribosomes are the smallest in size and weight among the various classes of ribosomes so far described (Borst 1972; Kroon, Agsteribbe and De Vries 1972; Borst and Grivell 1971), and for this reason they have been called "mini-ribosomes." However recent results indicate that this assumption is not valid. Three groups of workers (Sacchi et al. 1973; De Vries and Van der Koogh-Schuuring 1973; O'Brien, unpublished data) have independently established that the 55S ribosomes have a density of 1.42–1.45 g/cc in CsCl gradient, and this low density value can account for many of the anomolous properties ascribed to these ribosomes (Borst and Grivell 1971; Borst 1972; Kroon, Agsteribbe and De Vries 1972). Since great precautions were taken by these workers to ensure that their ribosomal preparations were not contaminated by membrane fragments or extraneous proteins, this density estimation probably reflects the true density of the 55S ribosomal particles. Based on this value, it can be calculated that the 55S ribosomes contain approximately 30% RNA and 70% protein. Assuming a molecular weight of 0.88×10^6 daltons for the ribosomal RNAs (Robberson et al. 1971; Dawid and Chase 1972), the molecular weight of the 55S ribosomes should be approximately 3.0×10^6 daltons, which is slightly heavier than that of *Escherichia coli* ribosomes. The large amounts of ribosomal proteins relative to ribosomal RNAs in the 55S ribosomes results in the low charge/mass ratio which is responsible for the slow mobility in polyacrylamide gel electrophoresis of these particles compared to rat liver 80S and *E. coli* 70S ribosomes (De Vries and Van der Koogh-Schuuring 1973).

BIOSYNTHESIS OF ORGANELLAR RIBOSOMAL RNAs

Wherever they have been tested, ribosomal RNAs of chloroplasts and mitochondria have been shown to hybridize with their organellar DNA. There are equal numbers of cistrons for large and small subunit RNAs, and for

522   N.-H. Chua and D. J. L. Luck

most of the cases analyzed, there appears to be only one set of ribosomal RNA cistrons per genome (the genome size was usually estimated by kinetic complexity). For mitochondria this is the situation in Neurospora (Schäfer and Küntzel 1972), Saccharomyces (Reijnders et al. 1972), Xenopus (Dawid 1972) and HeLa cells (Wu et al. 1972), and the same holds true for chloroplasts from Euglena (Rawson and Haselkorn 1973) and from leaves of several higher plants (Tewari and Wildman 1970; Ingle et al. 1970). The chloroplast of Chlamydomonas appears to be an exception, since hybridization studies (Bastia, Chiang and Swift 1971) and transcriptional mapping (Surzycki and Rochaix 1971) suggest the presence of two or three sets of ribosomal RNA cistrons. In any case, considering the multiplicity of genomes per organelle and the multiplicity of organelles (except for chloroplasts of Chlamydomonas) per cell, there appears to be considerable organellar ribosomal RNA gene redundancy for each nucleus.

Wherever they have been tested, organellar ribosomal RNAs have also shown a low level of hybridization to nuclear DNA (Reijnders et al. 1972; Wood and Luck 1969; Bastia, Chiang and Swift 1971; Ingle et al. 1970). Because the observed level of hybridization is low and can be substantially reduced by competition with cytoplasmic ribosomal RNAs, it has been interpreted as "nonspecific" hybridization or hybridization based on sequence homology with the cytoplasmic counterpart.

In *Neurospora crassa* a single high molecular weight precursor (32S) of large and small mitochondrial ribosomal RNAs has been identified by pulse labeling, pulse-chase, and hybridization competition studies (Kuriyama and Luck 1973). The model of ribosomal RNA production derived from these studies is shown in Figure 1. Although there is evidence that small and large subunit organellar ribosomal RNA genes are positioned in tandem in HeLa cell mitochondria (Wu et al. 1972) and Chlamydomonas chloroplasts (Surzycki and Rochaix 1971), there is yet no evidence of a discrete high molecular weight ribosomal RNA precursor molecule for any other organelle. Wilson and Chiang (1972) were unable to find such a precursor in short pulse-labeling studies of Chlamydomonas chloroplasts. On the basis of hybridization studies (Aloni and Attardi 1971) and of isolation of mitochondrial DNA-RNA complexes (Aloni and Attardi 1972), it has been postulated that in HeLa cells both strands of mitochondrial DNA are completely transcribed, but no discrete precursor of ribosomal RNA of intermediate size has yet been described in these mitochondria. It is of interest to note that in the Neurospora studies (Kuriyama and Luck 1973), pulse labeling under low temperature growth conditions, where the mass doubling time of the culture increases by sixfold, resulted in preferential incorporation of pulse label into the mitochondrial ribosomal RNA precursor. The application of this or any other method to slow ribosomal RNA maturation might be useful in the demonstration of a precursor for other organelles.

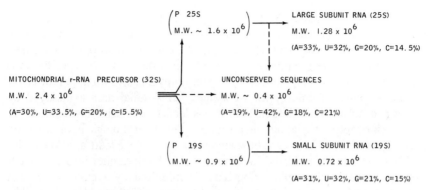

Figure 1   Model for biosynthesis of mitochondria ribosomal RNA in *Neurospora crassa* (adapted from Kuriyama and Luck 1973). Base compositions for 32S, 25S and 19S are shown in parentheses and are expressed as mole percent. The base composition of the unconserved sequence was calculated by difference (32S–25S and 19S). The hypothetical intermediates, P25S and P19S, were proposed on the basis of pulse-labeling experiments. They have not been isolated.

PROTEIN COMPONENTS OF ORGANELLE RIBOSOMES

Sites of Synthesis

Although ribosomal proteins of organelles have been extensively studied by one-dimensional gel electrophoresis (Hoober and Blobel 1969; Gualerzi and Cammarano 1969; Gualerzi and Cammarano 1970; Vasconcelos and Bogorad 1971; Janda and Wittmann 1968), only a few attempts have been made to characterize the various protein components by two-dimensional techniques as has been done for both *E. coli* (Kaltschmidt and Wittmann 1970) and rat liver ribosomes (Sherton and Wool 1972). In a limited number of cases in which two-dimensional techniques have been used, it was found that fungal mitochondrial ribosomes contain approximately 50–60 proteins (Lizardi and Luck 1972), whereas chloroplast ribosomes (Jones et al. 1972) and mitochondrial ribosomes of animal cells may contain up to 70–75 such proteins (O'Brien, unpublished data). Since both chloroplasts and mitochondria are able to carry out both transcription and translation, two questions can be raised with respect to the biogenesis of their ribosomes: (1) Where are the protein components of the organelle ribosomes synthesized within the cells? (2) Where are the structural genes for these proteins located?

The first question can be approached by the use of antibiotics that specifically block protein synthesis on either organelle or cytoplasmic ribosomes (Borst and Grivell 1971; Boulter, Ellis and Yarwood 1972; Mahler 1973). Several groups of workers have examined the effects of these inhibitors on the synthesis of either mitochondrial ribosomes (Davey, Yu and Linnane 1969) or mitochondrial ribosomal proteins (Küntzel 1969;

Lizardi and Luck 1972; Mahler 1973) in fungi. Their results show that most, if not all, of the ribosomal proteins are synthesized in the cell sap.

The situation is less clear, however, in the case of chloroplast ribosomes. Goodenough (1971) reported that in Chlamydomonas, 70S-specific antibiotics such as chloramphenicol and spectinomycin have no apparent effect on the amounts of chloroplast ribosomes per stromal area as assayed by electron microscopy, although both drugs lead to extensive disorganization of the chloroplast thylakoid membranes. This morphological observation is corroborated by the recent biochemical data of Honeycutt and Margulies (1973), who found that the incorporation of [³H]arginine into the structural proteins of chloroplast ribosomes is inhibited by cycloheximide but not chloramphenicol. It is not known, however, whether chloroplast ribosomes synthesized in the presence of chloramphenicol still contain their full complement of structural proteins, since these ribosomes have not been assayed for their function in vitro and no detailed analysis of their protein components has been attempted. From their studies Honeycutt and Margulies (1973) concluded that most of the chloroplast ribosomal proteins are made on the cytoplasmic ribosomes, but they also emphasized that their results did not rule out the possibility that a few of these may be made within the chloroplast. If the latter possibility were true, then these proteins must be added to the ribosomes relatively late during the assembly process, and their absence would not significantly affect the sedimentation coefficient of the chloroplast ribosomes.

In Ochromonas, on the other hand, the light-induced synthesis of chloroplast ribosomes is inhibited by chloramphenicol (Smith-Johannsen and Gibbs 1972), and similarly, the increase in the amounts of chloroplast ribosomes during greening of pea apices is also prevented by lincomycin, another specific inhibitor of 70S ribosomes (Ellis and Hartley 1971). Since lincomycin has no effect on the activity of the DNA-dependent RNA polymerase in chloroplasts (Ellis and Hartley 1971), these results clearly implicate chloroplast ribosomes in the synthesis of at least some of their own proteins which appear to be essential for the assembly of intact ribosomal particles.

At present it is not possible to resolve the apparently contradictory results obtained with Chlamydomonas (Goodenough 1971; Honeycutt and Margulies 1973) on the one hand and Ochromonas (Smith-Johannsen and Gibbs 1972), and similarly, the increase in the amounts of chloroplast species differences may account for the observed discrepancies, and the possibility remains that a few ribosomal proteins that are translated within the chloroplast are essential for the assembly of intact 70S ribosomes in pea and Ochromonas but not in Chlamydomonas. Clearly more detailed analysis of the ribosomal proteins similar to those that have been carried out for mitochondrial ribosomes (Lizardi and Luck 1972) is needed to resolve these problems.

Localization of Structural Genes

The identification and localization of structural genes for organellar ribosomal proteins present a more difficult problem than the elucidation of the sites of synthesis of these proteins. The compounds such as rifampicin (Surzycki 1969) and ethidium bromide (Mahler 1973), which specifically inhibit organelle DNA-dependent RNA polymerase, also block the synthesis of ribosomal RNA and thus prevent the assembly of ribosomes. Therefore no useful information could be obtained with the use of these drugs. The only approach is to isolate mutants that have altered organellar ribosomal proteins. In general, mutations in the nuclear genome are transmitted in a Mendelian pattern, whereas those in the organelle genome show uniparental (non-Mendelian) inheritance (Sager 1972). Genetic analysis of the mode of inheritance of such mutant strains should provide information as to the localization of the mutated genes.

Although the genetic approach appears promising, the problem is how to select for mutants that are specifically affected in their organellar ribosomal proteins. In *E. coli* and in other bacteria, it has been shown that mutants selected for resistance toward antibiotics inhibiting protein synthesis usually contain altered ribosomal protein components (Davies and Nomura 1972). Since organelle ribosomes are similar to bacterial ribosomes in many respects, including antibiotic sensitivity (Borst and Grivell 1971; Boulter, Ellis and Yarwood 1972), it seems feasible that a similar selection method may be applied to the isolation of organelle ribosome mutants. Indeed, this strategy has been used successfully by several laboratories, and many such antibiotic-resistant mutants showing both Mendelian and non-Mendelian inheritance have been isolated from *Chlamydomonas reinhardtii* (Sager 1954; Gillham 1969; Surzycki and Gillham 1971; Mets and Bogorad 1971; Sager 1972), *Saccharomyces cerevisiae* (yeast) (Thomas and Wilkie 1968; Coen et al. 1970), and *Paramecium aurelia* (Adoutte and Beisson 1970; Beale, Knowles and Tait 1972). Tables 1, 2 and 3 summarize the properties of only those mutant strains which have been characterized biochemically and in which a change in the organelle ribosome phenotype has been clearly demonstrated.

All the chloroplast ribosome mutants listed in Tables 1 and 2 have been isolated from the unicellular green alga, *C. reinhardtii*. This alga is isogamous and heterothallic, having two mating types designated as mt+ and mt−. The advantages of using this organism for studies on the genetics of chloroplast ribosomes have been discussed previously (Surzycki and Gillham 1971; Sager 1972). Sager (1954) was the first to isolate from this alga mutant strains that are resistant to the antibiotic streptomycin. These mutants fall into two genetic classes: mutants resistant to low levels of streptomycin (100 μg/ml) are inherited in Mendelian pattern, whereas those resistant to a higher level of the same drug (500 μg/ml) are transmitted by the mt+ parent only. Subsequent biochemical and genetic

Table 1 Non-Mendelian mutations affecting chloroplast ribosome phenotype in *C. reinhardtii*

Mutant	Resistance to antibiotic	Ribosome phenotype	Reference
Sr-2-60 *Sr-2-281*	streptomycin 500 μg/ml	Chloroplast ribosomes sediment at 66S instead of 70S; accumulation of 50S subunits	Gillham, Boynton and Burkholder (1970)
Sd-3-18	streptomycin-dependent 10 μg/ml	Reduced proportions of 70S ribosomes; accumulation of 50S subunits	Gillham, Boynton and Burkholder (1970)
Sr$_{35}$	streptomycin 500 μg/ml	70S ribosomes do not form dimers in presence of streptomycin	Boschetti and Bogdanov (1973)
Sm-2	streptomycin 500 μg/ml	One protein of 30S subunits altered In vitro synthesis of polyphenylalanine resistant to streptomycin; Resistance located in the 30S subunit	Ohta, Inouye and Sager (unpublished data) Schlanger and Sager (1974)
Spr-1-27 (*Sp 2-73*)	spectinomycin 100 μg/ml	70S ribosomes do not bind dihydrospectinomycin 30S ribosomal subunits lack one protein present in the WT 30S Chloroplast ribosomes sediment at 66S instead of 70S	Burton (1972) Boynton et al. (1973) Boynton et al. (1973) Boynton et al. (1973)

Spc	spectinomycin 50 μg/ml	In vitro incorporation of phenylalanine resistant to spectinomycin; resistance located in 30S subunits	Schlanger and Sager (1974)
nea	neamine	In vitro incorporation of phenylalanine resistant to neanine; resistance located in 30S subunits	Schlanger and Sager (1974)
car-r	carbomycin 150 μg/ml	In vitro incorporation of phenylalanine resistant to carbomycin; resistance located in 50S subunits	Schlanger, Sager and Ramanis (1972) Schlanger and Sager (1974)
cleo	cleocin 50 μg/ml	In vitro incorporation of phenylalanine resistant to cleocin; resistance located in 50S subunits	Schlanger and Sager (1974)
ery-U1	erythromycin 5×10^{-4}M (377 μg/ml)	50S subunits do not bind erythromycin One protein of 50S subunit replaced by series of new proteins	Mets and Bogorad (1971) Mets and Bogorad (1972)

Table 2 Mendelian mutations affecting chloroplast ribosome phenotype in *C. reinhardtii*

Mutant	Resistance to antibiotic	Ribosome phenotype	Reference
ac-20		70S ribosomes 10–20% WT level Low levels 66S ribosomes	Goodenough and Levine (1970) Boynton, Gillham and Chabot (1972)
cr-1		Low levels 70S ribosomes Accumulation of 50S subunits	Boynton, Gillham and Chabot (1972) Boynton, Gillham and Burkholder (1970)
ery-M1	erythromycin 5×10^{-4}M (377 μg/ml)	50S subunits do not bind erythromycin All 4 mutants of *ery*-M1 locus have same ribosomal protein altered	Mets and Bogorad (1971) Mets and Bogorad (1972) Bogorad et al. (unpublished data)
ery-M2d	erythromycin 5×10^{-4}M (377 μg/ml)	50S subunits do not bind erythromycin One ribosomal protein in 50S subunit altered	Mets and Bogorad (1971) Mets and Bogorad (1972)

Table 3 Mendelian and non-Mendelian mutations affecting mitochondrial ribosome phenotype

Organism	Mutant	Ribosome phenotype	Reference
S. cerevisiae	C^R_{321}, C^R_{323} resistant to chloramphenicol	In vitro protein synthesis by isolated mitochondria resistant to chloramphenicol; peptidyl transferase activity resistant to chloramphenicol	Grivell et al. (1973)
S. cerevisiae	E^R_{514}, E^R_{354} resistant to erythromycin	In vitro protein synthesis by isolated mitochondria resistant to erythromycin	Grivell et al. (1973)
S. cerevisiae	6–81c resistant to 4 $\mu g/ml$ erythromycin	Ribosomes partially resistant to linomycin in in vitro protein-synthesizing system	Grivell, Reijnders and De Vries (1971)
P. aurelia	$513E^R_1$ (EDIN) resistant to 250 $\mu g/ml$ erythromycin	Isolated mitochondrial ribosomes do not bind erythromycin	Tait (1972)
N. crassa	"poky" mi-1	Ribosomal protein profile different from wild type	Beale, Knowles and Tait (1972)
		Deficient in small ribosomal subunits	Rifkin and Luck (1971) Neupert, Massinger and Pfaller (1971)
		Deficient in rRNA methylation	Kuriyama and Luck (1974)

529

analysis by Gillham, Boynton and Burkholder (1970) of the chloroplast ribosomes in uniparental streptomycin-resistant and -dependent strains provides the first clue that there may be a correlation between uniparental mutations and changes in chloroplast ribosome phenotypes. Extension of these studies, principally by Gillham and Boynton's group, Sager's group, and Bogorad's group, have yielded additional information on uniparental mutants resistant to other antibiotics, and it is now known that at least seven such mutants have chloroplast ribosome phenotypes different from those of the wild-type strain (Table 1). In all these studies, the mutant chloroplast ribosomes have been shown to be altered by one or more of the following criteria:

(1) altered sedimentation coefficient of chloroplast ribosomes in sucrose gradients,
(2) accumulation of the 50S subunits,
(3) inability of isolated ribosomal subunits to bind antibiotics,
(4) resistance to antibiotics in an in vitro protein-synthesizing system and localization of the resistance to one of the two subunits by subunit-exchange experiments,
(5) alteration in ribosomal proteins as revealed by gel electrophoresis or column chromatography.

Among these various parameters, only the last one provides direct evidence linking the mutation to changes in the structure of ribosomal proteins. However extensive studies with antibiotic-resistant mutants in *E. coli* and other bacteria (Davies and Nomura 1972) have clearly shown that, with a few exceptions, the mutated genes code for the structural proteins of the ribosomes.

Since in *C. reinhardtii* the mutant ribosomes behave in a strikingly similar manner to those in *E. coli* with respect to the biochemical properties mentioned above, it can be surmised that the uniparental genes in *C. reinhardtii* code for the structural proteins of chloroplast ribosomes. Sager (1972) has obtained genetic evidence that the uniparental genes conferring resistance to spectinomycin, cleocin, carbomycin and streptomycin (cf. Table 1) are linked and are located on chloroplast DNA. Other antibiotic-resistant mutants described in Table 1 have not been mapped. If these mutated genes are linked to Sager's markers, then chloroplast DNA probably codes for at least seven chloroplast ribosomal proteins.

In addition to the uniparental mutants mentioned above, several antibiotic-resistant mutants showing Mendelian modes of inheritance have also been isolated (Sager 1954; Gillham and Levine 1962; Gillham 1969; Mets and Bogorad 1971). Among these Mendelian mutants, only the erythromycin-resistant strains have been investigated with respect to their chloroplast ribosome phenotype (Table 2). Mets and Bogorad (1971) have obtained a group of these mutants which have been mapped at two different loci, designated as *ery*-M1 and *ery*-M2. The *ery*-M1 gene is located between *pf*-2 and *ac*-7 on linkage group XI. The mapping of the *ery*-M2 locus has not been completed, but it is not linked to the *ery*-M1 locus. Four mutants (*ery*-M1a, *ery*-M1b, *ery*-M1c and *ery*-M1d), which are closely linked and are probably alleles of the *ery*-M1 locus, have been analyzed for their 50S ribosomal proteins by two-dimensional gel techniques. All these mutants have the same ribosomal protein altered, and in three of them the mutant forms of this protein could be distinguished from one another by the two-dimensional gel technique (Bogorad et al., unpublished results). These results strongly argue against the possibility that the protein in question is altered by a secondary modification. Rather, the primary structure of the affected ribosomal protein is probably coded for by the *ery*-M1 gene. Mutants of the *ery*-M2 series have not been studied in such detail; however, it has been shown that one of them (*ery*-M2d) also contains an altered 50S ribosomal protein (Mets and Bogorad 1972).

In higher plants (Bourque and Wildman 1973) and Acetabularia (Kloppstech and Schweiger 1973) there is also some evidence that some chloroplast ribosomal proteins may be coded for by nuclear genes. These results, taken together with those obtained by Bogorad's group, clearly implicate a role of the nucleus in the determination of chloroplast ribosome structure.

Non-Mendelian mutants resistant to 70S-specific antibiotics have also been isolated from *S. cerevisiae* (yeast) and *Paramecium aurelia*. In some of these mutants the resistance has been localized to the level of mitochondrial ribosomes on the basis of one or more of the following assays: (1) in vitro protein synthesis by isolated mitochondria, (2) peptidyl transferase activity of isolated mitochondrial ribosomes, and (3) binding of antibiotic to isolated mitochondrial ribosomes (cf. Table 3). In yeast, mutations conferring resistance to chloramphenicol and erythromycin have been shown to be localized in mitochondrial DNA (Michaelis, Petrochilo and Slonimski 1973). One-dimensional gel electrophoresis of mitochondrial ribosomal proteins at two different pH's did not reveal any significant difference between wild-type and the chloramphenicol-resistant and erythromycin-resistant mutants (Grivell et al. 1973). These results led Grivell et al. to suggest that the antibiotic resistance in these mutants is due to a change in mitochondrial ribosomal RNA. However, by analogy to the antibiotic-resistant mutants of *E. coli* (Davies and Nomura 1972) and the uniparental antibiotic mutants of *C. reinhardtii* (cf. Table 1), it is still possible that the mitochondrial genes which confer resistance to chloramphenicol and to erythromycin code for mitochondrial ribosomal proteins. In any case, differences in the ribosomal proteins between wild-type and mutant strains may only be detected by more sensitive methods of analysis, such as the two-dimensional gel technique.

Antibiotic-resistant mutants of *P. aurelia* have not been investigated in as great detail as those of yeast. Preliminary evidence suggests that the erythromycin-resistant strain may have altered mitochondrial ribosomes (Tait 1972; Beale, Knowles and Tait 1972). However, more work is required to establish the nature of this alteration.

COMPARISON OF STRUCTURAL PROTEINS BETWEEN ORGANELLE AND CELL SAP RIBOSOMES

The question of whether there are any similarities between proteins of organelle ribosomes and those of cytoplasmic ribosomes is important from the aspect of biogenesis of these ribosomes. It is conceivable that there may be a few common or structurally related proteins between the two ribosomal populations in the same cell, and some of these proteins eventually ending up in organelle ribosomes may be processed further, either by limited peptide cleavage or by secondary modification, upon their entry into the organelle (Lizardi and Luck 1972). Only a few workers have addressed themselves to this question and the available results are still inconclusive. Gualerzi and Cammarano (1969, 1970) found no resemblance between the protein patterns of chloroplast and cytoplasmic ribosomes of spinach, beet and lettuce by the split-gel technique. It is striking that chloroplast ribosomes contain 8–10 acidic proteins which are absent from the cytoplasmic ribosomes (Gualerzi and Cammarano 1969). Similar results have also been reported in *C. reinhardtii* (Hoober and Blobel 1969) and in other higher plants (Lyttleton 1968; Vasconcelos and Bogorad 1971). In contrast, Janda and Wittmann (1968) found good agreement between protein profiles of chloroplast and cytoplasmic ribosomes isolated from spinach, although this agreement is less striking in the case of bean. Furthermore, a rabbit antiserum prepared against cytoplasmic ribosomes of bean showed some cross-reaction with chloroplast ribosomes, indicating a weak immunological relationship between the two classes of ribosomes (Janda and Wittmann 1968; Wittmann 1972). It must be emphasized that in all these studies the ribosomal proteins were analyzed by one-dimensional gel electrophoresis, a technique which is not sensitive enough to detect a few common or structurally related proteins. In the case of mitochondrial ribosomes, fractionation of the proteins by either carboxymethyl cellulose chromatography (Küntzel 1969) or by a combination of isoelectric focusing and sodium dodecyl sulfate gel electrophoresis (Lizardi and Luck 1972) shows that they are significantly different from those of the cytoplasmic ribosomes. However these results also did not rule out limited structural similarities between some of these proteins.

ASSEMBLY OF ORGANELLE RIBOSOMES

Although much is known concerning the assembly of *E. coli* ribosomes (Nomura 1970) and eukaryotic ribosomes (Darnell 1968; Burdon 1971; Maden 1971), very little information is available as to where and how the organelle ribosomes are assembled. Since organellar ribosomal RNAs are transcription products of the corresponding organelle DNA, it is not unreasonable to assume that the assembly process occurs within the organelle. In principle it should be possible to use mutant strains that are perturbed in some steps of the assembly process to study aspects of this very complex phenomenon. However, only three mutant strains belonging to this category have been isolated so far (cf. Tables 2 and 3), and of these, only the "poky" mutant of *N. crassa* has been investigated in detail. The trait of the "poky" mutant is known to be transmitted in non-Mendelian fashion, and therefore the mutation is presumably localized in the mitochondrial DNA (Mitchell and Mitchell 1952). Analysis of the mitochondrial ribosomes showed that this mutant is deficient in the small ribosomal subunits (Rifkin and Luck 1971; Neupert, Massinger and Pfaller 1971; Kuriyama and Luck 1974). This deficiency is associated with under-methylation of the mitochondrial ribosomal RNAs. There is also alteration in the maturation of 19S RNA and failure of its assembly into small ribosomal subunits (Kuriyama and Luck 1974). In *C. reinhardtii* there is a Mendelian mutant, *cr*-1, which is deficient in the chloroplast small ribosomal subunits (Boynton, Gillham and Burkholder 1970; Boynton, Gillham and Chabot 1972), a phenotype reminiscent of that in the "poky" mutant. It would be interesting to see if this mutant is affected in the methylation process of the chloroplast ribosomal RNAs. Another Mendelian mutant of *C. reinhardtii, ac*-20, has been reported to have only 10–20% of chloroplast ribosomes compared to that of the wild-type strain (Goodenough and Levine 1970; Boynton, Gillham and Chabot 1972). However, the biochemical lesion of this strain has yet to be fully characterized.

CONCLUDING REMARKS

We have summarized the technical approaches that have been used to determine the number, site(s) of synthesis, and locations of structural genes for organellar ribosomal proteins. For no single organelle ribosome has the description yet been completed. Nonetheless, it is already clear in the case of the chloroplast of Chlamydomonas that genes for some ribosomal proteins are present in the nuclear genome while others are in the chloroplast. This kind of information has not yet been developed for mitochondria, but there is good evidence that the mitochondrial ribosomal proteins are synthesized on cytoplasmic ribosomes. Thus even before the present lines of investigation have been completed, we can formulate additional questions. How is transcription from two genetic sites regulated and coordinated? Does the control process also operate at the level of the ribosomal site of synthesis? Since this site may be external to the organelle, there is also the problem of transport of ribosomal proteins through membranes which normally show limited permeability to even small molecules. Here is another potential site for regulation. These questions, unique to organelle ribosomes, must be considered in addition to the kinds of problems now under investigation with prokaryotic and eukaryotic ribosomes.

References

Adoutte, A. and J. Beisson. 1970. Cytoplasmic inheritance of enthromycin-resistant mutations in *Paramecium aurelia. Mol. Gen. Genet.* **108**:70.

Aloni, Y. and G. Attardi. 1971. Expression of the mitochondrial genome in HeLa cells. II. Evidence for complete transcription of mitochondrial DNA. *J. Mol. Biol.* **55**:251.

———. 1972. Expression of the mitochondrial genome in HeLa cells. XI. Isolation and characterization of transcription complexes of mitochondrial DNA. *J. Mol. Biol.* **70**:363.

Attardi, G. and G. Amaldi. 1970. Structure and synthesis of ribosomal RNA. *Ann. Rev. Biochem.* **39**:183.

Bastia, D., K. S. Chiang and H. Swift. 1971. Studies on the ribosomal RNA cistrons of chloroplast and nucleus in *Chlamydomonas reinhardtii. Abstr. 11th Ann. Meet. Amer. Soc. Cell Biol.* **41**:25.

Beale, G. H., J. K. C. Knowles and A. Tait. 1972. Mitochondrial genetics in *Paramecium. Nature* **235**:396.

Borst, P. 1972. Mitochondrial nucleic acids. *Ann. Rev. Biochem.* **41**:333.

Borst, P. and L. A. Grivell. 1971. Mitochondrial ribosomes. *FEBS Letters* **13**:73.

Boschetti, A. and S. Bogdanov. 1973. Different effects of streptomycin on the ribosomes from sensitive and resistant mutants of *Chlamydomonas reinhardtii. Eur. J. Biochem.* **35**:482.

Boulter, D., R. J. Ellis and A. Yarwood. 1972. Biochemistry of protein synthesis in plants. *Biol. Rev.* **47**:113.

Bourque, D. P. and S. G. Wildman. 1973. Evidence that nuclear genes code for several chloroplast ribosomal proteins. *Biochem. Biophys. Res. Comm.* **50**:532.

Bourque, D. P., J. E. Boynton and N. W. Gillham. 1971. Studies on the structure and cellular location of various ribosome and ribosomal RNA species in the green alga *Chlamydomonas reinhardtii. J. Cell Biol.* **8**:153.

Boynton, J. E., N. W. Gillham and B. Burkholder. 1970. Mutations altering chloroplast ribosome phenotype in *Chlamydomonas.* II. A new Mendelian mutation. *Proc. Nat. Acad. Sci.* **67**:1505.

Boynton, J. E., N. W. Gillham and J. F. Chabot. 1972. Chloroplast ribosome deficient mutants in the green alga *Chlamydomonas reinhardtii* and the question of chloroplast ribosome function. *J. Cell Sci.* **10**:267.

Boynton, J. E., W. G. Burton, N. W. Gillham and E. H. Harris. 1973. Can a non-Mendelian mutation affect both chloroplast and mitochondrial ribosomes? *Proc. Nat. Acad. Sci.* **70**:3463.

Burdon, R. H. 1971. Ribonucleic acid maturation in animal cells. *Prog. Nucleic Acid Res. Mol. Biol.* **11**:33.

Burton, W. G. 1972. Dihydrospectinomycin binding to chloroplast ribosomes from antibiotic-sensitive and -resistant strains of *Chlamydomonas reinhardtii. Biochim. Biophys. Acta* **272**:305.

Chua, N.-H., G. Blobel and P. Siekevitz. 1973. Isolation of cytoplasmic and chloroplast ribosomes and their dissociation into active subunits from *Chlamydomonas reinhardtii*. *J. Cell Biol.* **57**:798.

Coen, D., J. Deutsch, P. Netter, E. Petrochilo and P. P. Slonimski. 1970. Mitochondrial genetics. I. Methodology and phenomenology. *Symp. Soc. Exp. Biol.* **24**:449.

Darnell, J. E., Jr. 1968. Ribonucleic acids from animal cells. *Bact. Rev.* **32**:262.

Davey, P. J., R. Yu and A. W. Linnane. 1969. The intracellular site of formation of the mitochondrial protein synthetic system. *Biochem. Biophys. Res. Comm.* **36**:30.

Davies, J. and M. Nomura. 1972. The genetics of bacterial ribosomes. *Ann. Rev. Gen.* **6**:203.

Dawid, I. B. 1972. Mitochondrial RNA in *Xenopus laevis*. I. The expression of the mitochondrial genome. *J. Mol. Biol.* **63**:201.

Dawid, I. and J. W. Chase. 1972. Mitochondrial RNA in *Xenopus laevis*. II. Molecular weight and other physical properties of mitochondrial ribosomal and 4S RNA. *J. Mol. Biol.* **55**:231.

De Vries, H. and R. Van der Koogh-Schuuring. 1973. Physiochemical characteristics of isolated 55S mitochondrial ribosomes from rat liver. *Biochem. Biophys. Res. Comm.* **54**:308.

Ellis, R. J. and M. R. Hartley. 1971. Sites of synthesis of chloroplast proteins. *Nature New Biol.* **233**:193.

Freeman, K. B., R. S. Mitra and B. Bartoov. 1973. Characteristics of the base composition of mitochondrial ribosomal RNA. *Sub-cell. Biochem.* **2**:183.

Galling, G. and B. R. Jordon. 1972. Isolation and characterization of cytoplasmic and chloroplastic 5S RNAs in the unicellular alga Chlorella. *Biochimie* **54**:1257.

Gillham, N. W. 1969. Uniparental inheritance in *Chlamydomonas reinhardtii*. *Amer. Naturalist* **103**:355.

Gillham, N. W. and R. P. Levine. 1962. Studies on the origin of streptomycin-resistant mutants in *Chlamydomonas reinhardtii*. *Genetics* **47**:1463.

Gillham, N. W., J. E. Boynton and B. Burkholder. 1970. Mutations altering chloroplast ribosome phenotype in *Chlamydomonas*. I. Non-Mendelian mutations. *Proc. Nat. Acad. Sci.* **67**:1026.

Goodenough, U. W. 1971. The effects of inhibitors of RNA and protein synthesis on chloroplast structure and function in wild-type *Chlamydomonas reinhardtii*. *J. Cell Biol.* **50**:35.

Goodenough, U. W. and R. P. Levine. 1970. Chloroplast structure and function in *ac-20*, a mutant strain of *Chlamydomonas reinhardtii*. III. Chloroplast ribosomes and membrane organization. *J. Cell Biol.* **44**:547.

Gray, P. N. and G. Attardi. 1973. An attempt to identify a presumptive 5S RNA-equivalent RNA species in mitochondrial ribosomes. *Abstr. 13th Ann. Meet. Amer. Soc. Cell Biol.* **59**:120a.

Greco, M., P. Cantatore, G. Pepe and C. Saccone. 1973. Isolation and characterization of rat liver mitochondrial ribosomes highly active in poly(U)-directed polyphenylalanine synthesis. *Eur. J. Biochem.* **37**:171.

Grivell, L. A. and G. S. P. Groot. 1972. Spinach chloroplast ribosomes active in protein synthesis. *FEBS Letters* **25**:21.

Grivell, L. A., L. Reijnders and P. Borst. 1971. Isolation of yeast mitochondrial ribosomes highly active in protein synthesis. *Biochim. Biophys. Acta* **247**:91.

Grivell, L. A., L. Reijnders and H. De Vries. 1971. Altered mitochondrial ribosomes in a cytoplasmic mutant of yeast. *FEBS Letters* **16**:159.

Grivell, L. A., P. Netter, P. Borst and P. P. Slonimski. 1973. Mitochondrial antibiotic resistance in yeast: Ribosomal mutants resistant to chloramphenicol, erythromycin and spiramycin. *Biochim. Biophys. Acta* **312**:358.

Gualerzi, C. and P. Cammarano. 1969. Comparative electrophoretic studies on the protein of chloroplast and cytoplasmic ribosomes of spinach leaves. *Biochim. Biophys. Acta* **190**:170.

———. 1970. Species specificity of ribosomal proteins from chloroplast and cytoplasmic ribosomes of higher plants. Electrophoretic studies. *Biochim. Biophys. Acta* **199**:203.

Honeycutt, R. C. and M. M. Margulies. 1973. Protein synthesis in *Chlamydomonas reinhardtii*. Evidence for synthesis of proteins of chloroplastic ribosomes on cytoplasmic ribosomes. *J. Biol. Chem.* **248**:6145.

Hoober, J. K. and G. Blobel. 1969. Characterization of the chloroplastic ribosomes of *Chlamydomonas reinhardtii*. *J. Mol. Biol.* **41**:121.

Ibrahim, N. G. and D. S. Beattie. 1973. Protein synthesis on ribosomes isolated from rat liver mitochondria: Sensitivity to erythromycin. *FEBS Letters* **36**:102.

Ingle, J., J. V. Possingham, R. Wells, C. J. Leaver and U. E. Loening. 1970. The properties of chloroplast ribosomal-RNA. *Symp. Soc. Exp. Biol.* **24**:303.

Janda, H. G. and H. G. Wittmann. 1968. Ribosomal proteins. V. Comparison of protein patterns of 70S and 80S ribosomes from various plants by polyacrylamide gel electrophoresis. *Mol. Gen. Genet.* **103**:238.

Jones, B. L., N. Nagabhushan, A. Gulyas and S. Zalik. 1972. Two-dimensional acrylamide gel electrophoresis of wheat leaf cytoplasmic and chloroplast ribosomal proteins. *FEBS Letters* **23**:167.

Kaltschmidt, E. and H. G. Wittmann. 1970. Ribosomal proteins. VII. Two-dimensional polyacrylamide gel electrophoresis for fingerprinting of ribosomal proteins. *Anal. Biochem.* **36**:401.

Kloppstech, K. and H. G. Schweiger. 1973. Nuclear genome codes for chloroplast ribosomal proteins in Acetabularia. II. Nuclear transplantation experiments. *Exp. Cell Res.* **80**:69.

Kroon, A. M., E. Agsteribbe and H. De Vries. 1972. Protein synthesis in mitochondria and chloroplasts. In *The mechanism of protein synthesis and its regulation* (ed. L. Bosch) pp. 539–582. North-Holland, Amsterdam.

Küntzel, H. 1969. Proteins of mitochondrial and cytoplasmic ribosomes from *Neurospora crassa*. *Nature* **222**:142.

———. 1971. The genetic apparatus of mitochondria from *Neurospora* and yeast. In *Current topics in microbiology and immunology*, vol. 54, pp. 94–118. Springer-Verlag, New York.

Kuriyama, Y. and D. J. L. Luck. 1973. Ribosomal RNA synthesis in mitochondria of *Neurospora crassa*. *J. Mol. Biol.* **73**:425.

———. 1974. Methylation and processing of mitochondrial ribosomal RNA's in "poky" and wild-type *Neurospora crassa*. *J. Mol. Biol.* **83**:253.

Leaver, C. J. and J. Ingle. 1971. The molecular integrity of chloroplast ribosomal ribonucleic acid. *Biochem. J.* **123**:235.

Lizardi, P. M. and D. J. L. Luck. 1971. Absence of a 5S RNA component in the mitochondrial ribosomes of *Neurospora crassa. Nature New Biol.* **229**:140.

————. 1972. The intracellular site of synthesis of mitochondrial ribosomal proteins in *Neuropsora crassa. J. Cell Biol.* **54**:56.

Lyttleton, J. W. 1962. Isolation of ribosomes from spinach chloroplasts. *Exp. Cell Res.* **26**:312.

————. 1968. Protein constituents of plant ribosomes. *Biochim. Biophys. Acta* **154**:145.

Maden, B. E. H. 1971. The structure and formation of ribosomes in animal cells. *Prog. Biophys. Mol. Biol.* **22**:127.

Mahler, H. R. 1973. Biogenetic autonomy of mitochondria. *CRC Critical Rev. Biochem.* **3**:381.

Mets, L. T. and L. Bogorad. 1971. Mendelian and uniparental alterations in erythromycin binding by plastid ribosomes. *Science* **174**:707.

————. 1972. Altered chloroplast ribosomal proteins associated with erythromycin-resistant mutants in two genetic systems of *Chlamydomonas reinhardtii. Proc. Nat. Acad. Sci.* **69**:3779.

Michaelis, G., E. Petrochilo and P. P. Slonimski. 1973. Mitochondrial genetics. III. Recombined molecules of mitochondrial DNA obtained from crosses between cytoplasmic petite mutants of *Saccharomyces cerevisiae.* Physical and genetic characterization. *Mol. Gen. Genet.* **123**:51.

Mitchell, M. B. and H. K. Mitchell. 1952. A case of "maternal" inheritance in *Neurospora crassa. Proc. Nat. Acad. Sci.* **38**:442.

Neupert, W., P. Massinger and A. Pfaller. 1971. Amino acid incorporation into mitochondrial ribosomes of *Neurospora crassa* wild type and *mi-1* mutant. In *Autonomy and biogenesis of mitochondria and chloroplasts* (ed. N. K. Boardman, A. W. Linnane and R. M. Smillie) pp. 328–338. North-Holland, Amsterdam.

Nomura, M. 1970. Bacterial ribosome. *Bact. Rev.* **34**:228.

Payne, P. I. and T. A. Dyer. 1971. Characterization of cytoplasmic and chloroplast 5S ribosomal ribonucleic acid from broad-bean leaves. *Biochem. J.* **124**:83.

————. 1972. Plant 5.8S RNA is a component of 80S but not 70S ribosomes. *Nature New Biol.* **235**:145.

Rawson, J. R. Y. and R. Haselkorn. 1973. Chloroplast ribosomal RNA genes in the chloroplast DNA of *Euglena gracilis. J. Mol. Biol.* **77**:125.

Rawson, J. R. and E. Stutz. 1969. Isolation and characterization of *Euglena gracilis* cytoplasmic and chloroplast ribosomes and their ribosomal RNA components. *Biochim. Biophys. Acta* **190**:368.

Reijnders, L., C. M. Keisen, L. A. Grivell and P. Borst. 1972. Hybridization studies with yeast mitochondrial RNA's. *Biochim. Biophys. Acta* **272**:396.

Rifkin, M. R. and D. J. L. Luck. 1971. Defective production of mitochondrial ribosomes in the poky mutant of *Neurospora crassa. Proc. Nat. Acad. Sci.* **68**:287.

Rijven, A. H. G. C. and J. A. Zwar. 1973. Methylation patterns of ribonucleic acids from chloroplasts and cytoplasm of fenugreek (*Trigorella foenumgrae-cum L.*) cotyledons. *Biochim. Biophys. Acta* **229**:564.

Robberson, D., Y. Aloni, G. Attardi and N. Davidson. 1971. Expression of the mitochondrial genome in HeLa cells. VI. Size determination of mitochondrial ribosomal RNA by electron microscopy. *J. Mol. Biol.* **60**:473.

Rossi, L. and C. Gualerzi. 1970. Non-random differences in the base composition of chloroplast and cytoplasmic ribosomal RNA from some higher plants. *Life Science* **9**:1401.

Sacchi, A., F. Cerbone, P. Cammarano and U. Ferrini. 1973. Physicochemical characterization of ribosome-like (55S) particles from rat liver mitochondria. *Biochim. Biophys. Acta* **308**:309.

Sager, R. 1954. Mendelian and non-Mendelian inheritance of streptomycin resistance in *Chlamydomonas reinhardtii*. *Proc. Nat. Acad. Sci.* **40**:356.

————. 1972. *Cytoplasmic genes and organelles*. Academic Press, New York.

Schäfer, K. P. and H. Küntzel. 1972. Mitochondrial genes in *Neurospora*. A single cistron for ribosomal RNA. *Biochem. Biophys. Res. Comm.* **46**:1312.

Schlanger, G. and R. Sager. 1974. Localization of five antibiotic resistances at the subunit level in chloroplast ribosomes of *Chlamydomonas*. *Proc. Nat. Acad. Sci.* (in press).

Schlanger, G., R. Sager and Z. Ramanis. 1972. Mutation of a cytoplasmic gene in *Chlamydomonas* alters chloroplast ribosome function. *Proc. Nat. Acad. Sci.* **69**:3551.

Sherton, C. C. and I. G. Wool. 1972. Determination of the number of proteins in liver ribosomes and ribosomal subunits by two-dimensional polyacrylamide gel electrophoresis. *J. Biol. Chem.* **247**:4460.

Smith-Johannsen, H. and S. P. Gibbs. 1972. Effects of chloramphenicol on chloroplast and mitochondrial ultrastructure in *Ochromonas danica*. *J. Cell Biol.* **52**:598.

Surzycki, S. J. 1969. Genetic functions of the chloroplast of *Chlamydomonas reinhardtii*: Effect of rifampicin on chloroplast DNA-dependent RNA polymerase. *Proc. Nat. Acad. Sci.* **63**:1327.

Surzycki, S. J. and N. W. Gillham. 1971. Organelle mutations and their expression in *Chlamydomonas reinhardtii*. *Proc. Nat. Acad. Sci.* **68**:1301.

Surzycki, S. J. and J. D. Rochaix. 1971. Transcriptional mapping of ribosomal RNA genes of the chloroplast and nucleus of *Chlamydomonas reinhardtii*. *J. Mol. Biol.* **62**:89.

Swanson, R. F. and I. B. Dawid. 1970. The mitochondrial ribosome of *Xenopus laevis*. *Proc. Nat. Acad. Sci.* **66**:117.

Tait, A. 1972. Altered mitochondrial ribosomes in an erythromycin-resistant mutant of *Paramecium*. *FEBS Letters* **24**:117.

Tewari, K. K. and S. G. Wildman. 1970. Information content in the chloroplast DNA. *Symp. Soc. Exp. Biol.* **24**:147.

Thomas, D. Y. and D. Wilkie. 1968. Recombination of mitochondrial drug-resistance factors in *Saccharomyces cerevisiae*. *Biochem. Biophys. Res. Comm.* **30**:368.

Vasconcelos, A. C. L. and L. Bogorad. 1971. Proteins of cytoplasmic, chloroplast, and mitochondrial ribosomes of some plants. *Biochim. Biophys. Acta* **228**:492.

Wilson, R. and K. S. Chiang. 1972. Absence of high molecular weight cyto-plasmic and chloroplast rRNA precursors in *Chlamydomonas reinhardtii. Abstr. 12th Ann. Meet. Amer. Soc. Cell Biol.* **55**:285a.

Wittmann, H. G. 1972. Ribosomal proteins from prokaryotes. *FEBS Symp.* **23**:3.

Wood, D. D. and D. J. L. Luck. 1969. Hybridization of mitochondrial ribo-somal RNA. *J. Mol. Biol.* **41**:211.

Wu, M., N. Davidson, G. Attardi and Y. Aloni. 1972. Expression of the mito-chondrial genome in HeLa cells. XIV. The relative position of the 4S RNA genes and of the ribosomal RNA genes in mitochondrial DNA. *J. Mol. Biol.* **71**:81.

PART II
Specific
Reviews

Ribosome Structure as Studied by Electron Microscopy

James A. Lake, David D. Sabatini
Department of Cell Biology
New York University Medical Center
New York, New York 10016

Yoshiaki Nonomura
Department of Pharmacology
Faculty of Medicine
University of Tokyo
Bunkyo-Ku Tokyo 113, Japan

INTRODUCTION

Much has been learned of the events occurring during ribosome function, but it is still difficult to correlate the sequence of macromolecular interactions involved in protein synthesis with the chemical composition of ribosomes. Presently individual functions are being assigned to ribosomal components, but our understanding of the operation of the translational machinery will only be complete when this information can be related to structural features. It is likely that this goal will be obtained through the use of ribosome crystals suitable for electron microscopy and X-ray diffraction, as well as by direct visualization of protein sites on ribosomes using specific antibodies.

We have studied biochemically and electron microscopically the ribosome crystal which can be induced by cooling chicken embryos in ovo (Byers 1966, 1967) and investigated the conditions leading to ribosome crystallization in vivo and in vitro. Our observations (Morimoto, Blobel and Sabatini 1972a,b) and those of other authors (Byers 1971; Carey 1970; Carey and Read 1971) indicate that crystals from chicken embryos are a promising subject for investigating the structure of ribosomes since they are made of normal 80S particles, which, although inactive, are potentially functional.

Here we present a summary of our electron microscopic studies on crystals of chicken embryo ribosomes. In addition, in light of the large amount of interest in *E. coli* ribosomes and in consideration of the relatively advanced state of our biochemical knowledge of them, we present the preliminary results of our comparative electron microscopic structural studies on eukaryotic (Nonomura, Blobel and Sabatini 1971) and *E. coli* ribosomes, as well as some observations (Lake, Kahan and Nomura, manuscript in preparation) on the localization of *E. coli* small subunit proteins obtained through direct visualization of specific bound antibodies.

The resolution in these studies has been sufficient to obtain a gross morphological characterization of the ribosomal subunits and to reveal details of the relationship between subunits. Some structural features, such as a partition in the small subunit, appear to be general characteristics of ribosomes of both eukaryotes and prokaryotes, probably related to the manner by which ribosomes associate with mRNA.

RAT LIVER RIBOSOMES

The model of the monomeric ribosomes in Figure 1 summarizes our interpretation of the images more frequently observed in samples of ribosomal subunits, their dimers and monomeric ribosomes (Nonomura, Blobel and Sabatini 1971). The most prominent features of the structure are the division of the small subunit (S) into two unequal parts, the asymmetric positioning of the small subunit upon the large, which thereby confers a definite handedness to the structure, and the notch in the profile of the large subunit (L) which is positioned below the partition of the small subunit.

Small Subunits

The shape of the small subunit is approximately that of a slightly curved and flattened prolate ellipsoid (230 Å × 140 Å × 115 Å). The profile of

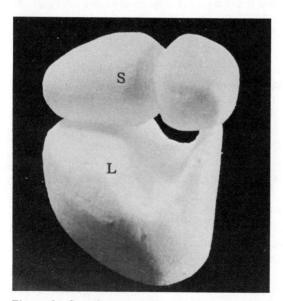

Figure 1 Styrofoam model of the eukaryotic ribosome. The small subunit is indicated by the letter *S* and the large subunit by *L*.

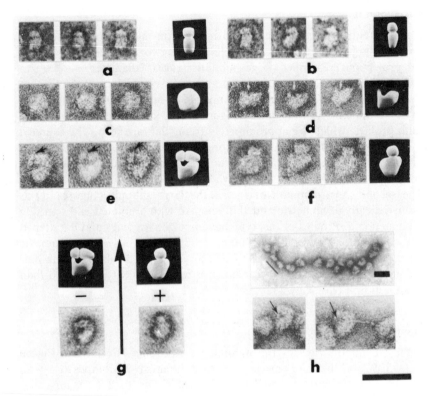

Figure 2 A gallery of rat liver ribosome monomers and subunits and corresponding views of the model shown in Figure 1. (*a,b*) Two approximately
orthogonal views of the small subunit; (*c,d*) two approximately orthogonal
views of the large subunit; (*e,f*) two views of the monomer; (*g*) views of a
monomer taken at two different tilt angles showing the conversion of one type
of monomer image into another; (*h*) polysomes showing the path of the strand
of mRNA. Bars = 500 Å.

the small subunit is divided into "one-third" and "two-thirds" regions by a
transversal partition seen in electron micrographs (Figure 2a,b) as a dense
line at 80 Å from one end of the subunit. Two types of elongated profiles
are common, slightly curved (Figure 2b) or straight (Figure 2a), which
would represent two approximately orthogonal views of the particle. "End-
on" views are rarely found in samples of isolated small subunits. Small
subunit dimers are formed when the ionic strength of a sample of isolated
small subunits is reduced from 500 mM to ~ 50 mM KCl. In most dimers
(not shown), individual small subunits bind to each other through a portion of their concave faces. The entire length of the same concave face of
the small subunit contacts the large subunit when native or reconstituted
monomeric ribosomes are observed (Figure 2e).

Large Subunits

The more frequent image types found in electron micrographs of samples of isolated large subunits can be explained by two approximately orthogonal projections. Rounded large subunit images (diameter $\sim$ 220 Å) (Figure 2c) are frequently seen in samples of isolated large subunits but are not recognized as components of images of large subunit dimers (not shown) or monomeric ribosomes (Figure 2e,f); in these cases, only large subunit profiles, which are approximately triangular, are seen. The rounded images of the large subunit are therefore likely to represent projections of the large subunits on a face of the particle on which the large subunits can dimerize or bind to the small subunit when a monomer is formed.

The second most common profile of the large subunit (Figure 2d) is characteristically asymmetric and skiff-shaped, with a pointed end opposite to a blunted end. Profiles of this type have two convex sides and a flattened side ($\sim$230 Å long) which has a notch ($\sim$30–40 Å) at $\sim$80 Å from the blunted end. The height of the skiff measured perpendicularly to the flattened side is $\sim$170 Å. The profile shown is more frequently observed than its enantiomorph (not shown).

Monomeric Ribosomes

The most common view of the monomeric ribosome (front view Figure 2e) ($\sim$50% of the monomeric ribosome images) is composed of an elongated, slightly curved, small subunit profile (up to 230 Å long and $\sim$120 Å wide) apposed through its concave side to the flattened side of the skiff-shaped large subunit profile (Figure 2d). Profiles of both subunits are separated by a linear dense region. In typical frontal views, the transverse partition of the small subunit profile is visible over a dense round spot ($\sim$40 to 50 Å in diameter), which is found to one side of the ribosome. This corresponds to the position of the notch on the large subunit profile and marks the confluence of the partition in the small subunit with the dense line separating both subunits.

The second most frequent view of the monomeric ribosome ($\sim$20% of the images) is the lateral view, which is characterized by two regions of unequal size separated by a thin and faint dense line (Figure 2f). The smaller region is approximately rectangular (height: 115–120 Å; width towards the base: $\sim$140 Å). This part of the profile is interpreted as resulting from an end view of the prolate-shaped small subunit, since the partition of the small subunit is not visible. In lateral views of monomers, the small subunit does not bind to the middle of the flattened side of the large subunit but is displaced towards one side of it. Most lateral views are such that if oriented with the flattened side of the large subunit profile horizontally and the small subunit towards the top, the small subunit profile appears displaced towards the right of the observer. An enantiomorph of this image is rarely observed.

The results of tilting experiments (Figure 2g) showed that frontal and lateral images are interconvertible and therefore correspond to different positions of the same particle on the supporting films. Because of the direction of the rotation (tilting) needed to convert a left-featured frontal view to a right lateral view, only a model like the one shown in Figure 1 is acceptable.

Polysomes

The ribosome model shown in Figure 1 is also consistent with images of ribosomes within polysomes. The strand of messenger RNA is seen in polysomes reaching the ribosome between subunits, in agreement with biochemical findings. Lateral views of the monomers and images intermediary between frontal and lateral views are frequently seen in negatively stained polysomes (Figure 2h), perhaps resulting from a restriction in the rotation of monomers by the strand of messenger joining them. The strand of mRNA could be followed through several ribosomes in a polysome. Its diameter (15–30 Å) suggests that there is considerable secondary structure in mRNA, or that mRNA is covered by protein, resulting in a larger diameter after negative staining. In some lateral views, the strand joined the ribosomes at both sides of the junction between subunits (Figure 2h), and in some frontal views, the strand could be followed up to the dense spot between subunits (Figure 2h), indicating that this marks a point of entrance or exit into the ribosome. These observations suggest that mRNA traverses the ribosome in a direction perpendicular to the long axis of the small subunit.

E. coli RIBOSOMES

In spite of the difference in particle size, which is reflected in the molecular weights, sedimentation values, number and sizes of proteins and size of RNAs, ribosomes from prokaryotes show interesting similarities in morphology to those of eukaryotes. In particular *E. coli* ribosomes closely resemble in several respects ribosomes derived from rat liver, a fact which suggests common architectural principles.

Small Subunits

The small subunit from *E. coli* is similar in shape to the rat liver small subunit, although somewhat smaller. The same characteristic views observed in preparations of rat liver (Figure 3b,d) are seen in electron micrographs

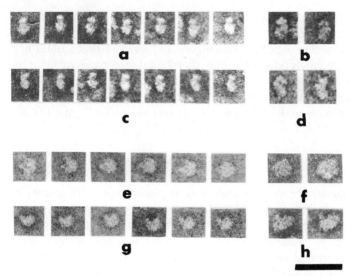

Figure 3   A gallery of *E. coli* and rat liver small and large subunits. (*a,c*) Two
views of the *E. coli* small subunit; (*e,g*) two views of the *E. coli* large subunit;
(*b,d,f,h*) views of rat liver subunits in comparable orientations to *E. coli*
subunits in *a,c,e* and *g,* respectively. Bar = 500 Å.

of *E. coli* small subunits (Figure 3a,c). One "one-third:two-thirds" partition is clearly present in both the "face-on" view (Figure 3a) and the side
view (Figure 3c). Both enantiomorphic types of curved side views are seen
in micrographs, indicating that the *E. coli* small subunit, like the rat liver
small subunit, assumes no marked preferential orientation on the supporting films.

Our observations show that the "one-third:two-thirds" transversal partition is characteristic of small ribosome subunits of both prokaryotes and
eukaryotes. The question of whether the small subunit can be approximated
by a prolate body, as suggested by the profiles observed by us, or whether
its shape corresponds to a flattened type or oblate, as proposed earlier
(Huxley and Zubay 1960), should be evaluated in the light of X-ray scattering data from solutions of small subunits (Hill, Thompson and Anderegg 1969). In preliminary calculations we have computed a theoretical pair
distribution function for a small subunit model similar to the one shown in
Figure 1 but derived from the actual shape and dimensions of the *E. coli*
small subunits shown in Figure 3a,c. This function is similar to that determined from X-ray data by Hill, Thompson and Anderegg (1969) but shows
three maxima, which presumably correspond to self vectors from the "one-
third" and the "two-thirds" regions and to cross vectors between both
regions. If, in solutions, both regions can flex relative to each other on a
hinge provided by the partition, the net effect on the pair distribution func-

tion would be to improve the agreement between the curves calculated from electron microscopy and from small-angle X-ray scattering. Thus these preliminary calculations indicate that the prolate, partitioned model may account for the X-ray diffraction data.

Large Subunits

The large subunit of *E. coli* shows some general similarities to the eukaryotic large subunit, but significant differences are present.

The view in Figure 3e corresponds to the circular profile (Figures 3f and 2c) of the eukaryotic large subunit. In this view, the *E. coli* large subunit profile is more angular and somewhat resembles the profile of a maple leaf, with the largest projection, described as a "nose" by Lubin (1968), located at its tip. This has been oriented towards the left in these galleries. The leaf profiles show two other smaller projections on each side of the tip which diverge at an angle (~60°) from the nose direction. Thus these profiles have an approximate line of mirror symmetry. It appears likely that in the three-dimensional structure the base of the nose lies near the notch region, which is apparent in the crescent-shaped side views of *E. coli* large subunits (see Figure 3g). In the leaf views the notch would then correspond to a dense patch of stain seen near the base of the nose. As expected, the *E. coli* large subunit is also smaller than the rat liver large subunit; although its largest diameter, which is in the direction of the nose, is about 200 Å, a value not too different from the 220 Å diameter observed for eukaryotes (Figure 3f).

As was just mentioned, a notch is present in crescent-shaped profiles of *E. coli* large subunits (Figure 3g) corresponding to the notch present in skiff-shaped profiles of rat liver large subunits shown in Figures 3h and 2d. Like its eukaryotic equivalent, the crescent profile corresponds to one of the two characteristic views which are orthogonal to the leaf profile ("circular" profile in rat liver).

Monomeric Ribosomes

The orientation of the small subunit with respect to the large subunit in the monomer is similar in both eukaryotes and prokaryotes. In the left frontal views in Figure 4a, the long axis of the small subunit is oriented horizontally across the top of the large subunit, and the "one-third:two-thirds" dividing line is apparent at the left. The notch on the large subunit which collects stain is frequently visible in this orientation and is located below the small subunit partition. The partition of the small subunit is observed with about equal frequency on the upper left side (Figure 4a) and on the upper right side (Figure 4c) of these images, indicating, in contradistinction to rat liver, that *E. coli* monomers display no preferred orientation on

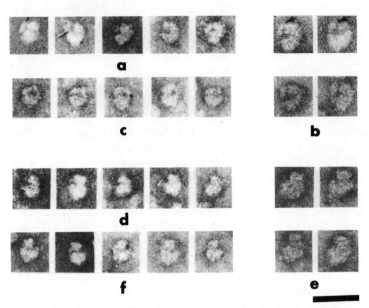

Figure 4 A gallery of *E. coli* and rat liver ribosome monomers. (*a*) Frontal view of *E. coli* monomers; (*c*) view of *E. coli* monomers enantiomorphic to that shown in *a;* (*d*) lateral view of *E. coli* monomers; (*f*) view of *E. coli* monomers enantiomorphic to that shown in *d;* (*b,e*) views of rat liver monomers in orientations comparable to those shown for *E. coli* in *a* and *d,* respectively. Bar = 500 Å.

the carbon-coated supporting films. Frontal images of rat liver monomers in their equivalent preferred orientation are shown in Figure 4b for comparison.

A gallery of lateral views of *E. coli* monomers demonstrating the asymmetric positioning of the small subunit on the large, similar to that in eukaryotic ribosomes, is shown in Figure 4d and f and should be compared with the examples from rat liver in Figure 4e. It should be noted that in *E. coli* ribosomes enantiomorphic forms of both frontal and lateral types of monomer images are all seen with approximately equal frequency.

Thus there are many similarities between 70S and 80S cytoplasmic ribosomes. The "one-third:two-thirds" partition of the small subunit is a general feature, as are the notch in side profiles of large subunits and the asymmetric attachment of the small subunit upon the large subunit. Images of polysomes suggest that the alignment of the "one-third:two-thirds" partition of the small subunit with the notch of the large subunit is a constant aspect likely to be related to a common spatial organization providing for specific interactions of mRNA and ribosome components directly participating in translation.

Antibody Labeling Studies on Small Subunits

Specific antibodies prepared against purified small subunit proteins can be used as markers to map the location of proteins by direct visualization of bound antibody in negatively stained subunit preparations. By adjusting the IgG and the 30S subunit concentrations to approximately equivalence, numerous dimers of small subunits, linked by single IgG antibodies, can be produced. An analysis of the images obtained at different subunit orientations shows that antibodies against protein S14 bind at a specific single site (Lake, Kahan and Nomura, manuscript in preparation; Lake 1974). Figure 5 shows a gallery of views of monomers and dimers of small subunits joined by antibodies against protein S14. Subunits in row *a* can be seen in the symmetrical (or "face on") view, while the lower subunit in each frame of row *b* is seen in the asymmetrical (or "side") view. Monomers, with a single antibody attached, are present in row *c* in both the symmetrical (left) and asymmetrical (right) views. All these views show that protein S14 is located at a single site in the upper one-third, or head, of the small subunit profile. At present, we have examined the location of approximately one-half of the small subunit proteins by this method (S3, S4, S5, S7, S9, S11, S13, S14, S16, S19 and S20) (Lake, Kahan and Nomura, unpublished results). Many of these, such as S13 and S19, are also located in the head of the small subunit, while others have their locations in the larger portion (such as S5 which is found slightly below the partition). An interesting observation concerns protein S4, which appears to be extended in conformation and has antigenic sites exposed at three different regions along the profile. The location of S4 seems to be consistent with the sequence of protein association suggested by the assembly map (Mizushima and Nomura 1970).

RIBOSOME TETRAMERS, p4 ARRAYS AND P422 CRYSTALS

Ribosomes crystallize in nature in a number of systems. Although presently available ribosome crystals are much too small for single-crystal X-ray diffraction, some of them are well suited for quantitative analysis using electron microscopy (Lake and Slayter 1970, 1972). At present, some of the most suitable crystals are those obtained by hypothermic treatment of chicken embryos (Byers, 1967).

Chain termination appears to be necessary for crystal formation in chicken embryos. If it is prevented, either by rapid cooling of the embryos or by a slow cooling of the eggs in the presence of cycloheximide, then the formation of a pool of monomers, which is necessary for crystallization, is almost completely suppressed (Morimoto, Blobel and Sabatini 1972b). Upon slow cooling, such as may occur when the hen leaves her nest, initiation of protein synthesis is blocked, but translation continues and crystallization occurs from the increased concentration of ribosome monomers.

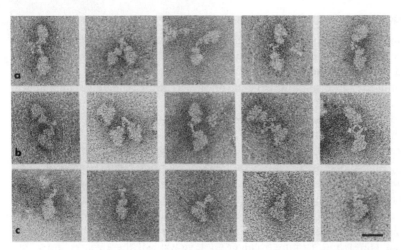

Figure 5 Gallery of electron micrographs of *E. coli* small subunits reacted with AS14 antibodies. Both F_{ab} and F_c regions of the antibody molecules are visible. (*a*) Dimers in the symmetrical view; (*b*) dimers in the asymmetrical view; (*c*) monomers in the symmetrical (left) and asymmetrical views (right). Bar = 200 Å.

Ribosome Tetramers

Several authors (Carey 1970; Byers 1971; Morimoto, Blobel and Sabatini 1972a) have isolated ribonucleoprotein particles with a sedimentation coefficient of 166S from hypothermic chicken embryos. These particles, which were first thought to be polysomes, constitute ~30% of the total ribonucleoprotein in 5-day-old embryos cooled for 24 hr. Electron microscopy (Figure 6) revealed that the 166S particles are tetramers of rounded units, with the morphological properties of ribosomes (Bell et al. 1965; Carey 1970; Byers 1971; Sabatini et al. 1972). Tetramers can function in protein synthesis, accepting poly (U) and the necessary protein factors without previous dissociation into free subunits (Morimoto, Blobel and Sabatini 1972a; Byers 1971). Within the isolated tetramers, which almost always lie flat on the supporting film, the four component ribosomes can be seen in nearly typical left-featured frontal views.

A gallery of tetramer profiles is shown in Figure 6a. The long arrows indicate the division between the large and the small subunits within each ribosome of the tetramer, and the small arrows point to the "one-third:two-thirds" dividing line within the small subunit. Profiles of small subunits within tetramers make contact with profiles of large subunits of neighboring ribosomes. In all tetramers examined, the sequence of contacts between profiles of individual ribosomes within tetramers was small to large subunit, proceeding in a clockwise sense. This indicates that tetramers were almost exclusively preferentially oriented on films and suggests that intratetramer bonds result from contacts between large and small subunits (see Figure

Figure 6 A gallery of tetramers of chicken embryo ribosomes is shown in *a*. The longer arrows point to the dividing line between large and small subunits, and the shorter arrows point to the "one-third:two-thirds" division of the small subunit. A model for the packing of ribosomes within the tetramer is shown in *b*. Bar = 500 Å.

6b). However, Byers (1971) and Carey and Read (1971) have presented evidence suggesting that intratetramer bonds also result from large-to-large' subunit contacts. They have reported that by raising the ionic strength, small subunits can be selectively released from tetramers, thereby generating tetramers of large subunits.

Ribosome p4 Arrays

Arrays of tetramers can be studied by negatively staining aliquots of homogenates or cell fractions capable of sedimentation at low speeds. It is then seen (Figure 7c) that ribosome tetramers crystallize into sheets across the twofold axes of plane group p4 through large subunit to large subunit

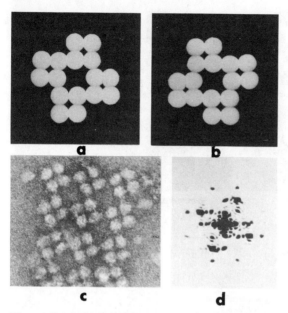

Figure 7 (*a,b*) Models representing the dextro and levo forms of the p4 ribosome sheets, respectively; (*c*) an electron micrograph of a levo crystal; and (*d*) the optical diffraction pattern of *c*.

bonds. Depending upon the orientation of the sheets on the support film during negative staining, either of two sides of these sheets are seen. Byers (1967) described them as "dextro" (Figure 7a) or "levo" (Figure 7b). Due to specific attachment of the crystals to the carbon film, we consistently observe the levo form, which is the one permitting the ribosomes to adopt an orientation close to that producing the left-featured frontal type of image. It is interesting to note that the left frontal view is the prevalent image seen in preparations of eukaryotic monomers.

It has been shown that tetramers (Morimoto, Blobel and Sabatini 1972a,b), and therefore the crystals, are formed by mature but inactive 80S ribosomes. These lack mRNA and peptidyl-tRNA and therefore can be disassembled into subunits by raising the ionic strength without the need of puromycin. By changing the ionic conditions, the interribosome binding in tetramers can be broken independently of the binding between subunits within monomeric ribosomes. At 1 mM Mg^{++} and in a medium of low ionic strength, tetramers are completely converted into monomers. The interribosome binding within tetramers can also be broken without previous dissociation into subunits by raising the ionic strength while maintaining the Mg^{++} concentration at 5 mM.

A levo crystal and its optical diffraction pattern are shown in Figure 7c and d, respectively. The optical diffraction pattern is formed by illuminating the original electron micrograph with parallel laser light and photographing the resulting Frauenhoffer diffraction. The pattern is analogous to the X-ray diffraction pattern that would be obtained if the crystal were large enough to use for single-crystal X-ray studies.

Ribosome P422 Crystals

Three-dimensional crystals which can be considered to result from a dextro sheet positioned above a levo sheet, so that the fourfold axes through the centers of overlapping tetramers coincide, can also be isolated from hypothermically treated chicken embryos (see Figures 8 and 9a). Furthermore, optical diffraction patterns of them (Figure 9c) show reflections to a Fourier resolution of nearly 60 Å.

Figure 8 shows a series of electron micrographs of one P422 crystal taken at different tilt angles on an axis which is approximately horizontal and lies at an angle of 13° with respect to the basic crystal repeat vector on the *a* axis. This number of views is sufficient to adequately utilize for the

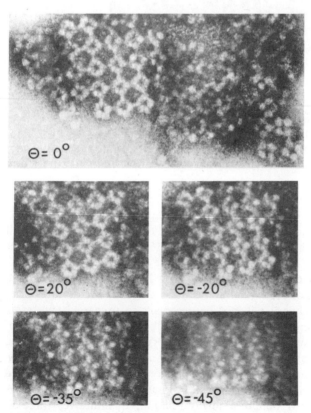

Figure 8   A tilt series of a P422 crystal of chicken embryo ribosomes.

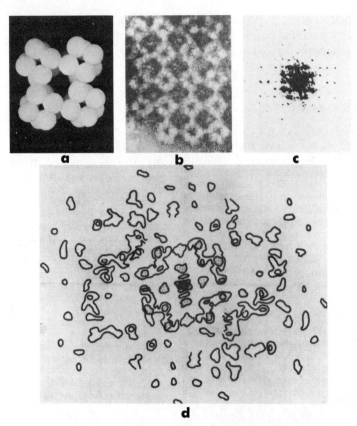

Figure 9   (*a*) A model showing the packing of ribosomes in the P422 ribosome crystal; (*b*) a region of a P422 crystal; (*c*) the optical diffraction pattern of *b;* (*d*) the intensity distribution of the computer-calculated Fourier transform calculated from *b.*

three-dimensional reconstruction the full resolution provided by the crystalline arrangement, although views perpendicular to the $\theta = 0$ direction will be associated with a higher noise level (Lake 1972a; Crowther, DeRosier and Klug 1970).

Because electrons can be focused by a lens, whereas X rays cannot, electron micrographs of the crystal also contain all of the necessary Fourier phase information. Thus it will not be necessary to prepare heavy atom derivatives of these crystals. Determining the phases from electron micrographs requires measuring their densities at regular points on a grid and calculating from this array a Fourier transform on a computer (for a review, see Lake 1972b). The intensity distribution from the computer-calculated Fourier transform of the crystal is shown in Figure 9d. By this process, we are now attempting to obtain a low resolution, quantitative, three-dimensional reconstruction of the structure of the chicken ribosome.

References

Bell, E., T. Humphreys, H. S. Slayter and C. E. Hall. 1965. Configuration of inactive and active polysomes of the developing down feather. *Science* **148**:1739.

Byers, B. 1966. Ribosome crystallization induced in chick embryo tissue by hypothermia. *J. Cell Biol.* **30**:Cl.

———. 1967. Structure and function of ribosome crystals in hypothermic chick embryo cells. *J. Mol. Biol.* **26**:155.

———. 1971. Chick embryo ribosome crystals: Analysis of bonding and functional activity *in vitro*. *Proc. Nat. Acad. Sci.* **68**:440.

Carey, N. H. 1970. Ribosomal aggregates in chick embryo tissues after exposure to low temperatures. *FEBS Letters* **6**:128.

Carey, N. H. and G. S. Read. 1971. The arrangement of ribosomes in ribosome tetramers from hypothermic chick embryos. *Biochem. J.* **121**:511.

Crowther, R. A., D. J. DeRosier and A. Klug 1970. The reconstruction of a three-dimensional structure from projections and its application to electron microscopy. *Proc. Roy. Soc.* (London) **A317**:319.

Hill, W. E., J. D. Thompson and J. W. Anderegg. 1969. X-ray scattering study of ribosomes from *Escherichia coli*. *J. Mol. Biol.* **44**:89.

Huxley, H. E. and G. Zubay. 1960. Electron microscope observations on the structure of microsomal particles from *Escherichia coli*. *J. Mol. Biol.* **2**:10.

Lake, J. A. 1972a. Reconstruction of three-dimensional structures from sectioned helices by deconvolution of partial data. *J. Mol. Biol.* **66**:255.

———. 1972b. Biological studies. In *Optical transforms* (ed. H. S. Lipson), p. 153. Academic Press, N.Y.

——— 1974. Ribosome structure: Three dimensional distribution of proteins S14 and S4. *Proc. 1974 Squaw Valley Conf. on Assembly Mechanisms*. Alan R. Liss, N.Y. (in press).

Lake, J. A. and H. S. Slayter. 1970. Three-dimensional structure of the chromatoid body of *Entamoeba invadens*. *Nature* **227**:1032.

———. 1972. Three-dimensional structure of the chromatoid body helix of *Entamoeba invadens*. *J. Mol. Biol.* **66**:271.

Lubin, M. 1968. Observations on the shape of the 50S ribosomal subunit. *Proc. Nat. Acad. Sci.* **61**:1454.

Mizushima, S. and M. Nomura. 1970. Assembly mapping of 30S ribosomal proteins from *E. coli*. *Nature* **226**:1214.

Morimoto, R., G. Blobel and D. D. Sabatini. 1972a. Ribosome crystallization in chicken embryos. I. Isolation, characterization and *in vitro* activity of ribosome tetramers. *J. Cell Biol.* **52**:338.

———. 1972b. Ribosome crystallization in chicken embryos. II. Conditions for the formation of ribosome tetramers *in vivo*. *J. Cell Biol.* **52**:355.

Nonomura, Y., G. Blobel and D. D. Sabatini. 1971. Structure of liver ribosomes studied by negative staining. *J. Mol. Biol.* **60**:303.

Sabatini, D. D., Y. Nonomura, T. Morimoto and G. Blobel. 1972. Structural studies on rat liver and chicken embryo ribosomes. *FEBS Symp.* **23**:147.

Chemical Approaches to the Analysis of Ribosome Architecture

Gary R. Craven, Brian Rigby and Li-Ming Changchien
The Laboratory of Molecular Biology
and the Department of Genetics
University of Wisconsin
Madison, Wisconsin 53706

INTRODUCTION

The present state of knowledge of ribosome structure bears a clear resemblance to the problem of enzyme structures as it developed prior to the impact of X-ray crystallography. During that period investigators utilized virtually any techniques they could imagine to uncover the fine details of protein architecture. One of the most widely applied methods to obtain significant information about protein structure has been the utilization of chemicals capable of selectively derivatizing polypeptide side chains. Over the years a substantial number of these reagents have been studied for their ability to react more or less selectively with amino acid functional groups. These reagents are used to determine which amino acid side groups are involved in the function of the enzyme and also to differentiate between so-called "buried" and "exposed" side chains. These data, combined with primary sequence information, have yielded considerable general knowledge of the three-dimensional structure of numerous proteins.

Application of Protein Modifying Reagents to the Problem of Ribosome Structure

A number of workers have attempted to extend the use of chemical reagents to determine some general features of ribosome structure (Acharya and Moore 1973; Craven and Gupta 1970; Ginzburg, Miskin and Azmir 1973; Huang and Cantor 1972; Hsiung and Cantor 1973; Kahan and Kaltschmidt 1972; Michalski, Sells and Morrison 1973; Miller and Sypherd 1973; Visentin, Yaguchi and Kaplan 1973). The initial effort with ribosomes has been to search for some gross reflections of protein organization within the ribosome architecture. Thus a fair number of different protein reagents have been screened for their reactivity with the bacterial ribosome. Two general conclusions can be inferred from these experiments. First, the individual proteins often show differential reactivity in the intact ribosome. That is, when the ribosome is treated with a specific reagent, different proteins incorporate the reagent to varying degrees. This is most extreme in the case of the 50S ribosome, in which a significant number of proteins have been found to be totally unreactive to all reagents so far examined (Hsiung and Cantor 1973; Michalski et al. 1973). The variation in protein reactivity has been interpreted to reflect general aspects of the ribosome's surface topography. It should be noted, however, that this interpretation has been challenged by several workers, at least for the 30S ribosome (Spitnik-Elson and Breiman 1971; Kurland 1972; Garrett and Wittmann 1973).

The second conclusion derived from these early studies is that the ribosomal proteins are in general more reactive to any given chemical reagent when exposed to that reagent in the "free state" (detached from the ribosome structure) than when included as members of the native particle. This is most probably due to the complex set of specific protein–protein and protein–RNA associations which compose the ribosome's structural network. Thus releasing the proteins from these specific interactions opens up many new sites for chemical modifications.

Identification of Specific Protein Influences on the Chemical Reactivity of Other Proteins in the 30S Ribosome

At least in theory, it should be possible to identify the individual protein–protein and protein–RNA interactions that operate to alter the pattern of chemical reactivity of each protein. For example, if a given protein has five sites reactive to a given chemical in the free state and none in the complete particle, then in principle, the macromolecular associations occurring during assembly which are responsible for this change in reactivity can be determined.

We have attempted to explore these alterations in the patterns of chemical reactivity as a possible means of detecting protein–protein interactions. It should be noted that we have also used this approach to examine protein–RNA interactions, as has W. Möller (personal communication). The approach rests on our ability to specify and catalog a set of functional groups for each protein which is reactive in the free state and then to construct in vitro a variety of subparticles, using the reconstitution technique devised by Traub and Nomura (1969), to determine the precise extent of influence all the other proteins exert to alter the catalog of reactivity. This is better understood by referring to Figure 1, which schematically illustrates the philosophy of this approach. In this example, it is our goal to determine the influence, if any, of protein #5 on the surface reactivity of proteins 1, 2, 3 and 4 when in complex with RNA. Thus we first react the four protein complexes with a radioactive reagent, isolate and purify the derivatized proteins, and then catalogue the pattern of modification by performing a peptide fingerprint analysis. We then carry out the same type of experiment with a new protein–RNA complex containing the original four proteins, numbered 1 through 4, and an additional fifth pro-

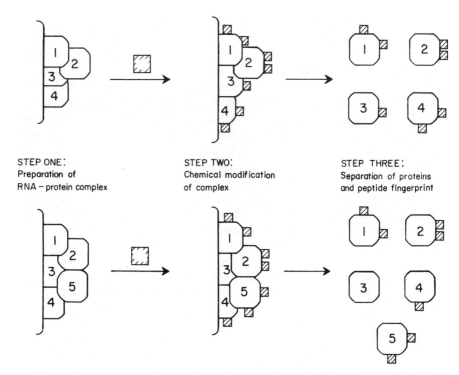

STEP ONE:
Preparation of
RNA – protein complex

STEP TWO:
Chemical modification
of complex

STEP THREE:
Separation of proteins
and peptide fingerprint

Figure 1 Schematic illustration of our method to determine protein influences on the chemical reactivity of other proteins. The line represents RNA, the numbered boxes represent individual proteins, and the shaded box refers to the chemical used for protein derivatization.

tein known from the assembly sequence of Mizushima and Nomura (1970) to be capable of specific binding to the complex. A comparison of the surface reactivity of proteins 1 through 4 is then made between the two particles. In the example of Figure 1, protein #5 directly covers a site of chemical modification on proteins 3 and 4, but has no influence on the surface reactivity of proteins 1 and 2. Thus the essential observation from the model experiment of Figure 1 would be a dramatic alteration in the peptide fingerprint pattern of proteins 3 and 4 without any substantial change in the profiles of 1 and 2. It should be noted that the cartoon of this model experiment includes a prejudice that the alteration in proteins 3 and 4 is a consequence of a site-specific interaction with protein 5 and not the result of any indirect involvement. This point will be discussed in more depth in a subsequent section.

Iodination as a Probe of Protein Topography

The success of our approach is entirely dependent on the efficacy of the reagent used in defining the surface topography of the ribosomal proteins.

We therefore initially surveyed a wide variety of chemicals and found that the iodination reaction, as described by Covelli and Wolff (1966), is especially convenient and appears to discriminate extremely well between the free state topography and the native ribosome topography of the individual proteins. This difference in reactivity to iodine, dependent on the state of association, is best illustrated by the experiments depicted in Figure 2. The figure is a comparison of two gel electrophoresis profiles. One (Figure 2B) is of the ribosomal proteins extracted from the iodinated native ribosome. This is contrasted with the pattern obtained when the 30S proteins are iodinated in the same buffer in the free state without 16S RNA (Figure 2A). All the proteins are extensively derivatized when reacted with iodine in free solution. However, when the 30S particle is hit, a number of proteins show little or no incorporation and the relative modification of those proteins open to reaction is significantly altered.

We were very encouraged by the possibility that iodine incorporation could be used as an effective probe of protein topography and proceeded to develop a method of cataloging the individual sites of iodination for each protein. We accomplished this goal by separation of the radioactive peptides produced by trypsin digestion, using Dowex 50 ion-exchange chromatography (Craven et al. 1969). The radioactivity profile of the iodinated peptides is in general highly reproducible and effectively reflects the surface reactivity of the individual proteins. Figure 3 shows four such fingerprints of iodinated proteins. The patterns for the two proteins shown (proteins S1 and S7) were obtained from protein iodinated in the free state and from protein purified from iodinated 30S particles. Examination of the gel pattern in Figure 2 reveals that these two proteins are much more extensively modified by iodine when members of an intact ribosome than when they are free in solution. However, the peptide patterns of these two proteins show that even in these instances, the intact ribosome structure protects much of the protein from reaction with iodine. We have completed this type of analysis for all the 30S proteins and have summarized our estimates of the number of sites available for reaction with iodine for each protein as it exists in the free state and as it exists within the three-dimensional restrictions of the native ribosome. The data indicate that with our procedure iodine derivatization can readily detect the surface topography of all the individual 30S proteins. In addition, the iodination patterns of all the proteins are significantly influenced by the macromolecular associations within the ribosome. In total, approximately 75% of the sites iodinated in the free state become buried or inaccessible when present in the 30S particle. It should be noted that this estimate is in sharp contrast to experiments recently published by Miller and Sypherd (1973). These authors iodinated 30S ribosomes and a free protein extract with ICl and found only a 1% increase in the number of tyrosines hit when the proteins were iodinated free in solution. With the lactoperiodase technique of iodination, they

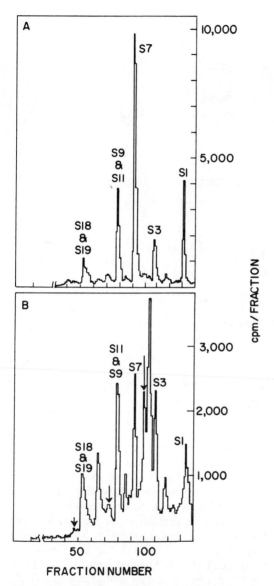

Figure 2 Polyacrylamide gel electrophoresis patterns of 30S ribosomal proteins. Proteins were iodinated with ^{125}I for 5 min at pH 7.8 and 28°C in a buffer solution containing 0.2 M Tris, 0.33 M KCl, 0.02 M MgCl$_2$. Iodination was carried out with a ratio of I$_3^-$ (KI + I$_2$) to ribosomes (or ribosome equivalents in the case of free proteins) of 25. Gels were sliced with a Gilson gel fractionator into 150 fractions. Radioactivity of the fractions was estimated by liquid scintillation.

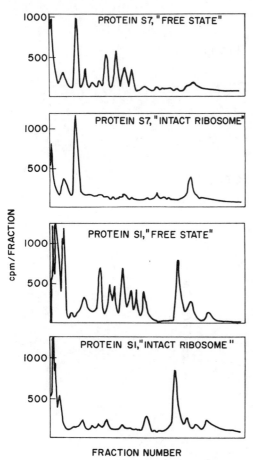

Figure 3   Dowex-50 ion-exchange chromatography of trypsin digests obtained from iodinated ribosomal proteins. Proteins were iodinated and separated as described in Figure 2. The aliquots of the gel fractions were counted and the regions corresponding to the various proteins were pooled and dialyzed against water. After dialysis the dialysis bag was opened, sliced into small pieces, and combined with the contents. The mixture of protein, water, acrylamide and dialysis bag was adjusted to 0.2 M NH_4HCO_3 and treated with 0.5 mg trypsin (CalBiochem, essentially chymotrypsin free) for 5hr at 37°C. The digest was centrifuged and the supernatant lyophilized. The product was applied to a Dowex-50 column equilibrated with pyridinium acetate and eluted according to Craven et al. (1969). The elutant was collected in fractions, which were lyophilized prior to estimation of radioactivity by liquid scintillation.

found approximately a 30% increase in tyrosine modification, yet this still is not comparable to the 75% increase in accessible sites we observe. There are a number of valid methodological differences which could explain this discrepancy. One major difference appears to be the degree of incorporation

Miller and Sypherd (1973) obtain with 30S ribosomes. Their gel patterns of protein extracted from iodinated 30S particles are strikingly different from ours as presented in Figure 2. They observe very high incorporation into all proteins, with only one protein peak showing a significant increase when the protein is iodinated in the free state. We have repeated our iodination of 30S ribosomes using the buffer conditions employed by Miller and Sypherd (1973) and find a dramatic increase in the incorporation of all proteins (as seen on gels) when compared to the buffer conditions we routinely use. We propose that use of the high ionic strength which we have employed is essential to maintain a compact ribosome structure that will yield the maximum detection of topological effects due to macromolecular interactions.

Determination of the Surface Topography of Reconstituted Subparticles

Satisfied that the iodination reaction could successfully distinguish the surface topography of the individual proteins dependent on their neighboring macromolecular associations, we proceeded to construct a number of complexes with RNA and isolated proteins, using the standard reconstruction procedure of Traub and Nomura (1969). The subparticles are separated from free protein by centrifugation through a layer of sucrose and then iodinated under our standard conditions. The iodinated proteins are separated and purified by preparative polyacrylamide gel electrophoresis and then digested with trypsin. The resultant preparation of iodinated peptides is fractionated on Dowex-50 and the radioactive profiles so obtained are taken as fingerprints of the surface reactivities of each protein. These fingerprints are compared for subparticles of different protein constitution.

Figure 4 is a selected example of the type of results this methodology yields. This figure shows reproductions of four chromatographic profiles of iodinated pepetides obtained from protein S13. In our hands, this protein forms a specific complex with 16S RNA (consistent with the observations of Mizushima and Nomura 1970). Comparison of the peptide pattern of the protein isolated from this complex with that of the same protein from a complex involving proteins S4, S8, S20, S15, S19, S7 and S13 shows a clear loss of incorporation in at least two major sites on protein S13. However, the addition of proteins S9 and S5 to the subparticles has no major effect on the surface reactivity of protein S13. Finally, when protein S10 is included in the complex, at least three major radioactive peptides disappear, indicating a substantial influence of this protein on the topography of protein S13.

Summary of Proposed Protein–Protein Associations

Table 1 summarizes most of the complexes for which we have completed the type of analysis illustrated by the experiments with protein S13. In

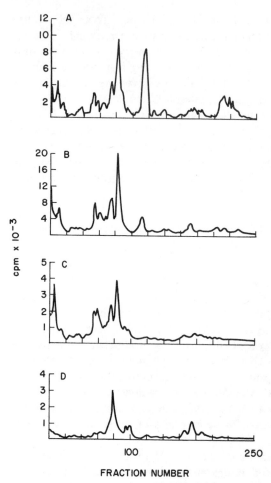

Figure 4 Ion-exchange chromatographic separation of peptides derived from protein S13. Protein S13 was incorporated as a component of various complexes. The complexes were iodinated and purified as described in Figure 3. (*A*) 16S RNA; (*B*) RNA and proteins S4, S8, S20, S15, S19, S13; (*C*) RNA and proteins S4, S8, S20, S15, S19, S13, S9, S5; (*D*) RNA and proteins S4, S8, S20, S15, S19, S13, S9, S5, S10.

general, the results of these experiments fall into one of two categories. Either the addition of a new protein to a complex of proteins and RNA has little or no effect on the iodination pattern of a given component protein, or the iodination pattern is observed to undergo a dramatic alteration. Thus of approximately 82 potential effects we might have seen in the experiments summarized by Table 1, only 15 are sufficiently dramatic to allow us to securely suggest specific changes in the surface topography of the proteins in question.

Table 1   Summary of complexes investigated

Proteins in complex	No. potential interactions examined	No. positive effects
S4, S20	1	0
S4, S15	1	1
S8, S20	1	1
S8, S15	1	1
S4, S8, S20	3	1
S4, S8, S20, S16 or S17	3	1
S4, S8, S20, S15, S7	7	1
S4, S8, S20, S15, S7, S19	5	1
S4, S8, S20, S15, S7, S19, S13	6	2
S4, S8, S20, S15, S7, S19, S13, S9	7	1
S4, S8, S20, S15, S7, S19, S13, S9, S5	8	0
S4, S8, S20, S15, S7, S19, S13, S9, S5, S10	9	3
S4, S8, S20, S15, S7, S19, S13, S9, S5, S10, S3	10	2
S7, S19, S14, S9	3	0
S7, S19, S14, S9, S10	4	0
S4, S8, S20, S15, S7, S19, S13, S6	7	0
S4, S8, S20, S15, S7, S19, S12	6	0
Total	82	15

It is impossible within the space limitations of this review to document all 15 "specific" protein–protein influences we have observed. However, we can at least summarize the identity of these influences (Figure 5). Note that from the experiments presented in Figure 4, we propose that one or more of the "early" proteins (S4, S8, S20, S7, S15, S19) has a pronounced effect on protein S13's surface topography (we have not as yet constructed the appropriate complexes to more precisely identify the protein(s) responsible). In addition, the dramatic effect of protein S10 on the iodination pattern of protein S13 is included in the summary as a specific protein influence. On the other hand, neither protein S9 nor protein S5 had any major effect on the reactivity of protein S13, and we therefore cannot propose any positive influence of these proteins. It is important to observe that we cannot rule out interactions among these proteins that may affect their surface topography in a way not detectable by iodine reactivity.

DISCUSSION

Throughout this presentation we have described the effect of a protein on the surface topography of another protein as an "influence." With particles as complex as the ribosome and its assembly intermediates, it is possible to interpret our experiments in a number of different ways. For example, the observed loss of specific iodination sites due to the presence of a new protein in the complex could be a consequence of a conformational change in the RNA induced by the protein, which in turn results in a new RNA–protein interaction with the second protein. Another possible explanation of our experiments could be that the protein in question undergoes a conformational change which is induced indirectly, either by the RNA chain or by another protein already present in the complex. However, the simplest and least contrived explanation is that the new protein in the complex becomes a tightly bound member of the complex, at least in part, due to specific protein–protein interactions, and that these direct protein associations cause topographical changes reflected in their surface chemical reactivity which are totally analogous to other protein systems composed of subunits. Unfortunately, there are no experiments at this time which allow us to convincingly argue that the alterations we observe in the surface reactivity of given proteins are due to direct protein–protein interaction, as opposed to an indirect alteration due to reshuffling of the component macromolecules. However, we do feel that indirect effects are not likely to produce the striking effects we have observed. In fact, we have often found minor influences on surface reactivity (i.e., incomplete loss or gain of reactive sites) which we feel can best be explained as due to indirect effects such as protein–protein–protein or protein–RNA–protein interactions. Thus

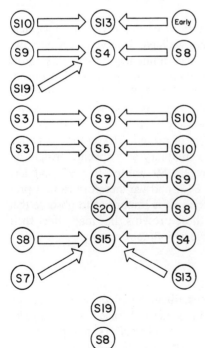

Figure 5 Summary of major protein–protein "influences." Large arrows indicate the effect of one protein on the surface topography of another protein.

we believe it is likely that most of the protein–protein effects we summarize in Figure 5 are a consequence of complex formation by direct intermolecular contacts among the component proteins. We are presently conducting experiments designed to test this hypothesis.

There are at present four other approaches to ribosome topography that have yielded significant information, two of which should be discussed here. The use of cross-linking reagents has begun to reveal the identity of some near neighbor proteins in the 30S ribosome (for example, see Lutter et al. 1972). One of these has been determined by Lutter (1973) to be composed of the proteins S7–S9. Referring to Figure 6, we have found an influence of protein S9 on S7. The other near neighbor relationships identified by these and other workers (with the exception of S5–S8) have involved proteins we have not as yet investigated. At least in this one case, we can conclude that two proteins which we propose to have a specific interaction are probably physically near one another in the ribosome.

Mizushima and Nomura (1970) have conducted an investigation which may have produced information about protein–protein relationships in the 30S ribosome. They have shown that a complex set of interdependent relationships is involved in the in vitro assembly of the 30S subunit. It has been suggested that when one protein requires the presence of another protein before it can become bound to the reconstitution complex, specific protein–protein interactions between the two proteins are responsible. Thus it is of

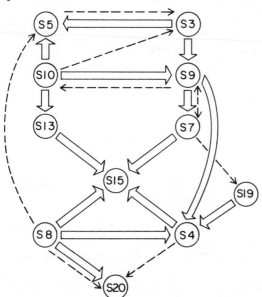

Figure 6   Two-dimensional map of our proposed protein–protein interactions (large arrows) and a comparison with the major interdependent assembly relationships (dotted arrows) described by Mizushima and Nomura (1970).

some interest to compare the protein–protein relationships we are suggesting with those observed to operate during in vitro assembly. Figure 6 is a redrawing essentially as a two-dimensional map of the information summarized in Figure 5. The large arrows represent our proposed interactions and the dashed arrows are taken from the data of Mizushima and Nomura (1970) and represent the major interdependent relationships that occur in the reconstitution reaction. Of the eight major assembly relationships involved with these proteins, four also show up as relationships producing alterations in surface topography. In addition, the two-dimensional summary of these data makes it clear that, assuming these relationships represent the relative positions of the proteins in the ribosome, there are no inconsistencies between the two kinds of experiments. This gives us added confidence that we are in fact detecting specific protein–protein interactions in the ribosome.

References

Acharya, A. S. and P. B. Moore. 1973. Reaction of ribosomal sulfhydryl groups with 5,5′-dithiobis(2-nitrobenzoic acid). *J. Mol. Biol.* **76**:207.

Covelli, I. and J. Wolff. 1966. Iodination of the normal and buried tyrosine residues of lysozyme. I. Chromatographic analysis. *Biochemistry* **5**:860.

Craven, G. R. and V. Gupta. 1970. Three dimensional organization of the 30S ribosomal proteins from *Escherichia coli*. I. Preliminary classification of the proteins. *Proc. Nat. Acad. Sci.* **67**:1329.

Craven, G. R., P. Voynow, S. J. S. Hardy and C. G. Kurland. 1969. The ribosomal proteins of *Escherichia coli*. II. Chemical and physical characterization of the 30S ribosomal proteins. *Biochemistry* **8**:2906.

Garrett, R. A. and H. G. Wittmann. 1973. Structure and function of the ribosome. *Endeavor* **32**:8.

Ginzburg, I., R. Miskin and A. Zamir. 1973. N-Ethyl maleimide as a probe for the study of functional sites and conformations of 30S ribosomal subunits. *J. Mol. Biol.* **79**:481.

Hsiung, N. and C. R. Cantor. 1973. Reaction of celite-bound fluorescein isothiocyanate with the 50S *E. coli* ribosomal subunit. *Arch. Biochem. Biophys.* **157**:125.

Huang, K. H. and C. R. Cantor. 1972. Surface topography of the 30S *Escherichia coli* ribosomal subunit: Reactivity towards fluorescein isothiocynate. *J. Mol. Biol.* **67**:265.

Kahan, L. and E. Kaltschmidt. 1972. Glutaraldehyde reactivity of the proteins of *Escherichia coli* ribosomes. *Biochemistry* **11**:2691.

Kurland, C. G. 1972. Structure and function of the bacterial ribosome. *Ann. Rev. Biochem.* **41**:377.

Lutter, L. C. 1973. Structure studies of the ribosome of *Escherichia coli*. Ph.D. thesis, University of Wisconsin, Madison.

Lutter, L. C., H. Zeichhardt, C. G. Kurland and G. Stöffler. 1972. Ribosomal protein neighborhoods. I. S18 and S21 as well as S5 and S8 are neighbors. *Mol. Gen. Genet.* **119**:357.

Michalski, C. J., B. H. Sells and M. Morrison. 1973. Molecular morphology of ribosomes, localization of ribosomal proteins in 50-S subunits. *Eur. J. Biochem.* **33**:481.

Miller, R. V. and P. S. Sypherd. 1973. Topography of the *Escherichia coli* 30S ribosome revealed by the modification of ribosome proteins. *J. Mol. Biol.* **78**:539.

Mizushima, S. and M. Nomura. 1970. Assembly mapping of 30S ribosomal proteins from *E. coli. Nature* **226**:1214.

Spitnik-Elson, P. and A. Breiman. 1971. The effect of trypsin on 30S and 50S ribosomal subunits of *Escherichia coli. Biochim. Biophys. Acta* **254**:457.

Traub, P. and M. Nomura. 1969. Structure and function of *Escherichia coli* ribosome. VI. Mechanism of assembly of 30S ribosomes studied *in vitro. J. Mol. Biol.* **40**:391.

Visentin, L. P., M. Yaguchi and H. Kaplan. 1973. Structure of the ribosome: Exposure of ribosomal proteins determined by competitive labelling. *Can. J. Biochem.* **51**:1487.

Affinity Labeling Techniques for Examining Functional Sites of Ribosomes

Charles R. Cantor, Maria Pellegrini and Helen Oen
Departments of Chemistry and Biological Sciences
Columbia University
New York, New York 10027

INTRODUCTION

No other technique can provide the detailed picture of macromolecular functional sites than the one potentially available from X-ray diffraction studies. However, a review of past studies on enzymes reveals that active site-directed chemical reagents have been particularly effective for identifying amino acid residues that are important for function. It seems natural to extend these successful techniques to attempt to identify ribosomal components which are located in various functional sites. Here attention will be restricted specifically to affinity labels. These are chemically reactive or activatable analogs of substrates, inhibitors or effectors. If properly designed, they will bind to a functional site in a manner that closely mimics that of the natural ligand. Once bound, the affinity label is capable of reacting covalently with residues of the binding site. Subsequent analysis of the covalent products permits identification of components of this site. The affinity labeling technique is well established. For a summary of the principles and recent applications, see the excellent reviews by Singer (1967), Shaw (1970) and Knowles (1972). In this brief report, we shall concentrate on some of the particular problems encountered when affinity labels are applied to the ribosome. A survey of successful results to date and possible future directions will be given.

Design of Affinity Labels

Two variables must be examined in constructing a useful affinity label: the choice of the reactive moiety and the site of its attachment to a substrate or ligand. With any affinity experiment, the goal is to optimize reaction at the functionally significant active site and minimize all other reactions of the label. If the chemically reactive moiety is too nonspecific or too labile, it may react very rapidly either with solvent or with random side chains on the proteins of the sample. In either case, a poor yield of specific active site modification will result. If the chemically reactive moiety is too specific or gentle, it may find itself incapable of reacting with any of the residues near its binding site. In this case, there will be no affinity labeling. Clearly the best choice lies in between these two extremes. Previous studies with enzymes have often used alkylating agents. Some well known examples are tosyl-phenylalanyl chloromethyl ketone (for chymotrypsin), tosyl-lysyl chloromethyl ketone and p-amidinophenacyl bromide (for trypsin). α-Bromocarbonyl or α-iodocarbonyl substrates have been the most widely used reactive moieties. They have seen success on a variety of enzymes (Shaw 1970). These compounds have moderate stability in water and are capable of reacting with a number of amino acid side chains including, cysteine, histidine, lysine and methionine (Means and Feeney 1971). Epoxides, diazo compounds, fluorophosphates and aromatic halides are others that have proved useful.

A second class of reactive moieties are photoactivatable species. These include arylazides and diazoesters and ketones which, upon irradiation with near ultraviolet light, yield nitrenes and carbenes, respectively (Knowles 1972). These photo products are extremely reactive, and it would be hard to imagine an active site that lacked a potential reactive side chain. On the other hand, this great reactivity means that frequently all that is observed is reaction with the solvent. Photoaffinity labels have the special potential advantage that their precursors are not protein reactive. The label can be prebound to an active site prior to photolysis. This minimizes the chance of nonspecific side reactions unless the photoexcited species lives long enough to leave the active site while still excited. Appropriate control experiments must be done to examine this possibility of "pseudo affinity labeling" (Ruoho et al. 1973). The main disadvantage of photoaffinity labels is the generally poor overall yield of reaction with any sites on the macromolecule.

A summary of many of the affinity labels which have been prepared for ribosome studies is given in Table 1. All of these labels have shown indications of covalent reaction with ribosomes. The pattern of reactivity and its possible biological significance have, in a few cases, been analyzed in considerable detail. The compounds shown in Table 1 fall into four classes: mRNA or GTP analogs, peptidyl-tRNA analogs and antibiotic analogs. In the last two categories both photoaffinity and normal reactive analogs have been studied; bromoacetyl and iodoacetyl moieties predominate.

Evidence That Affinity Labeling Has Occurred

An affinity label is prepared, incubated with ribosomes, and covalent attachment is observed. Now the problem is to show whether a specific affinity modification has occurred or whether what has happened is simply nonspecific or nonfunctional chemical modification. There are many tests, and a putative affinity label must pass at least some of them. The ribosome contains a multiplicity of potentially reactive groups. For example, most of the affinity labels used to date are principally sulfhydryl reagents, and the 70S particle contains many reactive sulfhdryls (Moore 1971; Acharya and Moore 1973). Therefore, there is an especially compelling need for an exhaustive set of tests to prove that specific affinity labeling has occurred.

Specificity of Reaction Pattern.   Since affinity labeling is presumed to take place at a binding site, one would expect a limited number of possible products. With ribosomes, a number of labels such as BrAcPhe-tRNA (Oen et al. 1973), *p*-nitrophenyl carbamyl Phe-tRNA (Czernilofsky et al. 1974), iodoamphenicol (Pongs, Bald and Erdmann 1973a) and bromamphenicol (Sonenberg, Wilchek and Zamir 1973) clearly pass this test. Most of the reactivity appears to occur with one, two or at most three different ribosomal proteins.

Alteration of Reaction Pattern.   The covalent products of the affinity label should be compared with any products from the reactive moiety alone. Differences are expected if the affinity label is selectively modifying an active site. For example, Pongs et al. (1973a) observed that iodoamphenicol incubated with 70S ribosomes reacts principally with protein L16, and to a lesser extent with S3. In contrast, iodoacetamide reacted with S18 and S21 under identical conditions. Sonenberg et al. (1973) showed that incubation of bromamphenicol with 50S particles selectively modified L2 and L27. BrAcPhe-methyl ester instead reacts much less specifically, and the major products include L1, L2, L10, L11 and L17. We have observed that the BrAcPhe-tRNA reaction with 70S or 50S particles is also much more specific than small molecular weight bromoacetyl compounds (Pellegrini et al., unpublished results; Pellegrini 1973).

Enhancement of Covalent Reaction.   The reaction of an affinity label can often show increased efficiency. If it is bound near an appropriate reactive moiety, a proximity rate enhancement should be seen. Under equivalent experimental conditions, total covalent reaction of BrAcPhe-tRNA with the 70S ribosome is a hundred times that of BrAcPhe or BrAcPhe-methyl ester (Pellegrini et al., unpublished results). This occurred under conditions in which ribosomes were present in great excess over reactive compounds. It is more difficult to see enhancements when the affinity label is present in excess. With a tenfold reagent excess, Pongs et al. (1973a) found comparable total reactivities of iodoacetamide and iodoamphenicol with the 70S particle.

Functional Binding of the Affinity Label.   If conditions can be found where the affinity label reacts slowly, it should be able to act as a normal substrate or at least as an inhibitor. BrAcPhe-tRNA binding to the 70S ribosome is poly(U) dependent (Pellegrini, Oen and Cantor 1972) and the bound analog is puromycin reactive (Pellegrini et al., unpublished results). Therefore it can bind correctly to the P site. The case is even more convincing with BrAcMet-tRNA$_f^{met}$, where specific messenger dependence, probably

Table 1 Structures of affinity labels reacted with *E. coli* ribosomes

Parent compound	Analog	Reference
I Chloramphenicol $R_1 = NO_2; R_2 = CHCl_2$	Ia: $R_1 = NO_2; R_2 = CH_2Br$ Ib: $R_1 = NO_2; R_2 = CH_2I$ Ic: $R_1 = N_2^+; R_2 = CHCl_2$ *Id: $R_1 = N_3$; $R_2 = CHCl_2$	Sonenberg et al. (1973) Pongs et al. (1973a) Leick, Votrin, Cooperman and Rich (unpublished results)
II Peptidyl-tRNA 	IIa: $R_1 = CH_2\phi$; $R_2 = CH_2Br$ *IIb: $R_1 = CH_2\phi$ $R_2 =$ IIc: $R_1 = CH_2\phi$ $R_2 =$ IId: $R_1 = CH_2CH_2-S-CH_3$ $R_2 = CH_2Br$ IIe(n): $R_1 = CH_2\phi$; $R_2 = (Gly)_n -\overset{O}{\overset{\|}{C}}-CH_2Br$ IIf: $R_1 = CH_2\phi$ $R_2 =$	Pellegrini et al. (1972) Hsiung, Reines & Cantor (un- published results) Czernilofsky & Kuechler (1972) Sopori, Pellegrini, Lengyel & Cantor (unpublished results) Eilat, Pellegrini, Oen, Lapidot & Cantor (unpublished results) Buchkareva et al. (1971)

III Puromycin

Me_2A

$R_1 = H; R_2 = H; R_3 = O-CH_3$

IIIa: $R_1 = H; R_2 = \overset{O}{\underset{}{C}}-CH_2I$ Pongs et al. (1973b)

$R_3 = O-CH_3$

IIIb: $R_1 =$ Harris et al. (1974)

$R_2 = H; R_3 = H$

*IIIc: $R_1 = R_2 = C-C-C-O-C_2H_5$ Leick, Votrin, Cooperman and Rich (unpublished results)

$R_3 = O-CH_3$

IV Streptomycin

IVa: 4-iodoacetamino-benzhydrazone of streptomycin   Pongs & Erdmann (unpublished results)

V UpUpUpU

$UpUpUp(nhr)^5U$ Lührmann et al. (1973)

$(nhr)^5 = C_5-H \longrightarrow C_5-NH-\overset{O}{\underset{}{C}}-CH_2I$

VI pppG

*VIa: N_3- $-O-pppG$ Möller & Maassen (unpublished results)

* Photoaffinity label

also requiring the presence of the correct initiation factors, can be demonstrated (Sopori et al., unpublished results). Here the bound analog is still capable of directing the correct factor-dependent binding of an aminoacyl-tRNA to the A site. The covalent reaction of another analog, p-nitrophenylcarbamyl Phe-tRNA, was markedly inhibited by the presence of puromycin (Czernilofsky et al. 1974). This also indicates P site binding. Bromamphenicol and iodoamphenicol both compete for binding with chloramphenicol. In addition, both affinity analogs inhibit peptidyl transferase activity (Pongs et al. 1973a; Sonenberg et al. 1973).

Binding: Prerequisite to Covalent Reaction.   Just because an affinity label can bind to the correct site, it does not necessarily follow that covalent reaction takes place from this site. The easiest way to check this is competition. For example, Harris et al. (1974) showed that puromycin and chloramphenicol could partially inhibit the covalent reaction of a bromoacetyl analog of puromycin (compound IIIb in Table 1) with 70S particles. We found that covalent reaction of BrAcPhe-tRNA at low magnesium concentration can be inhibited by excess unaminoacylated tRNA in a manner which exactly parallels inhibition of binding (Pellegrini et al., unpublished results). In addition, the covalent reaction itself is stimulated tenfold by poly(U). The inhibition of covalent reaction of p-nitrophenylcarbamyl Phe-tRNA by puromycin cited above is clear evidence that binding of this label is also necessary before appreciable covalent reaction can occur.

With BrAcPhe-tRNA, experimental conditions have been manipulated to cause binding either to the P site or, by competition with excess unaminoacylated tRNA at 30 mM Mg^{++}, to the A site (Eilat et al., unpublished results). The pattern of covalent reaction from each site is quite different: L2 and L27 are the major products from the P site, and L16 is reacted only from the A site.

Covalent Reaction before Work-up.   An excess of small molecule affinity label can easily be removed from a reaction mixture. However, tRNA analogs bind so tightly that ribosomal subunits must be separated to free any unreacted tRNA. One must be sure that the lengthy procedures needed for this do not promote unwanted nonspecific covalent reaction. With BrAcPhe-tRNA this was accomplished by incubating ribosomes with puromycin following the affinity labeling reaction (Pellegrini et al., unpublished results). The antibiotic releases most of the affinity label which has failed to react covalently. The effect of this puromycin treatment on the pattern of covalent reaction products is illuminating. The main products, L2 and L27, are unaltered. A number of minor products are eliminated. Hence they arise from nonspecific reaction during subunit separation. Czernilofsky and Kuechler (1972) used puromycin treatment in the same fashion to reduce side reaction of p-nitrophenylcarbamyl-Phe-tRNA.

Function of Covalently Attached Analog. In several cases it has been possible to obtain suggestive evidence that subsequent to covalent reaction, the ribosome-bound affinity analog can still participate in peptide bond formation. This is perhaps the strongest possible proof that labeling of a functional site has occurred. For example, 70S ribosomes preincubated with BrAcPhe-tRNA can still form a dipeptide with Phe of Phe-tRNA-Phe or puromycin (Oen et al. 1973). Similarly, a preincubated 70S ribosome BrAcMet-tRNA$_f^{met}$ complex will form a dipeptide with Ala-tRNA^{ala} (Sopori et al., unpublished results). The dipeptides in each case are found covalently attached to the same ribosomal proteins known to react with the affinity label alone. Another example occurs with the puromycin affinity analog IIIb developed by Harris et al. (1974). The analog was incubated with ribosomes for 26 hr at 26°C. Then the potential peptide donor UpAp-CpCpA-acetyl-Leu was added and incubation continued for 2 hr at 0°C. Peptide bond formation occurred with the covalently attached puromycin as detected by covalent incorporation of [³H]leucine. In this case, the affinity label reacts with the RNA of the 50S particle.

Other affinity analogs, such as chloramphenicol derivatives, would not be expected to participate in peptide bond formation. Instead, one can attempt to show that covalent attachment of these ribosomes causes irreversible inhibition of some function. Sonenberg et al. (1973) state that bromamphenicol reaction irreversibly inhibits the peptidyl transferase and causes loss of the ability to bind chloramphenicol and erythromycin. Iodoamphenicol also causes irreversible inhibition of polyphenylalanine synthesis (Pongs et al. 1973b). Similar irreversible effects on activity could be observed with another chloramphenicol derivative (Leick et al., unpublished results), and derivatives of puromycin (Pongs et al. 1973b; Leick et al., unpublished results) and streptomycin (Pongs and Erdmann, unpublished results). In most cases, the yield of covalent reaction with affinity labels has been far less than quantitative. This means that complete irreversible inhibition is not observed. Therefore in practice, one must compare the extent of irreversible inhibition of the reactive analog with that of the natural antibiotic.

Identification of Reaction Products

Experimental Procedures. All of the reaction products of ribosomal affinity labels that have been uniquely identified thus far have been proteins. For an unequivocal identification, electrophoresis of the proteins on two-dimensional acrylamide gels is required. The procedure of Kaltschmidt and Wittmann (1970), used by almost all workers, offers superb resolution. Protein spots are stained lightly and then cut out. Gels are dissolved and counted to find traces of radioactive affinity label. Frequently, sample oxidation is used to increase counting efficiency. However the two-dimensional gel procedure is accompanied by several potential difficulties. An appreciable amount of protein material remains in the first dimension disc gel. Therefore one cannot be sure that all radioactive affinity products have been identified. Also there is a possibility that by selective retention of some species in the first dimension, a distorted view of the quantitative distribution of products will be obtained. In some cases covalent derivatization of proteins by affinity labels may result in a marked alteration of electrophoretic mobility (Czernilofsky et al. 1974). It would be impractical, in routine experiments, to cut up an entire two-dimensional gel to search for products of unknown mobility. Therefore a risk exists that such products will be missed.

To circumvent these problems, we have found it convenient to analyze affinity label experiments by using one- and two-dimensional gel electrophoresis on aliquots of the same sample. The entire one-dimensional gel is cut up and counted. Usually there is little radioactivity found in the sample gel. Thus quantitative ratios of various products are obtained. The two-dimensional gel is used for identification of products. If these correspond quantitatively to the one-dimensional results, one can be fairly confident that all products have been detected and properly analyzed.

Most affinity labeling results thus far have yielded alkylated ribosomal proteins with little mobility change. The conspicuous exception is those different derivatives of L27. The products from reaction with BrAcPhe-tRNA (Pellegrini et al., unpublished results) and p-nitrophenylcarbamyl Phe-tRNA (Czernilofsky et al. 1974) all show such a marked change in mobility that identification from the positions on the gel is probable but not definitive. To confirm their identification of L27, Czernilofsky et al. made use of antibodies against specific ribosomal proteins. To circumvent the aggregation properties of ribosomal proteins, conditions of antibody excess were used and the resulting complexes analyzed by centrifugation. Anti-L27 specifically complexes a substantial part of the reaction products of the affinity-labeled ribosomes. This unequivocally identifies the product with mobility near L27 as authentic L27. It is clear that immunological techniques will be of considerable help in future affinity label experiments.

Results of Affinity Label Experiments.   A summary of some of the proteins identified in functional sites by affinity labeling is given in Table 2. A number of general conclusions can be drawn from these results. Most affinity reactions have occurred in low yield; however, controls to test the specificity of the reactions have usually been quite convincing. These two factors might appear to be contradictory, but they need not be. Few measurements exist on the extent of destruction of the label by reaction with solvent. The products of solvent reaction can still bind to ribosomes. With tRNA affinity labels, this will surely inhibit further covalent reaction since

Table 2 Results of affinity labeling studies

		Major reaction products											
Label*	Conditions	L2	L5	L11	L13–15	L16	L18	L24	L26, 27	L32, 33	5S or 23S RNA	S3 or S4	S14, 18, 19
Ia		++							++				
Ib						++							
IIa	direct reaction	++							++				
	via dipeptide	++				++			++				+
	in A site	+				++			+				
IIb	via dipeptide	++		++			+						
IIc		++			++	+							
IId	direct reaction	++							++				
	via dipeptide	++							++				
IIe(3)		+							++				+
IIe(6)									++				+
IIe(9)									++	++			+
IIe(12)								++	++	+			+
IIe(16)								++					+
IIIb											++		
IV							+						
VI			+	+			+					+	++

* Refer to Table 1 for chemical structure.

References are given in the text or in Table 1 for all results except IIa in A site, which is the unpublished work of Eilat, Pellegrini, Oen, de Groot, Lapidot and Cantor.

tRNA once bound is not easily displaced. A second unknown factor is the heterogeneity of ribosomes in both composition and activity.

When all of the potential difficulties of affinity labeling are considered, the number of successes to date and the extent of agreement between different laboratories are really impressive. All short peptidyl-tRNA derivatives bound to the P site react with L27 and L2. Other proteins labeled with these derivatives are L11 and L13 to L16. Bromamphenicol also reacts with proteins L2 and L27. BrAcPhe-tRNA at the A site and iodoamphenicol show a high specificity for protein L16. These findings are in good agreement with biochemical studies on the role of L16 (Nierhaus and Nierhaus 1973) and the importance of L11 (Nierhaus and Montejo 1973), L2 (Traugh and Traut 1972) and its *B. stearothermophilus* equivalent Fahnstock, Erdmann and Nomura 1973) for ribosome function. When all of these findings are integrated, it appears certain that protein L16 is at the A site. L2 and L27 appear to be P site proteins. L11 is nearby but its exact location is still unsure.

We have recently tried to extend the information available from affinity label studies by preparing a set of bromoacetyl-peptidyl-tRNAs of increasing chain length (Eilat et al., unpublished results). When the reaction pattern of these compounds is compared, a kind of chemical map results. As shown in Table 2, long peptidyl-tRNAs cease to react with L2 and react somewhat less with L27. Instead proteins L33 and L24 become major products. By tracing the quantitative reaction pattern as a function of length, these four proteins can be ordered relative to the 3' end of the peptidyl-tRNA. From closest to furthest this is L2, L27, L33 and L24.

One of the surprising features of all of these affinity label studies is the large number of proteins that have now been identified near the A and P sites. Many reagents react with more than one protein. This must be due to some flexibility in the binding sites or the mode of binding of the 3' end of tRNA. However, it also implies that many proteins are packed rather close together to form the peptidyl transferase center of the 50S particle. This is in good agreement with the recent neutron scattering results of Moore, Engleman and Schoenborn (1974), indicating that there is considerable clustering of the proteins of the 50S particle. A schematic illustration of the implications of affinity labels and other experiments on the 50S particle is shown in Figure 1. There is insufficient space in this brief review to attempt to justify many of the details of the figure; nor would such an attempt be sensible at the present time. Our knowledge of the structure of the 50S particle is still fragmentary. Figure 1 should be viewed as an abstract drawing with many details surely distorted. However, it shows the kind of information available from affinity labels, with all their advantages and considerable limitations.

The near future should see many more successful affinity label studies on ribosomes. The labels found for the 50S particle can now be used to

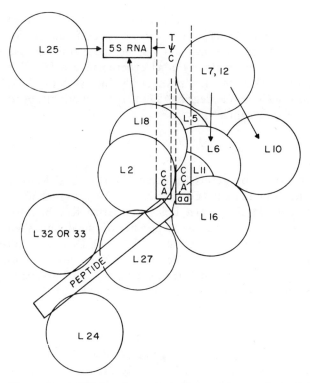

Figure 1   A highly schematic and approximate view of the peptidyl transferase center of the *E. coli* 50S ribosomal subunit. Proteins L2, L11, L16, L27 and L33 have all been localized near the center by more than one affinity label. L18 is modified by one tRNA analog and one GTP analog. The exact placement of any of these proteins, or others included because of known interactions with rRNAs, is totally hypothetical, except for L2, L27, L32–33 and L24 where some spatial information (as described in the text) is available.

study distortion of the A and P sites by various antibiotics. Conceivably they could also provide information about conformational changes during protein synthesis. Little work has reached a fruitful stage so far on the 30S particle. Messenger RNA analogs and various antibiotics seem ripe for further study. Many other derivatives of tRNA that could afford insight into the ribosomal binding site of the anticodon and other important regions seem feasible to make. Photoaffinity labels would seem to offer particularly attractive possibilities. These could be attached to positions on the 16S or 5S rRNA and reassembled bulk into ribosomes. A specific cross-link between the 16S rRNA and a 30S protein has already been reported (Kenner 1973). GTP and GDP analogs will be especially useful for locating proteins of functional significance (Maassen and Möller, unpublished results).

Affinity techniques will be useful for the specific introduction of cross-links or fluorescent labels. This is but a brief list of many likely future directions. Many initial attempts at affinity labeling will not be successful because there are so many places where an affinity label experiment can fail. For example, the first analog tried in our laboratory, chloroacetyl-Phe-tRNA, proved unreactive with ribosomes (Pellegrini 1973). It is important not to let initial failures be too discouraging.

Acknowledgments

Our work has been supported by grants from the USPHS (GM 19843) and the National Science Foundation (GB 34522X). We are grateful to Drs. Olaf Pongs, Theodore Möller, Vagn Leick and Ernst Kuechler for communicating results prior to publication. A number of our unpublished results cited here are the results of a collaborative effort with Dan Eilat and Yehuda Lapidot of the Hebrew University, Jerusalem. Others involve collaborative work with Mohan Sopori and Peter Lengyel of Yale University. We thank Dr. Patricia Cole for her assistance in constructing Figure 1.

References

Acharya, A. S. and P. B. Moore. 1973. Reaction of ribosomal sulfhydryl groups with 5,5'-dithiobis (2-nitrobenzoic acid). *J. Mol. Biol.* **76**:207.

Buchkareva, E. S., U. G. Budker, A. S. Girshovich, D. G. Knorre and N. M. Teplova. 1971. An approach to specific labeling of ribosome in the region of peptidyl transferase center using *N*-acylaminoacyl-tRNA with an active alkylating grouping. *FEBS Letters* **19**:121.

Czernilofsky, A. P. and E. Kuechler. 1972. Affinity label for the tRNA binding site on the *Escherichia coli* ribosome. *Biochim. Biophys. Acta* **272**:667.

Czernilofsky, A. P., E. E. Collatz, G. Stöffler and E. Kuechler. 1974. Proteins at the tRNA binding sites of *Escherichia coli* ribosomes. *Proc. Nat. Acad. Sci.* **71**:230.

Fahnestock, S., V. Erdmann and M. Nomura. 1973. Reconstitution of 50S ribosomal subunits from protein-free ribonucleic acid. *Biochemistry* **12**:220.

Harris, R. J., P. Greenwall and R. H. Symons. 1973. Affinity labeling of ribosomal peptidyl transferase by a puromycin analogue. *Biochem. Biophys. Res. Comm.* **55**:117.

Kaltschmidt, E. and H. G. Wittmann. 1970. Ribosomal proteins. VII. Two-dimensional polyacrylamide gel electrophoresis for fingerprinting of ribosomal proteins. *Anal. Biochem.* **36**:401.

Kenner, R. A. 1973. A protein-nucleic acid crosslink in 30S ribosomes. *Biochem. Biophys. Res. Comm.* **51**:932.

Knowles, J. R. 1972. Photogenerated reagents for biological receptor-site labeling. *Accts. Chem. Res.* **5**:155.

Lührmann, R., V. Schwarz and H. G. Gassen. 1973. Covalent binding of uridine-oligonucleotides to 70S *E. coli* ribosomes. *FEBS Letters* **32**:55.

Means, G. E. and R. E. Feeney. 1971. *Chemical modification of proteins.* Holden Day, San Francisco.

Moore, P. B. 1971. Reaction of *N*-ethyl maleimide with the ribosomes of *Escherichia coli. J. Mol. Biol.* **60**:169.

Moore, P. B., D. M. Engleman and B. P. Schoenborn. 1974. Asymmetry in the 50S ribosomal subunit of *E. coli. Proc. Nat. Acad. Sci.* **71**:172.

Nierhaus, D. and K. H. Nierhaus. 1973. Identification of the chloramphenicol-binding protein in *Escherichia coli* ribosomes by partial reconstitution. *Proc. Nat. Acad. Sci.* **70**:2224.

Nierhaus, K. and V. Montejo. 1973. A protein involved in the peptidyl transferase activity of *Escherichia coli* ribosomes. *Proc. Nat. Acad. Sci.* **70**:1931.

Oen, H., M. Pellegrini, E. Eilat and C. R. Cantor. 1973. Identification of 50S proteins at the peptidyl-tRNA binding site of *Escherichia coli* ribosomes. *Proc. Nat. Acad. Sci.* **70**:2799.

Pellegrini, M. 1973. Affinity label studies with bacterial ribosomes. *Ph.D. thesis,* Columbia University, New York.

Pellegrini, M., H. Oen and C. R. Cantor. 1972. Covalent attachment of a peptidyl-transfer RNA analog to the 50S subunit of *Escherichia coli* ribosomes. *Proc. Nat. Acad. Sci.* **69**:837.

Pongs, O., R. Bald and V. A. Erdmann. 1973a. Identification of chloramphenicol-binding protein in *Escherichia coli* ribosomes by affinity labeling. *Proc. Nat. Acad. Sci.* **70**:2229.

Pongs, O., R. Bald, T. Wagner and V. A. Erdmann. 1973b. Irreversible binding of N-iodoacetylpuromycin to *E. coli* ribosomes. *FEBS Letters* **35**:137.

Ruoho, A. E., H. Kiefer, P. E. Roeder and S. J. Singer. 1973. The mechanism of photoaffinity labeling. *Proc. Nat. Acad. Sci.* **70**:2567.

Shaw, E. 1970. Chemical modification by active-site-directed reagents. In *The enzymes,* 3rd ed. (ed. P. D. Boyer) vol. 1, pp. 91–146. Academic Press, New York.

Singer, S. J. 1967. Covalent labeling of active sites. *Advan. Protein Chem.* **22**:1.

Sonenberg, N., M. Wilchek and A. Zamir. 1973. Mapping of *Escherichia coli* ribosomal components involved in peptidyl transferase activity. *Proc. Nat. Acad. Sci.* **70**:1423.

Traugh, J. A. and R. R. Traut. 1972. Phosphorylation of ribosomal proteins of *Escherichia coli* by protein kinase from rabbit skeletal muscle. *Biochemistry* **11**:2503.

Fluorescence Spectroscopic Approaches to the Study of Three-dimensional Structure of Ribosomes

Charles R. Cantor, Kuei-huang Huang and Robert Fairclough
Departments of Chemistry and Biological Sciences
Columbia University
New York, New York 10027

INTRODUCTION

Fluorescence spectroscopy has a number of attractive features for the study of a system as large and complex as the ribosome. Because fluorescent probe molecules are individually or selectively introduced, one can concentrate attention on a small portion of the whole structure. The great sensitivity of the technique means that one can conveniently work with 10^{-6}–10^{-7} molar solutions in sample volumes as small as 0.2 ml. At least three different kinds of information can be obtained from fluorescence studies:

(1) Environmentally sensitive probes such as ethidium bromide (EB) and anilinonaphthalene sulfonate (ANS) are especially useful for monitoring conformational changes.
(2) Polarization or decay of fluorescence anisotropy allow motions and flexibility to be examined.
(3) Energy transfer between sets of attached fluorescent dyes enable the measurement of distances between fixed points.

All of these techniques are well established and have been developed and tested on simpler biochemical systems (for recent reviews, see Steiner and Weinryb 1971; Cantor and Tao 1971; Cantor and Timasheff 1974; Stryer 1968; Brand and Gohlke 1972). A few additional complications arise when a system the size of a ribosome is involved. Here we shall examine these difficulties, survey some of the results of previous fluorescence studies on ribosomes, and attempt to chart the course of possible future studies.

Choice of Fluorescent Probes

All fluorescence studies demand the presence of fluorescent moieties. Their selection and specific introduction is frequently the major obstacle to meaningful experiments. The natural fluorescence of ribosomes arises principally from aromatic amino acid residues and is very weak. Daya-Grosjean et al. (1972), however, have been able to use it to study some structural aspects of mutants of protein S4. Fluorescence measurements were performed on isolated single proteins. Mutants with altered binding affinity to the 16S rRNA were found also to have altered aromatic fluorescence. This suggests that tryptophan is important for the native binding conformation of S4 or may be directly involved in RNA binding. Direct experiments to look at S4 natural fluorescence in intact ribosomes would not be possible. The large number of aromatic amino acids in the whole particle means that the fluorescence of one site or one protein will be submerged in a static background from others. To see finer detail, it becomes necessary to introduce specific probes which can be selectively excited at longer wavelengths than protein absorption. In practice, the probe should have $\lambda_{max} \gtrsim 350$ nm. A summary of some of the probes and approaches that have been used to date is given in Table 1. Two general strategies exist. The easiest is to bind noncovalent probes. In most cases there are many binding sites. Although this leads to the same disadvantages cited above for natural protein fluorescence, some experiments can circumvent them. For example, EB is known to bind fairly selectively to double-strand nucleic acids. Bollen et al. (1970) used this fact to examine the nature of protein binding sites on 16S rRNA. They

Table 1   Flourescent labels for ribosomes

Accomplished*	Possible†
Ethidium: nonspecific RNA binding[1]	16S rRNA: 5′ end or random covalent
Tetracycline: binding[2]	5S rRNA: 3′ end, 5′ end or random
PBA-streptomycin: 30S[3]	covalent
G-factor: dansyl[4]	mRNA: 3′ end, 5′ end or specific
Natural tryptophan fluorescence[5]	internal
MBD: 50S binding[6]	aatRNA: on side chain
30S proteins: direct 70S reaction[7]	peptidyl tRNA: at N terminus
Single 30S proteins: reassembly[8]	tRNA internal sites: some odd base
50S proteins: direct reaction[9]	derivatives; e.g., Y base or Y base
Single 50S protein: L7, L12 readdition[10]	replacement
16S rRNA: 3′ end[11]	

*Varying degrees of success or usefulness.
† Varying degrees of difficulty.
[1] Bollen et al. 1970.
[2] White and Cantor 1971; Fey et al. 1973.
[3] Hall, Davis and Cantor, unpublished results.
[4] Gennis, Bodley and Cantor, unpublished results.
[5] Daya-Grosjean et al. 1972.
[6] Kenner and Aboderin 1971.
[7] Huang and Cantor 1972.
[8] Huang, Fairclough and Cantor, unpublished results.
[9] Hsiung and Cantor 1973.
[10] Litman, Lee and Cantor, unpublished results.
[11] Hsiung, Schreiber and Cantor, unpublished results.

observed that when proteins bound to rRNA pretreated with EB, about one-third of the bound EB was displaced. This could be monitored easily by a decrease in fluorescence. By examining the effect of individual 30S ribosomal proteins, a class which apparently recognized double-strand regions was identified.

It would be difficult to use a nonspecific probe like EB to look at one specific region in an intact particle. Occasionally noncovalent probes can be found with much higher specificity. For example, 7-(p-methoxybenzyl-amino)-4-nitrobenz-2-oxa-1,3-diazole (MBD) has a specific binding site on the 50S subunit (Kenner and Aboderin 1971). Nothing is known about the location or any functional significance, but the site must be nonpolar because of the high quantum yield of the bound MBD. The chances that a noncovalent probe will bind near a functional site can be increased by attaching it to a molecule with known function. For example, pyrene butyric acid-streptomycin binds to the streptomycin site on the 30S particle (Hall, Davis and Cantor, unpublished results). Tetracycline is a naturally fluorescent antibiotic with only a few strong binding sites (Fey et al. 1973), and by substituting Mn^{++} for Mg^{++}, it has been shown that tetracycline actually binds to ribosomes as a divalent ion complex (White and Cantor 1971). Ribosome-bound tetracycline was strongly fluorescent in the presence of Mg^{++} but almost totally quenched in Mn^{++}.

The second strategy is covalent attachment of fluorescent probes. This is more difficult experimentally, but one has more control in directing specific attachment. With ribosomal proteins the following approach has been reasonably successful in our laboratory. An individual, purified protein is reacted covalently with a celite-bound fluorescent dye (Gennis, Gennis and Cantor 1972; Huang and Cantor 1972). The covalently labeled protein is separated from celite by low-speed centrifugation and then eluted from two successive G25 Sephadex columns to remove any noncovalently attached fluorescent dye. In typical experiments, anywhere from 0.5 to more than 4 dyes are covalently attached per protein molecule. For 30S proteins the resulting product is reassembled back into 30S particles by adding 16S rRNA and the remainder of normal unmodified proteins, using the normal reconstitution conditions of Traub and Nomura (1969). This procedure has been successfully carried out with most of the 30S proteins and a variety of fluorescent reagents, including 7-chloro-4-nitrobenzoxadiazole (NBD-Cl), dansyl chloride, fluorescein isothiocyanate (FITC), rhodamine B isothiocyanate and N-iodoacetyl-N'-(5-sulfonic-1-naphthyl) ethylene diamine (1,5 I-AEDANS) (Hudson 1970). In the majority of cases tried, 30S particles result with appreciable biological activity. Work on 50S proteins has been more limited. I-AEDANS and FITC adducts of L7 and L12 have been prepared. These will successfully recombine with salt-ethanol-washed 50S particles with moderate recovery of G factor-dependent GTP binding activity (Litman, Lee and Cantor, unpublished results).

Other potential sites for specific covalent attachment of fluorescent dyes include the ends of ribosomal RNAs, several positions on tRNA (Yang 1973) or mRNAs, and the various initiation, elongation and termination factors. We have concentrated our attention on the 16S rRNA and tRNA. A large number of different fluorescent dyes can be attached conveniently to the 3' terminus of RNAs after periodate oxidation. Such derivatives of the 16S rRNA will reassemble to give functional 30S particles (Hsiung and Cantor, unpublished results). Similar tRNA derivatives will bind to 70S ribosomes; but with a molecule like tRNA, there is a real risk that addition of a dye will enhance nonspecific or at least nonfunctional binding. The ideal fluorescent derivative would appear to be a peptidyl analog still capable of peptide transfer. Although attempts to synthesize such molecules have not yet been successful, there would appear to be no fundamental obstacle.

Regardless of the mode of attachment of the fluorescent probe, one must also pay attention to its particular optical properties. Experimental sensitivity of normal static fluorescence measurements depends on ε (λ_{ex}) ϕ_f, the product of the extinction coefficient at the exciting wavelength and the fluorescence quantum yield. For commonly used probes, this product varies from around 40 for dansyl in a polar site to 40,000 for fluorescein. Studies on single proteins permit a sufficient range of concentrations to compensate for such variations in sensitivity. With ribosomes the high molecular weight and substantial Rayleigh scattering necessitate the use of probes with reasonably high sensitivity. Scattering can be overcome to some extent by choosing a probe with a large Stokes shift (roughly the difference between the wavelengths of maximum absorbance and maximum fluorescence). The scattered light will have the same wavelength as the exciting light, and the further this is from the emission wavelength, the less spectral contamination there will be. Direct measurements of fluorescence decay are useful in energy transfer or polarization measurements. For these it is usually necessary to choose a probe with a long singlet lifetime. NBD-Cl, EB, I-AEDANS and pyrene meet these criteria nicely. Pyrene is especially useful for polarization measurements on large systems, since its lifetime is so long, up to 200 nsec, that rather slow motions can be seen. There is one final important dye property: it must not be so photoreactive that a series of measurements results in marked destruction of the chromophore.

Use of Environmentally Sensitive Probes

Fluorescent groups like NBD, ANS, EB and dansyl show a marked sensitivity to their environment. In general, nonpolar or rigid media lead to high fluorescence quenching (Brand and Gohlke 1972). Such probes, covalently attached at particular ribosomal sites, provide an excellent monitor of the local environment. They are generally, but not necessarily, highly responsive to conformational changes.

We have used covalent NBD-Cl derivatives of individual purified 30S ribosomal proteins to monitor structural changes during reassembly. One would expect that as particles form, increased protein–protein and protein–nucleic acids contacts would occur. This would lead to rigid environments for attached dyes and therefore should show up as increases in fluorescence. Such an increase is just what we observe for most NBD proteins that have been examined. The changes are large, up to a factor of two in some cases. Thus one can directly monitor the reassembly kinetics by continuous recording from a sample at 38°C. A typical example is shown in Figure 1. For the four 30S proteins we have studied in detail, the kinetics of fluorescence enhancement are apparent first-order (neglecting a slow drift at long time). The rate constants fall in the range 0.13–0.22 min^{-1}. This is in excellent agreement with 0.15 min^{-1}, the value predicted for appearance of biological activity at 38°C from the rates and activation energies measured by Traub and Nomura (1969). These findings mean that most rigidification or protein–protein contacts seen by the fluorescence probes either accompany the RI to RI* transformation or follow it. The most likely in-

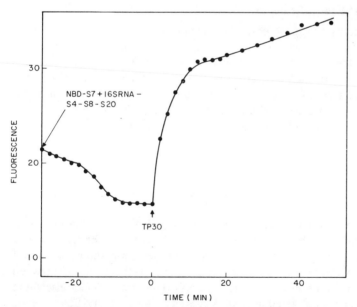

Figure 1 Direct monitoring of 30S ribosome reassembly using a covalent fluorescence probe. NBD–S7 was added to a preformed complex of S4, S8, S20–16S rRNA. Binding of NBD-S7 was accompanied by a decrease in fluorescence. Total 30S proteins were then added (at time zero). A rapid twofold fluorescence increase is observed. Throughout the experiment the sample was maintained at 38°C under normal optimal reassembly conditions (Traub and Nomura 1969).

terpretation is that this transformation involves the formation of a substantially condensed particle. This is in agreement with hydrodynamic data.

Environmentally sensitive probes can be used to search for specific protein–protein contacts. If a probe attached to a particular protein is enhanced by reassembly, one can ask which other proteins specifically induce this change. Our studies to date have centered on NBD-S5. The results are quite complex and indicate that a number of proteins, including S10, S21, S18, S6, S9, S14 and S19, may all be required for attainment of maximal fluorescence of NBD-S5. Probably not all of these act directly. Some may affect the correct placement of others which possibly form direct S5 contacts.

Chemical modification techniques and activity studies can yield much the same information as environmentally sensitive probes. The main advantage of the latter is that kinetics can be followed much more conveniently and, if necessary, on a much faster time scale. Since the measurement is nondestructive, the same particle can be used for further studies. For example, disassembly of the 30S particle or the rate of recombination with 50S could also be examined.

Singlet–singlet Energy Transfer

This is potentially the most powerful fluorescence technique for studying the three-dimensional structure of ribosomes. It is much more difficult to perform than the techniques mentioned previously. Two different fluorescent dyes, the donor and the acceptor, must be attached to specific ribosomal sites. If these are close enough, there is a probability, E, that if the donor is excited, instead of fluorescing it will transfer its energy to the acceptor. That moiety will then fluoresce. E, the energy transfer efficiency, can be related to the distance, R, between the two fluorescent dyes by the Förster theory for singlet energy transfer:

$$E = \frac{R_o^6}{R_o^6 + R^6}. \tag{1}$$

In this equation, R_o is the characteristic transfer distance. It can be calculated easily from the absorption spectrum of the acceptor, the fluorescence spectrum and quantum yield of the donor, and certain assumptions about the angular distribution of dyes and the nature of the medium between them. These assumptions have been discussed in great detail elsewhere (Dale and Eisinger 1974; Haugland, Yguerabide and Stryer 1969; Cantor and Timasheff 1974). In favorable cases R_o can be calculated to ± 20% or even better, so that distances can be measured to about that accuracy. The correctness of Equation 1 has been verified in a number of model systems (see, for example, Stryer and Haugland 1967). Studies on a number of different proteins and nucleic acids to date have been highly successful

(Wu and Stryer 1972; Bunting and Cathou 1973; Beardsley and Cantor 1970). Since experimentally realizable R_o's can be as large as 50 Å for some dye pairs, the method is capable of measuring distances as long as 80 Å.

For R_o to be large, the emission spectrum of the donor must overlap to a large extent the absorption spectrum of the acceptor. In addition, experimental measurements are simplest if donor and acceptor are found suitable for studies with ribosomes. In our experience to date, the most satisfactory dye pair is I-AEDANS (donor) and FITC (acceptor). This pair can have an R_o near 50 Å (Wu and Stryer 1972) and satisfies most of the other criteria mentioned above.

Energy transfer experiments with ribosomes will be more complex than in simpler systems. We shall consider a typical experiment to illustrate the problems that arise and how these can be dealt with. Two proteins are chosen. One is covalently conjugated with the donor, the other with the acceptor. Then three reassembled 30S particles are made. These are shown schematically in Figure 2. Consider, first, studies of the donor. Suppose both the sample with only donor and the sample with both donor and acceptor have exactly the same number of donors per particle. If no energy transfer occurs, these two samples will have exactly the same fluorescence. If there is transfer, the donor will be partially quenched. The ratio of donor fluorescence in the two samples will be

$$F^D_{D+A}/F^D_D = 1 - E. \qquad (2)$$

In practice, however, it is highly unlikely that two reassembled ribosome samples will contain exactly the same amount of donor-labeled protein.

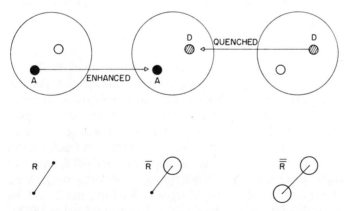

Figure 2 Schematics of a singlet–singlet energy transfer measurement. (*Top*) Samples that must be prepared and the optical differences expected between them. (*Bottom*) Distance measured, which can vary from a point-to-point estimate (R) to an average over two ensembles of distances ($\bar{R}$).

Therefore one must make an independent determination of the amount of donor and correct the ratio F_{D+A}^D/F_D^D. This cannot be done by fluorescence (because of possible energy transfer), nor will absorbance suffice (because most good donors in practice have small extinction coefficients). We have found it necessary to use radioactive I-AEDANS to determine concentrations. The problem of different donor stoichiometrics in two samples can be circumvented if fluorescence decay measurements are used. In this case, instead of Equation 2 one uses the analogous equation $\tau_{D+A}^D/\tau_D^D = 1 - E$. The lifetimes, τ, of the two samples are concentration independent. However if not all particles in the donor plus acceptor sample contain precisely one acceptor, the donor emission will show multiexponential decay kinetics. In favorable cases this can still be analyzed, but for particular sets of τ's the results can sometimes be quite unreliable.

Next consider studies of the acceptor. It is easy to show that if one excites at the donor λ_{max} and monitors acceptor fluorescence, the samples containing both donor and acceptor will have more intensity than the acceptor sample alone. In the simplest case the result is

$$F_{D+A}^A/F_A^A = 1 + E \frac{\varepsilon_D C_D}{\varepsilon_A C_A}. \qquad (3)$$

In Equation 3 $\varepsilon_D C_D/\varepsilon_A C_A$ is proportional to the ratio of the absorbance of donor and acceptor at the exciting wavelength. Once again it is hard to determine $\varepsilon_D C_D$ directly, but this can be calculated if radioactive donor is used. As before, if both samples in practice do not contain the same amount of acceptor, F_{D+A}^A/F_A^A must be corrected. Still more corrections are necessary if, as is frequently the case, some donor fluorescence occurring at the same wavelength is used to monitor the acceptor. These corrections make the actual equation employed far more complex than Equation 3, but in practice the procedure is still relatively straightforward.

If the reassembled ribosomes contained exactly one donor and one acceptor, measurement of E by Equations 2 and 3 could be used directly to compute a distance R in Equation 1. In practice, a situation this clean is rarely encountered. Certainly with ribosomes the stoichiometry of many of the proteins themselves is not unitary. Furthermore, dye conjugation procedures can lead to a complex mixture of derivatives containing no dye or 1, 2, 3, . . . dyes and on different sites. Without a laborious fractionation and analysis, all that one knows is the average number of donors per particle, μ_D, and the average number of acceptors, μ_A. It turns out that μ_D does not affect the results of calculating E from either Equation 2 or 3, so long as the correct value for $\varepsilon_D C_D$ is employed (Gennis and Cantor 1972). Therefore, if it is possible, the donor should be attached to the protein with the less certain stoichiometry. On the other hand, μ_A has a marked effect both on Equation 2 and 3 and on the proper value of R_o to be used in

Equation 1. Fortunately the effect on E is exactly the same for both equations. This effect can be accounted for and even turned to one's advantage. The details are complex and have been given elsewhere (Gennis and Cantor 1972). Here we shall just state the assumptions: Unless additional information is available, we assume that the stoichiometry of labeling is given by the Poisson distribution. If the donor or acceptor is an amine reagent, we assume it is on the surface of the protein but allow an equal probability for all surface locations. The protein is assumed to be a sphere. Fortunately the actual results derived are not particularly sensitive to the shape of the protein, or its size, or whether the dyes are at the surface or inside. They are not even that sensitive to placing all dyes at the surface or all at the center of the protein. Two variables are critical: the center-to-center distance between the donor and acceptor labeled proteins and μ_A.

Figure 3 illustrates the results expected for energy transfer between protein pairs in systems like ribosomes. By examining the figure and its caption, the reader should be convinced that even in the absence of any knowledge about the mechanism of labeling and the properties of the proteins, experimental measurements should provide a center-to-center distance with an accuracy of better than 25% of R_o. The more additional knowledge available, the more accurate one can be. Another feature of the method helps to increase accuracy. In practice what one wants to do is measure three protein–protein distances: A–B, B–C and A–C. Then by triangulation, the relative configuration of all three is obtained. From experiments on a four-protein model system, we have found that the final configuration is not all that sensitive to assumptions about R_o and the protein sizes, provided that the same donor-acceptor pair is used for all three measurements (Gennis and Cantor, unpublished results).

Our experimental results with ribosomes to date have been limited but encouraging. AEDANS–S4 and FITC–S20 are close enough to show transfer in the reassembled 30S particle. Preliminary analysis of the data suggests a distance between 40 and 60 Å, but more refinements in the experiments and calculations are needed to be certain. Energy transfer has also been seen between AEDANS–L12 and FITC–L7. Other preliminary experiments indicate that ribosome-bound pyrene butyric acid streptomycin is close enough to one of the 30S proteins, S1, S3, S4, S21, to show energy transfer (Hall, Davis and Cantor, unpublished results). More work will be needed to verify this. These experiments are not easy and the effects seen are small. However, the energy transfer technique does appear to be a realistic and useful strategy for obtaining information on the three-dimensional structure of the ribosome.

To gauge the future potential, let's compare singlet energy transfer with cross-linking, a technique which affords similar information. Cross-linking has two major advantages: (1) Per unit effort, far more distinct measurements can be made. In principle, a single sample could yield a number of

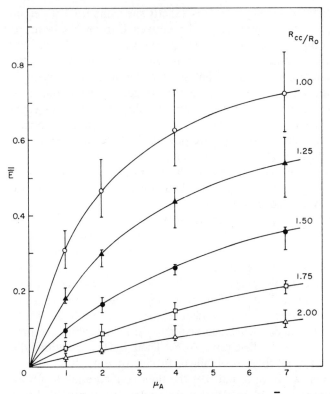

Figure 3 Predicted energy transfer efficiency, $\bar{\bar{E}}$, between two fluorescent-labeled spherical proteins. Each curve is for a specific center-to-center distance (R_{cc}) expressed in units of R_0, the characteristic transfer distance. Points shown are the results expected if proteins have a diameter 0.325 R_0. Vertical error bars show maximum range of predicted values for a wide variety of possible individual protein dimensions. μ_A is the mean number of acceptors located on one protein.

precise cross-links. (2) Very short distances can be measured more accurately by a cross-linker. Energy transfer has a number of real advantages once the difficulty of performing the experiments is accepted. (1) The results are actual distance measurements. (2) What is in between the donor and acceptor doesn't matter in any consequential way. Hence the distance between two proteins separated by part of the RNA or by another protein can still be measured. (3) Much longer distances can be measured. An average ribosomal protein, if a sphere, would have a diameter of around 32 Å. Energy transfer can be detected out to around 80 Å. Thus next nearest-neighbor close-packed proteins can surely be detected, or proteins separated by a few strands of RNA. (4) A negative energy transfer result can be meaningful, since it sets a lower limit on the distance between two proteins. In attempts to determine packing arrangements, this can often be as useful as knowing that two proteins are close. However it is clear that one must be skeptical of all negative results until a large body of positive results has been obtained. (5) With energy transfer, unlike cross-linking, there is less danger of covalent trapping of a functionally nonsignificant ribosome conformation.

The above discussion is in the context of protein-to-protein measurements. Both cross-linking and energy transfer can also be used to examine the location and configuration of functional sites. Dyes or bifunctional reagents can be attached to tRNAs, mRNAs or antibiotics. In practice, the relative ease of cross-linking and other chemical and enzymatic techniques suggests that these are the methods of choice to make a preliminary map of the connectivity of neighboring ribosomal proteins and sites. To refine this map, one could imagine an attempt to measure all protein–protein distances on the 30S particle by energy transfer. For the 30S particle alone, this would mean successfully reconstituting 210 different double-fluorescent-labeled ribosomes. The difficulty of this task is staggering, especially since no one dye will work for all proteins. More realistically, the role of energy transfer measurements will probably be to provide critical information about selected longer distances and regions of particular interest. This will be essential if some of the details of the three-dimensional organization of ribosomal proteins are to be revealed. Energy transfer should also be able to play a major role in examining conformational changes. It should be particularly interesting to follow distances between fluorescent-tagged L7, L12 and sites on the tRNA or other 50S ribosomal proteins. Also direct measurements of the distance between two bound tRNAs should be quite useful. These experiments seem reasonably practical now and could well afford a glimpse of any conformational changes involved in translocation.

One last question seems particularly amenable to energy transfer studies. Does the conformation of individual ribosomal proteins, tRNA and initiation and elongation factors alter upon attachment to the ribosome? Since high resolution structures are easier to obtain on small pieces, it is vital to know whether these findings will be directly applicable to ribosomes. This can be approached by specific internal distance measurements on a given protein or RNA both free and ribosome bound.

Acknowledgments

This work was supported by grants from the USPHS (GM-AI-19843) and the NSF (GB-34522X). We are very grateful to Bill Held and Masayasu Nomura for much helpful advice and direct assistance with reassembly techniques.

References

Beardsley, K. and C. R. Cantor. 1970. Studies of transfer RNA tertiary structure by singlet-singlet energy transfer. *Proc. Nat. Acad. Sci.* **65**:39.

Bollen, A., A. Herzog, A. Favre, J. Thibault and F. Gros. 1970. Fluorescence studies on the 30S ribosome assembly process. *FEBS Letters* **11**:49.

Brand, L. and J. R. Gohlke. 1972. Fluorescent probes for structure. *Annu. Rev. Biochem.* **41**:808.

Bunting, J. R. and R. E. Cathou. 1973. Energy transfer distance measurements in immunoglobulins. II. Localization of the hapten binding sites and the interheavy chain disulfide bond in rabbit antibody. *J. Mol. Biol.* **77**:223.

Cantor, C. R. and T. Tao. 1971. Application of fluorescence techniques to the study of nucleic acids. In *Procedures in nucleic acid research* (ed. G. L. Cantoni and D. R. Davies) vol. 2, pp. 31–93. Harper and Row, New York.

Cantor, C. R. and S. N. Timasheff. 1974. Optical spectroscopy of proteins. In *The proteins,* 3rd ed. (ed. H. Neurath and R. Hill). Academic Press, New York (in press).

Dale, R. E. and J. Eisinger. 1974. Polarized excitation energy transfer. In *Concepts in biochemical fluorescence* (ed. R. F. Chen and H. Edelhoch). Marcel Dekker, New York (in press).

Daya-Grosjean, L., R. A. Garrett, O. Pongs, G. Stöffler and H. G. Wittmann. 1972. Properties of the interaction of ribosomal proteins S4 and 16S RNA in *Escherichia coli* revertants from streptomycin dependence to independence. *Mol. Gen. Genet.* **119**:277.

Fey, G., M. Reiss and H. Kersten. 1973. Interaction of tetracyclines with ribosomal subunits from *Escherichia coli*. A fluorometric investigation. *Biochemistry* **12**:1160.

Gennis, L. S., R. B. Gennis and C. R. Cantor. 1972. Singlet energy transfer studies on associating protein systems. Distance measurements on trypsin, α-chymotrypsin and their protein inhibitors. *Biochemistry* **11**:2517.

Gennis, R. B. and C. R. Cantor. 1972. Use of nonspecific dye labeling for singlet energy transfer measurements in complex systems. A simple model. *Biochemistry* **11**:2509.

Haugland, R. P., J. Yguerabide and L. Stryer. 1969. Dependence of the kinetics of singlet-singlet energy transfer on spectral overlap. *Proc. Nat. Acad. Sci.* **63**:23.

Hsiung, N. and C. R. Cantor. 1973. Reaction of celite-bound fluorescein isothiocyanate with the 50S E. coli ribosomal subunit. *Arch. Biochem. Biophys.* **157**:132.

Huang, K.-h. and C. R. Cantor. 1972. Surface topography of the 30S *Escherichia coli* ribosomal subunit: Reactivity towards fluorescein isothiocyanate. *J. Mol. Biol.* **67**:265.

Hudson, E. N. 1970. Synthesis, characterization, and application of some covalent fluorescent probes for proteins. *Ph.D thesis,* University of Illinois, Urbana.

Kenner, R. A. and A. A. Aboderin. 1971. A new fluorescent probe for protein and nucleoprotein conformation. Binding of 7-(p-methoxybenzylamino)-4-

nitrobenzoxadiazole to bovine trypsinogen and bacterial ribosomes. *Biochemistry* **10**:4433.

Steiner, R. F. and I. Weinryb, ed. 1971. *Excited states of proteins and nucleic acids.* Plenum Press, New York.

Stryer, L. 1968. Fluorescence spectroscopy of proteins. *Science* **162**:526.

Stryer, L. and R. P. Haugland. 1967. Energy transfer: A spectroscopic ruler. *Proc. Nat. Acad. Sci.* **58**:719.

Traub, P. and M. Nomura. 1969. Structure and function of *Escherichia coli* ribosomes. VI. Mechanism of assembly of 30S ribosomes studies *in vitro*. *J. Mol. Biol.* **40**:391.

White, J. P. and C. R. Cantor. 1971. Role of magnesium in the binding of tetracycline to *Escherichia coli* ribosomes. *J. Mol. Biol.* **58**:397.

Wu, C. W. and L. Stryer. 1972. Proximity relationships in rhodopsin. *Proc. Nat. Acad. Sci.* **69**:1104.

Yang, C. H. 1973. Covalent attachment of fluorescent groups to tRNA and its application to structural studies of tRNA. *Ph.D. thesis,* Yale University, New Haven, Conn.

Neutron Scattering Studies of the *E. coli* Ribosome

P. B. Moore, D. M. Engelman
Department of Molecular Biophysics and Biochemistry
Yale University
New Haven, Connecticut 06520

B. P. Schoenborn
Department of Biology
Brookhaven National Laboratory
Upton, New York 11973

INTRODUCTION

It is obvious from the reports so far presented in this volume that increasing effort is being applied to understanding the three-dimensional structure of the ribosome. About a year ago it was suggested that the quaternary structure of the ribosome could be studied by the technique of low-angle scattering of thermal neutrons in solution (Engelman and Moore 1972). We would like to take this opportunity to report on the progress made over the last year towards bringing this proposal to fruition. In so doing, we will discuss some recent results on the global distribution of RNA and protein in the 50S ribosomal subunit of *E. coli*.

Background Information

To understand why neutron scattering is useful in this context, it is necessary to review some facts about how neutrons are generated and how they interact with matter (Schoenborn and Nunes 1972; Bacon 1962). At present, the only practical thermal neutron sources for the work under consideration are nuclear reactors. High energy neutrons are produced in abundance by fission in the reactor core. Some neutrons escape from the core and enter a surrounding blanket of material called the moderator, where they lose kinetic energy by collision events, rapidly reaching equilibrium with the moderator substance. These "thermalized" neutrons can be tapped from the reactor by the simple expedient of cutting a hole in the radiation shielding surrounding the reactor and providing a thin window in the moderator container through which the neutrons can pass.

The neutrons in the moderator region of the reactor have the Maxwellian distribution of energies characteristic of the moderator temperature. At 300°K, a usual operating temperature, the distribution has its peak at an energy corresponding to a DeBroglie wavelength of about 1.7 Å for neutrons and contains neutrons in the range 1–8 Å in experimentally useful amounts. Thus thermalized neutrons have wavelengths in the same range as the photons used in ordinary X-ray structural work. It is not surprising, therefore, that over the thirty years which have elapsed since the invention of the reactor, thermal neutrons have been used for structural work in much the same manner as X rays.

X rays mainly scatter from electrons. Accordingly, X-ray structural studies give information on the distribution of electron density in objects. Neutrons interact, instead, principally with atomic nuclei. The strength and nature of the interaction depends on nuclear parameters, such as spin and nuclear energy levels, and has no simple relationship to atomic mass or atomic number (see Marshall and Lovesey 1971). Table 1 gives the coherent scattering lengths of a number of biologically interesting atomic species. Scattering lengths measure the strength of the scattering interaction. In a similar table for X rays, the scattering lengths would be proportional to atomic number, which is not the case here. A striking fact obvious in Table 1 is that H has a negative scattering length. This means that neutrons scattered from H will be 180° out of phase compared to neutrons scattered from nuclei with the usual positive scattering lengths. In structural studies done with neutrons, H nuclei will therefore be represented as regions of negative density or "holes." The second fact of interest shown in Table 1 is that deuterium has a positive scattering length. In a word, these two isotopes are readily distinguished by neutrons. By way of comparison, X rays cannot detect such differences.

Biological macromolecules are rich in hydrogen, and therefore their average neutron scattering lengths per unit volume are sensitive to substitution of deuterium for hydrogen (see Table 2). In a low-angle scattering experiment, the parameter which determines the strength of the scattered

Table 1   Coherent neutron scattering lengths (b) for biologically interesting atomic species

Nucleus	$b \times 10^{13}$ cm
H	−3.74
D	+6.67
C	6.65
N	9.40
O	5.80
P	5.10

The scattering lengths given are taken from a compilation of C. G. Shull made at Massachusetts Institute of Technology, Cambridge, Massachusetts (1971) (personal communication).

Table 2   Average scattering length per unit volume (ρ) of some biologically important materials

Species	$\rho \times 10^{14} \text{ cm}/\text{Å}^3$
H protein	+3.14
D protein	+8.48
H RNA	+4.30
D RNA	+7.25
H_2O	−0.59
D_2O	+6.38

It is assumed that all exchangeable H in protein and RNA are fully exchanged with D. These figures are based on analysis of neutron scattering data obtained on H 50S subunits, D 50S subunits and H RNA polymerase, all derived from *E. coli*. It is assumed that approximately 20% of the hydrogen not bonded to carbon in the ribosome does not exchange with the solvent over a period of many days and that both protein and RNA behave in the same way in this respect. The RNA values are derived from the ribosome and RNA polymerase data by consideration of the relative content of RNA and protein in the ribosome (Moore, Engelman and Schoenborn 1974).

signal from a substance is its average scattering length per unit volume minus that of the medium in which it is immersed. Table 3 gives that difference for protein and RNA, fully H or fully D, immersed in D_2O. Deuterated RNA and protein are positive relative to D_2O; hydrogenated RNA and protein are negative.

E. coli will grow in media with any percentage of D up to 100%. By proper adjustment of growth conditions, *E. coli* can be grown so that its macromolecules include D to any predetermined extent. Thus the neutron scattering length per unit volume of these materials can be controlled to suit experimental needs, within the limits given in Table 3. For example, the scattering length of RNA and/or protein can be adjusted to equal that of D_2O. When this is done, the material so deuterated becomes undetectable by low-angle neutron scattering in D_2O medium. For all intents and purposes it is invisible. We call this condition "contrast matched." In the case of H-labeled protein or RNA, contrast matching can also be achieved by altering the medium. Since the scattering length of H_2O is very small, a judicious mixing of D_2O and H_2O can lead to a solvent whose scattering length is less than that of pure D_2O and equal to that of H-protein or H-RNA. The validity of both these approaches to contrast matching has been

Table 3   Average scattering length per
unit volume of protein and RNA
relative to D_2O

Species	$\Delta\rho \times 10^{14}$ cm/Å^3
H protein	-3.24
D protein	$+2.10$
H RNA	-2.08
D RNA	$+0.87$

ρ_{D_2O} has been subtracted from ρ_x to give
the values shown (see Table 2).

demonstrated experimentally (Moore, Engelman and Schoenborn, unpublished results).

The final point to make about neutron scattering lengths is that the differences produced by substituting D for H in macromolecules are large. In X-ray work it is customary to manipulate scattering lengths by adding high atomic number atoms to molecules in specific locations. By way of comparison, it might be pointed out that to produce a shift in X-ray scattering properties equivalent to the D–H difference we have been discussing for neutrons, you would have to label a protein with one uranium atom for every three amino acids. It would be naive to propose that substituting D for H in macromolecules or changing the D content of a solvent has no influence on the properties of macromolecular structures, but the viability of *E. coli* in 100% D medium and the activity of D-ribosomes in poly(U)-directed poly(Phe) incorporation in an H medium suggest that no drastic biochemical changes occur.

Outline of the Method for Obtaining Distances in Intrasubunit Macromolecular Complexes

The original suggestion for using neutron scattering as a tool for studying quaternary structure in ribosomes and other complex structures specifically exploited the size of the differences in scattering length generated by H/D substitution. Suppose one were to prepare contrast-matched 16S RNA and contrast-matched proteins S1, S2, . . . S21. Suppose fully D or fully H proteins S1, S2, . . . S21 were also prepared. Then by reconstitution it would be possible to make 30S particles which were all contrast-matched except for proteins X and Y inserted in all D or all H form. Incident neutrons would "see" two strongly scattering centers held in a constant spatial relationship by an invisible matrix. The scattering of a solution of such particles would resemble that of a diatomic gas, being dominated by a damped sinusoidal variation in intensity produced by interference between the neutrons scattered from the two labeled centers. The distance between two subunits can be obtained from the periodicity of the interference ripple to a good approximation and could be measured. It turns out that the experiment is feasible whether or not contrast matching is done on the rest of the particle, provided that the subunits of interest contrast with the remainder. A systematic series of pair separation measurements could provide enough intersubunit distances to allow one to locate them all in space by triangulation.

An analogous suggestion for studying molecular structure by solution scattering using X rays has been in the literature for some time (Kratky and Worthmann 1947). In this case, the parts of structures would be marked by heavy atom substitution, but otherwise the analysis would be identical. The possibility of applying the approach to the ribosome has been discussed recently by Hoppe (1972). The reason the X-ray version of this technique has not been used and appears infeasible in this case is that the signal due to the heavy atoms would be only 0.01–0.1% of the scattering produced by the rest of the ribosome (Hoppe 1972). In the neutron case the signal would be 1–5% of the total scatter without contrast matching and could be all of it with perfect contrast matching, an obvious advantage.

Theoretical Advances

The theoretical basis for this approach has been strengthened over the last year. It was originally suggested that one should measure four scattering curves to get each interprotein distance: those of (1) unlabeled subunits, (2) subunits with protein A labeled, (3) subunits with protein B labeled, and (4) subunits with both proteins labeled. To obtain the interference ripple needed for estimating the A–B distance, one would then subtract (curve 2 plus curve 3) from (curve 1 plus curve 4). As has been emphasized by Hoppe (1972, 1973), this approach works only if the particles in the samples do not interact with each other in a systematic way. Interaction effects, however, can be eliminated if one measures a mixture of samples 2 and 3 and subtracts their combined scattering profile from that of a mixture of samples 1 and 4. Operating in this way, the experiment will yield the correct result at any sample concentration provided that interactions are the same in both samples. The importance of this finding is that it may be necessary to use ribosome gels as samples for the interprotein spacing experiments.

As we pointed out earlier, the location of the nodes of the interference term measured in such an experiment will be influenced, albeit not too sensitively, by effects due to the combination of departures from spherical symmetry in the proteins in the ribosome and their relative orientation. It now appears that it may be possible to minimize the impact of shape problems on the final distances by proper processing of the data in each triangulation (Hoppe 1973).

Current Status

To the best of our knowledge, however, neither the neutron nor the X-ray form of this interference experiment has been achieved with macromolecular complexes. Until we have some experimental data in hand, an attempt at a critical assessment of its utility would be premature.

In order to put this idea to the test, a good deal of ground work is necessary, much of which has now reached a satisfactory state. In the first place, it has been necessary to determine how growth conditions affect the incorporation of D from the medium into RNA and protein in *E. coli*. It would be prohibitively expensive to grow cells on carbon sources such as deuterated algal hydrolysate in the quantity required. We have sought to determine whether cheaper deuterated carbon sources exist, what the consequences are of dispensing with deuterated carbon sources altogether, and how *E. coli* responds to varying concentrations of D_2O in the medium. A second, closely related area is the determination of how the scattering lengths per unit volume of ribosomal protein and ribosomal RNA vary with the extent of deuteration of the macromolecules themselves and with the D_2O concentration of the solvent. The third and final area of research has dealt with the development of neutron diffractometers suitable for low-angle solution studies. While it would be foolish to pretend that we have definitive answers for all these questions, nevertheless we seem to be in a position to go ahead. A complete account of the growth and scattering length determinations will appear in due course (Moore, Engelman and Schoenborn, manuscript in preparation).

The current scattering geometry is shown in Figure 1 and has been described elsewhere (Nunes 1973). It basically consists of two identical arrays of parallel vanes. The first array collimates the incident neutron beam so that the specimen is illuminated with about ten well-collimated, slit-shaped beams. The second set has vanes that absorb neutrons which enter the set in nonparallel directions. This set is aligned exactly with the beam-collimating slits at zero scattering angle. The detector slit and the detector itself move together as the scattering profile is scanned. The monochromator and detector positioned behind the second vane set effectively record the result of ten well-collimated slit source experiments done in parallel. The angular resolution of this arrangement has been tested by collecting data on the neutron scattering of a solution of the spherical virus R17. The data obtained clearly show several subsidiary maxima and compare favorably with X-ray data on this same virus.

Radius of Gyration Experiments

In the course of this work, it appeared that it might be possible to use the apparatus and experience obtained to measure the radius of gyration of the RNA and protein parts of a ribosomal subunit independent of each other.

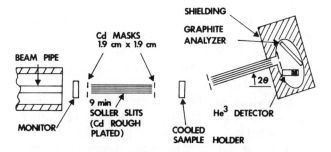

LOW ANGLE SPECTROMETER AT HFBR

Figure 1   The experimental apparatus. A beam of neutrons produced by the reactor emerges from the beampipe and passes through a monitor which measures the beam flux. The beam is defined by cadmium masks and collimated by the first Soller slit. It then passes through the thermally controlled specimen, and the neutrons scattered by the specimen are measured as a function of the angle 2θ. The angle of detection is defined by a second Soller slit. The pyrolytic graphite monochromator crystal is used to select a narrow wavelength band from the scattered neutrons by Bragg reflection into the ^{3}He detector (Nunes 1973). (Reprinted with permission from Moore, Engelman and Schoenborn 1974.)

These measurements were undertaken in part as a test of our equipment and understanding and in part for the information they would give about the ribosome. The results so far pertain to the 50S subunit. As an account of this work has already appeared (Moore, Engelman and Schoenborn 1974), we will confine ourselves to a summary of the experiments.

We grew *E. coli* under two conditions in D_2O-containing media. The first gave heavily D-protein and heavily H-labeled RNA. 50S subunits purified from these cells were suspended in an H_2O/D_2O solvent whose H_2O content was adjusted to contrast match the RNA. Under these conditions, the protein scattering length was highly positive relative to the solvent. The second growth condition gave less extensively deuterated protein, whose scattering length nearly equaled that of pure D_2O. The RNA formed was nearly 100% H and thus had a scattering length less than that of the solvent (100% D_2O). The first sample was designed to give neutron scattering only from its protein moiety. The second was intended to give appreciable scattering only from RNA.

The radius of gyration of a particle can be determined from the low-angle portion of its scattering curve. As the scattering angle becomes small, the scattering profile becomes Gaussian (to a good approximation) for all particles, regardless of shape (Beeman et al. 1957). As shown by Guinier (1939), the width of the low-angle Gaussian is inversely related to the radius of gyration of the particle squared. The narrower the Gaussian, the larger the radius of gyration of the particle. It is customary to plot such data as log (scattered intensity) versus scattering angle (2θ) squared. Plotted in this way, the Gaussian region of the scattering profile becomes a straight line whose slope is proportional to the radius of gyration squared. Figure 2 shows such a Guinier plot of low-angle neutron data for the two deuterated samples in question and for all H 50S subunits.

How do the radii of gyration of the RNA and protein of the subunit relate to the radius of gyration of the all H subunit to which both portions contribute? The answer to this question may be found by reference to the definition of the radius of gyration and the parallel axis theorem of classical mechanics. If the radius of gyration of the whole particle is R_T and that of the RNA and protein R_R and R_P, respectively, then

$$R_T{}^2 = f_P R_P{}^2 + f_R R_R{}^2 + f_P f_R \Delta^2, \qquad (1)$$

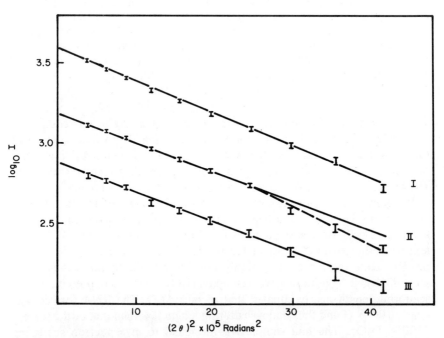

Figure 2 Guinier plots for samples of isotopically labeled 50S subunits. A plot of $(2\theta)^2$ vs. $\log_{10} I$ is shown for an example of each type of specimen so that the linear region for each sample can be estimated for use in subsequent analyses. (I) All H 50S subunits, linear to $(2\theta)^2 = 35.5 \times 10^{-5}$ radians. (II) 50S subunits labeled to emphasize protein, linear to $(2\theta)^2 = 24.7 \times 10^{-5}$ radians. (III) 50S subunits labeled to emphasize RNA, linear to $(2\theta)^2 = 35.5 \times 10^{-5}$ radians. The data points are shown as bars which represent in their height the statistical error of each measurement. Radii of gyration are determined from the slope of the linear portion of each curve. (Reprinted, with permission, from Moore, Engelman and Schoenborn 1974.)

where f_P and f_R are the mass fractions of protein and RNA ($f_R + f_P = 1.0$) and Δ is the distance between the centers of mass of the two distributions.

In these terms, the two deuterated samples represent efforts to achieve experimental situations where $f_P = 1.0$, $f_R = 0.0$ and $f_P = 0.0$, $f_R = 1.0$ so that R_P and R_R can be obtained directly. To the extent that contrast matching for the opposite species was not actually obtained, the radii measured in the deuterated samples will deviate from R_P and R_R. From the studies done on the dependence of scattering length on H/D ratio in RNA and protein and the measurement of the H/D ratios actually achieved in the samples used, the f_P and f_R values appropriate for each sample were estimated. The f's for all H subunits were also determined. Thus three equations were obtained which could be solved for R_P, R_R and Δ.

Table 4 gives the values actually obtained for the neutron radius of gyration of the three samples measured. These values were found by least squares extrapolation of all data collected to zero concentration. The "corrected" values are the extrapolated values corrected for slit aberration. The radii observed for the two deuterated samples are both less than that of the all H sample. Taking the data at face value, the only possible explanation is that Δ must be nonzero. In fact, if one simply assumes that the radii of the two deuterated samples R_P and R_R are perfect estimates, one finds a value for Δ of about 38 Å.

Table 5 gives the solution for R_P, R_R and Δ obtained using the values for f_P and f_R appropriate for each specimen. The error limits given reflect the statistical precision of the measurements and come primarily from consideration of the propogation of statistical errors in data collection through the data processing.

One question which may be asked about these values is how they compare with the radii of gyration data obtained by X-ray scattering meas-

Table 4 Apparent and corrected radii of gyration for three 50S subunit specimens

Specimen	Observed radius	Corrected radius
100% **H**	76.8 ± .95Å	78.0 ± .95Å
Protein dominant	74.9 ± .50Å	76.0 ± .50Å
RNA dominant	74.1 ± .60Å	75.0 ± .60Å

The observed radii of gyration were obtained by extrapolating the radii found for samples at different concentrations to zero concentration. In the case of the 100% H specimen, a variance-weighted least squares fitting was done; the error given is the variance of the extrapolated value. The slope found for the 100% H specimen was then imposed on the other two data sets and the extrapolated values found. The errors shown for these samples are effectively standard errors of the mean. The "corrected" radii are corrected for slit smearing as described elsewhere (Moore, Engelman and Schoenborn 1974).

Table 5   Final values for the parameters estimated in the neutron radius of gyration experiment

Parameter	Value
Rg of the total, all H 50S subunit	78.0 ± .95Å
Rg of the 50S protein component (in situ)	73.4 ± 2.0Å
Rg of the 50S RNA component (in situ)	72.5 ± 1.5Å
Separation of protein and RNA centers of mass	57.7 ± 10 Å

The values shown are derived from the data given in Table 3. The method for making these estimates is described elsewhere (Moore, Engelman and Schoenborn 1974).

urements. Equation 1 can be used with the values for R_P, R_R and Δ given in Table 5, and f's calculated using electron densities, to estimate the radius of gyration for the 50S subunit which should be obtained by X-ray low-angle scattering. The value we calculate, 76.0 Å, is in reasonable agreement with the values of 74.3 Å ± 1 (Serdyuk et al. 1970) and 76.0 Å ± 1 (Hill, Thompson and Anderegg 1969; Tolbert 1971) found earlier.

Critique of the Radius of Gyration Experiments

Natural questions to be asked about a result obtained by a novel technique are what level of reliance can be placed upon it and what are the problems in interpretation. We feel that the primary problem area in this work is the question of data correction. In order to collect perfect low-angle scattering data, specimens should be illuminated with a very fine circular beam of radiation and the scattered radiation measured with a point detector. In actual practice this ideal geometry is never used, because the low intensity of small beams does not produce enough scattered radiation to measure accurately, especially using very small detectors. The result is that both the source beams and the detectors used have appreciable area. The resulting degradation of the accuracy in the data collected, due to the geometric effects of finite beam and detector size, is called "slit smearing." A number of techniques have been proposed and employed in the past to correct data for these effects, none of which is fully satisfactory. In any case, the effect of slit smearing on the apparent radius of gyration in any experiment will be a function of the shape of the scattering profile observed, especially in its initial regions. This shape in turn reflects the overall shape of the mass distribution being examined.

From Figure 2 it is obvious that whereas the whole 50S subunit and the RNA sample give Gaussian data over the region shown, the protein sample does not. Thus we can assert with certainty that the shape of the mass distribution of protein in the 50S subunit is different from that of the RNA distribution or that of the whole particle; but what about the value for Δ? Due to limitations in neutron flux and the efficiency of the apparatus used, we do not know the scattering profiles of our samples over as wide a range of scattering angles as we would like. The problem is the following: If we had better data on these curves, would the correction process push up the value for protein radius of gyration relative to the other two values sufficiently to bring Δ to zero? (Since the total curve and the RNA curve seem similar, it is reasonable to suggest they will correct similarly.) If the other two values remained unchanged, this would imply that the proper correction is five times the correction we made. We feel that although this is improbable, it cannot be excluded altogether. It is hoped that improvements in data collecting equipment now being made will enable us to settle this issue in the near future.

In summation, what we have obtained is unambiguous evidence in favor of a difference in the shape of the protein and RNA distributions in the 50S subunit, from which we can conclude that the two types of macromolecules are not distributed homogeneously in the particle. Making the best analysis we currently can of the data in hand, it appears that a separation between the centers of mass of the two distributions is included as a component in this inhomogeneity. We will assume for the purposes of discussion that the value of Δ we obtain is approximately correct.

DISCUSSION

Anderegg and his coworkers have been able to investigate the X-ray solution scattering of the 50S subunits to far wider angles than we have been able to explore with neutrons to date. Using these scattering profiles, they have been able to suggest the approximate dimensions of the 50S subunit. The 50S subunit has a maximum linear dimension of about 230 Å. Thus the maximum possible separation between centers of mass of the protein and RNA is about 100 Å. The value we find for Δ, the separation between the RNA and protein centers of mass of 57.7 Å, is sizable, but not impossible. The only models for the 50S subunit that can accommodate this value are those in which the protein and RNA distributions are substantially displaced, giving a particle with a protein-rich side and an RNA-rich (or protein-poor) side.

Models of this type would appear to be consistent with much of the functional data on the 50S subunit. There is extensive data indicating a clustering of functional sites in the 50S subunit (see reviews by Traut et al., Kurland, and Möller, this volume). The A site for tRNA binding, the EF-G binding site and the EF-T site all appear to be overlapping functionally and probably physically as well. Recently Acharya, Moore and Richards (1973) showed that 50S-bound EF-G readily cross-links to about 100,000–120,000 daltons of 50S protein in the presence of a variety of bifunctional reagents. The 50S proteins involved include L7, L12 and possibly L2 as well (Acharya, personal communication). Many more 50S proteins have been identified as probable active site components that can be included in the approximately 100,000 daltons that cross-links to EF-G. Given that there are only about 475,000 daltons of protein in the structure, the implication is that this "active site region" is protein rich. It is also presumably peripheral, being available to protein factors in the environment. Thus it would appear reasonable to suggest that the active site region might correspond to the protein-rich surface of the 50S subunit that the neutron data indicate exists.

It is interesting that the data we currently have on the 30S subunit show that the uncorrected neutron radius of gyration of the all H particle is near 71 Å (the most recent published X-ray value is 72 ± 1.6 Å, Smith 1971). The protein and RNA values, for which our data are still quite weak, appear to deviate very little from this value. Thus Δ for the 30S subunit is probably much smaller than it is for the 50S subunit. Further data will have to be collected, however, before this conclusion can be drawn with certainty.

Finally, it should be pointed out that Equation 1 offers a possible alternative basis for finding distances between subunits in a structure, alternative to the interference technique discussed earlier. Following this strategy, three samples would have to be prepared; two containing labeled X or labeled Y only and one containing both labeled X and labeled Y, with all other components contrast matched in all cases. Radii of gyration would be measured on all three samples and the values combined according to Equation 1 to obtain Δ. The merit of this approach is that the value obtained for Δ would be independent of subunit shape, which would not be the case if interference data were used. The disadvantages of this method are that it puts stringent requirements on the contrast matching for the remaining components of the structure and would require measurements of each sample over a range of concentrations to correct for concentration effect. A similar idea has been applied to analysis of low-angle X-ray data on macromolecular complexes in the past (Damaschun et al. 1968).

Acknowledgments

This work was supported by Grants from the National Institutes of Health (A1-09167) and the National Science Foundation (GB-39275X). The research carried out at the Brookhaven National Laboratory was under the auspices of the U.S. Atomic Energy Commission.

References

Acharya, A. S., P. B. Moore and F. M. Richards. 1973. Crosslinking of elongation factor EF-G to the 50S ribosomal subunit of *E. coli. Biochemistry* **12**:3108.

Bacon, G. E. 1962. *Neutron diffraction.* Clarendon Press, Oxford.

Beeman, W. W., P. Kaesberg, J. W. Anderegg and M. B. Webb. 1957. Size of particles and lattice defects. *Handbuch der Physik* **32**:321.

Damaschun, G., P. Fichtner, H-V. Purschel and J. G. Reich. 1968. Untersuchung der Quarturstruktur von Proteinen mittels Rontgen-Kleinwinkelstreuung *Acta Biol. Med. Germ.* **21**:309.

Engleman, D. M. and P. B. Moore. 1972. A new method for the determination of biological quarternary structure by neutron scattering. *Proc. Nat. Acad. Sci.* **69**:1997.

Guinier, A. 1939. La diffraction des rayons X aux tres petits angles. *Ann. Phys.* 11e serie **12**:161.

Hill, W. E., J. D. Thompson and J. W. Anderegg. 1969. X-ray scattering study of ribosomes from *E. coli. J. Mol. Biol.* **44**:89.

Hoppe, W. 1972. A new x-ray method for the determination of the quaternary structure of protein complexes. *Israel J. Chem.* **10**:321.

————. 1973. The label triangulation method and the mixed isomorphous replacement principle. *J. Mol. Biol.* **78**:581.

Kratky, O. and W. Worthmann. 1947. Die Abhungigkeit der Rontgen-Kleinwinkelstreuung von Grosse und Form der kolloiden Teilchen in verdunnten Systemen. *Monatsch. Chem.* **76**:263.

Marshall, W. and S. W. Lovesey. 1971. *Theory of thermal neutron scattering,* pp. 1–16. Clarendon Press, Oxford.

Moore, P. B., D. M. Engelman and B. P. Schoenborn. 1974. Asymmetry in the 50S ribosomal subunit of *E. coli. Proc. Nat. Acad. Sci.* **71**:172.

Nunes, A. 1973. A simple neutron diffractometer for low angle biological studies. *Nucl. Inst. Meth.* **108**:189.

Schoenborn, B. P. and A. C. Nunes. 1972. Neutron scattering. *Annu. Rev. Biophys. Bioeng.* **1**:529.

Serdyuk, I. N., N. I. Smirnov, D. B. Ptitsyn and B. A. Fedorov. 1970. On the presence of a dense internal region in the 50S subparticle of *E. coli* ribosomes. *FEBS Letters* **9**:324.

Smith, W. S. 1971. An X-ray scattering study of the smaller ribosomal subunit of *E. coli. Ph.D. thesis,* University of Wisconsin, Madison.

Tolbert, W. R. 1971. A small angle X-ray scattering study of 50S ribosomal subunits for *E. coli. Ph.D. thesis,* University of Wisconsin, Madison.

Structure and Function of the *Escherichia coli* Ribosome: Immunochemical Analysis

Georg Stöffler
Max-Planck-Institut für Molekulare Genetik
Berlin-Dahlem, Germany

INTRODUCTION

The availability of specific antibodies has proven extremely useful in analyses of the structure and function of enzymes (Arnon 1971; Margoliash, Reichlin and Nisonoff 1967; Landsteiner 1945; Benjamini, Michaeli and Young 1972). The essential understanding of the nature of the antigenic determinants came from studies of the immunogenicity of synthetic copolymers of amino acids and from parallel studies with proteins of known primary sequences. The information that emerged was that antibodies can be directed against either amino acid sequences or against protein conformation and that the minimal immunogenic sequence was four contiguous amino acids, although as many as ten amino acids might be required to bind the antibody (Sela 1966, 1969; Sela et al. 1967; Benjamini, Michaeli and Young 1972).

It occurred to me seven or eight years ago that the same immunochemical methods that had been used to explore the structure of enzymes might well be applied to more complex assemblies and perhaps even to organelles. For this reason I began to raise antibodies against *Escherichia coli* ribosomes, ribosomal subunits and then, as they became available, against each of the 55 individual ribosomal proteins. The purpose of this article is to summarize the results, to show the contribution the immunochemical experiments have made to knowledge of the structure and function of bacterial ribosomes, and finally, to indicate possible further applications of this approach.

The earliest experiments were with antibodies against intact 70S ribosomes (Quash et al. 1962). Estrup and Santer (1966) immunized rabbits with *E. coli* 30S and 50S subunits and found that only one or two proteins appeared to be common to the two subunits; the remainder were immunologically distinct. In conformity with that finding, there was no cross-reaction of 30S or 50S core particles with antisera against 30S or 50S split proteins, respectively (Gupta and Singh 1972). Antibodies raised against 50S split proteins interfered with nonenzymatic binding of tRNA to ribosomes (Friedman, Olenick and Hahn 1968). The immunochemical experiments with antisera against eukaryotic ribosomes are discussed elsewhere (Wool and Stöffler, this volume).

Antibodies against Ribosomes and Ribosomal Subunits

Most of our experiments have been with antibodies to *Escherichia coli* ribosomes and a few with antisera to *Bacillus stearothermophilus* and *Bacillus subtilis* ribosomes. Bacterial ribosomes and their subunits are strongly immunogenic, more immunogenic than mixtures of the extracted ribosomal proteins (Zeichhardt 1974; Zeichhardt and Stöffler, unpublished data). Immunization with ribosome particles leads to the formation of precipitating antibodies directed primarily against the protein; formation of precipitating antibodies against the RNA is an exceptional occurrence (Zeichhardt 1974; Zeichhardt and Stöffler, unpublished data). The amount of cross-reaction between antisera against the 30S and 50S subunits was dependent on the purity of the particles (especially of the 50S subunit) used as immunogen. With 50S subunits, twice purified by zonal centrifugation, there was no cross-reaction. The precipitating antibodies against ribosomal subunits were mainly of the IgG class. Antisera raised against 70S ribosomes or 50S subunits were tested for their ability to react with purified ribosomal proteins; they precipitated approximately 70% of the individual proteins of the 50S subunit (22 out of 34 proteins). These results indicate that antibodies to a large proportion of the proteins are formed when animals are immunized with either 70S or 50S ribosomes. Each of eight different antisera against 70S ribosomes precipitated 28 individual proteins (purified from both 30S and 50S subunits); however, each individual serum also reacted with several additional proteins. Thus eight anti-70S ribosomes and two anti-50S sera together were capable of precipitating all 34 individual proteins. On the other hand, precipitating antibodies were not formed against all individual proteins of the 30S subunit, regardless of whether immunization was with 30S subunits or 70S ribosomes. Eight different antisera were tested (Zeichhardt 1974; Zeichhardt and Stöffler, unpublished data). It would appear that 30S subunits are weaker immunogens, but it is also possible that several proteins of the 30S subunit undergo a change in conformation upon extraction and are then poor antigens in the reaction with sera elicited against ribosomes. An antiserum against all of the proteins of the 30S subunit was tested for cross-reaction with eight pure 30S subunit proteins (Traut et al. 1969). From 28 pairs of proteins tested, 26 gave a reaction of nonidentity; two pairs of proteins, however, revealed partial identity. Unfortunately, the authors did not give the identity of these two proteins (Traut et al. 1969). It is of some importance that one 70S ribosome can bind at one time as many as 90–100 IgG molecules (Zeichhardt 1974). Thus it comes as no surprise that antibodies against ribosomal subunits and ribosomes, as well as their monovalent Fab fragments, inhibit the function of the particle (Hasenbank and Stöffler, unpublished data).

Only antisera against *Escherichia coli* ribosomes and their subunits have been investigated in detail. Although few experiments were done with antisera against other bacterial ribosomes, those that were carried out gave the same results as were obtained with antisera against *E. coli* ribosomes (Wittmann et al. 1969; Geisser 1971, 1973; Geisser, Tischendorf and Stöffler 1973; Geisser et al. 1973).

Antibodies against Individual Ribosomal Proteins

The purification of the proteins of *E. coli* 30S and 50S ribosomal subunits provided us with pure proteins to use as immunogens. In the initial experiments, antibodies were raised in rabbits against ten ribosomal proteins (Stöffler 1969; Wittmann et al. 1969). The immunological proof predated the biochemical proof that these proteins, later called S4, S5, L1, L3, L4, L5, L7, L15, L17 and L24, were not identical. After purification of the 21 proteins of the 30S subunit, we obtained antibodies against those proteins as well (Stöffler and Wittmann 1971a). At about the same time, 24 of the 34 proteins of the 50S subunit were isolated and purified and antibodies against them prepared (Stöffler and Wittmann 1971b). The remaining ten proteins were difficult to purify in sufficient amounts, and we have only recently succeeded in making antibodies against them (Stöffler et al. 1974a).

Extensive cross-reactivity was reported for several purified 30S subunit proteins (Fogel and Sypherd 1968), but this was probably due to the use of impure proteins for immunization. Antisera against three purified 30S proteins (30_5, 30_7, 30_{8a}) were found to react with the homologous proteins and not with any other 30S protein (Traut et al. 1969). The three proteins were later shown to be S9, S7 and S5 (Wittmann et al. 1971).

Antigenicity of Pure Ribosomal Proteins of *E. coli*

As briefly described above, precipitating antisera against all 21 30S and 34 50S proteins have been elicited, primarily in rabbits. The amount of single protein needed for the course of immunization varied between 2.8 and 15.0 mg; generally, 3–5 mg was sufficient. After 3–5 booster injections, the antibodies were mainly of the IgG class, with titers varying from 0.2 to 1.2 mg of specific antibody per ml of serum and usually between 0.4 and 0.8 mg per ml. Ribosomal proteins have proved to be relatively poor antigens, but the antisera possessed a high degree of specificity. In addition, antibodies against 12 proteins were elicited in sheep. There was no apparent difference in the immune response of rabbits and sheep, and the main advantage of the latter was that larger amounts of serum could be collected without requiring significantly more immunogen. Sheep sera do have one disadvantage: while they contain relatively high titers of several antibodies, including potent precipitins, the specifiicity is not nearly as great as in rabbit antisera. Thus sheep antisera were less useful for studies of inhibition of ribosome function. We have used proteins of the highest available purity. Nonetheless, some animals formed antibodies against traces of contaminating proteins. Since amounts of pure proteins sufficient for absorption of the contaminating antibodies were not available, such sera had to be discarded. In each case it was established that the reaction of an antiserum with a second protein was due to a contamination of the immunogen and not due to partial immunological identity (Stöffler and Wittmann 1971a,b; Stöffler et al. 1974a,b).

Antibodies against proteins can be directed against sequences or conformation or both (Celada and Strom 1972). A simple, reliable way to discriminate between the two types of antibodies is to add tryptic peptides derived from the immunogen to the precipitation reaction. Tryptic peptides have little tertiary structure and therefore they should inhibit precipitation of the corresponding intact antigen if the antibodies are directed against sequences but not if they are against conformation. By this criterion, antibodies against pure 30S proteins are mainly specific for sequences and not for conformation (Stöffler and Wittmann 1971a; Stöffler, unpublished data). With the exception of anti-L7 and -L12 the test has not yet been made with antisera against the 50S proteins. Precipitation of L7 and L12 by anti-L7 or -L12 was inhibited only 30 to 40% by their tryptic peptides, and therefore these two acidic proteins, in contrast to the others we have analyzed, induced appreciable numbers of antibodies against conformation as well as against sequences (Stöffler, unpublished data).

Lack of Immunological Identity among 30S Proteins. No immunological cross-reaction could be detected by immunodiffusion or by immunoprecipitation among the 21 proteins of the 30S subunits of *E. coli* ribosomes (Stöffler and Wittmann 1971a). Three different types of experiments were carried out. In one, total 30S proteins were placed in the center well and the antisera to each individual protein were placed in the peripheral wells on agar plates. Each protein-specific antiserum was tested adjacent to each of the others. Since these experiments do not require that the antigen be pure, they provide a quick and simple test for immunological identity. The procedure has another advantage: impure antisera are easily identified. Since all proteins are present in the antigen mixture (TP30), the occurrence of more than one line at antigen-antibody equivalence can be taken as an indication of contamination.

In a second series of experiments, a single protein (in the center well) was developed against antisera specific for each of the 30S proteins. A precipitation line was obtained only with the homologous antiserum. Each antiserum was next tested against pure proteins. Again the antisera precipitated only the specific antigen protein. Finally, quantitative immunoprecipitation suggested that each 30S ribosomal protein was a distinct entity, that is, they had no extensive common sequences (Stöffler and Wittmann 1971a).

50S Proteins Immunologically Distinct. The same methods that had been employed to test the immunological identity of the small subunit proteins were applied to those from the 50S particle. The proteins of the 50S subunit were, with the exception of L7 and L12, immunologically distinct. Anti-L7 precipitated proteins L12 and L7 equally well; anti-L12 reacted with L7 and L12 (Stöffler 1969; Stöffler and Wittmann 1971b; Figure 1a, 1b). If either antiserum was developed in Ouchterlony plates against the two proteins, the precipitation lines fused, indicating complete immunological identity. We concluded at the time that the proteins had similar, if not identical, primary structures (Stöffler 1969; Stöffler and Wittmann 1971b). Determination of the sequence of amino acids in the two proteins substantiated our surmise: the two proteins have identical sequences apart from an acetylated amino-terminal serine in protein L7 (Terhorst et al. 1972). An attempt was made to demonstrate the differences between L7 and L12 (e.g., the aminoacetylation of the amino-terminal serine) by immunological means but with no success. Antisera to the two proteins have been raised in ten animals, but none distinguished between them (Figure 1a,b). It therefore seems likely, although it is by no means proven, that the aminoacetylation does not appreciably alter the conformation and the immunogenicity of the protein (Stöffler et al. 1974a).

Proteins of 30S and 50S Subunits Are Different. We turned then to the question of whether the proteins of the two ribosomal subunits were immunologically distinct. The antisera against each of the 55 ribosomal proteins were screened by immunodiffusion on Ouchterlony plates for reaction with a mixture of all the proteins of the 30S or the 50S subunit. None of the pure antisera reacted with the mixture of proteins of the other subunit except for anti-S20 and anti-L26, where fusion of the precipitation lines indicated complete immunological identity (Stöffler and Wittmann 1971b; Tischendorf, Stöffler and Wittmann 1974; see Figure 1c). A large portion of the primary sequence of S20 has been determined (see Wittmann and Wittmann-Liebold, this volume), but nothing is yet known of the structure of L26. However, a mutant with an alteration in protein S20 has been isolated (Wittmann et al. 1974): the bacteria possess the same electrophoretic change in S20 and L26; hence the two proteins are very likely encoded by a single cistron. Proteins S5 and L11 of strain *E. coli* K, which migrate into the same position during two-dimensional electrophoresis, are immunologically unrelated (Tischendorf, Stöffler and Wittmann 1974; Figure 1d).

All these experiments were performed with ribosomal subunits that were purified by two additional zonal centrifugations. 50S subunits used without repurification contained appreciable amounts of proteins S5, S9, S11, S12, S14 and S20 (Morrison et al. 1973; Tischendorf, Stöffler and Wittmann 1974). It was inferred that the presence of small amounts of these proteins was compatible with their occurrence at the subunit interface (see below). It appears that repurification of subunits results in the removal of bound proteins rather than contaminating subunits (Tischendorf, Stöffler and Wittmann 1974).

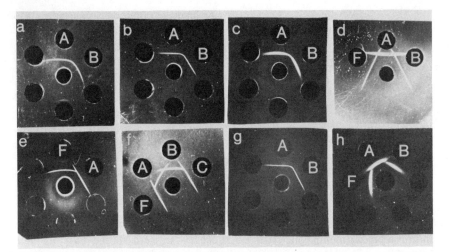

Figure 1 Precipitin patterns commonly observed in comparative double diffusion plate test in agarose gels. The procedure that followed for double immunodiffusion is described in detail elsewhere (Stöffler and Wittmann 1971a; Stöffler et al. 1971a).

(*a–c*) Pattern of identity (fusion; the compared antigens are serologically identical). (*a*) The center well contained 30 µl anti-L7 (serum) and the peripheral wells had in A: 1.5 µg protein L7 and in B: 1.5 µg L12. (*b*) Center well: 35 µl anti-L12 (serum); the peripheral wells contained proteins L7 (in A) and L12 (in B). (*c*) Center well: 90 µl anti-S20 (serum); peripheral wells had in A: 3 µg protein S20 and in B: 4 µg L26.

(*d*) Comparison of two immunologically unrelated antigens: pattern of nonidentity (intersection). The center well contained anti-S5 and anti-L11 γ-globulin (1:1 mixture, threefold enriched); the peripheral well had in A: 3 µg protein L11, in B: 2.2 µg S5, and in F: 30 µg TP30 (*E. coli*).

(*e–f*) Pattern of partial identity (partial intersection). Center well: 80 µl anti-S20 (γ-globulin, twofold enriched); the peripheral wells had in A: 20 µg TP30 *E. coli* A19 and in F: 25 µg TP30 (from mutant 64-2). N.B.: S20 was electrophoretically altered in this mutant (Wittmann et al. 1974). (*f*) Center well 60 µl anti-S4 (*E. coli* A19); the peripheral wells contained in A: 3.5 µg S4 (from mutant N425), in B: 3.5 µg S4 (from mutant N433) and in C and F: 2.8 µg S4 from *E. coli* A19 (Hasenbank et al. 1973). The precipitin bands of wild-type S4 and both mutants formed "spurs" (partial immunological identity). In addition, a "double spur" forms when two partially different antigens (mutants N425 and N433) are related with an antiserum to a third antigen which is different but related to both of them (anti-S4 elicited against S4 purified from *E. coli* A19). N.B.: The antiserum discriminated not only the antigenic differences between S4 from wild type and the two mutants with alterations in S4 (N425 and N433), in addition, it recognized antigenic differences among proteins S4 of the two mutants (see Hasenbank et al. 1973).

(*g*) Pattern of identity (despite electrophoretic differences between the two antigens). Center well: 90 µl anti-L22 (γ-globulin, twofold enriched); A: 60 µg TP70 A19; B: 65 µg TP70 (from mutant N281). N.B.: L22 (N281 is altered in two-dimensional electrophoresis but this was not detectable serologically) (Wittmann et al. 1973).

620

General Considerations and Conclusions. Cross-reaction proves sequence (or conformation) homology, but the absence of immunological cross-reaction does not prove lack of homology. An antigenic determinant comprises an absolute minimum of four amino acids; on the average, eight determinants were recognized on each 30S subunit protein (Stöffler and Wittmann 1971a). It is therefore reasonable to assume that on the average some 40 amino acids in each protein are recognized by antibodies. Thus it is conceivable that two proteins might hold in common three-fourths of their sequences and still not react. If a single immunogen is injected into different animals, they will form antibodies against different regions of the immunogen protein (individual specificity). Furthermore, the specificity of an antiserum changes during the course of immunization. Thus by testing antisera taken after various immunization times from different rabbits, a greater proportion of the amino acid sequence may be tested. We immunized two to eight animals with each single ribosomal protein, and none of these sera revealed any immunological similarity.

Amino acid sequence analyses of ribosomal proteins from *Escherichia coli* carried out so far have confirmed the conclusion previously drawn from immunological data. Thus among the sequences of the 30S subunit proteins that have been determined, there are only a few tetrapeptides common to two proteins (Wittmann and Wittmann-Liebold, this volume). Nonetheless, it must be pointed out that an antiserum against *Escherichia coli* S9 cross-reacted weakly (Higo et al. 1973) or not at all (Isono et al. 1973) with the homologous protein from *Bacillus stearothermophilus,* despite their having an identical nona- or undecapeptide near the amino terminus (Yaguchi et al. 1973; Higo, Loertscher and Nomura 1974).

Others have now prepared antibodies against single *Escherichia coli* ribosomal proteins. Antibodies against 14 purified 30S ribosomal proteins were raised in rabbits and said to react only with the immunogen protein (Higo et al. 1973). Antibodies against eight of the 30S proteins were also obtained by immunizing mice (Hawley and Slobin 1974). The advantage of using mice is that only 15 μg of each purified protein is required for immunization; the disadvantage is the small amount of sera obtained. For that reason the antisera were not characterized.

Application of Specific Antibodies

Correlation of Ribosomal Proteins of E. coli *Isolated in Different Laboratories.* The isolation, purification and characterization of the proteins of *Escherichia coli* was undertaken in a number of laboratories (see Witt-

---

Figure 1 (continued) (*h*) Partial homology between *E. coli* S19 and *B. stearothermophilus* BS-17 proteins. Center well: 80 μl anti-S19 (γ-globulin, threefold enriched); the peripheral well contained in A: 3 μg S19 from *E. coli* A19, in B: 5 μg *B. stearothermophilus* protein (BS-17), and in F: 35 μg TP30 of *B. stearothermophilus* (Isono et al. 1973).

mann, this volume). Different methods were employed for the separation and different systems of nomenclatures were used by the various groups to designate the proteins. The confusion that resulted was an impediment to further progress in research on the structure of the particle. It was of some importance to correlate the proteins prepared in the several laboratories and to adopt a common nomenclature (Wittmann et al. 1971). The plan that was agreed on was to co-opt the numbering system used by Kaltschmidt and Wittmann (1970) to specify the protein spots on a two-dimensional electropherogram of the ribosomal proteins. The two-dimensional electropherograms, together with a comparison of amino acid composition and peptide maps, were used for the correlation; molecular weights were of minimal value (Wittmann et al. 1971).

Specific antisera were also of great utility in the correlation, their principal advantages being in the rapidity with which the immunological test can be conducted and the small amount of protein antigen required (0.3–1.5 μg). The correlation of the proteins of the 30S subunit of *E. coli* isolated in Berlin, Uppsala, Madison and Geneva (Wittmann et al. 1971) and of the proteins of the 50S subunit purified in Uppsala and Berlin was by two-dimensional electrophoresis, amino acid composition, molecular weight and immunological identity (Kurland, Donner, Wittmann, Tischendorf and Stöffler, unpublished data). However, the same end was recently achieved more expeditiously with tritium-labeled 50S proteins using specific antisera only (Zimmermann and Stöffler 1974). A preliminary immunological correlation of 50S proteins, isolated in three different laboratories, is summarized in Table 1.

The correlation of the 55 proteins of *Escherichia coli* ribosomes isolated in several different laboratories, and the adoption of a uniform system of nomenclature is a unique example of a successful cooperative scientific enterprise that has benefited everyone.

Use of Specific Antibodies to Detect Mutational Alterations in Ribosomal Proteins. It was anticipated that mutations in ribosomal proteins would be of importance in analyzing the structure and function of the ribosome, provided of course that the mutants could be identified. An amino acid exchange that altered the net charge of the mutant would be detected by two-dimensional gel electrophoresis or column chromatography (see Wittmann and Wittmann-Liebold, and Davies and Nomura, this volume); however, a change in the mutant protein that did not alter the charge (e.g., replacement of one neutral amino acid by another) might go unrecognized. On the other hand, if the amino acid exchange occurred within an antigenic determinant, the result would be a reduction in the immunological reactivity of the mutant protein with the antiserum raised against the wild-type protein. The use of several separate antisera will, of course, increase the chances of detecting mutational alterations.

The method of choice in experiments of this type is, once again, Ouch-

Table 1   Correlation of ribosomal proteins of the 50S subunit of *Escherichia coli* and their interaction with 23S and 5S rRNA

Berlin	Geneva[a]	Uppsala[b]	Specific binding to 23S rRNA	5S rRNA
L1	11	7	+[d]	
L2	25	28	+[c,d]	
L3	10	11	+[d]	
L4	12	14	+[d]	
L5	7b	6II		(+)[g]
L6	9	6I	+[e,d]	(+)[e]
L7[h]	1a(1b)[h]	1(2)[h]		
L8	2b	3		
L9	2a	3		
L10	5	4		
L11	7a	9		
L12[h]	1b(1a)[h]	2(1)[h]		
L13	17	22	+[d]	
L14	13	16		
L15	20	25		
L16	27	26II		
L17	22	26I	+[c,d]	
L18	19	24	(+)[f]	+[e,f]
L19	16b	23	+[c,d]	
L20	31	36	+[c,d]	
L21	26	—		
L22	18	20		
L23	16a	18	+[c,d]	
L24	15	21	+[c,d]	
L25	11a	12		+[e,f]
L26 (S20)[h]	(25a)	—		
L27	24	30		
L28	28(25a)	34		
L29	8	10		
L30	14	19		
L31	—	—		
L32	—	—		
L33	21 .	27		
L34	—	35		

[a] Zimmermann and Stöffler 1974; [b] C. G. Kurland, D. Donner, G. W. Tischendorf, G. Stöffler and H. G. Wittmann, unpublished data; [c] Stöffler et al. 1971b; [d] Garrett et al. 1974; [e] Gray et al. 1972; [f] Gray et al. 1973; [g] Monier, this volume; [h] proteins immunologically not discernible.

terlony's double immunodiffusion (Figure 1). By this procedure, a partial homology between protein S7 from *E. coli* strain K and strain B was detected (Kaltschmidt et al. 1970).

Three separate genetic alterations in S5 are known: the first is between
E. coli strains B and K (an exchange in tryptic peptide T1 from alanine
to glutamic acid); the second, leading to spectinomycin resistance, is the
result of multiple exchanges in T10; and the third is an exchange in T2
known to confer a phenotypic change from streptomycin dependence to
independence (Wittmann-Liebold and Wittmann 1971; Funatsu, Schiltz
and Wittmann 1971; Funatsu, Nierhaus and Wittmann-Liebold 1972; Itoh
and Wittmann 1973). With anti-S5, only the last class of mutants could
be detected; the strain differences between *E. coli* B and *E. coli* K and the
alteration to spectinomycin resistance were immunologically silent (Stöffler
et al. 1971a; Kaltschmidt et al. 1970). Thus the mutation that leads to
streptomycin independence must occur outside of tryptic peptides T1 and
T10 (Hasenbank et al. 1973). Mutational alterations in protein L4
(erythromycin locus), in S20 (Figure 1e) and in S8 were also detected
with antisera (Wittmann et al. 1973; Wittmann et al. 1974; Wittmann,
Zubke, Stöffler and Apirion, unpublished data).

The value of the immunological approach is perhaps best illustrated by
the results of an investigation of a group of mutants in proteins S4 and S5
that suppressed streptomycin dependence (Deusser et al. 1970; Stöffler
et al. 1971a; Hasenbank et al. 1973). Of 100 mutants, 40 displayed
changes in two-dimensional polyacrylamide gel electrophoresis (24 in S4
and 16 in protein S5) (Hasenbank et al. 1973). Thirteen of these 100
mutants were investigated by immunochemical techniques and seven revealed an alteration in S4, three of which had not been uncovered by gel
electrophoresis. However, perhaps it is most instructive that six mutations
were silent by both methods of analysis (Deusser et al. 1970). Differences
between the alterations of S4 in two mutants (N425 and N433) were detected by both two-dimensional electrophoresis and by immunodiffusion
(Hasenbank et al. 1973; see Figure 1f). Thus immunochemical and electrophoretic techniques are mutually reinforcing for the detection of altered
ribosomal proteins (Figure 1g). Moreover, hemagglutination and complement fixation are likely to be even more sensitive than immunodiffusion
and hence increase the efficiency of immunological detection.

Ribosomal Proteins from Different Bacterial Species.   The ribosome is an
excellent object for the study of evolution. Ribosomes occur in pro- and
eukaryotic organisms and at all stages of evolution; they contain both
proteins and RNA and thus require not only conservation of function but
also conservation of essential protein–RNA and protein–protein interactions. Therefore a study of the homology between proteins of different
bacterial species was of interest.

Total reconstitution has been possible only with the large subunit of
B. stearothermophilus ribosomes (Nomura and Erdmann 1970). It was
important then to correlate the 50S proteins of that species with those of

E. coli, since a number of experiments could be done with the former, e.g., the sequence of assembly of the 50S particle and the function of individual proteins in subparticles lacking one protein.

Ribosomal proteins from seven species (*Aerobacter aerogenes, Erwinia carotovora, Serratia marcescens, Proteus vulgaris, Salmonella typhimurium* and *Shigella dispar*) belonging to the family Enterobacteriaceae were compared by two-dimensional electrophoresis and by Ouchterlony double diffusion and found to be very similar (Geisser et al. 1973). However, we did detect antigenic differences among homologous proteins which migrated identically in two-dimensional electrophoresis.

The sequence of amino acids at the amino termini of the proteins of *B. stearothermophilus* and *E. coli* ribosomes is very similar (Yaguchi et al. 1973; Higo, Loertscher and Nomura 1974) and the proteins are functionally interchangeable (Nomura, Traub and Bechmann 1968; Higo et al. 1973). Antisera specific for 15 of the 21 proteins of the 30S subunit of *E. coli* ribosomes cross-reacted with a mixture of the proteins of the 30S subunit (TP30) of *B. stearothermophilus;* the other six antisera gave no reaction (Geisser, Tischendorf and Stöffler 1973). The experiment indicated that there were at least 16 homologous proteins in the two bacterial species; however, it did not identify them. To accomplish the latter, 17 proteins purified from the 30S subunit of *B. stearothermophilus* were tested by immunodiffusion with *E. coli* antisera: 13 led to precipitation; the remaining four proteins were not precipitated by *E. coli* antisera. It is of importance that the correlation required, and was only possible with, several separate antisera against each of the *E. coli* proteins. Only five of 14 antisera against *E. coli* proteins tested by Higo et al. (1973) reacted with *B. stearothermophilus* TP30. However, the apparent discrepancy between the results of Geisser and of Higo is likely the result of the latter having tested only one antisera against each protein. Just as in earlier experiments (Wittmann et al. 1969) in which we had not detected cross-reaction between *B. stearothermophilus* TP70 and 10 single *E. coli* antisera against individual proteins, lack of cross-reactivity does not prove that structures are not related (Higo et al. 1973).

Antisera against 16 proteins of the 50S subunit revealed precipitation lines in immunodiffusion plates when reacted with a mixture of all the proteins of *B. stearothermophilus* ribosomes (Geisser, Tischendorf and Stöffler 1973). It is to be expected that with more sensitive immunological methods and with a variety of antisera against each protein, all the proteins of the two species may finally be correlated.

The ribosomal proteins from several other Bacillus species (*B. subtilis, B. licheniformis, B. pumilis, B. circulans, B. brevis, B. cereus* and *B. megaterium*) were also compared with each other and with *E. coli.* The ribosomal proteins from Bacilli showed little resemblances to those from *E. coli;* in addition, there was a great deal of immunological heterogeneity among the proteins of the ribosomes from the various Bacillus species. Only a small number of proteins could be correlated by two-dimensional gel electrophoresis. On the other hand, most of the antisera against 30S ribosomal proteins of *E. coli* precipitated their counterparts in Bacilli. Only anti-S20 and -S21 did not react with any Bacillus ribosomal protein. Approximately 16 of 32 antisera against *E. coli* 50S ribosomal proteins precipitated Bacillus protein. It is of interest that the number and the molecular weights of ribosomal proteins from all these bacterial species are conserved (Sun, Bickel and Traut 1972; Geisser, Tischendorf and Stöffler 1973; Geisser et al. 1973). Thus immunological methods are of far greater value in establishing structural homologies than two-dimensional electrophoretograms. From immunochemical experiments we were also able to determine that proteins of ribosomes from Hydrogenomonas, Rhodopseunomonas, Plesiomonas, Lactobacillus and Sarcina were more closely related to those of *E. coli* ribosomal proteins than to those of Bacillaceae, but not as close as those of *E. coli* to ribosomal proteins of Enterobacteriaceae (Wittmann et al. 1969; Geisser, Tischendorf and Stöffler 1973; Geisser et al. 1973; Geisser 1971, 1973). Ribosomal proteins from several species of Clostridium have also been compared with those from *E. coli;* their degree of homology is similar to that described for Bacilli (Katsaras, Geisser and Stöffler, unpublished data).

Thus by immunological methods, structural differences were detectable with ribosomal proteins from closely related organisms despite identical migration in two-dimensional electrophoresis, and homologies were still detectable among distantly related organisms although the two counterpart proteins migrated very differently in electrophoresis.

Protein–rRNA Interaction.   Shortly after Traub and Nomura (1968) achieved the physical and functional reconstitution of *E. coli* 30S subunits, the question was raised: Which ribosomal proteins can form specific complexes with rRNA independent of the presence of other proteins? Different methods established that five proteins of the 30S subunit bind independently to 16S rRNA (Mizushima and Nomura 1970; Schaup, Green and Kurland 1970; Schaup, Green and Kurland 1971; Garrett et al. 1971; Stöffler et al. 1971b; Zimmermann et al. 1972).

In the immunological approach, single proteins were incubated with rRNA under reconstitution conditions; rRNA was separated from unbound proteins by sieve chromatography. The bound protein could be detected and quantitated with specific antibodies. In accordance with results by other methods, S4, S7, S8, S15 and S20 were shown to bind specifically to 16S rRNA (Garrett et al. 1971; Stöffler et al. 1971b).

Similar investigations to determine which 50S proteins bind to 23S rRNA were, however, limited by the fact that 50S reconstitution was not practicable for *E. coli*. Experiments were therefore performed under 30S reconstitution conditions. Eight proteins, including L2, L6, L16, L17, L19,

L20, L23 and L24, were found to form specific complexes with 23S rRNA (Stöffler et al. 1971b,c). A more recent and more intensive investigation revealed specific binding of four additional proteins, L1, L3, L4 and L13, but, contrary to the previous report, no binding of L17 (Garrett et al. 1974). Weak but significant interaction of L18 with 23S rRNA was also observed (Gray et al. 1972).

Proteins S11 and S12 bind to 23S rRNA (Stöffler et al. 1971b; Morrison et al. 1973) and one preparation of L19 bound to 16S rRNA (Garrett et al. 1974), but only S11 formed a specific complex. For each of the three proteins, a role in subunit reassociation has been established independently by other methods (see section: Proteins at or near the Subunit Interface).

Using specific antibodies, proteins L18 and L25 were shown to attach specifically to 5S rRNA (Gray et al. 1972, 1973; Monier, this volume; for a summary, see Table 1).

I should emphasize that proteins radioactively labeled in vivo are more conveniently suited for these investigations than is the immunological approach. However, the latter method proved to be reliable and can be used when radioactive proteins are not available; it is superior to the use of proteins which are made radioactive in vitro, since protein–RNA complexes can be formed without chemically modifying the protein.

Use of Specific Antibodies to Identify Cross-linked Ribosomal Proteins. Specific antisera have been used to identify four protein pairs which can be cross-linked by bivalent reagents, namely, S5–S8, S18–S21, S13–S19 and S7–S9 (L. C. Lutter, U. Bode, H. Zeichhardt, G. Stöffler and C. G. Kurland, unpublished data; Lutter et al. 1974). These findings and their implications are discussed in detail in this volume by Kurland. Bollen et al. (personal communication) have made use of antibodies to detect the proteins cross-linked to the initiation factors IF-2 and IF-3 by bissuberimidate and confirmed the superiority of the immunochemical procedure to ammonolysis. It is instructive that when Traut and his associates reanalyzed the proteins of their 30S cross-linked pairs with specific antibodies, they were able to correct the earlier misidentification (Bickle, Hershey and Traut 1972; R. Traut, T. Sun, A. Bollen and L. Kahan, personal communication; see also Traut et al., this volume).

The Binding Site of Antibiotics.   It was considered possible that binding of an antibiotic to a ribosomal protein might be inhibited by its corresponding monovalent antibody fragment. We pursued such experiments with streptomycin, spectinomycin, tetracycline, thiostrepton and erythromycin (Lelong et al 1974; Highland et al. 1974c; Bollen, Maschler and Stöffler, unpublished data; Högenauer, Drews, Hasenbank and Stöffler, unpublished data; Wienen, Stöffler and Pestka, ·unpublished data). To our surprise, antibodies against as many as seven different proteins of the 30S subunit (anti-S1, -S10, -S11, -S18, -S19, -S20 and -S21) inhibited the binding of tritium-labeled dihydro-streptomycin, and the same antibodies, except anti-S11, inhibited the interaction of spectinomycin with the 30S subunit (Lelong et al. 1974; Bollen, Maschler and Stöffler, unpublished data). It is remarkable that anti-S12, -S4 and -S5 were not inhibitory for streptomycin binding, although mutational alterations in these proteins were shown to affect the phenotypic behavior of the ribosome towards these antibiotics. S3 and S5 were recently shown to be part of one binding site of dihydro-streptomycin and S4 bound a streptomycin analog in affinity labeling experiments (Schreiner and Nierhaus 1973; Pongs and Erdmann 1973).

Protein L11, which is required for the peptidyltransferase reaction (Nierhaus and Montejo 1973), is also necessary for the binding of thiostrepton (Highland et al. 1974c). Antibodies against L11 inhibited the peptidyltransferase activity of 50S subunits (Tate, Caskey and Stöffler 1974) but were inhibitory of thiostrepton binding only when 50S ribosomal cores were used in the assay (Highland et al. 1974c). The results showed that L11 is required for thiostrepton binding, but they also suggested that antibiotic binding and enzyme activity are contained in different portions of L11.

It would be easily conceivable that an antibiotic could still interact with the ribosome even though an antibody was attached to the target protein. However, it is very difficult to rationalize that the binding site for a drug with an average molecular weight of 500 is composed of six or seven proteins. Since it is known that binding of an antibody introduces a conformational change in the antigen (Celada and Strom 1972), the inhibitory effects could be induced by switching the 30S subunit into a nondrug-binding conformation. On the other hand, all those antibiotics interacted nonspecifically with Fab's (Wienen and Stöffler, unpublished data). More recent experiments provided evidence that unspecific inhibition can be eliminated by the use of specifically enriched antibodies; it was found that anti-L4 reduced erythromycin binding when purified antibodies were used (Wienen, Stöffler and Pestka, unpublished data). Prior results obtained by blocking antibiotic binding with antibodies, therefore, deserve a very cautious interpretation.

Accessibility of the Proteins of the Ribosome

Little is known about the arrangement of the molecular components in the ribosome. A large proportion of the rRNA must be naked since it binds dyes and is sensitive to ribonuclease digestion (Morgan and Rhoads 1965; Furano, Bradley and Childers 1966; Miall and Walker 1967; Santer 1963). The topography of the proteins is far less certain. The treatment of ribosomes with high concentrations of monovalent salts strips a subset of proteins (split proteins) from the ribosome, leaving behind a second group that are more firmly bound to the RNA (core particles) (Meselson et al. 1964). The then current interpretation was that the particle consisted of a core of RNA and protein, with an outer shell of a second set of proteins (split proteins). The construct has a certain heuristic value since it predicted that some proteins would not be accessible, a proposition eminently testable. Experiments were carried out on the sensitivity to trypsin digestion of the proteins of the intact ribosome (Chang and Flaks 1970; Craven and Gupta 1970; Crichton and Wittmann 1971; Spitnik-Elson and Breiman 1971), on the ease of H_2O-tritium water exchange (Page, Englander and Simpson 1967; Cotter and Gratzer 1971), and on the selective modification of amino acids by a series of reagents (Craven and Gupta 1970; Moore 1971; Kahan and Kaltschmidt 1971). We adopted another tact, that of testing whether antibodies against single proteins would react with intact ribosomes. Obviously if they did, at least one antigenic determinant for that protein must be accessible, presumably on the surface of the particle. The high specificity of the antibodies and their large size (40 × 50 × 70 Å for the monovalent fragment) were strong advantages of this approach. For example, the size made penetration of the antibody into the ribosome matrix unlikely; the same cannot be said of amino acid-specific reagents. A disadvantage is that only positive results carry weight since a protein might have nondeterminant carrying portions exposed. Moreover, during assembly of the ribosome, proteins may change conformation. The determinant region may still be on the surface of the ribosome but unable to react with the antibody because of the change in conformation. We have employed multiple immunological tests of ribosomal protein accessibility. In the end it turned out that a single method (immunodiffusion) would have sufficed.

Double Diffusion of Ribosomal Subunits against Antibodies Specific for Individual Proteins. The double diffusion method of Ouchterlony (1958) has the advantage that it is quick and simple. For that reason, the initial experiments were with antisera specific for single ribosomal proteins—they were developed against either 30S or 50S subunits by double diffusion in agar gels. Those experiments were with untreated subunits pretreated with glutaraldehyde. Glutaraldehyde fixation was used to prevent the stripping of proteins from the ribosome and the possibility that they reacted only then with antibody.

The 21 antisera against each of the individual 30S proteins all gave a precipitation line when tested against untreated 30S subunits. However, anti-S15 and anti-S17 did not react with glutaraldehyde-treated 30S subunits (Stöffler et al. 1973). We do not know whether glutaraldehyde treatment altered the antigenic determinants for those two proteins or whether the fixation simply prevented degradation of the subunit and the release of S15 and S17 during the relatively long time required for immunodiffusion.

Similar experiments were performed with 50S subunits. Antisera against 28 of the 34 proteins reacted strongly with 50S subunits regardless of whether the particles were fixed with glutaraldehyde or not. The remaining six antisera (anti-L13, -L15, -L24, -L26, -L28 and -L29) reacted only weakly (Hasenbank and Stöffler, unpublished data). Despite our initial concern about interpreting these results, later experiments with different methods (see below) proved that immunodiffusion, used properly, is actually a very reliable indicator of ribosomal protein accessibility.

Immunological "Sandwich Technique."   The formation of an immunoprecipitate requires that a minimum of two determinants on the antigen be accessible for antibody binding. The binding of a single antibody molecule to the ribosome leads to the formation of a soluble antigen-antibody complex; soluble complexes are also present if the antigen-antibody ratio is not optimal. The sandwich technique is a procedure for bringing about precipitation of the soluble complexes. To that end, we incubated radioactive ribosomal subunits with a specific rabbit IgG or Fab specific against a single ribosomal protein and then precipitated the complex with antibodies against rabbit IgG. The fraction of the total radioactivity precipitated is a measure of the extent of the formation of antigen-antibody complexes. Significant precipitation of the 30S subunit occurred with each of the antisera against the 21 subparticle proteins (Stöffler et al. 1973). The technique has not yet been used with 50S subunits.

Sedimentation Methods for Estimation of Antibody-Subunit Complex Formation

When a bivalent IgG interacts with ribosomal subunits, it can lead to the formation of dimers, larger aggregates, or precipitation of the particle. It should be possible, by sedimentation analysis, to determine which occurs and at the same time corroborate that separate proteins are accessible on the surface of the particle. For that purpose, ribosomal subunits were preincubated with specific IgGs and the reaction mixture centrifuged either in a sucrose gradient or in an ultracentrifuge. Sucrose gradient centrifugation allowed the simultaneous measurement of optical density and radioactivity, whereas analytical ultracentrifugation permitted a more accurate estimation of the sedimentation coefficients of the particles.

Sucrose Gradients.   IgGs directed against the 21 proteins of the 30S subunit, when preincubated with 30S ribosomal subunits, caused a reduction of the 30S peak during sucrose gradient centrifugation and, as a rule, formed only small amounts of dimers (Figure 2a–d; Stöffler et al. 1973). However, when 50S subunits were incubated with IgGs specific for each individual protein of the 50S subparticle, the subunits were also precipitated; but in general 50S subunit dimers[1] which were linked together by

---

[1] A complex of two 50S subunits or two 30S subunits coupled to each other by one divalent IgG molecule is referred to as a "dimer."

one divalent IgG could be resolved (Zeichhardt, Tischendorf and Stöffler, unpublished data; Figure 2 e–h). The results confirm that the subunit proteins are on the surface of the particle and exposed (Stöffler et al. 1973; Zeichhardt, Tischendorf and Stöffler, unpublished data). Several different IgG preparations of anti-L2, -L4, -L13, -L15, -L17, -L21, -L24, -L26 and -L29 prepared from separate animals had to be tested to obtain positive results. This was especially critical for anti-L2 and -L24. *B. stearothermophilus* L2 (and presumably *E. coli* L2 as well) undergoes a conformational change when it binds to 23S rRNA; the conformation change alters its immunogenicity (Tischendorf, Geisser and Stöffler 1973). L24 is resistant to trypsin digestion when present in the 50S subunit or in a complex with RNA, indicating that it has a different conformation in the free and the assembled state (Crichton and Wittmann 1971). The putative conformational changes in L2 and L24 may explain why more than one antisera against each was required to demonstrate that they are on the surface of the particle.

In general, IgGs against single 30S proteins caused precipitation of 30S subunits, whereas those directed against 50S proteins led to dimer formation (Figure 2). These subunit dimers were linked together by one divalent IgG molecule. Precipitation requires at least two accessible antigenic determinants, dimer formation only one. Thus the proteins of the 30S subunit may be more exposed (e.g., a greater portion of each protein molecule is accessible) than those of the 50S subparticles. The results are unlikely to be due to differences in the properties of the antibodies (titers, number of determinants, avidity, etc.) since these were very similar for IgGs against both 30S and 50S ribosomal proteins. When IgGs against a single 30S subunit protein were separated into subfractions specific for different antigenic regions of the protein, some IgG subfractions caused a greater amount of 30S dimers, indicating that the proteins did in fact have more than one accessible determinant (Figure 2d). There is however one inconsistency: IgGs against 50S subunit proteins caused dimer formation (one determinant) in sucrose gradient sedimentation tests but precipitation (two or more determinants) in immunoprecipitation tests (Ouchterlony or liquid medium). One rationalization is that more IgG molecules bind to a single 30S protein than to a 50S protein. That exegesis assumes that the binding of a single antibody to a protein alters its conformation and allows the binding of a second antibody to a second determinant; it implies that the change in conformation is more likely to occur with 30S than 50S proteins. The interpretation accords with the idea that the small subunit has a less rigid structure than the 50S particle and hence is less likely to be distorted by a bound antibody. Alternatively, 30S proteins may be arranged as a monolayer (Kurland, this volume) and therefore accessible to antibodies on both sides of the particle. The discrepancy between the formation of 50S dimers in sucrose gradients and precipitation in liquid

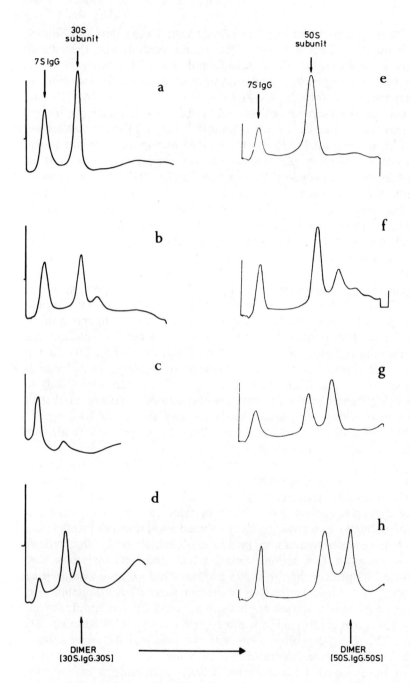

Figure 2 Sedimentation profiles of ribosomal subunit · IgG-antibody complexes in sucrose gradients.

(*a–d*) 30S subunits (1.0 A_{260}) were treated with IgGs prepared from non-immune sera (NIS) and from antisera to individual 30S subunit proteins.

media or on Ouchterlony plates can be explained in the following way: if two accessible determinants are close, only a dimer complex containing an antibody of high affinity would survive centrifugation. The binding constants between antigens and antibodies vary over a broad range (Eisen and Siskind 1964). Methods which employ sedimentation therefore select for the binding of those antibodies having an affinity sufficiently high to survive centrifugation. Thus if the antibody to one determinant was high affinity and the other low, one would expect dimers (afforded by the IgG with the high binding constant). In contrast, in the equilibrium conditions for antibody-antigen interactions that prevail in double immunodiffusion, one would expect precipitation since antibodies for both determinants would bind.

Analytical Ultracentrifugation.   Although analytical ultracentrifugation of subunit-IgG complexes was expected to give the same results as sucrose gradient sedimentation, we hoped that it would provide more accurate sedimentation coefficients for the complexes (2 subunits·1 IgG). There was another possible benefit to the exercise. An antibody that bound to a protein located at the "ends" of the subunit might form a dimer with a higher viscous drag and a lower sedimentation coefficient, whereas a "side protein-antibody-side protein" dimer would cause less viscous drag and

---

Figure 2 (*continued*) (*e–h*) 50S subunits (2.0 A_{260}) treated with IgGs specific for pure 50S subunit proteins. After preincubation of subunits with IgGs, the mixtures were layered on sucrose gradients and centrifuged (10–30% sucrose, w/w, in 10 mM Tris-HCl pH 7.8, 0.3 mM $MgCl_2$, 30 mM NH_4Cl, 6 mM 2-mercaptoethanol; 13 hr at 23.000 rpm, 4°C). A SW40 rotor and a L2-65 ultracentrifuge (Beckman) were used. The A_{260} was monitored in a continuous flow cell (Uvicord, LKB) and recorded with a W+W-Electronic Hi-speed Recorder 202; sedimentation direction was from left to right. IgG NIS did not reduce the 30S or 50S subunit peak as compared to controls without antibody addition (*a,e*).

(*b*) Anti-S11 · IgG (0.9 A_{280}) reduced the 30S monomer peak significantly and formed a small amount of 30S–IgG–30S complexes (dimers). A subfraction of this IgG (0.45 A_{280}) led however to a significant increase of dimer formation (*d*). Hence the unfractionated IgG contains antibodies specific for at least two different determinants, accessible on the 30S subunit in situ, thus allowing precipitation of 30S subunits.

(*c*) Anti-S2 · IgG (0.9 A_{280}) precipitates the 30S subunit almost completely. (*f*) Anti-L14 · IgG formed 50S–IgG–50S complexes (dimers) and higher aggregates. Anti-L1 · IgG (0.6 A_{280}) as well as anti-L19 both led to significant dimer formation (*g,h*).

N.B.: IgGs specific for individual 30S proteins in general precipitated 30S subunits, whereas antibodies against 50S proteins led almost exclusively to dimer formation. Further explanations are given in the text (Stöffler et al. 1973; sedimentation profiles from Zeichhardt, Tischendorf and Stöffler, unpublished data).

have a higher sedimentation coefficient. We thus hoped to be able to correlate protein position with sedimentation coefficients.

When the experiments were carried out, 19 separate IgGs against individual 30S proteins were found to react with the whole 30S particle (anti-S17 was not tested, and S19 which had given positive results during sucrose gradient centrifugation was now without effect). Surprisingly, ten of the IgGs caused significant dimer formation. The subunit-antibody equilibria may be different during equilibrium and zonal centrifugation. This result makes it more likely that dimer formation is related to antibody-ribosome equilibria than to the effects of the antibody on the conformation of the ribosome.

The sedimentation coefficients of the dimers, when they occurred, varied between 41 and 48. There was, however, no clear distribution into two different groups, and hence our attempt to define the position of the proteins on the particle was frustrated (Stöffler et al. 1973). Similar experiments performed by means of analytical ultracentrifugation with IgGs against 50S subunit proteins revealed similar results to those obtained with the other methods (Morrison 1974; Morrison, Garrett, Tischendorf and Stöffler, unpublished data).

Synopsis. Clearly a part of each ribosomal protein is on the surface of the particle and sufficiently accessible to be able to bind an antibody. To convincingly demonstrate this, it was necessary to use a battery of procedures and several antisera against the same protein, especially since a mistaken impression was sometimes produced by negative results with a single method.

That all ribosomal proteins are at least partly on the surface of the particle is of intrinsic importance for it provides a good deal of all we know of the ribosome. This information also bears on any theoretical models for the structure of the ribosome. Indeed, the model where the 30S subunit is an oblate ellipsoid of revolution consisting of a protein monolayer held together by RNA (Kurland, this volume) demands that each 30S protein be at least partly on the subunit surface. In fact, it would be surprising if the model did not accord with our observation, since the former is based on the latter. The important finding of Miller and Sypherd (1973) that all 30S proteins are iodinated by lactoperoxidase accords with our results. Discussion of whether some proteins are inaccessible in the 70S ribosome and hence likely to be located at the subunit interface is deferred to a later section.

Finally, the use of antibodies to study the role of individual proteins in ribosome function was predicated on the demonstration that antibodies bind to the proteins in the particle.

Inhibition of Poly(U)-dependent Polyphenlalanine Synthesis by IgGs and Fabs against Ribosomal Proteins. The synthesis of polyphenylalanine directed by polyuridylic acid requires the interaction of the ribosome with the template, with phenylalanyl-tRNA, and with the three elongation factors EF-Tu, EF-Ts and EF-G. We undertook to determine which ribosomal proteins were required for and engaged in polyphenylalanine polymerization. We assumed that many of the essential proteins might be at or near the binding site for one or another of the elongation factors. For these experiments we used IgG molecules or their monovalent Fab fragments. Monovalent Fabs have at least two advantages for experiments of this kind: first, they cannot cause precipitation or dimerization of ribosomes; second, the Fabs lack the acidic Fc part of the IgG molecule and hence will not interact nonspecifically with the ribosome.

The IgGs and the Fabs against 20 of the 30S subunit proteins (anti-S17 was not tested) inhibited phenylalanine polymerization (Stöffler et al. 1973; Maschler 1973; Lelong et al. 1974; Figure 3). The number of events required for the inactivation of the ribosome by each specific Fab has been found to be close to one (Maschler 1973; Lelong et al. 1974; Thammana et al. 1973; see also Figure 4). The extent of the inhibition,

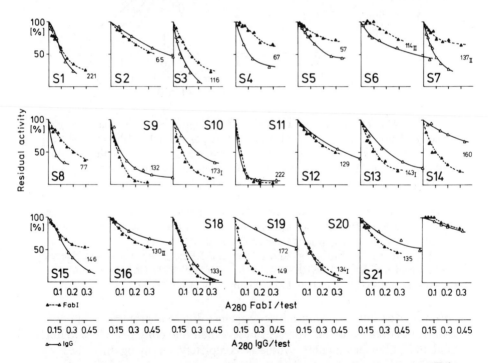

Figure 3 Concentration dependence of the inhibitory effects of IgGs and Fabs on in vitro poly(U)-dependent polyphenylalanine synthesis. Inhibition is expressed in percent cpm of the control; in the absence of IgG or Fab, 15,000 cpm (100%) of [14C]polyphenylalanine was synthesized (Maschler 1973; Lelong et al. 1974). ($\triangle$———$\triangle$) IgG; ($\blacktriangle$———$\blacktriangle$) Fab.

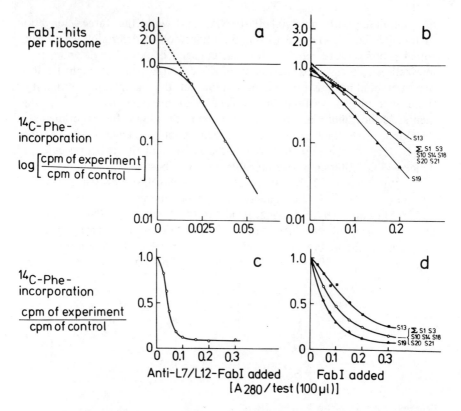

Figure 4 The effect of specific Fab I fragments on the protein synthetic capacity of ribosomes in a poly(U)-dependent polyphenyl synthesizing system. The activity at a given Fab concentration (expressed as a ratio of that obtained in the absence of Fab) is plotted as a function of the amount of Fab added (*c, d*). In (*a*) and (*b*), the plateau value obtained at a concentration of Fab where inhibition is complete is subtracted from each point in the inhibition curve. The intercept of the logarithmic plot is an index of the number of binding sites for a given Fab per ribosome; alternatively, it may reflect the number of partial steps of protein synthesis that were inhibited.

(*a*) (o⸺o) Inhibition by anti-L7 or anti-L12; (*b–d*) (o⸺o) average inhibition of anti-S1, -S3, -S10, -S14, -S18, -S20 and -S21; (■⸺■) anti-S13; (▲⸺▲) anti-S19 (Maschler 1973; Thammana et al. 1973).

however, was different with antisera against different proteins. Antibodies against fractional proteins (present in less than one copy per 30S subunit) inhibited polyphenylalanine synthesis completely, implying that they are essential for at least one step in the process.

The results with IgGs and Fabs against 50S proteins were very similar: 22 of the 34 completely inhibited polyphenylalanine synthesis, while the remainder inhibited synthesis by 60% or more, mostly by retarding the rate of protein synthesis. Preincubation of the antibodies with the individual antigen protein before adding them to the in vitro system abolished the inhibitory effects. Again, inhibition was described by a unimolecular inactivation curve (Lütgehaus and Stöffler, unpublished data). Among all those tested, only antibodies against L7 and L12 inactivated polyphenylalanine synthesis cooperatively (Figure 4 and Thammana et al. 1973). If we assume that inactivation is a Poisson process, the intercept of a logarithmic plot should yield a measure of the number of binding sites for the Fab against L7 and L12 on the ribosome. We found this intercept in each experiment to be greater than two and most frequently close to three (Figure 4). The implication of these findings for the function of L7 and L12 is discussed later.

It is important to distinguish these results from the foregoing demonstration of physical accessibility. It has been shown that only a relatively small fraction of ribosomes are active in in vitro protein synthesis (Gilbert 1963). Thus our previous experiments provided little information about the exposure of proteins during function, whereas the present ones bear on their accessibility in active ribosomes. Finally, these results were the basis for further experiments on the role of individual ribosomal proteins in the various partial reactions of protein synthesis.

Antibodies Inhibitory to Specific Functions of the 50S Subunit

Proteins L7 and L12.   Two acidic *E. coli* ribosomal proteins, L7 and L12, play an essential role in the function of the particle. They are required for the interaction of the ribosome with elongation factors EF-G and EF-Tu, with the initiation factor IF-2, and with release factors RF-1 and RF-2 (Hamel, Koka and Nakamoto 1972; Brot et al. 1972, 1974; Fakunding, Traut and Hershey 1973; for further references see Möller, this volume). Studies on the function of L7 and L12 were greatly facilitated by the discovery of a simple and convenient technique (Hamel, Koka and Nakamoto 1972) for the preparation of 50S subparticles lacking only those two proteins (PI particles). Nonetheless, the function of the proteins could not have been elucidated without specific antibodies. I shall summarize the experiments.

Immunological Identity of Proteins L7 and L12.   Antisera to L7 or L12 have been raised in ten animals; however, none of them distinguished between the two proteins (see above).[2]

We prepared the IgG fraction from antisera against L7 and L12 and found them to be strong inhibitors of poly(U)-directed polyphenylalanine

---

[2] Sera raised against a mixture of L7 and L12 are referred to as anti-L7/L12; similarly, L7/L12 is used for a mixture of the two proteins.

polymerization. Only insignificant differences in their inhibitory capacities were observed between the individual preparations. Since specific initiation does not occur during polyphenylalanine synthesis, we focused our attention primarily on the effects of antibodies against L7/L12 on the ribosome-catalyzed elongation reactions.

Interaction of Elongation Factor G with the Ribosome. The 50S core particle prepared during CsCl equilibrium centrifugation (Meselson et al. 1964) cannot carry out EF-G-dependent GTP hydrolysis. Addition of two acidic proteins, originally designated A_1 and A_2, restored activity (Kischa, Möller and Stöffler 1971). Proteins A_1 and A_2 actually are L7 and L12, and purified IgGs against L7 or L12 inhibited EF-G-dependent GTP hydrolysis by *E. coli* ribosomes (Kischa, Möller and Stöffler 1971). In contrast, an IgG prepared from an antiserum against a 30S subunit acidic protein (anti-S6) was without effect. To our knowledge, these experiments provided the first direct evidence for the involvement of an individual ribosomal protein in a specific ribosomal function.

The experiments do not exclude the possibility that proteins other than L7 and L12 are contained in the site to which EF-G binds. The availability of non-cross-reacting antibodies against all 55 ribosomal proteins provided us with the means to determine whether other proteins were necessary for the interaction of the ribosome and elongation factor G.

The interaction of EF-G with the ribosome consists of two partial reactions that can be studied separately. The first is the formation of a ribosome·EF-G·GDP complex on the 50S subunit; the reaction requires only the 50S subunit and is best observed in the presence of fusidic acid (Bodley and Lin 1970). The second reaction, the hydrolysis of GTP to GDP and P_i, requires both subunits (Nishizuka and Lipmann 1966).

We first tested whether any of 50 specific antibodies against separate ribosomal proteins would interfere with the formation of a 70S ribosome·EF-G·GDP complex. The results were decisive: only antibodies against proteins L7 and L12 were inhibitory (Highland et al. 1973; Figure 5). When the experiments were repeated with 50S subunits, again only anti-L7 and anti-L12 interfered with the formation of a 50S ribosome·EF-G·GDP complex (Highland et al. 1974a). In addition, five additional antibodies not tested earlier (anti-S17, -L26, -L31, -L32 and -L34) were now shown also to have no effect. The data are consistent with the view that the interaction between ribosomes and EF-G is limited to the site occupied by the acidic proteins L7 and L12.

We now proceeded to a study of the ribosome-dependent hydrolysis of GTP. The experiments were done in two ways. In the first series, antibodies against 50 separate proteins were tested for their ability to inhibit GTP hydrolysis by 70S ribosomes, and in the next series, subunits were treated with antibodies before they were combined for use in the assay. The latter

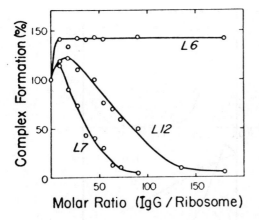

Figure 5 The effects of increasing concentrations of IgGs anti-L7 and anti-L12 on the formation of the ribosome EF-G · [³H]GDP complex. Conditions are as described by Highland et al. (1973). The results shown for anti-L6 are representative for IgGs specific for the remaining ribosomal proteins and IgG prepared from NIS.

experiments test the effect of the antibodies on subunit association; they also determine whether additional proteins become accessible after dissociation. Anti-L7 and anti-L12 were once again strong inhibitors of EF-G-dependent GTP hydrolysis by 70S ribosomes; however, there was also moderate but significant inhibition by IgGs and Fabs prepared from antisera against proteins L14 and L23 (Figure 6). Pretreatment of 50S subunits with anti-L14 and anti-L23 markedly increased their apparent inhibitory effects. Indeed, in the second series of tests when isolated subunits were pretreated with antibodies, the number of inhibitory antibodies was increased: they now included anti-L19 and anti-L27, and antibodies against two proteins of the 30S subunit, namely, anti-S9 and anti-S11 (Figure 6). It is noteworthy that the effect of anti-L7 and anti-L12 was just as strong when added to 70S ribosomes as when 50S subunits were pretreated before addition of 30S subunits. We presume that anti-L7 and anti-L12 inhibited GTP hydrolysis by preventing the interaction of EF-G with the ribosome, whereas the antibodies against S9 and S11 and against L14, L19, L23 and L27 merely prevented subunit reassociation which is a necessary prerequisite for GTPase activity (see also Marsh and Parmeggiani 1973). In support of our presumption, it was found that Fabs prepared from antisera against S9, S11, L14, L19, L23 and L27 did in fact prevent subunit reassociation (determined by sucrose density gradient sedimentation): at equal molar ratios reassociation and hydrolysis of GTP were inhibited para passu (Table 2). It is difficult to exclude that one or more of these proteins is not also directly involved in GTP hydrolysis.

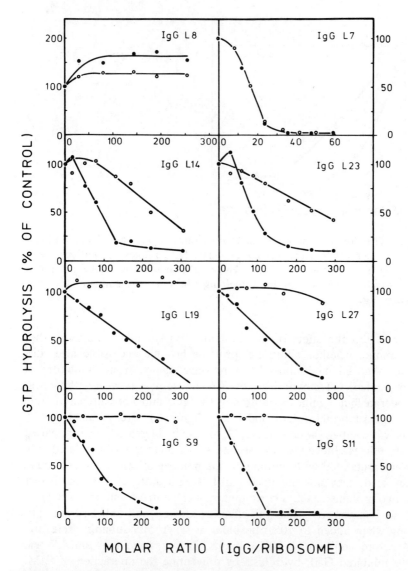

Figure 6 Concentration dependence of the effects of IgGs on the hydrolysis of GTP by ribosome and EF-G. Data from Highland et al. (1974a). (o———o) Treatment of 70S couples; (•———•) treatment of the appropriate subunit with IgGs. (N.B.: The inhibition by anti-L7 is the same with 70S ribosomes and pretreated 50S subunits.)

A complex that contains 5S RNA and proteins L18, L25 and either L5 or L6 or both will catalyze the hydrolysis of GTP and ATP (Horne and Erdmann 1973). The reaction does not require elongation factors but is inhibited by thiostrepton and, surprisingly, fusidic acid. Our results are not

Table 2   The effect of antibody binding on subunit association
and EF-G-directed GTP hydrolysis

Ribosomal protein specificity of the antibody preparation	% 70S ribosomes formed in sucrose gradients	GTP hydrolysis (% of control)
None	100	100
NIS	111	106
L15	107	102
S5	100	110
L14	34	25
L19	12	<1
L23	21	10
L27	40	34
S9	42	30
S11	1	3
L7	102	<1

The reaction conditions for measurement of GTP hydrolysis and for sucrose gradient analysis were as described (Highland et al. 1974a). Fab/ribosome ratios equal to 350 were used. When antibodies to proteins of the 50S subunit were used, each reaction contained 16 pmoles of 30S subunits and 6.4 pmoles of 50S subunits. The value reported for 100% was obtained under identical conditions but without Fab. When antibodies to the 30S subunit were tested, the reactions contained equal amounts (6.4 pmoles) of both subunits. Under these conditions, 100% association equalled 3.8 pmoles 70S and 100% GTP hydrolysis 220 pmoles. The values reported for the effects of antibodies to 30S proteins are in comparison to these (data taken from Highland et al. 1974a).

necessarily in conflict with those of Horne and Erdmann (1973). Anti-L7 and -L12 inhibit EF-G- and EF-Tu-mediated GTP hydrolysis by preventing the interaction of the elongation factors with the ribosome; the 5S rRNA complex does not require these factors. However, we should re-iterate that Fabs against L18, L25, L6 and L5 do not interfere with EF-G-mediated GTP hydrolysis (see above), although they inhibited poly(U)-dependent phenylalanine polymerization. It may be relevant that L6 and L18 (and L10 also) are required for the assembly of L7 and L12 into the 50S subunit (Stöffler et al. 1974b). 50S ribosomal particles lacking proteins L7 and L12 (PI-cores) were treated with monovalent antibodies specific for individual 50S ribosomal proteins. The subsequent ability of these Fabs to prevent reincorporation of L7/L12 into the particle was then assessed by testing for the binding of radioactively-labeled L7/L12 and for the restoration of EF-G-dependent [3H]GDP complex formation. Only antibodies against L10 completely inhibited reconstruction and recovery of function, but anti-L6, and to a lesser degree anti-L18, did partially in-hibit binding of L7/L12 (Stöffler et al. 1974b). The results conform with the findings of others using different methods (Schrier, Maassen and Möller 1973; Acharya, Moore and Richards 1973).

Interaction of Elongation Factor EF-Tu with the Ribosome. During one round of elongation two moles of GTP are hydrolyzed: the first hydrolysis is attendant on EF-Tu-mediated aminoacyl-tRNA binding and requires both subunits; the second occurs during EF-G-mediated translocation and only the 50S subunit is needed.

There is evidence that the two hydrolytic events are confined to a single site on the 50S subunit. They cannot occur simultaneously on a single ribosome (Cabrer, Vazquez and Modolell 1972; Miller 1972; Richter 1972; Richman and Bodley 1972), thiostrepton prevents both EF-G- and EF-Tu-dependent GTP hydrolysis (Modolell et al. 1971), and proteins L7/L12 are required for both reactions (Hamel, Koka and Nakamoto 1972; Brot et al. 1972).

The inhibition of poly(U)-directed polyphenylalanine synthesis by anti-L7/L12 is unique in that it is highly cooperative (Figure 4; Thammana et al. 1973), whereas inhibition by the other antibodies is unimolecular. The cooperativity of inhibition by anti-L7/L12 may be a reflection of the interaction between the three copies of L7/L12 contained in each 50S subunit (Thammana et al. 1973) or the interaction of L7/L12 with the two factors during elongation.

We have assessed in parallel the effect of antibodies (IgGs) on nonenzymatic and EF-Tu-dependent Phe-tRNA binding and on EF-Tu-catalyzed GTP hydrolysis. Inhibition of nonenzymatic Phe-tRNA binding can only occur at the aminoacyl-tRNA binding site, whereas inhibition of the elongation factor-catalyzed reactions might occur at the aminoacyl-tRNA or the factor binding sites. Antibodies that prevent subunit reassociation should inhibit both enzymatic and nonenzymatic Phe-tRNA binding, since both subunits are required. Anti-L7 and anti-L12 did inhibit completely and coordinately the two EF-Tu-mediated functions (Phe-tRNA binding and GTP hydrolysis), thereby indicating that EF-Tu interacts with those two ribosomal proteins. Anti-L7 and anti-L12 also inhibited (although to a lesser degree) nonenzymatic Phe-tRNA binding, suggesting that aminoacyl-tRNA interacts directly with proteins L7/L12 even in the absence of the elongation factor. The three activities were however also inhibited by anti-L4. That protein is then likely to constitute an imporant part of the aminoacyl-tRNA binding site. It is worth remarking that mutants resistant to erythromycin have an altered L4 (Wittmann et al. 1973). Antibodies to proteins L14 and L23 also inhibited the three activities, but since they are known to prevent subunit association (Highland et al. 1974a; Morrison et al. 1973), one cannot say whether or not they inhibit binding to the aminoacyl-tRNA site. We are in the process of trying to answer the question of whether the same set of antibodies that prevents the reassociation necessary for EF-G-dependent GTP hydrolysis also inhibits EF-Tu-catalyzed reactions. One impediment to the experiments is that the small subunit of prokaryotic ribosomes, unlike that from eukaryotes, will not bind aminoacyl-tRNA.

Peptide Chain Termination. The final step in protein biosynthesis is the discharge of the nascent polypeptide from tRNA bound to the ribosome. Three supernatant factors (release factors) are required: RF-1 and RF-2 recognize and respond to different termination codons (RF-1: UAA or UAG; RF-2: UAA or UGA); the third factor (RF-3 or S) is needed for the binding and release of RF-1 or RF-2 from the ribosome—RF-3 interacts with GTP and GDP. Termination has a final requisite: peptidyltransferase is necessary for the hydrolysis of the peptidyl-tRNA ester bond (Caskey et al. 1971).

The effects of antibodies specific for *E. coli* ribosomal proteins were tested on peptide chain termination. In the first instance, the effect of monovalent Fab fragments on the release of fMet from a fMet-tRNA·AUG·ribosome complex catalyzed by RF-1 and RF-2 was studied. Anti-L7 and anti-L12 completely inhibited the reaction (Tate, Caskey and Stöffler 1974). The inhibition by those two antibodies had been anticipated since 50S ribosomal core particles lacking proteins L7 and L12 are unable to interact with release factors (Brot et al. 1974). What was not certain was whether other proteins were also required. Indeed, we discovered that Fabs against L16 and L11, and also to a lesser extent those against S11, did inhibit the release reaction (Tate, Caskey and Stöffler 1974). L11 is necessary for peptidyltransferase activity (Nierhaus and Montejo 1973) and L16 for chloramphenicol binding—chloramphenicol inhibits peptidyltransferase (Pongs, Bald and Erdmann 1973; Nierhaus and Nierhaus 1973). Therefore, anti-L11 and -L16 are likely to inhibit release by interfering with peptidyltransferase. Anti-S11 is presumed to act indirectly by preventing subunit association which is necessary for termination and release.

The next experiments were of the nature of fine tuning, designed to determine the exact manner in which the various proteins participated in peptide chain termination by examining the effect of antibodies (Fabs) on partial reactions. Anti-L7 and -L12 inhibited the formation of a RF-2·UA [^{3}H]A·ribosome complex, anti-L11 and anti-L16 had no effect, and antibodies known to interfere with subunit association (e.g., S9 and S11) were only partially inhibitory (Tate, Caskey and Stöffler 1974). The hydrolysis of the peptidyl-tRNA ester bond assayed in ethanol without the termination codon was inhibited by anti-L7 and anti-L12 as well as by anti-L11 and anti-L16.

These experiments are then consistent. They indicate that L7 and L12 are part of the binding site for release factors, whereas L11 and L16 are part of the peptidyltransferase region. The other Fabs that inhibit peptide release do so by preventing subunit association. In full agreement with this, the latter were more inhibitory when subunits rather than 70S ribosomes were used for the assay.

Interchangeability of Proteins L7/L12 from Different Bacterial Ribosomes. Antisera against *E. coli* L7 and L12 gave a significant cross-reaction with ribosomal proteins from closely, as well as from distantly, related bacteria (Geisser, Tischendorf and Stöffler 1973; Geisser et al. 1973; Geisser 1973). They even precipitated an acidic protein isolated from yeast ribosomes and also reacted with two acidic proteins from the 60S subunit of rat liver (Wool and Stöffler, this volume).

We investigated whether *E. coli* anti-L7/L12 is capable of inhibiting poly(U)-directed phenylalanine synthesis as well as preventing EF-G·GDP·ribosome complex formation with other bacterial systems. Anti-L7/L12 inhibited strongly and completely both activities when incubated with ribosomes from *E. coli, Proteus vulgaris, Erwinia carotovora, Serratia marcescens, Plesiomonas shigelloides, Aeromonas punctata, Vibrio cuneatus, Bacillus pumilis, Bacillus stearothermophilus* and *Bacillus subtilis* (Geisser, Bodley, Hasenbank, Highland and Stöffler, unpublished data). In fact, the inhibitory activity of anti-L7 and -L12 was considerably stronger than their precipitating ability with their counterpart proteins purified from the various bacterial ribosomes. This could be due to a more rigorous conservation of the "active site" during evolution or a manifestation of the greater sensitivity of functional assays.

It was not surprising then that proteins L7 and L12 from species such as *Plesiomonas shigelloides, B. pumilis, B. subtilis* and *B. stearothermophilus* are active with *E. coli* core particles (Geisser, Bodley, Hasenbank, Highland and Stöffler, unpublished data). However, it was a surprise that the homologous protein from yeast could substitute for *E. coli* L7/L12 (Wool and Stöffler, this volume).

Affinity Label Experiments with 50S Proteins at the tRNA Binding Sites. N-substituted phenylalanyl-tRNA^{Phe} and tRNA$_F^{Met}$ analogs were synthesized and used as specific affinity labels for the 50S part of the ribosomal P site. The labeled proteins were identified by both two-dimensional electrophoresis and with specific antibodies; the latter were required since the introduction of the affinity label altered the electrophoretic mobility of the proteins significantly (Czernilofsky, Stöffler and Kuechler 1974; Czernilofsky et al. 1974; Hauptmann et al. 1974).

When *p*-nitrophenylcarbamyl-phenylalanyl-tRNA bound to ribosomes in response to poly(U), it reacted with proteins L27, L15, L2, L16 and L14 via covalent bond formation. It was concluded that these proteins are located at or near the tRNA binding sites; the most reactive protein was L27 (Czernilofsky et al. 1974).

When the *p*-nitrophenylcarbamyl-methionyl-tRNA$_F^{Met}$ analog was bound in the presence of R17–RNA and initiation factors, L27 was again very radioactive; L15 was also labeled but to a far lesser degree (Hauptmann et al. 1974; Czernilofsky, Stöffler and Kuechler 1974). Thus proteins L27 and L15 probably form at least part of the 50S ribosomal P site (see Cantor, this volume).

Antibodies Inhibitory to Specific Functions of the 30S Subunit

We had already shown that antibodies (Fabs) against each of the 30S ribosomal proteins inhibited polyphenylalanine synthesis, although the extent of the inhibition varied with antibodies against different proteins (see above and Figure 3). We were anxious to compare our results with those obtained with natural templates and with initiator tRNA. We proceeded then to study the effect of antibodies on the AUG-dependent binding of fMet-tRNA to 30S subunits catalyzed by initiation factors.

Inhibition by Specific Antibodies of Poly(AUG)-dependent fMet-tRNA binding Catalyzed by IF-1 and IF-2. The effect of 20 different Fabs on AUG-dependent binding of radioactive fMet-tRNA to purified 30S subunits catalyzed by initiation factors IF-1 and IF-2 were tested (Lelong et al. 1974). A number of Fabs were inhibitory. The degree of inhibition exerted by the various Fabs fell into three categories: (1) antibodies directed against S3, S10, S14, S19 and S21 were strong inhibitors; (2) antibodies against proteins S1, S2, S5, S6, S12, S13 and S20 were less inhibitory, although their effect was definite; (3) the remaining antibodies (S4, S7, S8, S9, S11, S15, S16, S18) were weak inhibitors or had no effect (Lelong et al. 1974). When fMet-tRNA binding to 70S ribosomes (rather than to 30S subunits) was measured with all three initiation factors, the results were much the same. At first we were surprised that Fabs against so many proteins interfered with the formation of an initiator–tRNA·30S complex. On reflection, it was apparent that antibodies could inhibit the formation of an initiation complex in a variety of ways. Formation of the complex requires the binding to the 30S subunit of four macromolecules—poly(AUG), IF-1, IF-2 and fMet-tRNA; interference with the binding of any one will result in the failure to form the complex.

Inhibition of EF-Tu-dependent Phe-tRNA Binding to 70S Ribosomes by Fabs. Correlative experiments to those just described were undertaken on the effect of antibodies (Fabs) on EF-Tu-catalyzed binding of Phe-tRNA to 70S ribosomes (Lelong et al. 1974). The overall reaction is comprised of a number of individual events: the binding of the template, of aminoacyl-tRNA, and of the elongation factor. Moreover, enzymatic Phe-tRNA binding is to 70S ribosomes only, and hence Fabs that prevent subunit reassociation will be inhibitory. The inhibitory antibodies were of two kinds: anti-S3, -S9, -S11, -S18, -S19 and -S21 were strong inhibitors of Phe-tRNA binding, while anti-S1, -S8, -S10, -S14 and -S20 had significant but less pronounced inhibitory effects (Lelong et al. 1974). Enzymatic Phe-tRNA binding was thus significantly inhibited by Fabs against one-half of the 30S subunit proteins. When 30S subunits were preincubated with antibody before 50S subunits were added and the assay conducted, anti-S3, -S8, -S9, -S11 and -S21 were now more strongly inhibitory than when 70S ribosomes were used. Anti-S8 and anti-S21 inhibited only when 30S subunits were preincubated with the Fabs (Figure 7), indicating they act at least in part by preventing subunit reassociation.

Identification of Ribosomal Proteins in the A and P Decoding Sites. Fifteen of the 21 proteins of the 30S ribosome are essential for the binding of initiator or aminoacyl-tRNA or both. We tried to deduce from our results the proteins located in the two decoding sites. Anti-S1, -S2, -S5, -S6, -S12 and -S13 were strong inhibitors of fMet-tRNA binding—e.g., of P-site binding (Table 3) and did not significantly inhibit Phe-tRNA binding. Of those, anti-S6 (and with some reservation anti-S12) had no effect on poly(U)-directed polyphenylalanine synthesis; therefore S6 and S12 are probably confined to the P site and involved specifically in initiation (Table 3). Similarly, anti-S1 might be inhibitory by preventing mRNA binding to the ribosome. A second group of antibodies (anti-S8, -S9, -S11, -S18) inhibited enzymatic Phe-tRNA binding but had little effect on binding of initiator-tRNA. The results suggest the four proteins (S8, S9, S11 and S18) are components of the A site (Table 3); however, antibodies against three of the four proteins (anti-S8, -S9 and -S11) also interfere with subunit reassociation, an observation which introduces a measure of uncertainty. Thus S18 is the only one of the four proteins that one can be confident is a part of the A site.

Several antibodies inhibited both fMet-tRNA and Phe-tRNA binding—anti-S3, -S10, -S14, -S19, -S20 and -S21; those proteins comprise an overlapping area impinging on both A and P sites.

The immunochemical results accord with the findings obtained in reconstitution (Nomura et al. 1969; Held, Mizushima and Nomura 1973) and exchange (Randall-Hazelbauer and Kurland 1972; Van Duin et al. 1972) experiments. In the latter, S2, S3 and S14 were found to be part of both the A and P sites. Our data confirm that S3 and S14 overlap the two sites. Bollen et al. (personal communication) reported that initiation factor IF-2 was cross-linked to a group of four 30S ribosomal proteins (S19, S13, S11, S1), and IF-3 could be cross-linked with S11, S12, S1 and S4. Antibodies against the cross-linked proteins (except S4) inhibited at least one reaction necessary for initiation. However, none of the antibodies against 30S proteins block IF-3 binding to 30S subunits significantly (Gualerzi and Pon 1973), thus supporting the viewpoint that IF-3 interacts with 16S rRNA and not with proteins. The results seem to contradict the findings just mentioned which were obtained in the cross-linking experiments (Bollen et al., personal communication). It is known, however, that IF-3 binds tenaciously

Table 3 The functional role of 30S subunit proteins

Classification	Protein synthesis directed by		tRNA-binding		Main inhibitors (Fab directed against)
	poly(U)[a]	natural[b] mRNA	phe-tRNA[a] (EF-Tu)	fMet-tRNA[a] (IF-1/IF-2)	
A site	+	+	+	−	S8, S9, S11, S18
P site	+	+	−	+	(S1), S2, S5, S12, S13
Proteins required for all functions (hybrid site)	+	+	+	+	(S3), S10, S14, S19, (S20)
Proteins specific for initiation	−	+	−	+	S6, (S12)
Natural mRNA recognition (± IF-3)	−	+	−	−	S4, S7, S15, (S16)
Subunit interface					S3, S8, S9, (S10), S11, S12, S14, S20, S21

[a] Maschler 1973; Lelong et al. 1974. [b] J. C. Lelong, E. Lazar, M. Crépin, I. Thibault, F. Gros, R. Maschler and G. Stöffler, unpublished data.

to the 30S subunit, whereas the association constant for the binding of an antibody to a ribosomal protein antigen is significantly lower. It is therefore possible that inhibition by Fabs was not detectable, since IF-3 binding was assayed by sucrose gradient centrifugation (Gualerzi and Pon 1973). It should be noted that an anti-S19 IgG inhibited the binding of IF-2 to 30S subunits (Bollen et al., personal communication).

Anti-S4, -S7, -S15 and -S16 have no appreciable effect on the binding of initiator or aminoacyl-tRNA to ribosomes (Lelong et al. 1974; Maschler 1973). It is of some importance that those proteins are necessary for sub-unit assembly (e.g., 30S reconstitution), that they interact directly with 16S RNA, and that they occur in one copy per 30S subunit (Nomura et al. 1969; Schaup, Green and Kurland 1971; Mizushima and Nomura 1970; Stöffler et al. 1971b; Voynow and Kurland 1971; Weber 1972). It seemed possible that S4, S7, S15 and S16 contribute directly only to the structure of the 30S subunit. On the other hand, proteins S8 and S20 are also essential for assembly and bind to rRNA, yet anti-S8 and anti-S20 inhibit the function of the particle (e.g., poly(U)-directed polyphenylalanine synthesis and aminoacyl-tRNA binding). Thus "structural" proteins are not necessarily functionally silent. Indeed, we now know that none of the proteins are unnecessary for function. Anti-S4, -S7, -S15 and -S16 are strong inhibitors of T4 mRNA-directed protein synthesis and λDNA-directed β-galacto-sidase synthesis (Lelong, Lazar, Crépin, Thibault, Gros, Maschler and Stöffler, unpublished data). These four proteins (S4, S7, S15 and S16) are thus probably essential for the recognition of natural mRNA (Table 3) and/or (at least in the case of S4) for IF-3 binding. IF-3 apparently functions in directing ribosomes to recognize initiation signals on mRNA. Further experiments are required to determine the exact role of the four proteins in initiation, be it mRNA binding or IF-3 binding.

When the results of all the assays are considered together, Fabs against all the 30S proteins interfered with one or another of the 30S subunit functions. Some antibodies specifically inhibited a single function, whereas others had pleiotropic effects. Certainly inhibition of a function by an antibody against a particular protein is not to be equated with a direct role of that particular protein in the function. The results make it difficult to credit the suggestion that proteins serve simply to enfold 16S rRNA and are not directly required for function of the 30S subunit (Kurland, this volume).

Proteins at or near the Subunit Interface

The ribosome is composed of two asymmetric subunits whose association, at least in vitro, is conditioned by the magnesium concentration (Tissières and Watson 1958). There is a great deal of evidence for an equilibrium between the associated and dissociated states of the subunits. The exact nature of the mode of subunit interaction (RNA–RNA, RNA–protein and protein–protein interaction) is not known. Two proteins of the 30S subunit

(S9 and S11) bind to 23S RNA (Stöffler et al. 1971b; Morrison et al. 1973), which suggests protein–RNA association may be important in the formation of 70S ribosomes. There is immunochemical evidence that S20 and L26 are identical; moreover, protein S5 is found in significant amounts with 50S subunits (Tischendorf, Stöffler and Wittmann 1974). S20/L26 and S5 may thus be shared by the subunits in the 70S ribosome and partitioned between them during their preparation. In accord with that interpretation, the 70S ribosome has one copy of S20/L26 and one of S5, whereas the 30S subunit has less than one copy of each (Voynow and Kurland 1971; Weber 1972). S20/L26 and S5 may be located at the subunit interface and indeed may condition the association of the subparticles.

It was of some importance to determine precisely which proteins occupied the interface area of the two subunits. The approach used was to preincubate one or the other of the subunits with an antibody (IgG or Fab) against a single ribosomal protein and determine whether it prevented the formation of 70S ribosomes when the other subparticle was added. The determination was then by analytical ultracentrifugation (Morrison et al. 1973). In another series of experiments the preparation was fixed with glutaraldehyde to prevent subsequent dissociation of subunits and discharge of low affinity antibodies and was analyzed by sedimentation in sucrose gradients. Various degrees of inhibition of reassociation were observed with antibodies against five 30S proteins (S9, S11, S12, S14, S20) and against eight proteins of the 50S subunit (L1, L6, L14, L15, L19, L20, L23, L27) (Morrison et al. 1973; Highland et al. 1974b; Zeichhardt, Tischendorf and Stöffler, unpublished data). As was to be expected, anti-S20 and anti-L26 inhibited reassociation when reincubated with either subunit (Morrison et al. 1973).

Functional assays can also be used to identify the proteins that occur at or near the subunit interface. The strategy is to determine the effect of an antibody on a function catalyzed by one subunit but dependent on or stimulated by the other. EF-G-dependent GTP hydrolysis, EF-Tu-mediated Phe-tRNA binding, the termination reactions, and the colicin-E3 effect are examples of applicable functions. It has already been pointed out (see above) that EF-G-mediated GTP hydrolysis is inhibited by antibodies against S9, S11, L14, L19, L23, L27 when the 30S or the 50S subunit is preincubated with those Fabs but not when the reaction is assayed with 70S ribosomes (Highland et al. 1974a,b; and Figure 6).

When 30S subunits were preincubated with Fabs against S3, S21, S8 and S9, EF-Tu-catalyzed Phe-tRNA binding was more strongly inhibited than when 70S ribosomes were treated (Lelong et al. 1974; Maschler 1973); presumably these enhanced effects resulted from hindrance of subunit association (Figure 7; Table 4). Only the Fabs directed against L14 and L23 were inhibitory for EF-G- as well as EF-Tu-dependent GTP hydrolysis and EF-Tu-mediated Phe-tRNA binding with 70S ribosomes (although only slightly), but they exerted stronger effects when preincubated with 50S subunits. The effects of anti-S9, -S11, -L19 and -L27 were absolutely de-

Table 4 Proteins at, or near, the subunit interface

Protein[a]	Prevention of subunit reassociation		Inhibition of EF-G-mediated GTP hydrolysis[d] at		Inhibition of EF-Tu-mediated Phe-tRNA Binding[e]	Termination	Protection of Phe-tRNA deacylation by RNase[g]	Binding to heterologous rRNA[h]
	Analytical ultracentrifugation[b]	Sucrose gradients[c]	1 mM Mg++	10 mM Mg++				
S2	–		–	–	–	+	–	–
S3	–		–	–	++		–	–
S8	–	++	–	–	+		–	–
S9	++	++	++	–	+	+	++	–
S10	–		–	–	–		++	–
S11	++	++	++	–	–	+	++	23S rRNA
S12	++	++	–	–	–		++	23S rRNA
S14	+	+	–	–	–		+	–
S20/L26	+	++	–	–	–		+	–

Protein							
S21	−			++		+	−
L1	+	+				(+)	−
L2	−					++	−
L4	−					−	−
L6	++	++				+	−
L10	−					++	−
L11	+	++		++		++	−
L14	++	++	+	++		++	−
L15	+					++	−
L16	−					++	−
L19	++	++	+	++		++	− (16S RNA)
L20	++	++				++	−
L23	++	++	+			++	−
L26/S20	+	++				+	−
L27	++	++				++	−

[a] Proteins without any effect are not recorded in the Table. [b] Morrison et al. 1973. [c] Zeichhardt, Tischendorf and Stöffler, unpublished data. [d] Highland et al. 1974a. [e] increase of inhibition by pretreatment of subunits with Fabs; Lelong et al. 1974; Maschler 1973. [f] Tate, Caskey and Stöffler 1974. [g] Lütgehaus and Stöffler, unpublished data. [h] Stöffler et al. 1971b; Morrison et al. 1973; Garrett et al. 1974.

(++) Complete inhibition; (+) significant but incomplete inhibition; (−) negative result; no symbol: experiment not done.

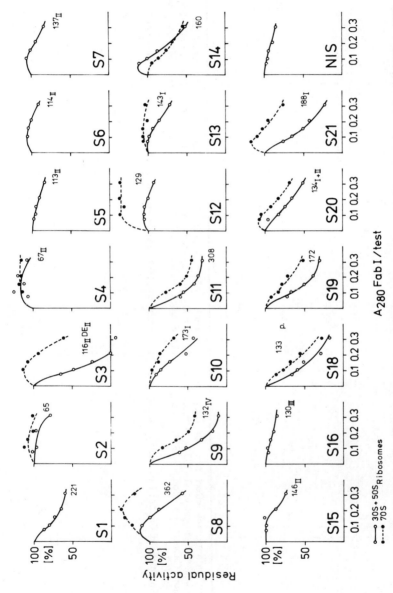

Figure 7 Inhibition of EF-Tu-dependent Phe-tRNA binding to 70S ribosomes (●——●) and 30S + 50S subunits (○——○). The activity at a given concentration of Fab (expressed as a ratio of that obtained in the absence of Fab) is plotted as a function of Fab concentration (Maschler 1973; Lelong et al. 1974).

pendent on preincubation, either with the separate subunits or with a mixture of subunits (Highland et al. 1974a,b; see also Figure 7 and Table 4).

Protection of the Aminoacyl-Ester Bond of Phe-tRNA from RNase Hydrolysis by Formation of 70S Ribosomes. The formation of 70S couples protects the aminoacyl-ester bond of Phe-tRNA from hydrolysis by RNase. 30S or 50S subunits were preincubated with Fabs against the individual proteins, the other subunit was then added, and Phe-tRNA binding was assayed by the method of Pestka (1968). To discriminate between an effect of a particular Fab on the binding of Phe-tRNA and on the formation of subunit couples, a sample from the reaction mixture was treated with pancreatic RNase. Seven Fabs directed against proteins of the 30S subunit (S9, S10, S11, S12, S14, S20 and S21) and eleven against 50S proteins (L2, L6, L10, L11, L14, L15, L16, L19, L20, L23 and L27) abolished protection from RNase digestion and hence may be presumed to have prevented the formation of 70S ribosomes.

The inhibition of subunit association (studied by sucrose gradient centrifugation) did not require that the subunits be preincubated with the Fab; inhibition occurred if the molar ratio of Fab to ribosome and the magnesium concentration were carefully adjusted (Figure 8). The kinetics of dissocia-

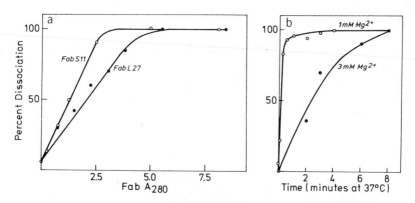

Figure 8 Inhibition of subunit reassociation by monovalent antibody fragments.

(*a*) Concentration dependence of Fab-induced inhibition of subunit reassociation with anti-S11 and anti-L27. Determination of subunit reassociation was by sucrose gradient centrifugation (Highland et al. 1974a; Noll, Bodley, Highland and Stöffler, unpublished data).

(*b*) Inhibition of subunit reassociation by anti-S11 (Fab) at 1 mM and 3 mM Mg^{++} as a function of the time of preincubation of Fabs with dissociated subunits (Noll, Bodley, Highland and Stöffler, unpublished data). At 10 mM, 14 mM and 20 mM Mg^{++}, no inhibitory effects were observed (data not shown).

tion were then also dependent on the magnesium concentration (Figure 8b). The simplest interpretation of the findings is that there is an equilibrium between 70S ribosomes and subunits and that antibodies to interface proteins bind to free subunits and prevent association rather than actually cause dissociation. For the purpose of description, it is convenient to refer to the process as dissociation. With anti-S11 at 1 mM Mg^{++}, dissociation was nearly complete in 20 seconds, whereas at 3 mM Mg^{++} it took 8 minutes (Figure 8b). At 10 mM Mg^{++} only a small amount of dissociation occurred in 30 minutes, and at 14 and 18 mM Mg^{++} no dissociation at all occurred (data not shown). It is important to note that 18 mM is the Mg^{++} concentration used to assay poly(U)-dependent polyphenylalanine synthesis, and in that test Fabs against all these "interface proteins" also strongly inhibit phenylalanine polymerization (Figure 9). Therefore during elongation, the antigenic determinants of those proteins must become available for antibody binding. Many of these antibodies revealed either little or no inhibition of partial reactions at high Mg^{++}. This implies that sub-

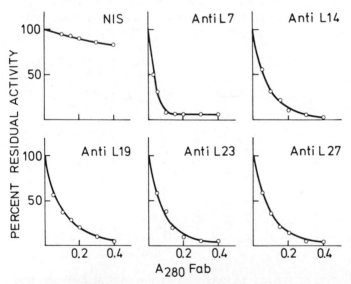

Figure 9 Inhibition of poly(U)-dependent polyphenylalanine synthesis by Fabs to individual 50S subunit proteins. Experimental details are described in the legends to Figures 3 and 4 (Maschler 1973; Lelong et al. 1974). N.B.: The Mg^{++} concentration in the in vitro protein synthesizing system is 18 mM and the Fabs against "interface proteins" L14, L19, L23, L27 lead to strong inhibition. However, these Fabs showed no effects on EF-G- and EF-Tu-mediated GTP hydrolysis and did not cause subunit "disassociation" in sucrose gradients (see Figures 6 and 8, Table 4 and text). Thus the two subunits should come apart during overall protein synthesis, most likely during translocation (Lütgehaus and Stöffler, unpublished data).

units separate (or open up) during one step of protein synthesis, perhaps during translocation.

If all the experiments are considered, then it seems that a relatively large number of proteins are located at or near the subunit interface (Table 4). The greatest number of interface proteins were identified with the amino-acyl-tRNA protection assay and in the reassociation experiments. Several proteins—S9, S11, L14, L19, L23, L27—were identified in every assay employed.

Although the proteins we have identified are likely to be located at the subunit interface, we do not know that they are involved in subunit interaction (e.g., protein–protein or protein–RNA interactions between the particles). Some of the antibodies might have inhibited reassociation indirectly by causing a conformational transition in one or the other subunit. A conformational change that prevents association could be induced by an antibody bound to a protein distant from the interface.

Electron Microscopic Localization of Ribosomal Proteins Using Specific Antibodies. Some 50S proteins located at the subunit interface have been visualized by electron microscopy: the proteins include L19 (Wabl 1973) and L14, L23 and L27 (Tischendorf, Zeichhardt and Stöffler 1974; Figure 10e). In each case a 50S dimer held together by a divalent IgG was seen.

Two well-defined shapes of 50S subunits were reproducibly encountered in these electron micrographs (Wabl 1973; Wabl, Barends and Nanninga 1973; Tischendorf, Zeichhardt and Stöffler 1974; see also Lake and Sabatini, this volume; Figure 10a,f). One projection resembled a pentagonal structure, or a globe carrying a crown with three crests (Figure 10a); the other projection showed particles with a half-circle image and one notch, asymmetrically located on the flat side of the 50S subunit (Figure 10f). Protein L1 could be localized in one of the peripheral crests of 50S subunits (Wabl 1973); L6 and L18 were localized at equivalent locations, as was L1 (Tischendorf, Zeichhardt and Stöffler 1974; Figure 10d). Only recently were we able to discriminate between antibodies bound to the "left" or the "right" of the two peripheral crests (Tischendorf, Zeichhardt and Stöffler, unpublished data). Finally, L17 and L22 were clearly shown to be at the more regular and convex side of the 50S subunit and therefore away from the interface region (Tischendorf, Zeichhardt and Stöffler 1974; Figure 10b,c). The procedure is certain to continue to be of great value in exploring the topography of the ribosomal subunit. The fact that L7 and L12 were assigned to the center crest by electron microscopy (Tischendorf, Zeichhardt and Stöffler, unpublished data), whereas anti-L7 and anti-L12 do not affect subunit reassociation and both L7 and L12 are accessible for antibody binding in 70S couples (see Figure 6), testifies to the importance of combining several methods in determining the identity of subunit interface proteins.

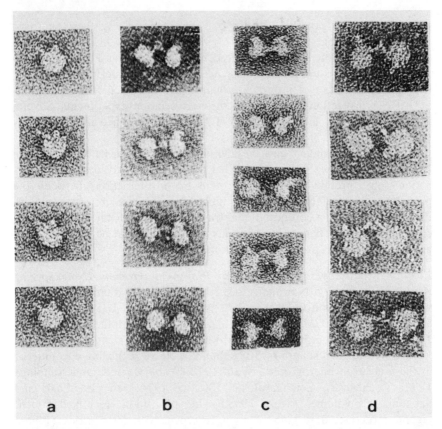

Figure 10   Electron micrographs of 50S subunits and 50S subunit dimers formed by specific IgGs to individual proteins. Samples were negatively stained with uranyl acetate; electron micrographs were taken with a Siemens Elmiscop a. A detailed description of the method is given elsewhere (Wabl 1973; Tischendorf, Zeichhardt and Stöffler 1974). The electron micrographs were kindly provided by G. W. Tischendorf. (*a*) Selected electron micrographs of 50S subunits, resembling a globe carrying a crown with three crests. 300,000 ×. (*b*) Selected electron micrographs of 50S dimers formed by anti-L22 · IgG. 260,000 ×. (*c*) 50S dimers held together by anti-L17 · IgG. 260,000 ×. (*d*) Selected pictures obtained with anti-L18 · IgG. Magnification was 300,000 × for the dimer in the upper line and 320,000 × for the remainder.

RETROSPECT AND PREVIEW

One of the most useful applications of the immunological approach was in establishing that all 55 ribosomal proteins are on the surface of the ribosome. This line of experiments can be promoted in the future and should provide a new series of important results at a deeper level of understanding. A few sequences of ribosomal proteins are already known (Wittmann and Wittmann-Liebold, this volume), and the next years should yield prolifera-

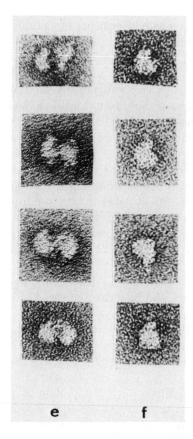

Figure 10 (continued) (*e*) Proteins visualized at the subunit interface. A dimer produced by anti-L14 is shown in the upper line. The three remaining electron micrographs demonstrate 50S dimers formed by anti-L23 · IgG. (*f*) Selected electron micrographs of 50S subunits showing the half-circular image with an asymmetrically located notch ("kidney form"). 300,000 ×.

tion of sequence characterizations. We will therefore be able to investigate which parts of the protein are actually located at the surface of the ribosome. Antibodies specific for single ribosomal proteins are mainly directed against primary sequences (Stöffler and Wittmann 1971b). Hence we might be able to show that particular antibody fractions purified from individual sera react with only certain fragments of the antigen protein (e.g., a tryptic peptide) but still with the intact subunit, whereas other fractions do not react with both. This could lead to the assignment of which parts of the sequences are really exposed and thus precede the more laborious determination by X-ray crystallographic investigations of the tertiary and quarternary structure of the proteins in the ribosome in situ.

We have shown that each of the 30S subunit proteins is involved in, if not also required for, one of the functions of the small subunit. We could also subdivide the proteins into groups which are involved in the 30S part of the A site, the P site or in a hybrid site. Furthermore, S6 and S12 appeared to be specifically required for initiation, whereas S4, S7, S15 and S16 might be necessary for the interaction of the ribosome with natural mRNA (Table 3).

A limited number of antibodies against individual proteins of the 50S subunit yielded inhibition effects when partial reactions of the 50S subunit were investigated. It is obvious that L7 and L12 are required for interaction with elongation factors EF-G and EF-Tu, the release factors RF-1 and RF-2 and the initiation factor IF-2. Although our experiments do not completely exclude possible extra interactions of these factors with rRNA, they make it unlikely that other 50S proteins than L7 and L12 are directly involved in the factor binding site, with the exception of L4 for EF-Tu binding.

L11 and L16 are part of the 50S A site, supporting the assumption (Caskey et al. 1971) that termination involves peptidyltransferase activity. L16 was among the proteins that could be affinity labeled with *p*-nitrophenylcarbamyl-phenylalanyl-tRNA. Thus our immunochemical investigations supported the previous chemical finding that L11 (peptidyltransferase) and L16 (chloramphenicol binding site) are at the A site (Pongs, Bald and Erdmann 1973; Nierhaus and Nierhaus 1973; Nierhaus and Montejo 1973; Pongs et al. 1974). Since anti-L4 inhibited both EF-Tu-dependent Phe-tRNA binding and GTP hydrolysis, L4 might also be located at the A site (Highland et al. 1974b).

A series of proteins (e.g., L11, L14, L19, L23, L27) are located at or near the subunit interface (see Table 4). Antibodies against this group of proteins exhibited their inhibitory effects predominantly by preventing the interaction of the 50S with the 30S subunit. Since L27 was also shown to react with N-substituted tRNA used as an affinity label, this protein may play more than a structural role at the subunit interface. In accordance with this view is the finding that L11, the peptidyltransferase, is also located at the interface (Tate, Caskey and Stöffler 1974).

It is worthwhile to stress that antibodies against interface proteins are not inhibitory to several partial functions when 70S ribosomes are used at 10 mM Mg^{++}, implying that at 10 mM Mg^{++} the parts of the proteins carrying the antigenic determinant are not accessible to the antibody. However, during polyphenylalanine synthesis performed at 18 mM Mg^{++}, all these antibodies lead to strong inhibition. This implies that the ribosome "breathes" at least during overall protein synthesis, or alternatively, that the subunits come apart (most likely during translocation).

Antibodies against all proteins which were shown to be involved in one of the partial functions strongly inhibited poly(U)-dependent polyphenylalanine synthesis. Anti-L1, -L17, -L20, -L24, -L28 and -L33 are strong inhibitors of overall synthesis, but they played no role in any of the partial reactions investigated. Antibodies against several other proteins only affected the rate of protein synthesis and did not lead to complete inhibition. It remains unsettled whether the latter group comprises "structural proteins" that are not directly involved in any in vitro function. They might, however, affect the translation of natural templates or participate in the interaction of the 50S subunit with the membrane. The first possibility is presently being investigated, whereas the second is approachable with methods similar to those used to study subunit interaction.

The availability of amino acid sequences will allow us to pose more detailed questions about the function of individual proteins. I would like to give one example in detail. L7 contains three methionines (in sequence positions 14, 17 and 26). If the protein is fragmented with cyanogen bromide, one obtains several pieces. The large fragment (27–120) is unable to bind to 50S ribosomal cores either alone or together with the small fragments (Stöffler, Kittler, Hasenbank and Pongs, unpublished data). However, antibodies raised against the fragment 27–120 inhibit the interaction of the ribosome with elongation factors as well as antibodies against the entire molecule. In support of this result, the inhibitory activity of anti-L7 and/or anti-L12 could be completely absorbed with the fragment 27–120 (Stöffler, Kittler, Hasenbank and Pongs, unpublished data). Similar experiments can also be performed with fragments 1–73 and 74–120 derived by selective tryptic cleavage at the single arginine in L7 and L12 at position 73. We have begun trying to assign the factor binding sites to one of the two fragments (Kittler, unpublished data). All kinds of immunochemical investigations requiring knowledge of the sequence of the antigen protein are no longer a distant prospect; they are the problem immediately at hand.

Other approaches have already been outlined above and, in particular, those dealing with the topography of ribosomal proteins. We hope soon to be able to establish a topographic ribosome model by the electron optical vizualization of individual ribosomal proteins. This, together with further knowledge about the functional contribution of each ribosomal protein, should end in a deeper insight into the structure of the ribosome and thereby allow a better understanding of the mechanism of protein synthesis.

Acknowledgments

The author is pleased to have the opportunity to thank Heinz-Günter Wittmann for his constant interest, support and encouragement during the course of this work. Thanks are due to my colleagues with whom I had the pleasure of working on problems described here; many of these rewarding collaborations have been supported by EMBO travel fellowships. I am thankful for the expert assistance of Karl-Heinz Rak from the very first. The author is especially indebted to Joseph Highland, Raimund Kittler, Reinhard Maschler, Gilbert Tischendorf and Heinz Zeichhardt for their permission to use results of unpublished experiments. The helpful criticisms and suggestions of Mark Achtmann and Gilbert Tischendorf during the preparation of this manuscript are gratefully acknowledged. Preparation of this article was rendered possible by innumerable discussions with my friends Ira Wool and Renate Hasenbank; I am pleased to thank them for their constant encouragement and their faithful help in many particulars, large and small.

References

Acharya, A., P. B. Moore and F. M. Richards. 1973. Cross-linking of elongation factor EF-G to the 50S ribosomal subunit of *Escherichia coli. Biochemistry* **12**:3108.

Arnon, R. 1971. Antibodies to enzyme—a tool in the study of antigenic specificity determinants. In *Current topics in microbiology and immunology,* vol. 54, pp. 47–93. Springer-Verlag, Berlin, Heidelberg, New York.

Benjamini, E., D. Michaeli and J. D. Young. 1972. Antigenic determinants of proteins of defined sequences. In *Current topics in microbiology and immunology,* vol. 58, p. 85. Springer-Verlag, Berlin, Heidelberg, New York.

Bickle, T. A., J. W. B. Hershey and R. R. Traut. 1972. Spatial arrangement of ribosomal proteins: Reaction of the *E. coli* 30S subunit with bis-imidoesters. *Proc. Nat. Acad. Sci.* **69**:1327.

Bodley, J. W. and L. Lin. 1970. Interaction of *E. coli* G factor with the 50S ribosomal subunit. *Nature* **227**:60.

Brot, N., W. P. Tate, C. T. Caskey and H. Weissbach. 1974. The requirement for ribosomal proteins L7 and L12 in peptide-chain termination. *Proc. Nat. Acad. Sci.* **71**:89.

Brot, N., E. Yamasaki, B. Redfield and H. Weissbach. 1972. The properties of an *E. coli* ribosomal protein required for the function of factor G. *Arch. Biochem. Biophys.* **148**:148.

Cabrer, B., D. Vazquez and J. Modolell. 1972. Inhibition by elongation factor EF-G of aminoacyl-tRNA binding to ribosomes. *Proc. Nat. Acad. Sci.* **69**:733.

Caskey, C. T., A. L. Beaudet, E. M. Scolnick and M. Rosman. 1971. Hydrolysis of fMet-tRNA by peptidyl transferase. *Proc. Nat. Acad. Sci.* **68**:3163.

Celada, F. and R. Strom. 1972. Antibody-induced conformational changes in proteins. *Quart. Rev. Biophys.* **5**:395.

Chang, F. N. and J. G. Flaks. 1970. Topography of the *Escherichia coli* 30S ribosomal subunit and streptomycin-binding. *Proc. Nat. Acad. Sci.* **67**:1321.

Cotter, R. I. and W. B. Gratzer. 1971. Accessibility of RNA and protein in the ribosome. *Eur. J. Biochem.* **23**:468.

Craven, G. R. and V. Gupta. 1970. Three-dimensional organization of the 30S ribosomal proteins from *Escherichia coli.* I. Preliminary classification of the proteins. *Proc. Nat. Acad. Sci.* **67**:1329.

Crichton, R. R. and H. G. Wittmann. 1971. Ribosomal proteins. XXIV. Trypsin digestion as a possible probe of the conformation of *Escherichia coli* ribosomes. *Mol. Gen. Genet.* **114**:95.

Czernilofsky, A. P., G. Stöffler and E. Kuechler. 1974. Messenger-RNA-abhängige Affinitätsmarkierung der 50S Untereinheit des *Escherichia-coli* Ribosoms. *Hoppe-Seyler's Z. Physiol. Chem.* **355**:89.

Czernilofsky, A. P., E. E. Collatz, G. Stöffler and E. Kuechler. 1974. Proteins at the tRNA binding sites of *Escherichia coli* ribosomes. *Proc. Nat. Acad. Sci.* **71**:230.

Deusser, E., G. Stöffler, H. G. Wittmann and D. Apirion. 1970. Ribosomal proteins. XVI. Altered S4 proteins in *Escherichia coli* revertants from streptomycin dependence to independence. *Mol. Gen. Genet.* **109**:298.

Eisen, H. N. and G. W. Siskind. 1964. Variations in affinities of antibodies during the immune response. *Biochemistry* **3**:996.

Estrup, F. and M. Santer. 1966. Immunological analysis of the proteins of *Escherichia coli* ribosomes. *J. Mol. Biol.* **20**:447.

Fakunding, J. L., R. R. Traut and J. W. B. Hershey. 1973. Dependence of initiation factor IF-2 activity on proteins L7 and L12 from *Escherichia coli* 50S ribosomes. *J. Biol. Chem.* **248**:8555.

Fogel, S. and P. S. Sypherd. 1968. Chemical basis for heterogeneity of ribosomal proteins. *Proc. Nat. Acad. Sci.* **59**:1329.

Friedman, D. I., J. G. Olenick and F. E. Hahn. 1968. A 50S ribosomal determinant: Immunological studies correlating function and structure. *J. Mol. Biol.* **32**:579.

Funatsu, G., K. Nierhaus and B. Wittmann-Liebold. 1972. Ribosomal proteins. XXII. Studies on the altered proteins S5 from a spectinomycin resistant mutant of *Escherichia coli*. *J. Mol. Biol.* **64**:201.

Funatsu, G., E. Schiltz and H. G. Wittmann. 1971. Ribosomal proteins. XXVII. Localization of the amino acid exchanges in protein S5 from two *E. coli* mutants resistant to spectinomycin. *Mol. Gen. Genet.* **114**:106.

Furano, A. V., D. F. Bradley and L. G. Childers. 1966. The conformation of the ribonucleic acid in ribosomes. Dye stacking studies. *Biochemistry* **5**:3044.

Garrett, R. A., S. Müller, P. Spierer and R. A. Zimmermann. 1974. Binding of 50S ribosomal subunit proteins to 23S RNA of *Escherichia coli*. *J. Mol. Biol.* (in press).

Garrett, R. A., K. H. Rak, L. Daya and G. Stöffler. 1971. Ribosomal proteins. XXIX. Specific protein binding sites on 16S rRNA of *Escherichia coli*. *Mol. Gen. Genet.* **114**:112.

Geisser, M. 1971. Vergleichende immunologische und elektrophoretische Untersuchungen an Ribosomen von *Enterobacteriaceen*. Diplom-Arbeit, Freie Universität Berlin.

―――. 1973. Vergleichende Untersuchungen an den ribosomalen Proteinen verschiedener Bakterienspezies. *Ph.D. thesis, Freie Universität Berlin*.

Geisser, M., G. W. Tischendorf and G. Stöffler. 1973. Comparative immunological and electrophoretic studies on ribosomal proteins of *Bacillaceae*. *Mol. Gen. Genet.* **127**:129.

Geisser, M., G. W. Tischendorf, G. Stöffler and H. G. Wittmann. 1973. Immunological and electrophoretical comparison of ribosomal proteins from eight species belonging to *Enterobacteriaceae*. *Mol. Gen. Genet.* **127**:111.

Gilbert, W. 1963. Polypeptide synthesis in *E. coli*. I. Ribosomes and the active complex. *J. Mol. Biol.* **6**:374.

Gray, P. N., R. A. Garrett, G. Stöffler and R. Monier. 1972. An attempt at the identification of the proteins involved in the incorporation of 5S RNA during 50S ribosomal subunit assembly. *Eur. J. Biochem.* **28**:412.

Gray, P. N., G. Bellemare, R. Monier, R. A. Garrett and G. Stöffler. 1973. Identification of the nucleotide sequences involved in the interaction between *Escherichia coli* 5S RNA and specific 50S subunit proteins. *J. Mol. Biol.* **77**:133.

Gualerzi, C. and C. L. Pon. 1973. Nature of the ribosomal binding site for initiation factor 3 (IF-3). *Biochem. Biophys. Res. Commun.* **52**:792.

Gupta, R. S. and U.N. Singh. 1972. Biogenesis of ribosomes: Free ribosomal protein pools in *Escherichia coli. J. Mol. Biol.* **69**:279.

Hamel, E., M. Koka and T. Nakamoto. 1972. Requirement of an *E. coli* 50S ribosomal protein component for effective interaction of the ribosome with T and G factors and with guanosine triphosphate. *J. Biol. Chem.* **247**:805.

Hasenbank, R., C. Guthrie, G. Stöffler, H. G. Wittmann, L. Rosen and D. Apirion. 1973. Electrophoretic and immunological studies on ribosomal proteins of 100 *Escherichia coli* revertants from streptomycin dependence. *Mol. Gen. Genet.* **127**:1.

Hauptmann, R., A. P. Czernilofsky, H. O. Voorma, G. Stöffler and E. Kuechler. 1974. Identification of a protein at the ribosomal donor site. *Biochem. Biophys. Res. Commun.* **56**:331.

Hawley, D. A. and I. Slobin. 1974. The immunological reactivity of 30S ribosomal proteins in 70S ribosomes from *Escherichia coli. Biochem. Biophys. Res. Commun.* (in press).

Held, W. A., S. Mizushima and M. Nomura. 1973. Reconstitution of *Escherichia coli* 30S ribosomal subunits from purified molecular components. *J. Biol. Chem.* **248**:5720.

Highland, J. H., J. W. Bodley, J. Gordon, R. Hasenbank and G. Stöffler. 1973. Identity of the ribosomal proteins involved in the interaction with elongation factor G. *Proc. Nat. Acad. Sci.* **70**:142.

Highland, J. H., E. Ochsner, J. Gordon, J. Bodley, R. Hasenbank and G. Stöffler. 1974a. Coordinate inhibition of elongation factor G function and ribosomal subunit association by antibodies to several ribosomal proteins. *Proc. Nat. Acad. Sci.* **71**:627.

Highland, J. H., E. Ochsner, J. Gordon, R. Hasenbank and G. Stöffler. 1974b. Inhibition of phenylalanyl-tRNA binding and elongation factor Tu dependent GTP hydrolysis by antibodies specific for several ribosomal proteins. *J. Mol. Biol.* **86**:175.

Highland, J. H., G. A. Howard, E. Ochsner, G. Stöffler, R. Hasenbank and J. Gordon. 1974c. Identification of the ribosomal protein responsible for the binding of thiostrepton to *E. coli* ribosomes. *J. Biol. Chem.* (in press).

Higo, K. J., K. Loertscher and M. Nomura. 1974. Amino terminal sequences of some *Escherichia coli* 30S ribosomal proteins and functionally corresponding *Bacillus stearothermophilus* ribosomal proteins. *J. Bact.* **118**:180.

Higo, K. J., W. Held, L. Kahan and M. Nomura. 1973. Functional correspondence between 30S ribosomal proteins of *Escherichia coli* and *Bacillus stearothermophilus. Proc. Nat. Acad. Sci.* **70**:944.

Horne, J. R. and V. A. Erdmann. 1973. ATPase and GTPase activities associated with a specific 5S RNA-protein complex. *Proc. Nat. Acad. Sci.* **70**:2870.

Isono, K., S. Isono, G. Stöffler, L. P. Visentin, M. Yaguchi and A. T. Matheson. 1973. Correlation between 30S ribosomal proteins of *Bacillus stearothermophilus* and *E. coli. Mol. Gen. Genet.* **127**:191.

Itoh, T. and H. G. Wittmann. 1973. Amino acid replacement in proteins S5 and S12 from streptomycin dependence to independence. *Mol. Gen. Genet.* **127**:19.

Kahan, L. and E. Kaltschmidt. 1972. Glutaraldehyde reactivity of *E. coli* ribosomes. *Biochemistry* **11**:2691.

Kaltschmidt, E. and H. G. Wittmann. 1970. Ribosomal proteins. XII. Number of proteins in small and large ribosomal subunits of *Escherichia coli* as determined by two-dimensional gel electrophoresis. *Proc. Nat. Acad. Sci.* **67**:1276.

Kaltschmidt, E., G. Stöffler, M. Dzionara and H. G. Wittmann. 1970. Ribosomal proteins. XVII. Comparative studies on ribosomal proteins of four strains of *Escherichia coli. Mol. Gen. Genet.* **109**:303.

Kischa, K., W. Möller and G. Stöffler. 1971. Reconstitution of a GTPase activity by a 50S ribosomal protein from *E. coli. Nature* **233**:62.

Landsteiner, K. 1945. *The specificity of serological reactions* (revised edition). Harvard University Press, Cambridge, Massachusetts.

Lelong, J. C., D. Gros, F. Gros, A. Bollen, R. Maschler and G. Stöffler. 1974. Function of individual 30S subunit proteins of *E. coli*. The effect of specific immunoglobulin fragments (Fab) on the activities of ribosomal decoding sites. *Proc. Nat. Acad. Sci.* **71**:248.

Lutter, L. C., U. Bode, C. G. Kurland and G. Stöffler. 1974. Ribosomal protein neighborhoods. III. Cooperativity of ribosomal assembly. *Mol. Gen. Genet.* **129**:167.

Margoliash, E., M. Reichlin and A. Nisonoff. 1967. The relation of immunological activity and primary structure in cytochrome C. In *Conformation of biopolymers* (ed. G. N. Ramachandran). Academic Press, New York.

Marsh, R. C. and A. Parmeggiani. 1973. Requirement of proteins S5 and S9 from 30S subunits for the ribosome-dependent GTPase activity of elongation factor G. *Proc. Nat. Acad. Sci.* **70**:151.

Maschler, R. 1973. Immunochemische Untersuchungen zur Funktion der Einzelproteine der 30S-Untereinheit von *Escherichia coli* Ribosomen. *Ph.D. thesis,* Freie Universität Berlin.

Meselson, M., M. Nomura, S. Brenner, C. Davern and D. Schlessinger. 1964. Conservation of ribosomes during bacterial growth. *J. Mol. Biol.* **9**:696.

Miall, S. H. and I. O. Walker. 1967. Structural studies on ribosomes. I. The binding of proflavine to *E. coli* ribosomes. *Biochim. Biophys. Acta* **145**:82.

Miller, D. L. 1972. Elongation factors EF-Tu and EF-G interact at related sites on the ribosome. *Proc. Nat. Acad. Sci.* **69**:752.

Miller, R. V. and P. S. Sypherd. 1973. Topography of the *Escherichia coli* 30S ribosome revealed by the modification of ribosomal proteins. *J. Mol. Biol.* **78**:539.

Mizushima, S. and M. Nomura. 1970. Assembly mapping of 30S ribosomal proteins from *E. coli. Nature* **226**:1214.

Modolell, J., B. Cabrer, A. Parmeggiani and D. Vazquez. 1971. Inhibition by siomycin and thiostrepton of both aminoacyl-tRNA and factor G binding to ribosomes. *Proc. Nat. Acad. Sci.* **68**:1796.

Moore, P. B. 1971. Reaction of *N*-ethyl maleimide with the ribosomes of *Escherichia coli. J. Mol. Biol.* **60**:169.

Morgan, P. S. and D. G. Rhoads. 1965. Binding of acridine orange to yeast ribosomes. *Biochim. Biophys. Acta* **102**:311.

Morrison, C. 1974. A sedimentation approach to the topography of the proteins in ribosomes of *Escherichia coli* using protein specific antibodies. *Ph.D. thesis,* Freie Universität Berlin.

Morrison, C. A., R. A. Garrett, H. Zeichhardt and G. Stöffler. 1973. Proteins occurring at, or near the subunit interface of *E. coli* ribosomes. *Mol. Gen. Genet.* **127**:359.

Nierhaus, D. and K. H. Nierhaus. 1973. Identification of the chloramphenicol-binding protein in *Escherichia coli* ribosomes by partial reconstitution. *Proc. Nat. Acad. Sci.* **70**:2224.

Nierhaus, K. H. and V. Montejo. 1973. A protein involved in the peptidyltransferase activity of *Escherichia coli* ribosomes. *Proc. Nat. Acad. Sci.* **70**:1931.

Nishizuka, Y. and F. Lipmann. 1966. The interrelationship between guanosine triphosphatase and amino acid polymerization. *Arch. Biochem. Biophys.* **116**:344.

Nomura, M. and V. A. Erdmann. 1970. Reconstitution of 50S ribosomal subunits from dissociated molecular components. *Nature* **228**:744.

Nomura, M., P. Traub and H. Bechmann. 1968. Hybrid 30S ribosomal particles reconstituted from components of different bacterial origins. *Nature* **219**:793.

Nomura, M., S. Mizushima, M. Ozaki, P. Traub and C. V. Lowry. 1969. Structure and function of ribosomes and their molecular components. *Cold Spring Harbor Symp. Quant. Biol.* **34**:49.

Ouchterlony, Ö. 1958. Diffusion-in-gel methods for immunological analysis. *Progr. Allergy,* vol. 5, pp. 1–78. Karger, Basel/New York.

Page, L. A., S. W. Englander and M. V. Simpson. 1967. Hydrogen exchange studies on ribosomes. *Biochemistry* **6**:968.

Pestka, S. 1968. Studies on the formation of transfer ribonucleic acid-ribosome complexes. IV. A new assay for codon recognition and interaction of transfer ribonucleic acid with 50S subunits. *J. Biol. Chem.* **243**:4038.

Pongs, O. and V. A. Erdmann. 1973. Affinity labeling of *E. coli* ribosomes with a streptomycin-analogue. *FEBS Letters* **37**:47.

Pongs, O., R. Bald and V. A. Erdmann. 1973. Identification of chloramphenicol-binding protein in *Escherichia coli* ribosomes by affinity labeling. *Proc. Nat. Acad. Sci.* **70**:2229.

Pongs, O., K. H. Nierhaus, V. A. Erdmann and H. G. Wittmann. 1974. Active sites in *Escherichia coli* ribosomes. *FEBS Letters* **40** (Suppl.):S28.

Quash, G., J. P. Dandeu, E. Barbu and J. Panijel. 1962. Recherches préliminaires sur les antigènes des ribosomes. *Ann. Inst. Pasteur* **103**:3.

Randall-Hazelbauer, L. L. and C. G. Kurland. 1972. Identification of three 30S proteins contributing to the ribosomal A-site. *Mol. Gen. Genet.* **115**:234.

Richman, N. and J. W. Bodley. 1972. Ribosomes cannot interact simultaneously with elongation factors EF-Tu and EF-G. *Proc. Nat. Acad. Sci.* **69**:686.

Richter, D. 1972. Inability of *E. coli* ribosomes to interact simultaneously with the bacterial elongation factors EF-Tu and EF-G. *Biochem. Biophys. Res. Commun.* **46**:1850.

Santer, M. 1963. Ribosomal RNA on the surface of ribosomes. *Science* **141**:1049.

Schaup, H. W., M. Green and C. G. Kurland. 1970. Molecular interactions of ribosomal components. I. Identification of RNA binding sites for individual 30S ribosomal proteins. *Mol. Gen. Genet.* **109**:193.

————. 1971. Molecular interactions of ribosomal components. II. Site-specific complex formation between 30S proteins and ribosomal RNA. *Mol. Gen. Genet.* **112**:1.

Schreiner, G. and K. H. Nierhaus. 1973. Protein involved in the binding of dihydrostreptomycin to ribosomes of *Escherichia coli. J. Mol. Biol.* **81**:71.

Schrier, P., J. Maassen and W. Möller. 1973. Involvement of 50S ribosomal proteins L6 and L10 in ribosome dependent GTPase activity of elongation factor G. *Biochem. Biophys. Res. Commun.* **53**:90.

Sela, M. 1966. Immunological studies with synthetic polypeptides. *Adv. in Immunol.* **5**:29.

————. 1969. Antigenicity: Some molecular aspects. *Science* **166**:1365.

Sela, M., B. Schechter, J. Schechter and F. Borek. 1967. Antibodies to sequential and conformational determinants. *Cold Spring Harbor Symp. Quant. Biol.* **32**:537.

Spitnik-Elson, P. and A. Breiman. 1971. The effect of trypsin on 30S and 50S subunits of *E. coli. Biochim. Biophys. Acta* **254**:457.

Stöffler, G. 1969. Immunologische Untersuchungen mit Antiseren gegen isolierte ribosomale Proteine aus 30S und 50S Untereinheiten aus *Escherichia coli. Hoppe Seyler's Z. Physiol. Chem.* **350**:1166.

Stöffler, G. and H. G. Wittmann. 1971a. Sequence differences of *Escherichia coli* 30S ribosomal proteins as determined by immunochemical methods. *Proc. Nat. Acad. Sci.* **68**:2283.

————. 1971b. Ribosomal proteins. XXV. Immunological studies on *Escherichia coli* ribosomal proteins. *J. Mol. Biol.* **62**:407.

Stöffler, G., E. Deusser, H. G. Wittmann and D. Apirion. 1971a. Ribosomal proteins. XIX. Altered S5 ribosomal protein in *Escherichia coli* revertant from streptomycin dependence to independence. *Mol. Gen. Genet.* **111**:334.

Stöffler, G., L. Daya, K. H. Rak and R. A. Garrett. 1971b. Ribosomal proteins XXVI. The number of specific protein binding sites on 16S RNA and 23S RNA of *Escherichia coli. J. Mol. Biol.* **62**:411.

————. 1971c. Ribosomal proteins. XXX. Specific binding sites on 23S RNA of *Escherichia coli. Mol. Gen. Genet.* **114**:125.

Stöffler, G., G. W. Tischendorf, R. Hasenbank and H. G. Wittmann. 1974a. The number of individual proteins of the 50S ribosomal subunit from *Escherichia coli* as determined by immunochemical methods. *Eur. J. Biochem.* (in press).

Stöffler, G., R. Hasenbank, J. W. Bodley and J. H. Highland. 1974b. Inhibition of protein L7/L12 binding to 50S ribosomal cores by antibodies specific for proteins L6, L10 and L18. *J. Mol. Biol.* **86**:171.

Stöffler, G., R. Hasenbank, M. Lütgehaus, R. Maschler, C. A. Morrison, H. Zeichhardt and R. A. Garrett. 1973. The accessibility of proteins of the *Escherichia coli* 30S ribosomal subunit to antibody binding. *Mol. Gen. Genet.* **127**:89.

Sun, T., T. A. Bickle and R. R. Traut. 1972. Similarity in size and number of ribosomal proteins from different prokaryotes. *J. Bact.* **111**:474.

Tate, W. P., C. T. Caskey and G. Stöffler. 1974. Inhibition of peptide chain termination by antibodies specific for ribosomal proteins. *J. Mol. Biol.* (in press).

Terhorst, C. P., W. Möller, R. Laursen and B. Wittmann-Liebold. 1972. Amino acid sequence of a 50S ribosomal protein involved in both EF-G and EF-T dependent GTP-hydrolysis. *FEBS Letters* **28**:325.

Thammana, P., C. G. Kurland, E. Deusser, J. Weber, R. Maschler, G. Stöffler and H. G. Wittmann. 1973. Structural and functional evidence for a repeated 50S subunit ribosomal protein. *Nature* **242**:47.

Tischendorf, G. W., M. Geisser and G. Stöffler. 1973. Comparison of ribosomal RNA-binding protein "L2" isolated from different bacterial species. *Mol. Gen. Genet.* **127**:147.

Tischendorf, G. W., G. Stöffler and H. G. Wittmann. 1974. The comparative characterization of the proteins of *Escherichia coli* 30S and 50S ribosomal subunits. The identity of 20S and L26. *Eur. J. Biochem.* (in press).

Tischendorf, G. W., H. Zeichhardt and G. Stöffler. 1974. The localization of ribosomal proteins on the surface of the 50S subunit by electron optical vizualization of antibody-ribosome-complexes. *Mol. Gen. Genet.* (in press).

Tissières, A. and J. D. Watson. 1958. Ribonucleoprotein particles from *Escherichia coli*. *Nature* **182**:778.

Traub, P. and M. Nomura. 1968. Structure and function of *E. coli* ribosomes. V. Reconstitution of functionally active 30S ribosomal particles from RNA and protein. *Proc. Nat. Acad. Sci.* **59**:777.

Traut, R. R., H. Delius, C. Ahmad-Zadeh, T. A. Bickle, P. Pearson and A. Tissières. 1969. Ribosomal proteins of *E. coli:* Stoichiometry and implications for ribosome structure. *Cold Spring Harbor Symp. Quant. Biol.* **36**:25.

Van Duin, J., P. H. van Knippenberg, M. Dieben and C. G. Kurland. 1972. Functional heterogeneity of the 30S ribosomal subunit of *Escherichia coli*. II. Effect of S21 on initiation. *Mol. Gen. Genet.* **116**:181.

Voynow, P. and C. G. Kurland. 1971. The ribosomal proteins of *E. coli*. III. Stoichiometry of the 30S ribosomal proteins. *Biochemistry* **10**:517.

Wabl, M. 1973. Elektronenmikroskopische Lokalisierung von Proteinen auf der Oberfläche ribosomaler Untereinheiten von *Escherichia coli* mittels spezifischer Antikörper. *Ph.D. thesis*, Freie Universität Berlin.

Wabl, M. R., P. J. Barends and N. Nanninga. 1973. Tilting experiments with negatively stained *E. coli* ribosomal subunits. An electron microscopic study. *Cytobiologie* **7**:1.

Weber, H. J. 1972. Stoichiometric measurements of 30S and 50S ribosomal proteins from *Escherichia coli*. *Mol. Gen. Genet.* **119**:233.

Wittmann, H. G., G. Stöffler, W. Piepersberg, P. Buckel, D. Ruffler and A. Böck. 1974. Altered S5 and S20 ribosomal proteins in revertants of an alanyl-tRNA synthetase mutant of *Escherichia coli*. *Mol. Gen. Genet.* (in press).

Wittmann, H. G., G. Stöffler, D. Apirion, L. Rosen, K. Tanaka, M. Tamaki, R. Takata, S. Dekio, E. Otaka and S. Osawa. 1973. Biochemical and genetic studies on two different types of erythromycin resistant mutants of *Escherichia coli* with altered ribosomal proteins. *Mol. Gen. Genet.* **127**:175.

Wittmann, H. G., G. Stöffler, I. Hindennach, C. G. Kurland, L. Randall-Hazelbauer, E. A. Birge, M. Nomura, E. Kaltschmidt, S. Mizushima, R. R. Traut and T. A. Bickle. 1971. Correlation of 30S ribosomal proteins of *Escherichia coli* isolated in different laboratories. *Mol. Gen. Genet.* **111**:327.

Wittmann, H. G., G. Stöffler, E. Kaltschmidt, V. Rudloff, H. G. Janda, M. Dzionara, D. Donner, K. Nierhaus, M. Cech, I. Hindennach and B. Wittmann. 1969. Protein-chemical and serological studies on ribosomes of bacteria, yeast and plants. *FEBS Symposium* **21**:33.

Wittmann-Liebold, B. and H. G. Wittmann. 1971. Ribosomal proteins. XX. Isolation and analysis of the tryptic peptides of proteins S5 from strain K and B of *Escherichia coli*. *Biochim. Biophys. Acta* **251**:44.

Yaguchi, M., C. Roy, A. T. Matheson and L. P. Visentin. 1973. The amino sequence of the N-terminal region of some 30S ribosomal proteins from *Escherichia coli* and *Bacillus stearothermophilus:* Homologies in ribosomal proteins. *Canad. J. Biochem.* **51**:1215.

Zeichhardt, H. 1974. Die Antigenität von Ribosomen aus *Escherichia coli—* quantitative Untersuchungen. *Diplom-Arbeit,* Freie Universität Berlin.

Zimmermann, R. A. and G. Stöffler. 1974. Purification of proteins from the 50S ribosomal subunit of *Escherichia coli* by ion-exchange chromatography. *J. Biol. Chem.* (in press).

Zimmermann, R. A., A. Muto, P. Fellner, C. Ehresmann and C. Branlant. 1972. Location of ribosomal protein binding sites on 16S ribosomal RNA. *Proc. Nat. Acad. Sci.* **69**:1282.

Ribosome Genetics Revealed by Hybrid Bacteria

Paul S. Sypherd
Department of Medical Microbiology
College of Medicine
University of California
Irvine, California 92650

Syozo Osawa
Research Institute for Nuclear Medicine and Biology
Hiroshima University
Hiroshima, Japan

INTRODUCTION

Genetic mosaicism in plants and animals has made it possible to study problems of chromosome linkage and gene action that otherwise could not be approached experimentally. An extension of this natural phenomenon has been the production of hybrids between the somatic cells of different animals. Certain hybrid cell lines have made it possible to assign specific genes to a given chromosome or chromosome group (Weiss and Green 1967; Matsuya, Green and Basilico 1968). As a general technique, hybrid cells may be useful for mapping the genes which control essential functions, especially where there are no apparent methods for selecting conditional (i.e., temperature-sensitive) mutations.

The possibility that genetic hybrids could be used to map the genes for ribosomal proteins and RNA was first suggested by the studies of Leboy, Cox and Flaks (1964). In studying the differences that occur naturally between the ribosomal proteins of *Escherichia coli* strains, they showed that one protein had a different electrophoretic mobility in strain K12 from its counterpart in strain B. The genetic determinant for this protein (the "K character") was shown to be linked to the locus controlling streptomycin resistance (*str*A) when crosses between the two strains were performed.

Mapping the Genes for Ribosomal Proteins

Naturally occurring differences in the chromatographic behavior of ribosomal proteins from different *E. coli* strains were investigated further by Osawa and his colleagues. Takata, Dekio and Osawa (1969) found that in addition to the "K character," or protein S7, the 30S protein S5 of *E. coli* K12 could be distinguished from its counterpart in strain B on carboxymethyl-cellulose (CMC). Similarly, Takata, Dekio and Osawa (1969) and

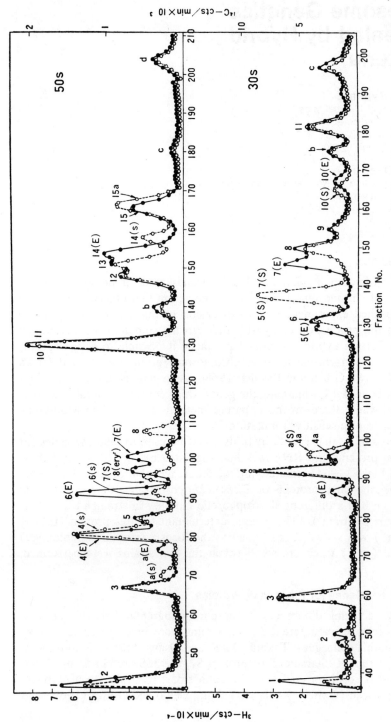

Figure 1 Chromatography of ribosomal proteins on carboxymethyl-cellulose. The proteins of *S. typhimurium* were labeled with [³H]lysine (o – – – o) and those of *E. coli* JC411EO1 were labeled with [¹⁴C]lysine (•——•). The proteins were mixed and run simultaneously. (Reprinted from Dekio, Takata and Osawa, 1970b, with permission of Springer-Verlag.)

670

Dekio (1971) described differences in protein S4 between strains K12 and W3637 and in S17 between JC411 and AT2472. They showed that the genes specifying these two proteins were cotransduced with the $strA$ marker by phage P_1.

The real floodgate of mapping data was opened when attention was turned to comparisons between *E. coli* and other enteric bacteria. O'Neil (1970) found differences in as many as four 30S ribosomal proteins between *E. coli* and *Shigella boydii, Salmonella typhimurium, Salmonella typhosa, Erwinia amylorora* and *Protens mirabilis*. Similar analyses of both 30S and 50S ribosomal proteins on various enteric bacteria were performed by Dekio, Takata and Osawa (1970a) and Osawa, Itoh and Otaka (1971). In the case of *S. typhimurium*, Dekio, Takata and Osawa (1970b) showed that four 30S and six 50S ribosomal proteins were distinguishable from *E. coli* by CMC chromatography. These differences can be seen in Figure 1.

On the basis of these findings, attempts to map several genes for ribosomal proteins by intergeneric crosses were made by O'Neil, Baron and Sypherd (1969). They used *S. typhosa* as a recipient in conjugation crosses with Hfr strains of *E. coli* K12 and produced both diploid and haploid hybrids containing varying portions of the *E. coli* chromosome. Electrophoretic analyses performed on the ribosomes of the hybrids, together with detailed genetic studies to show the extent of donor chromosome present in the Salmonella recipient, showed that the genes for three proteins of the 30S ribosome were clustered near the $strA$ locus. One of the hybrids was diploid for the $strA$ region, as indicated by the segregation of Str^r and Str^s clones from the sensitive hybrid. This hybrid also contained the alleles from both parents for the three proteins, since the proteins characteristic of both parents were present in the hybrid. Figure 2A shows the separation of the 30S ribosomal proteins from the *E. coli* donor (E), the diploid hybrid (H), and the *S. typhosa* recipient (S). When selection was made for integration of the Str^r allele, a stable haploid recombinant was formed. As Figure 2B shows, in this haploid hybrid (H) the ribosomal proteins which can be distinguished are characteristic of the *E. coli* donor (E).

When constructing hybrids for use in mapping ribosomal proteins, one is forced to select for the incorporation of nutritional or drug-resistance markers to identify the donor segments of chromosome. In its initial stages, the mapping effort is aided by the fact that relatively large segments of chromosome can be included in the hybrid cell. However, once the ribosome genes are limited to a certain region of the chromosome, the large segments become a handicap. This is because close linkage between ribosome and nutritional gene cannot be established, since large numbers of recombinants cannot be analyzed by chromatography or electrophoresis of their ribosomal proteins.

Closer linkage between ribosome genes and nutritional markers can be established by employing a transducing phage which is known to carry only 1–1.5% of the bacterial chromosome. The problem here is that transducing

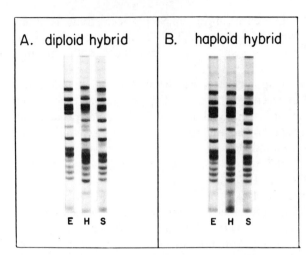

A. diploid hybrid   B. haploid hybrid

E H S   E H S

Figure 2 Electrophoretic analysis of 30S ribosomal proteins from an *E. coli-S. typhosa* hybrid (H). The proteins of both parents are shown for comparison: (E), *E. coli* Hfr; (S), *S. typhosa* recipient. (*A*) Proteins of a hybrid which is diploid for a region including *str*A to *pro*B. (*B*) Proteins of a hybrid which is haploid and contains *E. coli* genes from *str*A to *xyl*. Acrylamide gels were run at pH 4.5 in 8 M urea. *See* Sypherd, O'Neil and Taylor (1969) and O'Neil, Baron and Sypherd (1969) for genetic details.

phages have a restricted species range and the very analysis depends on gene transfers between species. The restricted species range of the transducing phage was circumvented by Dekio, Takata and Osawa (1970b) and by O'Neil and Sypherd (1971) in the following way. A large chromosome segment was transferred from an Hfr *S. typhimurium* to *E. coli* by conjugation followed by the selection of appropriate markers close to *str*A (e.g., *mal*A, *arg*G and *aro*E). Recombinants for the nutritional markers were then screened by isolating their ribosomes and performing an electrophoretic or chromatographic analysis of the proteins. Some recipients were found to contain the donor (Salmonella) ribosomal proteins. These hybrids, which contained variable and undetermined lengths of the donor chromosome integrated into them, were then used for the preparation of the transducing phage P_1. The recipient for transduction was a second *E. coli* strain, leading ultimately to the transfer of ribosomal protein genes from *S. typhimurium* to *E. coli* and comprising a DNA segment on the order of 1% the length of the *E. coli* chromosome.

By the combination of conjugation and transduction for forming *E. coli*–Salmonella hybrids, several new ribosomal protein genes could be placed near the *str*A locus. In fact, all of the distinguishable differences in the proteins of these organisms were shown to be controlled by genes in that region. The possibility that ribosomal protein genes were located elsewhere could not be excluded since at most only 13 or 14 proteins out of a total of 55 could be examined in crosses of this type.

In an effort to expand this map even more, Osawa and his colleagues pursued the species differences in ribosomal proteins by examining other enteric bacteria. Because phage P_1 can also be propagated on *Shigella dysenteriae,* Dekio (1971) compared the proteins of this organism with *E. coli.* As with *S. typhimurium,* genes for the distinguishable proteins could be transferred by transduction between the two species and were linked to *str*A.

In a comparison of *E. coli* and *Serratia marcescens,* at least nine 30S and nine 50S proteins could be distinguished between the species on columns of carboxymethyl-cellulose. The formation of hybrids between *E. coli* and *S. marcescens* is generally difficult to achieve for chromosomal markers. However episomes are known to be more easily transferred across this species barrier. Therefore Takata (1972) prepared episomes of various lengths in *E. coli* which covered the region of the chromosome around *str*A. One such episome, JCH 13, extending from *arg*G to *aro*B, could be transferred to *arg*G$^-$ *S. marcescens* recipients. The resulting hybrid cells maintained the episome as an autonomous element, as evidenced by the fact that the cells could be cured of their donor state and the *arg*$^+$ character by acridine. Chromatographic analyses of the ribosomal proteins revealed that the genes for ribosomal proteins of both parents were present in the hybrid cells. In addition, it was clearly seen that genes for all nine distinguishable *E. coli* 30S ribosomal proteins were present on the episome. An analysis of the 50S ribosomal proteins showed that six of the distinguishable *E. coli* proteins were encoded on the episome. The remaining three distinguishable 50S proteins were usually not observed. The failure to observe all nine 50S proteins could have been due to the difficulty with which these *E. coli* proteins incorporate into hybrid ribosomes or could serve as an indication that genes for these proteins lie in another region of the chromosome. This latter possibility seems remote since crosses involving other species (Dekio, Takata and Osawa 1970b) showed that two of these three proteins were indeed linked to *str*A.

As shown in Figure 3, the combination of interstrain, interspecies and intergenus crosses resulted in mapping 20 proteins of both 30S and 50S ribosomes near the *str*A region. The fact that proteins mapped by this procedure were chosen essentially randomly, that is, by virtue of their natural differences, makes a strong case for all ribosomal protein genes residing as a cluster in that region. In apparent support of this conclusion is the demonstration that several ribosomal protein genes lie in a single transcriptional unit (Nomura and Engbaek 1973). However at least one 30S ribosomal protein, S18, has been shown to be controlled by a gene which maps at a considerable distance from the 64-min location of *str*A (Bollen, et al. 1973). This protein, which maps between 74 and 80 min,

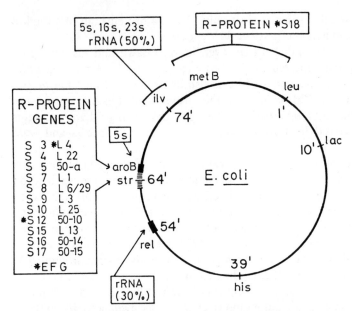

Figure 3 Chromosomal location of ribosome genes. These sites were determined from a combination of genetic hybrids and other methods. The proteins marked with (*) were determined by analysis of drug-resistant or temperature-sensitive mutants. Proteins 50-a, 50-10, 50-14 and 50-15 have not been identified in the nomenclature of Wittmann et al. (1971). Modified from Osawa et al. (1972).

is required for full activity of the 30S particle, as judged from the reconstitution work of Nomura et al. (1969). Whether it represents a unique class of ribosomal proteins that constitutes a separate organized genetic function must await further work.

The use of genetic hybrids for mapping ribosomal proteins has not been restricted to *E. coli* and its relatives. Smith, Goldthwaite and Dubnau (1969) and Tanaka et al. (1973) have shown that the technique can also be applied to *Bacillus subtilis,* where DNA from one strain can be used to transform a second strain which shows some differences in the electrophoretic mobility of certain ribosomal proteins.

Mapping the Genes for Ribosomal RNAs

The finding that many ribosomal protein genes are clustered and comprise a single transcriptional unit raises the possibility that the genes for all ribosomal components, including the RNAs, might reside in the same region of the chromosome. In addition, for some time there has been the notion that rRNA might act as the messenger for ribosomal proteins. Despite the rather compelling reasons for locating rRNA genes on the genetic map, the technical difficulties were great. The possibility of employing conditional lethal mutations was remote since there are multiple copies of the genes (Yankofsky and Spiegelman 1962). In an attempt to locate rRNA genes, Cutler and Evans (1967) employed synchronized cultures of *E. coli* and DNA-RNA hybridization, but with ambiguous results. However, their data did suggest that rRNA genes fell into two distinct and separable regions of the chromosome.

An early application of genetic hybrids to this problem was carried out by Sypherd, O'Neil and Taylor (1969). They first showed that the 23S RNA of *S. typhosa* contained the pancreatic oligonucleotide A_4GC not found in the 23S RNA of *E. coli*. Conversely, *E. coli* contained an A_5U sequence not found in *S. typhosa*. By employing hybrids composed of *E. coli* chromosome segments in *S. typhosa,* they limited the gene for 23S RNA to a region between *xyl* and *proB*. This was in agreement with an earlier finding by Vermeulen (1966), who used RNA-DNA hybridization in various Hfr strains to locate the rRNA genes near the *ilv* region.

The gene locations for rRNA were refined by Osawa and his associates (Muto, Takata and Osawa 1971; Matsubara, Takata and Osawa 1972) by taking advantge of differences in the 16S RNA of the K12 and B strains of *E. coli*. In this case, the K12 strain contained a decanucleotide, generated by pancreatic ribonuclease, which was not evident in strain B. Conjugation between the two strains and selection for nutritional recombinants produced a hybrid line of K12 containing the *met*B gene from strain B (strain B being the donor). This hybrid lost its K12 characteristic decanucleotide, presumably replaced by the B-strain sequence.

The genetic hybrid experiments pointed to a region in the upper left-hand portion of the chromosome as the site for 23S and 16S RNA. It was presumed that the genes for both RNAs were adjacent, since other studies indicated that these genes, as well as those for 5S RNA, are linked (Pato and von Meyenberg 1970; Doolittle and Pace 1970; Bremer and Berry 1971). Examining genetic hybrids for oligonucleotide differences did not, however, reveal whether or not several sites might exist for rRNA genes, as the early work by Cutler and Evans (1967) had suggested. Birnbaum and Kaplan (1971) constructed merodiploid hybrids by transferring *E. coli* episomes into *Protens mirabilis*. The DNA of the hybrids was then used to perform RNA-DNA hybridization studies. Their results showed that only one-half of the RNA genes could lie in the region previously described (Yu, Vermeulen and Attwood 1970; Sypherd, O'Neil and Taylor 1969). In subsequent studies Kaplan and his colleagues (Unger et al. 1972) have shown that a second cluster comprising about 30% of the rRNA genes lies in the 55–59-min region of the *E. coli* chromosome.

The existence of several rRNA gene clusters separated on the chromosome was further supported by the mapping of 5S RNA. This small RNA, found in the 50S subunit (Rosset and Monier 1963) and required for its function (Erdmann et al. 1971), shows some heterogeneity in its sequence (Jarry and Rosset 1971). Such heterogeneity, reflecting the redundancy of the genes controlling the RNA structure, was used by Jarry and Rosset (1973) to map the 5S RNA genes. Among the oligonucleotides produced by ribonuclease T_1 are several sequences unique to the K12 and MRE600 strains, respectively, of *E. coli*. Transfer of a gene coding for an RNA with the unique oligonucleotide could be assessed by carrying out crosses between the two strains. In this way Jarry and Rosset (1973) demonstrated three separate gene clusters (Figure 3); the *aro*B–*mal*A region (66 min), the *rbs–ilv* region (74 min) and the *met*E–*rha* region (75–76 min). A fourth possible region could not be determined.

Since there is experimental evidence that the genes for 23S, 16S and 5S RNA are contiguous (Doolittle and Pace 1971; Pato and von Meyenberg 1970; Bremer and Berry 1971), the gene locations for 5S RNA should coincide with those for the two larger RNAs. At the present time the agreement is only approximate, but surprisingly good when one considers the different approaches that have been used.

Genetic hybrids, produced by interstrain, interspecies or intergenus crosses, have been invaluable in deducing chromosome map positions for the genes controlling ribosomal protein and ribosomal RNA structure. The mapping data, together with other experimental results, show that many of the ribosomal protein genes comprise a tight cluster in the *str*A region and could be considered to make up an operon (Nomura and Engbaek 1973). In addition, the clear separation in the map locations for the proteins and rRNAs argue forcefully against the notion that the rRNAs are messengers for the proteins. Finally, these map locations provide genetic "landmarks" for further studies on genes that control the structure, function and assembly of ribosomes.

References

Birnbaum, L. and S. Kaplan. 1971. Localization of a portion of the ribosomal RNA genes in *Escherichia coli*. *Proc. Nat. Acad. Sci.* **68**:925.

Bollen, A., M. Faelen, J. Lecocq, A. Herzog, J. Zengel, L. Kahn and M. Nomura. 1973. The structural gene for the ribosomal protein S18 in *Escherichia coli*. I. Genetic studies on a mutant having an alteration in the protein S18. *J. Mol. Biol.* **76**:463.

Bremer, H. and L. Berry. 1971. Cotranscription of 16S and 23S ribosomal RNA in *Escherichia coli*. *Nature New Biol.* **234**:81.

Cutler, R. and J. Evans. 1967. Relative transcription activity of different segments of the genome throughout the cell division cycle of *Escherichia coli*. The mapping of ribosomal and transfer RNA and the determination of the direction of replication. *J. Mol. Biol.* **26**:91.

Dekio, S. 1971. Genetic studies of the ribosomal proteins in *Escherichia coli*.

VII. Mapping of several ribosomal protein components by transduction experiments between *Shigella dysenteriae* and *Escherichia coli. Mol. Gen. Genet.* **113**:20.

Dekio, S., R. Takata and S. Osawa. 1970a. Genetic studies of the ribosomal proteins in *Escherichia coli*. III. Compositions of ribosomal proteins in various strains of *Escherichia coli. Mol. Gen. Genet.* **107**:32.

―――. 1970b. Genetic studies of the ribosomal proteins in *Escherichia coli*. VI. Determination of chromosomal loci for several ribosomal protein components using a hybrid strain between *Escherichia coli.* and *Salmonella typhimurium. Mol. Gen. Genet.* **109**:131.

Doolittle, F. and N. Pace. 1970. Synthesis of 5S ribosomal RNA in *Escherichia coli* after rifampicin treatment. *Nature* **228**:125.

―――. 1971. Transcriptional organization of the ribosomal RNA cistrons in *Escherichia coli. Proc. Nat. Acad. Sci.* **68**:1786.

Erdmann, V., S. Fahnestock, K. Higo and M. Nomura. 1971. Role of 5S RNA in the function of 50S ribosomal subunits. *Proc. Nat. Acad. Sci.* **68**:2932.

Jarry, B. and R. Rosset. 1971. Heterogeneity of 5S RNA in *Escherichia coli. Mol. Gen. Genet.* **113**:43.

―――. 1973. Localization of some 5S RNA cistrons on *Escherichia coli* chromosome. *Mol. Gen. Genet.* **121**:151.

Leboy, P., E. Cox and J. Flaks. 1964. The chromosomal site specifying a ribosomal protein in *Escherichia coli. Proc. Nat. Acad. Sci.* **52**:1367.

Matsubara, M., R. Takata and S. Osawa. 1972. Chromosomal loci for 16S ribosomal RNA in *Escherichia coli. Mol. Gen. Genet.* **117**:311.

Matsuya, Y., H. Green and C. Basilico. 1968. Properties and uses of human-mouse hybrid cell lines. *Nature* **220**:1199.

Muto, A., R. Takata and S. Osawa. 1971. Chemical and genetic analysis of 16S ribosomal RNA in *Escherichia coli. Mol. Gen. Genet.* **111**:15.

Nomura, M. and F. Engbaek. 1973. Expression of ribosomal protein genes as analyzed by bacteriophage Mu-induced mutations. *Proc. Nat. Acad. Sci.* **69**:1526.

Nomura, M., S. Mizushima, M. Ozaki, P. Traub and C. Lowry. 1969. Structure and function of ribosomes and their molecular components. *Cold Spring Harbor Symp. Quant. Biol.* **34**:49.

O'Neil, D. 1970. Genetic mapping of 30S and 50S ribosomal protein cistrons by intergeneric transduction and conjugation. *Ph.D. thesis,* University of Illinois, Urbana.

O'Neil, D. and P. Sypherd. 1971. Cotransduction of *str*A and ribosomal protein cistrons in *Escherichia coli–Salmonella typhimurium* hybrids. *J. Bact.* **105**:947.

O'Neil, D., L. Baron and P. Sypherd. 1969. Chromosomal location of ribosomal protein cistrons determined by intergeneric bacterial mating. *J. Bact.* **99**:242.

Osawa, S., T. Itoh and E. Otaka. 1971. Differentiation of the ribosomal protein compositions in the genus *Escherichia* and its related bacteria. *J. Bact.* **107**:168.

Osawa, S., E. Otaka, R. Takata, S. Dekio, M. Matsubara, T. Itoh, A. Muto, K. Tamaka, H. Teraoka and M. Tamaki. 1972. Ribosomal protein genes in bacteria. *FEBS Symp.* **23**:313.

Pato, M. and K. von Meyenberg. 1970. Residual RNA synthesis in *Escherichia coli* after inhibition of initiation of transcription by rifampicin. *Cold Spring Harbor Symp. Quant. Biol.* **35**:497.

Rosset, R. and R. Monier. 1963. A propos de la présence d'acide ribonucléique de faible poids moléculaire dans les ribosomes d'*Escherichia coli*. *Biochim. Biophys. Acta* **68**:653.

Smith, I., C. Goldthwaite and D. Dubnau. 1969. The genetics of ribosomes in *Bacillus subtilis*. *Cold Spring Harbor Symp. Quant. Biol.* **34**:85.

Sypherd, P., D. O'Neil and M. Taylor. 1969. The chemical and genetic structure of bacterial ribosomes. *Cold Spring Harbor Symp. Quant. Biol.* **34**:77.

Takata, R. 1972. Genetic studies of the ribosomal proteins in *Escherichia coli*. VIII. Mapping of ribosomal protein components by intergeneric mating experiments between *Serratia marcescens* and *Escherichia coli*. *Mol. Gen. Genet.* **118**:363.

Takata, R., S. Dekio and S. Osawa. 1969. Genetic studies of the ribosomal proteins in *Escherichia coli*. I. Mutants and strains having 30S ribosomal subunit with altered protein components. *Mol. Gen. Genet.* **105**:113.

Tanaka, K., M. Tamaki, S. Osawa, A. Kimura and R. Takata. 1973. Erythromycin-resistant mutants of *Bacillus subtilis*. *Mol. Gen. Genet.* **127**:157.

Unger, M., L. S. Birnbaum, S. Kaplan and A. Pfister. 1972. Location of the ribosomal RNA cistrons of *Escherichia coli:* A second site. *Mol. Gen. Genet.* **119**:377.

Vermeulen, C. 1966. The genetic mapping of the ribosomal RNA loci in *E. coli* and the direction of chromosome replication in Hfr strains. *Ph.D. thesis*, University of Illinois, Urbana.

Weiss, M. and H. Green. 1967. Human-mouse hybrid cell lines containing partial complements of human chromosomes and functioning human genes. *Proc. Nat. Acad. Sci.* **58**:1104.

Wittmann, H., G. Stöffler, I. Hindennach, C. Kurland, L. Randall-Hazelbauer, E. Birge, M. Nomura, E. Kaltschmidt, S. Mizushima, R. Traut and T. Bickle. 1971. Correlation of 30S ribosomal proteins of *Escherichia coli* isolated in different laboratories. *Mol. Gen. Genet.* **111**:327.

Yankofsky, S. and S. Spiegelman. 1962. The identification of the ribosomal RNA cistron by sequence complementarity. II. Saturation of and competitive interaction of the RNA cistron. *Proc. Nat. Acad. Sci.* **48**:1466.

Yu, M., C. Vermeulen and K. Attwood. 1970. Location of the genes for 16S and 23S ribosomal RNA in the genetic map of *Escherichia coli*. *Proc. Nat. Acad. Sci.* **67**:26.

The Ribosome Cycle

Raymond Kaempfer
The Biological Laboratories
Harvard University
Cambridge, Massachusetts 02138

INTRODUCTION

Three forms of ribosomal particles are generally observed in extracts of both prokaryotic and eukaryotic cells: polysomes, single ribosomes and ribosomal subunits. Analysis of the functional relationship between these three classes and their role in the ribosome cycle in protein synthesis has been complicated chiefly for two reasons. First, a complete turn of the ribosome cycle—repeated passage of a ribosome over messenger RNA—requires the participation of so many components that faithful reproduction in a purified, reconstituted cell-free system is virtually impossible. Individual steps can be analyzed more precisely, but their placement within the cycle is often problematical. Second, the three observed states of ribosomal particles are stable enough to be recovered in extracts, but the existence of additional, less stable states can only be revealed by special or indirect methods. Moreover, particles in different states may not always be separated by ultracentrifugation. These complications at times have led to apparently contradictory results.

To simplify the analysis of available facts, I shall present a general scheme for the ribosome cycle, discuss the evidence in support of its individual features, show how the available observations in bacterial and eukaryotic systems can be accommodated within this scheme, and finally present the alternative interpretations to which the scheme remains open.

THE RIBOSOME CYCLE—A GENERAL SCHEME

Several ribosomes are engaged in protein synthesis as they move along a messenger RNA molecule, forming a *polysome* (depicted at the top of Figure 1). At the termination codon (t), the completed polypeptide chain is released and the ribosome leaves messenger RNA to become a *termination ribosome.* This undissociated intermediate of the ribosome cycle is unstable and readily dissociates, yielding a pair of *free ribosomal subunits.* These free ribosomal subunits are themselves short-lived intermediates that can proceed along two different paths. Ordinarily, the small ribosomal subunit binds a molecule of initiation factor IF-3, and a pair of relatively stable, *native ribosomal subunits* is formed. Alternatively, when insufficient IF-3 is available, free ribosomal subunits associate with high affinity to form a relatively stable *single ribosome,* which by itself does not function as

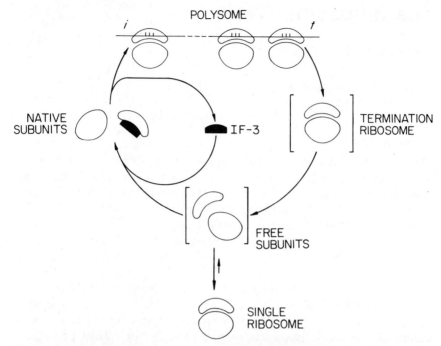

Figure 1   General scheme for the ribosome cycle. See text for description. Only components essential to the cycle are depicted. Intermediates thought to be unstable are in brackets. An arrow within this simplified diagram may represent more than one step. Binding of IF-3 to small subunits is reversible in vitro. IF-3 and its cycle are probably more complex in eukaryotic systems.

an intermediate of the ribosome cycle, but can return to it when dissociation into free subunits occurs. Native subunits are at once incapable of forming single ribosomes and are able to reenter polysomes. During one of the steps leading to formation of a new initiation complex on messenger RNA (*i*), initiation factor IF-3 is released and recycled.

Specific aspects of this cycle have been proposed before, as we shall see below. Some apparently diverging proposals, however, can be accommodated within the scheme of Figure 1.

RIBOSOMAL SUBUNIT EXCHANGE

Demonstration in Intact Cells

Cyclic dissociation of ribosomes has been demonstrated in growing bacterial cells by analysis of the density of ribosomes and ribosomal subunits after transfer of a culture uniformly labeled with heavy isotopes to a medium containing only light isotopes (Kaempfer, Meselson and Raskas

1968). Upon transfer, heavy ribosomes are progressively replaced by two species of hybrid density, one containing a heavy 50S subunit and a light 30S subunit, and vice versa for the other hybrid. Within the limits of detection, the two ribosomal subunit species remain intact during growth and are continuously recycled through polysomes.

A similar demonstration of ribosomal subunit exchange in eukaryotic cells (Kaempfer 1969) was made possible by the fact that the yeast *Candida krusei* is capable of rapid growth in heavy isotopes. Immediately upon transfer to light medium, the cells contain only fully heavy 80S cytoplasmic ribosomes, derived from polysomes by digestion with ribonuclease (Figure 2a). One generation after transfer, however, approximately half of the label originally present in heavy ribosomes is seen to sediment in the region intermediate between heavy and light, where hybrid ribosomes are expected (Figure 2b). Four generations after transfer, this label is found almost exclusively in ribosomes sedimenting as hybrids (Figure 2c). Since heavy subunits retain their original density for at least four generations after transfer, subunit exchange in ribosomes must have occurred during this period, yielding two reciprocal species of hybrid ribosomes. Though not

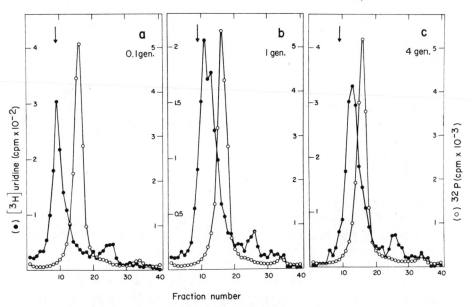

Figure 2   Sedimentation distribution of cytoplasmic ribosomes of yeast after transfer from heavy to light medium. A culture of *C. krusei* was grown in heavy [³H]uridine-containing medium and transferred to unlabeled light medium. After the indicated periods of growth in light medium, the cells were lysed and the extracts were treated with ribonuclease and layered on isokinetic sucrose gradients. ³²P-labeled light cell extract was included in each gradient. Arrow indicates expected position of heavy 80S ribosomes (Kaempfer 1969).

resolved in the experiment, the existence of two species is indicated by the broadening of the ^{3}H peak in (c). The proportion of ^{3}H label in heavy ribosomal subunits does not increase after transfer, indicating that subunits fully retain their capacity to recycle through polysomes.

These experiments provided direct evidence that ribosomes in prokaryotic and eukaryotic cells frequently undergo subunit exchange, apparently by dissociation into their two subunits and re-formation from a pool of free subunits that continuously recycle through polysomes. Indirect evidence consistent with subunit exchange came from studies of the rate of equilibration of newly synthesized ribosomal subunits with polysomes in bacteria (Mangiarotti and Schlessinger 1967) and mammalian cells (Joklik and Becker 1965; Hogan and Korner 1968b).

Role in Protein Synthesis

To determine how often a ribosome undergoes subunit exchange and to show that the exchange observed in vivo is connected to protein synthesis, a cell-free system was necessary. To this end, a polysome-containing extract from bacteria grown in heavy isotopes was allowed to synthesize protein in the presence of a large excess of similar extract from light cells (Kaempfer 1968). Not only did extensive ribosomal subunit exchange occur, but it was absolutely dependent upon protein synthesis. Moreover, exchange of subunits was 50% complete within 2.5 min, a time span not much greater than that expected for a complete cycle of translation in vivo. The kinetics of exchange thus is compatible with an exchange of subunits for each round of protein synthesis.

These findings led to the conclusion that ribosomes dissociate into their two subunits upon termination of protein synthesis or shortly thereafter, and are reformed from subunits at initiation. Subsequent in vitro studies of bacteriophage RNA translation by Grubman and Nakada (1969) further strengthened this conclusion.

Evidence for a similar dissociation step in mammalian cell-free protein synthesis was provided by Falvey and Staehelin (1970), who used labeled 40S ribosomal subunits from liver to demonstrate a rapid exchange of subunits in unlabeled ribosomes upon their release from polysomes, and by Howard, Adamson and Herbert (1970), who used labeled polysomes to demonstrate their rapid and complete exchange with subunits in an unlabeled rabbit reticulocyte lysate capable of multiple rounds of protein synthesis.

Ribosomal subunits of membrane-bound polysomes and free polysomes appear to share the same ribosome cycle, for upon their release from membrane-bound polysomes, 40S subunits exchange rapidly with added subunits derived from free polysomes; 60S subunits of membrane-bound ribosomes, on the other hand, do not exchange, most likely because they remain attached to the membrane (Borgese, Blobel and Sabatini 1973).

Two kinds of experiments indicate that a ribosome undergoes subunit exchange between individual rounds of translation, but not during polypeptide synthesis on messenger RNA. In the first, the rate of exchange in vitro was studied in the presence of puromycin. If exchange of subunits normally occurs after termination, but not during polypeptide synthesis, puromycin, by causing premature release of nascent chains, might be expected to induce an abnormally high rate of subunit exchange. Indeed, in the presence of puromycin, full ribosomal subunit exchange occurs instantaneously (Kaempfer and Meselson 1969), whereas normally it requires 15 min to reach completion (Kaempfer 1968). In the second experiment, a mixture of heavy and light polysome extracts was allowed to synthesize proteins under conditions permitting little, if any, chain initiation. After more than half of the ribosomes had left polysomes, the remaining polysomes were found to contain only fully heavy and fully light ribosomes (Kaempfer 1970).

Function of Ribosomal Subunits

The need for ribosome dissociation in protein synthesis had not been perceived in early studies of cell-free polypeptide formation because these were done almost exclusively with synthetic polyribonucleotides, to which 70S ribosomes could attach at random sites. This need became clear when it was shown that the binding of N-formylmethionyl-tRNA$_F$ to the initiation codon AUG in natural (Nomura and Lowry 1967; Nomura, Lowry and Guthrie 1967; Guthrie and Nomura 1968) or synthetic messenger RNA (Ghosh and Khorana 1967) uniquely requires 30S and 50S ribosomal subunits, not 70S ribosomes.

These studies and the data on ribosomal subunit exchange led to the concept that initiation of protein synthesis on natural messenger RNA requires ribosomal subunits, with attachment of 30S subunits and N-formylmethionyl-tRNA$_F$ to AUG initiation codons preceding the entry of 50S ribosomal subunits. Many details of the formation of new initiation complexes on messenger RNA are now known (see Haselkorn and Rothman-Denes 1973).

SINGLE RIBOSOMES

Constancy in Number of Ribosomal Subunits

When protein synthesis slows down under a variety of metabolically adverse conditions, single ribosomes accumulate at the expense of polysomes in both prokaryotic and eukaryotic cells. Joklik and Becker (1965) first noted that whereas the level of single ribosomes may vary widely depending on the metabolic state of the cell, the level of ribosomal subunits remains essentially constant. Working with HeLa cells, they showed that this was also true after treatment with puromycin. This constancy in the number of ribosomal subunits was further documented for ascites cells (Hogan and Korner 1968a) and for bacteria (Subramanian, Davis and Beller 1969).

Properties of Single Ribosomes

Single ribosomes lack messenger RNA or growing polypeptide chains and thus are not engaged in protein synthesis. Transfer RNA also appears to be absent from bacterial single ribosomes (Tai and Davis 1972).

Single ribosomes can be distinguished from translating ribosomes by several criteria. Bacterial single ribosomes dissociate more readily than polysomal ribosomes when the Mg^{++} ion concentration is lowered (Ron, Kohler and Davis 1968; Kelley and Schaechter 1969); they are sensitive to dissociation by pancreatic ribonuclease, whereas polysomal ribosomes are resistant (Kaempfer 1970); in the presence of initiation factor IF-3 and sufficiently low Mg^{++} concentrations, they are converted into subunits, whereas polysomal ribosomes are not (Subramanian, Davis and Beller 1969); they can be dissociated selectively by Na^+ ions (Kaempfer 1970; Beller and Davis 1971); and they appear to sediment more slowly in sucrose gradients (Kaempfer 1970).

The latter property is explained by pressure-induced dissociation, a phenomenon first described by Infante and Baierlein (1971), who used sea urchin single ribosomes, and confirmed for bacterial single ribosomes (Spirin 1971; Van Diggelen, Oostrom and Bosch 1971; Subramanian and Davis 1971).

The early in vitro studies of ribosomal subunit exchange with heavy isotopes (Kaempfer 1968, 1970; Kaempfer and Meselson 1969) preceded the discovery of pressure-induced dissociation, and in contrast to the in vivo studies described above (Kaempfer, Meselson and Raskas 1968; Kaempfer 1969), the ribosomes analyzed in the former experiments contained significant numbers of single ribosomes. It was thus appropriate to ask if the ribosomes sedimenting as density hybrids in velocity gradients were in fact hybrids or were products of pressure-induced dissociation generated from heavy ribosomes as they advanced ahead of the light ribosomes; their hybrid nature was confirmed by Subramanian and Davis (1971), who used fixation with glutaraldehyde to eliminate the pressure-induced dissociation.

Mammalian single ribosomes are less compact than polysomal ribosomes and also dissociate more readily in salt solutions (Vournakis and Rich 1971).

Slow Equilibration of Single Ribosomes
with Subunits and Polysomes In Vivo

In intact mammalian cells, label in newly synthesized ribosomal subunits enters polysomes extensively and without much delay; by contrast, the rise in specific radioactivity of single ribosomes lags far behind that of the ribosomal subunits and only approaches their specific activity after several hours (Joklik and Becker 1965; Girard et al. 1965; Hogan and Korner 1968b; Kabat and Rich 1969). Mammalian ribosomal subunits, therefore, equilibrate much more slowly with single ribosomes than with polysomes. Since rapid and extensive protein synthesis occurred in these experiments, and since the majority of ribosomes continued to be associated with polysomes, this finding implies that the single ribosomes observed cannot be obligatory intermediates of the polysome-subunit cycle.

This conclusion receives additional support from in vitro studies (Howard, Adamson and Herbert 1970; Falvey and Staehelin 1970; Kaempfer 1970, 1971, 1972; Kaempfer and Kaufman 1972) that will be discussed in later sections.

Although they do not behave as intermediates of the cycle, single ribosomes can return to polysomes when the rate of overall protein synthesis increases (Staehelin, Verney and Sidransky 1967; Hogan and Korner 1968a).

THE DISSOCIATION STEP IN THE RIBOSOME CYCLE

Dissociation after Termination

As we have seen, the kinetics of ribosomal subunit exchange in vitro (Kaempfer 1968) and the requirement for ribosomal subunits in initiation complex formation (Guthrie and Nomura 1968) provide strong evidence that ribosomes must undergo dissociation upon termination of polypeptide synthesis or shortly thereafter. These experiments, however, do not pinpoint the precise step at which dissociation occurs. Such knowledge is particularly important for understanding the relationship of single ribosomes to the ribosome cycle. Because of the complexity of the experimental evidence, I shall examine it in some detail.

Rapid Exchange of Ribosomal Subunits after Termination

An essential observation is that accumulation of single ribosomes at the expense of polysomes, during slowing protein synthesis, is accompanied by rapid and extensive exchange of subunits in single ribosomes. This exchange can be demonstrated when single ribosomes are allowed to accumulate during polypeptide synthesis in a $1:10^3$ mixture of extracts from isotopically heavy and light bacteria. The single ribosomes originating from heavy polysomes are found to be of hybrid density, indicating that they contain one heavy and one light subunit (Kaempfer 1970; Subramanian and Davis 1971). Similarly, rapid exchange of added, labeled 40S subunits with mammalian single ribosomes occurs when the latter accumulate at the expense of polysomes (Falvey and Staehelin 1970).

These experiments eliminated the possibility that single ribosomes are stable intermediates of the ribosome cycle that accumulate during slowing protein synthesis because they fail to dissociate (Kohler, Ron and Davis 1968; Colombo, Vesco and Baglioni 1968; Algranati, Gonzalez and Bade 1969). Indeed, in a bacterial extract greatly diluted with respect to polysomes, polysome runoff results in the accumulation of ribosomal subunits, not single ribosomes (Kaempfer 1970; see Figure 3).

The rapid exchange of subunits after termination could be explained in two possible ways (Kaempfer 1970). Either ribosomes on completing polypeptide chains dissociate into ribosomal subunits (possibly via a short-lived undissociated intermediate) which combine readily to form stable single ribosomes that, once formed, equilibrate only slowly with the pool of subunits, or single ribosomes are released directly from polysomes but then equilibrate rapidly and continuously with the pool of subunits, thus remaining unstable. To distinguish between these alternatives, knowledge of the rate of equilibration of single ribosomes with ribosomal subunits is essential.

The Equilibrium between Single Ribosomes and Ribosomal Subunits

We have already seen that single ribosomes in mammalian cells equilibrate slowly with subunits and polysomes. The rate of equilibration of bacterial single ribosomes with subunits was studied in a dilute polysome extract from ^{3}H-labeled cells supplemented with S-100 from unlabeled cells. As

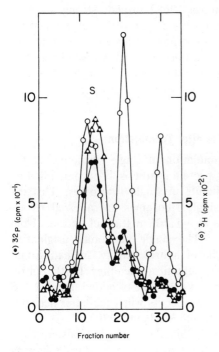

Figure 3 Stability of single ribosomes during accumulation of ribosomal subunits at the expense of polysomes. ^{32}P-labeled *E. coli* single ribosomes (●) were incubated with dilute ^{3}H-labeled polysome extract (○) in conditions for protein synthesis, with S-100 from RNase I⁻ cells. After incubation, the sample was centrifuged directly; no ribonuclease was added. Profile of unincubated single ribosomes (△), sedimented in a parallel gradient, is superimposed. *S*, position of single ribosomes (Kaempfer 1970).

seen in Figure 3, which illustrates the sedimentation distribution of ribosomes and ribosomal subunits, protein synthesis in these conditions results
in the accumulation of extensive amounts of 30S and 50S subunits at the
expense of polysomes (open circles). The rationale of this procedure was
to attain low subunit concentrations that limit the rate of subunit association. Although ^{3}H-labeled polysomes were converted extensively into ribosomal subunits, a fifteenfold smaller number of ^{32}P-labeled single ribosomes
included in the same incubation mixture was not (Figure 3, filled circles),
but instead sedimented essentially as before incubation (triangles). This
demonstrates the essential stability of single ribosomes in conditions of
protein synthesis leading to the conversion of polysomal ribosomes into
subunits. The ^{32}P-labeled single ribosomes in Figure 3 had been prepared
freshly and were taken from a sucrose gradient. Although they did not dissociate significantly under conditions permitting rapid cell-free polypeptide
synthesis, single ribosomes prepared by this method dissociate readily at
lower Mg^{++} ion concentrations (Kaempfer 1972). The result of Figure 3,
therefore, offers support for the view that single ribosomes, unlike ribosomes leaving polysomes, equilibrate very slowly with the pool of ribosomal
subunits. Independent support for this conclusion is offered by the data
of Figure 6, as we shall see below.

The same conclusion was reached for mammalian single ribosomes.
During repeated rounds of protein synthesis in a reticulocyte lysate, label
in polysomal ribosomes exchanges rapidly and completely with ribosomal
subunits, but not noticeably with single ribosomes (Howard, Adamson and
Herbert 1970). Moreover, Falvey and Staehelin (1970) found that in
contrast to the rapid subunit exchange occurring upon termination of protein synthesis in liver extracts, post-termination exchange in single ribosomes is extremely slow. This was later confirmed by Henshaw, Guiney
and Hirsch (1973).

A different conclusion was reached by Subramanian and Davis (1971),
who observed that isotopically heavy and light single ribosomes, generated
separately in bacterial polysome extracts, exchanged subunits readily
when the preincubated extracts were mixed. The discrepancy, however,
could well be due to the higher concentration of phosphoenolpyruvate used
in this experiment, which acts to lower the effective Mg^{++} ion concentration. As we shall see later, the rate of exchange of single ribosomes with
subunits depends critically on the effective concentration of divalent cations, being slow at concentrations permitting optimal protein synthesis,
but increasing at lower concentrations.

When comparing the results of various studies on the rate of equilibration of bacterial single ribosomes with subunits (Kaempfer 1970, 1971,
1972; Subramanian and Davis 1971, 1973), it is important to take into
account the concentrations of Mg^{++}-chelating agents present in the reaction mixture. Nucleoside triphosphates in particular bind Mg^{++} ions

tightly and stoichiometrically (Burton 1959; O'Sullivan and Perrin 1961). Corrections for chelation were applied by Kaempfer (1972). When effective, rather than added, Mg^{++} ion concentration is considered, then the equilibration rates reported in these studies agree quite well.

An Undissociated Intermediate Released from Polysomes

As already mentioned, the earlier studies of ribosomal subunit exchange left open the possibility that an undissociated intermediate intervenes between polysomes and the pool of ribosomal subunits. Such an intermediate, however, would have to be so unstable that its presence would not be detectable by the techniques employed previously. Indeed, by carefully controlling conditions, Subramanian and Davis (1973) were able to show that undissociated ribosomes are released from polysomes during normal termination or by treatment with puromycin. This was done by mixing preisolated heavy and light polysomes and incubating them at a higher Mg^{++} concentration than usual or at 25°C instead of 37°C. The ribosomes released under these conditions were found to be fully heavy and fully light. Thus, bacterial ribosomes can leave polysomes in undissociated form, at least under certain experimental conditions.

The release of ribosomes appears to require the participation of elongation factor G and a release factor present in the supernatant (Hirashima and Kaji 1972). The mechanism of release is not yet known.

The High Affinity of Ribosomal Subunits

The strong tendency of ribosomal subunits to form single ribosomes is already apparent from the fact that single ribosomes, not subunits, accumulate in response to fluctuations in the level of polysomes. How strongly the equilibrium between bacterial ribosomal subunits and single ribosomes favors the latter can be seen from Figure 4. Here labeled polysomes (open circles) were allowed to run off in the presence of an excess of unlabeled 30S subunits. Protein synthesis results in the accumulation of label in single ribosomes and 30S subunits, but no label accumulates in 50S subunits (filled circles). Evidently after they had left polysomes, the labeled 50S subunits all associated with 30S subunits into single ribosomes, but due to the excess of unlabeled 30S subunits, far fewer labeled 30S subunits were able to do so.

Mammalian ribosomal subunits likewise possess a high affinity for each other, with the equilibrium strongly favoring single ribosomes (Falvey and Staehelin 1970). In fact, compared to bacterial ribosomes, even lower Mg^{++} ion concentrations are required before mammalian single ribosomes dissociate completely.

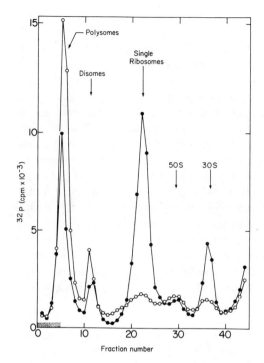

Figure 4   Asymmetric transfer of subunits from polysomes to single ribosomes. A ³²P-labeled *E. coli* polysome lysate was incubated in conditions for
protein synthesis with unlabeled S-100 containing an excess of 30S subunits,
in the presence (o) or absence (•) of 10^{-4}M sparsomycin. After cooling, each
sample was centrifuged through an isokinetic sucrose gradient supported by a
cushion of 75% (w/v) sucrose (shaded area). The two sedimentation profiles
are superimposed (Kaempfer 1971).

The high affinity of the two species of ribosomal subunits is not surprising, for they must function together in protein synthesis. Indeed, the ability
of subunits to associate is closely correlated with their ability to initiate
translation (Noll, Hapke and Noll 1973).

INITIATION FACTOR IF-3: INHIBITOR OF
RIBOSOMAL SUBUNIT ASSOCIATION

The observed constancy of ribosomal subunits during fluctuations in the
levels of polysomes and single ribosomes led to the postulate that the
supply of subunits is regulated by a factor, present in limiting amounts,
that forms a complex with one of the ribosomal subunits (Kohler, Ron
and Davis 1968). Indeed, it was possible to recover from bacterial 30S

subunits a factor capable of converting single ribosomes into ribosomal subunits, provided the Mg^{++} concentration was sufficiently low (Subramanian, Ron and Davis 1968; Subramanian, Davis and Beller 1969). This factor proved to be identical with initiation factor IF-3 (Sabol et al. 1970; Subramanian and Davis 1970); stoichiometric amounts of IF-3 are needed to effect the net dissociation of ribosomes.

The effect of IF-3 on the nature of the products accumulating at the expense of bacterial polysomes during cell-free protein synthesis is illustrated in Figure 5. The extract had been prepared from rapidly growing cells and contained polysomes, disomes and subunits but essentially no single ribosomes (filled circles: A, D). Protein synthesis results in the accumulation of single ribosomes (B, F); this process is accompanied by the extensive conversion of added 50S subunits (open circles: A, B) or 30S subunits (open circles: D, E) into single ribosomes. In the presence of added IF-3, however, polysomes disappear while 50S and 30S ribosomal subunits accumulate (C, F), and few, if any, added subunits are seen to enter single ribosomes. Thus IF-3 inhibits the accumulation of single ribosomes at the expense of polysomes and instead promotes the accumulation of ribosomal subunits.

These effects of IF-3 on preformed single ribosomes and on the ribosomal particles generated from polysomes during protein synthesis can be interpreted in two ways. Initiation factor IF-3 could act to dissociate single ribosomes into subunits. Alternatively, IF-3 could prevent the association of ribosomal subunits into single ribosomes, without necessarily promoting dissociation of single ribosomes.

The Role of Magnesium Ions

Sensitivity of Ribosome Dissociation Effect of IF-3.   At first, the net dissociation of single ribosomes induced by IF-3 was interpreted to mean that the dissociation step in the ribosome cycle does not occur spontaneously but requires the interaction of IF-3 with a ribosome released from polysomes, to yield a 50S subunit and a 30S·IF-3 complex (Subramanian and Davis 1970). An essential property of IF-3, however, is its inability to effect the conversion of single ribosomes into ribosomal subunits unless the Mg^{++} ion concentration is sufficiently low (Subramanian, Ron and Davis 1968; Subramanian and Davis 1970). The dissociation effect is detectable only at Mg^{++} ion concentrations below 5 mM, maximally at 1.5–2 mM, but is essentially undetectable at the higher Mg^{++} concentrations optimal for in vitro protein synthesis with natural messenger RNA (7 mM and above).

Inhibition of Ribosomal Subunit Association by IF-3.   The preceding results suggested that the dissociation effect of IF-3 depends absolutely on

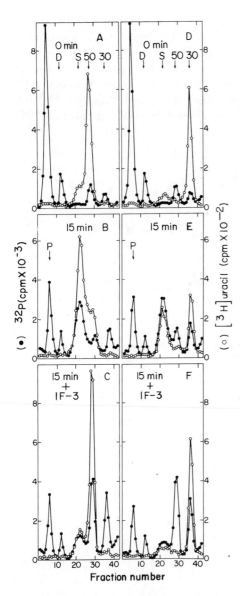

Figure 5 Inhibition of single ribosome accumulation during translation in the presence of added IF-3. ^{32}P-labeled *E. coli* polysomes (●) were incubated for the indicated time in conditions for protein synthesis, in the presence of ^3H-labeled (○) 50S subunits (*A, B, C*) or 30S subunits (*D, E, F*). IF-3 was present in samples *C* and *F*. Cushion gradient analysis as in Figure 4. Peaks from left to right represent polysomes (*P*), disomes (*D*), single ribosomes (*S*), 50S and 30S ribosomal subunits (Kaempfer 1971).

conditions that permit the spontaneous dissociation of single ribosomes. Indeed, it can be shown that IF-3 prevents the association of ribosomal subunits into single ribosomes but does not actively promote their dissociation (Kaempfer 1971, 1972).

To study the mechanism by which IF-3 controls the distribution between single ribosomes and ribosomal subunits, an experimental approach was chosen that allows independent determination of the extent of ribosome dissociation and of ribosomal subunit association in one and the same reaction mixture (Kaempfer 1971, 1972). ^{3}H-labeled single ribosomes, ^{32}P-labeled 50S subunits, and unlabeled S-100 containing an excess of 30S subunits are incubated under optimal conditions for protein synthesis, at 6.4 mM Mg^{++}. As seen in Figure 6a, labeled 50S subunits associate with unlabeled 30S subunits to single ribosomes. When 2–3 molecules of IF-3 per 30S subunit are present (Figure 6b), most ^{32}P label remains at 50S,

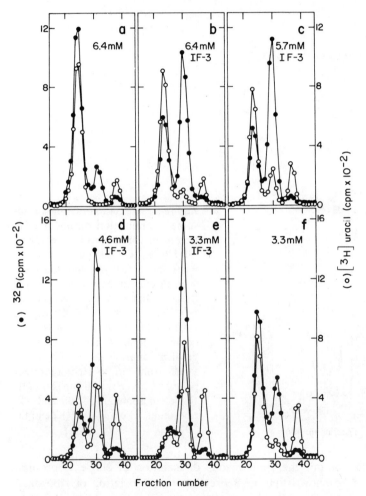

Figure 6 Inhibition of ribosomal subunit association by IF-3 and its independence of magnesium ion concentration. ^{32}P-labeled *E. coli* 50S ribosomal subunits (•) and ^{3}H-labeled single ribosomes (o) were incubated with unlabeled S-100 containing an excess of 30S subunits, at the indicated Mg^{++} concentrations. IF-3 was present where shown. Sucrose gradient peaks from left to right represent single ribosomes, 50S and 30S subunits (Kaempfer 1972).

yet this amount of IF-3 (250–350 molecules per single ribosome) does not cause any significant dissociation of single ribosomes. When the Mg^{++} concentration is progressively lowered, however, as in Figures 6c, d and e, the added IF-3 causes increasing amounts of ^{3}H-labeled 30S and 50S subunits to accumulate; few single ribosomes are left after incubation at 3.3 mM Mg^{++}.

The simplest explanation of these findings is that IF-3 inhibits the association of 30S and 50S subunits over the entire range of Mg^{++} concentration shown. Lowering the Mg^{++} concentration results in the spontaneous dissociation of single ribosomes; the subunits so generated are unable to return to single ribosomes because IF-3 blocks their association. If this explanation is correct, we should expect single ribosomes to dissociate at low Mg^{++} concentration even in the absence of IF-3; their ^{3}H-labeled 50S subunits should reassociate with unlabeled 30S subunits present in excess over labeled ones, but most of their ^{3}H-labeled 30S subunits should be excluded from this interaction and therefore should accumulate. That this is indeed the case can be seen in Figure 6f: at 3.3 mM Mg^{++}, a condition in which most ^{32}P-labeled 50S subunits still enter single ribosomes, the ^{3}H label is found mainly in two species, i.e., single ribosomes and 30S subunits. As expected, the level of ^{3}H-labeled 30S subunits in Figure 6f is significantly greater than in (a), and essentially the same as in the sample that contained IF-3 (E). In spite of the rapid equilibration occurring at 3.3 mM Mg^{++}, when IF-3 is absent ribosomal subunits are still found predominantly in single ribosomes (Figure 6f).

Single ribosomes thus equilibrate slowly with ribosomal subunits at effective Mg^{++} concentrations of 6 mM or above and progressively more rapidly at lower concentrations of Mg^{++}. On the other hand, the inhibition of ribosomal subunit association by IF-3 is largely independent of Mg^{++} in the concentration range studied. The apparent dissociation of single ribosomes in the presence of IF-3, seen at 3.3 mM Mg^{++} but not at 6.4 mM (Figure 6), is accounted for by the difference in spontaneous equilibration rate between the two conditions. In agreement with this conclusion, Noll and Noll (1972) reported that the rate of spontaneous dissociation of single ribosomes increases fiftyfold between 5 and 2 mM Mg^{++} and is not affected by IF-3.

Thus IF-3 does not actively promote dissociation of single ribosomes. A preliminary search for a ribosome dissociation factor, active in the presence of IF-3, has been negative (Kaempfer 1972), although Noll and Noll (1972) report that saturating amounts of IF-1 slightly stimulate the rate of dissociation.

Location, Amount and Recycling of IF-3

The association of IF-3 with 30S subunits, but not with 50S subunits or single ribosomes, has been demonstrated directly by the use of radioactive factor (Sabol and Ochoa 1971; Pon, Friedman and Gualerzi 1972; Thibault et al. 1972; Jay and Kaempfer 1974). IF-3 is released and recycled upon formation of a 70S initiation complex on messenger RNA (Sabol and Ochoa 1971; Benne et al. 1973).

When single ribosomes are incubated at low Mg^{++} with labeled IF-3, the label is associated exclusively with 30S particles (Sabol and Ochoa 1971; Pon, Friedman and Gualerzi 1972; Thibault et al. 1972). In the presence of streptomycin or neomycin, moreover, spontaneous dissociation of single ribosomes is inhibited, and no labeled IF-3 is found on 30S subunits (Thibault et al. 1972).

From the yield of IF-3 (Sabol et al. 1970), it can be calculated that a bacterial cell contains less than 0.1 molecule of this factor per ribosome (Subramanian and Davis 1970); this amount of IF-3 corresponds to about one molecule per native 30S subunit. Indeed, Sabol and Ochoa (1971) find that a 30S subunit binds close to one molecule of IF-3.

As would be expected for a recycling factor, IF-3 does not appear to bind very tightly to 30S subunits. As seen in Figure 6B, considerable escape formation of single ribosomes from subunits can occur even when active IF-3 is present in two- to three-fold excess over 30S subunits. Thus 50S subunits appear to compete with IF-3 for 30S subunits. Indeed, the ability of IF-3 to hold 30S and 50S subunits apart can be overcome completely by high concentrations of divalent cations, such as spermidine (Kaempfer 1972); IF-3 is released from 30S subunits (Sabol and Ochoa 1971) when association with 50S subunits is forced by raising the Mg^{++} concentration (Subramanian, Davis and Beller 1969).

Mammalian IF-3-like Factor

A factor activity analogous to bacterial IF-3, in that it blocks formation of single ribosomes from subunits and promotes recycling of ribosomal subunits through polysomes, can be isolated from rabbit reticulocytes. When initiation of protein synthesis is blocked with aurintricarboxylate, addition of this factor results in the accumulation of ribosomal subunits, rather than single ribosomes, at the expense of polysomes (Kaempfer and Kaufman 1972; Mizuno and Rabinovitz 1973). Labeled 60S subunits (Figure 7, curve a), incubated with unlabeled reticulocyte lysate in conditions of slowing protein synthesis, are converted extensively into single ribosomes (b), apparently by association with unlabeled 40S subunits generated from the polysomes; no label enters polysomes. In the presence of increasing amounts of reticulocyte IF-3, however, fewer subunits are converted into single ribosomes (c), but instead they efficiently enter polysomes (d).

This factor is much more efficient in recycling ribosomal subunits through polysomes than in recruiting single ribosomes, once they have been

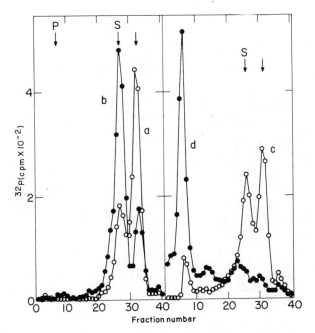

Figure 7 Effect of reticulocyte initiation factor on the fate of 60S ribosomal subunits during cell-free protein synthesis. ^{32}P-labeled rabbit reticulocyte 60S subunits were incubated in a reaction mixture containing unlabeled polysome lysate and incubated in conditions for protein synthesis (without hemin) for (*a*) 0 min; (*b*) 15 min; 15 min in the presence of (*c*) 1× or (*d*) 3× initiation factor. Cushion gradient analysis as in Figure 4; sedimentation profiles are super-imposed. *P*, polysomes; *S*, single ribosomes (Kaempfer and Kaufman 1972).

allowed to accumulate (Kaempfer and Kaufman 1972). If the factor is added to a polysome lysate after incubation for 12 min, by which time most polysomes have disappeared and single ribosomes have accumulated (Figure 8a), and incubation is continued for another 12 min, many single ribosomes fail to return to polysomes (c). Yet the amount of factor added was clearly sufficient to recycle all ribosomes among polysomes between 12 and 24 min (Figures 8b, d). Since in the sample first incubated without factor (a), fewer ribosomes were leaving polysomes at 12 min than could be recycled by the amount of factor added at that time, we would have expected that if the factor actively dissociated them, all single ribosomes should have entered polysomes by 24 min. This is not the case; therefore, the rate of dissociation of single ribosomes into subunits must remain limiting, even in the presence of sufficient factor to recycle an equal amount of ribosomes in polysomes. This supports the concept that the factor acts to inhibit the formation of single ribosomes from 60S and 40S subunits, but does not actively dissociate them. The slow entry of

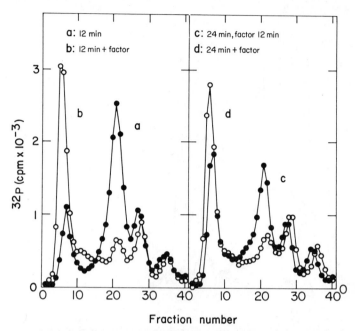

Figure 8 Kinetics of polysome recycling and single ribosome mobilization by reticulocyte factor. Four samples (*a–d*), containing ³²P-labeled rabbit reticulocyte polysome lysate, were incubated in conditions for protein synthesis (without hemin) for the times indicated. The same amount of factor was present in *b* and *d* throughout incubation, and in *c* was added after 12 min. Cushion gradient analysis as in Figure 4; sedimentation profiles are superimposed. Peaks from right to left represent 40S, 60S, single ribosome and polysome species (Kaempfer and Kaufman 1972).

single ribosomes into polysomes, even in the presence of factor, can then be accounted for by the observation that at the Mg^{++} ion concentration permitting optimal protein synthesis, 2 mM, reticulocyte single ribosomes exchange very slowly with 40S and 60S subunits (Howard, Adamson and Herbert 1970).

The reticulocyte IF-3 of Kaempfer and Kaufman (1972, 1973) shares with another initiation factor, IF-E3 (Schreier and Staehelin 1973), the property of promoting binding of globin messenger RNA to ribosomes. Both factors chromatograph similarly to IF-M3, an initiation factor first described by Prichard et al. (1970, 1971) that is required for translation of natural mRNA, and a similar factor from chick muscle (Heywood 1970). These factors are an order of magnitude larger in molecular size than IF-3 from *E. coli* (21,000 daltons; Sabol et al. 1970).

The activity of a reticulocyte factor (DF, Lubsen and Davis 1972; or EIF-3, Nakaya, Ranu and Wool 1973) capable of converting single ribosomes into subunits (provided the Mg^{++} ion concentration is sufficiently low) appears to depend on spontaneous ribosome dissociation and thus may well reflect the above-mentioned activity of IF-3 to block association of ribosomal subunits at the higher Mg^{++} ion concentration needed for protein synthesis. Mizuno and Rabinovitz (1973) report that their factor (DF) not only keeps subunits apart, but also dissociates single ribosomes in an energy-dependent process. The latter activity, however, was only observed upon addition of 15 mM creatine phosphate, so that it cannot be excluded that the resulting lower effective Mg^{++} concentration promoted spontaneous ribosome dissociation.

Whether IF-3, IF-E3, IF-M3 and DF function as one protein or several is not yet clear. Merrick, Lubsen and Anderson (1973) found that the IF-M3 and DF activities are separated by chromatography on phosphocellulose. However, while DF so prepared lacks IF-M3 activity, the IF-M3 still contains DF activity. Since phosphocellulose has a tendency to disrupt protein complexes, it is not excluded that DF in fact is a subunit of IF-M3 (IF-3, IF-E3) that retains its ability to hold subunits apart. The answer will come with further knowledge of the structure of these proteins. In this context, it is worth noting also that two species of native 40S subunits can be detected (Baglioni, Vesco and Jacobs-Lorena 1969), containing about 10^5 and 7×10^5 daltons of accessory protein, respectively (Hirsch et al. 1973); these subunits, as well as native 60S subunits, have lost their affinity for salt-washed complementary subunits (Henshaw, Guiney and Hirsch 1973).

THE RIBOSOME CYCLE: ANALYSIS

Returning to the general scheme of the ribosome cycle presented in Figure 1, we can now evaluate the evidence in support of its individual steps and consider possible alternative interpretations for certain features of the cycle in bacteria or eukaryotic cells.

The Bacterial Cycle

The dissociation of ribosomes between individual rounds of translation has been established firmly by the demonstration of ribosomal subunit exchange in vivo and its characteristics in vitro and by the requirement of subunits for initiation codon recognition.

The existence of an unstable pair of free subunits is demonstrated by the rapid exchange of subunits accompanying single ribosome accumulation, by the high affinity of subunits for each other and their overriding tendency to form single ribosomes as soon as they have left polysomes, and by the evidence that IF-3 must act on free subunits, not undissociated ribosomes, to generate a pair of stable subunits.

The action of IF-3 is to prevent the association of subunits into single ribosomes, without actively promoting dissociation of ribosomes. Since it is present in limiting amounts, binds stoichiometrically to 30S subunits, and is released during initiation complex formation, IF-3 must recycle as shown in the scheme.

Native subunits, defined here as 50S particles and 30S·IF-3 complexes, are also known to contain other factors needed for initiation, but only IF-3 prevents their association and thus stabilizes them; the limited quantity of IF-3 most likely determines the observed constancy in the level of ribosomal subunits.

The existence of a termination ribosome has been demonstrated under experimental conditions chosen to prevent ribosome dissociation; this particle is released from polysomes. Since normally it may dissociate so rapidly that it cannot be detected, the termination ribosome must be an unstable intermediate between polysomes and free subunits.

Single ribosomes accumulate at the expense of polysomes, as just mentioned, in a process accompanied by rapid subunit exchange. The rate of equilibration of single ribosomes with free subunits is extremely sensitive to the concentration of divalent cations (see Figure 6), being rapid at Mg^{++} ion concentrations below those optimal for in vitro protein synthesis on natural messenger RNA, but very slow at the higher effective Mg^{++} concentrations (7–10 mM Mg^{++}) that permit protein synthesis. This Mg^{++} dependence explains the discrepancy in rates of equilibration observed earlier (Kaempfer 1970; Subramanian and Davis 1971). It is not certain, however, if in the bacterial cell this rate is slow or rapid as compared to the time it takes to complete a polypeptide chain, although indirect arguments favor a slow rate (Kaempfer 1972). These are that the tendency of free subunits to associate in vitro must be even greater at the higher concentration of ribosomal particles in the cell, since higher concentrations favor association (second order in subunit concentration) over dissociation (first order in ribosome concentration), and that the effective mono- and divalent cationic conditions in vivo, although unknown, must approximate the ionic conditions for optimal synthesis in vitro in order to insure both maximal rate and fidelity of protein synthesis.

It has not been shown directly if single ribosomes seen in bacterial cell extracts are side-track products or intermediates of the cycle. Although the fact that single ribosomes are regularly observed in polysome lysates has been taken as evidence for their role as intermediates (Davis 1971), that fact could equally support their role as side-track products of the cycle. It is possible to prepare extracts lacking single ribosomes by rapid lysis of a relatively small culture of bacteria growing at a maximal rate (see Figure 5A, D); the same lysis technique yields single ribosomes unless these precautions are taken. This result supports the concept that during maximal rates of protein synthesis the cell contains virtually only polysomes and subunits, as first suggested by Mangiarotti and Schlessinger (1966), while single ribosomes are formed only during slowing synthesis (Kaempfer 1970). Unlike for mammalian cells, there are no direct data available on the in vivo equilibrium between bacterial single ribosomes with subunits and polysomes. One reason for this may be that while fluctuations in the rate of synthesis of protein and, correspondingly, the level of single ribosomes, occur commonly in eukaryotic cells and are not necessarily related to growth, bacteria generally divide rapidly and can respond rapidly to deteriorating environmental conditions by slowing the rate of ribosome synthesis and diluting the existing number of ribosomes by an increase in cell mass. Thus in growing bacteria, the accumulation of single ribosomes probably is only transitory.

The foregoing considerations lead to the following alternatives for the bacterial cycle:

(1) The termination ribosome may form only in special circumstances and may not be an obligatory intermediate.

(2) Both termination ribosome and single ribosome may exist, the former an unstable intermediate of the cycle, the latter a relatively stable particle outside the cycle. This yields the cycle depicted in the scheme of Figure 1. The difference in stability between single ribosomes and termination ribosomes could reflect a difference in structure, perhaps through an alteration in one or both subunits. Whether structural processing of subunits occurs at individual steps of the ribosome cycle is not yet established.

(3) Only termination ribosomes may exist. In that case, single ribosomes in fact would be termination ribosomes and thus intermediates of the ribosome cycle. They would have to be in rapid equilibrium with the subunit pool to explain the rapid subunit exchange accompanying accumulation of single ribosomes. This is the view offered by Subramanian and Davis (1973). As pointed out above, however, ionic conditions for efficient protein synthesis in vitro tend to favor a slow equilibrium between single ribosomes and the pool of subunits. In addition, the bacterial ribosome cycle according to this view must differ fundamentally from the eukaryotic ribosome cycle, whereas by contrast the two cycles would be quite similar under alternatives (1) or (2).

The Eukaryotic Cycle

The cyclic dissociation of ribosomes has been established in vivo and in vitro; repeated recycling of subunits through polysomes has been demonstrated directly in reticulocyte lysates.

The existence of an unstable pair of subunits is supported by the rapid exchange of subunits after termination and the high affinity of subunits for each other.

The action of IF-3 appears similar to that of bacterial IF-3: preventing the association of subunits into single ribosomes while promoting recycling of subunits through polysomes. Whether these functions are carried out by several distinct proteins or a single, complex protein remains to be established firmly. The apparent inefficiency of reticulocyte factor in recruiting single ribosomes, compared to its efficiency in recycling subunits through polysomes, indicates that this factor, too, may not actively promote ribosome dissociation in conditions of protein synthesis.

Native subunits are stable in cell extracts and can rapidly recycle through polysomes. Both 40S and 60S subunits appear to be modified after termination; two species of native 40S subunits have been detected, differing in amount of associated protein. The exact structural requirements for the stability of these intermediates in the cycle are not yet known.

The existence of a termination ribosome so far has not been demonstrated. If it exists, this particle must be unstable.

Single ribosome accumulation is accompanied by rapid subunit exchange, but once they have formed, single ribosomes equilibrate only very slowly with subunits cycling through polysomes. As in bacteria, the equilibrium between subunits and single ribosomes strongly favors the latter. Single ribosomes can return to polysomes when the rate of protein synthesis increases, but their mobilization is slow compared to the average translation time.

The most likely alternatives for the eukaryotic cycle thus are:

(1) The termination ribosome may not exist.
(2) Termination ribosomes and single ribosomes both may exist.

These two alternatives correspond directly to (1) and (2) of the bacterial cycle, but alternative (3) clearly is excluded for the eukaryotic cycle.

Finally, whichever alternative applies, the scheme of Figure 1 can accommodate the cycling of subunits through membrane-bound polysomes without need for additional assumptions.

PHYSIOLOGICAL SIGNIFICANCE OF THE RIBOSOME CYCLE

The cycle as depicted in Figure 1 provides the cell with a means of regulating the number of ribosomes active in protein synthesis in response to physiological changes.

These aspects have been discussed elsewhere in more detail (Kaempfer 1971). Briefly, during maximal rates of protein synthesis, the small subunits generated from polysomes will combine immediately with IF-3 and subsequently enter new initiation complexes while IF-3 is recycled. Though present in limiting quantity, the amount of IF-3 suffices to maintain the flow of subunits between polysomes. However, as soon as the number of subunits leaving polysomes begins to exceed the number of subunits entering polysomes, as during slowing protein synthesis, an imbalance is created; while more small subunits accumulate, IF-3 is recycled less frequently. Free subunits not intercepted by IF-3 will associate with high affinity to form single ribosomes, and this process will continue until the number of small subunits leaving polysomes returns to the number of available IF-3 molecules. The net effect is a reduction in the number of ribosomes translating messenger RNA. Conversely, an increase in the rate of initiation will lead to a temporary excess of recycled IF-3 over small subunits leaving polysomes; free IF-3 then can effect a net shift in equilibrium between single ribosomes and subunits in favor of the latter, until a new steady state is reached.

References

Algranati, I., N. Gonzalez and E. Bade. 1969. Physiological role of 70S ribosomes in bacteria. *Proc. Nat. Acad. Sci.* **62**:574.

Baglioni, C., C. Vesco and M. Jacobs-Lorena. 1969. The role of ribosomal subunits in mammalian cells. *Cold Spring Harbor Symp. Quant. Biol.* **34**:555.

Beller, R. and B. Davis. 1971. Selective dissociation of free ribosomes of *Escherichia coli* by sodium ions. *J. Mol. Biol.* **55**:477.

Benne, R., N. Naaktgeboren, J. Gubbens and H. Voorma. 1973. Recycling of initiation factors IF-1, IF-2 and IF-3. *Eur. J. Biochem.* **32**:372.

Borgese, D., G. Blobel and D. Sabatini. 1973. *In vitro* exchange of ribosomal subunits between free and membrane-bound ribosomes. *J. Mol. Biol.* **74**:415.

Burton, K. 1959. Formation constants for the complexes of adenosine di- or triphosphate with magnesium or calcium ions. *Biochem. J.* **71**:388.

Colombo, B., C. Vesco and C. Baglioni. 1968. Role of ribosomal subunits in protein synthesis in mammalian cells. *Proc. Nat. Acad. Sci.* **61**:651.

Davis, B. 1971. Role of subunits in the ribosome cycle. *Nature* **231**:153.

Falvey, A. and T. Staehelin. 1970. Structure and function of mammalian ribosomes. II. Exchange of ribosomal subunits at various stages of *in vitro* polypeptide synthesis. *J. Mol. Biol.* **53**:21.

Ghosh, H. and H. Khorana. 1967. Studies on polynucleotides. LXXXIV. On the role of ribosomal subunits in protein synthesis. *Proc. Nat. Acad. Sci.* **58**:2455.

Girard, M., H. Latham, S. Penman and J. Darnell. 1965. Entrance of newly formed messenger RNA and ribosomes into HeLa cell cytoplasm. *J. Mol. Biol.* **11**:187.

Grubman, M. and D. Nakada. 1969. Ribosomal subunits and MS2 phage RNA-directed protein synthesis. *Nature* **223**:1242.

Guthrie, C. and M. Nomura. 1968. Initiation of protein synthesis: A critical test of the 30S subunit model. *Nature* **219**:232.

Haselkorn, R. and L. Rothman-Denes. 1973. Protein synthesis. *Ann. Rev. Biochem.* **42**:397.

Henshaw, E., D. Guiney and C. Hirsch. 1973. The ribosome cycle in mammalian protein synthesis. I. The place of monomeric ribosomes and ribosomal subunits in the cycle. *J. Biol. Chem.* **248**:4367.

702   R. Kaempfer

Heywood, S. 1970. Specificity of mRNA binding factor in eukaryotes. *Proc. Nat. Acad. Sci.* **67**:1782.

Hirashima, A. and A. Kaji. 1972. Factor-dependent release of ribosomes from messenger RNA. Requirement for two heat-stable factors. *J. Mol. Biol.* **65**:43.

Hirsch, C., M. Cox, W. van Venrooy and E. Henshaw. 1973. The ribosome cycle in mammalian protein synthesis. II. Association of the native smaller ribosomal subunit with protein factors. *J. Biol. Chem.* **248**:4377.

Hogan, B. and A. Korner. 1968a. Ribosomal subunits of Landschutz ascites cells during changes in polysome distribution. *Biochim. Biophys. Acta* **169**:129.

———. 1968b. The role of ribosomal subunits and 80S monomers in polysome formation in an ascites tumour cell. *Biochim. Biophys. Acta* **169**:139.

Howard, G., S. Adamson and E. Herbert. 1970. Subunit recycling during translation in a reticulocyte cell-free system. *J. Biol. Chem.* **245**:6237.

Infante, A. and R. Baierlein. 1971. Pressure-induced dissociation of sedimenting ribosomes: Effect on sedimentation patterns. *Proc. Nat. Acad. Sci.* **68**:1780.

Jay, G. and R. Kaempfer. 1974. Host interference with viral gene expression: Mode of action of bacterial factor i. *J. Mol. Biol.* **82**:193.

Joklik, W. and Y. Becker. 1965. Studies on the genesis of polyribosomes. I. Origin and significance of subribosomal particles. *J. Mol. Biol.* **13**:496.

Kabat, D. and A. Rich. 1969. The ribosomal subunit-polyribosome cycle in protein synthesis in embryonic skeletal muscle. *Biochemistry* **8**:3742.

Kaempfer, R. 1968. Ribosomal subunit exchange during protein synthesis. *Proc. Nat. Acad. Sci.* **61**:106.

———. 1969. Ribosomal subunit exchange in the cytoplasm of a eukaryote. *Nature* **222**:950.

———. 1970. Dissociation of ribosomes on polypeptide chain termination and origin of single ribosomes. *Nature* **228**:534.

———. 1971. Control of single ribosome formation by an initiation factor for protein synthesis. *Proc. Nat. Acad. Sci.* **68**:2458.

———. 1972. Initiation factor IF-3: A specific inhibitor of ribosomal subunit association. *J. Mol. Biol.* **71**:583.

Kaempfer, R. and J. Kaufman. 1972. Translational control of hemoglobin synthesis by an initiation factor required for recycling of ribosomes and for their binding to messenger RNA. *Proc. Nat. Acad. Sci.* **69**:3317.

———. 1973. Inhibition of cellular protein synthesis by double-stranded RNA: Inactivation of an initiation factor. *Proc. Nat. Acad. Sci.* **70**:1222.

Kaempfer, R. and M. Meselson. 1969. Studies of ribosomal subunit exchange. *Cold Spring Harbor Symp. Quant. Biol.* **34**:209.

Kaempfer, R., M. Meselson and H. Raskas. 1968. Cyclic dissociation into stable subunits and re-formation of ribosomes during bacterial growth. *J. Mol. Biol.* **31**:277.

Kelley, W. and M. Schaechter. 1969. Magnesium ion-dependent dissociation of polysomes and free ribosomes in *Bacillus megaterium*. *J. Mol. Biol.* **42**:599.

Kohler, R., E. Ron and B. Davis. 1968. Significance of the free 70S ribosomes in *E. coli* extracts. *J. Mol. Biol.* **36**:71.

Lubsen, N. and B. Davis. 1972. A ribosome dissociation factor from rabbit reticulocytes. *Proc. Nat. Acad. Sci.* **69**:353.

Mangiarotti, G. and D. Schlessinger. 1966. Extraction of polyribosomes and ribosomal subunits from fragile, growing *Escherichia coli. J. Mol. Biol.* **20**:123.

———. 1967. Polyribosome metabolism in *Escherichia coli.* II. Formation and lifetime of RNA molecules, ribosomal subunit couples and polyribosomes. *J. Mol. Biol.* **29**:395.

Merrick, W., N. Lubsen and W. Anderson. 1973. A ribosome dissociation factor from rabbit reticulocytes distinct from initiation factor M3. *Proc. Nat. Acad. Sci.* **70**:2220.

Mizuno, S. and M. Rabinovitz. 1973. Factor-promoted dissociation of free ribosomes in a rabbit reticulocyte lysate system: Inhibition and requirement for an energy source. *Proc. Nat. Acad. Sci.* **70**:787.

Nakaya, K., R. Ranu and I. Wool. 1973. Dissociation of eukaryotic ribosomes by purified initiation factor EIF-3. *Biochem. Biophys. Res. Comm.* **54**:246.

Noll, M. and H. Noll. 1972. Mechanism and control of initiation in the translation of R17 RNA. *Nature New Biol.* **238**:225.

Noll, M., B. Hapke and H. Noll. 1973. Structural dynamics of bacterial ribosomes. II. Preparation and characterization of ribosomes and subunits active in the translation of natural messenger RNA. *J. Mol. Biol.* **80**:519.

Nomura, M. and C. Lowry. 1967. Phage f2 RNA-directed binding of formyl-methionyl-tRNA to ribosomes and the role of 30S ribosomal subunits in the initiation of protein synthesis. *Proc. Nat. Acad. Sci.* **58**:946.

Nomura, M., C. Lowry and C. Guthrie. 1967. The initiation of protein synthesis: Joining of the 50S ribosomal subunit to the initiation complex. *Proc. Nat. Acad. Sci.* **58**:1487.

O'Sullivan, W. and D. Perrin. 1961. The stability constant of $MgATP^=$. *Biochim. Biophys. Acta* **52**:612.

Pon, C., S. Friedman and C. Gualerzi. 1972. Studies on the interaction between ribosomes and $^{14}CH_3$-F3 initiation factor. *Mol. Gen. Genet.* **116**:192.

Prichard, P., J. Gilbert, D. Shafritz and W. Anderson. 1970. Factors for the initiation of hemoglobin synthesis by rabbit reticulocyte ribosomes. *Nature* **226**:511.

Prichard, P., D. Picciano, D. Laycock and W. Anderson. 1971. Translation of exogenous messenger RNA for hemoglobin on reticulocyte and liver ribosomes. *Proc. Nat. Acad. Sci.* **68**:2752.

Ron, E., R. Kohler and B. Davis. 1968. Magnesium ion dependence of free and polysomal ribosomes from *Escherichia coli. J. Mol. Biol.* **36**:83.

Sabol, S. and S. Ochoa. 1971. Ribosomal binding of labelled initiation factor F3. *Nature New Biol.* **234**:233.

Sabol, S., M. Sillero, K. Iwasaki and S. Ochoa. 1970. Purification and properties of initiation factor F3. *Nature* **228**:1269.

Schreier, M. and T. Staehelin. 1973. Initiation of eukaryotic protein synthesis: [Met-tRNA$_f$ · 40S ribosome] initiation complex catalysed by purified initiation factors in the absence of mRNA. *Nature New Biol.* **242**:35.

Spirin, S. 1971. On the equilibrium of the association-dissociation reaction of ribosomal subparticles and on the existence of the so-called "60S intermediate" ("swollen 70S") during centrifugation of the equilibrium mixture. *FEBS Letters* **14**:349.

Staehelin, T., E. Verney and H. Sidransky. 1967. The influence of nutritional change on polyribosomes of the liver. *Biochim. Biophys. Acta* **145**:105.

Subramanian, A. and B. Davis. 1970. Activity of initiation factor F3 in dissociating *Escherichia coli* ribosomes. *Nature* **228**:1273.

————. 1971. Rapid exchange of subunits between free ribosomes in extracts of *Escherichia coli*. *Proc. Nat. Acad. Sci.* **68**:2453.

————. 1973. Release of 70S ribosomes from polysomes in *Escherichia coli*. *J. Mol. Biol.* **74**:45.

Subramanian, A., B. Davis and R. Beller. 1969. The ribosome dissociation factor and the ribosome-polysome cycle. *Cold Spring Harbor Symp. Quant. Biol.* **34**:223.

Subramanian, A., E. Ron and B. Davis. 1968. A factor required for ribosome dissociation in *Escherichia coli*. *Proc. Nat. Acad. Sci.* **61**:761.

Tai, P. and B. Davis. 1972. Transfer RNA content of runoff and complexed ribosomes of *Escherichia coli*. *J. Mol. Biol.* **67**:219.

Thibault, J., A. Chestier, D. Vidal and F. Gros. 1972. Interaction of the radioactive translation initiation factor IF3 with ribosomes. *Biochimie* **54**:829.

Van Diggelen, O., H. Oostrom and L. Bosch. 1971. Association products of native and derived ribosomal subunits of *E. coli* and their stability during centrifugation. *FEBS Letters* **19**:115.

Vournakis, J. and A. Rich. 1971. Size changes in eukaryotic ribosomes. *Proc. Nat. Acad. Sci.* **68**:3021.

Alternative Views on the Ribosome Cycle

Bernard D. Davis
Bacterial Physiology Unit
Harvard Medical School
Boston, Massachusetts 02115

INTRODUCTION

While earlier studies of organisms, which were limited to their visible features, emphasized diversity, studies at a molecular level have increasingly revealed the underlying unity. This development has been exceedingly gratifying, for it has converted a mass of empirical biochemical data into direct evidence for the evolutionary continuity of the living world. Yet at the molecular level diversity is also encountered, especially in regulatory mechanisms, and it is just as inevitable a consequence of evolution, and just as interesting, as is uniformity. Hence the existence of extensive biochemical uniformity should not tempt us to try to force all metabolic processes into a Procrustean bed fitted for *E. coli*.

Protein synthesis illustrates this point. It exhibits striking unity between prokaryotes and eukaryotes in the details of the microcycle of chain elongation, but the process of initiation seems to be less uniform (for example, in formylation)—perhaps because it is a site of regulation. In particular, there are grounds for questioning Kaempfer's proposal (this volume) that eukaryotes and prokaryotes have the same ribosomal macrocycle. His model, which involves unstable termination ribosomes and stable reserve ribosomes, fits the data so far available for eukaryotes; but with bacteria, there is considerable evidence for a simpler scheme in which there is only one kind of free runoff ribosome, and it is in rapid spontaneous equilibrium with free subunits. This paper will review that evidence and will comment briefly on certain historical aspects of the ribosome cycle.

Subunit Exchange between Ribosomes after Runoff

The key evidence was provided by experiments designed for another purpose: to determine whether the 70S runoff ribosomes that accumulate have been released as such or as subunits which then reassociate. To answer this question, it was necessary to slow post-runoff exchange by elevating the $[Mg^{++}]$ or by lowering the temperature (Subramanian and Davis 1973). It was also necessary to fix the particles before analyzing them by zonal centrifugation, since otherwise the effects of pressure on sedimentation patterns (Infante and Baierlein 1971) mimic the presence of hybrid ribosomes (Subramanian and Davis 1971). In fact, this pressure effect invalidates all earlier demonstrations of hybrid ribosomes performed with unfixed free ribosomes; it does not apply to the classical demonstration of subunit exchange by Kaempfer, Meselson and Raskas (1968), in which the hybrids were measured as complexed ribosomes.

With the elimination of these complications, it could be shown that the immediate product of runoff is a 30S·50S couple (Subramanian and Davis 1973). Kaempfer (this volume) agrees with this conclusion but not with our interpretation of an additional important finding, i.e., that subunit exchange at 37°C was rapid up to 10 mM Mg^{++} (Subramanian and Davis 1973). He suggests that the free Mg^{++} in these experiments may have been below the physiological range because of the presence of 5 mM phosphoenolpyruvate. However, other studies in our laboratory showed that in a similar protein-synthesizing system, also containing 5 mM phosphoenolpyruvate, 10 mM Mg^{++} was *above* the optimum for protein synthesis by initiating ribosomes, and it was in the optimal range for pure chain elongation (Tai et al. 1973). It thus seems clear that runoff bacterial ribosomes exchange subunits rapidly under ionic conditions that we have found optimal for in vitro protein synthesis.

There is also a disagreement over experimental findings. Kaempfer's (this volume) main evidence for two classes of free ribosomes in bacteria is that accumulated free ribosomes isolated from a sucrose gradient were not nearly as readily dissociated by IF-3, or as active in subunit exchange, as ribosomes being released from polysomes in the same mixture. We have been unable to confirm this difference. In similar experiments in our hands, the two sets of ribosomes exchanged subunits equally rapidly at a Mg^{++} concentration (10 mM) optimal for protein synthesis (Subramanian and Davis, unpublished; Gottlieb and Davis, unpublished). (We may further note that if termination ribosomes and accumulated ribosomes have the same exchange rates, it is very unlikely that their dissociation constants— the true index of stability—will differ.)

The discrepancy could arise from alterations in some preparations of runoff ribosomes. Alternatively, it remains conceivable that the ribosome in the act of release may go through a transient conformation with an increased probability of dissociating, compared to normal runoff ribosomes; the two laboratories may differ in the suitability of their conditions for detecting such an effect. But even if such a difference in the ribosome during release can be confirmed, it would not justify designating the more stable, accumulated ribosomes as a reserve form since they dissociate rapidly under physiological conditions (Subramanian and Davis 1973).

Reversible Interaction with the Dissociation Factor (IF-3)

The dissociation of free ribosomes was further characterized by demonstrating the rapid reversibility of their reaction with IF-3 (Beller, Davis and Gottlieb 1974), using ribosomes freed of ligands either by runoff in cells or by washing with 1 M NH_4Cl. After equilibration with a limited amount of IF-3, a shift in the concentration of either IF-3, ribosomes or Mg^{++} caused a rapid increase or decrease in the dissociation (i.e., within 1 min or so) to the value expected. Moreover when labeled ribosomes were added to an equilibrated mixture, they partly dissociated, and the initial ribosomes partly reassociated, so that both sets rapidly reached the same equilibrium value.

These experiments encountered an interesting complication, which illustrates how significant artificial alterations in ribosomal particles may be for studies of dissociation. With a preparation of NH_4Cl-washed ribosomes that appeared to respond normally to IF-3 under the usual assay conditions, an increase in ribosome concentration unexpectedly caused a decrease, rather than an increase (as predicted from the mass law), in the number of ribosomes dissociated by a given amount of IF-3. The effect was evidently due to the presence of damaged particles (30S or 70S) which bound IF-3 without contributing additional dissociation, since less manipulated runoff ribosomes yielded the predicted results.

It is thus clear that IF-3 is in equilibrium with free ribosomes and 30S·IF-3 complexes. Moreover since free ribosomes are in equilibrium with free 30S subunits, as we have seen, and since these subunits can complex reversibly with IF-3 (Sabol and Ochoa 1971), it is clear that IF-3 can cause net dissociation by shifting the spontaneous equilibrium between free ribosomes and free subunits. However, it is also possible that IF-3 can directly displace 50S subunits from 30S·50S couples, just as allosteric effectors can both stabilize a spontaneous inactive conformation of an enzyme (Monod model) and induce that conformation (Koshland model) (Koshland and Neet 1968). Kaempfer (this volume) considers a direct interaction with free ribosomes excluded because IF-3 acted preferentially on the products of termination, competing with presumably stable, previously accumulated ribosomes. We could not confirm this difference, as noted above, but Kaempfer's conclusion appears to be correct, since added IF-3 failed to increase the rate of exchange of ribosomes with added labeled subunits (Noll and Noll 1972; Gottlieb and Davis, unpublished). An increase in rate would be expected if IF-3 interacted directly with 70S particles as well as with 30S particles.

Terminology and Mammalian DF

When a specific protein was found to be required for net dissociation of runoff ribosomes, it was naturally called a dissociation factor. Kaempfer later proposed (see Kaempfer, this volume) that a true dissociation factor should be defined, instead, as something that acts directly on the ribosome rather than on the subunits in equilibrium with it, and he has renamed IF-3 an anti-association factor. However, just as with allosteric effectors which may either induce or stabilize a conformational change, a definition based on overall action seems logical, and it remains so whatever the detailed mechanism. I therefore see no advantage in the proposed change in terminology. We might also note that the term "free" seems to distinguish runoff ribosomes from complexed (polysomal) ribosomes more clearly than the term "single." In addition it seems quite descriptive, since runoff ribosomes (or NH_4Cl-washed ribosomes) are now known to lack bound tRNA (Tai and Davis 1972) as well as mRNA and nascent polypeptide.

Matters of terminology are not necessarily trivial: they may have conceptual consequences and may influence the reading of history, as is illustrated by Kaempfer's description of a factor from reticulocytes that he labels reticulocyte IF-3. In the presence of this fraction, runoff yielded an accumulation of subunits rather than 80S ribosomes. This anti-association activity was considered novel and central because the same fraction was relatively inactive in dissociating accumulated runoff ribosomes (Kaempfer and Kaufman 1972). However, there is no reason to believe that this factor is distinct from a reticulocyte dissociation factor (DF) previously described by Lubsen and Davis (1972). This factor had been assayed by its net dissociation of free ribosomes prepared in other ways, which apparently yielded runoff ribosomes rather than reserve ribosomes.

Although reticulocyte DF has been separated chromatographically from all known initiation factors (Merrick, Lubsen and Anderson (1973), Kaempfer (this volume) further proposes that reticulocytes have an IF-3 quite homologous to bacterial IF-3 in its multiple functions. He suggests that the separated DF might be a protein subunit artificially dissociated from the IF-3. This is an interesting idea, but the attempted unification of eukaryotes and prokaryotes seems premature. There is no evidence for the proposed fragmentation, and the mammalian initiation factors separated in various laboratories have not yet been clearly identified with each other or related functionally to homologous bacterial factors. Indeed, their number is not yet certain, and some alleged factors may be true ribosomal proteins that have been washed off along with the initiation factors (Lubsen and Davis 1974a). Moreover, whereas bacterial DF is found only on the small native subunits, reticulocytes appear to have a specific DF on each kind of native subunit (Lubsen and Davis 1974b); the function of this double set is not clear.

CONCLUSIONS

The evidence seems to me to show that in bacteria, polysomes release intact, free 70S ribosomes, which are in rapid spontaneous equilibrium with their free subunits, and these in turn are in equilibrium with 30S subunits stabilized by complexing with IF-3. When protein synthesis is slowed by various methods that cause net loss of polysomes, the level of these runoff ribosomes rises, but they do not become a special class of inactive, reserve ribosomes. This model for the ribosomal macrocycle is supported by considerable evidence for rapid equilibration among the various components. It is not consistent with Kaempfer's evidence (this volume) that accumulated free ribosomes exchange subunits sluggishly, but we have been unable to confirm that finding.

This experimental discrepancy may be due to persistent, unphysiological alterations in ribosomal conformation occurring during the processing of the cell extract. This interpretation is difficult to establish directly but it is consistent with more general observations on ribosomal stabiliy. Thus exposure of subunits to abnormal ionic environments impairs their subsequent specific reactions (including reassociation): reactivation in normal environments is very slow, and a normal ligand (fMet-tRNA) can strikingly accelerate the reactivation (Miskin, Zamir and Elson 1970; Zamir, Miskin and Elson 1971). Moreover, it is well known that ribosomes prepared by usual procedures become more active in protein synthesis if heated briefly to 50°C prior to testing; and, as we noted above, ribosome preparations differed in the effect of concentration variations on dissociation by IF-3 (Beller, Davis and Gottlieb 1974). The recent history of the ribosomes may affect their capacity for subunit exchange more than their other properties, since the exchange does not involve any ligands. Hence undetected, probably unphysiological, conformational alterations of free ribosomes could be a major source of the controversies that have plagued the study of the ribosome cycle.

The model proposed here, like Kaempfer's, is probably oversimplified in at least one respect, i.e., the interpretation of the "native" subunits. By definition, these are the particles in a lysate that remain unassociated even at high Mg^{++} concentrations. Some of these are undoubtedly immature particles (Lindahl 1973). However, subunits made from free ribosomes in vitro vary in their ability to reassociate (Van Diggelen and Bosch 1973; Noll, Hapke and Noll 1973). In our hands (Beller, Davis and Gottlieb 1974), subunits made either with pure IF-3 or with a crude preparation of initiation factors have readily reassociated at 10 mM Mg^{++} and 0°C, the conditions usual for analyzing the particles in lysates. Hence the initiating native subunits in lysates, which contain the initiation factors, appear to have some unknown additional features.

References

Beller, R. J., B. D. Davis and M. Gottlieb. 1974. Reversible equilibrium in the reaction between ribosomes and the dissociation factor of *Escherichia coli*. *Biochemistry* **13**:939.

Infante, A. and R. Baierlein. 1971. Pressure-induced dissociation of sedimenting ribosomes: Effect on sedimentation patterns. *Proc. Nat. Acad. Sci.* **68**:1780.

Kaempfer, R., M. Meselson and H. Raskas. 1968. Cyclic dissociation into stable subunits and re-formation of ribosomes during bacterial growth. *J. Mol. Biol.* **31**:277.

Kaempfer, R. and J. Kaufman. 1972. Translational control of hemoglobin synthesis by an initiation factor required for recycling of ribosomes and for their binding to messenger RNA. *Proc. Nat. Acad. Sci.* **69**:3317.

Koshland, D. E., Jr. and K. E. Neet. 1968. The catalytic and regulatory properties of enzymes. *Ann. Rev. Biochem.* **37**:359.

Lindahl, L. 1973. Two new ribosomal particles in *Escherichia coli*. *Nature New Biol.* **243**:170.

Lubsen, N. H. and B. D. Davis. 1972. A ribosome dissociation factor from rabbit reticulocytes. *Proc. Nat. Acad. Sci.* **69**:353.

―――. 1974a. Use of purified polysomes from rabbit reticulocytes in a specific test for initiation factors. *Proc. Nat. Acad. Sci.* **71**:68.

―――. 1974b. A ribosome dissociation factor on both native subunits in rabbit reticulocytes. *Biochim. Biophys. Acta* **335**:196.

Merrick, W., N. H. Lubsen and W. F. Anderson. 1973. A ribosome dissociation factor from rabbit reticulocytes distinct from initiation factor M3. *Proc. Nat. Acad. Sci.* **70**:2220.

Miskin, R., A. Zamir and D. Elson. 1970. Inactivation and reactivation of ribosomal subunits: The peptidyl transferase activity of the 50S subunit of *Escherichia coli*. *J. Mol. Biol.* **54**:355.

Noll, M. and H. Noll. 1972. Mechanism and control of initiation in the translation of R17 RNA. *Nature New Biol.* **238**:225.

Noll, M., B. Hapke and H. Noll. 1973. Structural dynamics of bacterial ribosomes. II. Preparation and characterization of ribosomes and subunits active in the translation of natural messenger RNA. *J. Mol. Biol.* **80**:519.

Sabol, S. and S. Ochoa. 1971. Ribosomal binding of labeled initiation factor F3. *Nature New Biol.* **234**:233.

Subramanian, A. and B. D. Davis. 1971. Rapid exchange of subunits between free ribosomes in extracts of *Escherichia coli*. *Proc. Nat. Acad. Sci.* **68**:2453.

―――. 1973. Release of 70S ribosomes from polysomes in *Escherichia coli*. *J. Mol. Biol.* **74**:45.

Tai, P.-C. and B. D. Davis. 1972. Transfer RNA content of runoff and complexed ribosomes of *Escherichia coli*. *J. Mol. Biol.* **67**:219.

Tai, P.-C., B. J. Wallace, E. L. Herzog and B. D. Davis. 1973. Properties of initiation-free polysomes of *Escherichia coli*. *Biochemistry* **12**:609.

Van Diggelen, O. P. and L. Bosch. 1973. The association of ribosomal subunits of *Escherichia coli*. *Eur. J. Biochem.* **39**:499.

Zamir, A., R. Miskin and D. Elson. 1971. Inactivation and reactivation of ribosomal subunits: Amino acyl-transfer RNA binding activity of the 30S subunit of *Escherichia coli*. *J. Mol. Biol.* **60**:347.

The Ribosomal Components Involved in EF-G- and EF-Tu-dependent GTP Hydrolysis

W. Möller
Laboratory for Physiological Chemistry
State University of Leiden
The Netherlands

INTRODUCTION

This review concerns recent developments in the determination of the ribosomal components involved in GTP hydrolysis, which is dependent on the peptide chain elongation factors Tu (EF-Tu) and G (EF-G).

EF-Tu participates with GTP in the enzymatic binding of aminoacyl-tRNA to the ribosomal A site, whereas EF-G is necessary for translocation of peptidyl-tRNA from the A site to the P site, messenger movement, and ejection of "spent" tRNA. There is general agreement that EF-G and EF-Tu bind to a common region on the ribosome and that this region promotes the cleavage of GTP. Convincing evidence has been accumulated that two ribosomal proteins of the large ribosomal subunit play an equally important part in the transient attachment of the elongation factors EF-G and EF-Tu to the ribosome and thus in the induction of GTPase activity. The two proteins, known as L7 and L12, have structural properties that are unique among the ribosomal proteins and are also identical proteins except for the presence of an acetyl group on the N-terminal serine of L12.

The L7 and L12 requirement for GTP hydrolysis in a number of partial reactions in protein synthesis are described here, and examples are given to support proximity of the peptidyl transferase center and the GTPase center in terms of common ribosomal components. This proximity raises the question of how the ribosome controls the proper temporal succession by which EF-Tu and EF-G exert their function in tRNA binding and translocation. The functional state of the ribosome, as determined by the occupancy of the A site or P site by peptidyl-tRNA, may be decisive in this process. Conformational changes in a restricted region of the ribosome are considered, tRNAs and the elongation factors being seen as active participants. In this context, GTP hydrolysis is viewed as a mechanism by which GTP and GDP could act as allosteric regulators for the dynamic attachment of the elongation factors.

This chapter has no pretension to completeness in any of the areas it deals with. It may serve as a plea for the development of methods for the investigation of changes in neighbor relations and conformations of individual ribosomal components during the different phases of elongation.

711

With respect to this stage of "comparative" ribosome structure, we still know virtually nothing.

Role of L7/L12 in Elongation Factor-dependent Hydrolysis of GTP

The role of L7 and L12 in elongation factor-dependent hydrolysis of GTP was independently discovered in two laboratories. Kischa, Möller and Stöffler (1971) showed that after CsCl centrifugation at 0.04 M Mg^{++}, 50S ribosomes lost their ability to catalyze EF-G-dependent hydrolysis, but this activity could be regained by addition of either purified L7 (A1) or L12 (A2) protein. It was found that L7 was less active than L12 with CsCl-prepared cores. Moreover, antibodies prepared against pure L7/L12 specifically inhibited the EF-G-dependent GTPase specificity.

Hamel, Koka and Nakamoto (1972) demonstrated that washing of 50S ribosomes with a mixture of ethanol and ammonium chloride resulted in a marked loss of both EF-Tu and EF-G-dependent GTPase activity. Addition of the ethanol-ammonium chloride wash (PI protein) to the 50S ribosomal cores (PI ribosomes) restored the EF-Tu and EF-G-dependent hydrolysis of GTP (see Figures 1 and 2). Their PI protein fraction moved as two bands in gel electrophoresis and corresponded mostly to the 50S acidic proteins of Traub and Nomura (1968) and to the "A protein" (Möller and Castleman 1967; Möller, Castleman and Terhorst 1970).

Within a year, a number of groups had confirmed and extended the observation of an L7/L12 requirement in the partial reactions needing EF-G and EF-Tu (Brot et al. 1972; Sopori and Lengyel 1972; Weissbach et al. 1972; Sander, Marsh and Parmeggiani 1972; Hamel and Nakamoto 1972). Concurrent studies by other groups had shown that EF-Tu and EF-G interact with a common ribosomal region on the ribosome (Richman and Bodley 1972; Cabrer, Vazquez and Modolell 1972; Miller 1972; Richter 1972; Modolell and Vazquez 1973). The combined evidence strongly suggested that L7/L12 play a key role in the function of both of the elongation factors EF-G and EF-Tu. These observations led to an attempt to prove directly that EF-G reacts at a site containing L7/L12. Cross-linking of EF-G to the 50S ribosomal subunit indicates that L7/L12 are indeed located at or near the EF-G binding site (Acharya, Moore and Richards 1973).

More detailed data on the role of L7 and L12 in the various partial reactions of protein synthesis are presented in the next section.

Role of L7/L12 in a Number of Partial Reactions of Protein Synthesis

Hamel, Koka and Nakamoto (1972) showed that L7/L12-deficient ribosomes, prepared by extracting ribosomes at 0°C with ethanol-NH$_4$Cl, have a decreased ability to (1) bind aminoacyl-tRNA dependent on EF-Tu, (2) hydrolyze GTP dependent on EF-G and EF-Tu, (3) carry out transloca-

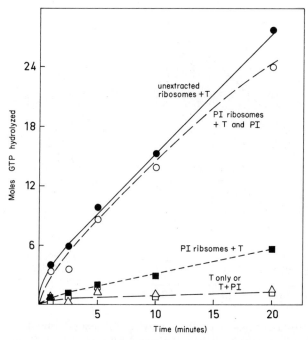

Figure 1 Kinetics of EF-T-dependent GTPase activity. All assays contained, in a volume of 0.10 ml, 50 pmoles [γ-^{32}P]GTP, 27 μg aminoacyl-tRNA, 125 nmoles fusidic acid, 1.9 μg EF-T preparation, and 100 μg ribosomes and 0.010 ml PI as indicated. Incubation was at 37°C for the indicated times. There were five sets of reactions: EF-T only ($\triangle$ --- $\triangle$), T factor and PI ($\square$ --- $\square$), unextracted ribosomes and EF-T factor ($\bullet$ --- $\bullet$), PI ribosomes and EF-T ($\blacksquare$ --- $\blacksquare$), and PI ribosomes, PI, and EF-T ($\circ$ --- $\circ$). (Reprinted, with permission, from Hamel, Koka and Nakamoto 1972.)

tion, and (4) synthesize polyphenylalanine. These reactions could all be restored by adding L7/L12 to the reaction mixture.

The results of Brot et al. (1973) show that both L7 and L12 are equally effective in restoring the ability of NH$_4$Cl-ethanol-depleted ribosomes to hydrolyze GTP on addition of either EF-Tu or EF-G. In addition, it was observed that L7 and L12 equally restore polyphenylalanine synthesis (Brot et al. 1973) or the in vitro synthesis of β-galactosidase in a DNA-dependent protein synthesis system using L7/L12-depleted ribosomes (Kung et al. 1973). Under specific conditions, therefore, L7 and L12 are equally active in all aspects of polypeptide synthesis.

A requirement for L7/L12 has also been found in the initiation and termination reactions of protein synthesis. Three initiation factors have been shown to be required for the initiation of polypeptide synthesis in *Escherichia coli*. IF2, similar to EF-Tu and EF-G, displays an "uncoupled"

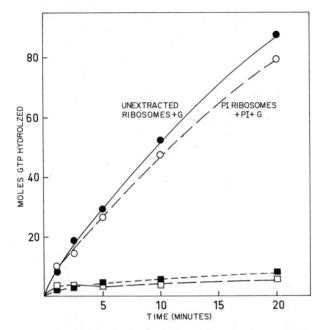

Figure 2 Kinetics of EF-G-dependent GTPase activity. All assays contained, in a volume of 0.05 ml, 150 nmoles [γ-³²P]GTP 17 μg EF-G, as well as 50 μg ribosomes and 0.010 ml PI as indicated: Incubation was at 37°C for the indicated times. There were four sets of reactions: EF-G factor only (□ – – – □), unextracted ribosomes and EF-G (• – – – •), PI ribosomes and EF-G (■ – – – ■), and PI ribosomes, PI, and EF-G factor (○ – – – ○). (Reprinted, with permission, from Hamel, Koka and Nakamoto 1972.)

GTPase activity that is lost on removal of L7 and L12 but can be restored by the addition of ribosomal protein L7 and L12 to the extracted subunits (Kay, Sander and Grunberg-Manago 1973; Lockwood et al. 1974). Moreover, L7 and L12 function equally well in stimulating the IF2-uncoupled hydrolysis of GTP (see Figure 3).

Since the uncoupled IF2 GTP hydrolysis proceeds in the absence of initiation, IF2-dependent GTP hydrolysis was also measured in an initiation system containing AUG, IF1, IF2, IF3, GTP and fMet-tRNA (Lockwood et al. 1974). Again L7 and L12 were required for the coupled hydrolysis of GTP as well as for the 50S ribosomal subunit-dependent stimulation of fMet-tRNA binding. In this coupled system, L7 and L12 were needed in restoring both GTP hydrolysis and fMet-tRNA binding. In addition, L7/L12 were required for the ability of IF2 to function catalytically. It should be noted that L7/L12-depleted 50S subunits can still combine with 30S subunits to form 70S particles (Sander, Marsh and Parmeggiani 1972).

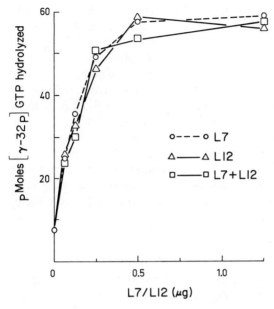

Figure 3 Effect of L7 and L12 concentration on IF2-dependent GTP hydrolysis. Reaction mixtures (0.1 ml) contained 20 mM Tris-HCl pH 7.4, 20 mM NH$_4$Cl, 15 mM Mg acetate, 2 mM 2-mercaptoethanol, 2.2 nmoles [γ-^{32}P]GTP and, as indicated, 0.2 μg IF2, 24 pmoles 30S subunits, 5 pmoles of 50S or extracted 50S subunits. The incubation was for 4 min at 37°C and the ^{32}P$_i$ released was assayed. The ribosomal proteins L7 and L12 were extracted (Hamel, Koka and Nakamoto 1972; Brot et al. 1972) from the 50S ribosomal subunits with an ethanol-NH$_4$Cl mixture and separated as described by Möller et al. (1972). (Reprinted, with permission, from Lockwood et al. 1974.)

A significant difference between the termination and the other reactions described above is that there is no GTP hydrolysis associated with termination in *Escherichia coli.*

The termination reaction is conveniently assayed by measuring release of fMet from a fMet-tRNA-terminator codon-ribosome complex (Caskey et al. 1968). Very little fMet is released from L7/L12-depleted ribosomes, and addition of these proteins to the reaction mixtures almost completely restores the activity. This effect is also totally dependent on the presence of the release factors RF1 or RF2 and the terminator codon UAA (Brot et al. 1974).

It had previously been demonstrated that the binding of the terminator codon to ribosomes is directly related to the formation of a RF-UAA-ribosome complex, and the amount of UAA bound can be used to assay RF binding to the ribosome (Scolnick and Caskey 1969). The amount of UAA bound to L7/L12-depleted ribosomes in the presence of RF2 is greatly decreased, and addition of L7/L12 restores the capacity of the depleted ribosomes to bind RF (Brot et al. 1974). Since binding of RF and hydrolysis of nascent peptidyl-tRNA do not require GTP hydrolysis, the need for L7/L12 may be concerned only with the proper binding of soluble factors rather than with the hydrolysis of GTP. The partial release by methanol of the requirement for L7/L12 in EF-G and EF-Tu dependent GTP hydrolysis supports this notion (Ballesta and Vazquez 1972).

In summary, the requirement for L7/L12 in the initiation, elongation and termination phase of protein synthesis strongly indicates that L7/L12 function in the recognition process of all the soluble factors that interact with the 50S subunit. Structural studies on the attachment site of EF-G to the 50S subunit are consistent with this view (see section on Possible Function for GTP Hydrolysis in Elongation).

GTPase and ATPase Activity of a 5S RNA–Protein Complex

A few years ago, reconstitution experiments done by Nomura's group indicated that 5S RNA-free "50S ribosomes" from *Bacillus stearothermophilus* show greatly diminished activities in poly(U)-dependent polyphenylalanine synthesis. This activity could be restored by the addition of 5S RNA to the reconstitution mixture (Erdmann et al. 1971). Analysis of the 5S RNA-free "50S ribosomes" showed them to be deficient in about four 50S proteins. Apparently these proteins were not incorporated in the absence of 5S RNA. On closer examination, the 5S RNA–protein-deficient particles had lost a large number of functional properties, including G factor-dependent GTP binding (Erdmann et al. 1971).

These findings led to the isolation of a native 5S RNA–protein complex from *Bacillus stearothermophilus* with the major proteins B-L5 and B-L22. A similar complex could be reconstituted in the case of *Escherichia coli* and consisted of 5S RNA in association with the major proteins E-L18 and E-L25 and the minor proteins E-L5, E-L20 and E-L30 (Horne and Erdmann 1972). Such an in vitro complex in *Escherichia coli* had been anticipated to some extent by the observation of Gray et al. (1972) that only the proteins L6, L18 and L25 participate in complex formation between 23S RNA and 5S RNA, and among the three proteins only L6 binds to 23S RNA (Stöffler et al. 1971). Furthermore, binding studies indicated that L18 and L25 bind to 5S RNA, whereas L6 did not (Gray et al. 1972).

On further investigation of the functional properties of the complex, Horne and Erdmann (1973) found that the 5S RNA–protein complex from *Bacillus stearothermophilus* was able to hydrolyze GTP as well as ATP in the absence of elongation factor EF-G.

Furthermore, thiostreptone and fusidic acid were found to inhibit GTPase activity, although the inhibiting concentrations were rather high, especially for thiostreptone (50% activity at 0.1 mM of antibiotic). The finding of GTPase activity in a 5S RNA–protein complex led Horne and Erdmann to propose that the GTPase center for protein synthesis is located on one or more of the proteins of this complex. However, the fact that complete restoration of the EF-G-dependent GTPase activity can occur without L25 (see below) argues against this protein being the "GTPase enzyme" (L25 is a split protein in 0.04 M Mg^{++} CsCl equilibrium centrifugation). This leaves L18 as one of the most likely candidates for direct participation in GTP hydrolysis.

The significance of the interesting findings of Horne and Erdmann seems to us insufficiently clear at present. Evaluation of ATPase activity comparable to that of GTPase on the 5S RNA–protein complex is difficult, because no comparable activities seem to exist in EF-G-dependent ribosomes. Closer analysis of these authors' kinetic data also indicates that a ribosomal system from *Escherichia coli,* similar especially with respect to molar equivalence of GTP and ribosomes, splits about one hundred times more GTP in the presence of EF-G (Sander, Marsh and Parmeggiani 1972; Schrier, Maassen and Möller 1973).

Work on the relationship of the 5S RNA–protein complex to EF-G-dependent hydrolysis on the ribosome will have to be done before the general implications of this 5S RNA system can be judged. In particular, the nature of the interaction of the elongation factors with the putative GTPase center of protein synthesis should be better understood. Certainly it would be interesting to know which proteins are neighbors of the 5S RNA–protein complex in situ and to see whether they correlate with the ones required for elongation factor-dependent GTP hydrolysis. The presence of L6 on a 23S RNA–5S RNA–protein complex (Gray et al. 1972) and its involvement in EF-G-dependent GTP hydrolysis (Schrier, Maassen and Möller 1973) tend to support the view of Erdmann.

Proximity of GTPase and Peptidyl Transferase Centers

To define other ribosomal proteins possibly participating in GTP hydrolysis, Schrier et al. (1973) used an approach originally developed by Meselson and coworkers (1964). 50S ribosomes were stripped gradually with 5 M CsCl solutions of decreasing Mg^{++} concentrations, and the resultant core preparations were tested for GTPase activity in the presence of 30S ribosomes, EF-G, L7/L12 and single proteins isolated from these split fractions.

At 0.04 M Mg^{++}, 50S CsCl cores lacking the proteins L1, L7, L8, L10, L12, L16, L25, L28, L31 and L33 were obtained. Such cores are completely inactive in factor-dependent GTP hydrolysis but can be reactivated by L7 or L12. To restore full GTPase activity, a large molar excess of L12 to 50S cores is needed, whereas the same excess of L7 restores only part of

the original GTPase activity. For the 50S CsCl cores, the reduced activity
of L7 proved to depend on the presence of L10 in the reconstitution mixture. When L10 was added to 50S cores together with L7 or L12, both
GTPase activities became equal (Figure 4). Moreover, in the presence of
one mole L10/mole 50S, only two moles of L7 and L12 per mole 50S
"cores" were needed to reconstitute a GTPase activity comparable to that
of the intact particle. The steep rise in the concentration activity curves
(see Figure 4) on addition of L10 suggests a role for this protein in the
physical assembly process of L7 and L12; presumably in the absence of
L10, the acetylated form L7 (Möller et al. 1972; Terhorst, Wittmann-Liebold and Möller 1972) binds less effectively to the 50S subunit. In this
connection it should be mentioned that L7 and L12 seem equally active in
all reactions using ethanol-ammonium chloride washed cores (Brot et al.
1973). However, using similar cores, Parmeggiani's group found L7 to be
less active than L12 in both EF-G- and EF-Tu-dependent GTPase (Sander,
Marsh and Parmeggiani 1972). Recently similar results have been obtained

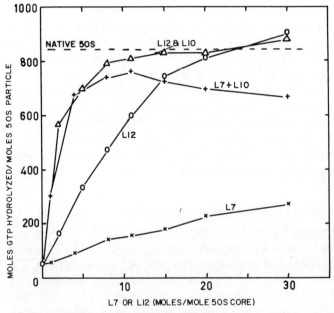

Figure 4   Effect of L10 on EF-G-dependent GTP hydrolysis. Reaction mixture
contained 26 pmoles 50S CsCl cores (prepared at 40 mM Mg++), 90 pmoles
30S ribosomes, 5 μg EF-G and 50 nmoles (0.5 mM) of [γ-^{32}P] GTP in 0.100 ml.
Reactions were done with varying concentrations of L7 (x – – – x), L12
(o – – – o), L7 and 0.56 μg L10 (1.0 mole/mole), 50S core (+ – – – +), and
L12 and 0.56 μg L10 (△ – – – △). For further experimental conditions and
work-up of the GTPase assay see Schrier, Maassen and Möller 1973.

in our laboratory; the most obvious explanation would be partial extraction of L10 due to a somewhat different washing procedure (Hamel, Koka and Nakamoto 1972).

Recently a direct approach for investigating the proteins involved in GDP binding on the ribosome was used (Maassen and Möller 1974). On irradiation, the azidophenylester of guanosine-5'-diphosphate (APh-GDP), in which the β-phosphate is esterified to azidophenol, gives a reactive nitrene (Figure 5). This nitrene attacks the bond of a ribosomal component in its vicinity by an insertion reaction, which may be followed by breakage of that bond. Figure 6 shows that APh-GDP is an effective inhibitor of GDP binding to the ribosome. The assay described by Bodley et al. (1970) uses fusidic acid to stabilize the ribosome-GDP-fusidic acid complex. The inhibitory effect of APh-GDP having been demonstrated, *Escherichia coli* 50S ribosomes were incubated with a thousandfold excess of the photo-affinity label for 5 min at 0°C. Next the mixture was irradiated for 30 min at 0°C with UV light ($\lambda > 320$ nm), which does not damage the ribosome. To increase the fractional yield of reacted ribosomes, the whole procedure was repeated four times. After recentrifugation of the ribosomes, no L16 was detected in the supernatant fluid. On analysis of the proteins of the ribosomes by two-dimensional gel electrophoresis, most of the protein L16 had disappeared. The gradual disappearance of L16 was observed by visual inspection of the dye-colored spots; moreover two spots, corresponding with L11 and L5, had become diffuse. A completely normal pattern was observed in the control, which contained the APh-GDP compound without fusidic acid. This control was subjected to an identical procedure of irradiation. When APh-GDP tritiated in the guanosine moiety was used, the most strongly labeled spots were those corresponding to L11, L5, L30 and L18, and no counts were detected at the position of L16. The 50S fusidic acid-dependent labeling of L5, L30 and L18 by APh-GDP correlates well with the GTPase activity of the 5S RNA–protein complex with the major proteins E-L18 and E-L25 and the minor proteins E-L5, E-L20 and E-L30 (Horne and Erdmann 1972, 1973). The most likely interpretation of the disappearance of L16 is that the APh-GDP had cleaved L16 or that the APh-GDP/L16 reaction product has an electrophoretic mobility different from free L16. On the other hand, the reduced amount of L16 may be an

Figure 5 Formula of the 4-azidophenylester of guanosine-5'-diphosphate. On irradiation at $\lambda > 310$ nm N_2 splits off under formation of a reactive nitrene.

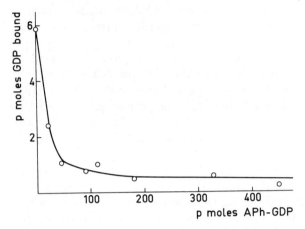

Figure 6   Inhibition of [³H]GDP binding on 50S ribosomes by APh-GDP in the presence of fusidic acid. Reaction mixture contained 10 mM Tris-HCl, 10mM Mg acetate, 10 mM NH₄Cl, 3 mM fusidic acid, 1 mM DTT, 28 pmoles 50S ribosomes, 70 pmoles [³H]GDP, 7 μg EF-G and varying amounts (0–450 pmoles) of APh-GDP. 50S ribosomes were incubated for 5 min at 0°C in the presence of APh-GDP and fusidic acid in a reaction volume of 47 μl. Next 70 pmoles [³H]GDP were added and the mixture incubated for another 5 min at 0°C. The work-up of the samples, including Millipore filtration, was done according to Bodley et al. (1970). Values are averages of several experiments (Maassen and Möller 1974).

indirect effect, e.g., a weakening of the L16 binding to the ribosome cannot be ruled out rigidly.

In recent reconstitution experiments L16 was mapped as the component that binds chloroamphenicol and therefore is possibly the aminoacyl binding protein (Nierhaus and Nierhaus 1973). A similar conclusion was reached independently on the basis of affinity labeling with the chloroamphenicol analog, monoiodoamphenicol (Pongs, Bald and Erdmann 1973). Moreover, L16 stimulates the peptidyl transferase protein L11 (Nierhaus and Montejo 1973). Interestingly the protein L11 was strongly labeled by APh-GDP. This suggests that the peptidyl transferase and the binding site for GDP are close neighbors. A similar situation may exist for L6, which is essential for GTPase activity (Schrier, Maassen and Möller 1973) and the inhibition of the CACCA-binding by chloroamphenicol (K. H. Nierhaus, personal communication), although allosteric effects might be considered here. The findings with L11 point to a hitherto unexpected proximity of the peptidyl transferase and GTPase centers. Antibiotic studies did not suggest a mutually functional interdependency of the two centers (Modolell, Vazquez and Monro 1971; Nierhaus and Nierhaus 1973).

Role of 30S Proteins in EF-G- and EF-Tu-dependent
GTPase Activity of Ribosomes

The elongation factors EF-G and EF-Tu require the presence of both 30S and 50S subunits for expression. Although the dominant role of the 50S subunit has been established, less information is available on the role of the 30S subunit. Recently Marsh and Parmeggiani (1973) reported that the 30S proteins S5 and S9 were required for the ribosome-dependent GTPase activity of EF-G. On addition to 30S cores prepared by CsCl centrifugation (30S CsCl cores), these two proteins individually contributed only a small fraction to the reconstituted GTPase activity as compared to their combined effect. A similar cooperative effect was noted in the capacity of the reconstituted 30S particles to associate with 50S ribosomes when S5 and S9 were added to CsCl cores. Omission of S5 from the total split protein mixture added to CsCl cores gave a reconstituted 30S particle that could not associate with 50S ribosomes (Marsh and Parmeggiani 1973). These experiments suggest a role for S5 and S9 in the association reaction and indicate the indirect mode of action of the 30S split proteins in the EF-G-dependent GTPase reaction. They are also in line with the observation that omission of S9 from a total reconstitution mixture of 30S ribosomes affects polypeptide synthesis and tRNA binding in a drastic way (Nomura et al. 1969). An effect of S9 on subunit association and possibly thereby on EF-G-dependent GTPase activity was confirmed by immunological experiments. When added to 30S ribosomes, an inhibitory effect of anti-S9 on the GTPase activity was shown. No such effect was observed when the antibodies against S9 were added to the intact 30S–50S couple, which would confirm that S9 is at a locus only available after dissociation (Highland and Stöffler 1973, unpublished data). Interestingly, S9 shows almost no reaction with N-ethylmaleimide in the 70S ribosome but reacts strongly after dissociation to free 30S and 50S subunits (Chang 1973); S9 is also shielded by the 50S subunit against modification by fluorescein isothiocyanate (Huang and Cantor 1972).

Recently six ribosomal proteins—S4, S7, S8, S9, S11 and S15—were reported to be required in the EF-G-dependent GTPase reaction (Cohlberg 1974). Several others, including S5, were required for polyphenylalanine synthesis but were dispensable in the GTPase reaction. These results were obtained on 30S particles lacking a single protein as a result of using a partial reconstitution system (core particles plus a mixture of split proteins minus one) and a total reconstitution system (16S RNA plus total mixture of proteins minus one). Although the addition of S5 alone to 30S cores stimulates GTPase activity, omission from 30S particles in partial and total reconstitution experiments produced no loss of activity (Cohlberg 1974). In contrast, a loss of GTPase activity in partial reconstitution was found by Marsh and Parmeggiani (1973).

Using a similar approach to study the role of the 30S ribosomal subunit in EF-Tu-dependent GTPase, Sander, Marsh and Parmeggiani (1973) found that in this case, S2 and S9 were the major components of the required split proteins and S5 the minor one. Addition of S2, S5 and S9 restored additively 90% of the activity of the native 30S, whereas the 30S core alone was inactive. When added alone, S2 also showed activity with EF-Tu (Marsh and Parmeggiani 1973); the effect of S2 in EF-G-dependent hydrolysis was somewhat less pronounced.

In this connection it is interesting to note that upon irradiation of 70S ribosomes with azidophenyl-GDP in the presence of fusidic acid, S2 had disappeared from the ribosomes (Maassen and Möller, unpublished data). A similar disappearance of L16 was observed in the case of 50S ribosomes. This draws attention to the fact that both S2 and L16 are reported to be involved in aminoacyl-tRNA binding to the A site (Randall-Hazelbauer and Kurland 1972; Nierhaus and Nierhaus 1973).

Alternate Recognition of EF-Tu and EF-G by the Ribosome

Several groups reported independently that the polypeptide chain elongation factors EF-Tu and EF-G interact with a common or overlapping site on the large ribosomal subunit (Richman and Bodley 1972; Cabrer, Vazquez and Modolell 1972; Miller 1972; Richter 1972; Modolell and Vazquez 1973). The evidence is based mainly on the fact that EF-Tu and EF-G cannot interact simultaneously with the ribosome. Therefore there must be a mechanism that regulates the alternate binding of EF-Tu and EF-G to the ribosome. One mechanism to effect this regulation could be a conformational change in a specific part of the ribosome. Our own preference would locate this part near or at the binding site of the elongation factors (EF-binding site). On the other hand, the EF-recognition sites may be regulated by the occupants of the A site and P site, namely, peptidyl-tRNA, aminoacyl-tRNA and deacetylated tRNA. As a result, the relative arrangement of the different tRNAs could control the proper alternation of the elongation factors. Support for the latter possibility was provided by two studies in which the binding of EF-G and EF-Tu to ribosomes was compared in the so-called pre- and post-translocation phases (see Modolell, Cabrer and Vazquez 1973 for the prokaryotes and Nombela and Ochoa 1973 for the eukaryotes). Pretranslocation ribosomes carrying peptidyl-tRNA at the A site interact preferentially with EF-G (or EF2). Post-translocation ribosomes with peptidyl-tRNA at the P site show decreased interaction with EF-G (or EF2). Conversely, ribosomes in the post-translocation phase reacted preferentially with EF1 (the counterpart of EF-Tu in prokaryotes). Pretranslocation ribosomes show decreased interaction with EF1 (Nombela and Ochoa 1973). Nombela and Ochoa prefer a conformational change of the ribosome as a consequence of the shift of peptidyl-tRNA following translocation or peptide bond formation. Their suggestion invites detailed studies of the possible effects of relative displacements of peptidyl-tRNA on specified components of the ribosome. Modolell and coworkers arrived at the same conclusion as Nombela and Ochoa but leave open the possibility that the nature of the control of EF-G interaction with the ribosome resides in steric hindrance of EF-G by peptidyl-tRNA when in the P site, as opposed to the entry of EF-G when in the A site.

A Possible Function for GTP Hydrolysis in Elongation

We shall now turn briefly to a discussion of the manner in which the ribosome may use GTP hydrolysis for elongation. Lipmann (1969), on the basis of a specific requirement for GTP hydrolysis in translocation, favored an energy coupling mechanism in which the energy released on GTP hydrolysis is used to pull the messenger "one triplet forward." Gupta et al. (1971) proved that the actual movement of messenger RNA along the ribosome is indeed triggered by GTP and EF-G, "with one step of messenger movement as expected three nucleotides long." This left the problem of how the energy of hydrolysis is transmitted through EF-G into the ribosomal subcomponents that effect translocation. Moreover, GTP hydrolysis is promoted by two different elongation factors at a common site; aminoacyl-tRNA binding occurs with one factor (EF-Tu) and tRNA release and translocation by another (EF-G). There is no doubt a formal analogy between ATP-linked stepwise interaction of actin and myosin molecules and the GTP-linked stepwise interaction of EF-G with its EF-binding site; also, the structural properties of L7/L12 including their multimeric occurrence (Möller et al. 1972; Deusser 1972; Weber 1972; Thammana et al. 1973) are reminiscent of parts of the contractile system (Kischa et al. 1971). Accordingly, among a large number tested, only antibodies raised against L7 or L12 could block EF-G interaction (Highland et al. 1973), and cross-linking of EF-G to ribosomes mapped L7 and L12 at or near the EF-G binding site (Acharya, Moore and Richards 1973).

In principle, two possible functions for GTP hydrolysis can be considered. First, it could be argued that the energy released by GTP hydrolysis is primarily conserved in a change of conformational energy of ribosomal components located at or near the center of GTP cleavage. These components could be responsible either directly or indirectly through a process of subcomponent interactions for the triggering of translocation and tRNA binding.

Second, the possibility should be left in mind that GTP hydrolysis primarily serves as a mechanism which can convert GTP-EF-ribosome complexes into GDP-EF-ribosome complexes. The advantage would be that each of the two nucleotides may impose its own conformation on the ribosome-bound elongation factors and thus ensure binding and release of the elongation factors.[1] Necessarily the actual site of GTP hydrolysis would not be on the elongation factors. For instance, it would be conceivable to have the locus which binds the nucleotide moiety situated on the elongation factors and the ribosomal GTPase enzyme in proximity of the γ-phosphate of GTP. On the other hand, the site of binding and hydrolysis may be separated, with the GTPase enzyme and the elongation factors at opposite ends. Therefore it seems of interest to know whether GTP hydrolysis is simply diffusion controlled. Perhaps the type of hindrance could be found from comparative rate studies using reconstituted 50S ribosomes which lack specified subcomponents.

This view of GTP hydrolysis is illustrated in a model indicating how GTP and GDP could modulate the stability of EF–ribosome complexes (Figure 7). EF-GTP complexes are presumed to bind to the ribosome but not EF-GDP complexes. According to this model, hydrolysis would be concerned with the binding and release of the factors rather than with subsequent events in protein synthesis (cf. tRNA binding, translocation of peptidyl-tRNA and messenger movement). GTP and GDP each may impose a different conformation on EF-Tu; similarly when GDP replaces GTP in EF-G-ribosome interaction, a change in EF-G conformation may occur. Here we should also add that EF2-nucleotide complexes have been found in eukaryotes (Skogerson and Moldave 1968), and that no release of IF-2 from the ribosome occurs when GTP hydrolysis is blocked (Benne and Voorma 1972; Lockwood et al. 1974).

Support for this model is provided by the recent demonstration that EF-Tu·GTP complexes more easily exchange protein hydrogens than do EF-Tu·GDP (Printz and Miller 1973). Moreover, EF-Tu·GTP seems to possess a binding site for a hydrophobic dye whereas EF-Tu·GDP does not (Printz and Miller 1973). These findings support our contention that EF binding may be controlled by nucleotides. The hydrophobic site exposed by GTP may in turn interact with L7 and L12, which are rather hydrophobic proteins (Terhorst et al. 1973).

The proposed scheme does little to explain the multiple occurrence of L7/L12 in the ribosome. Perhaps a comparative study of their tertiary and quaternary structure both free and on the ribosome may provide more insight. In this respect, reductive methylation may be a helpful method because with retention of its ribosome function, specific lysine residues of L7 and L12 are more reactive in situ than in isolation (Amons and Möller 1974). These results indicate that in situ the proteins L7/L12 unfold or

---

[1] Recently Y. Kaziro presented independent evidence that the role of GTP hydrolysis in elongation is the removal of the elongation factors from ribosomes (Kaziro 1973).

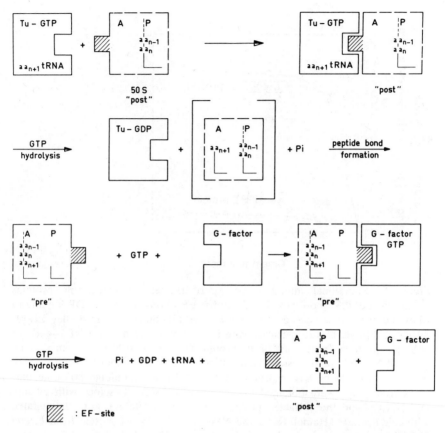

Figure 7   Possible role of GTP hydrolysis in the alternate binding of EF-Tu and EF-G elongation factor binding site. EF site comprises a region where L7, L12 and L10 are located.

are more exposed to the solvent than the "free" dimeric form (Möller et al. 1972).

In conclusion it may be said that the hydrophobic properties and multimeric occurrence of L7/L12 may be more than coincidental in the interactions of the elongation factors with the ribosome.

Occurrence of L7 and L12 in Eukaryotes

The importance of L7 and L12 in bacterial polypeptide synthesis has stimulated research on the occurrence of similar proteins in eukaryotes. Recently Wool and coworkers presented evidence that L7 and L12 seem to be conserved in eukaryotes. The universal occurrence of these proteins would be striking, particularly because only L7 and L12 seem to have been

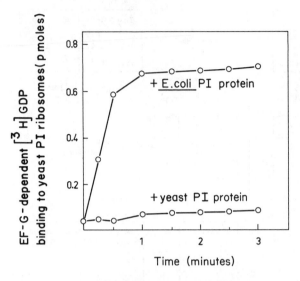

Figure 8 Functional interchangeability of the ethanol-extractable proteins from 80S and 70S ribosomes. The assay for EF-G-directed [³H]GDP binding to ribosomes was the same as described elsewhere (Richter, Lin and Bodley 1971). 80 µg of 80S yeast core particles were incubated in the presence of 2–4 µg of 80S split proteins from yeast or PI protein from *E. coli* 50S ribosomes, 5 µg EF-G, [³H]GDP, and fusidic acid (Richter, Lin and Bodley 1971). GDP binding to ribosomes was measured by the Millipore technique. 80S core particles from yeast were obtained by extraction of 80S ribosomes with ethanol and NH₄Cl, and the extracted proteins (80S split proteins) were precipitated with cold acetone (Hamel, Koka and Nakamoto 1972). PI protein from *E. coli* 50S ribosomes was obtained in a similar way (Hamel, Koka and Nakamoto 1972). The 80S split protein fraction consisted of, among others, two acidic ribosomal proteins that migrated on two-dimensional electrophoresis in the acidic region of L7/L12 from *Escherichia coli*. A similar fraction has been detected in 80S reticulocyte ribosomes, the two most acidic bands having isoelectric points close to pH 4.5 (D. Richter and W. Möller, unpublished data).

uniquely conserved, as judged on the basis of immunological data (see Wool and Stöffler, this volume). Independent work by Richter and Möller, using a different approach, also support that L7 and L12 seem conserved in eukaryotes and functionally interchangeable. When supplemented with L7 and L12 from *Escherichia coli* ribosomes, the bacterial EF-G substituted in 80S yeast cores for the corresponding yeast EF2 in GTP hydrolysis and GDP binding (D. Richter and W. Möller, unpublished data). The functional interchangeability of the ethanol-extractable proteins from 80S and 70S ribosomes with respect to GDP binding is demonstrated in Figure 8.

SUMMARY

On the basis of the preceding discussion the following key points may be summarized:

(1) The 50S proteins L7 and L12 play a key role in the binding of the elongation factors EF-Tu and EF-G. Their presence is required for elongation factor-dependent GTP hydrolysis, translocation and enzymatic tRNA binding. In addition, L7/L12 are essential for the IF2-dependent initiation reactions and the binding of the release factor RF2.

(2) The peptidyl transferase and the GTPase center seem to be close neighbors on the 50S ribosomal subunit.

(3) A 5S RNA–protein complex having both ATPase and GTPase activity has been isolated from *Escherichia coli* and *Bacillus stearothermophilus*.

(4) Alternating selectivity of EF-G and EF-Tu binding to the ribosome may be related to the occupancy by peptidyl-tRNA of either the A site or the P site. Activation of the EF-G binding site may occur through a conformational change on the ribosome following the formation of the peptide bond.

(5) The precise role of GTP hydrolysis in ribosomal polypeptide synthesis remains to be determined. The importance of differences in conformation between GTP and GDP elongation factor complexes is stressed.

(6) Proteins similar to L7 and L12 also occur in eukaryotic ribosomes.

Acknowledgments

I thank Dr. H. Weissbach and Dr. N. Brot for kindly furnishing a manuscript of their work before publication. I thank my colleagues for friendly advice. The helpful criticism of Dr. L. I. Slobin is gratefully acknowledged, and I am indebted to Dr. V. A. Erdmann, Dr. M. Nomura, Dr. S. Ochoa and Dr. A. Parmeggiani for making available their work before publication.

References

Acharya, A. S., P. B. Moore and F. M. Richards. 1973. Cross-linking of elongation factor EF-G to the 50S ribosomal subunit of *Escherichia coli*. *Biochemistry* **12**:3108.

Amons, R. and W. Möller. 1974. Reductive methylation studies on L7 and L12 ribosomal proteins. *Eur. J. Biochem.* **44**:97.

Ballesta, J. P. G. and D. Vazquez. 1972. Effects of thiostrepton and syomycin on elongation factors G- and T-dependent GTP hydrolysis. *Proc. 8th Meet. Fed. Eur. Biochem. Soc.*, Vol. 27, RNA Viruses and Ribosomes, p. 281. North Holland, Amsterdam.

Benne, R. and H. O. Voorma. 1972. Entry site of formylmethionyl-tRNA. *FEBS Letters* **20**:347.

Bodley, J. W., F. J. Zieve, L. Lin and S. T. Zieve. 1970. Studies on transloca-tion. III. Conditions necessary for the formation and detection of a stable ribosome-G factor-guanosine diphosphate complex in the presence of fusidic acid. *J. Biol. Chem.* **245**:5656.

Brot, N., R. Marcel, E. Yamasaki and H. Weissbach. 1973. Further studies on the role of 50S ribosomal proteins in protein synthesis. *J. Biol. Chem.* **248**:6952.

Brot, N., W. P. Tate, C. T. Caskey and H. Weissbach. 1974. The requirement of ribosomal proteins L7 and L12 for peptide chain termination. *Proc. Nat. Acad. Sci.* **71**:89.

Brot, N., E. Yamasaki, B. Redfield and H. Weissbach. 1972. The properties of an *E. coli* ribosomal protein required for the function of factor G. *Arch. Biochem. Biophys.* **148**:148.

Cabrer, B., D. Vazquez and J. Modolell. 1972. Inhibition by elongation factor EF-G of aminoacyl-tRNA binding to ribosomes. *Proc. Nat. Acad. Sci.* **69**:733.

Caskey, C. T., R. Tomkins, E. Scolnick, T. Caryk and M. Nirenberg. 1968. Sequential translation of trinucleotide codons for initiation and termination of protein synthesis. *Science* **162**:135.

Chang, F. N. 1973. Conformational changes in ribosomal subunits following dissociation of the *Escherichia coli* 70S ribosome. *J. Mol. Biol.* **78**:563.

Cohlberg, J. A. 1974. Activity of protein-deficient 30S ribosomal subunits in elongation factor G-dependent GTPase. *Biochem. Biophys. Res. Comm.* **57**:225.

Deusser, E. 1972. Heterogeneity of ribosomal populations in *Escherichia coli* cells grown in different media. *Mol. Gen. Genet.* **119**:249.

Erdmann, V. A., S. Fahnestock, K. Higo and M. Nomura. 1971. Role of 5S RNA in the functions of 50S ribosomal subunits. *Proc. Nat. Acad. Sci.* **68**:2932.

Gray, P. N., R. A. Garrett, G. Stöffler and R. Monier. 1972. An attempt at the identification of the proteins involved in the incorporation of 5-S RNA during 50-S ribosomal subunit assembly. *Eur. J. Biochem.* **28**:412.

Gupta, S. L., J. Waterson, M. L. Sopori, S. M. Weissman and P. Lengyel. 1971. Movement of the ribosome along the messenger ribonucleic acid during protein synthesis. *Biochemistry* **10**:4410.

Hamel, E. and T. Nakamoto. 1972. Studies on the role of an *Escherichia coli* 50S ribosomal component in polypeptide chain elongation. *J. Biol. Chem.* **247**:6810.

Hamel, E., M. Koka and T. Nakamoto. 1972. Requirement of an *Escherichia coli* 50S ribosomal protein component for effective interaction of the ribosome with T and G factors and with guanosine triphosphate. *J. Biol. Chem.* **247**:805.

Highland, J. H., J. W. Bodley, J. Gordon, R. Hasenbank and G. Stöffler. 1973. Identity of the ribosomal proteins involved in the interaction with elongation factor G. *Proc. Nat. Acad. Sci.* **70**:147.

Horne, J. R. and V. A. Erdmann. 1972. Isolation and characterization of 5S RNA-protein complexes from *Bacillus stearothermophilis* and *Escherichia coli. Mol. Gen. Genet.* **119**:337.

————. 1973. ATPase and GTPase activities associated with a specific 5S RNA-protein complex. *Proc. Nat. Acad. Sci.* **70**:2870.

Huang, K. H. and C. R. Cantor. 1972. Surface topography of the 30S *Escherichia coli* ribosomal subunit: Reactivity towards fluorescein isothiocyanate. *J. Mol. Biol.* **67**:265.

Kay, A., G. Sander and M. Grunberg-Manago. 1973. Effect of ribosomal protein L12 upon initiation factor IF-2 activities. *Biochem. Biophys. Res. Comm.* **51**:979.

Kaziro, Y. 1973. The role of guanosine triphosphates in polypeptide elongation reaction in *Escherichia coli*. In *Organization of energy transducing membranes* (ed. M. Nakao and L. Packer) p. 187. University Park Press, Baltimore.

Kischa, K., W. Möller and G. Stöffler. 1971. Reconstitution of a GTPase activity by a 50S ribosomal protein from *E. coli. Nature New Biol.* **233**:62.

Kung, H. F., J. E. Fox, C. Spears, N. Brot and H. Weissbach. 1973. Studies on the role of ribosomal proteins L7 and L12 in the *in vitro* synthesis of β-galactosidase. *J. Biol. Chem.* **248**:5012.

Lipmann, F. 1969. Polypeptide chain elongation in protein biosynthesis. *Science* **164**:1024.

Lockwood, A. H., U. Maitra, N. Brot and H. Weissbach. 1974. The role of ribosomal proteins L7 and L12 in polypeptide initiation in *Escherichia coli. J. Biol. Chem.* **249**:1213.

Maassen, J. A. and W. Möller. 1974. Identification by photo-affinity labeling of the proteins in *Escherichia coli* ribosomes involved in EF-G-dependent binding. *Proc. Nat. Acad. Sci.* **71**:1277.

Marsh, R. C. and A. Parmeggiani. 1973. Requirements of 30S proteins S5 and S9 for the ribosome-dependent GTPase activity of elongation factor G. *Proc. Nat. Acad. Sci.* **70**:151.

Meselson, M., M. Nomura, S. Brenner, C. Davern and D. Schlessinger. 1964. Conservation of ribosomes during bacterial growth. *J. Mol. Biol.* **9**:696.

Miller, D. L. 1972. Elongation factors EF-Tu and EF-G interact at related sites on ribosomes. *Proc. Nat. Acad. Sci.* **69**:752.

Modolell, J. and D. Vazquez. 1973. Inhibition by aminoacyl transfer ribonucleic acid of elongation factor G-dependent binding of guanosine nucleotide to ribosomes. *J. Biol. Chem.* **248**:488.

Modolell, J., B. Cabrer and D. Vazquez. 1973. The interaction of elongation factor G with N-acetylphenylalanyl transfer ribonucleic acid ribosome complexes. *Proc. Nat. Acad. Sci.* **70**:3561.

Modolell, J., D. Vazquez and R. E. Monro. 1971. Ribosomes. G-factor and siomycin. *Nature New Biol.* **230**:109.

Möller, W. and H. Castleman. 1967. Primary structure heterogeneity in ribosomal proteins from *Escherichia coli. Nature* **215**:1293.

Möller, W., H. Castleman and C. P. Terhorst. 1970. Characterization of an acidic protein in 50S ribosomes of *E. coli. FEBS Letters* **8**:192.

Möller, W., A. Groene, C. Terhorst and R. Amons. 1972. Purification and partial characterization of two acidic proteins, A_1 and A_2, isolated from 50-S ribosomes of *Escherichia coli*. *Eur. J. Biochem.* **25**:5.

Nierhaus, D. and K. H. Nierhaus. 1973. Identification of the chloramphenicol-binding protein in *Escherichia coli* ribosomes by partial reconstitution. *Proc. Nat. Acad. Sci.* **70**:2224.

Nierhaus, K. H. and V. Montejo. 1973. A protein involved in the peptidyl transferase activity of *Escherichia coli* ribosomes. *Proc. Nat. Acad. Sci.* **70**:1931.

Nombela, C. and S. Ochoa. 1973. Conformational control of the interaction of eucaryotic elongation factors EF_1 and EF_2 with ribosomes. *Proc. Nat. Acad. Sci.* **70**:3556.

Nomura, M., S. Mizushima, M. Ozaki, P. Traub and C. V. Lowry. 1969. Structure and function of ribosomes and their molecular components. *Cold Spring Harbor Symp. Quant. Biol.* **34**:49.

Pongs, O., R. Bald and V. A. Erdmann. 1973. Identification of chloroamphenicol-binding protein in *Escherichia coli* ribosomes by affinity labeling. *Proc. Nat. Acad. Sci.* **70**:2229.

Printz, M. P. and D. L. Miller. 1973. Evidence for conformational changes in elongation factor Tu induced by GTP and GDP. *Biochem. Biophys. Res. Comm.* **53**:149.

Randall-Hazelbauer, L. L. and C. G. Kurland. 1972. Identification of three 30S proteins contributing to the ribosomal A-site. *Mol. Gen. Genet.* **115**:234.

Richman, N. and J. W. Bodley. 1972. Ribosomes cannot interact simultaneously with elongation factor EF-Tu and EF-G. *Proc. Nat. Acad. Sci.* **69**:686.

Richter, D. 1972. Inability of *E. coli* ribosomes to interact simultaneously with the bacterial elongation factors EF-Tu and EF-G. *Biochem. Biophys. Res. Comm.* **46**:1850.

Richter, D., L. Lin and J. W. Bodley. 1971. Studies on translocation. IX. The pattern of action of antibiotic translocation inhibition in eucaryotic and procaryotic systems. *Arch. Biochem. Biophys.* **147**:186.

Sander, G., R. C. Marsh and A. Parmeggiani. 1972. Isolation and characterization of two acidic proteins from the 50S subunit required for GTPase activities of both EF-G and EF-T. *Biochem. Biophys. Res. Comm.* **47**:866.

———. 1973. Role of split proteins from 30S subunits in the ribosome-EF-T GTPase reaction. *FEBS Letters* **33**:132.

Schrier, P. I., J. A. Maassen and W. Möller. 1973. Involvement of 50S ribosomal proteins L6 and L10 in the ribosome dependent GTPase activity of elongation factor G. *Biochem. Biophys. Res. Comm.* **53**:90.

Scolnick, E. and T. Caskey. 1969. Peptide chain termination. V. The role of release factors in mRNA terminator codon recognition. *Proc. Nat. Acad. Sci.* **64**:1235.

Skogerson, L. and K. Moldave. 1968. Characterization of the interaction of aminoacyl transferase II with ribosomes. *J. Biol. Chem.* **243**:5454.

Sopori, M. L. and P. Lengyel. 1972. Components of the 50S ribosomal subunit involved in GTP cleavage. *Biochem. Biophys. Res. Comm.* **46**:238.

Stöffler, G., L. Daya, K. H. Rak and R. A. Garrett. 1971. Specific protein binding sites on 23S RNA of *Escherichia coli*. *Mol. Gen. Genet.* **114**:125.

Thammana, P., C. G. Kurland, E. Deusser, J. Weber, R. Maschler, G. Stöffler and H. G. Wittmann. 1973. Structural and functional evidence for a repeated 50S subunit ribosomal protein. *Nature New Biol.* **242**:47.

Terhorst, C., B. Wittmann-Liebold and W. Möller. 1972. 50S ribosomal proteins. Peptide studies on two acidic proteins, A₁ and A₂, isolated from 50S ribosomes of *Escherichia coli. Eur. J. Biochem.* **25**:13.

Terhorst, C., W. Möller, R. Laursen and B. Wittmann-Liebold. 1973. The primary structure of an acidic protein from 50S ribosomes of *Escherichia coli* which is involved in GTP hydrolysis dependent on elongation factors G and T. *Eur. J. Biochem.* **34**:138.

Traub, P. and M. Nomura. 1968. Structure and function of *Escherichia coli* ribosomes. I. Partial fractionation of the functionally active ribosomal proteins and reconstitution of artificial subribosomal particles. *J. Mol. Biol.* **34**:575.

Weber, H. J. 1972. Stoichiometric measurements of 30S and 50S ribosomal proteins from *Escherichia coli. Mol. Gen. Genet.* **119**:233.

Weissbach, H., B. Redfield, E. Yamasaki, R. C. Davis, Jr., S. Pestka and N. Brot. 1972. Studies on the ribosomal sites involved in factors Tu- and G-dependent reactions. *Arch. Biochem. Biophys.* **149**:110.

Cellular Regulation of Guanosine Tetraphosphate and Guanosine Pentaphosphate

Michael Cashel
Laboratory of Molecular Genetics
National Institutes of Health
Bethesda, Maryland 20014

Jonathan Gallant
Department of Genetics
University of Washington
Seattle, Washington 98105

The Pleiotropic Cellular Response to Amino Acid Availability

Most cells possess the capacity to coordinately regulate a rich variety of physiological activities in response to nutritional abundance. Perhaps the most widely studied of these mechanisms is the bacterial response to limitation of *any* amino acid for protein synthesis. This response was first noted as a stringent dependence of RNA accumulation on amino acid availability (Sands and Roberts 1952; Pardee and Prestidge 1956; Gros and Gros 1958) and later termed the "stringent response" (Stent and Brenner 1961). A mutant strain in which this amino acid dependence was relaxed was encountered (Borek, Rockenbach and Ryan 1956). Stent and coworkers mapped the mutant allele at the "RNA Control" (RC) or *rel*A locus (Alföldi, Stent and Clowes 1962). Many more mutants of this locus have since been isolated by Fiil and Friesen (1968), providing isogenic strain pairs in which the functions of the *rel* gene product may be experimentally isolated. Fiil (1969) has shown that the *rel*$^-$ allele is recessive to the wild type (*rel*$^+$). Much less common are mutations in at least two other loci, *rel*B (Lavallé 1965) and *rel*C (Friesen, unpublished). Since little has been reported about these unusual classes of relaxed mutants, in this review we will deal exclusively with the *rel*A locus.

Several immediate effects of withdrawing an amino acid can be anticipated, such as insufficient free amino acid for cognate tRNA aminoacylation and derepression of specific amino acid biosynthetic pathways (Neidhardt 1966). However, Table 1 summarizes a growing list of unrelated cellular activities which rather unexpectedly also respond to amino acid starvation, but only in *rel*$^+$ cells and not in *rel*$^-$ strains. This variety of processes regulated by a single gene product led to a search for the pleiotropic effector mediating these alterations. In *rel*$^+$ cells, amino acid starvation led to reduced ^{32}P labeling of most phosphorylated compounds and occasioned a striking accumulation of two novel nucleotides, which were designated

733

Table 1 Amino acid dependent activities controlled by the *rel* gene

Activity	Effect	Reference	ppGpp in vitro
rRNA transcription	Inhibition	Lazzarini & Dahlberg (1971)	Travers (this volume)
tRNA transcription	Inhibition	Primakoff & Berg (1970); Ikemura & Dahlberg (1973)	
mRNA transcription	Inhibition	Lazzarini & Dahlberg (1971); Gallant & Margason (1972)	
mRNA transcription	Stimulation	Lavallé & deHauwer (1968)	Aboud & Pastan (1973)
mRNA transcription	No effect	Primakoff & Berg (1970)	
Protein stability	Inhibition	Sussman & Gilvarg (1969); Goldberg (1971) Hansen, Bennett & von Meyenberg (1973)	
Coding fidelity	Inhibition(*rel⁻*)	Hall & Gallant (1972)	
Polysome reassembly	Inhibition	Cozzone & Donini (1973)	
Protein synthesis	Inhibition		Arai et al. (1972); Legault, Jeantet & Gros (1972); Yoshida, Travers & Clark (1972)

Phospholipid synthesis	Inhibition	Merlie & Pizer (1973)
Lipid synthesis	Inhibition	Polakis, Guchhait & Lane (1973)
Nucleotide synthesis	Inhibition	Cashel & Gallant (1968)
Pyrimidine transport	Inhibition	Edlin & Neuhard (1967)
Purine transport	Inhibition	Nierlich (1968)
α-Methyl glucoside transport	Inhibition	Sokawa & Kaziro (1969)
Glycolysis	Inhibition	Irr & Gallant (1969)
Glucose respiration	Inhibition	Sokawa, Nakao-Sato & Kaziro (1970)
Cellular outgrowth	Inhibition(*rel*−)	Alföldi et al. (1963); Sokawa, Sokawa & Kaziro (1971)

Sokawa, Sokawa & Kaziro (1972)
Sokawa, Nakao & Kaziro (1968)

Gallant, Irr & Cashel (1971)
Hochstadt-Ozer & Cashel (1972)
Hochstadt-Ozer & Cashel (1972)

magic spots (MS) I and II; in *rel⁻* cells neither nucleotide accumulated (Cashel and Gallant 1969; Cashel 1969). Table 1 also summarizes in vitro evidence of the inhibitory activities of ppGpp.

rel-Dependent Synthesis of ppGpp

A survey with several *rel⁻* strains confirmed that the accumulation of MS I was invariably associated with the *rel⁺* function (Cashel 1969). Further studies of other *rel⁻* mutants in *E. coli, B. subtilis,* and *S. typhimurium* have confirmed this association (Swanton and Edlin 1972; Raué 1971).

MS I has been isolated from amino acid starved *rel⁺* cells and its structure determined to be guanosine 5'-diphosphate, 2'- or 3'-diphosphate (ppGpp) (Cashel and Kalbacher 1970). Cellular MS II is a guanosine pentaphosphate (Cashel and Kalbacher 1970) and was determined from in vitro preparations to have the analogous structure, pppGpp (Haseltine et al. 1972). Assignment of the 2'- or 3'-pyrophosphate residue of ppGpp to the 3' position in the in vitro product by Sy and Lipmann (1973) has been confirmed by ¹³C NMR spectra (Que et al. 1973).

Usually *rel⁺*-dependent ppGpp accumulation is evoked by starvation for any amino acid. Inhibitors of formyl transfer produce the same effect, which was originally interpreted in terms of involvement of protein synthesis initiation (Khan and Yamazaki 1972; Lund and Kjeldgaard 1972a). More recently, the complex effects of these inhibitors on ppGpp can be shown to be largely reversed by mixtures of amino acids (Kjeldgaard, unpublished; Cashel, unpublished), suggesting that they may merely produce amino acid starvation. Similarly, nitrogen starvation evokes *rel⁺*-dependent ppGpp accumulation (Edlin and Donini 1971; Irr 1972). In view of what is now known of the *rel* gene product, any treatment which provokes ppGpp accumulation in *rel⁺* but not in *rel⁻* cells can very probably be attributed to limited aminoacylation of one or more tRNA species (see below).

Some cellular activities show an amino acid specificity which depends on the frequency of an amino acid in proteins, such as polysome reaggregation and proteolysis (Ron 1971; Cozzone and Donini 1973; Pine 1973). The formation of ppGpp, like RNA Control (Stent and Brenner 1961), is independent of such specificity but instead responds in a virtually equivalent fashion to starvation for any amino acid (Cashel 1969). The *rel*-dependent synthesis of ppGpp is also quite independent of steady-state growth conditions occurring after amino acid starvation with cells grown on glucose, glycerol or succinate, and either aerobically or anaerobically (Lund and Kjeldgaard 1972a; Cashel, unpublished).

Dependence on tRNA Aminoacylation

Like stringent control of RNA synthesis, the accumulation of ppGpp is induced by restricting the activity of various tRNA synthetases even in the presence of all 20 free amino acids (Cashel 1969; deBoer et al. 1971; Fiil, von Meyenberg and Friesen 1972; Lund and Kjeldgaard 1972a). On the other hand, inhibition of protein synthesis at the ribosomal level evokes neither ppGpp accumulation (see below) nor inhibition of RNA accumulation. These two observations show that the *rel*+ gene product transacts its business somewhere between the amino acid charging of tRNA and the subsequent discharging of aminoacyl-tRNA onto growing peptide chains. In fact, direct inhibition of peptide chain growth antagonizes the stringent response (Kurland and Maaløe 1962) and blocks the production of ppGpp (Cashel 1969; deBoer et al. 1971; Lund and Kjeldgaard 1972a). The trickle of amino acids produced by protein turnover in amino acid starved cells is continuously consumed by residual protein synthesis, leaving the cognate tRNA pool uncharged. When this consumption is directly inhibited, recharging occurs to near normal levels and the stringent response vanishes (Kaplan, Atherly and Barret 1973). Most conditions which block ribosomal function probably operate through this sort of "trickle-charging" mechanism (Sokawa, Sokawa and Kaziro 1972; Atherly 1973; Lund and Kjeldgaard 1972a; Rabbani and Srinivasan 1973).

A singular exception is tetracycline, which blocks the stringent response and ppGpp formation when trickle-charging is completely abolished (Kaplan, Atherly and Barret 1973). Thus the drug which inhibits both enzyme-dependent aminoacyl-tRNA binding as well as enzyme-independent binding of uncharged tRNA (Levin 1970) emerges as a direct inhibitor of ppGpp synthesis. In retrospect, these observations suggest that uncharged tRNA binding at a tetracycline-sensitive site triggers ppGpp synthesis.

Kinetics

Since ppGpp formation invariably accompanies the stringent response, it is conceivable that this accumulation is a secondary consequence of blocked RNA accumulation rather than a primary expression of *rel*+ gene product activity. Yet no formation of ppGpp occurs when RNA synthesis is inhibited directly (Cashel and Gallant 1969; Gallant et al. 1970; Lazzarini, Cashel and Gallant 1971). Stronger evidence along these lines follows from the fact that ppGpp accumulation begins within a few seconds after imposing amino acid starvation, preceding the onset of RNA Control by at least a minute (Cashel 1969; Gallant et al. 1970; Fiil, von Meyenberg and Friesen 1972). Evidently a functional ppGpp regulatory mechanism exists in growing cells; this conclusion is also drawn by correlations between growth rate and ppGpp levels during exponential growth (Lazzarini, Cashel and Gallant 1971; Fiil, von Meyenberg and Friesen 1972). After 5–10 min of amino acid starvation, the level of ppGpp increases by at least an order of magnitude to plateau at a level which can persist for hours. This plateau represents a metabolically active steady state, since both exogenous phosphate and guanine continue to label ppGpp (Cashel 1969; Gallant et al.

1970; Fiil, von Meyenberg and Friesen 1972). Readdition of amino acids to starved cells occasions a first-order decay (Cashel 1969; Gallant, Margason and Finch 1972). The first-order rate constants vary from 3/min to 0.1–0.2/min, depending on genetic factors which will be discussed later. The fact that *rel+* is the dominant allele (Fiil 1969) implies that the *rel* gene product mediates an increase in the rate of synthesis. Calculation of the rates of ppGpp synthesis and breakdown by and large bears this out, although some change in degradation rates occurs (Gallant, Margason and Finch 1972; Fiil, von Meyenberg and Friesen 1972).

Typically the level of ppGpp rises to about one-third that of ATP, the most abundant nucleotide, although it has been reported to exceed ATP in some strains (Lund and Kjeldgaard 1972a; Swanton and Edlin 1972). The pppGpp levels range from virtually undetectable in some *E. coli* strains to typically one-third the ppGpp; in *B. subtilis* pppGpp can exceed the ppGpp levels (Swanton and Edlin 1972; Gallant and Margason 1972). In any event, a very substantial portion of cellular energies can be directed towards ppGpp formation. For example, amino acid starved K12 cells in the steady state form and turnover about 10^6 molecules/min/cell (Lund and Kjeldgaard 1972a). It is unknown whether energy retrieval occurs by the degradation reactions.

Dependence of ppGpp Synthesis on mRNA

Addition of rifampicin to growing cells leads to a now familiar chain of events ultimately restricting protein synthesis for lack of messenger RNA. Rifampicin also inhibits ppGpp formation in amino acid starved cells: the longer the rifampicin pretreatment, the more severe the inhibition (Wong and Nazar 1970; Lund and Kjeldgaard 1972b). Since message depletion in rifampicin-treated cells leads to restricted protein synthesis, trickle-charging is undoubtedly largely responsible for this effect. When trickle-charging is reduced through the use of a temperature-sensitive activating enzyme mutant, then ppGpp formation is not inhibited by rifampicin for about 15 min (Erlich, Laffler and Gallant 1971; de Boer et al. 1973). Later ppGpp synthesis does appear to be progressively inhibited under these conditions (de Boer et al. 1973; Erlich, unpublished). This delayed effect might mean that the presence of mRNA is required for ppGpp synthesis (de Boer et al. 1973). However, since the temperature-sensitive mutant employed in these studies is leaky (Kaplan, Atherly and Barret 1973), trickle-charging still cannot be ruled out.

It is notable that infection with phage T_7, whose RNA polymerase is rifampicin resistant, permits ppGpp accumulation as well as protein synthesis in the presence of rifampicin, indicating that the relationship between ppGpp synthesis and mRNA availability is not host specific (Lund and Kjeldgaard 1972b. In cells infected with phage R_{17}, whose major coat protein product contains no histidine, starvation for histidine does not elicit ppGpp accumulation, whereas starvation for another amino acid (iso-leucine) does (Khan and Yamazaki 1970; Watson and Yamazaki 1972). This might mean that ppGpp formation involves a codon-specific interaction, but again trickle-charging is a sufficient explanation.

Deductions from *rel*-Dependent ppGpp Synthesis

Thus the presence of the *rel* gene product is a necessary but not sufficient condition for ppGpp synthesis and the stringent response. It is clear that the sufficient conditions also include restricted aminoacylation of any tRNA species and the target of tetracycline inhibition, which is presumably a functional ribosomal acceptor site. The effect of types of protein synthesis inhibition other than tetracycline are consistent with the idea that a functional protein synthetic apparatus is required; the effect of rifampicin is consistent with a requirement for mRNA on the ribosomes. However, both of the latter notions are seriously qualified by an alternative explanation, namely, the indirect trickle-charging mechanism alluded to earlier. These considerations place certain constraints on the kinds of mechanisms which could account for ppGpp synthesis. For example, mechanisms can be ruled out which respond only to charged or uncharged tRNA on the one hand, or only to restriction of peptide chain growth on the other. Instead an idling reaction of protein synthesis involving both functional ribosomes and tRNA is the simplest explanation (Cashel and Gallant 1969; Lund and Kjeld-gaard 1972).

rel-Independent Accumulation of ppGpp

Neidhardt (1963) noted that during slowing of growth rates accompanying transition from an amino acid rich medium to an amino acid poor medium, relaxed mutants overproduced RNA much as during simple amino acid starvation. In marked contrast, Neidhardt also noted that upon adjustment to media containing poor major carbon and energy sources, the RNA Control during the transition was identical in relaxed and stringent strains. Evidently *rel*⁻ mutants retain a second mechanism for controlling RNA accumulation in response to some stimulus other than amino acid starvation.

This mechanism also governs the level of ppGpp. For example, *rel*⁻ cells accumulate large amounts of ppGpp during the diauxic lag accompanying downshifts from glucose to poorer carbon sources, as well as downshifts from glucose to nothing (Lazzarini, Cashel and Gallant 1971; Harshman and Yamazaki 1971; Winslow 1971). Another condition in which similar accumulations of ppGpp are evoked even in *rel*⁻ cells is plasmolysis in hypertonic sodium chloride (Harshman and Yamazaki 1972) or sucrose (Cashel, unpublished). We believe that leakage of intermediary metabolites in such maltreated cells makes this condition physiologically equivalent to downshift in growing cells.

In both cases, net RNA synthesis is restricted as it is during the stringent response. Thus more than one mechanism can give rise to ppGpp accumulation, and in each case RNA accumulation is shut off.

The downshift mechanism displays another peculiarity which further distinguishes it from the stringent response: pppGpp accumulation does not occur in parallel with ppGpp during downshift as it does during the stringent response in most strains. On the contrary, downshift occasioned by high salt addition leads to a concomitant increase in ppGpp and disappearance of pppGpp (Harshman and Yamazaki 1972). This surprising observation implies that the downshift mechanism has to do with interconversion of the two nucleotides.

We mentioned earlier that the stringent response operates primarily by increasing the rate of synthesis of ppGpp in rel^+ cells. There is evidence that downshift affects the steady state level of ppGpp differently, primarily reducing the turnover of ppGpp. Chloramphenicol inhibition studies (Gallant, Margason and Finch 1972) showed that the rate of ppGpp turnover decreased 4- to 10-fold during the course of downshift in both members of a rel^+/rel^- strain pair. More recently, similar observations have been made on cells subjected to salt plasmolysis (Laffler and Gallant, unpublished). In both cases, kinetic calculations indicate that decreased turnover is the major cause of the accumulation of ppGpp, although a relatively small increase in the nucleotide's rate of synthesis may also occur.

The decreased rate of ppGpp turnover and the absence of pppGpp characteristic of the downshift mechanism rather resemble the phenotype of the recently identified $spoT^-$ mutation (Laffler and Gallant 1974). This spontaneous mutant allele has the following effects on rel^+ strains: (1) amino acid starvation is associated with the appearance of ppGpp but not of pppGpp; (2) the levels of ppGpp are increased; and (3) the rate of decay of ppGpp after amino acid resupplementation is reduced from 3/min to 0.1–0.2/min. $SpoT^-$ is recessive to the wild-type allele in regard to all three phenotypic effects. This implies that the mutant is defective in phosphorylating ppGpp to pppGpp rather than having an increased capacity to hydrolyze pppGpp to ppGpp.

The decreased ppGpp turnover in $spoT^+$ strains during downshift evidently is a phenocopy of the effects of the $spoT^-$ mutation. This suggests that the availability of a glycolytic intermediate regulates the phosphorylation of ppGpp to pppGpp mediated by the $spoT^+$ gene product (Stamminger and Lazzarini 1974).

CONCLUSIONS

Studies with the protagon-like (Posner and Gies 1905) magic spots have been clarified to the point where regulation of intracellular levels of ppGpp and pppGpp emerges as central to understanding the stringent response to

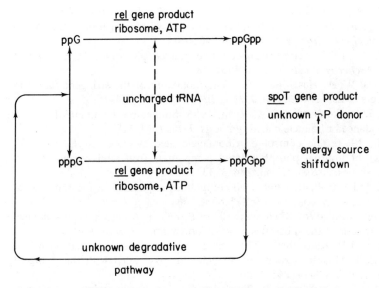

Figure 1   ppGpp and pppGpp reactions in *Escherichia coli.*

amino acid starvation as well as the more general cellular response to nutritional abundance.

Whole cell studies suggest that the intracellular ppGpp levels are controlled by independent adjustment of two valves (Figure 1). One valve, the conversion of ppG to ppGpp by the *rel*A gene product on the ribosome, is controlled by the level of uncharged tRNA and is sensitive to tetracycline (see Block and Haseltine, this volume). The second valve, the phosphorylation of ppGpp to pppGpp, involves the *Spo*T gene product and is coupled to energy source metabolism.

Thus the existence of *rel*A gene mutants has allowed the dissection of the complex regulatory events dependent upon this locus as well as the biochemical basis of ppGpp synthesis on the ribosome (Haseltine et al. 1972; Block and Haseltine 1973). The availability of other mutants, such as *rel*B, *rel*C and *spo*T, is similarly expected to facilitate approaches to both modes of ppGpp regulation.

References

Aboud, M. and I. Pastan. 1973. Stimulation of *lac* transcription by guanosine 5'-diphosphate 2'- (or 3') diphosphate and transfer ribonucleic acid. *J. Biol. Chem.* **248**:3356.

Alföldi, L., G. S. Stent and R. C. Clowes. 1962. The chromosomal site for the RNA Control (R.C.) locus in Escherichia. *J. Mol. Biol.* **5**:348.

Alföldi, L., G. S. Stent, M. Hoogs and R. Hill. 1963. Physiological effects of the RNA Control (R.C.) gene in *E. coli. Z. Vererbungs.* **94**:285.

Arai, K., N. Arai, M. Kawakita and Y. Kaziro. 1972. Interaction of guanosine 5'-diphosphate, 2'- (or 3') diphosphate (ppGpp) with elongation factors from *E. coli. Biochem. Biophys. Res. Comm.* **48**:190.

Atherly, A. G. 1973. Temperature-sensitive relaxed phenotype in a stringent strain of *Escherichia coli. J. Bact.* **113**:178.

Block, R. and W. A. Haseltine. 1973. Thermolability of the stringent factor in *rel* mutants of *Escherichia coli. J. Mol. Biol.* **77**:625.

Borek, E., J. Rockenbach and A. Ryan. 1956. Studies on a mutant of *E. coli* with unbalanced ribonucleic acid synthesis. *J. Bact.* **71**:318.

Cashel, M., 1969. The control of ribonucleic acid synthesis in *E. coli.* IV. Relevance of unusual phosphorylated compounds from amino acid starved stringent strains. *J. Biol. Chem.* **244**:3133.

Cashel, M. and J. Gallant. 1969. Two compounds implicated in the function of the R.C. gene of *E. coli. Nature* **221**:838.

————. 1968. Control of RNA synthesis in *E. coli.* I. Amino acid dependence on the synthesis of the substrates of RNA polymerase. *J. Mol. Biol.* **34**:317.

Cashel, M. and B. Kalbacher. 1970. The control of ribonucleic acid synthesis in *E. coli.* V. Characterization of a nucleotide associated with the stringent response. *J. Biol. Chem.* **245**:2309.

Cozzone, A. and P. Donini. 1973. Turnover of polysomes in amino acid starved *Escherichia coli. J. Mol. Biol.* **76**:149.

De Boer, H. A., H. A. Raué, G. Ab and M. Gruber. 1971. Role of the ribosome in stringent control of bacterial RNA synthesis. *Biochim. Biophys. Acta* **246**:157.

DeBoer, H. A., A. J. J. van Ooyen, G. Ab and M. Gruber. 1973. The role of messenger RNA and peptidyl-tRNA in the synthesis of the guanine nucleotides MS I and MS II by ribosomes *in vivo. FEBS Letters* **30**:335.

Edlin G. and P. Donini. 1971. Synthesis of guanosine 5'-diphosphate, 2'- (or 3') diphosphate and related nucleotides in a variety of physiological conditions. *J. Biol. Chem.* **246**:4371.

Edlin G. and J. Neuhard. 1967. Regulation of nucleoside triphosphate pools in *Escherichia coli. J. Mol. Biol.* **24**:225.

Erlich, H., T. Laffler and J. Gallant. 1971. ppGpp formation in *Escherichia coli* treated with rifampicin. *J. Biol. Chem.* **246**:6121.

Fiil, N. 1969. A functional analysis of the *rel* gene in *Escherichia coli. J. Mol. Biol.* **45**:195.

Fiil, N. and J. D. Friesen. 1968. Isolation of "relaxed" mutants of *E. coli. J. Bact.* **95**:729.

Fiil, N., K. von Meyenburg and J. D. Friesen. 1972. Accumulation and turnover of guanosine tetraphosphate in *Escherichia coli. J. Mol. Biol.* **71**:769.

Gallant, J. and G. Margason. 1972. Amino acid control of messenger ribonucleic acid synthesis in *Bacillus subtilis. J. Biol. Chem.* **247**:2289.

Gallant, J., J. Irr and M. Cashel. 1971. The mechanism of amino acid control of guanylate and adenylate biosynthesis. *J. Biol. Chem.* **246**:5812.

Gallant, J., G. Margason and B. Finch. 1972. On the turnover of ppGpp in *Escherichia coli. J. Biol. Chem.* **247**:6055.

Gallant, J., H. Erlich, B. Hall and T. Laffler. 1970. Analysis of the RC function. *Cold Spring Harbor Symp. Quant. Biol.* **35**:397.

Goldberg, A. L. 1971. A role of amino acyl-tRNA in the regulation of protein breakdown in *Escherichia coli. Proc. Nat. Acad. Sci.* **68**:362.

Gros, F. and F. Gros. 1958. Role des acides amines dans la synthese des acides nucleiques chez *Escherichia coli. Exp. Cell Res.* **14**:104.

Hall, B. and J. Gallant. 1972. Defective translation in RC⁻ cells. *Nature New Biol.* **237**:131.

Hansen, M. T., P. M. Bennett and K. von Meyenburg. 1973. Intracistronic polarity during dissociation of translation from transcription in *Escherichia coli. J. Mol. Biol.* **77**:589.

Harshman, R. B. and H. Yamazaki. 1971. Formation of ppGpp in a relaxed and stringent strain of *Escherichia coli* during diauxic lag. *Biochemistry* **10**:3980.

———. 1972. MS I accumulation induced by sodium chloride. *Biochemistry* **11**:615.

Haseltine, W. A., R. Block, W. Gilbert and K. Weber. 1972. MS I and MS II made on ribosome in idling step of protein synthesis. *Nature* **238**:381.

Hochstadt-Ozer, J. and M. Cashel. 1972. The regulation of purine utilization in bacteria. V. Inhibition of purine phosphoribosyltransferase activities and purine uptake in isolated membrane vesicles by guanosine tetraphosphate. *J. Biol. Chem.* **246**:7067.

Ikemura, T. and J. E. Dahlberg. 1973. Small ribonucleic acids of *E. coli*. II. Noncoordinate accumulation during stringent control. *J. Biol. Chem.* **248**:5033.

Irr, J. 1972. Control of nucleotide metabolism and ribosomal ribonucleic acid synthesis during nitrogen starvation of *E. coli. J. Bact.* **110**:554.

Irr, J. and J. Gallant. 1969. The control of RNA synthesis in *E. coli*. II. Stringent control of energy metabolism. *J. Biol. Chem.* **244**:2233.

Kaplan, S., A. G. Atherly and A. Barret. 1973. Synthesis of stable RNA in stringent *E. coli* cells in the absence of changed transfer RNA. *Proc. Nat. Acad. Sci.* **70**:689.

Khan, S. R. and H. Yamazaki. 1970. Continued expression of the ribonucleic acid control gene during inhibition of *E. coli* ribonucleic acid and protein synthesis. *J. Bact.* **102**:702.

———. 1972. Trimethoprim-induced accumulation of guanosine tetraphosphate (ppGpp) in *E. coli. Biochem. Biophys. Res. Comm.* **48**:169.

Kurland, C. G. and O. Maaløe. 1962. Regulation of ribosomal and transfer RNA synthesis. *J. Mol Biol* **4**:193.

Laffler, T. and J. Gallant. 1974. *Spo*T, a new genetic locus involved in the stringent response in *E. coli. The Cell* **1**:27.

Lavallé, R. 1965. Noveaux mutants de regulation de la synthese de l'ARN. *Bull. Soc. Chim. Biol.* **47**:1567.

Lavallé, R. and G. DeHauwer. 1968. Messenger RNA synthesis during amino acid starvation in *Escherichia coli. J. Mol. Biol.* **37**:269.

Lazzarini, R. A. and A. E. Dahlberg. 1971. The control of ribonucleic acid synthesis during amino acid deprivation in *Escherichia coli. J. Biol. Chem.* **246**:420.

Lazzarini, R. A., M. Cashel and J. Gallant. 1971. On the regulation of guanosine tetraphosphate levels in stringent and relaxed strains of *Escherichia coli. J. Biol. Chem.* **246**:4381.

Legault, L., C. Jeantet and F. Gros. 1972. Inhibition of *in vitro* protein synthesis by ppGpp. *FEBS Letters* **27**:71.

Levin, J. G. 1970. Codon-specific binding of deacylated transfer ribonucleic acid to ribosomes. *J. Biol. Chem.* **245**:3195.

Lund, E., and N. O. Kjeldgaard. 1972a. Metabolism of guanosine tetraphosphate in *Escherichia coli. Eur. J. Biochem.* **28**:316.

———. 1972b. Protein synthesis and formation of guanosine tetraphosphate. *FEBS. Letters* **26**:306.

Merlie, J. P. and L. I. Pizer. 1973. Regulation of phospholipid synthesis in *Escherichia coli* by guanosine tetraphosphate. *J. Bact.* **116**:355.

Neidhardt, F. C. 1963. Properties of a bacterial mutant lacking amino acid control of RNA synthesis. *Biochim. Biophys. Acta* **68**:365.

———. 1966. Role of amino acid activating enzymes in cellular physiology. *Bact. Rev.* **30**:701.

Nierlich, D. P. 1968. Amino acid control over RNA synthesis, a re-evaluation. *Proc. Nat. Acad. Sci.* **60**:1345.

Pardee, A. and L. Prestidge. 1956. The dependence of nucleic acid synthesis on the presence of amino acids in *Escherichia coli. J. Bact.* **71**:677.

Pine, M. J. 1973. Regulation of intracellular proteolysis in *Escherichia coli. J. Bact.* **115**:107.

Polakis, S. E., R. B. Guchhait and M. D. Lane. 1973. Stringent control of fatty acid synthesis in *Escherichia coli:* Possible regulation of acetyl CoA carboxylase by ppGpp. *J. Biol. Chem.* **248**:7957.

Posner, E. R. and W. J. Gies. 1905. Is protagon a mechanical mixture of substances or a definite chemical compound? *J. Biol. Chem.* **1**:59.

Primakoff, P. and P. Berg. 1970. Stringent control of transcription of phage φ80psu₃. *Cold Spring Harbor Symp. Quant. Biol.* **35**:391.

Que, L., G. R. Willie, M. Cashel, J. W. Bodley and G. R. Gray. 1973. Guanosine 5'-diphosphate, 3'-diphosphate; assignment of structure by ^{13}C nuclear magnetic resonance spectroscopy. *Proc. Nat. Acad. Sci.* **70**:2563.

Rabbani E. and P. R. Srinivasan. 1973. Role of the translocation factor G in the regulation of ribonucleic acid synthesis. *J. Bact.* **113**:1177.

Raué, H. A. 1971. *Ph.D thesis.* Rijksuniversiteit te groningen. The Netherlands.

Ron, E. Z. 1971. Polysome turnover during amino acid starvation in *Escherichia coli. J. Bact.* **108**:263.

Sands, M. K. and R. B. Roberts. 1952. The effects of a tryptophan-histidine deficiency in a mutant of *Escherichia coli. J. Bact.* **63**:505.

Sokawa, J., Y. Sokawa and Y. Kaziro. 1972. Stringent control in *E. coli. Nature New Biol.* **240**:242.

Sokawa, Y. and Y. Kaziro. 1969. Amino acid-dependent control of the transport of α-methyl glucoside in *E. coli. Biochem. Biophys. Res. Comm.* **34**:99.

Sokawa, Y., E. Nakao-Sato and Y. Kaziro. 1968. On the nature of the control by RC gene in *Escherichia coli;* amino acid dependent control of lipid synthesis. *Biochem. Biophys. Res. Comm.* **33**:108.

———. 1970. RC gene control in *Escherichia coli* is not restricted to RNA synthesis. *Biochim. Biophys. Acta* **199**:256.

Sokawa, Y., J. Sokawa and Y. Kaziro. 1971. Function of the *rel* gene in *E. coli. Nature New Biol.* **234**:7.

Stamminger, G. and R. A. Lazzarini. 1974. Altered metabolism of the guanosine tetraphosphate, ppGpp, in mutants of *E. coli. The Cell* **1**:85.

Stent, G. S. and S. Brenner. 1961. A genetic locus for the regulation of ribonucleic acid synthesis. *Proc. Nat. Acad. Sci.* **47**:2005.

Sussman, A. J. and C. Gilvarg. 1969. Protein turnover in amino acid-starved strains of *Escherichia coli* K-12 differing in their ribonucleic acid control. *J. Biol. Chem.* **244**:6304.

Swanton, M. and G. Edlin. 1972. Isolation and characterization of an RNA relaxed mutant of *B. subtilis. Biochem. Biophys. Res. Comm.* **46**:583.

Sy, J. and F. Lipmann. 1973. Identification of the synthesis of guanosine tetraphosphate (MS I) as insertion of a pyrophosphoryl group into the 3′-position in guanosine 5′-diphosphate. *Proc. Nat. Acad. Sci.* **70**:306.

Watson, R. and H. Yamazaki. 1972. Expression of the *rel* gene during R17 infection. *Biochemistry* **11**:611.

Winslow, R. M. 1971. A consequence of the *rel* gene during a glucose to lactate downshift in *Escherichia coli. J. Biol. Chem.* **246**:4872.

Wong, J. F.-T. and R. N. Nazar. 1970. Relationship of the MS nucleotides to the regulation of ribonucleic acid synthesis in *Escherichia coli. J. Biol. Chem.* **245**:4591.

Yoshida, M., A. Travers and B. F. C. Clark. 1972. Inhibition of translation initiation complex formation by MS I. *FEBS Letters* **23**:163.

In Vitro Synthesis of ppGpp and pppGpp

Ricardo Block
Committee on Higher Degrees in Biophysics
Harvard University
Boston, Massachusetts 02138

William A. Haseltine
Biology Department
Massachusetts Institute of Technology
Cambridge, Massachusetts 02139

INTRODUCTION

When wild-type or certain strains of *E. coli* suffer an amino acid "starvation," they undergo the "stringent" response. The main characteristic of this response is a sharp curtailment in the rate of synthesis of stable RNA species. Mutants that continue to accumulate RNA during amino acid starvation are called *relaxed*. The classical genetic locus for this effect is the *rel*A gene. Cashel and Gallant (1969) found that amino acid starvation in stringent strains causes the accumulation of two unusual guanine nucleotides, ppGpp (magic spot I, MSI) and pppGpp (magic spot II, MSII), which did not appear in relaxed strains. They then postulated that high intracellular concentrations of these compounds lead to a cessation of RNA accumulation and to the other physiological changes characteristic of the stringent response. The preceding chapter by Cashel and Gallant reviews the in vivo metabolism of ppGpp and pppGpp and the experiments that implicate these nucleotides as mediators of the stringent response.

In vivo experiments strongly suggest that some aspect of the protein synthetic machinery is involved in the synthesis of ppGpp and pppGpp. Cashel and Gallant (1969) speculated that during the stringent response a reaction normally involved in protein synthesis idles and produces ppGpp and pppGpp. This notion is supported by the following evidence: the synthesis of ppGpp and pppGpp during amino acid starvation is dependent upon a high intracellular concentration of uncharged tRNA (Fangman and Neidhardt 1964), the presence of functioning ribosomes (Lund and Kjeldgaard 1972a), and a pool of messenger RNA (Lund and Kjeldgaard 1972b). Furthermore, most antibiotics that inhibit protein synthesis also inhibit the synthesis of the unusual nucleotides.

In vitro, ppGpp and pppGpp can be synthesized in a ribosome-dependent reaction. This reaction was first demonstrated on crude ribosomes, synthesizing the guanosine tetra- and pentaphosphates from GTP and ATP using ribosomes from stringent, but not relaxed, strains (see Table 1 for reaction conditions). When the ribosomes are washed with 0.5 M NH₄Cl, both the washed ribosomes and the ribosomal wash are required for the reaction (Haseltine et al. 1972).

The 0.5 M NH₄Cl ribosomal wash contains the "stringent factor," a protein which has been purified to near homogeneity (Block, Siev and Haseltine, unpublished; Cashel, unpublished). This protein is the product of the *rel*A gene (Block and Haseltine 1973). ATP is the donor of a pyrophosphate group to GTP or GDP in the reaction (Haseltine et al. 1972; Sy and Lipmann 1973).

The nucleotides are synthesized on a ribosome–messenger-tRNA complex. The acceptor site of the ribosome must be occupied by an uncharged tRNA which recognizes a codon in that site (Haseltine and Block 1973; Pedersen, Lund and Kjeldgaard 1973). ppGpp and pppGpp are not made by ribosomes actively engaged in protein synthesis or by those in which a charged tRNA is enzymatically bound to the acceptor site (Haseltine and Block 1973).

The stringent factor itself is the enzyme catalyzing the synthetic reaction, since it can catalyze the pyrophosphate transfer in a ribosome-independent reaction when purified to near homogeneity (Sy, Ogawa and Lipmann 1973; Block, unpublished).

The Pyrophosphate Transfer

The synthesis of ppGpp and pppGpp has an absolute requirement for ATP and its function is that of pyrophosphate donor to either GTP or GDP (Haseltine et al. 1972). The reaction diagramed below is probably an attack by the 3'-OH group of the guanosine ribose on the alpha-beta bond of the ATP.

$$pp\text{-}5'\text{-}G + pp^*pA \longrightarrow pp\text{-}5'\text{-}G\text{-}3'\text{-}p^*p + pA$$
$$ppp\text{-}5'\text{-}G + pp^*pA \longrightarrow ppp\text{-}5'\text{-}G\text{-}3'\text{-}p^*p + pA$$

Using ATP labeled with ^{32}P in either the beta or gamma positions of ATP, Sy and Lipmann (1973) found that the beta label produced ppGp*p and that the gamma label produced ppGpp*. Furthermore, a reaction between [³H]GDP and [$\gamma - ^{32}P$]ATP yielded ppGpp with a ³H/³²P ratio of 1/1 (also Cashel, unpublished). Thus the terminal pyrophosphate group of ATP is transferred as a unit. Sy and Lipmann (1973) also showed that the pyrophosphate was transferred to the 3'-OH group. They made pGp* by digesting ppGp*p with Zn^{++}-dependent yeast inorganic pyrophosphatase. Practically 100% of the ^{32}P was released by treating the pGp* with a 3'-nucleotidase prepared from rye grass. It is not known whether the pyrophosphate group is first transferred to an acceptor group on a protein before it is transferred to the 3'-OH group, but high concentrations of inorganic phosphate do not inhibit the reaction, suggesting a direct attack by

Table 1   In vitro reaction conditions for synthesis
of ppGpp and pppGpp by crude ribosomes

Component	Amount/50 µl reaction mixture	
Tris-acetate, pH 7.8	42	mM
Dithiothreitol	1.4	mM
Magnesium-acetate	11.4	mM
Ammonium-acetate	27.0	mM
GTP	0.55	mM
ATP	2.2	mM
Potassium-acetate	10.0	mM
Ribosomes	100	µg
[α-^{32}P]GTP	1	µCi

Ribosomes are prepared as described in Haseltine et al.
(1972). Reactions are incubated for 30 min at 37°C and terminated by addition of 1 µl 88% formic acid; the precipitate is
discarded. The conversion of the substrates to ppGpp and
pppGpp is followed by chromatography of the clear supernatant
on thin-layer polyethyleneimine (PEI) plates, which separates
the nucleotides according to the number of phosphates when
developed in 1.5 M KH_2PO_4 pH 3.4 (Cashel et al. 1969).

the 3'-OH on the alpha-beta bond of the ATP (Cashel, unpublished;
Haseltine and Block, unpublished).

The GTP analog, beta-gamma methylene guanosine 5'-triphosphate
(GMPPCP), can also accept the pyrophosphate group (Haseltine et al.
1972), and dATP can serve as a donor, albeit at reduced efficiency
(Cashel, unpublished). The beta-gamma adenyl amido-diphosphate
(AMPPNP) cannot serve as a pyrophosphate donor but can act as a
competitor for ATP in the synthetic reaction (Richter 1973).

The primary product of the reaction using low-salt ribosomes and GTP
as a substrate is ppGpp. This is not true in the more purified systems described below and is probably due to a degradation of GTP to GDP and of
pppGpp to ppGpp by beta-gamma GTPases cosedimenting with the ribosomes.

The Ribosomal Reaction

The ribosome-mediated synthesis of ppGpp and pppGpp requires the
stringent factor present in the 0.5 M NH₄Cl wash, the 30S and 50S ribosomal subunits, an RNA to serve as a message, and an uncharged tRNA
capable of recognizing a codon carried by the message (Pedersen et al.
1973; Haseltine and Block 1973) (see Table 2). The requirement for
tRNA and mRNA can only be shown if the high salt-washed ribosomes

Table 2   Requirements for the ribosomal reaction

	Percent conversion of GTP to MS			
Ribosome concentration	Ribo-somes alone	$+ rel^+$ factor	$+ rel^+$ factor $+$ tRNA	$+ rel^+$ factor $+$ tRNA $+$ poly(AUG)
3 × 1M NH$_4$Cl-washed (10 mg/ml)	0.7	50.3	54.3	52.6
4 × 1M NH$_4$Cl-washed (7.3 mg/ml)	0.9	39.4	39.7	47.8
Low Mg++ dialyzed (5.0 mg/ml)	0.8	10.0	26.9	59.5
30S subunits (2.0 mg/ml)	1.0	0.2	1.8	0.5
50S subunits (4.2 mg/ml)	0.7	0.7	4.4	11.6
30S + 50S subunits (6.2 mg/ml)	0.8	2.2	13.0	48.5
Rifampicin-treated, low Mg++ dialyzed (3.5 mg/ml)	2.0	1.6	1.3	52.2

Ribosomes and ribosomal subunits were prepared as described in Pedersen et al. (1973). The rel^+ factor was precipitated from the 0.5 M NH$_4$Cl ribosomal wash with 33% (NH$_4$)$_2$SO$_4$ and passed through a DEAE-Sephadex column equilibrated with 0.2 M NH$_4$Cl. The rel^+ factor when present was at a concentration of 25 μg/ml. tRNA and poly(AUG) were 0.6 mg/ml and 0.3 mg/ml, respectively. (Table reproduced, with permission, from Pedersen et al. 1973.)

are freed of both endogenous tRNA and mRNA. This can be done by preparing ribosomal subunits from ribosomes extracted from cells treated with rifampicin. The rifampicin treatment prevents initiation of new RNA chains and allows the ribosomes to run off the existing mRNA molecules. The ribosomal subunits are prepared by dialysis of these "rifampicin-treated" ribosomes against solutions of very low magnesium ion concentration followed by centrifugation through sucrose gradients. Using such ribosomal subunits, one can demonstrate that the synthesis of ppGpp and pppGpp requires the specific recognition of the mRNA by the tRNA, as is shown in Figure 1. For example, purified tRNA^{phe}, which recognizes the codons UUU and UUC, stimulated the synthetic reaction when the template was poly(U) but not when poly(AG) or poly(AUG) was used.

The uncharged tRNA must occupy the acceptor (A) site of the ribosome; if the acceptor site is empty, ppGpp and pppGpp are not made. This conclusion is drawn from experiments using the natural message from the phage R17. The ribosomes were positioned at the natural start sequences by the formation of an initiation complex. This complex did not make the

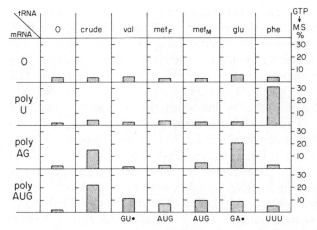

Figure 1   Effect of different combinations of polynucleotides and tRNAs. The conversion of GTP to ppGpp and pppGpp was measured as described in Table 1 with AS19 low Mg^{++} ribosomes (1.8 mg/ml) and stringent factor from rifampicin-treated AS19 (50 μg/ml). The polynucleotides were added at a concentration of 0.3 mg/ml, crude tRNA at 0.6 mg/ml, and the purified tRNAs at 0.4 mg/ml. The percent conversion of GTP into MS is given by the bars. The codons for the amino acids are indicated at the bottom of the figure. (Reprinted with permission from Pedersen et al. 1973).

MS nucleotides unless a tRNA capable of recognizing the codon adjacent to the start codon was added to the reaction. The amino acid sequence for the coat protein of R17 begins fMet-Ala-Ser and that for the replicase begins fMet-Ser-Lys (Steitz, unpublished). Addition of purified tRNA[Ala] or tRNA[Ser] stimulated the synthesis of ppGpp and pppGpp only if the initiation complex had been formed; other uncharged tRNA species, such as tRNA[Glu], tRNA[Val] and tRNA[Phe], which could not recognize the codon in the acceptor site, did not stimulate the reaction (Haseltine and Block 1973).

Charged tRNA[Ala] and tRNA[Ser] do not stimulate the synthesis of the magic spots; neither do they inhibit the stimulation by the homologous uncharged tRNA species when the elongation factors EF-Tu and EF-Ts are not present. However, if charged tRNA is enzymatically bound, ppGpp and pppGpp are not synthesized even in the presence of the homologous uncharged tRNA species. That is, in the presence of EF-Tu and EF-Ts, a mixture of charged and uncharged tRNA[Ala] added to the R17 initiation complex does not stimulate the synthesis of ppGpp and pppGpp. Whether or not the ribosome will synthesize ppGpp and pppGpp depends on whether the A site is occupied by a charged or an uncharged tRNA.

In vitro studies with antibiotic inhibitors of protein synthesis also suggest that the A site of the ribosome plays a key role in the synthesis of the magic spots (Pedersen et al. 1973; Haseltine et al. 1972). Of all the antibiotics tested in vitro that inhibit ribosome function in vivo, only tetracycline and thiostrepton strongly inhibit the ppGpp and pppGpp synthetic reaction. Others, such as chloramphenicol, kasugamycin, streptomycin and lincomycin, do not inhibit the synthetic reaction in vitro; moreover, they probably antagonize the stringent response in vivo through amino acid sparing (see Cashel and Gallant, this volume).

Tetracycline probably works by inhibiting the binding of uncharged tRNA to the A site (Levin 1970). However, tetracycline may inhibit directly the stringent factor activity for it inhibits the nonribosomal reaction (see below) mediated by stringent factor (Cashel, unpublished; Lund and Kjeldgaard, unpublished). Thiostrepton probably acts by also preventing the binding of the uncharged tRNA to the A site; it interacts with the 50S subunit and is known to block the binding of charged tRNA·EF-Tu·GTP complex to the ribosome (Cundliffe 1971). Fusidic acid inhibits the synthetic reaction at large concentrations, but this may be due to the detergent effect of this drug. Ramagopal and Davis (unpublished) have shown that stringent factor activity is present on native 50S subunits, on polysomes from gently lysed cells, and on 70S ribosomes from alumina-ground cells harvested either during exponential growth or after runoff. They detected no activity in native 30S subunits.

The "Idling" Reaction

The guanine nucleotides do not accumulate in a reaction where the ribosomes are actively engaged in protein synthesis. This means that either they are not made at all or that they are consumed at the same rate as they are synthesized. In the coupled, DNA-directed, protein-synthesizing system described by Zubay, Chambers and Cheong (1970), the magic spots can be synthesized if the amino acids are left out of the reaction mixture (mimicking the in vivo situation of amino acid starvation), and in this case, no synthesis of β-galactosidase dependent on the presence of an exogenous DNA template λplac5 can be detected. If a full complement of amino acids is present, then β-galactosidase is synthesized but the magic spots do not accumulate (Haseltine et al. 1972). During poly(U)-dependent synthesis of polyphenylalanine in a mixture that contains uncharged and charged tRNA[Phe], 30S and 50S subunits, the stringent factor, EF-Tu, EF-Ts and EF-G, ppGpp and pppGpp are not made until the supply of charged tRNA is exhausted, at which point the synthesis of pppGpp and ppGpp begins (see Figure 2). Once synthesized, ppGpp is not degraded in a reaction in which protein synthesis occurs. However, pppGpp is hydrolyzed to ppGpp. The conversion of pppGpp to ppGpp can be catalyzed by EF-G in a ribosome-dependent hydrolysis as well as by EF-Tu and IF2 in the ribosomal binding of aminoacyl-tRNA and fMet-tRNA, respectively (Haseltine and Block, unpublished; Hamel and Cashel, unpublished).

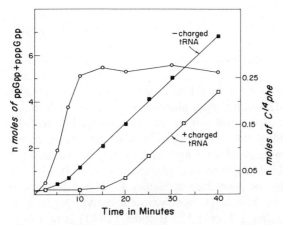

Figure 2   ppGpp and pppGpp are not synthesized by ribosomes actively engaged in protein synthesis. The reaction contains per 50 μl: 30 μg 50S and 70 μg 30S ribosomal subunits, 5 μg poly(U), 5 μg charged [^{14}C]Phe-tRNA^{Phe} where indicated, 5 μg uncharged tRNA^{Phe} and 1 μg stringent factor purified free of EF-Tu, EF-Ts, and 0.2 μg EF-G. Open circles represent nmoles of [^{14}C]Phe incorporated into acid insoluble material. Squares represent nmoles of ppGpp and pppGpp synthesized.

Functional Studies on the Ribosome and on Uncharged tRNA

The ribosome-dependent, uncharged tRNA-stimulated synthesis of ppGpp and pppGpp can be used as a model system to study the functional binding of uncharged tRNA to ribosomes. For example, tRNA modified by periodate oxidation of the 3′-OH or by removal of the 3′-terminal adenosine by digestion with snake venom phosphodiesterase no longer stimulates the synthesis of the guanine nucleotides (Pedersen et al. 1973; Haseltine and Block 1973). Furthermore, the tetranucleotide TφCG excised from *E. coli* tRNA inhibits the tRNA-dependent ribosomal synthesis of ppGpp and pppGpp (Richter, unpublished), apparently by competing for a site on the 5S RNA (Erdmann, Horne, Pongs, Sprinzl and Zimmermann, unpublished). A submethylated tRNA isolated by Bjork and Isaksson (1970) from *E. coli,* −m^{5}U (5-methyluracil deficient), possesses a normal capacity to function in synthesis of the magic spots. Another submethylated tRNA, −m$_2^2$G (*N*^{2}-dimethylguanine deficient), isolated from yeast by Phillips and Kjellin-Stroby (1967), is unable to stimulate synthesis of the magic spots (Lund, Pederson and Kjeldgaard, unpublished). Wild-type yeast tRNA shows the same stimulating activity as *E. coli* tRNA (see below).

Synthesis of the guanine nucleotides requires that both the 16S and 5S ribosomal RNA species be intact on the ribosome. Colicin E3 inhibits the synthesis of the guanine nucleotides both in vivo and in vitro (Lund, Pedersen and Kjeldgaard, unpublished). This colicin specifically cleaves 16S ribosomal RNA (Senior and Holland 1971; Bowman et al. 1971).

Ribosomes deficient in 5S RNA are unable to synthesize ppGpp and pppGpp unless 5S RNA is restored (Lund, Pedersen and Kjeldgaard, unpublished).

The 50S subunit can be inactivated with respect to the synthesis of ppGpp and pppGpp by removal of proteins either by treatment with CsCl (to prepare gamma cores) or by a 1 M LiCl wash (Lund, Pedersen and Kjeldgaard, unpublished; Haseltine, unpublished; Richter and Nierhaus, unpublished). The activity of the 50S subunit can be restored by reconstituting the ribosomes by adding the split proteins back to the washed cores. The acidic proteins of the 50S subunit L7 and/or L12 may be required for the reaction. Block (unpublished) finds that ribosomes washed free of the 50S proteins L7 and L12 by the procedure of Hamel, Koka and Nakamoto (1972) can be stimulated in their magic spot synthetic activity up to fivefold by the addition of either L7 or L12. Richter (1973) and Lund, Pedersen and Kjeldgaard (unpublished) find no requirement for the L7 or L12 acidic proteins. However, the stringent factor used in these experiments was not as pure as that used by Block and may contain sufficient amounts of these proteins as contaminants.

These results indicate that this reaction will be a fruitful probe for those studying the relationship between ribosomal structure and function.

The Stringent Factor

The stringent factor has been purified from the ribosomal wash to near homogeneity, a 180-fold purification (Block, Siev and Haseltine, unpublished; Cashel, unpublished). This purification is outlined in Table 3. The

Table 3   Purification of stringent factor from the ribosomal wash

	Concentration	Total no. units	Units/mg
(1) 0.5 M NH_4Cl wash	3.1 mg/ml	104,397	3.4
(2) 0–35% $(NH_4)_2SO_4$	2.7 μg/ml	43,716	73
(3) 0–20% $(NH_4)_2SO_4$	600 μg/ml	7840	262
(4) Hydroxyapatite column, 50 mM KPO_4 pH 7.1 step	250 μg/ml	5413	542

Assays were done using the conditions described in Table 1. The reaction mixture contained per 50 μl: 120 μg ethanol-extracted ribosomes (Hamel et al. 1972), 1 μl stringent factor preparation, and 2 μg purified proteins L7, L12. Reactions were incubated at 37°C and points taken at different times to calculate specific activities, i.e., 1 unit = 250 pmoles of GTP converted to the magic spots per minute. After steps (1), (2) and (3), the protein suspensions in 0.5 M NH_4Cl were dialyzed against a buffer with no NH_4Cl, and the precipitate formed was spun down and resuspended in high salt again. These precipitates contained most of the activity, and the supernatants were discarded. The low-salt precipitations were a purification step in themselves and were the source of material for the next step in the purification.

stringent factor is a monomeric protein with molecular weight of 75,000 on SDS-polyacrylamide gels. It sediments, in a 0.5 M NH_4Cl glycerol gradient, at 4.5S. It is not identical to any of the previously described factors associated with protein synthesis, such as IF1, IF2, IF3, EF-Tu, EF-Ts, EF-G and the termination factors. This was established by testing these active factors for their ability to promote ppGpp synthesis when added to the ribosome (Haseltine and Block 1973 and unpublished).

The stringent factor is routinely isolated from the 0.5 M NH_4Cl ribosomal wash; however, it is present in significant amounts in the S100 fraction of extracts (Cashel, unpublished). It may also be present in the pellet of an S30, the so-called membrane fraction (Cashel, unpublished; Pedersen and Haseltine, unpublished). Purification of the stringent factor to near homogeneity leads to very low yields of about 1 mg per kilogram of cells (Block, Siev and Haseltine, unpublished; Cashel, unpublished). Extrapolating for losses in purification (see Table 3) gives a number of 10 to 20 molecules per cell. Possible extraction of 10 times more stringent factor activity out of an S30 pellet made from gently lysed cells in low salt and no magnesium (Cashel, unpublished) would still give a number of 100 to 200 molecules per cell, which is much less than one per ribosome.

The *rel* Mutations: *relA*

The stringent factor is the product of the *rel*A gene, the classical "relaxed" locus described by Stent and Brenner (1961). None of the *rel*A mutants originally examined in vitro contained stringent factor activity as measured in the standard assay described in Table 1 (Haseltine et al. 1972). However, stringent factor activity was found in other partially relaxed strains (Block and Haseltine 1973). Fiil and Friesen (1968) isolated a set of 21 independent rel^- mutants in this cistron. These mutants fell into two categories with respect to RNA accumulation following amino acid starvation: a "high relaxed" class, which continues to synthesize RNA at approximately the same rate after amino acid starvation as does the classical W6 strain, and the "low relaxed," which accumulates substantially less RNA than does W6. Many of these latter strains possess measurable stringent factor activity, albeit at lower levels than the rel^+ strain. Moreover, the stringent factor activity of some of the mutants is more thermosensitive in vitro than that of the rel^+ parental strain. This thermosensitivity varies from strain to strain. Figure 3 shows the denaturation curves of the stringent factor activity on ribosomes of the rel^+ parent and four different *rel*A mutations. This physical variation in the stringent factor isolated from strains differing only by changes in the *rel*A gene strongly suggests that the stringent factor is the direct product of the *rel*A gene. It is likely that the *rel*A mutants isolated to date are missense mutants. No amber mutants have been found. Furthermore, the alteration in the stringent factor may still leave a physiologically functional protein.

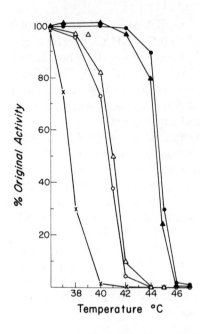

Figure 3 Denaturation curves of the stringent factor activity on the ribosomes. Low-salt washed ribosomes were held at the temperatures indicated for 5 min each. The heated ribosomes were assayed as described in Table 1 for their ability to synthesize ppGpp and pppGpp at 30°C for 60 min. (●) Strain CP78 (*rel*⁺); (▲) strain CP90; (△) strain CP84; (○) strain CP141; (×) strain CP80.

Gallant and Harada (unpublished) have been able to detect stringent factor activity in the 0.5 M NH₄Cl wash from a "high relaxed" strain, CP79. In a reaction containing high salt-washed ribosomes and the high-salt wash extracted from strain CP79, ppGpp and pppGpp can be synthesized if, in the presence of poly(U), there is a substantial excess of tRNA^{Phe} over ribosomes. This strain does not synthesize ppGpp when assayed in the standard reaction described in Table 1. Furthermore, Pedersen and Haseltine (unpublished) observe ppGpp and pppGpp synthesis in gently lysed extracts containing the membrane fraction made from all the relaxed strains including W6. The activity varies from mutant to mutant.

*rel*B

Lavalle (1965) described a second relaxed locus. This mutant expresses its relaxed phenotype only after a delay of some ten minutes following the onset of amino acid starvation. Initially RNA accumulation is sharply restricted. The intracellular concentration of ppGpp rises sharply following amino acid starvation and then decays to the normal basal level after ten minutes (Lavalle, unpublished). Biochemical analysis of the mutant is now in progress.

*rel*C

Recently a new class of *rel* mutants, *rel*C, has been discovered (Friesen, Fiil and Haseltine, unpublished). This mutation maps at a different site from the classical *rel*A site. Following amino acid starvation, these mutants continue to accumulate RNA and the intracellular concentration of ppGpp and pppGpp remains at the low but normal basal level. Furthermore, ppGpp does reach high concentrations during carbon-source downshift, as it also does in the *rel*A mutants (Fiil and Haseltine, unpublished).

The defect in the *rel*C mutant is in the 50S ribosomal subunit and not in the stringent factor. The stringent factor isolated from this mutant appears normal. The defect is in the 50S particle, since, as shown in Table 4, wild-type 30S subunits and *rel*C 50S subunits make ppGpp and pppGpp at one-tenth the normal rate, whereas a reaction mixture containing *rel*C 30S and wild-type 50S is normal. Furthermore, the mutant ribosomes can be made active in the synthesis of the magic spots by the addition of the acidic ribosomal proteins split from wild-type 50S subunits (Friesen, Fiil and Haseltine, unpublished). This indicates that the *rel*C locus either codes for the structural gene of one of the acidic proteins of the 50S subunit or for an enzyme which modifies one or more of the 50S ribosomal proteins. The two-dimensional gel pattern of the mutant is the same as that of the wild-type parental strain (Mollin, unpublished), and the mutant ribosomes are not defective in poly(U)-directed polyphenylalanine synthesis.

The *rel*C gene maps at 79 minutes on the new Taylor and Trotter map (1972), very close to *rts,* a gene which renders the 50S subunit temperature sensitive for protein synthesis. The *rel*C locus is also covered by the episome F112, the same episome which carries the structural gene for the 30S ribosomal protein S18.

The Nonribosomal Reaction

The stringent factor is probably itself the enzyme that catalyzes the magic spot synthetic reaction. Sy et al. (1973) showed that the synthesis of ppGpp and pppGpp could be catalyzed by the 0.5 M NH_4Cl ribosomal wash alone upon the addition of 20% methanol. There are no detectable ribosomes or ribosomal subunits in their ribosomal wash preparation, and the reaction is resistant to thiostrepton, an antibiotic which binds to the

Table 4   The *rel*C mutation: A defect in the 50S ribosomal subunit

30S wild type	+	−	+	−	−	−	+	−
50S wild type	−	+	+	−	−	−	−	+
30S mutant	−	−	−	+	−	+	−	+
50S mutant	−	−	−	−	+	+	+	−
nmoles ppGpp + pppGpp	0.25	0.27	5.2	0.1	0.1	0.2	0.3	5.4

The reaction mixtures (50 μl) contained the salts and buffer at the concentrations described in Table 1 and 30 μg of 50S and 20 μg of 30S subunits where indicated. The subunits were isolated from strain CP78 (wild type) and *rel*C mutant C305, 5 μg poly(U), 5 μg uncharged tRNA[Phe] and 1 μg partially purified stringent factor (step 2 of Table 3). The reaction was incubated at 37°C for 30 min.

50S subunit and inhibits the ribosome-mediated reaction. The nonribosomal synthetic reaction is sensitive to small concentrations (20 μg/ml) of oxytetracycline even when the most highly purified preparation of stringent factor is used (Cashel, unpublished). The rate of magic spot synthesis of the ribosome-independent reaction is at least 20-fold less than that mediated by the ribosome. The reaction conditions optima differ from those of the ribosome-mediated reaction. The Mg^{++} optimum is lower and the NH_4Cl optimum of 100 mM is higher and inhibitory in the ribosomal reaction. The temperature optimum is 25–30°C and drastically falls at 37°C. Twenty-percent methanol (Sy et al. 1973), DMSO (Kjeldgaard, unpublished) and Triton X (Cashel, unpublished) stimulate the rate of synthesis of the ribosome-independent reaction.

The minimum requirements for this reaction have not been definitively established. Nevertheless, the most highly purified preparation of stringent factor available (greater than 95% pure) can synthesize pppGpp in the presence of only the substrates, salts and buffer. Here the product is exclusively pppGpp when GTP is used as a substrate (Block, unpublished). This synthesis can be observed with small concentrations of stringent factor alone (4 μg/ml) at temperatures below 14°C. Thus the active site must reside on the stringent factor itself. The rate of synthesis of pppGpp is 100–200 times slower than that of the ribosome-dependent reaction at 37°C. The addition of 20% methanol and/or several acidic proteins, such as casein and the L7 or L12 ribosomal proteins from the 50S subunit, can stimulate the rate of pppGpp synthesis up to 20-fold. Richter and Stöffler (unpublished) observe that addition of purified antibodies to the ribosomal proteins L7 and L12 totally inhibits the ribosome-independent reaction mediated by the 0.5 M NH_4Cl ribosomal wash. Antibodies to proteins L1, L6, L10, L16 and L18 inhibit the reaction slightly, but antibodies to the remaining 50S proteins or a mixture of antibodies to the 30S subunit have no effect.

The synthesis of ppGpp by the pellet of an S30, the membrane fraction, may also be, at least in part, ribosome independent. A fraction of the activity is resistant to high concentrations of thiostrepton (Pedersen and Haseltine, unpublished) and to extensive RNase digestion (Cashel, unpublished). This reaction is sensitive to deoxycholate and phospholipase (Cashel, unpublished).

Heterologous Systems

Ribosomes isolated from sources other than *E. coli* are capable of catalyzing magic spot synthesis when supplied with the stringent factor of *E. coli*. Ribosomes extracted from *B. stearothermophilus*, *Proteus vulgaris* and bean chloroplasts are active. However yeast cytoplasmic and mitochondrial ribosomes, as well as mouse ascites tumor cell ribosomes, were completely devoid of activity (Lund, Pedersen and Kjeldgaard, unpublished).

Hybrid ribosomes made from ribosomal subunits of *E. coli* and calf brain ribosomes did not complement *E. coli* stringent factor in magic spot synthetic activity (Richter 1973).

Crude yeast tRNA as well as purified yeast tRNA[Phe] show the same stimulating activity in the ribosomal system as does tRNA extracted from *E. coli* (Lund, Pedersen and Kjeldgaard, unpublished).

CONCLUSIONS

The in vitro experiments suggest a model for the synthesis of ppGpp and pppGpp during the stringent response.

The stimuli that induce the stringent response in vivo, such as starvation for a required amino acid or inactivation of an aminoacyl-tRNA synthetase, give rise to the accumulation of an uncharged tRNA species counterpointed by the depletion of the homologous charged tRNA population. Thus when a ribosome traveling along a messenger RNA molecule reaches a codon for which there is no cognate-charged tRNA, it receives an uncharged tRNA molecule which recognizes the codon, occupies the acceptor site and activates the ribosome–stringent factor complex. The stringent factor catalyzes the transfer of the terminal phosphate from ATP to GTP and GDP and may also be involved in binding the uncharged tRNA to the acceptor site. When a charged tRNA is available to be enzymatically bound to the acceptor site, ppGpp and pppGpp are not made. Thus one would expect the rate of ppGpp and pppGpp synthesis to be determined by the fraction of the ribosomes having an uncharged tRNA in the acceptor site, which would be determined by the ratio of charged to uncharged tRNA.

This model explains the in vivo requirements for ppGpp and pppGpp synthesis during the stringent response: functioning ribosomes, uncharged tRNA and messenger RNA. The model also explains why the synthesis of the nucleotides is greatly reduced when protein synthesis resumes. The failure of the *rel*A mutants to increase the rate of ppGpp synthesis upon amino acid starvation may be due either to the inability of the ribosome–tRNA complex to activate the mutant factor or to a stringent factor damaged so that the catalytic site is altered or so that it has a lower affinity for uncharged tRNA.

It is likely that ppGpp is the signal which causes the stringent response and the cessation of stable RNA accumulation because accumulation and disappearance of the magic spots are always followed closely by onset and disappearance of the stringent response (see Cashel and Gallant, this volume). Also, mutations in the *rel*A and *rel*C genes damage proteins which are necessary for the synthesis of ppGpp, and this might be a sufficient explanation for the effect of these lesions on RNA metabolism.

The association of the translocation machinery with the synthesis of ppGpp and pppGpp raises the possibility that the protagon-like (Posner and Gies 1905) magic spots may be normal intermediates in protein synthesis. Hamel and Cashel (unpublished) find that pppGpp can substitute in vitro for GTP in reactions catalyzed by IF2 and EF-Tu, EF-Ts, but that it very poorly supports poly(U)-dependent polyphenylalanine synthesis. On the other hand, the magic spot compounds may be products of a side reaction which only accumulate in response to crisis conditions. If the relA product is needed for translation in vivo, then the existing relA mutants must retain the ability to function in protein synthesis, while losing the ability to synthesize ppGpp and pppGpp at high rates during amino acid starvation. In this context, it is interesting that all the relA mutants isolated to date are probably missense mutations and retain some stringent factor activity in vitro. Recall also that relA strains accumulate ppGpp during carbon-source shiftdown experiments and that they do have normal basal levels of ppGpp. The accumulation of ppGpp in relA mutants is partly due to a decrease in the turnover rate of ppGpp (Gallant, Margason and Finch 1972), but the mechanism of synthesis is not yet fully understood. Either the stringent factor in relA mutants is altered in such a way that although it does not respond to the signal present during amino acid starvation, it can respond to other or stronger signals, or else there must be another mechanism for synthesis of ppGpp.

Acknowledgments

The authors would like to thank Mike Cashel, John Gallant and Niels Ole Kjeldgaard for communicating their results prior to publication and to John Hershey, Herb Weissbach and Nat Brot for kind gifts of enzymes. We are also grateful to Walter Gilbert for helpful criticism of this manuscript. This work was supported in part by NIH grant GM-09541.

References

Bjork, G. R. and L. A. Isaksson. 1970. Isolation of mutants of *Escherichia coli* lacking 5-methyluracil in transfer ribonucleic acid or 1-methylguanine in ribosomal RNA. *J. Mol. Biol.* **51**:83.

Block, R. and W. A. Haseltine. 1973. Thermolability of the stringent factor in *rel* mutants of *Escherichia coli. J. Mol. Biol.* **77**:625.

Bowman, C. M., J. E. Dahlberg, T. Ikemura, J. Konisky and M. Nomura. 1971. Specific inactivation of 16S ribosomal RNA induced by colicin E3 in vivo. *Proc. Nat. Acad. Sci.* **68**:964.

Cashel, M. and J. Gallant. 1969. Two compounds implicated in the function of the RC gene of *Escherichia coli. Nature* **221**:838.

Cashel, M., R. A. Lazzarini and B. Kalbacher. 1969. An improved method for thin-layer chromatography of nucleotide mixtures containing ^{32}P-labeled orthophosphate. *J. Chromatog.* **40**:103.

Cundliffe, E. 1971. The mode of action of thiostrepton *in vivo. Biochem. Biophys. Res. Comm.* **44**:912.

Fangman, W. L. and F. C. Neidhardt. 1964. Demonstration of an altered aminoacyl ribonucleic acid synthetase in a mutant of *Escherichia coli. J. Biol. Chem.* **239**:1839.

Fiil, N. and J. D. Friesen. 1968. Isolation of "relaxed" mutants of *Escherichia coli. J. Bact.* **95**:729.

Gallant, J., G. Margason and B. Finch. 1972. On the turnover of ppGpp in *Escherichia coli. J. Biol. Chem.* **247**:6055.

Hamel, E., M. Koka and T. Nakamoto. 1972. Requirement of an *Escherichia coli* 50S ribosomal protein component for effective interaction of the ribosome with T and G factors and with guanosine triphosphate. *J. Biol. Chem.* **247**:805.

Haseltine, W. A. and R. Block. 1973. Synthesis of guanosine tetra- and pentaphosphate requires the presence of a codon-specific, uncharged transfer ribonucleic acid in the acceptor site of ribosomes. *Proc. Nat. Acad. Sci.* **70**:1564.

Haseltine, W. A., R. Block, W. Gilbert and K. Weber. 1972. MSI and MSII made on ribosome in idling step of protein synthesis. *Nature* **238**:381.

Lavalle, R. 1965. Nouveaux mutants de regulation de la synthese de l'ARN. *Bull. Soc. Chim. Biol.* **47**:1567.

Levin, J. 1970. Codon specific binding of deacylated tRNA to ribosomes. *J. Biol. Chem.* **245**:3195.

Lund, E. and N. O. Kjeldgaard. 1972a. Metabolism of ppGpp in *Escherichia coli. Eur. J. Biochem.* **28**:316.

————. 1972b. Protein synthesis and formation of guanosine tetraphosphate. *FEBS Letters* **26**:306.

Pedersen, F. S., E. Lund and N. O. Kjeldgaard. 1973. Codon specific tRNA dependent *in vitro* synthesis of ppGpp and pppGpp. *Nature New Biol.* **243**:13.

Phillips, J. H. and K. Kjellin-Stroby. 1967. Studies on microbial ribonucleic acid. IV. Two mutants of *Saccharomyces cerevisiae* lacking N²-dimethylguanine in soluble ribonucleic acid. *J. Mol. Biol.* **26**:509.

Posner, E. R. and W. J. Gies. 1905. Is protagon a mechanical mixture of substances or a definite chemical compound? *J. Biol. Chem.* **1**:59.

Richter, D. 1973. Formation of guanosine tetraphosphate (magic spot I) in homologous and heterologous systems. *FEBS Letters* **34**:291.

Senior, B. W. and I. B. Holland. 1971. Effect of colicin E3 upon the 30S ribosomal subunit of *Escherichia coli. Proc. Nat. Acad. Sci.* **68**:959.

Stent, G. S. and S. Brenner. 1961. A genetic locus for the regulation of ribonucleic acid synthesis. *Proc. Nat. Acad. Sci.* **47**:2005.

Sy, J., Y. Ogawa and F. Lipmann. 1973. Non-ribosomal synthesis of guanosine tetraphosphate (MSI) as insertion of a pyrophosphoryl group into the 3′-position in guanosine 5′-diphosphate. *Proc. Nat. Acad. Sci.* **70**:306.

Sy, J., Y. Ogawa and F. Lipmann. 1973. Non-ribosomal synthesis of guanosine 5′, 3′-polyphosphates by the ribosomal wash of stringent *Escherichia coli. Proc. Nat. Acad. Sci.* **70**:2145.

Taylor, A. L. and C. D. Trotter. 1972. Linkage map of *Escherichia coli* strain K-12. *Bact. Rev.* **36**:504.

Zubay, G., D. A. Chambers and L. C. Cheong. 1970. Cell free studies on the regulation of the *lac* operon. In *The lactose operon* (ed. J. R. Beckwith and D. Zipser) pp. 375–391. Cold Spring Harbor Laboratory, Cold Spring Harbor, New York.

Ribosomal RNA Synthesis In Vitro

Andrew Travers
Medical Research Council
Laboratory of Molecular Biology
Cambridge, England, CB2 2QH

INTRODUCTION

In vitro studies of ribosomal RNA transcription aim to reproduce as closely as possible the in vivo synthesis of rRNA and hence to provide a molecular basis for the phenomena of regulation of rRNA synthesis in vivo.

An extreme example of the regulation of rRNA accumulation in vivo is the stringent response, defined genetically by the *rel* locus. The product of the *rel* gene is a protein which is involved in the synthesis of the nucleotide guanosine 5'-diphosphate 3'-diphosphate or ppGpp (see Block and Haseltine, this volume). When *rel*+ strains are functionally starved for an amino acid, the rate of net rRNA synthesis is drastically reduced and the intracellular concentration of ppGpp rapidly rises, attaining and usually exceeding that of guanosine 5'-triphosphate (Cashel 1969; Gallant et al. 1970). Two classes of mutant fail to restrict net rRNA synthesis in a normal manner. One, a mutation to *rel*−, fails to synthesize ppGpp on amino acid starvation (Cashel 1969) and is recessive to *rel*+ (Fiil 1969). This observation has led to the suggestion that ppGpp acts as a direct inhibitor of rRNA synthesis in vivo (Cashel 1969). A second class of mutant, of which ts103 is an example, contains an altered RNA polymerase (Jacobson and Gillespie, unpublished observations) which is defective in vitro for pppA initiations (Jacobson and Gillespie 1970). This mutation is dominant to *rel*+. This observation suggests that RNA polymerase itself is the target of stringent control and that the net synthesis of rRNA is controlled primarily at the level of synthesis. Nevertheless, the evidence on the point is not conclusive (see Kjeldgaard, this volume), and consequently, in vitro experiments have attempted to demonstrate directly selective inhibition of rRNA synthesis by ppGpp.

Any gross control on the rate of rRNA synthesis in vivo is presumably a consequence of controlling the rate of initiation of rRNA synthesis (Stomato and Pettijohn 1971). Thus in this paper, particular attention will be paid to methods of altering the initiation specificity of transcription in vitro.

IN VITRO SYSTEMS FOR rRNA SYNTHESIS

Two approaches have been used to study rRNA synthesis in vitro. In the first, the transcription machinery is reconstituted from its highly purified components. An alternative approach is to isolate in toto the transcription machinery of bacterial cells, either in the form of a relatively crude preparation associated with membranes (Winsten and Huang 1972; Murooka and Lazzarini 1973) or in the form of a preparation containing only RNA polymerase and supercoiled bacterial DNA (Stonington and Pettijohn 1971; Worcel and Burgi 1972). With such transcription complexes it is hoped to preserve as far as possible the characteristics of in vivo transcription.

Reconstituted Systems

The simplest reconstituted system comprises purified bacterial DNA and bacterial RNA polymerase holoenzyme. Initial studies showed that little or no detectable rRNA was synthesized in such a system (Pettijohn et al. 1970; Travers, Kamen and Schleif 1970). However, subsequent work indicated that in different experimental conditions RNA polymerase could initiate rRNA synthesis efficiently (Hussey et al. 1972; Haseltine 1972; Pettijohn 1972; Birnbaum and Kaplan 1973), such that rRNA comprised ~ 10% of the total in vitro transcript. The differences between these results are thus dependent on the ability of the hybridization assay to distinguish between ≤ 2% rRNA and 10% rRNA. In the subsequent discussion I shall assume that the assay can so distinguish.

The standard procedure in experiments of this type is to mix all the components of the reaction mixture at 0°C and subsequently to incubate the mixture at 37°C. This procedure results in a lag in the synthesis of rRNA relative to total RNA synthesis, a lag which is especially apparent at high KCl concentrations (Pettijohn 1972; Travers, Baillie and Pedersen 1973). This lag can be abolished by preincubating the DNA template at 37°C. More extensive studies have demonstrated that the rate of rRNA synthesis on purified *E. coli* DNA by holoenzyme is strongly dependent on the temperature of preincubation of the DNA template. Between 34 and 36°C there is an abrupt fivefold change in the absolute rate of rRNA synthesis, although there is little change in total RNA synthesis over this temperature range (Travers, Baillie and Pedersen 1973). Above 36°C and below 34°C, the rate of rRNA synthesis doubles for every 10° rise in temperature. A similar, although smaller, abrupt change occurs in the rate of total RNA synthesis between 30 and 32°C. One consequence of these changes in the rates of total and rRNA synthesis is that the proportion of rRNA in the transcript is variable, rRNA comprising 4–7% below 30°C, ~2% between 32 and 34°C, and 10–14% between 36 and 45°C.

The activation of the *E. coli* DNA template is reversible (Travers, Baillie and Pedersen 1973). This, coupled with the striking resemblance of the change in rRNA synthesis over a narrow temperature range to a nucleic acid melting curve, has resulted in the suggestion that the phenomenon is a consequence of cooperative structural changes in the DNA template occurring at or close to the promoter sites for rRNA synthesis. The promoter conformations above and below the transition temperature are termed *open* and *closed,* respectively (Travers, Baillie and Pedersen 1973). Nevertheless, because in these experiments RNA synthesis was used as an indirect indicator of DNA structure, it remains possible that the observed characteristics of the conformational change are partly determined by the interaction between RNA polymerase and the promoter.

A further complication is that the activation of *E. coli* DNA for rRNA synthesis is readily apparent only at KCl concentration $> \sim 0.075$ M. At 0.01 M KCl the temperature dependence of rRNA synthesis differs strikingly from that at 0.1 M KCl. Whereas at 0.1 M KCl the rate of rRNA synthesis increases sevenfold between 28 and 38°C, at 0.01 M KCl rRNA synthesis increases by only 50% over the same temperature range and comprises a constant $\sim 10\%$ of the transcript. Moreover at all temperatures tested, the absolute rate of rRNA synthesis at 0.01 M KCl is considerably higher than that at 0.1 M KCl. These KCl-dependent changes in the proportional and absolute rates of rRNA synthesis correlate with changes in the capacity of polymerase holoenzyme to transcribe different DNA templates (Matsukage 1972).

Is rRNA synthesis in reconstituted systems correct, i.e., is it initiated at the same sites as in vivo? Since the 5′ termini of neither the in vivo nor the in vitro rRNA transcript have yet been characterized, it is impossible to answer this question directly. Nevertheless it is clear that rRNA sequences are preferentially transcribed from *E. coli* DNA, for although the rRNA genes comprise only $\sim 0.4\%$ of the *E. coli* genome (Yankofsky and Spiegelman 1962; Kennell 1968), rRNA can comprise 10% of the in vitro transcript. Moreover this preferential transcription of rRNA requires the polymerase σ factor (Haseltine 1972). A further indication that rRNA synthesis is initiated in vitro at specific sites identical with or close to the natural promoter sites is the observation that the 16S and 23S rRNA sequences are transcribed sequentially in vitro in the same order as in vivo (Pettijohn 1972).

The pattern of in vitro rRNA transcription can be altered by substituting the mutant ts103 polymerase for the wild-type polymerase holoenzyme. Compared to the parent enzyme, the ts103 polymerase synthesizes rRNA twice as efficiently from the closed rRNA promoter and much less efficiently from the open rRNA promoter. The relaxed phenotype of the mutant is thus quantitatively duplicated in vitro only by RNA synthesis from the closed rRNA promoter (Travers, unpublished observations). This observation again suggests that rRNA synthesis in reconstituted systems is accurate.

Protein–DNA Complexes

The crude membrane-associated protein–DNA complex differs from simple reconstituted systems in supporting a very high proportional rate of initiation of rRNA synthesis. The total transcript from such a complex can contain about 30% rRNA (Murooka and Lazzarini 1973; Winsten and Huang 1972). Nevertheless the proportional rate of rifampicin-sensitive rRNA synthesis in such a complex can increase from an initial value of ~12% to a final value of ~40% (Murooka and Lazzarini 1973). The molecular basis of this phenomenon has not been explained, and consequently it is unclear whether the initial or the final rate of rRNA initiation is more representative of the in vivo situation.

The supercoiled form of the *E. coli* genome is unable to support detectable initiation of rRNA by added RNA polymerase, although this form of *E. coli* DNA is otherwise a good template for the polymerase (Pettijohn et al. 1970; Pettijohn 1972). Thus either rRNA promoters must be blocked in this template, or, conceivably, the supercoiling could restrain the conformation of the rRNA promoter so that it remains in the closed configuration at temperatures at least up to 37°C (Travers, Baillie and Pedersen 1973).

EFFECT OF TRANSCRIPTION FACTORS

H₁ Protein

H_1 is a neutral, thermostable DNA binding protein, normally isolated as a tetramer with a molecular weight of ~30,000 (Cukier-Kahn, Jacquet and Gros 1972). This protein was initially characterized as a factor which stimulated RNA synthesis by RNA polymerase holoenzyme on T4 DNA. With *E. coli* DNA as template, H_1 again stimulates total RNA synthesis slightly at 38°C and 0.1 M KCl (Travers and Cukier-Kahn, unpublished results). However, H_1 concomitantly reduces the proportional and absolute amount of rRNA in the transcript by five- and threefold, respectively. Further experiments show that at a DNA:protein weight ratio of 1:1, H_1 shifts the transition temperature between the open and closed forms of the rRNA promoter from 35 to 41°C. Thus below 35°C and above 41°C H_1 appears to have no significant effect on rRNA synthesis in vitro (Travers and Cukier-Kahn, unpublished observations).

Polyamines

Another class of molecules which exists in vivo in high concentration and might also bind directly to the DNA template are the polyamines, putrescine and spermidine. Initial studies on the effect of polyamines on rRNA synthesis show that at 35°C and 0.05 M KCl—conditions resulting in an intermediate level of rRNA synthesis—spermidine increases the proportion of rRNA in the transcript, whereas putrescine decreases this proportion (Travers, unpublished observations). Thus both polyamines alter the quality of the transcript from *E. coli* DNA.

Protein Synthesis Elongation Factor EF-T

The initiation specificity of RNA polymerase holoenzyme can be charged by the protein factor ψ (Travers, Kamen and Schleif 1970), since shown to be identical with the protein synthesis elongation factor EF-T (Blumenthal, Landers and Weber 1972). Preparations of EF-T purified from $Q\beta$ replicase preferentially stimulate RNA synthesis from the closed rRNA promoter, the proportion of rRNA in the *E. coli* DNA transcript being increased two- to threefold at 0.01 M KCl and up to fivefold at 0.1 M KCl (Travers 1973). Thus in the presence of EF-T, rRNA can comprise ~20% of the transcript under the former conditions. The stimulation of rRNA synthesis by EF-T correlates with a selective increase in pppG initiations. In contrast, EF-T has little effect on synthesis from the open rRNA promoter at 0.1 M KCl and even inhibits this synthesis at 0.01 M KCl. The effect of EF-T on in vitro transcription is not restricted to rRNA synthesis, for the factor also strongly inhibits T4 DNA transcription below 0.1 M KCl but again has little effect at high KCl concentrations (Travers 1973). Thus the effects of EF-T on T4 DNA transcription and RNA synthesis from the open rRNA promoter are very similar but differ strikingly from the effect of the factor on RNA synthesis from the closed rRNA promoter.

CONTROL OF rRNA SYNTHESIS IN VITRO

Attempts to demonstrate the control of rRNA synthesis have concentrated on the question of whether ppGpp directly inhibits rRNA synthesis relative to the synthesis of other RNA species. This approach depends on the assumption that, with a given template, rRNA is the major or perhaps the only RNA species synthesized in vitro whose synthesis is also subject to stringent control in vivo.

With purified *E. coli* DNA as template, rRNA synthesis by holoenzyme alone from either the open or the closed form of the rRNA promoter is not selectively inhibited by ppGpp (Haseltine 1972; Travers 1973), although ppGpp does inhibit total RNA synthesis by ~40% (Cashel 1970). However, the selective enhancement of synthesis from the closed rRNA promoter by EF-T preparations is abolished by ppGpp (Travers 1973). The K_i for this preferential inhibition of rRNA synthesis by ppGpp is ~0.1–0.2 mM and thus is similar to the calculated K_i for the inhibition of net rRNA synthesis by ppGpp in vivo (Fiil, von Meyenburg and Friesen 1972). In contrast, ppGpp does not preferentially reduce synthesis from the open rRNA promoter in the presence of EF-T. Thus in a highly purified in vitro system, both the closed rRNA promoter and EF-T appear to be necessary for selective control by ppGpp. In addition to the effect of ppGpp, guanosine 5′-diphosphate also preferentially inhibits the stimulation of RNA synthesis from the closed rRNA promoter by EF-T (Travers, unpublished observations). Significantly, EF-T binds both ppGpp and ppG (Blumenthal, Landers and Weber 1972).

Further evidence that ppGpp can selectively inhibit a specific class of RNA synthesis has been derived from studies of crude RNA polymerase preparations. When RNA polymerase is purified by zone sedimentation from an S30 or S100 bacterial extract, two to three distinct polymerase activities can be distinguished (Snyder 1973; Travers and Buckland 1973) on the basis of physical and functional differences. Depending on the method of isolation, these activities sediment at either 15S, 21S and 27S (Travers and Buckland 1973) or at 14S, 16S and 18S (Snyder 1973), respectively. The latter sedimentation coefficients are probably a more accurate reflection of the molecular weights of these polymerase forms. The slowest sedimenting species synthesizes rRNA efficiently from the closed promoter and its activity on all templates tested is inhibited by both ppGpp and ppG (Travers and Buckland 1973). This contrasts with purified RNA polymerase holoenzyme, which is inhibited by ppGpp but not by ppG (Cashel 1970). The central polymerase activity synthesizes no detectable rRNA from the closed promoter and is insensitive to both ppGpp and ppG. The fastest sedimenting polymerase supports an intermediate level ($\sim$5% of transcript as rRNA) of rRNA synthesis from the closed promoter. Thus only those polymerase activities that synthesize rRNA from the closed rRNA promoter are sensitive to ppGpp. These data suggest that both the initiation specificity and the sensitivity to ppGpp of RNA polymerase can be varied in vitro.

Total rifampicin-sensitive RNA synthesis from a membrane-associated protein–DNA complex isolated from exponentially growing *E. coli* is strongly inhibited by both ppGpp and by ppG but not by ppA (Travers, unpublished observations). However, Murooka and Lazzarini (1973) have shown that the rifampicin-sensitive rRNA synthesis in such a complex is only decreased from $\sim$30% to $\sim$17% of the total transcript by 0.1–0.2 mM ppGpp in conditions where the rate of rRNA synthesis relative to total RNA synthesis steadily increases during the course of the incubation.

RELEVANCE OF IN VITRO RESULTS

To what extent does the in vitro transcription of rRNA reproduce the phenomenon observed in vivo? In the reconstituted system containing highly purified DNA and RNA polymerase holoenzyme, the relative rate of rRNA synthesis is dependent on the DNA conformation and also on the addition of transcription factors. Similarly, the sensitivity of this synthesis to ppGpp is also dependent on the same parameters. In vitro, the in vivo phenotype of the ts103 RNA polymerase is duplicated only from the closed form of the rRNA promoter. This suggests first, that in vivo the rRNA promoter normally is in a closed conformation, and second, that the normal rate of rRNA synthesis from this promoter conformation at high ionic strength represents a basal level of synthesis.

If the rRNA promoter were normally closed in vivo, we would expect that this conformation would need to be stabilized by interaction with DNA-binding proteins and/or polyamines, especially since there are no indications of unusual temperature dependence of rRNA synthesis in vivo (e.g., Fiil, von Meyenburg and Friesen 1972). A suggestion that such stabilization may occur is apparent from the observation that the DNA-binding H_1 protein raises the transition temperature between the open and closed conformations of the rRNA promoters.

Although synthesis from the closed rRNA promoter adequately reproduces both the phenotype of ts103 and the stringent response in vitro, this in itself does not constitute proof that this mechanism is the one operative in vivo. Such proof clearly requires the isolation of other bacterial mutants defective in rRNA synthesis.

References

Birnbaum, L. S. and S. Kaplan. 1973. *In vitro* synthesis of *Escherichia coli* ribosomal RNA. *J. Mol. Biol.* **75**:73.

Blumenthal, T., T. A. Landers and K. Weber. 1972. Bacteriophage $O\beta$ replicase contains the protein biosynthesis elongation factors EF-Tu and EF-Ts. *Proc. Nat. Acad. Sci.* **69**:1313.

Cashel, M. 1969. The control of ribonucleic acid synthesis in *Escherichia coli*. IV. Relevance of unusual phosphorylated compounds from amino acid starved strains. *J. Biol. Chem.* **244**:3133.

———. 1970. Inhibition of RNA polymerase by ppGpp, a nucleotide accumulated during the stringent response to amino acid starvation in *E. coli*. *Cold Spring Harbor Symp. Quant. Biol.* **35**:407.

Cukier-Kahn, R., M. Jacquet and F. Gros. 1972. Two heat-resistant low molecular weight proteins from *Escherichia coli* that stimulate DNA-directed RNA synthesis. *Proc. Nat. Acad. Sci.* **69**:3643.

Fiil, N. 1969. A functional analysis of the *rel* gene in *Escherichia coli*. *J. Mol. Biol.* **45**:195.

Fiil, N. P., K. von Meyenburg and J. D. Friesen. 1972. Accumulation and turnover of guanosine tetraphosphate in *Escherichia coli*. *J. Mol. Biol.* **71**:769.

Gallant, J., H. Erlich, B. Hall and T. Laffler. 1970. Analysis of the RC function. *Cold Spring Harbor Symp. Quant. Biol.* **35**:397.

Haseltine, W. A. 1972. *In vitro* transcription of *Escherichia coli* ribosomal RNA genes. *Nature* **235**:329.

Hussey, C., J. Pero, R. G. Shorenstein and R. Losick. 1972. *In vitro* synthesis of ribosomal RNA by *Bacillus subtilis* RNA polymerase. *Proc. Nat. Acad. Sci.* **69**:407.

Jacobson, A. and D. Gillespie. 1970. An RNA polymerase mutant defective in ATP initiations. *Cold Spring Harbor Symp. Quant. Biol.* **35**:85.

Kennell, D. 1968. Titration of the gene sites on DNA by DNA-RNA hybridization. II. The *Escherichia coli* chromosome. *J. Mol. Biol.* **34**:85.

Matsukage, A. 1972. The effect of KCl concentration on the transcription by

E. coli RNA polymerase. I. Specific effect of the combination of nucleoside triphosphates. *Mol. Gen. Genet.* **118**:11.

Murooka, Y. and R. A. Lazzarini. 1973. *In vitro* synthesis of ribosomal RNA by a DNA-protein complex isolated from *Escherichia coli. J. Biol. Chem.* **248**:6248.

Pettijohn, D. E. 1972. Ordered and preferential initiation of ribosomal RNA synthesis in vitro. *Nature New Biol.* **235**:204.

Pettijohn, D. E., K. Clarkson, C. R. Kossman and O. G. Stonington. 1970. Synthesis of ribosomal RNA on a protein-DNA complex isolated from bacteria: A comparison of ribosomal RNA synthesis *in vitro* and *in vivo. J. Mol. Biol.* **52**:281.

Snyder, L. 1973. Change in RNA polymerase associated with the shut-off of host transcription by T4. *Nature New Biol.* **243**:131.

Stomato, T. D. and D. E. Pettijohn. 1971. Regulation of ribosomal RNA synthesis in stringent bacteria. *Nature New Biol.* **234**:99.

Stonington, O.G. and D. E. Pettijohn. 1971. The folded genome of *Escherichia coli* isolated in a protein-DNA complex. *Proc. Nat. Acad. Sci.* **68**:6.

Travers, A. 1973. Control of ribosomal RNA synthesis *in vitro. Nature* **244**:15.

Travers, A. and R. Buckland. 1973. Heterogeneity of *E. coli* RNA polymerase *Nature New Biol.* **243**:257.

Travers, A., D. L. Baillie and S. Pedersen. 1973. The effect of DNA conformation on ribosomal RNA synthesis *in vitro. Nature New Biol.* **243**:161.

Travers, A. A., R. I. Kamen and R. F. Schleif. 1970. Factor necessary for ribosomal RNA synthesis. *Nature* **228**:748.

Winsten, J. A. and P. C. Huang. 1972. Ribosomal RNA synthesis *in vitro:* A protein-DNA complex from *Bacillus subtilis* active in the initiation of transcription. *Proc. Nat. Acad. Sci.* **69**:1387.

Worcel, A. and E. Burgi. 1972. On the structure of the folded chromosome of *Escherichia coli. J. Mol. Biol.* **71**:127.

Yankofsky, S. A. and S. Spiegelman. 1962. The identification of the ribosomal RNA cistron by sequence complementarity. II. Saturation of and competitive interaction at the ribosomal RNA locus. *Proc. Nat. Acad. Sci.* **48**:1466.

Complex Interactions of Antibiotics with the Ribosome

Bernard D. Davis, Phang-Cheng Tai and Brian J. Wallace*
Bacterial Physiology Unit
Harvard Medical School
Boston, Massachusetts 02115

INTRODUCTION

The search for antibiotics has yielded not only many life-saving drugs, but also an even larger number of valuable reagents for the study of cell physiology. The specificity and variety of their inhibitory actions have reaffirmed Claude Bernard's dictum: "Poisons are delicate instruments that dissect vital units"; and the ribosome has turned out to be the vital unit most frequently affected.

Antibiotics have contributed to our knowledge of the ribosome both through study of their mode of action and through their use in the isolation of mutants with altered ribosomes. However we shall not review the latter literature, except to note a particularly ingenious recent use of heterozygotes carrying multiple mutations to resistance: the polarity effects in these strains showed that the genes for several ribosomal proteins (and for factor EF-G) are linked in a single operon (Nomura and Engbaek 1972). It might also be noted that in studies of eukaryotic cells the selective action of cycloheximide on cytoplasmic ribosomes, and that of chloramphenicol on mitochondrial ribosomes, has been indispensable for recognizing that the latter make only a fraction of the mitochondrial proteins (Ashwell and Work 1970).

The early studies of protein synthesis with synthetic messengers revealed the microcycle of chain elongation but not the macrocycle of initiation and release. Hence the first anti-ribosomal antibiotics to be understood were those that can act on elongating ribosomes. Many others yielded equivocal results until the availability of a natural messenger, viral RNA, made it possible to detect specific interactions with initiating ribosomes. We shall consider a few particularly fruitful observations with members of the first class and shall then concentrate on the second class.

Dedicated to the memory of Selman A. Waksman (1888–1973), who not only discovered streptomycin but thereby demonstrated the value of a systematic search for antibiotics.
* Present address: Department of Biochemistry, John Curtin School of Medical Research, Canberra, Australia.

We shall see that the specific interactions with initiating ribosomes do not always involve a simple block of some step in initiation: some antibiotics must bind to the ribosome during initiation in order to exert their inhibitory effect on a later step. Moreover, the resultant blocked complex is unstable, and cyclic reinitiation and blockade by the released ribosomes leads to the formation of abnormal polysomes (polyinitiation complexes). Among the antibiotics that act in this way, the aminoglycosides have a particularly complex pattern, for in addition to blocking initiating ribosomes they have a different effect on ribosomes already engaged in chain elongation. These findings support the suggestion of Pestka (1971) that interactions of the ribosome with antibiotics, like its interactions with normal ligands, can be markedly influenced by its cyclic changes in conformation.

This brief review will necessarily be highly selective. For more comprehensive surveys of the large literature on antibiotic action on the ribosome, the reader is referred to Weisblum and Davies (1968), Pestka (1971) and Cundliffe (1972a).

Antibiotic Effects on Elongating Ribosomes

Puromycin, an analog of acceptor aminoacyl-tRNA, inhibits further growth of the peptide chain by causing its premature release as peptidyl-puromycin. This "puromycin reaction" has served as a very useful model system in studies of the peptidyl transfer reaction: it has shown that the large subunit alone can mediate peptidyl transfer and that no supernatant factor or energy source is required (Traut and Monro 1964; Monro 1967b; Monro et al. 1969).

The puromycin reaction has also been useful for determining whether various antibiotics or factors cause a charged tRNA to be bound in the A or in the P site. (Indeed, the P site has become operationally defined as the puromycin-reactive site, although this reactivity may well not specify a unique position of the tRNA.) However, when an antibiotic is found to inhibit the puromycin reaction, its site of action is still ambiguous: the effect could be due either to inhibition of peptidyl transferase (with the peptidyl-tRNA in the P site but unreactive) or to inhibition of translocation (with the peptidyl-tRNA fixed in the A site). The problem was solved for certain antibiotics by the discovery of the "fragment reaction," in which, in the presence of alcohol, a short 3'-terminal fragment of peptidyl-tRNA (or of fMet-tRNA) can bind to the large subunit and react with puromycin (Monro 1967a). Since this reaction does not involve translocation, it can presumably be blocked only by inhibitors of peptidyl transfer.

Chloramphenicol, sparsomycin and several other antibiotics have been shown in this way to block peptidyl transfer directly. Chloramphenicol and sparsomycin appear to block the reaction in different ways since the former makes the binding of an fMet-tRNA fragment to the large subunit more labile and the latter makes it more stable (Monro, Celma and Vazquez 1969). These effects may be useful in further dissecting the peptidyl transfer reaction into component steps.

Fusidic acid, a steroid antibiotic, blocks chain elongation by acting on EF-G rather than directly on the ribosome: it binds to EF-G in solution and selects for resistant mutants with an altered EF-G (Kinoshita, Kuwano and Tanaka 1968; Tocchini-Valentini, Felicetti and Rinaldi 1969). This interaction with a definable macromolecular component of the system has made possible a particularly sophisticated analysis, showing that fusidic acid blocks the recycling of EF-G, rather than its total function. In the presence of the antibiotic, the EF-G can bind to the ribosome and catalyze one round of GTP hydrolysis, associated with a translocation; but release of the resulting EF-G·GDP complex is then inhibited (Bodley, Zieve and Lin 1970).

Further studies with fusidic acid revealed an unexpected feature of the ribosome cycle: mutually exclusive binding of EF-G and EF-Tu. Thus if the single translocation that occurs in the presence of this antibiotic were followed by recognition, as was expected, the peptide should then be transferred to the A site and it should be held there since the next translocation would be blocked. In fact, however, in fusidic-blocked ribosomes with excess EF-G, the polypeptide is puromycin-reactive and hence evidently in the P site (Modolell and Davis 1969; Cundliffe 1972b). The explanation came with the finding that bound EF-G, blocked by fusidic acid from its normal release, inhibits the binding of the EF-Tu·GTP·aminoacyl-tRNA complex (Cabrer, Vazquez and Modolell 1972; Miller 1972; Richman and Bodley 1972; Richter 1972). Fusidic acid is thus primarily, though indirectly, an inhibitor of recognition.

Further evidence on the interaction of EF-G and EF-Tu is provided by the finding that the binding of both factors is blocked by certain other antibiotics: thiostrepton, siomycin and thiopeptin (Bodley, Lin and Highland 1970; Weisblum and Demohn 1970; Kinoshita, Liou and Tanaka 1971; Modolell, Vazquez and Monro 1971; Modolell et al. 1971; Watanabe and Tanaka 1971; Pestka 1972a). These observations, and the mutual exclusion of EF-G and EF-Tu (shown by the effect of fusidic acid), suggest that these factors bind to overlapping sites, which may involve a common GTPase.

Some features of the action of fusidic acid still require explanation. If it allows one translocation and then blocks the product in the P site, one might expect an initiating system to accumulate a dipeptide in its presence, but in fact, a tripeptide accumulates (Beller and Lubsen 1972). Moreover, in protoplasts, after an initial blockade of the polypeptides in the P site, there is slow, "leaky" protein synthesis (even with high antibiotic concentrations); but instead of the expected continual rapid recycling to a block in the P site, a new steady state is reached in which one-third of the peptides are not puromycin-reactive (Cundliffe 1972b).

Pleiotropic Action of Streptomycin

We have seen that a single antibiotic (e.g., thiostrepton) can affect the interaction of the ribosome with two different normal ligands, EF-G and EF-Tu. Multiple effects are even more conspicuous with streptomycin and related aminoglycosides (e.g., kanamycin, gentamycin, neomycin, paromomycin). Indeed, it is partly because of this pleiotropy that "the" mechanism of action of aminoglycosides, intensively pursued in many laboratories for decades, has been so elusive.

The most striking pleiotropy involves a paradox. Sublethal concentrations of streptomycin, allowing continued synthesis, increase the frequence of errors and thus cause phenotypic suppression (Gorini and Kataja 1964). Moreover, in the translation of synthetic homopolynucleotides in lysates, even high streptomycin concentrations cause extensive misreading, with only partial inhibition. But though these observations first revealed the ability of alterations in the ribosome to affect the accuracy of translation (see Gorini, this volume), it was not clear how the same high concentrations could inhibit protein synthesis completely in the cell (and very rapidly with mutants that lack the normal permeability barrier: Turnock 1970). This problem languished until the use of viral RNA made it possible to reproduce in extracts the complete inhibition observed in cells. This development will be discussed in the following section.

Streptomycin also antagonizes the dissociation of ribosomes into subunits, in response either to the dissociation factor (Garcia-Patrone et al. 1971; Herzog, Ghysen and Bollen 1971; Wallace, Tai and Davis 1973) or to lowering of the Mg^{++} concentration (Herzog 1964). In addition, with ribosomes that have been inactivated by exposure to abnormal ionic conditions, streptomycin impairs reactivation in normal media (Miskin and Zamir 1972). This effect presumably involves multiple ion-binding sites; hence it has been ascribed to an increased rigidity of the particle, which may be viewed as an additional pleiotropic action or may underlie some of the effects already noted.

An additional effect of streptomycin, still unexplained after many years, is damage to the cell membrane (Anand and Davis 1960). This effect appears at moderate streptomycin concentrations, and almost as soon as the inhibition of protein synthesis (Dubin, Hancock and Davis 1963). Mutations to ribosomal resistance eliminate this effect (along with the ribosomal effects described above), suggesting that alterations in the bacterial ribosome may influence the cytoplasmic membrane.

On the basis of these findings and others to be detailed below, the known pleiotropic effects of streptomycin can be summarized as: complete inhibition of initiating ribosomes, destabilization of the initiation complex, partial inhibition of preformed polysomes, increased misreading on these polysomal ribosomes, impaired dissociation of free ribosomes into subunits, increased rigidity of the ribosome, and membrane damage.

Differential Action of Streptomycin on Initiating Ribosomes and on Preformed Polysomes

With viral RNA as messenger, streptomycin was found to cause complete inhibition of initiating ribosomes (Anderson, Davies and Davis 1967). In addition, when the antibiotic was added after synthesis had reached a steady state, it still caused rapid and complete inhibition (Modolell and Davis 1968). Since this system was then believed to engage in little reinitiation, streptomycin appeared to be blocking elongating as well as initiating ribosomes. However, it was found to cause only partial inhibition of endogenous incorporation in a crude cell extract (Modolell and Davis 1968; Luzzatto, Apirion and Schlessinger 1969a), which suggested that polysomal ribosomes might not be completely inhibited. More direct evidence was obtained with purified polysomes (either endogenous or formed in vitro on viral RNA) which could complete their nascent chains but could not reinitiate (Tai et al. 1973). As Figure 1 shows, with such preparations

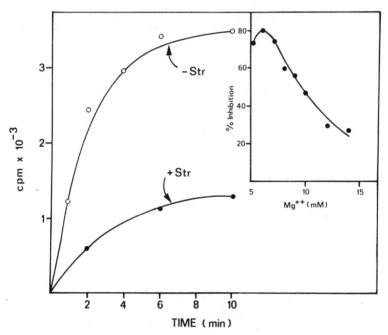

Figure 1   Inhibition by streptomycin of peptide synthesis on purified endogenous polysomes of *E. coli*. Reaction mixtures contained [^{14}C]valine (70 Ci/mole), 19 other amino acids, 50 μg purified endogenous polysomes (in 0.1 ml), and 8 mM Mg^{++}. After incubation at 34°C, with or without streptomycin (20 μg/ml = 34 μM), samples were taken at intervals and the incorporated radioactivity was measured. For the effect of Mg^{++} on streptomycin inhibition (*inset*), reaction mixtures, with or without streptomycin at the Mg^{++} concentration indicated, were incubated for 6 min. (Wallace et al. 1973).

addition of streptomycin caused protein synthesis to shift immediately to a new, lower rate (Wallace et al. 1973). This partial inhibition of elongating ribosomes evidently reflects a different action of streptomycin from that observed with initiating ribosomes, rather than a lower affinity, since increasing concentrations of streptomycin, above a low, saturating value, do not further increase the degree of inhibition. This inhibition of elongating ribosomes differs from the inhibition of initiating ribosomes in another respect: it is strongly antagonized by elevated Mg^{++} (Figure 1; also Luzzatto, Apirion and Schlessinger 1969a).

Further investigation showed that although streptomycin prevents initiating ribosomes from synthesizing protein, it does not prevent them from forming initiation complexes; rather, it blocks them at some step shortly after initiation. In addition, the blocked complex is unstable. The intermediate complexes, made with 30S subunits (Figure 2) or with 70S ribosomes plus GMPPCP, are stable; but when the complex is completed by the hydrolysis of GTP on the 70S ribosome (Figure 2), the fMet-tRNA is released, with a half-life of about 5 min at 37°C (Modolell and Davis 1970; Lelong et al. 1971). This instability, as we shall see, provided the

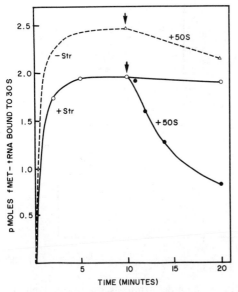

Figure 2   Breakdown of 30S initiation complexes on addition of 50S subunits in the presence of streptomycin. A 200-μl reaction mixture, containing 1.5 mg/ml 30S subunits, 0.42 mg/ml f-[^{14}C]Met-tRNA, poly(AUG), and other components, was incubated for 10 min with streptomycin and then further with or without addition of an equal number of 50S subunits. Parallel reaction mixtures were incubated without streptomycin. Samples taken at intervals were analyzed for radioactivity retained by a nitrocellulose filter. (Modolell and Davis 1970).

key to an old problem: the dominance of sensitivity to streptomycin in heterozygotes.

Curiously, the cleavage of a 3'-terminal fragment of 16S RNA on the ribosome by colicin E3 (Nomura et al., this volume) resembles the binding of streptomycin in causing a partial, Mg^{++}-dependent inhibition of elongating ribosomes and a more complete inhibition of initiating ribosomes (Tai and Davis 1974). Moreover, the binding of streptomycin and the action of the colicin antagonize each other (Dahlberg et al. 1973; Tai and Davis 1974). However, alteration of the ribosome by the colicin does not induce misreading (Tai and Davis 1974).

Streptomycin-induced Misreading of Natural Messenger

Since streptomycin slows elongating ribosomes, it evidently affects their function while allowing them to continue to synthesize protein. These ribosomes might therefore provide the circumstances required for misreading. The paradox of inhibition versus misreading could then be explained. Thus although streptomycin at high concentrations would soon block every initiating ribosome in a cell, at low concentrations it would hit only an occasional initiating ribosome and the rest of the ribosomes would continue their cycles. During this process an occasional elongating ribosome would bind a molecule of streptomycin, which could increase its misreading. This model would fit the earlier finding that a very narrow concentration range of streptomycin was required for stimulation of the formation of an immunologically cross-reacting, inactive enzyme in cells (Bissell 1965).

This postulated mechanism was tested by observing the effect of streptomycin on the ability of preformed polysomes to incorporate a mixture of only 15 amino acids, all labeled. (The synthesizing system was evidently contaminated with the missing amino acids, for in the absence of streptomycin it incorporated about one-third as much of this mixture as when it was supplemented with the five missing amino acids, unlabeled.) As Table 1 shows, streptomycin markedly reduced incorporation in the control complete system, as expected, but it stimulated incorporation in the incomplete system. In fact, in the presence of streptomycin the two systems incorporated at the same level: i.e., under the test conditions (with a limited supply of the labeled amino acids) any codon calling for a missing amino acid could apparently be filled as effectively as though that amino acid were present. It thus seems that the high frequency of misreading in the presence of streptomycin, observed earlier with synthetic messengers, also occurs with natural messenger under conditions that bypass the block after initiation. This conclusion is consistent with the results of an early study with phage RNA (Schwartz 1965), in which low concentrations of streptomycin stimulated the incorporation of single-labeled amino acids in a system where asparagine was limiting.

Table 1   Effect of streptomycin on peptide synthesis by polysomes with an incomplete set of amino acids

Incubation (min)	cpm incorporated*		% Inhibition (−) or stimulation (+)
	−Str	+Str	
A 5	33,722	16,422	−51
10	45,569	19,968	−56
B 5	10,642	16,694	+56
10	12,778	20,205	+59

In experiment A, 56 μg purified *E. coli* polysomes were incubated in 100 μl of a peptide-synthesizing system at 8 mM Mg^{++} (Tai et al. 1973) containing 0.3 μCi of a mixture of 15 ^{14}C-labeled amino acids (New England Nuclear, NEC-445, and neutralized with Tris), in addition to unlabeled cysteine, methionine, glutamine, asparagine and tryptophan (each at 100 μM). In experiment B, the 5 unlabeled amino acids were omitted. Streptomycin (str) (20 μg/ml) was added where indicated at 0 time. After incubation at 34°C for 5 and 10 min, the polypeptide formed was precipitated by 1 ml 5% trichloroacetic acid, heated at 90°C for 20 min, collected on a Millipore filter, and counted in a Nuclear-Chicago gas-flow counter.

* 236 cpm in the control tube (at 0 time) has been subtracted. (Data of B. J. Wallace, P.-C. Tai, E. L. Herzog and B. D. Davis, unpublished.)

Cyclic Reinitiation by Streptomycin Ribosomes

Though the streptomycin ribosomes released from the unstable blocked initiation complexes are irreversibly stopped from future synthesis of protein, they are not inert: they can reinitiate, at the expense of energy supplied by GTP. Moreover, several such reinitiating ribosomes can form unstable blocked complexes on a single molecule of messenger. Hence cells killed by streptomycin maintain a substantial level of short polysomes, and these turn over rapidly despite the absence of protein synthesis (Wallace and Davis 1973).

This unusual ribosome cycle in streptomycin-inhibited cells has been demonstrated by several lines of evidence. Thus despite the lack of protein synthesis in these cells, pulse-labeling with uracil showed that the mRNA in their polysomes turns over as rapidly as that in the polysomes of growing cells (see Luzzatto, Apirion and Schlessinger 1969b). Moreover, the polysomes in the treated cells can be pulse-labeled with methionine but not with valine (as shown below in Figure 5 for another antibiotic; see also Lennette and Apirion 1970). Finally, when reinitiation is blocked directly by trimethoprim or hydroxylamine, or indirectly by using rifampicin to prevent mRNA renewal (Figure 3), the polysome level falls rapidly. It is thus clear that in cells the initiating streptomycin ribosome not only is blocked shortly after initiation and falls off after a few minutes, as previously demonstrated in extracts, but it also reinitiates. The rate of reinitiation, and hence the polysome level, varies from one strain to another (Figure 3).

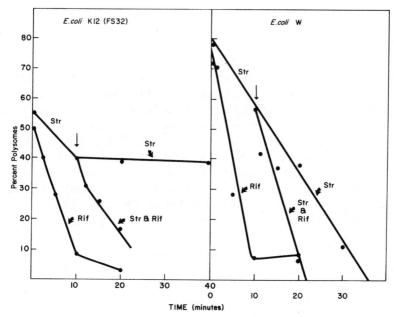

Figure 3   Effect of rifampicin on polysome level in streptomycin-treated cells. To growing cultures of sensitive strains FS32 (*left*) and W (*right*), streptomycin was added, or rifampicin (200 μg/ml), or both sequentially. Samples were removed at intervals and analyzed for polysomes. (Wallace and Davis 1973; Wallace, Tai and Davis 1973).

The released streptomycin ribosomes, as noted above, are more stable to dissociation than are free ribosomes from untreated cells (Wallace, Tai and Davis 1973). This finding explains the earlier observation of unusually stable 70S particles in cells treated with streptomycin (Luzzatto, Apirion and Schlessinger 1968, 1969b). These particles were interpreted as blocked initiation complexes; but though such complexes are indeed formed in the presence of streptomycin, they are found, as we have seen, in polysomes.

Dominance of Sensitivity over Resistance to Streptomycin

The complex effect of streptomycin on initiating ribosomes provides an explanation for an old problem: the dominance of sensitivity in str^R/str^S heterozygotes (Lederberg 1951). The puzzle was increased by the further finding that such cells contain a mixture of sensitive and resistant ribosomes (Sparling et al. 1968) and that they are not merely inhibited but are killed by streptomycin, like sensitive haploid cells (Sparling and Davis 1972). Why are the resistant ribosomes inactive in protein synthesis, and why do the cells not resume growth when plated in the absence of streptomycin?

The explanation is based on the fact that streptomycin not only causes sensitive ribosomes to block initiation sites, but also causes rapid turnover of the blocked complexes (Wallace and Davis 1973). As we have seen, the block after initiation lasts, on the average, for several minutes. This duration is long, compared with the normal interval between initiations on an initiation site (ca. 1 sec); hence the blocked sensitive ribosomes will exclude the resistant ribosomes in a heterozygote very effectively. (The competition is especially effective since the ratio of sensitive to resistant ribosomes in heterozygotes is high: about 2:1 in one K12 strain [Sparling et al. 1968] and even higher in another [Chang et al. 1973].) On the other hand, the block is not quite permanent, and the eventual release of the streptomycin ribosomes, followed by reinitiation, allows them to block new mRNA (and to cease protecting old mRNA from breakdown). This cyclic abortive re-initiation, observed in sensitive cells, can account for the dominance observed in heterozygotes.

This mechanism was confirmed by observations on lysates from heterozygous cells that had been treated with streptomycin until they had ceased to synthesize protein. Though these lysates had no endogenous activity, the resistant fraction of their ribosomes resumed activity when an excess of viral messenger was supplied (Wallace and Davis 1973). Evidently the lack of available messenger can account for the cessation of protein synthesis in these cells.

Specific Effects of Other Antibiotics on Initiating Ribosomes

The specific inhibition of ribosomes shortly after initiation, first observed with aminoglycosides, does not appear to be a rare pattern. A very limited search has revealed a similar pattern, including an unstable block after initiation, for two additional antibiotics, spectinomycin and erythromycin. Moreover, in parallel studies, Pestka (1972b), observing the puromycin reaction instead of protein synthesis, found that a number of antibiotics (including these two) do not block the release of polypeptide from endogenous polysomes in lysates of *E. coli,* though they do block the reaction with various synthetic model systems.

It is noteworthy that streptomycin, spectinomycin and erythromycin share the novel pattern of an unstable block after initiation, since their actions differ in other fundamental respects. Spectinomycin and erythromycin, unlike streptomycin, are bacteristatic (i.e., have a reversible effect) rather than bactericidal; they do not cause misreading; and erythromycin acts on the large subunit of the ribosome (Wilhelm and Corcoran 1967) and the other two on the small subunit.

Spectinomycin, like streptomycin, selects for one-step, highly resistant mutants (Anderson, Davies and Davis 1967), with an altered protein (S5) in the small subunit (Bollen et al. 1969; Dekio and Takata 1969). Comparison of preformed polysomes with initiating ribosomes showed that low concentrations of spectinomycin inhibited protein synthesis only with the latter (Figure 4). Moreover, cells inhibited by spectinomycin maintain polysomes at a substantial level: these polysomes can be rapidly pulse-labeled with uracil, they can be pulse-labeled with methionine but not with valine (Figure 5), and they disappear rapidly when RNA renewal is blocked by rifampicin (Wallace, Tai and Davis 1974). Hence spectinomycin blocks ribosomes after, rather than before, initiation; the blocked complex is unstable (though it has a somewhat longer life than the streptomycin complex), and the released ribosomes reinitiate. This cyclic blockade of initiation sites can account for the previously observed dominance of sensitivity over resistance to spectinomycin (Davies, Anderson and Davis 1965; Sparling and Davis 1972).

Though the action of spectinomycin thus resembles in several respects that of streptomycin its effect is evidently less drastic: in the usual concentration range its interaction with initiating ribosomes is reversible (and hence nonlethal), and with elongating ribosomes it has no detectable effect, rather than causing slowing (Figure 4).

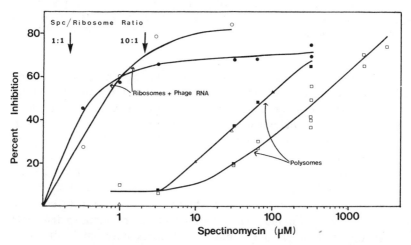

Figure 4 Effect of spectinomycin on peptide synthesis in initiating systems (free ribosomes plus initiation factors plus phage RNA) and in noninitiating systems (purified polysomes). The former systems were incubated for 30 min at 34°C and the latter for 10 min (note early completion of synthesis by purified polysomes in Figure 1). The free ribosomes were obtained either by washing with 1 M NH₄Cl (o——o) or by runoff in slow-cooled cells (•——•); the IF-free polysomes were either endogenous (□——□; ■——■) or prepared on phage R17 RNA (△——△). Spectinomycin was added at 0 time except for one preparation of polysomes. (■——■), which was preincubated with the antibiotic (Wallace, Tai and Davis 1974).

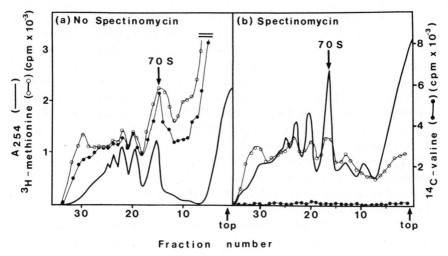

Figure 5 Effect of spectinomycin on pulse-labeling of polysomes with [³H]-methionine and [¹⁴C]valine. A growing culture of *E. coli* MRE600 was incubated with or without spectinomycin (200 μg/ml) for 20 min and then was incubated for 1 min with the further addition of the two labeled amino acids (plus unlabeled isoleucine). After rapid cooling the cells were harvested and lysed, and the ribosomal particles were fractionated by zonal centrifugation and analyzed for A_{254} and for ³H and ¹⁴C. (Wallace, Tai and Davis 1974).

With erythromycin, selective inhibition of initiating ribosomes was suggested earlier by the finding of only a very limited effect on endogenous synthesis in lysates of *Staphylococcus aureus* (Mao and Putterman 1968) and *Bacillus subtilis* (Oleinick and Corcoran 1969). This selectivity was subsequently established with purified polysomes of *E. coli:* these are completely unresponsive to erythromycin, up to high concentrations, whereas initiating ribosomes are very sensitive. (Tai, Wallace and Davis 1974). With this antibiotic it has been further shown that the difference in responsiveness is due to a difference in affinity: at low erythromycin concentrations, sufficient for maximal inhibition of an initiating system, free ribosomes reversibly bind one molecule, whereas polysomal ribosomes bind little or none (Oleinick and Corcoran 1969; Tai, Wallace and Davis 1974).

The complexes blocked after initiation by erythromycin are evidently unstable, like those blocked by streptomycin or spectinomycin, for in erythromycin-treated cells they are also found in polysomes that turn over rapidly without protein synthesis. The evidence is much the same as that described above for streptomycin and spectinomycin (Tai, Wallace and Davis 1974); it includes pulse-labeling with methionine but not with valine. This cyclic blockade of initiation sites explains the finding that sensitivity to erythromycin is dominant over resistance (Nomura and Engbaek 1972).

Among antibiotics that selectively inhibit initiating ribosomes, one might expect that some would interfere with the intricate process of forming the initiation complex. The best-documented examples are the antibiotic kasugamycin (Okuyama et al. 1971) and the synthetic inhibitor aurintricarboxylate (Grollman and Stewart 1968): these have been found to block formation of initiation complexes in extracts. Selective interference with initiation was inferred from the observation that these agents, at low concentrations, did not inhibit endogenous incorporation in S30 extracts as completely as they inhibited the translation of added messenger. Studies with purified polysomes extended this conclusion: initiating ribosomes were inhibited, without appreciable inhibition of elongating ribosomes, over a broad concentration range with kasugamycin and a narrow range with aurintricarboxylate (Tai, Wallace and Davis 1973).

Basis for Differential Interactions with Free and Elongating Ribosomes

We have seen that streptomycin, spectinomycin and erythromycin each specifically block some stage in the transition between initiation and the repeated microcycles of chain elongation. The specificity might depend on a conformation peculiar to this transition. However, observations on binding suggest that the two-stage action of these antibiotics on initiating ribosomes involves binding to the free ribosome, followed by retention of the bound antibiotic through the process of initiation until the susceptible reaction is blocked. Thus streptomycin binds tightly to free ribosomes (in a 1:1 molar ratio), and this binding is evidently irreversible under conditions prevailing in the cell. Erythromycin also binds with high affinity (though reversibly) to free ribosomes, and it has only low affinity for polysomal ribosomes. The binding of spectinomycin has not been studied.

Unlike erythromycin (and possibly spectinomycin), streptomycin binds with high affinity to polysomal ribosomes (as well as to free ribosomes). Both kinds of binding probably involve the same site, since mutations to resistance strikingly decrease the affinity of the ribosome in both states. Nevertheless, after binding, streptomycin polysomal ribosomes can continue to synthesize protein but free ribosomes cannot. These findings suggest that on the polysomal ribosomes the normal ligands permit only a part of the streptomycin-binding site to be exposed, and the resultant looser binding would have a less drastic effect on ribosomal function.

The differential action of streptomycin on free and on elongating ribosomes also provides a basis for explaining another unsolved problem in this field: the effect of various reversible inhibitors of protein synthesis on the irreversible, lethal action of streptomycin. This action is prevented by chloramphenicol (Jawetz, Gunnison and Speck 1951; Plotz and Davis

1962), which stabilizes polysomes, but not by puromycin (White and White 1964), which causes rapid cycling of ribosomes on and off messenger. We know that the binding of streptomycin by free ribosomes is irreversible. If the binding to polysomal ribosomes, with its less drastic effect on function, is reversible under conditions in the cell, it would explain why killing is antagonized by inhibitors that stabilize polysomes.

Reactions Blocked after Initiation

While it is clear that spectinomycin and erythromycin at the low concentrations required to stop protein synthesis each block some step shortly after initiation, these steps have not been established. Preliminary studies with initiation complexes formed in the presence of spectinomycin (Wallace, Tai and Davis 1974) or erythromycin (Tai, Wallace and Davis, manuscript in preparation) have shown that these complexes are not impaired in the first recognition or in the first peptidyl transfer after initiation. These findings suggest a block in the first translocation, but they are also compatible with a block in some step in a subsequent round of translation.

Earlier studies (reviewed in Pestka 1971; Cundliffe 1972a) have indicated that spectinomycin blocks translocation, whereas for erythromycin there is contradictory evidence both for a similar block and for a block in peptidyl transfer. However, it now becomes difficult to interpret these experiments, for they were not designed to take into account the difference in interactions of the antibiotic with initiating and with elongating ribosomes, or to use antibiotic concentrations low enough to ensure selective inhibition of the former.

Particularly cogent and interesting evidence was provided by Cundliffe and McQuillen (1967), who observed that the release of nascent polypeptide from polysomes by puromycin in protoplasts of *Bacillus megaterium* was inhibited by erythromycin alone, but not by erythromycin following pretreatment with a tetracycline. Since the latter blocks the A site (Hierowski 1965; Suarez and Nathans 1965; Gottesman 1967; Suzuka, Kaji and Kaji 1966) and hence should fix the peptidyl-tRNA in the P site, it seemed clear that erythromycin does not affect the puromycin reaction of peptidyl-tRNA already in the P site; hence the inhibition of the reaction by erythromycin alone would imply a block in translocation. However this interpretation no longer seems certain, for we have seen that erythromycin at low concentrations does not bind to polysomal ribosomes (of *E. coli*) or inhibit their reaction with puromycin (Pestka 1972b; Tai, Wallace and Davis 1974). Hence tetracycline might be antagonizing the action of erythromycin not because it specifically blocks the A site, but because it fixes the ribosomes on polysomes or because it limits the conformational mobility of the polysomal ribosomes and hence might impair their binding of erythromycin.

Similar considerations apply to later experiments in which protoplasts were pretreated with fusidic acid instead of tetracycline (Burns and Cundliffe 1973). Erythromycin and spectinomycin caused a slow decrease in reactivity of the accumulated polypeptides with puromycin, which suggested slow leakage past the fusidic block in recognition, followed by a block in translocation by erythromycin or spectinomycin. However, the leakage could be slowly yielding the conformation required for binding of spectinomycin or erythromycin, which might then block any subsequent reaction. These ingenious experiments with mixtures of antibiotics thus do not appear to establish translocation as the step blocked by erythromycin or by spectinomycin; a direct block in peptidyl transfer still remains possible.

With streptomycin, misreading and inhibition may represent different degrees of distortion of the same region of the ribosome, for, as we have noted, mutations to resistance have a large effect on both these actions of the antibiotic. Since the misreading on elongating ribosomes implies an effect on recognition, it seems quite possible that the blockade of initiating ribosomes results from a more severe interference with the same process. However, this indirect evidence is not decisive, and since streptomycin has a pleiotropic action, including an increase in ribosomal rigidity (Miskin and Zamir 1972), it might block more than one step.

CONCLUSIONS

While some antibiotics inhibit a particular reaction on the ribosome in a relatively straightforward manner, several others must bind to free or to initiating ribosomes in order to exert their characteristic inhibition at some step shortly after initiation. It is thus clear that the ligands on elongating ribosomes prevent the kind of binding of streptomycin, spectinomycin and erythromycin that occurs with free ribosomes; this interference may well result from reduced conformational flexibility of the ribosome. It is striking that all three blocked complexes are unstable, for the three antibiotics differ markedly in structure and in action; if a common feature is responsible for this instability, it remains to be determined. But whatever the mechanism, the ability of the released ribosomes to recycle, thus blocking initiation sites for long periods and then becoming free to block new mRNA, can account for the dominance of sensitivity over resistance to each of these antibiotics.

References

Anand, N. and B. D. Davis. 1960. Damage by streptomycin to the cell membrane of *Escherichia coli. Nature* **185**:22.

Anderson, P., J. E. Davies and B. D. Davis. 1967. The effect of spectinomycin on polypeptide synthesis in extracts of *Escherichia coli. J. Mol. Biol.* **29**:203.

Ashwell, M. and T. S. Work. 1970. The biogenesis of mitochondria. *Ann. Rev. Biochem.* **39**:251.

Beller, R. J. and N. H. Lubsen. 1972. Effect of polypeptide chain length on dissociation of ribosomal complexes. *Biochemistry* **11**:3271.

Bissell, D. M. 1965. Formation of an altered enzyme by *Escherichia coli* in the presence of neomycin. *J. Mol. Biol.* **14**:619.

Bodley, J. W., L. Lin and J. H. Highland. 1970. Studies on translation VI. Thiostrepton prevents the formation of a ribosome-G factor-guanine nucleotide complex. *Biochem. Biophys. Res. Comm.* **41**:1406.

Bodley, J. W., F. J. Zieve and L. Lin. 1970. Studies on translocation IV. The hydrolysis of a single round of guanosine triphosphate in the presence of fusidic acid. *J. Biol. Chem.* **245**:5662.

Bollen, A., J. Davies, M. Ozaki and S. Mizushima. 1969. Ribosomal protein conferring sensitivity to the antibiotic spectinomycin in *Escherichia coli.* *Science* **165**:85.

Burns, D. J. W. and E. Cundliffe. 1973. Bacterial protein synthesis: A novel system for studying antibiotic action in vivo. *Eur. J. Biochem.* **37**:570.

Cabrer, B., D. Vazquez and J. Modolell. 1972. Inhibition by elongation factor EFG of aminoacyl-tRNA binding to ribosomes. *Proc. Nat. Acad. Sci.* **69**:733.

Chang, F. N., Y. J. Wang, C. J. Fetterolf and J. G. Flaks. 1973. Unequal contribution to ribosomal assembly of both Str alleles in *Escherichia coli* merodiploids and its relationship to the dominance phenomenon. *J. Mol. Biol.* **82**:273.

Cundliffe, E. 1972a. Antibiotic inhibitors of ribosome function. In *Molecular basis of antibiotic action* (ed. E. F. Gale at el.) p. 278. Wiley-Interscience, New York.

————. 1972b. The mode of action of fusidic acid. *Biochem. Biophys. Res. Comm.* **46**:1794.

Cundliffe, E. and K. McQuillen. 1967. Bacterial protein synthesis: The effects of antibiotics. *J. Mol. Biol.* **30**:137.

Dahlberg, A. E., E. Lund, N. O. Kjeldgaard, C. M. Bowman and M. Nomura. 1973. Colicin E$_3$ induced cleavage of 16S ribosomal ribonucleic acid; blocking effects of certain antibiotics. *Biochemistry* **12**:948.

Davies, J., P. Anderson and B. D. Davis. 1965. Inhibition of protein synthesis by spectinomycin. *Science* **149**:1096.

Davis, B. D. 1971. Role of subunits in the ribosome cycle. *Nature* **231**:153.

Dekio, S. and R. Takata. 1969. Genetic studies of the ribosomal proteins in *Escherichia coli*. II. Altered 30S ribosomal protein component specific to spectinomycin resistant mutants. *Mol. Gen. Genet.* **105**:219.

Dubin, D., R. Hancock and B. D. Davis. 1963. The sequence of some streptomycin-induced changes in *E. coli. Biochim. Biophys. Acta* **74**:476.

Garcia-Patrone, M., C. A. Perazzolo, F. Baralle, N. S. Gonzalez and I. D. Algranati. 1971. Studies on dissociation factor of bacterial ribosomes: Effect of antibiotics. *Biochim. Biophys. Acta* **246**:291.

Gorini, L. and E. Kataja. 1964. Phenotypic repair by streptomycin of defective genotypes in *E. coli. Proc. Nat. Acad. Sci.* **51**:487.

Gottesman, M. E. 1967. Reaction of ribosome-bound peptidyl transfer ribonucleic acid with aminoacyl transfer ribonucleic acid or puromycin. *J. Biol. Chem.* **242**:5564.

Grollman, A. P. and M. L. Stewart. 1968. Inhibition of the attachment of messenger ribonucleic acid to ribosomes. *Proc. Nat. Acad. Sci.* **61**:719.

Herzog, A. 1964. An effect of streptomycin on the dissociation of *Escherichia coli* 70S ribosomes. *Biochem. Biophys. Res. Comm.* **15**:172.

Herzog, A., A. Ghysen and A. Bollen. 1971. Sensitivity and resistance to streptomycin in relation with factor-mediated dissociation of ribosomes. *FEBS Letters* **15**:291.

Hierowski, M. 1965. Inhibition of protein synthesis by chlortetracycline in the *E. coli in vitro* system. *Proc. Nat. Acad. Sci.* **53**:594.

Jawetz, E., J. B. Gunnison and R. S. Speck. 1951. Studies on antibiotic synergism and antagonism: The interference of aureomycins, chloramphenicol and terramycin with the action of streptomycin. *Amer. J. Med. Sci.* **222**:404.

Kinoshita, T., G. Kuwano and N. Tanaka. 1968. Association of fusidic acid sensitivity with G factor in a protein-synthesizing system. *Biochem. Biophys. Res. Comm.* **33**:769.

Kinoshita, T., Y. Liou and N. Tanaka. 1971. Inhibition by thiopeptin of ribosomal functions associated with T and G factors. *Biochem. Biophys. Res. Comm.* **44**:859.

Lederberg, J. 1951. Streptomycin-resistance: A genetically recessive mutation. *J. Bact.* **61**:549.

Lelong, J. C., M. A. Cousin, D. Gros, M. Grunberg-Manago and F. Gros. 1971. Streptomycin induced release of fmet-tRNA from the ribosomal initiation complex. *Biochem. Biophys. Res. Comm.* **42**:530.

Lennette, E. and D. Apirion. 1970. The level of fmet-tRNA on ribosomes from streptomycin treated cells. *Biochem. Biophys. Res. Comm.* **41**:804.

Luzzatto, L., D. Apirion and D. Schlessinger. 1968. Mechanism of action of streptomycin in *E. coli:* Interruption of the ribosome cycle at the initiation of protein synthesis. *Proc. Nat. Acad. Sci.* **60**:873.

———. 1969a. Streptomycin action: Greater inhibition of *Escherichia coli* ribosome function with exogenous than with endogenous messenger ribonucleic acid. *J. Bact.* **99**:206.

———. 1969b. Polysome depletion and blockage of the ribosome cycle by streptomycin in *Escherichia coli. J. Mol. Biol.* **42**:315.

Mao, J. C. H. and M. Putterman. 1968. Accumulation in gram-positive and gram-negative bacteria as a mechanism of resistance to erythromycin. *J. Bact.* **95**:1111.

Miller, D. L. 1972. Elongation factors EFTu and EFG interact at related sites on ribosomes. *Proc. Nat. Acad. Sci.* **69**:752.

Miskin, R. and A. Zamir. 1972. Effect of streptomycin on ribosome interconversion, a possible basis for the action of the antibiotic. *Nature New Biol.* **238**:78.

Modolell, J. and B. D. Davis. 1968. Rapid inhibition of polypeptide chain extension by streptomycin. *Proc. Nat. Acad. Sci.* **61**:1279.

———. 1969. Significance of the effects of streptomycin and fusidic acid on polysome stability. In *Progress in antimicrobial and anticancer chemotherapy* (Proc. 6th Int. Congress Chemotherapy) pp. 464–467. University of Tokyo Press, Tokyo, Japan.

———. 1970. Breakdown by streptomycin of initiation complexes formed on ribosomes of *Escherichia coli. Proc. Nat. Acad. Sci.* **67**:1148.

Modolell, J. W., D. Vazquez and R. E. Monro. 1971. Ribosomes, G-factor, and siomycin. *Nature New Biol.* **230**:109.

Modolell, J., B. Cabrer, A. Parmeggiani and D. Vasquez. 1971. Inhibition by siomycin and thiostrepton of both aminoacyl-tRNA and factor G binding to ribosomes. *Proc. Nat. Acad. Sci.* **68**:1796.

Monro, R. E. 1967a. Ribosome-catalyzed reaction of puromycin with a formylmethionine-containing oligonucleotide. *J. Mol. Biol.* **25**:347.

――――. 1967b. Catalysis of peptide bond formation by 50S ribosomal subunits from *Escherichia coli. J. Mol. Biol.* **26**:147.

Monro, R. E., M. L. Celma and D. Vazquez. 1969. Action of sparsomycin on ribosome-catalyzed peptidyl transfer. *Nature* **222**:356.

Monro, R. E., T. Staehelin, M. L. Celma and D. Vazquez. 1969. The peptidyl transfer activity of ribosomes. *Cold Spring Harbor Symp. Quant. Biol.* **34**:357.

Nomura, M. and F. Engbaek. 1972. Expression of ribosomal protein genes as analyzed by bacteriophage Mu-induced mutations. *Proc. Nat. Acad. Sci.* **69**:1526.

Okuyama, A., N. Machiyama, T. Kinoshita and N. Tanaka. 1971. Inhibition by kasugamycin of initiation complex formation on 30S ribosomes. *Biochem. Biophys. Res. Comm.* **43**:196.

Oleinick, N. L. and J. W. Corcoran. 1969. Evidence of a limited access of erythromycin A to functional polysomes and its action on bacterial translocation. In *Progress in antimicrobial and anticancer chemotherapy* (Proc. 6th Int. Congress Chemotherapy) pp. 202–208. University of Tokyo Press, Tokyo, Japan.

Pestka, S. 1971. Inhibitors of ribosome functions. *Ann. Rev. Microbiol.* **25**:487.

――――. 1972a. Thiostrepton: A ribosomal inhibitor of translocation. *Biochem. Biophys. Res. Comm.* **40**:667.

――――. 1972b. Effect of antibiotics on peptidyl puromycin synthesis on polyribosomes from *Escherichia coli. J. Biol. Chem.* **247**:4669.

Plotz, P. H. and B. D. Davis. 1962. Absence of a chloramphenicol-insensitive phase of streptomycin action. *J. Bact.* **83**:802.

Richman, N. and J. W. Bodley. 1972. Ribosomes cannot interact simultaneously with elongation factors EFTu and EFG. *Proc. Nat. Acad. Sci.* **69**:686.

Richter, D. 1972. Inability of *E. coli* ribosomes to interact simultaneously with the bacterial elongation factors EFTu and EFG. *Biochem. Biophys. Res. Comm.* **46**:1850.

Schwartz, J. H. 1965. An effect of streptomycin on the biosynthesis of coat protein of coliphage f2 by extracts of *E. coli. Proc. Nat. Acad. Sci.* **53**:1133.

Sparling, P. F. and B. D. Davis. 1972. Bacterial action of streptomycin and comparison with spectinomycin in heterozygotes of *Escherichia coli. Antimicrob. Agents Chemother.* **1**:252.

Sparling, P. F., J. Modolell, Y. Takeda and B. D. Davis. 1968. Ribosomes from *Eschericia coli* merodiploids heterozygous for resistance to streptomycin and to spectinomycin. *J. Mol. Biol.* **37**:407.

Suarez, G. and D. Nathans. 1965. Inhibition of aminoacyl-sRNA binding to ribosomes by tetracycline. *Biochem. Biophys. Res. Comm.* **18**:743.

Suzuka, I., H. Kaji and A. Kaji. 1966. Binding of specific sRNA to 30S ribosomal subunits: Effect of 50S ribosomal subunits. *Proc. Nat. Acad. Sci.* **55**:1483.

Tai, P.-C. and B. D. Davis. 1974. Activity of colicin E3-treated ribosomes in initiation and in chain elongation. *Proc. Nat. Acad. Sci.* **71**:1021.

Tai, P.-C., B. J. Wallace and B. D. Davis. 1973. Actions of aurintricarboxylate, kasugamycin, and pactamycin. *Biochemistry* **12**:616.

———. 1974. Selective action of erythromycin on initiating ribosomes. *Biochemistry* (in press).

Tai, P.-C., B. J. Wallace, E. L. Herzog and B. D. Davis. 1973. Properties of initiation-free polysomes of *Escherichia coli*. *Biochemistry* **12**:609.

Tocchini-Valentini, G. P., L. Felicetti and G. M. Rinaldi. 1969. Mutants of *Escherichia coli* blocked in protein synthesis: Mutant with an altered G factor. *Cold Spring Harbor Symp. Quant. Biol.* **34**:463.

Traut, R. R. and R. E. Monro. 1964. The puromycin reaction and its relation to protein synthesis. *J. Mol. Biol.* **10**:63.

Turnock, G. 1970. The action of streptomycin in a mutant of *Escherichia coli* with increased sensitivity to the antibiotic. *Biochem. J.* **118**:659.

Wallace, B. J. and B. D. Davis. 1973. Cyclic blockade of initiation sites by streptomycin-damaged ribosomes in *Escherichia coli:* An explanation for dominance of sensitivity. *J. Mol. Biol.* **75**:377.

Wallace, B. J., P.-C. Tai and B. D. Davis. 1973. Effect of streptomycin on the response of *Escherichia coli* ribosomes to the dissociation factor. *J. Mol. Biol.* **75**:391.

———. 1974. Selective inhibition of initiating ribosomes by spectinomycin. *Proc. Nat. Acad. Sci.* **71**:1634.

Wallace, B. J., P.-C. Tai, E. L. Herzog and B. D. Davis. 1973. Partial inhibition of polysomal ribosomes of *Escherichia coli* by streptomycin. *Proc. Nat. Acad. Sci.* **70**:1234.

Watanabe, S. and K. Tanaka. 1971. Effect of siomycin on the G factor dependent GTP hydrolysis by *Escherichia coli* ribosomes. *FEBS Letters* **13**:267.

Weisblum, B. and J. Davies. 1968. Antibiotic inhibitors of the bacterial ribosome. *Bact. Rev.* **32**:493.

Weisblum, B. and V. Demohn. 1970. Inhibition by thiostrepton of the formation of a ribosome-bound guanine nucleotide complex. *FEBS Letters* **11**:149.

White, J. R. and H. L. White. 1964. Streptomycinoid antibiotics: Synergism by puromycin. *Science* **146**:772.

Wilhelm, J. M. and J. W. Corcoran. 1967. Antibiotic glycosides. VI. Definition of the 50S ribosomal subunit of *Bacillus subtilis* 168 as a major determinant of sensitivity to erythromycin A. *Biochemistry* **6**:2578.

Streptomycin and Misreading of the Genetic Code

Luigi Gorini
Department of Microbiology and Molecular Genetics
Harvard Medical School
Boston, Massachusetts 02115

Historical Background

In 1961 we accidentally isolated a streptomycin-resistant auxotroph whose arginine requirement could be satisfied by streptomycin (Gorini, Gundersen and Burger 1961). It was shown a few years later (Gorini and Kataja 1964a), using the same streptomycin-resistant parent and by selecting for streptomycin dependence in minimal medium but not in broth, that a class of mutants defective in a number of unrelated metabolic pathways could be easily obtained. The members of this class share the property of "conditional streptomycin dependence (CSD)." It could be shown in extracts that the enzyme normally missing in the mutant was formed during growth in the presence of streptomycin. We suggested that streptomycin might induce mistakes in the transmission of information from DNA to protein, compensating for the effect of the mistake encoded in the mutant DNA. Such a mechanism is analogous to that of "informational suppression" (Gorini 1970), by which the effect of a mutation persists at the level of the gene but is suppressed at the level of translation. However, informational suppression is found to be the consequence of a mutation in one of the molecules involved in the translation process (most commonly in a tRNA), whereas suppression in the case of CSD mutants appeared to depend on the environment, i.e., the presence or absence of streptomycin in the growth medium. We have suggested the designation of "phenotypic suppression" for this streptomycin effect.

If streptomycin produces phenotypic suppression by inducing mistakes in translation, then the drug could be expected to induce mistakes also in cell-free protein synthesis directed by defined messengers. Indeed, when tested (Davies, Gilbert and Gorini 1964), it was found that streptomycin and other amino-glucoside antibiotics, in addition to inhibiting incorporation of the correct amino acid coded for by a synthetic homopolymer, caused extensive incorporation of other incorrect amino acids not usually coded for by that polymer. This phenomenon was called "misreading" and was found to be relatively specific in that with any given homopolymer, streptomycin causes misincorporation of only a small number of amino acid species. Further studies (Davies, Gorini and Davis 1965) of this specificity have revealed some precise rules. For instance, UUU, the codon for phenylalanine, can only be misread as isoleucine, serine, tyrosine, or leucine. From this it may be concluded that (1) streptomycin can cause misreading of only one base at a time; (2) the misread base is in the 5'-terminal or the internal position in the triplet; and (3) U residues are misread as either C or A. Furthermore, by using defined heteropolymers, it was shown (Davies, Jones and Khorana 1966) that misreading of individual triplets may be influenced by the nature of the neighboring bases. This neighboring-base effect decreases the misreading possibilities, since a number of coding triplets are not misread except in a given reading context. By studying the poly(U)-directed incorporation of either phenylalanine (the correct amino acid), isoleucine (one of the misread amino acids), or arginine (which is not misread and is used as a negative control) when only one of these amino acids at a time is supplied to the incorporating system, one can detect an ambiguity which is intrinsic to the code itself (Gorini 1967). Under these conditions, either phenylalanine or isoleucine are equally well incorporated, but not arginine. However, when the competition between the different aminoacyl-tRNA molecules is reestablished by adding all the amino acids to the incorporating system, then only phenylalanine is incorporated, unless streptomycin is added, in which case isoleucine is again misincorporated, but not arginine. A simple interpretation of these results is that streptomycin magnifies an ambiguity which is intrinsic to the code and which follows precise rules, of which the "wobbling" (Crick 1966) at the 3' position is only one possibility. There is no experimental evidence that streptomycin induces new misreadings of its own.

Misreading and phenotypic suppression are influenced by mutations at the site (strA) controlling sensitivity to streptomycin. This was easy to demonstrate for misreading because the effect of streptomycin can be tested in vitro (Davies, Gilbert and Gorini 1964) without interference by the killing effect of the drug. For phenotypic suppression, which is performed in vivo, sublethal doses of streptomycin must be used when testing streptomycin-sensitive cells (Gorini and Kataja 1965). Both phenotypic suppression and misreading show maximal drug responses in streptomycin-sensitive wild-type cells. The presence of a strA mutation in the strain leads not only to streptomycin resistance, but also to a simultaneous restriction of streptomycin-induced misreading and of streptomycin-induced suppression. Restriction of misreading is the more drastic effect; in fact, no measurable misreading is found with any of a number of streptomycin-resistant mutants tested. The restriction of phenotypic suppression is less drastic since suppression is lowered to different extents in different streptomycin-resistant mutants. Thus several classes of streptomycin-resistant mutants could be distinguished on the basis of the amount of streptomycin-induced suppression they allow (Breckenridge and Gorini 1970; Gorini and Kataja 1964a; Strigini and Gorini 1970). It is known (Ozaki, Mizushima and Nomura 1969) that the streptomycin sensitivity site strA defines protein S12 of the 30S ribosomal subunit. Hence, on studying misreading in an in vitro system reconstituted from mixed components of streptomycin-sensitive and resistant cells, it was found that the restriction of misreading was a property exclusively associated with only the 30S subunit component from the resistant cells.

The findings described above furnished the most convincing evidence that the ribosome intervenes directly in the decoding process by maintaining and perhaps influencing fidelity of translation. Our studies also suggested a possible explanation for the bactericidal action of streptomycin. In the following sections these two questions will be analyzed separately.

Role of the Ribosome in Translational Fidelity

The transfer of information from DNA to protein is based on the mechanism of pairing between complementary bases, i.e., A (adenine) pairs with T (thymine) or U (uracil), and G (guanine) pairs with C (cytosine). Other decoding possibilities do however exist, especially at the level of codon-anticodon pairing, because of the "triplet" nature of the interaction between the mRNA and the tRNA. These possibilities can be most easily analyzed in the case of mutations giving rise to a nonsense codon in the translatable section of a gene. Usually such a mutation leads to premature chain termination and the production of a truncated peptide because no tRNA in the cell has an anticodon matching the nonsense codon. It was found (Gorini, Jacoby and Breckenridge 1966), however, that nonsense mutants can have a "leaky" phenotype, indicating that translation has some chance to continue through the nonsense block with production of a few completed molecules of enzyme. This "translational leakiness" should not be confused with "structural leakiness" due to a normal amount of an altered enzyme with low activity resulting from a missense mutation. Translational leakiness can only be accounted for by assuming that nonsense codons are actually read with low efficiency by some of the tRNAs present in the cell. Several observations about misreading in vitro and in vivo support this assumption:

(1) In vitro the isolated nonsense codon UGA (the most "leaky" of the three nonsenses) is found to bind six different aminoacyl-tRNAs (Söll et al. 1965), albeit with low efficiency.

(2) When the phenylalanine incorporation directed by poly(U) is studied, it can be shown that a defined set of amino acids other than phenylalanine can be misincorporated (Anderson, Gorini and Breckenridge 1965). This misreading occurs even without the addition of streptomycin, although at a much lower level (see preceding section).

(3) Misreading has been studied in vivo (Strigini and Brickman 1973) by checking the effect of streptomycin on the ability of any one nonsense suppressor tRNA to suppress the other two nonsense codons. To

avoid any side effect, a set of mutants was derived from a single parent in a strictly comparable way, the nonsense mutations being carried at the same genetic site. The following precise pattern was found:

(a) The UAG suppressor is able to misread the UAA codon at a low but significant level, whereas the UGA codon is not misread at all;

(b) the UAA suppressor reads UAG very well, as expected by wobble (Crick 1966), but also misreads UGA at a low level;

(c) some misreading of the UAA codon, but not the UAG codon, is observed with the UGA suppressor.

Streptomycin magnifies these misreadings several times but does not induce any new ones, a situation identical to that found in vitro. An even more significant analogy between misreading in vitro and in vivo is drawn from the fact that in both cases misreading follows identical rules, i.e., misreading of only one base occurs at a time and only at the 5′ and central positions (see preceding section).

All the above observations suggest as a reasonable hypothesis that an appreciable amount of ambiguity, intrinsic to the codon-anticodon pairing, does indeed exist. The term "ambiguity" does not, however, represent a completely random phenomenon but signifies translation by a defined number of alternative decoding possibilities at a level very low in comparison to that reached by normal reading (of the order of 0.1%).

In addition to being translated at low efficiency by wild-type tRNAs pairing ambiguously, a nonsense codon can be read specifically and with high efficiency by a mutated tRNA whose anticoding specificity has changed by mutation. These mutated tRNAs are called informational suppressors (Gorini 1970) and were assumed initially to always carry a mutated anticodon specifically matching by conventional base-pairing the codon they suppress. From recent work, however, it has become clear that such a mutation does not necessarily occur in the anticodon. Indeed, a well-characterized UGA suppressor was shown (Hirsh 1971; Hirsh and Gold 1971) to be a *trp*-tRNA with a normal *trp* anticodon but carrying a mutation in another site of the tRNA molecule. This suppressor does not carry the UGA anticodon. Nevertheless it is specific for UGA, and its efficiency of suppression is as high as that of well-characterized UAG suppressor *su*3 (Goodman et al. 1968) which does carry the UAG anticodon. This stresses the role of the entire tRNA molecule in establishing anticodon specificity for a given codon since it appears that a "correct" anticodon obtained by mutation can be out of context in the resulting tRNA suppressor molecule. Its ability to bind the codon can therefore be equal or even worse than that of a tRNA which retains an "incorrect" anticodon but has mutated to alter the molecular context of the anticodon triplet. A clear-cut distinction between suppression by tRNA suppressors and translational leakiness, based on conventional versus unconventional base-pairing, is not tenable anymore. The mechanism underlying both phenomena is the same, namely, the availability of alternate decoding possibilities with widely different strengths of codon-anticodon binding.

We are now in a position to describe more correctly the conditional streptomycin-dependent (CSD) phenotype. The defect carried by the CSD mutants was found (Gorini, Jacoby and Breckenridge 1966) to be a nonsense mutation which is leaky in the $strA^+$ parent but appears negative in its $strA$ mutated derivative. This is because the ribosomal $strA$ mutation restricts translational leakiness. Streptomycin antagonizes this ribosomal restriction and the leakiness appears again. The antagonistic effect of streptomycin varies in degree and is sometimes not detectable at all, depending on degree (and reversibility) of restriction exerted by different $strA$ mutations. As expected, since translational leakiness is basically just a weak suppression, $strA$ mutations also restrict translation by nonsense and missense tRNA suppressors (Biswas and Gorini 1972a). Streptomycin also antagonizes this effect. The same quantitative relationship between differently restrictive $strA$ mutations holds for both leakiness and suppression. A unitary interpretation of these phenomena is that the ribosome interferes (through its $strA$ component) with the translation efficiency of at least some tRNAs and that streptomycin antagonizes this ribosomal interference. Such a model poses two questions. One concerns the site of action of streptomycin since its antagonism to $strA$ does not necessarily indicate a direct action of the drug at the $strA$ site or even on the ribosome. The second concerns the need for the ribosome to discriminate tRNAs, since restriction does not generally affect cellular growth and therefore does not seem to affect all tRNAs to the same extent.

Very strong evidence for streptomycin's exerting its antagonistic effect directly on the ribosome came from the isolation (Rosset and Gorini 1969) of the ribosomal mutation *ram,* which alters protein S4 of the 30S subunit (Zimmermann, Garvin and Gorini 1971). This mutation was shown to affect translational leakiness and tRNA suppression in the same way as the addition of streptomycin. It has even been found (Biswas and Gorini 1972a) that the presence of a *ram* mutation in the genome plus the addition of streptomycin to the growth medium have an additive effect in antagonizing restriction. Nevertheless, *ram* and streptomycin are not completely equivalent, since *ram* is not a lethal mutation, even in the $strA^+$ context. This point will be elaborated in the next section. The *ram* mutants were selected by looking for a secondary mutation that would restore translational leakiness in a restrictive $strA$ genetic context. The analogy between *ram* and the misreading effect of streptomycin holds in all situations tested and in particular in CSD ($strA40$) and in streptomycin-dependent ($strA^D$) mutants. In fact, the double mutants $strA40$ *ram*1 (Rosset and Gorini 1969) and $strA^D$ *ram*1 (Bjare and Gorini 1971) do not require streptomycin for leakiness or growth, respectively, whereas their parents strA40 ram+ and strAᴰ ram+ do. As ram is a ribosomal mutation, we conclude that streptomycin acts directly on the ribosome. The fact that an altered protein S4 influences the function of protein S12 (or vice versa) also indicates a strong interaction of the ribosomal components.

The postulated (Gorini 1971) role of the ribosome in discriminating between wild-type and suppressor tRNAs can be extended to include discrimination between a wild-type tRNA in a normal translation situation and that same tRNA in a misreading situation. The hypothesis that discrimination only affects tRNAs exhibiting a structural peculiarity as a consequence of a mutation seems inadequate; the alternative suggestion that the discriminating factor is the strength of codon-anticodon binding fits all the data better. In fact, as pointed out earlier, misreading by ambiguity and tRNA suppression represent a spectrum of alternative decoding possibilities with increasing strength in the codon-anticodon binding. We suggest that the tRNA-ribosomal binding has some weak but definite capacity to interfere with codon-anticodon binding and that this can be increased or decreased by ribosomal mutations. Actually the mechanism by which only one tRNA, out of several which bind the ribosome, reaches the codon position can be visualized as a ribosomal ability to prevent incoming tRNA from lining up properly with the codon unless the attractive forces of complementary base-pairing between codon and anticodon are sufficiently strong. In general, this requires that all three bases of the codon complement those of the anticodon. However, this complete complementation does not occur in translational leakiness or in suppression by tRNA mutated outside the anticodon. On the other hand, the strength of codon-anticodon attraction for suppressors mutated in the anticodon may be lower than that normally found in perfectly matching triplets if the resulting mutant anticodon is out of context with the rest of the tRNA molecule. In a restrictive strA ribosome, the screening force is much stronger than that in the wild-type strA+ and magnifies the difference between weakly binding and strongly binding tRNAs. This again suggests a discriminatory power of ribosomes.

The presence of a screening force in the strA+ wild-type ribosome can be best demonstrated by the effect of the mutation ram: in ram mutants most, if not all, of the ribosomal screen is abolished so that weakly binding tRNA can now bind indiscriminately and can burden the cell with faulty proteins (Biswas and Gorini 1972a). This situation is evident in strains carrying missense suppressors because they tend to insert an amino acid which is "good" at the mutated codon site but is "wrong" wherever else that same codon is present. This should cause generalized mistranslation which is lethal. However, in cells with wild-type ribosomes the generalized mistranslation is kept under control, since the growth rate even in strains carrying two missense suppressors is close to normal. It might be that those nonlethal missense suppressors one is able to select affect only those tRNAs whose specificity includes the bases surrounding the codon, i.e., the mRNA reading context. The influence of the reading context on the efficiency of suppression has been described already (Salser 1969; Strigini and Gorini 1970). It is thus conceivable that nonlethal missense suppressors bind preferentially only to mutated codons which are now "out of context" and are therefore not found elsewhere in the cell's genetic material in an efficiently recognizable form. On the other hand, it was found (Biswas and Gorini 1972a) that by introducing a *ram* mutation into a missense suppressor strain, growth becomes severely inhibited and production of substantial amounts of faulty proteins can be demonstrated by determination of cross-reacting material (CRM) replacing active enzymes. It was also shown that by superimposing a restrictive *str*A mutation on the *ram* mutant parent, mistranslation is lowered back to the barely detectable level exhibited by wild type. In spite of the supposed extension of specificity of missense suppressors to the reading context, a strain carrying such suppressors is very sensitive to *ram* and *str*A ribosomal mutations because the possibilities for mistranslation appear to be greatly extended if they are not kept under ribosomal discriminative control.

It is known that by exposure to a high concentration of streptomycin (500 μg/ml), a gamut of resistant mutants is obtained distinguishable either by their indifference to, growth-stimulation by, or dependence on streptomycin. The mutations are all alleles of the *str*A gene, as demonstrated for several of them genetically (Breckenridge and Gorini 1970; Momose and Gorini 1971) and biochemically (Funatsu and Wittmann 1972). These alleles impose different degrees of restriction. It is tempting to suggest that their behavior toward streptomycin is related to the degree of selectivity of their ribosomal screen (Gorini, Rosset and Zimmermann 1967; Gorini 1971). At one end of the scale are the streptomycin-indifferent mutants in which the screen is discriminatory against alternative decoding possibilities but does not disturb normal decoding; at the other end are the streptomycin-dependent mutants in which the screen is so selective that even some normal tRNAs are no longer allowed to read their codons, so that growth does not occur unless streptomycin is present to reverse the restriction. This hypothesis for streptomycin dependence is supported by the following facts:

(1) The streptomycin requirement for growth can be satisfied by other aminoglycosides, like paromomycin (Pm), or by ethyl alcohol (Eth), whose only known mode of action linked to streptomycin is their ability to enhance misreading (Gorini, Rosset and Zimmermann 1967).

(2) Four classes of dependent mutants were selected from *E. coli* strain B: Sm, Pm, or Eth dependent; Sm or Eth dependent; Pm dependent;

and Sm dependent (Momose and Gorini 1971). All were found to be *str*A restrictive mutants (Momose and Gorini 1971).

(3) In all of the above classes of dependent mutants the introduction of a *ram* mutation eliminates the drug dependence (Bjare and Gorini 1971; Dabbs and Gorini, unpublished data).

These facts are consistent with the idea that the growth of these strains only requires a factor which will reverse their restriction, irrespective of how this is achieved.

Misreading and Bactericidal Action

The most dramatic effect of streptomycin is the rapid killing of a large spectrum of bacteria. When the misreading effect of the drug was also discovered, it was hinted that perhaps misreading might cause killing. Up to now, however, the question as to how these two effects are linked is still unsettled. In *E. coli* at least, killing must always be accompanied by misreading in the sense that all strains sensitive to killing contain ribosomes which misread extensively when tested against streptomycin in vitro. By testing different aminoglycoside antibiotics inducing misreading, it was found (Davies, Gilbert and Gorini 1964) that in mutants resistant to one drug, misreading persists for all other drugs to which the mutant is still sensitive. This is clearly illustrated by the behavior of a peculiar class of mutants that is resistant to streptomycin and paromomycin if the drugs are used separately but sensitive to killing by either drug if the cells are previously grown in the presence of the other. This property, called "phenotypic masking" (Gorini, Rosset and Zimmermann 1967), is intriguing in itself and will be discussed later. However, it is pertinent to note here that the ribosomes extracted from cells of a phenotypically maskable mutant grown either in streptomycin or in paromomycin are stimulated to misread in vitro either by paromomycin or streptomycin, respectively, exactly paralleling the shift in sensitivity to killing (Zimmermann, Rosset and Gorini 1971).

By contrast, misreading is not necessarily accompanied by killing. In fact, streptomycin-induced misreading can cause growth inhibition in vivo under conditions where killing does not occur. The mechanism of this inhibition is as follows: An increase in the level of misreading, caused by the presence of streptomycin, leads to increased synthesis of faulty proteins in the cell (Gorini and Kataja 1964b). The presence of faulty proteins can either limit or completely inhibit growth. The concentration of streptomycin used can determine the extent of this misreading since the CSD effect is streptomycin-concentration dependent (Gorini and Kataja 1965). Thus in cells particularly sensitive to the increase in the level of misreading, a drug concentration can be reached at which growth is completely inhibited.

The strain is said to be "sensitive," although this has a meaning quite different from that currently used, i.e., this sensitivity, in contrast to killing, is fully reversible when the drug is removed. Such an effect is especially noticeable in *ram* mutant strains which already have a level of misreading higher than the wild-type, even in the absence of streptomycin. The following relevant situations will be analyzed:

(1) *E. coli* strain B wild-type (*str*A$^+$) is killed when the concentration of streptomycin reaches 5 μg/ml, irrespective of the presence of a *ram* mutation. It was shown (Rosset and Gorini 1969), however, that below that concentration streptomycin inhibits growth of the *ram* mutant (slow growth with 1 μg and no growth at all with 2 μg of drug per ml), whereas the growth of the *ram*$^+$ parent is unaffected by these sublethal drug concentrations.

(2) Streptomycin-resistant mutants carrying *str*A40 (a moderately restrictive *str*A allele) are unaffected by any drug concentration tested, i.e., up to 2000 μg/ml. The growth of their *ram*1 derivative, however, is completely inhibited when the concentration of streptomycin reaches more than 40 μg/ml. In contrast, the growth of the *ram*1 derivative of a streptomycin-resistant mutant carrying *str*A1 (a highly restrictive *str*A allele) remains unaffected by any drug concentration tested, i.e., up to 1000 μg/ml (Rosset and Gorini 1969).

(3) Even in the presence of the highly restrictive *str*A1 allele, growth is inhibited by relatively low concentrations of streptomycin if the strain carries a missense suppressor in addition to *ram*1 (Biswas and Gorini 1972a). For this effect the minimal inhibitory concentration of the drug varies from 50 to 200 μg/ml depending on the missense suppressor carried by the strain. This is consistent with the idea that although streptomycin's reversal of *str*A1 restriction is so small that it is generally not noticeable in other systems, it is, however, large enough to produce a visible effect in the *ram*1-missense suppressor system.

In conclusion, the examples discussed above show that the effects of *ram* and of streptomycin are cumulative and that the resulting growth inhibition only reflects a magnification of the level of misreading beyond the limit acceptable by the cell.

The above analysis and several other observations listed below indicate that the killing effect of streptomycin cannot be seen as an extrapolation of the misreading effect but seems to be a different aspect of streptomycin action. The killing action of streptomycin requires cellular growth (no killing is seen with resting cells) and is very fast; the viable count drops by a factor of 10^5 in less than 5 minutes. Accumulation of faulty proteins could eventually become irreversible but not that fast. Furthermore, in *str*A$^+$ cells the threshold drug concentration for killing is the same for the *ram*$^+$ parent as for the *ram*1 mutant, although the mutant produces more misreading as indicated by its 20% slower rate of growth. Note also that when the *ram* mutation is in a suitable genetic context, such as a *str*A$^+$ strain carrying missense suppressors, high levels of misreading [up to 60% of β-galactosidase CRM (Biswas and Gorini 1972a)] are seen and the growth rate slows drastically; yet the situation is not lethal. Finally, in a partial diploid *str*A$^+$/*str*A1 the growth-inhibition effect of streptomycin is dominant, i.e., growth stops in the presence of streptomycin, but the killing effect is not; after washing out streptomycin, all cells appear viable (Breckenridge and Gorini 1969). This result is difficult to interpret without attributing two actions to streptomycin: one accounting for irreversible inactivation of sensitive ribosomes (a *str*A$^+$/*str*A$^+$ diploid is killed as expected) and a second accounting for reversible inhibition of resistant ribosomes in the presence of inactivated sensitive ribosomes (a *str*A1/*str*A1 diploid is not inhibited).

If killing is not a consequence of misreading, then a different mechanism for killing should be sought. Assuming that this effect is also at the ribosomal level (as misreading is), a quick and irreversible effect of streptomycin on ribosomal activity in vitro has been looked for. The experimental results obtained by different investigators are diverse and point in two directions, i.e., prevention of formation of the initiation complex (Luzzatto, Apirion and Schlessinger 1968) or stoppage of chain elongation (Modolell and Davis 1969). A thorough analysis of these lines of research is outside the purpose of this paper which deals with misreading, an event which unequivocally operates at the chain elongation step (leakiness occurs at any site inside a cistron). One point, however, seems worth analyzing further here, namely, whether streptomycin acts on the same ribosomal site for misreading as for killing. In spite of the superficial evidence that mutations at the *str*A site control both misreading and killing, a dual role for this site is not necessarily inferred. Concerning misreading, evidence is presented in the preceding section indicating that protein S12, the product of the *str*A gene, is responsible for ribosomal restriction. Streptomycin antagonizes restriction, but given the extensive interaction between ribosomal components, this does not necessarily imply that the drug acts directly on S12. It has been shown (Ozaki, Mizushima and Nomura 1969), in fact, that isolated S12 protein does not bind streptomycin and moreover that the *ram* mutation, which also antagonizes restriction by S12, is an alteration of protein S4, another component of the 30S subunit.

It is known (Cox, White and Flaks 1964; Davies 1964) that streptomycin binds to intact 30S subunits derived from *str*A$^+$-sensitive cells but not to those derived from *str*A mutated strains. However, it was recently found (Biswas and Gorini 1972b) that the drug binds specifically to naked 16S RNA isolated from 30S ribosomal subunits whatever their origin. Presumably the streptomycin attachment site(s) located on the 16S RNA are masked when S12 is altered, showing that the site where the mutation to antibiotic resistance occurs does not necessarily define the binding site of the drug. It was also found (Biswas and Gorini 1972b) that it is impossible to reconstitute in vitro functional ribosomes from the 16S RNA–streptomycin complex, suggesting that the drug could act on ribosomal assembly in vivo. The following experiment (Garvin, Rosset and Gorini 1973) supported this idea. Cells carrying the *str*A40 allele (moderately restrictive but streptomycin resistant and with 30S subunits unable to bind streptomycin in vitro) were pregrown in the presence of the antibiotic and then, after washing out the streptomycin, the level of misreading in these cells was assayed by measuring their ability to support growth of T4 phages carrying different nonsense mutations. This misreading ability appears maximal after at least four generations of pregrowth in the presence of antibiotic and therefore seems related to the newly synthesized ribosomes. Appropriate controls show that some carry-over of streptomycin to the fresh medium cannot account for the result observed. Ribosomes extracted from these cells, avoiding exposure to high salt concentration, were able to misread in a poly(U) system, but this ability was lost after washing with 1 M NH_4Cl. Presence of streptomycin is not detectable in the unwashed ribosomal preparation or in the high salt wash; therefore, we tentatively suggest that a labile conformational change is induced in the ribosome by streptomycin during pregrowth which is lost in vitro at high salt concentrations. Similarly, ribosomal preparations from phenotypically masked mutants misread either with streptomycin or with paromomycin depending upon pregrowth in either paromomycin or streptomycin, respectively. Again none of the pregrowth antibiotic could be detected as carry over into the in vitro system. Moreover, these ribosomes also lose their misreading ability when washed with 1 M NH_4Cl (Zimmermann, Rosset and Gorini 1971). We conclude that the possibility of streptomycin disrupting ribosomal assembly in vivo is still another aspect of the mode of action of streptomycin.

References

Anderson, W. F., L. Gorini and L. Breckenridge. 1965. Role of ribosomes in streptomycin-activated suppression. *Proc. Nat. Acad. Sci.* **54**:1076.

Biswas, D. K. and L. Gorini. 1972a. Restriction, de-restriction and mistranslation in missense suppression. Ribosomal discrimination of transfer RNAs. *J. Mol. Biol.* **64**:119.

———. 1972b. The attachment site of streptomycin to the 30S ribosomal subunit. *Proc. Nat. Acad. Sci.* **69**:2141.

Bjare, U. and L. Gorini. 1971. Drug dependence reversed by a ribosomal ambiguity mutation, *ram*, in *Escherichia coli. J. Mol. Biol.* **57**:423.

Breckenridge, L. and L. Gorini. 1969. The dominance of streptomycin sensitivity re-examined. *Proc. Nat. Acad. Sci.* **62**:979.

————. 1970. Genetic analysis of streptomycin resistance in *Escherichia coli*. *Genetics* **65**:9.

Cox, E. C., J. R. White and J. G. Flaks. 1964. Streptomycin action and the ribosome. *Proc. Nat. Acad. Sci.* **51**:703.

Crick, F. H. C. 1966. Codon-anticodon pairing: The wobble hypothesis. *J. Mol. Biol.* **19**:548.

Davies, J. 1964. Studies on the ribosomes of streptomycin-sensitive and re-sistant strains of *Escherichia coli*. *Proc. Nat. Acad. Sci.* **51**:659.

Davies, J., W. Gilbert and L. Gorini. 1964. Streptomycin, suppression, and the code. *Proc. Nat. Acad. Sci.* **51**:883.

Davies, J., L. Gorini and B. D. Davis. 1965. Misreading of RNA codewords induced by aminoglycoside antibiotics. *Mol. Pharmacol.* **1**:93.

Davies, J., D. S. Jones and H. G. Khorana. 1966. A further study of mis-reading of codons induced by streptomycin and neomycin using ribo-polynucleotides containing two nucleotides in alternating sequence as tem-plates. *J. Mol. Biol.* **18**:48.

Funatsu, G. and H. G. Wittmann. 1972. Ribosomal proteins. XXXIII. Loca-tion of amino acid replacements in protein S12 isolated from *Escherichia coli* mutants resistant to streptomycin. *J. Mol. Biol.* **68**:547.

Garvin, R. T., R. Rosset and L. Gorini. 1973. Ribosomal assembly influenced by growth in the presence of streptomycin. *Proc. Nat. Acad. Sci.* **70**:2762.

Goodman, H. M., J. Abelson, A. Landy, S. Brenner and J. D. Smith. 1968. Amber suppression: A nucleotide change in the anticodon of a tyrosine transfer RNA. *Nature* **217**:1019.

Gorini, L. 1967. Induction of code ambiguity by aminoglycoside antibiotics. *Fed. Proc.* **26**:5.

————. 1970. Informational suppression. *Ann. Rev. Genet.* **4**:107.

————. 1971. Ribosomal discrimination of tRNAs. *Nature New Biol.* **234**:261.

Gorini, L. and E. Kataja. 1964a. Phenotypic repair by streptomycin of defective genotypes in *E. coli*. *Proc. Nat. Acad. Sci.* **51**:487.

————. 1964b. Streptomycin-induced oversuppression in *E. coli*. *Proc. Nat. Acad. Sci.* **51**:995.

————. 1965. Suppression activated by streptomycin and related antibiotics in drug sensitive strains. *Biochem. Biophys. Res. Comm.* **18**:656.

Gorini, L., W. Gundersen and M. Burger. 1961. Genetics of regulation of enzyme synthesis in the arginine biosynthetic pathway of *Escherichia coli*. *Cold Spring Harbor Symp. Quant. Biol.* **26**:173.

Gorini, L., G. Jacoby and L. Breckenridge. 1966. Ribosomal ambiguity. *Cold Spring Harbor Symp. Quant. Biol.* **31**:657.

Gorini, L., R. Rosset and R. A. Zimmermann. 1967. Phenotypic masking and streptomycin dependence. *Science* **157**:1314.

Hirsh, D. 1971. Tryptophan transfer RNA as the UGA suppressor. *J. Mol. Biol.* **58**:439.

Hirsh, D. and L. Gold. 1971. Translation of the UGA triplet *in vitro* by trypto-phan transfer RNA's. *J. Mol. Biol.* **58**:459.

Luzzatto, L., D. Apirion and D. Schlessinger. 1968. Mechanism of action of streptomycin in *E. coli*: Interruption of the ribosome cycle at the initiation of protein synthesis. *Proc. Nat. Acad. Sci.* **60**:873.

Modolell, J. and B. D. Davis. 1969. Mechanism of inhibition of ribosomes by streptomycin. *Nature* **224**:345.

Momose, H. and L. Gorini. 1971. Genetic analysis of streptomycin dependence in *Escherichia coli*. *Genetics* **67**:19.

Ozaki, M., S. Mizushima and M. Nomura. 1969. Identification and functional characterization of the protein controlled by the streptomycin-resistant locus in *E. coli*. *Nature* **222**:333.

Rosset, R. and L. Gorini. 1969. A ribosomal ambiguity mutation. *J. Mol. Biol.* **39**:95.

Salser, W. 1969. The influence of the reading context upon the suppression of nonsense codons. *Mol. Gen. Genet.* **105**:125.

Söll, D., E. Ohtsuka, D. S. Jones, R. Lohrmann, H. Hayatsu, S. Nishimura and H. G. Khorana. 1965. Studies on polynucleotides. XLIX. Stimulation of the binding of aminoacyl-sRNAs to ribosomes by ribotrinucleotides and a survey of codon assignments for 20 amino acids. *Proc. Nat. Acad. Sci.* **54**:1378.

Strigini, P. and E. Brickman. 1973. Analysis of specific misreading in *Escherichia coli*. *J. Mol. Biol.* **75**:659.

Strigini, P. and L. Gorini. 1970. Ribosomal mutations affecting efficiency of amber suppression. *J. Mol. Biol.* **47**:517.

Zimmermann, R. A., R. T. Garvin and L. Gorini. 1971. Alteration of a 30S ribosomal protein accompanying the *ram* mutation in *Escherichia coli*. *Proc. Nat. Acad. Sci.* **68**:2263.

Zimmermann, R. A., R. Rosset and L. Gorini. 1971. Nature of phenotypic masking exhibited by drug-dependent streptomycin A mutants of *Escherichia coli*. *J. Mol. Biol.* **57**:403.

Effects of Colicin E3
on Bacterial Ribosomes

M. Nomura and J. Sidikaro
Institute for Enzyme Research
University of Wisconsin
Madison, Wisconsin 53706

K. Jakes and N. Zinder
The Rockefeller University
New York, New York 10021

INTRODUCTION

Studies of the relation of the structure to function of ribosomes can be approached in several ways. For example, the role of the various components of ribosomes can be directly analyzed by reconstitution experiments (see Nomura and Held, this volume). Another approach is modification of these components in situ by the use of specific reagents. In this article, the inactivation of ribosomes by colicin E3 will be discussed in the light of our developing understanding of its specific mode of action.

COLICIN E3

The bacteriocins are bactericidal substances which are synthesized by certain strains of bacteria and have killing activity against some other strains of the same or closely related species. Colicins are one class of such bacteriocins. Several colicins have been studied with respect to their chemical nature, genetic determinants, and mode of action. Colicins are proteins and are produced by cells which carry a plasmid called colicinogenic factor or Col factor. The presence of a Col factor in a cell confers on it both the ability to produce the colicin as well as immunity against the corresponding colicin (Fredericq 1958). In killing sensitive cells, colicins first adsorb to specific receptors on the cell surface. Colicin E3 shares its adsorption specificity with colicins E1 and E2 and with bacteriophage BF23. Cells which lack such receptors are resistant to the colicins. For example, most of the mutants of *E. coli* that are resistant to BF23 are also resistant to colicins E1, E2, and E3; the resistance is due to failure of the colicins or phage to adsorb to the cells (Fredericq 1958; Nomura 1967).

The specific biochemical events that take place after the adsorption of the various colicins to sensitive cells differ depending on the type of colicin (Nomura 1963). As will be described below, ribosome inactivation takes place in the case of colicin E3. (For reviews of earlier work on colicins in general, see Fredericq 1958; Nomura 1967; for more recent developments, see Hager 1973).

Colicin E3 is a protein with a molecular weight of 60,000 (Herschman and Helinski 1967; Glick et al. 1972). The genetic determinant for E3, Col E3 factor, is a small double-stranded circular DNA with a molecular weight of about 5×10^6 (Roth and Helinski 1967; Inselburg 1973).

Mode of Action of Colicin E3: In Vivo Studies

Colicin E3 causes specific inhibition of protein synthesis in treated cells without affecting RNA and DNA synthesis (Nomura 1963; Nomura and Maeda 1965). In this respect the mode of action of E3 is different from other related colicins, such as E1 and E2. The inhibition of protein synthesis by E3 is not due to the inhibition of mRNA synthesis (Nomura and Maeda 1965). Respiration and energy-producing metabolism also continue without inhibition.

Ribosomes from E3-treated cells have very little activity in in vitro polypeptide synthesis (Konisky and Nomura 1967). The supernatant from E3-treated cells functions normally. Furthermore, it was found that the 30S subunits are inactive; the 50S subunits retain activity. The altered molecular component responsible for the inactivity of the 30S subunits was identified as the 16S RNA using the reconstitution technique (Bowman et al. 1971). The 30S ribosomal proteins from inactive subunits are active in the reconstitution of 30S subunits. No alteration in ribosomal protein composition has been detected (Senior and Holland 1971; Bowman et al. 1971). A chemical alteration of the 16S RNA from E3-inactivated 30S subunits was found by Senior and Holland (1971) and by Bowman et al. (1971). Inactive RNA ("E3–16S RNA") sediments somewhat more slowly in sucrose gradient centrifugation (Senior and Holland 1971) and migrates somewhat faster on polyacrylamide gels (Bowman et al. 1971) than control 16S RNA. This difference is caused by the loss of a 50-nucleotide fragment (the "E3 fragment," Bowman et al. 1971) from the 3' terminus of the 16S RNA molecule. The sequence of the E3 fragment was studied using preparations obtained both in vivo and in vitro (Bowman et al. 1971; Santer and Santer 1972; see also Fellner, this volume; Zimmermann, this volume). The most likely sequence of the E3 fragment is shown in Figure 1.

E3–16S RNA, although missing about 50 nucleotides at the 3' end, does not have any hidden breaks in the molecule. The size of the molecule does not appear to change even after destruction of secondary structure by denaturing agents. All of the oligonucleotides found to be missing from E3–16S RNA are present in the E3 fragment. These results strongly indicate that E3 causes a single (or a very few) nucleolytic cleavage (Bowman et al. 1971; Bowman, Sidikaro and Nomura 1973).

The question of whether all the E3 fragments are retained by the inactivated 30S subunits in vivo is not completely settled. Earlier experiments showed that some E3 fragment is found in the soluble protein fraction as

GUCGmUAACAAGGUAACCGUAG$_{3-5}$m$^6_2$Am$^6_2$ACCUGCGGUUGGAUC(CUCACUUC)A$_{OH}$

Figure 1   The most probable sequence of the E3 fragment. The sequence is based on the data obtained from Bowman et al. (1971), Bowman (1972b), Santer and Santer (1972) and Ehresmann et al. (1972). Recent work by Shrine and Dalgarno (1974) has shown that the 3′-terminal sequence of *E. coli* 16S RNA is (pyd)-ACCUCCUUA$_{OH}$. They suggested that the sequence ACCUCC could recognize a conserved sequence found in the ribosome binding sites of various coliphage mRNAs.

well as the ribosome fraction (Bowman et al. 1971). However, Samson, Senior and Holland (1972) have reported that all of the E3 fragment remains associated with ribosomes (see also the discussion of in vitro experiments below). It is likely that inactivation of 30S subunits is due to the cleavage of 16S RNA rather than the loss of the fragment. The loss may be a secondary step which takes place more slowly after the cleavage reaction. In this connection, it should be pointed out that experiments done with E3-inactivated 30S subunits might give different results depending on the degree of loss of the E3 fragment. The kinetics of cleavage of 16S RNA in vivo parallels the kinetics of inhibition of protein synthesis (Samson, Senior and Holland 1972) and indicates that this specific cleavage is the primary effect of colicin E3 in vivo.

The next question then, is, which step in protein synthesis in vivo is inhibited by this cleavage reaction. Earlier experiments showed that ribosomes isolated from E3-treated cells are inactive in poly(U)-dependent Phe-tRNA binding (Konisky and Nomura 1967) and in poly(AUG)-dependent fMet-tRNA binding (Konisky 1968). Senior, Kwasniak and Holland (1970) followed the fate of pulse-labeled nascent proteins on polysomes during partial inhibition of protein synthesis by E3. Their results clearly indicate that a step in chain elongation or termination is blocked. However, the possibility that E3-treated cells may also be defective in initiation was not ruled out. The fMet-tRNA binding experiments mentioned above and the in vitro studies (Tai and Davis 1974) described below suggest that in addition to an elongation step, an initiation step is also blocked.

In contrast to the finding by Senior, Kwasniak and Holland (1970), who reported increasing instability of polysomes, Dahlberg (personal communication) has found that polysomes are initially stabilized. Stabilization of polysomes after E3 treatment has also been observed by Tai and Davis (1974) in in vitro studies.

Inactivation of Ribosomes by E3 In Vitro

Although early experiments failed to show ribosome inactivation by E3 in vitro (Konisky and Nomura 1967), it is now clear that this failure was due to the use of crude E3 preparations containing both E3 and the E3 immunity substance. Incubation of 70S ribosomes from *E. coli* with purified E3 results in inactivation of the ribosomes (Boon 1971; Bowman, Sidikaro and Nomura 1971). The inactivation is caused by cleavage of the 16S RNA in a manner similar to inactivation in vivo. The in vitro cleavage of the 16S RNA leads to recovery of the "E3 fragment" with the 30S subunit in nearly theoretical yield (Boon 1971); therefore the in vitro results are in agreement with the in vivo findings of Samson, Senior and Holland (1972). The E3 fragment produced in vitro is identical to the one produced in vivo. This was shown by isolating both fragments and comparing oligonucleotide patterns obtained after RNase T_1 digestion (Bowman, Sidikaro and Nomura 1971).

E3 acts catalytically in vitro (Boon 1971; Bowman, Sidikaro and Nomura 1971). In this respect E3 behaves like an enzyme. However, E3 has stringent requirements for its action. 16S RNA itself is not cleaved by E3 (Boon 1971; Bowman et al. 1971). The simultaneous presence of both 30S and 50S subunits is needed for E3-induced cleavage (Boon 1972; Bowman 1972a). The cleavage does not occur if 30S subunits are incubated with E3, separated from the colicin (or the colicin is inactivated by antibody), and then the 30S subunits are mixed with 50S subunits. Therefore the colicin does not induce a stable change in the 30S subunit, such as activation of a masked RNase associated with the ribosomes, which can later be activated by the addition of the 50S subunit. The reverse possibility was also eliminated by the experiments establishing the requirement for the simultaneous presence of both 30S and 50S subunits (Boon 1972; Bowman 1972a).

The activity of E3 in vitro is strongly dependent on the Mg^{++} concentration; this dependence is similar to that of the activity of the ribosomes in an f2-RNA-directed polypeptide synthesizing system (Jakes and Zinder 1974).

The last two observations, i.e., the requirement for both 30S and 50S subunits and the Mg^{++} dependence, indicate that the formation of 70S ribosomes is probably required and that there is a certain structure of 70S ribosomes which is favorable as a "substrate" for E3. The observation that unfolded ribosomes are resistant to the E3 action (Meyhack, Meyhack and Apirion 1973) is also consistent with this view.

There are several other indications that the structure of the ribosomes is important for E3 action. First, when E3-induced inactivation of ribosomes was studied in a crude cell-free extract suitable for protein synthesis, it was found that the presence of f2-RNA stimulates the rate and the extent of inactivation of ribosomes. Moreover, message-like nucleic acids containing natural start signals, such as phage f1 single-stranded DNA, which form initiation complexes with ribosomes and fMet-tRNA but do not direct the synthesis of polypeptides in this crude cell-free system, also stimulated the inactivation of ribosomes by E3 (Jakes and Zinder, manuscript in preparation). Poly(U), which lacks natural initiation signals, did not stimulate ribosome inactivation by E3 in this system. Endogenous message in these crude extracts, however, is not responsible for the ribosome inactivation seen in the absence of any added mRNA. Ribosomes in extracts prepared from rifampicin-treated cells are still sensitive to E3, although there is no measurable endogenous message on those ribosomes (Jakes and Zinder, manuscript in preparation). Also since 30S ribosomes prepared from individually purified 30S proteins and 16S RNA can be inactivated by E3 in the presence of highly purified 50S subunits (Sidikaro and Nomura, unpublished experiments), the presence of mRNA is probably not obligatory for inactivation. Thus the above observations may indicate that under the conditions mentioned, the presence of f2-RNA or other nucleic acids containing natural initiation signals increases the number of ribosomes with a structure suitable for E3-induced cleavage.

The second indication of the importance of ribosome structure for E3 action is the finding that streptomycin, gentamycin and tetracyclin (but not erythromycin and kasugamycin) protect ribosomes against E3-induced cleavage of 16S RNA both in vivo and in vitro (Dahlberg et al. 1973). Similarly, it was found that poly(U)-directed binding of Phe-tRNA to the A site, and probably to the P site, protects ribosomes from E3-induced inactivation (Kaufmann and Zamir 1973). It is probably unlikely that bindings of these antibiotics (streptomycin, gentamycin and tetracyclin) and Phe-tRNA all take place at the same site. In addition, ribosomes participating in chain elongation, including the poly(U)-dependent polyphenylalanine synthesizing system, are susceptible to E3-induced ribosome inactivation (Tai and Davis 1974; A. Zamir, personal communication; A. Dahlberg, personal communication). Thus it would be difficult to explain the observed protection of free ribosomes by the above-mentioned antibiotics or Phe-tRNA, together with poly(U), on the basis of steric interference with E3 action. Rather, it is more likely that the binding of antibiotics or Phe-tRNA affects the overall conformation of the ribosome and renders it resistant to the E3 action. However, interference by simple steric hindrance has not been rigorously excluded.

From all the evidence presented above, it is clear that E3 has stringent requirements for its action. Yet it was found that E3 is capable of specific in vitro inactivation of not only *E. coli* ribosomes but also ribosomes from different bacterial species, such as *Bacillus stearothermophilus* and *Azotobacter vinelandii* (Sidikaro and Nomura 1973). These species are resistant to E3 in vivo, probably because they fail to adsorb the colicin. (Ribosomes from E3-resistant mutants of *E. coli* are sensitive to the action of E3 in vitro [Bowman, Sidikaro and Nomura 1971].) The E3 fragment obtained from the 3′ end of *B. stearothermophilus* 16S RNA is slightly different in size from the *E. coli* fragment and is different in sequence, as judged from the difference in oligonucleotide patterns after T_1 digestion (Sidikaro and Nomura 1973). Presumably, the ribosomes from these distantly related organisms have retained a common structural feature required for susceptibility to E3-induced cleavage. Furthermore, preliminary experiments using protein-deficient "30S" particles reconstituted from purified 30S proteins and 16S RNA have shown that some 30S proteins are not required for E3-induced 16S RNA cleavage to take place (Sidikaro and Nomura, unpublished experiments). Further studies may give information regarding minimal structural features required for E3 action.

Immunity Substance

As mentioned above, a substance that inhibits E3 action in vitro has been found in crude E3 preparations as well as in whole cell extracts (Boon 1971; Bowman, Sidikaro and Nomura 1971). This substance has been purified and characterized (Jakes, Zinder and Boon 1974; Sidikaro and Nomura 1974). It is a small protein with a molecular weight of about 10,000 and can be found only in E3-colicinogenic cells. This protein can account for the immunity property of E3-colicinogenic cells, and we call it "E3 immunity substance."

E3 immunity substance inhibits E3-induced ribosome inactivation in vitro. The inhibition is due to the direct interaction of the immunity substance with E3 rather than its interaction with the ribosomes (Jakes, Zinder and Boon 1974; Sidikaro and Nomura 1974). However, the interaction between E3 and the immunity substance observed in vitro is reversible. Completely active E3 can be purified from a mixture of E3 and excess immunity substance which has no activity in ribosome inactivation in vitro. However, E3 immunity substance does not induce any noticeable effect on the killing action of E3 on living sensitive cells.

Purified E3 immunity substance is useful in examining the details of E3-induced ribosome inactivation in vitro. It is also useful in showing that an observed effect in vitro is colicin specific if the effect is prevented by purified E3 immunity substance (e.g., Sidikaro and Nomura 1973).

Use of Colicin E3 to Analyze Ribosome Structure and Function

Studies on the mechanism of E3-induced ribosome inactivation in vitro have already given some useful information about ribosome structure and function. Ribosomes with the defective 16S RNA (cleavage and/or loss of the E3 fragment) can neither carry out polypeptide chain initiation nor chain elongation (Tai and Davis 1974). This is consistent with earlier observations mentioned above that E3-inactivated ribosomes (obtained from cells treated with E3) show a greatly reduced activity in poly(U)-directed Phe-tRNA binding (Konisky and Nomura 1967) and in poly-

(AUG)-directed fMet-tRNA binding (Konisky 1968). However, Tai and Davis (1974) have found that E3-inactivated ribosomes can carry out chain elongation if the Mg^{++} concentration is elevated during the assay of activity, but that the same Mg^{++} concentration increase cannot relieve the defect in chain initiation with natural mRNA. Thus it appears that at high Mg^{++} concentrations, the E3-induced nucleolytic cleavage of 16S RNA affects only the initiation function of the 30S subunits.

However, the identity of the step affected by colicin E3 is not entirely clear. Recently Turnowsky and Högenauer (1973) reported that the ribosomes treated with E3 in vitro have defects in the Phe-tRNA binding function, assayed in the presence of EF-T, but are able to carry out the AUG-dependent fMet-tRNA binding reaction. The results imply that the step affected by colicin E3 is *not* the initiation step, but a step which requires the binding of tRNA to the "acceptor site" on the ribosome. As mentioned above, the earlier experiments indicated that the 30S subunits isolated from E3-treated *E. coli* cells are inactive in the poly(AUG)-directed fMet-tRNA binding reaction (Konisky 1968). Possibly the inactivated ribosomes (purified and assayed as 30S subunits) obtained in the earlier in vivo experiments may have differed from those used by Turnowsky and Högenauer (1973). For example, most of the E3 fragment might have been lost from the 30S subunits inactivated in vivo. Further studies are needed to clarify these discrepancies.

As already described above, E3 requires a specific ribosome structure for its action in vitro. Thus sensitivity to E3 might be used as a probe to detect certain alterations of ribosome structure. The fact that E3 cannot act on free poly(U)–Phe-tRNA–ribosome complexes and yet is able to inhibit chain elongation in poly(U)-directed phenylalanine synthesis (Kaufmann and Zamir 1973; A. Zamir, personal communication) may suggest the presence of a stage in the elongation cycle when ribosomes having poly(U) and Phe-tRNA take E3-sensitive structure.

Another use of E3 is the production of modified RNAs, that is, the E3–16S RNA and the E3 fragment. These RNA molecules have been used for 16S RNA sequence work (Santer and Santer 1972). In addition, E3–16S RNA has been examined for its ability to bind ribosomal proteins. It was found that E3–16S RNA is able to bind all the 30S ribosomal proteins, except S21, and form (inactive) particles which sediment at 30S (Bowman et al. 1971; Bowman 1972b). It should be noted that ribosomes isolated from E3-treated cells retain S21 as well as all other proteins, even though 16S RNA is cleaved, perhaps because most of the E3 fragment is still present in the ribosomes. Although further studies are required, failure of S21 to bind E3–16S RNA-containing particles in reconstitution experiments suggests the importance of the 3' portion of 16S RNA in S21 binding.

Finally, we note that there are some other bacteriocins which have been reported to inhibit protein synthesis without inhibiting the synthesis of other macromolecules, as does E3. Examples are colicin D (Timmis and Hedges 1972) and cloacin DF13 (de Graaf, Planta and Stouthamer 1971). In the latter case, ribosomes isolated from cloacin-treated cells have been reported to be inactive in a cell-free protein synthesizing system. It would be interesting and useful for ribosome research if we could find other bacteriocins which cause different chemical alterations of the ribosome.

Acknowledgments

The work done by M. Nomura and J. Sidikaro was supported in part by grant No. GB-31086X2 from the National Science Foundation and grant No. GM-20427-01 from the National Institutes of Health. The work done by K. Jakes and N. Zinder was supported in part by grant No. GB-29626-38343X from the National Science Foundation.

References

Boon, T. 1971. Inactivation of ribosomes *in vitro* by colicin E3. *Proc. Nat. Acad. Sci.* **68**:2421.

————. 1972. Inactivation of ribosomes *in vitro* by colicin E3 and its mechanism of action. *Proc. Nat. Acad. Sci.* **69**:549.

Bowman, C. M. 1972a. Inactivation of ribosomes by colicin E3 *in vitro*: Requirement for 50S ribosomal subunits. *FEBS Letters* **22**:73.

————. 1972b. Studies on the biochemical action of colicin E3. *Ph.D. thesis*, University of Wisconsin, Madison.

Bowman, C. M., J. Sidikaro and M. Nomura. 1971. Specific inactivation of ribosomes by colicin E3 *in vitro* and mechanism of immunity in colicinogenic cells. *Nature New Biol.* **234**:133.

————. 1973. Mode of action of colicin E3. In *Chemistry and function of colicins* (ed. L. Hager) pp. 87–106. Academic Press, New York.

Bowman, C. M., J. E. Dahlberg, T. Ikemura, J. Konisky and M. Nomura. 1971. Specific inactivation of 16S ribosomal RNA induced by colicin E3 *in vivo*. *Proc. Nat. Acad. Sci.* **68**:964.

Dahlberg, A., E. Lund, N. O. Kjeldgaard, C. M. Bowman and M. Nomura. 1973. Colicin E3-induced cleavage of 16S ribosomal ribonucleic acid: Blocking effects of certain antibiotics. *Biochemistry* **12**:948.

Ehresmann, C., P. Stiegler, P. Fellner and J-P. Ebel. 1972. The determination of the primary structure of the 16S ribosomal RNA of *Escherichia coli*. II. Nucleotide sequences of products from partial enzymatic hydrolysis. *Biochimie* **54**:901.

Fredericq, P. 1958. Colicins and colicinogenic factors. *Symp. Soc. Exp. Biol.* **12**:104.

Glick, J. E., S. J. Kerr, A. M. Gold and D. Shemin. 1972. Multiple forms of colicin E3 from *Escherichia coli* CA-38. *Biochemistry* **11**:1183.

de Graaf, F. K., R. J. Planta and A. H. Stouthamer. 1971. Effect of a bacteriocin produced by *Enterobacter cloacoe* on protein biosynthesis. *Biochim. Biophys. Acta* **240**:122.

Hager, L. P., ed. 1973. *Chemistry and function of colicins.* Academic Press, New York.

Herschman, H. R. and D. R. Helinski. 1967. Purification and characterization of colicin E2 and colicin E3. *J. Biol. Chem.* **242**:5360.

Inselberg, J. 1973. Colicin factor DNA: A single nonhomologous region in Col E2-E3 heteroduplex molecules. *Nature New Biol.* **241**:234.

Jakes, K. and N. D. Zinder. 1974. Stimulation on in vitro colicin E3 action by messenger RNA and message analogs. *J. Biol. Chem.* (in press).

Jakes, K., N. D. Zinder and T. Boon. 1974. Purification and properties of colicin E3 immunity protein. *J. Biol. Chem.* **249**:438.

Kaufmann, Y. and A. Zamir. 1973. Protection of *E. coli* ribosomes against colicin E3-induced inactivation by bound aminoacyl-tRNA. *FEBS Letters* **36**:277.

Konisky, J. 1968. Biochemical effects of colicins on the bacterium *Escherichia coli.* I. Alteration of ribosomes induced by colicin E3 *in vivo.* II. Action of colicin I on non-colicinogenic and colicinogenic cells. *Ph.D. thesis,* University of Wisconsin, Madison.

Konisky, J. and M. Nomura. 1967. Interaction of colicins with bacterial cells. II. Specific alteration of *Escherichia coli* ribosomes induced by colicin E3 *in vivo. J. Mol. Biol.* **26**:181.

Meyhack, B., I. Meyhack and D. Apirion. 1973. Colicin E3: A unique endoribonuclease. *Proc. Nat. Acad. Sci.* **70**:156.

Nomura, M. 1963. Mode of action of colicins. *Cold Spring Harbor Symp. Quant. Biol.* **28**:315.

———. 1967. Colicins and related bacteriocins. *Ann. Rev. Microbiol.* **21**:257.

Nomura, M. and A. Maeda. 1965. Mechanism of action of colicins. *Zentr. Bakteriol. Parasitenk. Abt. I. Orig.* **196**:216.

Roth, T. F. and D. R. Helinski. 1967. Evidence for circular DNA forms of a bacterial plasmid. *Proc. Nat. Acad. Sci.* **58**:650.

Samson, A. C. R., B. W. Senior and I. B. Holland. 1972. The kinetics of colicin E3 induced fragmentation of *Escherichia coli* 16S ribosomal RNA in vivo. *J. Supramol. Struc.* **1**:135.

Santer, U. V. and M. Santer. 1972. The sequence of the 3′-OH end of the 16S RNA of *Escherichia coli. FEBS Letters* **21**:311.

Senior, B. W. and I. B. Holland. 1971. Effect of colicin E3 upon the 30S ribosomal subunit of *Escherichia coli. Proc. Nat. Acad. Sci.* **68**:959.

Senior, B. W., J. Kwasniak and I. B. Holland. 1970. Colicin E3-directed changes in ribosome function and polyribosome metabolism in *Escherichia coli* K12. *J. Mol. Biol.* **53**:205.

Shine, J. and L. Dalgarno. 1974. The 3′-terminal sequence of *Escherichia coli* 16S ribosomal RNA: Complementarity to nonsense triplets and ribosome binding sites. *Proc. Nat. Acad. Sci.* **71**:1342.

Sidikaro, J. and M. Nomura. 1973. Colicin E3-induced *in vitro* inactivation of ribosomes from colicin-insensitive bacterial species. *FEBS Letters* **29**:15.

———. 1974. E3-immunity substance: A protein from E3-colicinogenic cells that accounts for their immunity to colicin E3. *J. Biol. Chem.* **249**:445.

Tai, P. C. and B. D. Davis. 1974. Activity of colicin E3-treated ribosomes in initiation and in chain elongation. *Proc. Nat. Acad. Sci.* **71**:1021.

Timmis, K. and A. J. Hedges. 1972. The killing of sensitive cells by colicin D. *Biochim. Biophys. Acta.* **262**:200.

Turnowsky, F. and G. Högenauer. 1973. Colicin E3, an inactivating agent of the ribosomal A-site. *Biochem. Biophys. Res. Comm.* **55**:1246.

Yeast Ribosomes: Genetics

Calvin S. McLaughlin
Department of Molecular Biology and Biochemistry
University of California, Irvine
Irvine, California 92664

INTRODUCTION

The yeasts represent an important class of eukaryotic organisms both from the point of view of their evolutionary position and from their potential importance in molecular biology. The yeast cell contains a nuclear membrane, a nucleolus, a centriole, mitrochrondia, an endoplasmic reticulum, lysosomes, a Golgi body, vacuoles, a plasma membrane and a cell wall. Thus their cellular structure is that of a eukaryotic cell. The genetic information for all of this cellular complexity is contained in the smallest amount of DNA of any eukaryotic cell. Haploid spores of *Saccharomyces cerevisiae* contain only 0.84×10^{10} daltons of DNA (Darland 1969), which is equivalent to about three times the DNA content of the *E. coli* chromosome. The kinetics of thermal renaturation of denatured DNA are consistent with a genome of this size (Bicknell and Douglas 1970). The life cycle in yeast is similar to that found in higher eukaryotic organisms, with a restricted period of DNA synthesis and mitotic and meiotic cell division cycles (Hartwell 1970).

For problems in molecular and cellular biology which are common to both prokaryotic and eukaryotic organisms, yeast offers little advantage over the well-studied prokaryotic organisms except that complementatio. testing is straightforward. However for certain problems on the cellular level which are peculiar to eukaryotic organisms, yeast is the choice organism. In particular, yeast offers a number of distinct advantages for the study of those problems of translation and transcription which are peculiar to eukaryotic organisms. The genetics of yeast is well studied. Genetic analysis in *Saccharomyces cerevisiae* has demonstrated the existence of at least 17 chromosomes, and over 150 genetic loci have been mapped (Figure 1) (Mortimer and Hawthorne 1966, 1969, 1973). A number of well-marked strains are available. One of the chief advantages of yeast in any study of the biochemical genetics of indispensable cellular processes, such as transcription and translation, lies in the fact that both the haploid phase and the diploid phase of the life cycle are stable. This makes complementation testing straightforward. Since approaches to the genetics of ribosomes involve the isolation of a large number of mutants of similar phenotypes, a simple procedure for complementation testing is essential if one is to deter-

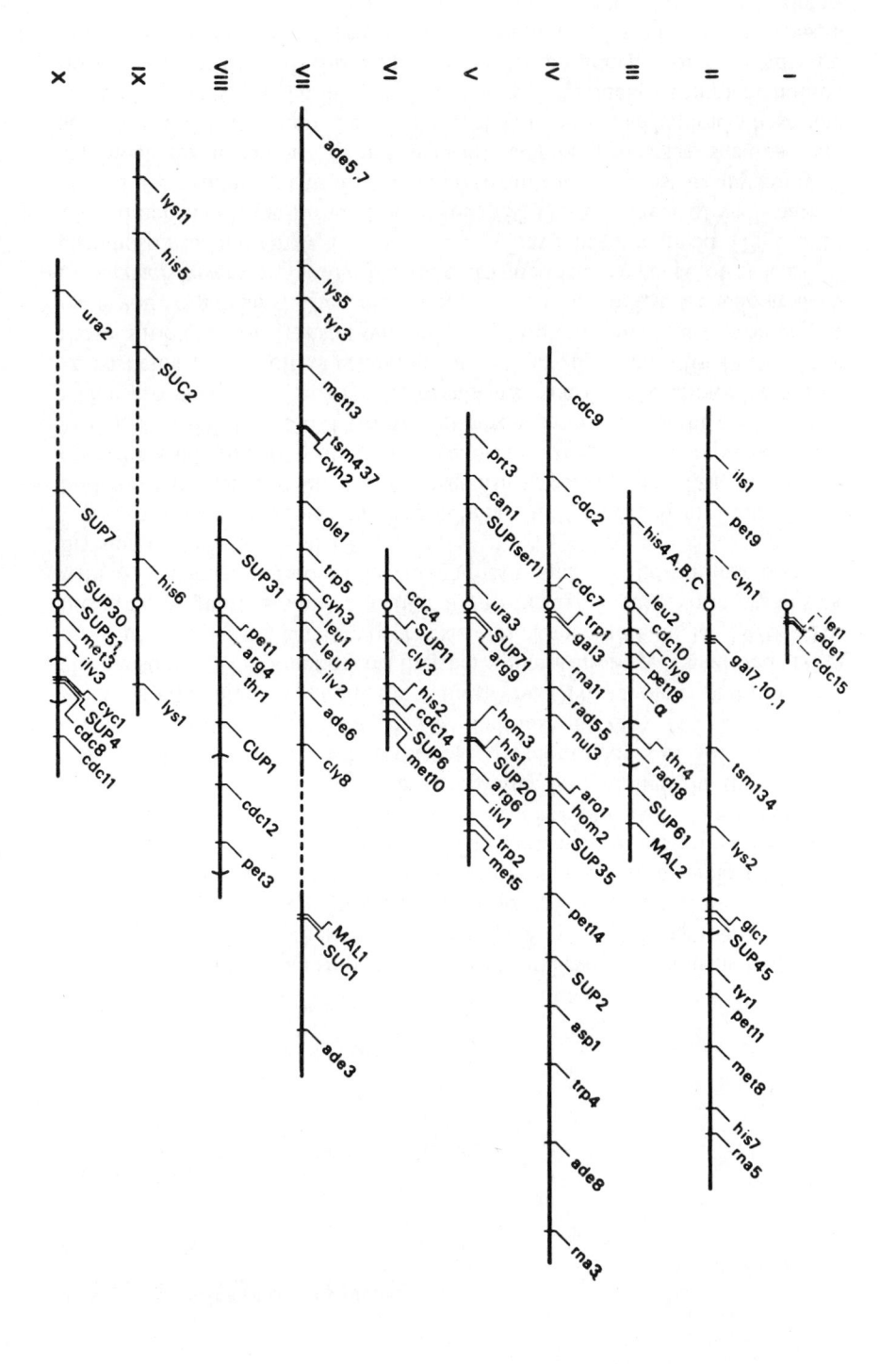

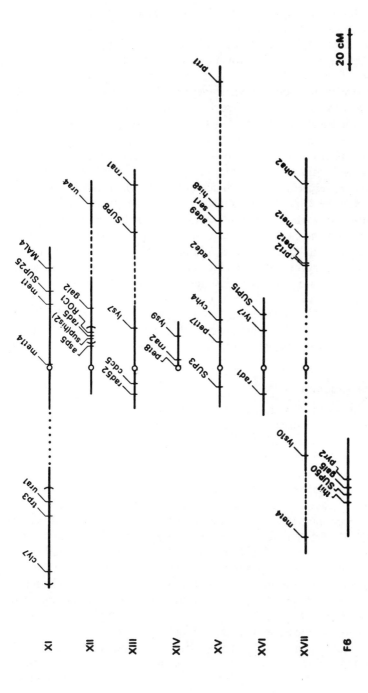

Figure 1 Genetic map of Saccharomyces. Linkages established by tetrad or random spore analysis are represented by solid lines; those determined by mitotic and trisomic analysis are indicated by dashed and dotted lines, respectively. The sequence of genes within parentheses has not been determined relative to outside markers. (Reprinted, with permission, from Mortimer and Hawthorne 1973.)

817

mine the number of genetic loci involved before detailed and laborious biochemical studies are begun for each mutant. Because of the short generation time and mode of growth, the techniques developed for the biochemical genetics of bacteria, such as replica plating, are directly applicable to yeast. Large quantities of yeast can be grown easily and inexpensively for enzyme purifications.

However, one reservation as to the general utility of yeast for the elucidation of the macromolecular processes involving the eukaryotic ribosome must be recognized at the outset. Yeast and other fungi occupy a unique position in the evolutionary tree, considerably diverged from other eukaryotic organisms and certainly diverged from man. Thus one runs the risk that the results found in yeast may not be typical of eukaryotic cells in general. The importance of this is a difficult question which can be answered only after extensive studies have shown whether the processes involving yeast ribosomes do resemble those in other eukaryotic organisms. However, the results to date indicate that in a number of areas directly concerned with the structure, function and biosynthesis of ribosomes, yeasts have a typical eukaryotic organization. On the basis of current work, the yeasts most studied in the area of this review include the closely related species *Saccharomyces cerevisiae* and *Saccharomyces carlsbergensis,* both of which are budding yeasts. The fission yeast *Schizosaccharomyces pombe* has been used in a number of studies. This review will focus on the 80S cellular, as contrasted to the mitochondrial, ribosome.

Chromosome Structure

The genetic behavior of yeast chromosomes closely resembles that of the chromosomes of higher organisms. Genetic analysis of *S. cerevisiae* indicates that the genetic material is divided into at least 17 chromosomes (Mortimer and Hawthorne 1973) and that each chromosome has a centromere which behaves as it does in higher organisms (Mortimer and Hawthorne 1966). At another level, biochemical studies on isolated yeast chromatin indicate that it has a structure similar to that found in other higher organisms. The nucleoprotein complex isolated from *S. cerevisiae* is composed of about 36% DNA, 4% RNA and 60% protein (Van der Vliet, Tonino and Rozijn 1969). The protein is acid extractable to the extent of about 70%. This DNA-associated protein stabilizes the melting profile of the DNA in the chromosome as the histones do in higher eukaryotic organisms (Van der Vliet, Tonino and Rozijn 1969). Further studies have indicated that the chromatin-like material from *S. cerevisiae* contains a complement of yeast histones, resembling those of other higher eukaryotic cells except for the almost complete absence of the histone F1 (Wintersberger, Smith and Letnansky 1973). Yeast chromatin is only poorly transcribed by isolated yeast polymerases. All of these properties are strikingly similar to those previously described for chromosomes from animal cells. Electron micrographs of the yeast nucleohistone fibers indicate that the yeast fibers have an average diameter of about 175 Å. This is similar to the fiber diameters observed with other eukaryotes which contain the full complement of histones (Grey, Peterson and Ris 1973).

RNA Polymerases

The presence of multiple DNA-dependent RNA polymerases from yeast has been reported (Adman, Schultz and Hall 1972; Ponta, Ponta and Wintersberger 1972). The yeast RNA polymerases have many properties similar to those isolated from various multicellular eukaryotic organisms. The yeast polymerases 1b, 2 and 3 resemble the equivalent enzymes from sea urchin with respect to salt optima, column elution position, α-amanitin sensitivity and preference for a native rather than a denatured template (Adman, Schultz and Hall 1972). Yeast cells also contain, in common with other eukaryotic cells, a poly(A) polymerase (Twu and Bretthauer 1971). This poly(A) polymerase activity has been shown to be due to two polymerases which differ in their primer requirements. Poly(A) polymerase 1 requires poly(A) sequences as a primer (Haff and Keller 1973).

mRNA Metabolism

The messenger RNA in yeast is predominantly monocistronic in the sense that each mRNA apparently codes for one polypeptide chain (Petersen and McLaughlin 1973). This monocistronic arrangement of mRNA contrasts with that observed in prokaryotic cells and is consistent with observations on mammalian cells in tissue culture (Kuff and Roberts 1967). Yeast mRNAs also contain poly(A) sequences (McLaughlin et al. 1973). These sequences, while smaller than those reported for mammalian mRNAs, average about 50 residues in length, suggesting that poly(A) sequences on mRNA may be a general feature of eukaryotes. Messenger RNA in *S. cerevisiae* is unstable, with a half-life of around 20 min (Hutchinson, Hartwell and McLaughlin 1969; Tonnesen and Friesen 1973). This half-life is considerably longer than that reported for prokaryotic cells (Levinthal, Keynan and Higa 1962) but shorter than that reported for mammalian cells (Perry 1973), both in absolute time and as a fraction of the cell generation time.

rRNA Structure

The RNA components of yeast ribosomes are similar to those found in other eukaryotic cells. The 80S ribosome dissociates under conditions of high salt or low Mg^{++} concentration into two subunits of about 60S and

40S (Walters and Van Os 1970; Van der Zeijst, Kool and Bloemers 1972). The large subunit contains a 5S RNA molecule and a 25S RNA molecule. The latter is bound noncovalently to a 5.8S RNA molecule. The 40S subunit contains a single 18S RNA molecule (Warner, this volume). The sequence of the 5.8S RNA from *S. cerevisiae* has been determined (Rubin 1973). Since there are about 150 copies of DNA homologous to the 27S rRNA in a haploid cell, and the 27S RNA appears to be the direct precursor of the 5.8S rRNA, it is significant that no sequence heterogeneity was detected in the 5.8S rRNA. Evidence of the conserved nature of this 5.8S rRNA was derived by comparing the RNase T₁ fingerprints of 5.8S RNA from four other species of Saccharomyces: *S. italicus, S. uvarum, S. schevalieri* and *S. carlsbergensis*. The similarities observed suggest that the 5.8S sequence may be subject to severe evolutionary constraints. Interestingly enough, the 5.8S RNA contains a pseudouridine residue. The 5S rRNA from *S. cerevisiae* has been sequenced and shows some homology with the sequence of the 5S RNA from *E. coli* and the sequence from KB cells. The sequence homologies with the 5S RNA isolated from KB cells are particularly striking (Hindley and Page 1972).

Chromosomal Location of the rRNA Cistrons

About 1.3 and 0.7% of the cellular DNA is found to be homologous to the 18S and 25S ribosomal RNA components, respectively, in different species of yeast (Schweizer, MacKechnie and Halvorson 1969; de Kloet 1970). In a mercury-cesium chloride gradient, a satellite DNA which accounts for 10–12% of the total DNA appears on the light side of the major nuclear DNA peak and contains all of the rRNA cistrons, including those required for the 5S RNA (Retel and Planta 1972). Rubin and Sulston (1973) demonstrated that the 5S cistrons are closely linked to the 18S and 25S ribosomal RNA cistrons in *S. cerevisiae*. Using DNA of 3.5 × 10⁶ molecular weight, they found that the cistrons for the 18S RNA and 25S RNA cosedimented with the cistrons for the 5S RNA. The number of cistrons coding for 5S RNA is the same as that coding for the 18S and 25S RNA. Their data is consistent with an alternating arrangement of these cistrons similar to that demonstrated for bacteria (Jaskunas, Davies and Nomura, this volume). In the other eukaryotic organisms that have been examined, the 18S and 28S cistrons are not linked to the 5S cistrons (Reeder, this volume). Interestingly, the 5S RNA in yeast appears to be under a distinctly different pattern of control than the 18S and 25S RNAs (see below).

Two approaches have been used to localize the ribosomal RNA cistrons on the yeast chromosome. The first approach takes advantage of the fact that the amount of DNA per chromosome in yeast is less than that present in the chromosome of a bacteria. It has been possible to isolate and separate the yeast chromosomes by low speed centrifugation in a sucrose gradient.

The chromosomes sediment as a heterogeneous material of very high molecular weight (Petes and Fangman 1972; Petes, Byers and Fangman 1973; Blamire et al. 1972). The average molecular weight of the chromosomal DNA is distributed between 400 and 700 × 10⁶ daltons. DNA-rRNA hybridizations through such a sucrose gradient indicate that the rRNA cistrons are located on two distinct size classes of chromosomal DNA. The minor peak, representing about 30% of the ribosomal RNA cistrons, appears to have a molecular weight of about 400–450 × 10⁶ daltons. The majority of the rRNA cistrons appear to be located on chromosomes having a size of about 700 × 10⁶ daltons (Finkelstein, Blamire and Marmur 1972). Thus the ribosomal RNA genes in *S. cerevisiae* could be located on as few as two (or as many as 17) chromosomes. Similar, although not identical, data has been obtained in *S. carlsbergensis*. There, too, the ribosomal RNA cistrons appear to be located on chromosomes of two distinct sizes. The first, containing most of the cistrons coding for rRNA, is the size of an average chromosome. The rRNA cistrons located on chromosome(s) of smaller than average size contain the minority of the rRNA cistrons, in contrast to the situation observed in *S. cerevisiae* (de Kloet 1973).

The genetic approach to the localization of the ribosomal RNA cistrons has taken advantage of the fact that certain monosomic (2n − 1) and disomic (1n + 1) strains of yeast are stable. Halvorson and coworkers used disomic strains to determine that the ribosomal RNA cistrons were not located on all of the yeast chromosomes (Goldberg et al. 1972). Available data indicates that ribosomal RNA genes are not located on chromosomes 3, 5, 6, 7, 9, 11, 12 or 13 (Goldberg et al. 1972; Gimmler and Schweizer 1972). Using monosomic strains, Marmur and coworkers have shown that 70% of the total rRNA cistrons are located on chromosome 1 (Finkelstein, Blamire and Marmur 1972). Few other known genes are located on chromosome 1. Thus these workers speculate that the genes for the ribosomal proteins could also be located on chromosome 1. However, the distribution of the antibiotic-resistant mutants that affect ribosome structure suggests that the ribosomal protein genes are scattered throughout the genome. This would be consistent with the lack of clustering of other functionally related structural genes (Figure 1). There appears to be very little clustering of genes for related processes in yeast (Mortimer and Hawthorne 1973).

Mutations Directly Affecting the Ribosome

The number of mutations known to affect the ribosomes directly is relatively small. Although some temperature-sensitive mutants might be expected to harbor lesions in the genes for ribosomal proteins, no such mutant has been found to date; however, a number of mutations have been isolated which do affect other areas involved in transcription and translation (Hartwell and

McLaughlin 1968; Hutchinson, Hartwell and McLaughlin 1969; Klyce and McLaughlin 1973). Temperature-sensitive mutants defective in one of ten genes that control ribosome formation in yeast have been described (Hartwell, McLaughlin and Warner 1970). These mutants have been extensively studied and will be discussed later in the section on ribosomal biosynthesis. The search for genes that directly affect ribosome structure through the selection of antibiotic-resistant mutants has been more successful. Cycloheximide has been identified in numerous studies as an antibiotic that affects ribosome function (Gale et al. 1972). Early experiments indicated that one type of cycloheximide resistance in *S. cerevisiae* was mediated through cycloheximide-resistant ribosomes (Siegel and Sisler 1965; Cooper, Banthorpe and Wilkie 1967). The 60S ribosomal subunit was subsequently shown to be altered in the resistant ribosomes (Rao and Grollman 1967). The genetic locus involved in the alteration of the 60S ribosomal subunit is *cyh2* on chromosome 7 (Figure 1) (C. McLaughlin, unpublished observations). There is no direct evidence to indicate whether the alteration involves a protein or an rRNA species. However, the hybridization studies described above indicate that chromosome 7 is not a likely site for any of the rRNA cistrons. A diploid heterozygous for cycloheximide resistance has been shown to have 50% resistant ribosomes, which suggests that the gene is not concerned with a modifying enzyme. Resistant ribosomes can be washed in high concentrations of monovalent salts without altering their resistance to cycloheximide (C. McLaughlin, unpublished observations).

The antibiotic cryptopleurine blocks a step in the elongation process of protein synthesis and acts on the ribosome. Resistant mutants of *S. cerevisiae* have been isolated, and their resistance has been shown to be due to modified ribosomes that are resistant to the drug. The gene involved maps very close to the mating type locus on chromosome 3 (Figure 1). To date, no evidence is available to indicate whether the resistant ribosome has an altered protein or RNA component (Skogerson, McLaughlin and Wakatama 1973).

Trichodermin is an antibiotic that has been shown to act on the termination phase of protein synthesis in yeast (Stafford and McLaughlin 1973). It acts at the ribosomal level by directly binding to ribosomes with an affinity constant of 9.2×10^5 M^{-1} (C. M. Wei, personal communication). Resistant mutants have been isolated, and the mutation conferring the ribosomal resistance results in an altered 60S ribosomal subunit (Schindler, Grant and Davies 1974). Although this genetic alteration has not been mapped, it does not appear to be linked either to the locus for cryptopleurine or to that for cycloheximide resistance.

An interesting approach to the isolation of mutants affecting the ribosome has been developed by Bayliss and Vinopal (1971). They isolated a mutant that conferred streptomycin sensitivity and cold sensitivity apparently by affecting ribosome formation and function. Yeasts are not normally sensitive to streptomycin; however, it proved possible to isolate a sensitive mutant by selecting from a histidine auxotroph for mutants which could grow without histidine only in the presence of streptomycin. The ribosomes from these mutants are sensitive to inhibition of in vitro protein synthesis by streptomycin. They are also cold sensitive and accumulate a 28S ribonucleoprotein particle when grown at low temperatures (Bayliss and Ingraham 1974).

Singh and Manney (1974) have isolated a number of cold-sensitive mutants of *S. cerevisiae* that are unable to grow at temperatures of less than 10°C. Nine different complementation groups are involved in this phenotype. Five of the loci have been mapped. Mutations at two of the loci generally confer cycloheximide sensitivity along with cold sensitivity. The critical in vitro experiments have not yet been done, but the authors point out that these loci may affect the ribosome.

Mutations Affecting Ribosome Biosynthesis

The pattern of synthesis and processing of rRNA in *Saccharomyces cerevisiae* has been established through the work of Udem and Warner (1972, 1973) and others (Sillevis Smitt et al. 1972). The basic pattern resembles that in other eukaryotic cells and is reviewed elsewhere in this volume by Warner. In yeast, a 35S precursor is cleaved to two fragments of 27S and 20S precursor RNA. The 20S precursor is converted into an 18S RNA found in the small ribosomal subunit, and the 27S precursor is converted into a 25S and a 5.8S RNA found in the large subunit. Planta and coworkers and Rozijn and coworkers have established that the basic process of ribosomal RNA maturation is similar in *Saccharomyces carlsbergensis* and added a number of details (Retel and Planta 1970; Van den Bos, Retel and Planta 1971; Van den Bos, Klootwijk and Planta 1972; Van den Bos and Planta 1973; Sillevis Smitt et al. 1972). Taber and Vincent (1969) have indicated that the fission yeast *Schizosaccharomyces pombe* follows a similar precursor product pattern for rRNA biosynthesis. Thus all the yeast examined follow the same basic pattern of synthesis and processing. The range of sizes reported for the precursors in the different species may not be real since a careful comparison of the molecular sizes has not been undertaken.

During sporulation the rRNA processing pathway is apparently altered, and a 20S RNA accumulates to a level of approximately 4% of the total cellular RNA in *S. cerevisiae*. Hybridization experiments and pulse-chase experiments in the presence and absence of cycloheximide indicate that the 20S sporulation-specific RNA appears to be a variant of the 20S ribosomal RNA precursor normally observed (Sogin, Haber and Halvorson 1972). The function of this RNA during sporulation is not known.

A number of temperature-sensitive mutants defective in ribosome biosynthesis have been isolated and characterized (Hartwell, McLaughlin and Warner 1970; Warner and Udem 1972). A detailed analysis of the ten genes involved in this defect failed to assign specific defects to any of the mutants. Each appeared to be simultaneously defective in rRNA synthesis and processing. Presumably this effect is due to the very tight control over synthesis and processing in eukaryotic cells (discussed elsewhere in this volume by Warner). Recent attempts to elucidate the molecular defects in these mutants have been partially successful, and at least one of the genes has been shown to be involved in the transformation of the 27S RNA precursor to the 25S:5.8S product (T. Helser, personal communication). However, studies with these mutants suggest that the eukaryotic cell has a very tight regulatory system, which is used to maintain coordinate processing of the rRNA molecules.

The exception to this rule of coordinate synthesis and processing occurs with 5S RNA. Neither its synthesis nor its processing, if any, is affected by the temperature-sensitive lesions that block the synthesis and processing of the other ribosomal RNAs in the temperature-sensitive mutants described above. Thus 5S RNA continues to be made at near normal rates for a substantial period of time after the processing of the other rRNA species is severely inhibited (T. Helser, personal communication). Lomofugin severely inhibits the synthesis of 18S and 25S RNA without affecting the rate of synthesis of 5S RNA (Fraser, Creanor and Mitchison 1973). This indicates that although the cistrons for 5S RNA and 18S and 25S RNA are closely linked, the synthesis and processing of 5S RNA is controlled separately from that of the other rRNA species.

The results discussed in this article clearly represent the first, and as yet rather primitive, stage in the development of yeast as a major organism for the biochemical genetics of the eukaryotic ribosome. Nonetheless, the inherent experimental potential of the organism should guarantee a rapidly increasing rate of progress over the next few years.

Acknowledgments

The work from the author's laboratory has been supported by a grant from the National Cancer Institute CA 10628. I am grateful to my colleagues who read and criticized an earlier draft of this article.

References

Adman, R., L. D. Schultz and B. D. Hall. 1972. Transcription in yeast: Separation and properties of multiple RNA polymerases. *Proc. Nat. Acad. Sci.* **69**:1702.

Bayliss, F. T. and J. L. Ingraham. 1974. A mutation in *S. cerevisiae* conferring streptomycin and cold-sensitivity by affecting ribosome formation and function. *J. Bact.* **118**:319.

Bayliss, F. T. and R. T. Vinopal. 1971. Selection of ribosomal mutants by antibiotic suppression in yeast. *Science* **174**:1339.

Bicknell, J. N. and H. C. Douglas. 1970. Nucleic acid homologies among species of *Saccharomyces*. *J. Bact.* **101**:505.

Blamire, J., D. R. Cryer, D. B. Finkelstein and J. Marmur. 1972. Sedimentation properties of yeast nuclear and mitochondrial DNA. *J. Mol. Biol.* **67**:11.

Cooper, D., D. V. Banthorpe and D. Wilkie. 1967. Modified ribosomes conferring resistance to cycloheximide in mutants of *S. cerevisiae*. *J. Mol. Biol.* **26**:347.

Darland, G. K. 1969. The physiology of sporulation in *S. cerevisiae*. *Ph.D. thesis,* University of Washington.

De Kloet, S. 1970. The formation of RNA in yeast: Hybridization of high molecular weight RNA species to yeast DNA. *Arch Biochem. Biophys.* **136**:402.

————. 1973. Distribution of ribosomal RNA cistrons among yeast chromosomes. *J. Bact.* **114**:1034.

Finkelstein, D. B., J. Blamire and J. Marmur. 1972. Location of ribosomal RNA cistrons in yeast. *Nature New Biol.* **240**:279.

Fraser, R. S. S., J. Creanor and J. M. Mitchison. 1973. Rapid and selective inhibition of the synthesis of high molecular weight RNA in yeast by lomofugin. *Nature* **244**:222.

Gale, E. F., E. Cundliffe, P. E. Reynolds, N. H. Richmond and M. J. Waring. 1972. Cycloheximide and related glutarmide antibiotics. In *The molecular basis of antibiotic action,* pp. 357–361. John Wiley and Sons, London.

Gimmler, G. M. and E. Schweizer. 1972. Use of disomics in mapping ribosomal RNA genes in *S. cerevisiae. J. Mol. Biol.* **72**:811.

Goldberg, S., T. Oyen, J. M. Idriss and H. O. Halvorson. 1972. Use of disomic strains to study the arrangement of ribosomal cistrons in *Saccharomyces. Mol. Gen. Genet.* **116**:139.

Grey, R. H., J. B. Peterson and H. Ris. 1973. The organization of yeast nucleohistone fibers. *J. Cell Biol.* **58**:244.

Haff, L. A. and E. B. Keller. 1973. Two distinct poly(A) polymerases in yeast nuclei. *Biochem. Biophys. Res. Comm.* **51**:704.

Hartwell, L. H. 1970. Biochemical genetics of yeast. *Ann. Rev. Gen.* **4**:373.

Hartwell, L. H. and C. S. McLaughlin. 1968. Temperature-sensitive mutants of yeast exhibiting a rapid inhibition of protein synthesis. *J. Bact.* **96**:1664.

Hartwell, L. H., C. S. McLaughlin and J. R. Warner. 1970. Identification of 10 genes that control ribosome formation in yeast. *Mol. Gen. Genet.* **109**:42.

Hindley, J. and S. M. Page. 1972. Nucleotide sequence of yeast 5S ribosomal RNA. *FEBS Letters* **26**:157.

Hutchinson, H. T., L. H. Hartwell and C. S. McLaughlin. 1969. Temperature-sensitive yeast mutant defective in RNA production. *J. Bact.* **99**:807.

Klyce, H. R. and C. S. McLaughlin. 1973. Characterization of temperature-sensitive mutants of yeast by a photomicrographic procedure. *Exp. Cell Res.* **82**:47.

Kuff, E. and N. E. Roberts. 1967. In vivo labeling pattern of free polysomes: Relationship to tape theory of mRNA function. *J. Mol. Biol.* **26**:211.

Levinthal, C., A. Keynan and A. Higa. 1962. Messenger RNA turnover and protein synthesis in *B. subtilis* inhibited by actinomycin D. *Proc. Nat. Acad. Sci.* **48**:1631.

McLaughlin, C. S., J. R. Warner, M. Edmunds, H. Nakazato and M. Vaughan. 1973. Poly(A) sequences in yeast messenger ribonucleic acid. *J. Biol. Chem.* **248**:1466.

Mortimer, R. K. and D. C. Hawthorne. 1966. Yeast genetics. *Ann. Rev. Microbiol.* **20**:151.

―――. 1969. Yeast genetics. In *The yeasts* (ed. A. H. Rose and J. S. Harrison) vol. 1, pp. 386–543. Academic Press, London.

―――. 1973. Genetic mapping in *Saccharomyces*. IV. Mapping of temperature-sensitive genes and use of disomic strains. *Genetics* **74**:33.

Perry, R. P. 1973. On the role of poly(A) sequences in mRNA metabolism. In *Molecular cytogenetics* (ed. B. A. Hamkalo and A. J. Papaconstantinou) pp. 133–145. Plenum Press, New York.

Petersen, N. S. and C. S. McLaughlin. 1973. Monocistronic mRNA in yeast. *J. Mol. Biol.* **81**:33.

Petes, T. D. and W. L. Fangman. 1972. Sedimentation properties of yeast chromosomal DNA. *Proc. Nat. Acad. Sci.* **69**:1188.

Petes, T. D., B. Byers and W. L. Fangman. 1973. Size and structure of yeast chromosomal DNA. *Proc. Nat. Acad. Sci.* **70**:3072.

Ponta, H., U. Ponta and E. Wintersberger. 1972. Purification and properties of DNA-dependent RNA polymerases from yeast. *Eur. J. Biochem.* **29**:110.

Rao, S. S. and A. P. Grollman. 1967. Cycloheximide resistance in yeast: A property of 60S ribosomal subunits. *Biochim. Biophys. Res. Comm.* **29**:696.

Retel, J. and J. Planta. 1970. On the mechanism of the biosynthesis of ribosomal RNA in yeast. *Biochim. Biophys. Acta* **224**:458.

―――. 1972. Nuclear satellite DNAs of yeast. *Biochim. Biophys. Acta* **281**:299.

Rubin, G. M. 1973. The nucleotide sequence of *S. cerevisiae* 5.8S ribosomal RNA. *J. Biol. Chem.* **248**:3860.

Rubin, G. M. and J. E. Sulston. 1973. Physical linkage of the 5S cistron to the 18S and 28S ribosomal RNA cistrons in *S. cerevisiae*. *J. Mol. Biol.* **79**:521.

Schindler, D., P. Grant and J. Davies. 1974. Trichodermin resistance, a mutation affecting eucaryotic ribosomes. *Nature New Biol.* (in press).

Schweizer, E., C. MacKechnie and H. O. Halvorson. 1969. The redundancy of ribosomal and transfer RNA genes in *S. cerevisiae*. *J. Mol. Biol.* **40**:261.

Siegel, M. R. and H. D. Sisler. 1965. Site of action of cycloheximide in cells of *Saccharomyces pastorianus*. III. Further studies on the mechanism of action and the mechanism of resistance in Saccharomyces species. *Biochim. Biophys. Acta* **103**:558.

Sillevis Smitt, W. W., J. M. Vlak, R. Schiphof and T. H. Rozijn. 1972. Precursors of ribosomal RNA in yeast nucleus. *Exp. Cell Res.* **71**:33.

Singh, A. and T. R. Manney. 1974. Genetic analysis of mutations affecting growth of *Saccharomyces cerevisiae* at low temperature. *Genetics* (in press).

Skogerson, L., C. S. McLaughlin and E. Wakatama. 1973. Modification of ribosomes in cryptopleurine-resistant mutants of yeast. *J. Bact.* **116**:818.

Sogin, S. J., J. E. Haber and H. O. Halvorson. 1972. Relationship between sporulation-specific 20S ribonucleic acid and ribosomal ribonucleic acid processing in *Saccharomyces cerevisiae*. *J. Bact.* **112**:806.

Stafford, M. E. and C. S. McLaughlin. 1973. Trichodermin, a possible inhibitor of the termination process of protein synthesis. *J. Cell. Physiol.* **82**:121.

Taber, R. L. and W. S. Vincent. 1969. The synthesis and processing of ribosomal RNA precursor molecules in yeast. *Biochim. Biophys. Acta* **186**:317.

Tonnesen, T. and J. D. Friesen. 1973. Inhibitors of ribonucleic acid synthesis in *Saccharomyces cerevisiae:* Decay rate of messenger ribonucleic acid. *J. Bact.* **115**:889.

Twu, J. S. and R. K. Bretthauer. 1971. Properties of a polyriboadenylate polymerase isolated from yeast ribosomes. *Biochemistry* **10**:1576.

Udem, S. A. and J. R. Warner. 1972. Ribosomal RNA synthesis in *Saccharomyces cerevisiae. J. Mol. Biol.* **65**:227.

————. 1973. The cytoplasmic maturation of a ribosomal precursor ribonucleic acid in yeast. *J. Biol. Chem.* **218**:1412.

Van den Bos, R. C. and R. J. Planta. 1973. Structural comparison of 37S and 32S ribosomal precursor RNA in yeast with the mature ribosomal RNA components. *Biochim. Biophys. Acta* **294**:465.

Van den Bos, R. C., J. Klootwijk and R. J. Planta. 1972. Structural comparison of 17S ribosomal RNA of yeast and its immediate precursor, 18S RNA. *FEBS Letters* **24**:93.

Van den Bos, R. C., J. Retel and R. J. Planta. 1971. The size and the location of the ribosomal RNA segments in ribosomal precursor RNA of yeast. *Biochim. Biophys. Acta.* **232**:494.

Van der Vliet, P. C., G. J. M. Tonino and T. H. Rozijn. 1969. Studies on the yeast nucleus. III. Properties of a deoxyribonucleoprotein complex derived from yeast. *Biochim. Biophys. Acta* **195**:473.

Van der Zeijst, B. A. M., A. J. Kool and H. P. J. Bloemers. 1972. Isolation of active ribosomal subunits from yeast. *Eur. J. Biochem.* **30**:15.

Walters, J. A. L. I. and G. A. J. Van Os. 1970. The dissociation and association behavior of yeast ribosomes. *Biochim. Biophys. Acta* **199**:453.

Warner, J. R. and S. A. Udem. 1972. Temperature-sensitive mutations affecting ribosome synthesis in *Saccharomyces cerevisiae. J. Mol. Biol.* **65**:243.

Wintersberger, U., P. Smith and K. Letnansky. 1973. Yeast chromatin. Preparation from isolated nuclei, histone composition and transcription capacity. *Eur. J. Biochem.* **33**:123.

Progress in the Structural Analysis of Mammalian 45S and Ribosomal RNA

B. E. H. Maden, M. Salim and J. S. Robertson
Department of Biochemistry
University of Glasgow
Glasgow, Scotland

INTRODUCTION

Several features distinguish eukaryotic from prokaryotic ribosomes with respect to both their structure and their formation. Eukaryotic ribosomes are larger than prokaryotic ones, their RNA is more highly methylated, and their formation takes place in a specialized organelle, the nucleolus, and involves transcription of a single large RNA precursor molecule (45S RNA in mammalian cells) (Maden 1971). The latter contains, in addition to the two ribosomal RNA sequences, considerable stretches of nonconserved material that are eliminated during maturation.

For a clearer insight into these phenomena, detailed chemical information on rRNA and its precursors is essential. The work described here has been directed towards this end. It relates in particular to two of the distinctive eukaryotic features mentioned above: rRNA methylation and the extensive nonconserved sequences in the precursor molecules. The work was carried out with HeLa cells. However, a comparable analysis has been carried out for methylation on yeast rRNA (Klootwijk and Planta 1973), and the results obtained, though different in detail, were strikingly similar in several general respects.

Methylation Patterns of HeLa Cell rRNA and Precursors

Sequence analysis of the methylated regions within rRNA and its precursors promised to be interesting in light of several previous findings with HeLa cells. These earlier (pre–1970) findings may be summarized as follows:

(1) Nearly all methylation occurs rapidly in the nucleolus, commencing on nascent 45S RNA (Greenberg and Penman 1966; Zimmerman and Holler 1967). There appeared originally to be one exception, late methylation of dimethyl A on 18S RNA (Zimmerman 1968).

(2) Most or all early methylation seemed to occur on ribosomal sequences and not on nonconserved sequences of 45S RNA (Wagner, Penman and Ingram 1967; Weinberg et al. 1967; Weinberg and Penman 1970), though evidence on this point was not conclusive.

(3) At least 80% of methylation appeared to be ribose methylation (Wagner, Penman and Ingram 1967; Vaughan et al. 1967). Most of the possible alkali-stable dinucleotides were present in alkaline hydrolysates of rRNA, as also were two 28S specific products which were tentatively identified as trinucleotides (Wagner, Penman and Ingram 1967). This suggested that many chemically different sequences within rRNA are ribose-methylated.

(4) Methionine starvation experiments indicated that RNA methylation is essential for ribosome maturation (Vaughan et al. 1967).

In our first experiments using sequencing techniques, methyl-labeled RNA was prepared to high specific activity and was "fingerprinted" (Sanger, Brownlee and Barrell 1965) after digestion with T_1 ribonuclease (Maden, Salim and Summers 1972). Many methylated products were resolved (Figure 1a). The 28S and 18S patterns were different from each other, and these differences permitted the presence of one or both rRNA sequences to be unambiguously detected in the various nucleolar precursor molecules. In this way we confirmed by direct chemical means the "maturation pathway" for rRNA, which had previously been inferred from less direct evidence by Weinberg and Penman (1970):

$$45S \longrightarrow 41S \begin{array}{c} \nearrow 32S \longrightarrow 28S \\ \\ \searrow 20S \longrightarrow 18S \end{array}$$

(See Maden, Salim and Summers [1972] for the fingerprints.)

Two other points were noticed in these experiments. First, there were no extra well-resolved T_1 products in fingerprints of the precursors as compared with mature rRNA, with a minor exception in 45S RNA. This supported the earlier inference that practically all methylation of 45S RNA is within the ribosomal sequences. Second, three 18S-specific products were absent from the 45S, 41S and 20S precursor RNAs, and one of these 18S products contained dimethyl A, in agreement with Zimmerman's (1968) earlier observation.

To pursue these points and also to obtain sequence and quantitative information on the methylated oligonucleotides, we carried out a series of further experiments (Salim and Maden 1973; Maden, Lees and Salim 1972; Maden and Salim 1974). Late methylation was unambiguously distinguished from early methylation by methyl labeling after a brief (5 min) preincubation of cells in actinomycin D (Figure 1b). Long "T_1 plus phosphatase" fingerprints were prepared in order to separate all hitherto unresolved T_1 products (Figure 1c). A variety of procedures for sequence analysis and quantitation were then used, from which the following fairly complete data were obtained (see Table 1).

28S RNA contains some 71 methyl groups. All of these are added early (i.e., to 45S RNA) with one very minor exception. All except five early

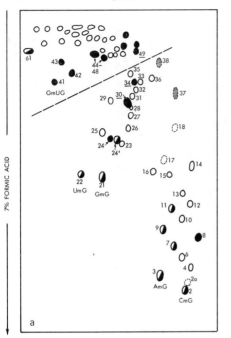

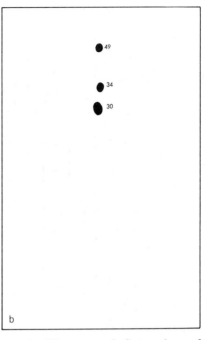

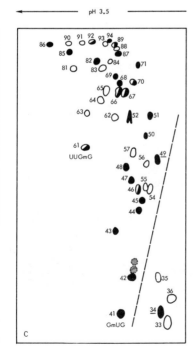

Figure 1 Diagrams of fingerprints of methyl-labeled HeLa cell rRNA. (*a*) 28S + 18S RNA, standard T$_1$ fingerprint. (Reprinted, with permission, from Maden, Salim and Summers 1972.) (*b*) Late methylations on 18S RNA, shown by labeling after actinomycin D (see text), T$_1$ fingerprint. (*c*) 28S + 18S RNA, T$_1$ plus alkaline phosphatase fingerprint with long separations in both dimensions. (*b,c:* Reprinted, with permission, from Salim and Maden 1973.) Products above the dotted line in (*a*) and (*c*) were generally fractionated by the T$_1$ plus phosphatase method. Clear spots, 28S specific products; black spots, 18S specific products; half-black spots, common to 28S and 18S RNA (but not necessarily identical sequences in the two rRNA species). Fingerprints of 45S RNA are almost indistinguishable from those of 28S + 18S RNA, except for absence of known late-methylated products and presence of two weak and variable extra spots.

Table 1 Summary of the methylated sequences from HeLa cell ribosomal RNA

Product number	Sequence or methylated component	CH₃ groups mol⁻¹ observed	CH₃ groups mol⁻¹ suggested	Comments
28S				
2a	mC-G	faint	0.25	*late*, base
14	...mA and ..Cm-A or Am-C...	1.60	2	base & ribose
28	U-(A,mA)-C-G	0.97	1	base
29	A-A-ψm-G	0.88	1	doubly modified component (ψm)
33	(mC,C,A-A-Am-U)-G	2.22	2	base & ribose
36	A-C$_{3^-5}$-Gm-A-Am-A-G	1.76	2	
62	Cm-A-Gm-U-U-G	2.21	2	
63	mU-U-ψ-A-G	1.18	1	base
66	A-A-U-U-Gm-C-A-G	0.95	1	"5.8S"
67	..mC...,2Am-U...	2.51	3	base & ribose
83	...Um-Gm-U...	1.93	2	alkali-stable trinucleotide
90	...Gm-U,Um-C,Gm-A....	2.53	3	alkali-stable trinucleotide
92a	...Um-Gm-ψ....	1.83	2	alkali-stable trinucleotide
93	...Am-Gm-Cm-A....	1.63	3?	alkali-stable tetranucleotide

all other 28S products			46–47	all singly ribose methylated
Total 28S CH₃ groups			71	
18S	m$_2^6$A-m$_2^6$A-C-U-G	3.67	4	*late*, base
	... mA	0.75	1	*late*, base
	... m^{7}G?	0.95	1	*late*, base
	.. Cm-A,Um-A ...	1.82	2	
	.. Um-C,Um-G	2.22	2	
all other 18S products			36	probably all singly ribose methylated
Total 18S CH₃ groups			46	
Total 28S + 18S			117	(111 early, 6 late)
Total 45S			111	(all except 5 or 6 on ribose groups)

Most T_1 products are singly ribose methylated and undergo methylation on 45S RNA. All products specifically itemized are "unusual" in some respect, such as multiple modification, base methylation, and/or late methylation (see comments). In several cases where two or more methyl groups are found within a single "spot," it is clear that these methyl groups exist with a single T_1 product (for example 28S products 36 and 62). However, for the following spots we cannot rule out the possibility of unresolved mixtures of products: 14, 33, 67 and 90. Most of these data first appeared in Salim and Maden (1973).

methylations are on ribose groups, the other five apparently being on bases. There are instances of methyl-group clustering: two doubly methylated trinucleotides occur, one triply methylated tetranucleotide, and a few sequences in which two methyl groups are on nearby but not adjacent nucleotides.

18S RNA contains some 46 methyl groups, of which 40 are added early and six late (Figure 1b). Most 18S T_1 products are different from the 28S products. All early 18S methylations seem to be on ribose groups, with one possible exception which is still uncharacterized. However, the six late methylations are all base methylations, and four of these occur within a single product, $m_2^6Am_2^6ACUG$. Most interestingly, a very similar late-methylated sequence occurs in yeast 17S RNA (Klootwijk, van den Bos and Planta 1972a) and towards the 3' end of *E. coli* 16S RNA (Ehresmann, Fellner and Ebel 1971; Hayes et al. 1971). A similarly modified sequence may therefore be widespread in the smaller rRNA of both eukaryotes and prokaryotes. By contrast, early ribose methylation seems to be a peculiarly eukaryotic phenomenon (cf. Retel, van den Bos and Planta 1969).

In both 28S and 18S RNA, most of the methylated products occur approximately once per mole, though there are a few exceptions. Some of the smallest T_1 products occur more than once per mole, and a few products occur in clear fractional yields. Nevertheless, the fact that most products occur in near integral molar yields suggests a fairly high degree of sequence homogeneity in HeLa cell rRNA, at least with respect to the methylated sequences (see also Klootwijk and Planta 1973).

The methylation pattern of 45S RNA is almost indistinguishable from that of 28S + 18S RNA, apart from absence of known late methylations. Two weak and variable extra 45S spots are sometimes seen, whose significance is still unknown. Apart from these minor exceptions, the data provide the strongest evidence yet obtained that methylation of 45S RNA is essentially confined to its ribosomal sequences. We shall consider the possible implications of this finding later.

Nonconserved Sequences in the Precursor Molecules

What of the nonconserved regions of the precursor molecules? These regions are remarkable for their size, amounting to almost half of the total molecule in mammalian 45S RNA (McConkey and Hopkins 1969; Weinberg and Penman 1970). The nonconserved material is GC rich (Willems et al. 1968; Jeanteur, Amaldi and Attardi 1968) and unmethylated, but otherwise little is known about it.

As a first approach to this problem, we sought for extra oligonucleotides in T_1 digests of $^{32}PO_4$-labeled 32S RNA as compared with 28S RNA (Robertson and Maden 1973). Some 13 such oligonucleotides were found

(Figure 2a; Table 2). Most of these are rich in C, none are chemically modified, and most of the larger ones appear to occur once per mole, again suggesting sequence homogeneity. The oligonucleotides probably represent some 10% of the total nonconserved material in 32S RNA, the remainder presumably being degraded to small T_1 products which are indistinguishable from 28S ribosomal products.

45S RNA yields even more complex fingerprints than 32S RNA and reveals yet further T_1 products which are not present in either the 28S, 18S or 32S nonconserved sequences. These products are currently being characterized (Robertson and Maden, unpublished observations).

The presence of such distinctive products in digests of 32S and 45S RNA shows clearly that these molecules contain, within their nonconserved re-

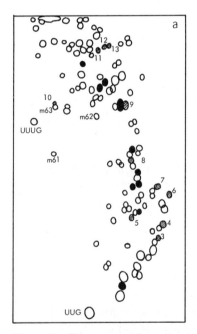
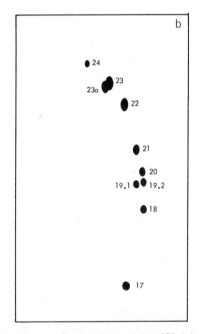

Figure 2   Diagrams of long T_1 plus phosphatase fingerprints of (*a*) ^{32}P-labeled 32S RNA and (*b*) 5.8S RNA. Most of the products shown contain two or more uridylate residues, other products (which contribute less information) having run off the ends of the fingerprints. In (*a*), the shaded products 3–13 are absent from 28S fingerprints and are presumed to be derived from the nonconserved part of 32S RNA. Products marked black in (*a*) are recognizable by various qualitative and quantitative criteria as being derived from the 5.8S sequence (compare *b*). Products m61, m62 and m63 are well-resolved 28S methylated products, which occur once per mole, and were used for quantitation of other products. (These products are numbered 61, 62 and 63 in Figure 1).

Table 2   Nucleotide sequences from the nonconserved region
of HeLa cell 32S ribosomal precursor RNA

Product number	Sequence	Molar yield	
		observed	suggested
1	U-C-C-C-G	5–8	5–8
2	C-(C$_3$,U)-G	3–6	3–6
			(1 in 28S?)
3	C-(C,U)-C-U-C-G	1.04	1
4	C-(C$_2$,U)-C-U-C-G	1.12	1
5	A-U-U-C-C-G	1.47	1–2?
6	C-(C$_2$,U)-C-C-U-C-C-G	0.67	1
7	C-(C$_3$,U)-C-U-A-A-G	0.85	1
8	A-(A,U-C)-A-U-C-G	0.88	1
9	U-(C$_2$,U)-C-C-U-C-G	0.89	1
10	U-(A,U)-U-G	0.46	?
11	U-(C,U$_2$)-C-C-C-U-C-G	0.86	1
12	(C$_3$,U)-C-U-C-U-U-G	0.71	1
13	C-(C$_5$,U$_4$)-C-G	0.61	1

Products 1 and 2 have run off the end in Figure 2a. Product 10 is always weakly labeled, and it is not clear whether it is a genuine "fractional" product or a contaminant. Partial sequence analyses and quantitation were carried out as described elsewhere (Robertson and Maden 1973).

gions, sequences that are chemically different from the ribosomal sequences. RNA-DNA hybridization studies also point to this conclusion (Jeanteur and Attardi 1969). Moreover, knowledge of an array of distinctive oligonucleotides from rRNA and its precursors should be useful in a wide range of further structural studies. One situation where such knowledge has recently proved useful concerns "5.8S" RNA.

"5.8S" RNA

When 28S rRNA is briefly heated or otherwise denatured, a small noncovalently attached component is released. This was termed 7S RNA by its discoverers (Pene, Knight and Darnell 1968) but has more recently been recognized to be approximately 5.8S. Strong circumstantial evidence suggested that the 5.8S sequence is contained in covalently linked form within the 32S and 45S precursor RNAs and is generated as a separate entity by nucleolytic cleavage during the final 32S to 28S maturation step (Pene, Knight and Darnell 1968). We have confirmed that this is the case by oligonucleotide analysis (see Figure 2). A number of characteristic "5.8S" oligonucleotides are clearly recognizable in "T$_1$ plus phosphatase" fingerprints of 28S RNA and are recognizably absent from heat-treated 28S RNA (which lacks the 5.8S fragment; fingerprint not shown). Other 5.8S spots are not qualitatively unique in 28S RNA, but their presence or absence can be recognized by quantitation, and the identities of all the respective oligonucleotides can be confirmed by enzymic digestion procedures, carried out in parallel with the corresponding oligonucleotides from pure 5.8S RNA. By these criteria all of the oligonucleotides in Figure 2b can be identified in fingerprints of 32S RNA (Figure 2a), whether heat treated or not. Moreover, one of the products is methylated (23a in Figure 2b, identical to product 66.28 in Table 1). This product is therefore recognizable in fingerprints of [^{14}C] methyl-labeled 45S RNA, confirming that the 5.8S sequence is present within the transcription unit (Maden and Robertson 1974).

IMPLICATIONS AND PROSPECTS

The information which we have obtained about methylation is of considerable interest, particularly at the level of 45S RNA. The 45S molecule undergoes rapid methylation at some 110 points within its ribosomal sequences. All but five or six of these early methylations are on ribose groups, the great majority are on regions with different primary sequences (though some methyl-group clustering occurs), and one ribose methyl group is situated within the 5.8S sequence.

The many different primary sequences that undergo ribose methylation suggest that this type of modification is specified by secondary structure, or conformation, rather than purely by primary structure within 45S RNA (Maden, Salim and Shepherd 1974; Salim and Maden 1973). A similar view has been expressed for methylation of yeast ribosomal precursor RNA (Klootwijk, van den Bos and Planta 1972b). If this view of ribose methylation is correct, then the ribosomal sequences must contain, at many points, conformational features that are absent from the nonconserved regions of 45S RNA, since the latter are unmethylated, at least in HeLa cells.

The known requirement for methylation in HeLa cell ribosome maturation (Vaughan et al. 1967) has led one of us to speculate (Maden 1971) that the role of ribose methylation might be to protect certain exposed phosphodiester bonds against accidental nucleolytic cleavage (via 2′–3′ cyclic intermediates) during maturation. The present results do not yet permit critical appraisal of this suggestion, but it would clearly be of interest to learn more about the spatial arrangement of the methylated regions within rRNA and the ribosome, as well as to pursue analysis of the nonconserved regions to a point where some clue emerges as to their biological role. In this respect, 5.8S RNA may prove to be of special interest since it appears to be an "island" of ribosomal material that is recognizable to the methylating enzyme(s) and hydrogen bonded to the main 28S sequence, although probably separated from the latter in terms of its primary sequence by a region of nonconserved material.

Finally, it is possible that by extension of this work, information might be obtained that could be correlated with results from other methods, such as electron microscopy of partly denatured RNA molecules, whereby the gross arrangement of the 28S, 18S and nonconserved regions in 45S RNA has recently been visualized (5′–28S–nonconserved–18S–nonconserved–3′: Wellauer and Dawid 1973).

Acknowledgments

This work was supported by a grant from the Medical Research Council. The skilled technical assistance of Mr. J. Forbes is gratefully acknowledged.

References

Ehresmann, C., P. Fellner and J-P. Ebel. 1971. The 3′-terminal nucleotide sequence of the 16S ribosomal RNA from *Escherichia coli*. *FEBS Letters* **13**:325.

Greenberg, H. and S. Penman. 1966. Methylation and processing of ribosomal RNA in HeLa cells. *J. Mol. Biol.* **21**:527.

Hayes, F., D. Hayes, P. Fellner and C. Ehresmann. 1971. Additional nucleotide sequences in precursor 16S ribosomal RNA from *Escherichia coli*. *Nature New Biol.* **232**:54.

Jeanteur, Ph. and G. Attardi. 1969. Relationship between HeLa cell ribosomal RNA and its precursors studied by high resolution RNA-DNA hybridization. *J. Mol. Biol.* **45**:305.

Jeanteur, Ph., F. Amaldi and G. Attardi. 1968. Partial sequence analysis of ribosomal RNA from HeLa cells. II. Evidence for sequences of nonribosomal type in 45S and 32S ribosomal RNA precursors. *J. Mol. Biol.* **33**:757.

Klootwijk, J. and R. J. Planta. 1973. Analysis of the methylation sites in yeast ribosomal RNA. *Eur. J. Biochem.* **39**:325.

Klootwijk, J., R. C. van den Bos and R. J. Planta. 1972a. Secondary methylation of yeast ribosomal RNA. *FEBS Letters* **27**:102.

———. 1972b. Analysis of the methylation pattern of yeast ribosomal RNA. *Abstr. Commun. Meet. Fed. Eur. Biochem. Soc.* **8**:492.

Maden, B. E. H. 1971. The structure and formation of ribosomes in animal cells. *Prog. Biophys. Mol. Biol.* **22**:127.

Maden, B. E. H. and J. S. Robertson. 1974. Demonstration of the "5.8S" ribosomal sequence in HeLa cell ribosomal precursor RNA. *J. Mol. Biol.* (in press).

Maden, B. E. H. and M. Salim. 1974. The methylated nucleotide sequences in HeLa cell ribosomal RNA and its precursors. *J. Mol. Biol.* (in press).

Maden, B. E. H., C. D. Lees and M. Salim. 1972. Some methylated sequences and the numbers of methyl-groups in HeLa cell rRNA. *FEBS Letters* **28**:293.

Maden, B. E. H., M. Salim and J. Shepherd. 1974. Ribosomal RNA and ribosome formation in HeLa cells. *Biochem. Soc. Symp. Eukaryotic ribosomes, structure and formation* **37**:23.

Maden, B. E. H., M. Salim and D. F. Summers. 1972. Maturation pathway for ribosomal RNA in the HeLa cell nucleolus. *Nature New Biol.* **237**:5.

McConkey, E. H. and J. W. Hopkins. 1969. Molecular weights of some HeLa ribosomal RNAs. *J. Mol. Biol.* **39**:545.

Pene, J. J., E. Knight and J. E. Darnell. 1968. Characterization of a new low molecular weight RNA in HeLa cell ribosomes. *J. Mol. Biol.* **33**:609.

Retel, J., R. C. van den Bos and R. J. Planta. 1969. Characteristics of the methylation *in vivo* of ribosomal RNA in yeast. *Biochim. Biophys. Acta* **195**:370.

Robertson, J. S. and B. E. H. Maden. 1973. Nucleotide sequences from the non-conserved region of HeLa cell 32S ribosomal precursor RNA. *Biochim. Biophys. Acta* **331**:61.

Salim, M. and B. E. H. Maden. 1973. Early and late methylations in HeLa cell ribosome maturation. *Nature* **244**:334.

Sanger, F., G. G. Brownlee and B. G. Barrell. 1965. A two-dimensional fractionation procedure for radioactive nucleotides. *J. Mol. Biol.* **13**:373.

Vaughan, M. H., R. Soeiro, J. R. Warner and J. E. Darnell. 1967. The effects of methionine deprivation on ribosome synthesis in HeLa cells. *Proc. Nat. Acad. Sci.* **58**:1527.

Wagner, E. K., S. Penman and V. Ingram. 1967. Methylation patterns of HeLa cell ribosomal RNA and its nucleolar precursors. *J. Mol. Biol.* **29**:371.

Weinberg, R. A. and S. Penman. 1970. Processing of 45S nucleolar RNA. *J. Mol. Biol.* **47**:169.

Weinberg, R. A., U. Loening, M. Willems and S. Penman. 1967. Acrylamide gel electrophoresis of HeLa cell nucleolar RNA. *Proc. Nat. Acad. Sci.* **58**:1088.

Wellauer, P. K. and I. B. Dawid. 1973. Secondary structure maps of RNA; processing of HeLa ribosomal RNA. *Proc. Nat. Acad. Sci.* **70**:2827.

Willems, M., E. Wagner, R. Laing and S. Penman. 1968. Base composition of ribosomal RNA precursors in the HeLa cell nucleolus: Further evidence of nonconservative processing. *J. Mol. Biol.* **32**:211.

Zimmerman, E. F. 1968. Secondary methylation of ribosomal ribonucleic acid in HeLa cells. *Biochemistry* **7**:3156.

Zimmerman, E. F. and B. W. Holler. 1967. Methylation of 45S ribosomal RNA precursor in HeLa cells. *J. Mol. Biol.* **23**:149.

Differential Replication of Ribosomal RNA Genes in Eukaryotes

Brian B. Spear*
Department of Biology
Yale University
New Haven, Connecticut 06520

General Occurrence of Differential rDNA Replication

Several lines of evidence have demonstrated that the amount of DNA in the cell nucleus, and hence the nuclear information content of the cell, remains constant in development (Alfert 1954). From this fact, it has been concluded that cellular differentiation takes place by selective expression of the genes rather than by the selective loss or multiplication of specific genes. While DNA constancy has repeatably been demonstrated to be a sound general rule, it has some striking exceptions. The most carefully studied of these exceptions is the variability in the number of copies of the genes for the ribosomal RNAs.

In the oocytes of a great variety of organisms, especially the Amphibia, there is a selective extrachromosomal replication of the rDNA, the DNA sequences that code for 28S and 18S rRNA (Brown and Dawid 1968; Gall 1968, 1969). The rDNA amplification is generally accompanied by an increase in the number of nucleoli in the germinal vesicle. The amount of extrachromosomal rDNA varies from a few times to several thousand times the amount of rDNA in the somatic nucleolus organizer. The subject of ribosomal gene amplification in oocytes is treated more completely by Reeder in this volume.

There have also been several reports of the differential replication of ribosomal RNA genes in cells other than those of the germ line. Koch and Cruceanu (1971) examined DNA from cultured human liver cells which had been treated with a hormone, triiodothyronine. This hormone is known to induce the formation of extra nucleoli. Hybridization of 28S and 18S rRNA to DNA from cells treated with triiodothyronine demonstrated that the number of rRNA genes had increased to 1.8 times the number in untreated cells. A similar increase in ribosomal gene number occurs in the cells of regenerating newt iris (Collins 1972). After removal of the lens, iris cells show great nucleolar activity, and the DNA from these cells hybridizes to rRNA at a level that is 1.6 times that in the DNA

* Present address: Department of Molecular, Cellular and Developmental Biology, University of Colorado, Boulder, Colorado 80302

from normal irises. In both of these instances, the differential replication of rDNA may be due to extrachromosomal amplification, but the results might also be interpreted as premature replication of the nucleolus oraganizer during the cell cycle. Engberg, Mowat and Pearlman (1972) concluded that a 30% increase in the rDNA content of Tetrahymena which had been well fed, as compared to those which had been starved, was the result of the rDNA entering S phase before the rest of the nuclear DNA.

A recent report by Howells (1972) indicates that there is a selective replication and degradation of the rDNA during the cell cycle in Chlamydomonas. By hybridizing rRNA to DNA from algae at different points in the cell cycle, Howells showed that the number of rRNA genes per cell varied by a factor of three. The lowest number of rRNA genes was found at the time of nucleolar breakdown during mitosis and the highest number during S phase. Both the threefold change in the number of rRNA genes and the timing of the rDNA increase and decrease were indications that the differential rDNA synthesis was not simply due to asynchronous replication of nucleolar DNA. In the onion *Allium cepa,* there is a variation of the rDNA content from cell to cell (Avanzi, Maggini and Innocenti 1973). In the metaxylem cell lines of the root tip there is considerable DNA synthesis, and much of this DNA is complementary to rRNA, as shown by in situ and in vitro rRNA hybridization. Avanzi and coworkers concluded that the rRNA genes in the metaxylem cells had been replicated to several thousand times the amount measured in meristematic cells.

rDNA in Drosophila

The genes for ribosomal RNA have been extensively studied in the fruit fly, Drosophila. The ease with which Drosophila can be manipulated genetically has attracted a number of researchers, and so the differential rDNA replication in this organism has been the subject of detailed investigation. For this reason, the rest of this article will concentrate on the genes for rRNA in Drosophila.

In *Drosophila melanogaster* the genes for rRNA are present in multiple copies and in double-stranded form make up 0.5–1.0% of the wild-type genome (Vermuelen and Atwood 1965; Ritossa and Spiegelman 1965; Tartof 1971; Spear and Gall 1973). This fraction of the Drosophila genome, assuming a diploid DNA content of 0.36×10^{-12}g (Rasch, Barr and Rasch 1971), represents about 285–550 copies of the sequence coding for 18S and 28S rRNA.

Analyses by genetic mapping and by RNA-DNA hybridization have shown that the rRNA genes are clustered and that the rDNA is located at or near the nucleolus organizer (Ritossa and Spiegelman 1965). The nucleolus organizer is a secondary constriction in the heterochromatin of the X and Y chromosomes (Kaufmann 1934; Cooper 1959), shown schematically in Figure 1. Ritossa and Spiegelman (1965) hybridized rRNA to DNA from flies whose genomes had one, two, three or four doses of a region of the heterochromatin that included the nucleolus organizer. The levels to which the rRNA bound to the different DNAs were in proportion to the number of nucleolus organizers present, thereby demonstrating the linkage of this chromosomal region with the rDNA.

The only gene known to map in the region of the nucleolus organizer is bobbed (*bb*) (Lindsley and Grell 1967). Bobbed is a pleiotropic gene whose phenotype includes shortened bristles, reduced fertility, slow development, and abnormal formation of the abdominal cuticle. The bobbed mutation is hypomorphic, that is, the mutation functions the same as a normal allele but at a lower level. A fly with two bobbed mutations has a phenotype closer to wild type than a fly with one bobbed mutation and a deletion of the bobbed locus. The linkage of bobbed to the ribosomal genes, and therefore to the nucleolus organizer, was established by the RNA-DNA hybridization studies of Ritossa, Atwood and Spiegelman (1966). These researchers found that in four bobbed alleles tested, the DNA from bobbed flies had an rDNA level strikingly lower than DNA from wild-type flies. Therefore it appears that the bobbed mutations are partial deletions of the rDNA, or in cytological terms, partial deletions of the nucleolus organizer. A detailed review of the genetics and biochemistry of the bobbed locus has recently been written by Ritossa (1974).

Magnification of rDNA

Male Drosophila with a bobbed mutation on one sex chromosome and a severely deficient or deleted bobbed locus on the other pass on to their progeny a bobbed locus with extra rDNA. If the male progeny (G_1) have sex chromosomes identical to those of their fathers, their bobbed phenotype is less severe than that of their fathers. The G_1 males in turn pass on extra rDNA to their progeny. The males of the next generation (G_2) are non-bobbed, both in phenotype and in rDNA content, despite having the

Figure 1   The diploid X and Y chromosomes of *Drosophila melanogaster,* after Cooper (1959). The clear area is the euchromatin and the blackened area is the heterochromatin. The stippled region marked *NO* is the nucleolus organizer, which is also the location of the rDNA and of the gene for bobbed.

same sex chromosomes as their severely bobbed grandfathers. This process of phenotypic reversion and rDNA accumulation is known as magnification (Ritossa 1968; Ritossa and Scala 1969).

Recent experiments by Harford (unpublished) have confirmed that the magnification of bobbed is a true chromosomal change. First, alteration of the genotypic background had no effect on the magnified phenotype, thus ruling out the possibility of accumulation of modifiers affecting the phenotype. Second, a magnified and an unmagnified bobbed locus segregated out of a female in the manner that would be expected of alleles. This would indicate that magnification cannot be attributed to transferable factors, such as rDNA episomes. Third, the magnified bobbed locus was mapped by recombination at the same location on the chromosome as the wild-type allele.

In certain cases, a magnified bobbed locus can return to its original bobbed condition if it is associated with a chromosome bearing a wild-type (bb^+) allele. The return to bobbed depends on how many generations the bobbed mutation has been in a magnifying condition, that is, in a male opposite a chromosome deleted for bobbed, and how many generations the magnified locus has subsequently been opposite a wild-type allele. In general, the number of flies which return to bobbed is inversely proportional to the number of generations of magnification and directly proportional to the subsequent number of generations with a bb^+ allele. After about four generations of magnification, none of the magnified alleles will return to bobbed (Henderson and Ritossa 1970).

Ritossa and his coworkers have studied the rDNA increase that accompanies the phenotypic reversion during magnification (Ritossa 1968; Ritossa and Scala 1969; Ritossa et al. 1971). In one of the bobbed loci studied, the percent rDNA attributable to that locus increased from 0.09% to 0.25% during the process of magnification. This increase proceeded by more or less equal steps with each generation (Ritossa 1968; Ritossa et al. 1971). In the first generation that the bobbed mutation is associated with an rDNA-deficient chromosome, there is a significant increase in rDNA content, but with no marked effect on the phenotype. This was interpreted as "nonfunctional" rDNA (Ritossa et al. 1971). The synthesis of this nonfunctional rDNA was believed to be a necessary first step in magnification, but it might also be a separate phenomenon such as "rDNA compensation," an rDNA increase somewhat different from magnification which will be discussed in a later section.

The molecular process which brings about rDNA magnification is not well understood. Clearly it must involve some extremely sensitive cellular sensing and control mechanisms. Several mechanisms for increasing the number of rRNA genes have been proposed, but all have drawbacks. One possibility is that there is strong selection for those flies in the population that have higher rDNA contents. However, the work of Atwood (1969) indicates that all the males in the population are equally capable of magnifying. Furthermore, Ritossa and Scala (1969) found that there was no selection against those males in the population which had lower rDNA levels, since the ratio of male offspring to female offspring was close to one.

A mechanism that involves selection for those sperm with higher rDNA contents is attractive, especially since magnification seems to occur only in males. However, sperm activity is independent of the nuclear contents of the spermatocytes (Lindsley and Grell 1969). Therefore, sperm would have no differences in function depending on the number of rRNA genes they contained. There could be selection for spermatogonia which have increased rDNA content due to unequal mitotic exchange.

Ritossa (1972) has proposed a model that involves the specific extra-chromosomal synthesis of circular rDNA molecules in germ cells of magnifying males. These circles would then integrate into the chromosomes by recombination, thus increasing the chromosomal rDNA content. There is evidence in Xenopus oocytes of extrachromosomal rDNA in the form of circles (Hourcade, Dressler and Wolfson 1973; Bird, Rochaix and Bakken 1973; Miller 1966) which have been synthesized on a chromosomal template (Brown and Blackler 1972). However, in Drosophila there is no evidence for extrachromosomal rDNA synthesis in ovaries, male genitalia or eggs (Mohan and Ritossa 1970; Boncinelli et al. 1972; Polan et al. 1973), nor is there evidence for rDNA circles. The increased recombination observed in males that are undergoing magnification may be due to the process of circle integration, as suggested by Ritossa (1973). However, these data support equally well other mechanisms that involve recombination, such as unequal crossing over of homologs or unequal mitotic exchange.

rDNA Compensation

Male Drosophila which are lacking the Y chromosome, whose sex chromosome constitution is termed XO, have only one nucleolus organizer and therefore would be expected to have one-half the rDNA content of their XX female mothers. This is not the case. The hybridization of rRNA to DNA from XO males is about 0.8 times the hybridization to DNA from XX females (Tartof 1971, 1973). Experimentally, Tartof found the DNA of XO flies to be 0.35% rDNA and the DNA of XX flies to be 0.43% rDNA. The experiments of Tartof have been repeated, using the methods of Spear and Gall (1973) with whole flies (shown in Figure 2). The results of these tests were consistent with those of Tartof; XO is 0.32% rDNA and XX is 0.43% rDNA.

rDNA compensation occurs not only in XO males, but also in females in which one of the X chromosomes has a deletion for the nucleolus or-

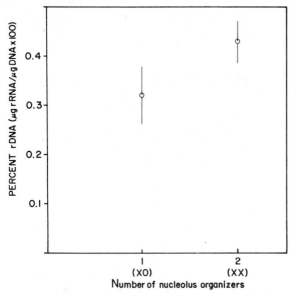

Figure 2 Hybridization of Drosophila [³H]rRNA to DNA extracted from whole Drosophila adults by the methods of Spear and Gall (1973). XO flies have one nucleolus organizer per diploid genome and XX flies have two.

ganizing region (Xno⁻) (Tartof 1971, 1973). In these females, rDNA compensation will occur whether the rDNA-bearing chromosome has a wild-type or mutant bobbed allele. XO and Xno⁻/X flies are phenotypically wild, whereas X*bb*/O and X*bb*/Xno⁻ flies are bobbed, often severely. When the X chromosome carries a bobbed mutation, the degree of rDNA compensation is proportional to the original amount of rDNA carried by the bobbed locus (Tartof 1973). Curiously, rDNA compensation does not seem to occur in males which are Xno⁻/Y (Tartof 1971, 1973).

The rDNA increase in rDNA compensation seems to be of a purely somatic nature. The additional rDNA is not inherited, that is, it is not passed on to the next generation. Tartof (1973) noted a slight increase over four generations in the rDNA content of a bobbed locus associated with the Xno⁻ chromosome. However, when the fourth generation X*bb*/Xno⁻ females were mated to XY males, the rDNA content of the X*bb* chromosome in the X*bb*/Y progeny was back to its original level. Therefore, it seems that even over several generations, there is no inheritance of the extra rDNA in rDNA compensation.

Both rDNA compensation and rDNA magnification involve an increase in rDNA levels in flies which have a deficiency for rDNA. For this reason, the two phenomena appear to be very similar if not the same. However, there are some major differences between magnification and rDNA compensation. First, magnification is restricted to males, whereas rDNA compensation can take place in either males or females (Tartof 1971). Second, magnification has a marked effect on the bobbed phenotype (Ritossa 1968), but rDNA compensation seems to have none (Tartof 1973; Stern 1929). Third, the changes in rDNA content due to magnification are inherited (Ritossa 1968; Henderson and Ritossa 1970), whereas those due to rDNA compensation are not (Tartof 1971, 1973). It has been suggested that magnification and rDNA compensation may involve the same mechanisms but that the latter takes place only in the somatic cells. However, recent investigations (Spear and Gall 1973) have suggested that rDNA compensation may be restricted to only those cells which have polytene chromosomes.

Independent Replication of rDNA in Polytene Chromosomes

In many tissues of Drosophila, especially those of the larva, cells undergo chromosome replication without cell division or the separation of chromatids. This endoreplication results in giant polytene chromosomes which, in the larval salivary gland, have as many as 1024 or 2048 chromosome strands, the result of nine or ten doublings of the original diploid chromosome (Rasch 1970). During the process of polyteny, not all of the DNA is replicated to the same extent. While the euchromatin is fully and evenly replicated, the heterochromatic regions (see Figure 1) are replicated no more than once or twice (Rudkin 1965; Berendes and Keyl 1967; Mulder, van Duijn and Gloor 1968; Gall, Cohen and Polan 1971).

Some studies have shown that the ribosomal RNA genes which are located in the heterochromatin (Figure 1) are also underreplicated with respect to the euchromatic DNA. Hennig and Meer (1971) found that the DNA from polytene salivary glands of *Drosophila hydei* larvae was about 0.1% rDNA, whereas DNA from whole larvae or adults was 0.2% rDNA and DNA from embryos was about 0.4% rDNA. From these results, Hennig and Meer estimated that the rDNA in polytene chromosomes of *D. hydei* was three rounds of replication behind the euchromatic DNA. Sibatani (1971) found a somewhat comparable case in *Drosophila melanogaster*. In his experiments, DNA from larvae, which have a high proportion of polytene tissue, had a lower rDNA percent than did the DNA from adults. In *Rhynchosciara angelae,* rRNA hybridizes to salivary gland DNA to about one-half the level at which it binds to ovarian DNA (Gambarini and Meneghini 1972).

Recent experiments on DNA from diploid and polytene cells of *D. melanogaster* (Spear and Gall 1973) have confirmed the underreplication of rDNA in polytene chromosomes. Furthermore, the results indicate that the percent rDNA in the salivary gland DNA is independent of the genotype of the larva. DNA from the salivary glands of wild-type female larvae was found to be about 0.1% rDNA. DNA extracted from the diploid brains and imaginal discs was found to be 0.47% for the Oregon R strain and 0.37% for the Urbana S strain. When the tissues were obtained from XO male larvae, the diploid DNA was found to have an rDNA content about one-half that of the XX diploid DNA. However, the percent rDNA in polytene DNA from XO larvae was agɛin 0.1%, the same as in XX polytene DNA.

A replotting of the data of Spear and Gall is shown in Figure 3. The diploid DNA from XO larvae is about 0.29% rDNA, whereas the diploid DNA from XX larvae is about 0.51% rDNA. In the DNA from polytene cells, the two genotypes have the same rDNA content, about 0.08%.

Similar studies have been carried out with DNA from larvae of other genotypes, such as XY and Xno$^-$/Y, and those with three or four nucleolus organizers per diploid cell. In these experiments, as in those with XO and XX larvae, the rDNA content of DNA from diploid cells was the simple sum of the rDNA contents of the combined nucleolus organizers.

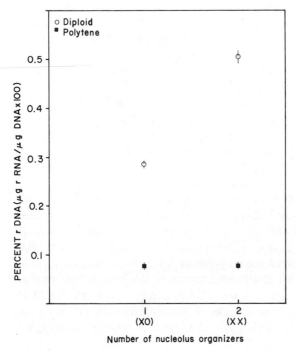

Figure 3   The rDNA content of DNA from diploid and polytene cells of XO and XX *D. melanogaster* larvae. The data have been replotted from Spear and Gall (1973).

In the DNA from polytene cells, however, the rDNA content was consistently close to 0.1% regardless of the number of nucleolus organizers or of the rDNA content of the diploid cells from the same larvae.

These results led to the conclusion that, in polytene cells, the control of replication of rDNA is to some extent independent of the control of replication of euchromatic DNA (Spear and Gall 1973). This independent replication of the rDNA allows the polytene cells to compensate for variations in the number of nucleolus organizers in the genotype. It is not yet known whether independent polytenization can compensate for deficiencies in the number of rRNA genes in the bobbed mutants.

Conclusions on Differential rDNA Replication in Drosophila

In order for rDNA magnification and rDNA compensation to be clarified, these phenomena must be more fully understood at the cellular level. Drosophila has a very complex variety of cell types, involving polyploid and polytene cells as well as the "normal" diploid cells. For this reason, the results obtained by hybridization of rRNA to DNA extracted from whole flies can give only a rough approximation of the events at the chromosomal level.

Analysis of DNA extracted from tissues with defined chromosome types allows a possible explanation for the process of rDNA compensation. A comparison of the results in Figure 2 with those in Figure 3 indicates that the non-additivity of rDNA in DNA from whole flies is not a property of DNA from diploid cells. Non-additivity of rDNA, however, does occur in the polytene cells of the larval salivary gland. Therefore, it might be reasonable to view rDNA compensation as a phenomenon that is restricted to those cells of the Drosophila adult which have polytene chromosomes. Like rDNA compensation, differential polyteny takes place in both males and females and is restricted to the somatic tissues. It is possible that the bobbed phenotype is the result of rDNA deficiencies in the nonpolytene cells, thus explaining the lack of any phenotypic effect of rDNA compensation.

Magnification, on the other hand, cannot be explained by a differential replication of rRNA genes in polytene tissues. Since magnification is an inherited phenomenon, the variation in rRNA gene number must take place in the germ line cells, which are not polytene, as well as in the somatic cells. It seems at this point that rDNA magnification can only be viewed as a change in the tandem number of rRNA genes within the nucleolus organizer. One might expect that this variability of rRNA gene number may be occurring in many other organisms and will become apparent when the genetics of rDNA in these organisms is better known.

Summary: The Terminology of Differential rDNA Replication

The differential replication of rDNA can take place in a wide variety of cellular systems by what appear to be numerous controlling and replicating mechanisms. Since each new mechanism tends to acquire its own name, it seems useful to briefly distinguish between them so that the names might have a more general value.

The term "amplification" as used by Brown and Dawid (1968) refers to the specific extrachromosomal replication of the rDNA in oocytes. This term might also be applicable to the specific increase in rRNA gene number in other cells if the increase is demonstrated to be extrachromosomal. "Magnification" (Ritossa 1968) refers to the inherited increase in rDNA in Drosophila that accompanies the return of the bobbed phenotype to wild type. As yet, the term magnification does not imply a specific chromosomal mechanism. "rDNA compensation" is the name for the noninherited rDNA increase in Drosophila (Tartof, personal communication) best typified by the increased rDNA multiplicity in XO males (Tartof 1971). Like magnification, rDNA compensation refers to a phenomenon rather than a mechanism. The mechanism of rDNA compensation may be that of "independent polytenization" (Spear and Gall 1973), which leads to a variability in the number of lateral copies of the nucleolus organizer in polytene chromosomes.

There are several ways in which the measurable rDNA content of cells can vary in response to genetic, developmental or environmental factors. Among the possible mechanisms are extrachromosomal rDNA synthesis, change in the number of tandem copies of the rRNA gene within the nucleolus organizer, change in the number of lateral rDNA copies to form a polystranded nucleolus organizer, or simple replication of the rDNA either earlier or later than the bulk of the nuclear DNA. As it becomes possible to divide the cases of differential rDNA replication into these and other classes, the terminology will become more specific and descriptive of the basic controlling and replicating systems involved.

Acknowledgments

The author would like to thank Dr. J. G. Gall for his advice and encouragement. Critical review of the manuscript by Drs. C. I. Davern, G. Herrick and D. M. Prescott is also greatly appreciated. This paper was written with the support of Grant VC 85 from the American Cancer Society to J. G. Gall and USPHS Training Grant HD 00032.

References

Alfert, M. 1954. Composition and structure of giant chromosomes. *Int. Rev. Cytol.* **3**:131.

Atwood, K. C. 1969. Some aspects of the bobbed problem in *Drosophila*. *Genetics* (suppl.) **61**:319.

Avanzi, S., F. Maggini and A. M. Innocenti. 1973. Amplification of ribosomal cistrons during the maturation of metaxylem in the root of *Allium cepa*. *Protoplasma* **76**:197.

Berendes, H. D. and H. G. Keyl. 1967. Distribution of DNA in hetero-chromatin and euchromatin of polytene nuclei of *Drosophila hydei*. *Genetics* **57**:1.

Bird, A. P., J. D. Rochaix and A. H. Bakken. 1973. The mechanism of gene amplification in *Xenopus laevis* oocytes. In *Molecular cytogenetics* (ed. B. A. Hamkalo and J. Papaconstantinou) pp. 49–58. Plenum Press, New York.

Boncinelli, E., F. Graziani, L. Polito, C. Malva and F. Ritossa. 1972. rDNA magnification at the bobbed locus of the Y chromosome in *Drosophila melanogaster*. *Cell Differentiation* **1**:133.

Brown, D. D. and A. W. Blackler. 1972. Gene amplification proceeds by a chromosome copy mechanism. *J. Mol. Biol.* **63**:75.

Brown, D. D. and I. Dawid. 1968. Specific gene amplification in oocytes. *Science* **160**:272.

Collins, J. M. 1972. Amplification of ribosomal ribonucleic acid cistrons in regenerating lens of *Triturus*. *Biochemistry* **11**:1259.

Cooper, K. W. 1959. Cytogenetic analysis of major heterochromatic elements (especially Xh and Y) in *Drosophila melanogaster*, and the theory of "heterochromatin." *Chromosoma* **10**:535.

Engberg, J., D. Mowat and R. E. Pearlman. 1972. Preferential replication of the ribosomal RNA genes during a nutritional shift-up in *Tetrahymena pyriformis*. *Biochim. Biophys. Acta* **272**:312.

Gall, J. G. 1968. Differential synthesis of the genes for ribosomal RNA during amphibian oogenesis. *Proc. Nat. Acad. Sci.* **60**:553.

———. 1969. The genes for ribosomal RNA during oogenesis. *Genetics* (suppl.) **61**:121.

Gall, J. G., E. H. Cohen and M. L. Polan. 1971. Repetitive DNA sequences in *Drosophila*. *Chromosoma* **33**:319.

Gambarini, A. G. and R. Meneghini. 1972. Ribosomal genes in salivary gland and ovary of *Rhynchosciara angelae*. *J. Cell Biol.* **54**:421.

Henderson, A. and F. Ritossa. 1970. On the inheritance of rDNA of mag-nified bobbed loci in *D. melanogaster*. *Genetics* **66**:463.

Hennig, W. and B. Meer. 1971. Reduced polyteny of ribosomal RNA cistrons in giant chromosomes of *Drosophila hydei*. *Nature New Biol.* **233**:70.

Hourcade, D., D. Dressler and J. Wolfson. 1973. The amplification of ribosomal RNA genes involves a rolling circle intermediate. *Proc. Nat. Acad. Sci.* **70**:2926.

Howells, S. H. 1972. The differential synthesis and degradation of ribosomal DNA during the vegetative cell cycle in *Chlamydomonas reinhardi*. *Nature New Biol.* **240**:264.

Kaufman, B. P. 1934. Somatic mitoses of *Drosophila melanogaster*. *J. Morph.* **56**:125.

Koch, J. and A. Cruceanu. 1971. Hormone induced gene amplification in somatic cells. *Hoppe-Seyler's Z. Physiol. Chem.* **352**:137.

Lindsley, D. L. and E. H. Grell. 1967. Genetic variations of *Drosophila melanogaster*. Carnegie Institute of Washington, Publ. 627.

————. 1969. Spermiogenesis without chromosomes in *Drosophila melanogaster*. *Genetics* (suppl.) **61**:69.

Miller, O. L. 1966. Structure and composition of peripheral nucleoli of salamander oocytes. *Nat. Cancer Monogr.* **23**:53.

Mohan, J. and F. Ritossa. 1970. Regulation of ribosomal RNA synthesis and its bearing on the bobbed phenotype in *Drosophila melanogaster*. *Develop. Biol.* **22**:495.

Mulder, M. P., P. van Duijn and H. J. Gloor. 1968. The replicative organization of DNA in polytene chromosomes of *Drosophila hydei*. *Genetica* **39**:385.

Polan, M. L., S. Freidman, J. G. Gall and W. Gehring. 1973. Isolation and characterization of mitochondrial DNA from *Drosophila melanogaster*. *J. Cell Biol.* **56**:580.

Rasch, E. 1970. DNA cytophotometry of salivary gland nuclei and other tissue systems in *Dipteran* larvae. In *Introduction to quantitative cytophotometry—II* (ed. G. L. Wied and G. F. Bahr). Academic Press, New York.

Rasch, E. M., H. J. Barr and R. W. Rasch. 1971. The DNA content of sperm of *Drosophila melanogaster*. *Chromosoma* **33**:1.

Ritossa, F. M. 1968. Unstable redundancy of genes for ribosomal RNA. *Proc. Nat. Acad. Sci.* **60**:509.

————. 1972. Procedure for magnification of lethal deletions of genes for ribosomal RNA. *Nature New Biol.* **240**:109.

————. 1973. Crossing over between X and Y chromosomes during ribosomal DNA magnification in *Drosophila melanogaster*. *Proc. Nat. Acad. Sci.* **70**:1950.

————. 1974. The bobbed locus. In *Genetics and biology of Drosophila* (ed. M. Ashburner and E. Novitski). Academic Press, New York (in press).

Ritossa, F. M. and G. Scala. 1969. Equilibrium variations in the redundancy of rDNA in *Drosophila melanogaster*. *Genetics* (suppl.) **61**:305.

Ritossa, F. M. and S. Spiegelman. 1965. Localization of DNA complementary to ribosomal RNA in the nucleolus organizer region of *Drosophila melanogaster*. *Proc. Nat. Acad. Sci.* **53**:737.

Ritossa, F. M., K. C. Atwood and S. Spiegelman. 1966. A molecular explanation of the bobbed mutants of *Drosophila* as partial deficiencies of "ribosomal" DNA. *Genetics* **54**:819.

Ritossa, F., C. Malva, E. Boncinelli, F. Graziani and L. Polito. 1971. The first steps of magnification of the DNA complementary to ribosomal RNA in *Drosophila melanogaster*. *Proc. Nat. Acad. Sci.* **68**:1580.

Rudkin, G. T. 1965. Structure and function of heterochromatin. In *Genetics today* (ed. S. J. Geerts) pp. 359–374. Pergamon Press, New York.

Sibatani, A. 1971. Difference in the proportion of the DNA specific to ribosomal RNA between adults and larvae of *Drosophila melanogaster*. *Mol. Gen. Genet.* **114**:177.

Spear, B. B. and J. G. Gall. 1973. Independent control of ribosomal gene replication in polytene chromosomes of *Drosophila melanogaster*. *Proc. Nat. Acad. Sci.* **70**:1359.

Stern, C. 1929. Uber die additive wirkung multipler allele. *Biol. Zbl.* **49**:261.

Tartof, K. D. 1971. Increasing the multiplicity of ribosomal RNA genes in *Drosophila melanogaster. Science* **171**:294.

————. 1973. Regulation of ribosomal RNA gene multiplicity in *Drosophila melanogaster. Genetics* **73**:57.

Vermeulen, C. W. and K. C. Atwood. 1965. The proportion of DNA complementary to ribosomal RNA in *Drosophila melanogaster. Biochem. Biophys. Res. Comm.* **19**:221.

Phosphorylation of Ribosomal Proteins in Eukaryotes

**Alphonse Krystosek, Lawrence F. Bitte,
M. Lawrence Cawthon and David Kabat**
Department of Biochemistry
University of Oregon Medical School
Portland, Oregon 97201

INTRODUCTION

Approximately four years ago, we first obtained evidence that several of the proteins in rabbit reticulocyte ribosomes are phosphoproteins rather than simple proteins. Their phosphoryl groups turn over intracellularly with a half-life of approximately 25 min, they become radioactive when the cells are incubated with [^{32}P]orthophosphate, and they occur in o-phosphoserine and in o-phosphothreonine residues (Kabat 1970, 1972). Thus the phosphoproteins are sites of an active and continuous metabolism. In addition, phosphorylation of ribosomal proteins occurs in other tissues (Blat and Loeb 1971; Loeb and Blat 1970; Majumder and Turkington 1972; Barden and Labrie 1973; Correze, Pinell and Nunez 1972; Bitte and Kabat 1972), in higher plants (Trewavas 1973) and in yeasts (J. Warner, personal communication), and it may therefore be ubiquitous in eukaryotes. However, ribosome phosphorylation is apparently absent from bacteria (Gordon 1971).

In this paper we review the evidence which indicates that ribosomal phosphoproteins exist intracellularly and describe some of the physiological variables, including cyclic AMP, that influence this metabolism. Also described are some of our preliminary attempts to analyze the functions of ribosome phosphorylation. In addition, we have briefly reviewed the studies of ribosomal protein phosphorylation in vitro which have employed [γ-^{32}P]ATP and several different kinds of protein kinase.

Evidence for Ribosomal Phosphoproteins

When rabbit reticulocytes are incubated in a nutritionally rich medium with [^{32}P]orthophosphate at 37°C, several of their ribosomal proteins become highly radioactive. Furthermore, acid hydrolysates of these proteins contain radioactive o-phosphoserine and o-phosphothreonine (Kabat 1970, 1972; Bitte and Kabat 1974). Typical electrophoretic fractionations of ^{32}P-labeled ribosomal proteins are shown in Figure 1. The electrophoresis is according to molecular weight in polyacrylamide gels which contain sodium

40S 60S

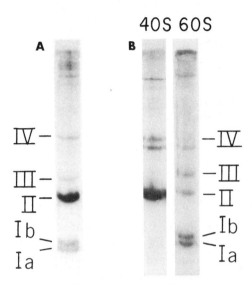

Figure 1   Electrophoresis of ^{32}P-labeled reticulocyte ribosomal proteins in 8% sodium dodecyl sulfate-polyacrylamide gels. Migration is toward the bottom and the radioactive components are visualized by autoradiography. The cells were incubated with [^{32}P]orthophosphate for 60 min. Each gel was used to analyze 30 μg of ribosomal preparation. *A*, Undissociated ribosomes, prepared by washing in a buffer containing 0.25 M KCl, 0.01 M MgCl$_2$, 0.01 M Tris-HCl (pH 7.4), and 0.5% sodium deoxycholate. *B*, Analysis of 60S and 40S ribosomal subunits, prepared without sodium deoxycholate (Blobel and Sabatini 1971). There is a slight contamination of 60S subunits by 40S subunits. Frames *A* and *B* are independent experiments. The component which migrates just ahead of band IV (B) is not reproducibly found in ribosome preparations. (Reprinted, with permission, from Cawthon et al. 1973.)

dodecyl sulfate, and the radioactive components are visualized by autoradiography. Among the approximately 70 proteins in highly purified ribosomes, we reproducibly find the five phosphoproteins labeled Ia, Ib, II, III and IV. These have molecular weights, based on their electrophoretic mobilities, of approximately 18,200; 19,500; 27,500; 33,000 and 53,000, respectively. The ribosomes in frame A were washed with a high ionic strength buffer and with sodium deoxycholate, a detergent that removes the proteins that bind directly to hemoglobin messenger RNA (Lebleu et al. 1971). Consequently, the phosphoproteins I–IV are very likely constituents of the ribosomal subunits. In agreement with this conclusion, frame B shows an analysis of purified ribosomal subunits. The ribosomes were dissociated with puromycin in a high ionic strength buffer (Blobel and Sabatini 1971), and the subunits were purified by sedimentation in a sucrose gradient. Each of the phosphoproteins reproducibly found in whole ribosomes is also found in purified ribosomal subunits. Components Ia, Ib and III occur in large subunits, whereas II occurs in small subunits. Components having the mobility of IV are often found on both subunits. Additional phosphoproteins occasionally contaminate even highly purified ribosomal subunits; however, they contain only a small fraction of the radioactivity which is incorporated into ribosome preparations.

Data supporting these conclusions have been obtained with mouse sarcoma-180 (S-180) tumor cells. Ribosomes from S-180 cells also reproducibly contain five phosphoproteins, and these coelectrophorese in sodium dodecyl sulfate–polyacrylamide gels with the phosphoproteins from rabbit reticulocyte ribosomes (Bitte and Kabat 1972, 1974). Accordingly, we have concluded that similar ribosomal phosphoproteins occur in these two very different eukaryote cells. As shown in Figure 2, the S-180 phosphoproteins also have the same ribosomal subunit localizations as those in reticulocytes. These extensive similarities support the conclusion that the phosphoproteins are true ribosomal constituents rather than contaminants; they also suggest that homologous phosphoproteins occur in ribosomes from different mammalian species.

Quantitative analysis of [^{32}P]orthophosphate incorporation into ribosomes in reticulocytes has indicated that the phosphoproteins I–IV become radioactive because their phosphoryl groups are turning over with a half-life of approximately 25 min at 37°C (Kabat 1972). Furthermore, each reticulocyte ribosome contains an average of 7–11 of these rapidly turning over phosphoryl groups. Direct chemical analyses of phosphate content of purified ribosomal proteins have suggested that there are approximately

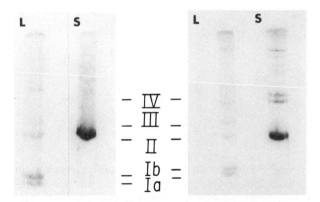

Figure 2   Electrophoresis of ^{32}P-labeled ribosomal subunit proteins from mouse S-180 cells (*left*) and from rabbit reticulocytes (*right*). The ribosomes were labeled intracellularly and were analyzed as in Figure 1. *L* is large subunits, *S* is small subunits. The reticulocyte data are the same experiment as in Figure 1. The ribosomal phosphoproteins from these two sources are very similar.

14–20 phosphoryl groups in ribosomes from rat liver (Loeb and Blat 1970), from reticulocytes and S-180 cells (Bitte and Kabat 1974), and from adrenal tumor cells (Walton and Gill 1973). The chemical analyses may be overestimates because some proteins lacking phosphate contribute to the phosphomolybdate color reaction. However, it is also possible that several phosphoryl groups may be turning over only slowly in ribosomal proteins; these would have escaped detection by the ^{32}P-labeling studies (i.e., in Figures 1 and 2). In any case, these analyses support the conclusion that ribosomal proteins from different eukaryotic cells are phosphorylated intracellularly to a stoichiometrically significant extent.

Chemical Heterogeneity of Mammalian Ribosomes

Evidence also indicates that eukaryotic ribosomes are heterogeneous with respect to their phosphorylation at specific sites. As illustrated in Figure 3, there are striking and reproducible differences between the phosphorylation patterns of polyribosomes and of single ribosomes. The figure shows the electrophoretic patterns on 4% polyacrylamide gels of ^{32}P-labeled phospho-

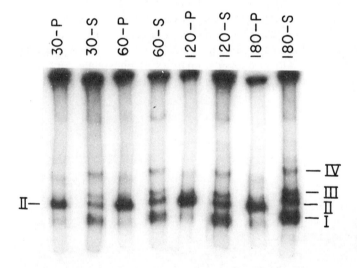

Figure 3   Electrophoresis of ^{32}P-labeled reticulocyte ribosomal constituents in sodium dodecyl sulfate-polyacrylamide gels. The gels are 4%, and the radioactive components are visualized by autoradiography. The polysomes (*P*) and single ribosomes (*S*) were from cells labeled with [^{32}P]orthophosphate for 30, 60, 120 or 180 min. Bands Ia and Ib are not resolved on these 4% gels; however, other experiments have indicated that both of these phosphoproteins are more highly labeled in single ribosomes than in polysomes. (Reprinted, with permission, from Kabat 1972).

proteins from these classes of ribosomes. The ribosomes were isolated after incubating reticulocytes with [^{32}P]orthophosphate for different times. We reproducibly find that proteins I, III and IV are more heavily phosphorylated in single ribosomes than in polyribosomes, whereas protein II is more heavily phosphorylated in polysomes. Similar data have been obtained with S-180 cells (Bitte and Kabat 1972). These data strongly support other evidence (e.g., Adamson, Howard and Herbert 1969; Kabat and Rich 1969; Hogan and Korner 1968) indicating that single ribosomes are not participating in the ribosomal subunit-polyribosome cycle of protein synthesis. Rather, single ribosomes constitute a stagnant inactive pool of ribosomes in animal cells. The subunits of single ribosomes reenter the active pool of native subunits only very slowly and thus may be excluded from protein synthesis for long periods of time. The phosphorylation differences between polysomes and single ribosomes have been analyzed in more detail (Kabat 1972), and it has been concluded that they are caused by a configurational difference between these two classes of ribosomes. Ultracentrifugal evidence for such a configurational difference has been obtained by other workers (Vournakis and Rich 1971). In other words, we believe that the arrangement of ribosomal components differs in polysomes and in single ribosomes, and that this difference of architecture causes the differences in the steady-state ^{32}P-labeling levels for the various phosphoproteins. Most interestingly, these data show that phosphorylation renders ribosomes chemically heterogeneous and that phosphorylation patterns may be used as an intracellular probe for analyzing structure-function interrelationships among mammalian ribosomes.

Role of Cyclic AMP

When reticulocytes are incubated with cyclic AMP or with dibutyryl cyclic AMP, there occurs a selective stimulation of protein II phosphorylation (Cawthon et al. 1973). The slight increase in the labeling of the other phosphoproteins is caused by a cyclic AMP-dependent increase in ATP-specific activity, rather than by any change in extent of phosphorylation. As can be seen in the experiment in Figure 4, protein-II labeling is increased approximately fivefold by the added cyclic AMP, whereas the labeling of the other phosphoproteins is relatively unaffected. This and other evidence (Kabat 1970) has shown that protein-II phosphorylation can be regulated independently of the other ribosomal phosphoproteins. In agreement with this conclusion, protein-II labeling is a variable proportion of the total ribosomal labeling in different preparations of reticulocytes (e.g., compare Figures 1a and 4) and the extent of its stimulation by cyclic AMP is correspondingly variable.

In S-180 cells, the labeling of protein II appears to always be extensive, and exogenously added cyclic AMP or dibutyryl cyclic AMP do not further

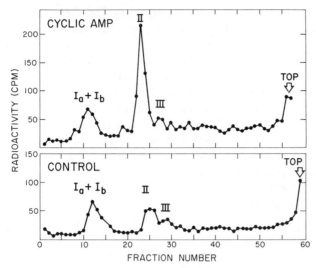

Figure 4 Electrophoresis of ^{32}P-labeled ribosomal proteins in 8% sodium dodecyl sulfate-polyacrylamide gels. Migration is toward the left. The cell incubations with [^{32}P]orthophosphate were for 100 min, in the absence or presence of 4 mM cyclic AMP. Gel sections (1 mm) were assayed for radioactivity as described elsewhere (Bitte and Kabat 1974). (Adapted from Cawthon et al. 1973.)

stimulate its phosphorylation. However, these additions do cause an increase in [^{32}P]orthophosphate incorporation into ATP, and there is an attendant increase in the labeling of all of the phosphoproteins. Nevertheless, these increases do not reflect an actual enhancement of ribosomal protein phosphorylation.

Studies of the Function of Ribosomal Protein Phosphorylation

In approaching the problem of the functions of ribosome phosphorylation, it is essential to realize that there are several different phosphoproteins and that ribosomes have a complex metabolism. For example, they have active and inactive forms. They are turning over intracellularly and must be recognized by degradative enzymes. They also dimerize and interact with membranes and with other cellular components. Obviously not all of the functions of ribosomal proteins are related solely to protein synthetic activity. Furthermore, it is important to understand that the assay conditions may alter the phosphoproteins. For example, ribosomes contain bound protein kinase (Majumder and Turkington 1972; Kabat 1971; Jergil 1972; Traugh, Mumby and Traut 1973) and protein synthesis is routinely assayed in high ATP concentrations. Furthermore, ribosomes also contain bound phosphoprotein phosphatase (Kabat 1971) which can dephosphorylate the ribosomal proteins.

Eil and Wool (1973b) have recently attempted to analyze the functions of ribosomal protein phosphorylation by comparing the protein synthetic activities of isolated liver ribosomal subunits ("control") with the same subunit preparations after they had been phosphorylated in vitro by incubation with ATP and with protein kinase ("phosphorylated"). Although no evidence for functional differences was obtained, these studies were inconclusive for several reasons. First, the control liver ribosomes may actually have been fully phosphorylated at their physiologically relevant sites; no measurements were made of their endogenous phosphorylation. Second, as described below, in vitro phosphorylation has not been shown to result in the modification of any of the specific serine and threonine residues which are phosphorylated intracellularly. Actually the available evidence indicates that at least some, and perhaps the majority, of the sites ^{32}P-labeled in vitro with [γ-^{32}P]ATP and various protein kinases are negligibly labeled when the cells are incubated with [^{32}P]orthophosphate.

An approach we have used to analyze for possible functions of ribosomal protein phosphoryl groups has been to compare the protein synthetic activity of isolated reticulocyte ribosomes with the same ribosomes dephosphorylated in vitro by *Escherichia coli* alkaline phosphatase (orthophosphoric monoester phosphohydrolase, E.C. 3.1.3.1). The ribosomes used for our initial studies were unfractionated and contained both polyribosomes and single ribosomes.

Highly purified alkaline phosphatase can dephosphorylate all of the sites on ribosomes which become radioactive when the reticulocytes are incubated with [^{32}P]orthophosphate. As seen in Figure 5, no radioactive proteins are fully resistant to the enzyme treatment. However, the dephosphorylation reaction requires high concentrations of enzyme and exhibits unexpected kinetics (Figure 6). At any phosphatase concentration, a fraction of the ^{32}P-labeled sites appear relatively resistant to dephosphorylation. However, analysis of these relatively resistant sites by electrophoresis indicates that they occur in all of the five phosphoproteins: Ia, Ib, II, III and IV. Furthermore, the number of apparently resistant sites depends on the phosphatase concentration. These studies indicate that the affinity of phosphatase for the ribosomal phosphoryl groups is low and that the reaction may be influenced by the heterogeneity of ribosomal configurations. The dephosphorylation of polysomes, for example, may depend upon whether peptidyl-tRNA is in the A or D sites. In this context, it is interesting that efficient dephosphorylation of trout protamine also requires high concentrations and prolonged incubations with *E. coli* alkaline phosphatase (Marushige, Ling and Dixon 1969). Figure 6 also indicates that there is endogenous phosphatase activity in the reticulocyte ribosome preparation; this endogenous enzyme can be removed by washing the ribosomes with

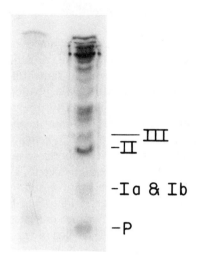

Figure 5   Dephosphorylation of ^{32}P-labeled ribosomal proteins with *E. coli* alkaline phosphatase. The proteins from control (*right*) and from enzyme-treated ribosomes are compared by electrophoresis in sodium dodecyl sulfate-polyacrylamide gels and the radioactive components are visualized by auto-radiography. The ^{32}P-labeled reticulocyte ribosomes were purified using a low ionic strength buffer (0.01 M KCl, 0.01 M Tris-HCl pH 7.35, 0.0015 M $MgCl_2$) and are consequently contaminated by putative nonribosomal phospho-proteins (Bitte and Kabat 1974). The Worthington (type BAPF) phosphatase concentration was 0.1 mg/ml, and the dephosphorylation reaction was for 15 min. All of the phosphoproteins are efficiently dephosphorylated. The band labeled *P* occurs specifically on polysomes and is extracted by buffers which contain sodium deoxycholate (Kabat 1970).

solutions containing 0.5% sodium deoxycholate. Control experiments also indicated that the integrity of reticulocyte polysomes and of ribosomal proteins was unaffected by the phosphatase treatments.

Following dephosphorylation, the alkaline phosphatase can be rapidly and completely removed from reticulocyte ribosomes. Gel filtration on an agarose A-5m column is a very effective method (Figure 7). The ribosomes are recovered in the excluded volume within 30 min of application to the column and they are diluted only 2- to 3-fold.

By these methods, we prepared control and dephosphorylated reticu-locyte ribosomes. Their globin synthesizing activities were assayed in a modification of the messenger RNA-dependent, cell-free system which has been described by Anderson and his colleagues (Crystal et al. 1972; Schafritz, Drysdale and Isselbacher 1973). The major modification was to use the reticulocyte ribosome preparations (containing polysomes plus

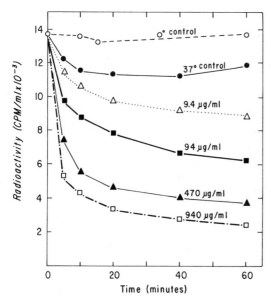

Figure 6 Dephosphorylation of ³²P-labeled ribosomal proteins with different concentrations of *E. coli* alkaline phosphatase. The reaction mixtures were at 37°C in 0.25 M KCl, 0.01 M MgCl₂, 0.01 M Tris-HCl pH 7.9, 0.005 M 2-mercaptoethanol.

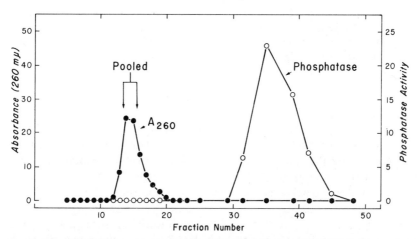

Figure 7 Separation of *E. coli* alkaline phosphatase from ribosome preparations by gel filtration on an agarose A-5m column (BioRad Laboratories, Richmond, California). The column dimensions are 30 cm × 1.2 cm diameter. The eluting buffer was 0.05 M Tris-HCl pH 7.4, 0.01 M KCl, 0.005 M MgCl₂, 0.0017 M DTT, and the elution was at 0°C.

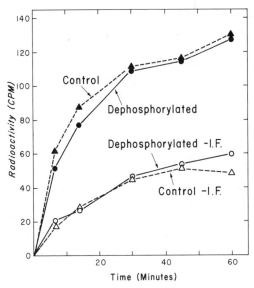

Figure 8   Assay of protein synthetic activity of control and of dephosphorylated reticulocyte ribosomes. The ribosomes were a mixture of polysomes and of single ribosomes, and the dephosphorylation was with 0.8 mg/ml alkaline phosphatase at 37°C for 10 min. in the conditions described in Figure 6. The control ribosomes were not treated with phosphatase, were maintained at 0°C, and were passed through the agarose columns. The assay system is described in the text and contained 0.2 A_{260} units of ribosomes per 100 μl solution. Incorporation of [^{14}C]leucine was linearly dependent upon ribosome concentration in this concentration range. The data points are the hot trichloroacetic acid-precipitable radioactivity in 10 μl of reaction mixture. The samples denoted by solid points contained a saturating amount of initiation factors.

single ribosomes) instead of separately adding globin messenger RNA plus messenger-depleted ribosomes. The modified system was strongly stimulated by initiation factors and was also stimulated approximately twofold by addition of excess rabbit globin messenger RNA. The extent of [^{14}C]leucine incorporation by the maximally stimulated system (i.e., containing ribosomes plus initiation factors plus globin messenger RNA) was approximately 15 moles of leucine incorporation/mole ribosomes.

The control and dephosphorylated ribosomes were approximately equally active in the cell-free system (Figure 8), both in the presence and absence of a saturating amount of initiation factors. Similarly, these two ribosome preparations were equally active in the fully optimized system containing additional globin messenger RNA.

Although these results are negative, there are four major reasons why we believe it is premature to conclude that the phosphoryl groups on ribosomal proteins lack essential functions. First, as described above, their functions might possibly be unrelated to protein synthetic activity per se. Second, it is conceivable that endogenous protein kinase or phosphoprotein phosphatase in the assay system could alter the two ribosome preparations so rapidly that they become functionally equivalent; however, we believe that this is unlikely because we have observed no significant functional differences even at very early times of incorporation. Third, it is important to realize that isolated eukaryotic ribosomes function in vitro at only a small fraction of their intracellular rate. There seems to be no valid reason for assuming that the rate-limiting steps for ribosome function in vitro are the same ones which control ribosome function in vivo. The fourth point also seems to be important, namely, the conditions selected as being optimal for assay of ribosome function may be precisely those which negate the role of the phosphoryl groups. For example, an effect of initiation factors on poly-uridylic acid-dependent polyphenylalanine synthesis can only be revealed at low Mg^{++} concentrations (Schafritz et al. 1970; Schafritz and Anderson 1970). Similarly, muscle glycogen phosphorylase a and b differ by one phosphoryl group per polypeptide chain; however, the inactive dephos-phorylated enzyme can be activated by various means besides addition of a phosphoryl group, e.g., by changes in the concentrations of glucose-6-phosphate, AMP, ATP and phosphate ion (Morgan and Parmeggiani 1964). Likewise the activity of the mammalian pyruvate dehydrogenase complex is subject to product inhibition by acetyl-CoA and by NADH, in-dependent of its regulatory phosphorylation-dephosphorylation cycle (Reed et al. 1974).

In Vitro Phosphorylation of Ribosomal Proteins

Certain eukaryotic ribosomal proteins can be phosphorylated in vitro by incubating ribosomes or ribosomal subunits with $[\gamma\text{-}^{32}P]ATP$ and with protein kinases (Loeb and Blat 1970; Kabat 1971; Majumder and Turk-ington 1972; Eil and Wool 1973a; Li and Amos 1971; Walton and Gill 1973; Delaunay et al. 1973; Yamamura et al. 1972; Fontana, Picciano and Lovenberg 1972; Stahl, Welfle and Bielka 1972; Jergil 1972; Traugh, Mumby and Traut 1973; Traugh and Traut 1974; Correze, Pinell and Nunez 1972; Barden and Labrie 1973). The kinases utilized were either bound to the isolated ribosomes (Kabat 1971; Li and Amos 1971; Eil and Wool 1973a; Jergil 1972; Fontana, Picciano and Lovenberg 1972; Correze, Pinell and Nunez 1972; Barden and Labrie 1973) or present in the cytosol. The reaction with ribosome-associated kinases has been found to be either unaffected or weakly stimulated by cyclic AMP. Similarly, both cyclic AMP-dependent and -independent protein kinases from the cytosol have been employed. However, the site specificity of in vitro ribosome phos-phorylation has not been reported to be affected by cyclic AMP.

The physiological significance of these in vitro studies remains uncertain.

A major problem is that the number of ribosomal proteins which can be phosphorylated by these enzymes in vitro appears to be considerably larger than the number which are phosphorylated in vivo. Data supporting this conclusion are available for ribosomes from pituitary (Barden and Labrie 1973) and mammary glands (Majumder and Turkington 1972), rat liver (Loeb and Blat 1970; Stahl, Welfle and Bielka 1972; Correze, Pinell and Nunez 1972; Eil and Wool 1973a; Wool and Stöffler, this volume), and reticulocytes (Kabat 1971; Delaunay et al. 1973; Traugh, Mumby and Traut 1973). At least 14 proteins can be phosphorylated in vitro when separated ribosomal subunits are utilized as the substrate. However, the number of proteins phosphorylated is fewer when undissociated 80S ribosomes are employed (Eil and Wool 1973a; Stahl, Welfle and Bielka 1972). Preliminary evidence suggests that at least several of the ribosomal proteins phosphorylated in vivo are among the proteins which can be phosphorylated in vitro (Majumder and Turkington 1972; Kabat 1971). Apparently, an appreciable fraction of the protein sites which can be phosphorylated in vitro are either not modified intracellularly or are only phosphorylated at certain stages of the ribosome's life cycle.

Current Problems Concerning
Ribosome Phosphorylation

Several precautions should be considered in evaluating experiments on the intracellular labeling of ribosomal phosphoproteins with [^{32}P]orthophosphate. First, cyclic AMP and other physiological variables can alter phosphate transport into cells and ATP specific activity (Weber and Edlin 1971; De Venanzi, Altares and Forero 1964; Zahlten et al. 1972). Consequently, induced changes of ^{32}P-incorporation into ribosomes can occur without any change in ribosome phosphorylation. For example, exogenously added cyclic AMP causes an increase in specific activity of phosphoryl groups incorporated into ribosomal proteins in reticulocytes (Cawthon et al. 1973). Second, phosphoprotein phosphatase adsorbs to ribosomes (e.g., see Figure 6; Kabat 1971). This enzyme can partially dephosphorylate ribosomes during their preparation, and every effort must be made to maintain the solutions at 0°C. The loss of radioactivity occurs more rapidly from proteins Ia and Ib than from protein II. A third problem also arises due to loss of ^{32}P label from ribosomes during their isolation. Although ribosomal subunits contain each of the phosphoproteins present in undissociated ribosomes (Figure 1), the specific radioactivity of subunits is generally only approximately 30–50% as high as that of intact ribosomes. Our evidence suggests that this radioactivity loss can only be partially attributed to dephosphorylation by phosphatase during subunit preparation. It is possible that some of the ribosomal phosphoproteins are partially extracted from subunits during their preparation. In support of this idea, Walton and Gill

(1973) have recently presented evidence that phosphorylation of ribosomal proteins in vitro may cause their partial dissociation from ribosomes.

The number of phosphoproteins in eukaryotic ribosomes also remains somewhat uncertain. Our evidence suggests that at least five proteins are phosphorylated to some extent intracellularly in reticulocytes and in S-180 cells. Conceivably there may be additional proteins which are phosphorylated relatively weakly or in which the phosphoryl groups are quiescent and turn over relatively slowly; our experiments would have failed to detect such phosphoproteins. Majumder and Turkington (1972) have reported that there are eight ribosomal phosphoproteins which become labeled after 4 hr of [^{32}P]orthophosphate incorporation by mouse mammary glands. However, the criteria used for establishing ribosome purity were not described and purified ribosomal subunits were not examined. It has also been reported that there is one major and several minor radioactive ribosomal phosphoproteins in rat liver (Loeb and Blat 1970; Blat and Loeb 1971). Our data are consistent with the latter conclusion since protein II is clearly the major radioactive phosphoprotein in S-180 cells and in reticulocytes which have been treated with cyclic AMP (Figures 1, 2, 4). Protein II is also unique among the phosphoproteins in that it is localized selectively in polysomes (Figure 3), it occurs in the smaller ribosomal subunits (Figure 4), and its labeling is increased by cyclic AMP (Figure 4). A failure to observe the clear labeling of other phosphoproteins could readily occur in a tissue that contained predominantly polyribosomes and relatively few single ribosomes, especially if the intracellular cyclic AMP concentration were sufficiently high. Thus, while it is hard to compare work from different laboratories, it would seem that there is no firm evidence for tissue- or species-specific patterns of ribosomal phosphoproteins. In fact, our own work (Bitte and Kabat 1972) suggests that ribosomal phosphoproteins in very different cells are likely to be very similar.

The most important remaining problem concerns the function(s) of ribosomal protein phosphorylation. There are good reasons for considering the available negative evidence (Eil and Wool 1973b; Krystosek, Bitte and Kabat 1974) (see also Figure 8) on this point as inconclusive. Indeed, we believe that the ribosomal phosphoryl groups must have essential functions because they have apparently been retained during eukaryotic evolution. Furthermore, the pattern of ribosomal protein phosphorylation appears to be subject to physiological control. We therefore conclude that ribosomal protein phosphorylation plays a functional role which remains unknown.

Acknowledgments

This research was supported by grants from the U.S. Public Health Service (HL14960-04) and from the National Science Foundation (GB-39815). L. B. and A. K. were supported by a predoctoral training grant (GM-01200) from the U.S. Public Health Service.

References

Adamson, S. D., G. A. Howard and E. Herbert. 1969. The ribosome cycle in a reconstituted cell-free system from reticulocytes. *Cold Spring Harbor Symp. Quant. Biol.* **34**:547.

Barden, N. and F. Labrie. 1973. Cyclic adenosine 3′, 5′-monophosphate-dependent phosphorylation of ribosomal proteins from bovine anterior gland. *Biochemistry* **12**:3096.

Bitte, L. and D. Kabat. 1972. Phosphorylation of ribosomal proteins in sarcoma 180 tumor cells. *J. Biol. Chem.* **247**:5345.

———. 1974. Isotopic labeling and analysis of phosphoproteins from mammalian ribosomes. In *Methods in enzymology* (ed. K. Moldave) vol. 30, pp. 563. Academic Press, New York.

Blat, C. and J. Loeb. 1971. Effect of glucagon on phosphorylation of some rat liver ribosomal proteins in vivo. *Fed. Eur. Biochem. Soc. Letters* **18**:125.

Blobel, G. and D. Sabatini. 1971. Dissociation of mammalian polyribosomes into subunits by puromycin. *Proc. Nat. Acad. Sci.* **68**:390.

Cawthon, M. L., L. F. Bitte, A. Krystosek and D. Kabat. 1973. Effects of cyclic adenosine 3′, 5′-monophosphate on ribosomal protein phosphorylation in reticulocytes. *J. Biol. Chem.* **249**:275.

Correze, C., P. Pinell and J. Nunez. 1972. Effects of thyroid hormones on phosphorylation of liver ribosomal proteins and on protein phosphokinase activity. *FEBS Letters* **23**:87.

Crystal, R. G., A. W. Nienhuis, N. A. Elson and W. F. Anderson. 1972. Initiation of globin synthesis. Preparation and use of reticulocyte ribosomes retaining initiation region messenger ribonucleic acid fragments. *J. Biol. Chem.* **247**:5357.

Delaunay, J., J. E. Loeb, M. Pierre and G. Schapira. 1973. Mammalian ribosomal proteins: Studies on the *in vitro* phosphorylation patterns of ribosomal proteins from rabbit liver and reticulocytes. *Biochim. Biophys. Acta* **312**:147.

De Venanzi, F., C. D. Altares and J. Forero. 1964. Organ distribution of radioactive orthophosphate and total phosphate in the rat. The influence of glucagon and insulin. *Diabetes* **13**:609.

Eil, C. and I. G. Wool. 1973a. Phosphorylation of liver ribosomal proteins. Characteristics of the protein kinase reaction and studies of the structure of phosphorylated ribosomes. *J. Biol. Chem.* **248**:5122.

———. 1973b. Function of phosphorylated ribosomes. The activity of ribosomal subunits phosphorylated *in vitro* by protein kinase. *J. Biol. Chem.* **248**:5130.

Fontana, J. A., D. Picciano and W. Lovenberg. 1972. The identification and characterization of a cyclic AMP-dependent protein kinase on rabbit reticulocyte ribosomes. *Biochem. Biophys. Res. Comm.* **49**:1225.

Gordon, J. 1971. Determination of an upper limit to the phosphorus content of polypeptide chain elongation factor and ribosomal proteins in *Escherichia coli. Biochem. Biophys. Res. Comm.* **44**:579.

Hogan, B. L. M. and A. Korner. 1968. The role of ribosomal subunits and 80-S monomers in polysome formation in an ascites tumor cell. *Biochim. Biophys. Acta* **169**:139.

Jergil, B. 1972. Protein kinase from rainbow trout testis ribosomes. Partial purification and characterization. *Eur. J. Biochem.* **28**:546.

Kabat, D. 1970. Phosphorylation of ribosomal proteins in rabbit reticulocytes. Characterization and regulatory aspects. *Biochemistry* **9**:4160.

———. 1971. Phosphorylation of ribosomal proteins in rabbit reticulocytes. A cell-free system with ribosomal protein kinase activity. *Biochemistry* **10**:197.

———. 1972. Turnover of phosphoryl groups in reticulocyte ribosomal phosphoproteins. *J. Biol. Chem.* **247**:5338.

Kabat, D. and A. Rich. 1969. The ribosomal subunit-polyribosome cycle in protein synthesis of embryonic skeletal muscle. *Biochemistry* **8**:3742.

Krystosek, A., L. F. Bitte and D. Kabat. 1974. Phosphorylation if ribosomal proteins in higher organisms. In *Proc. 3rd Int. Symp., Metabolic Interconversion of Enzymes* (ed. E. H. Fischer et al.) p. 165. Springer-Verlag, Berlin and New York.

Lebleu, B., G. Marbaix, G. Huez, J. Temmerman, A. Burny and H. Chantrenne. 1971. Characterization of the messenger ribonucleoprotein released from reticulocyte polyribosomes by EDTA treatment. *Eur. J. Biochem.* **19**:264.

Li, C. and H. Amos. 1971. Alteration of phosphorylation of ribosomal proteins as a function of variation of growth conditions of primary cells. *Biochem. Biophys. Res. Comm.* **45**:1398.

Loeb, J. E. and C. Blat. 1970. Phosphorylation of some rat liver ribosomal proteins and its activation by cyclic AMP. *Fed. Eur. Biochem. Soc. Letters* **10**:105.

Majumder, G. C. and R. W. Turkington. 1972. Hormone-dependent phosphorylation of ribosomal and plasma membrane proteins in mouse mammary gland in vitro. *J. Biol. Chem.* **247**:7207.

Marushige, K., V. Ling and G. H. Dixon. 1969. Phosphorylation of chromosomal basic proteins in maturing trout testis. *J. Biol. Chem.* **244**:5953.

Morgan, H. E. and A. Parmeggiani. 1964. Regulation of glycogenolysis in muscle. III. Control of muscle glycogen phosphorylase activity. *J. Biol. Chem.* **239**:2440.

Reed, L. J., F. H. Pettit, T. E. Roche, P. J. Butterworth, C. R. Barrera and C. S. Tsai. 1974. Structure, function and regulation of the mammalian pyruvate dehydrogenase complex. In *Proc. 3rd Int. Symp., Metabolic Interconversion of Enzymes* (ed. E. H. Fischer et al.) p. 199. Springer-Verlag, Berlin and New York.

Schafritz, D. A. and W. F. Anderson. 1970. Isolation and partial characterization of reticulocyte factors M$_1$ and M$_2$. *J. Biol. Chem.* **245**:5553.

Schafritz, D. A., J. W. Drysdale and K. J. Isselbacher. 1973. Translation of liver messenger ribonucleic acid in a messenger-dependent reticulocyte cell-free system. Properties of the system and identification of ferritin in the product. *J. Biol. Chem.* **248**:3220.

Schafritz, D. A., P. M. Prichard, J. M. Gilbert and W. F. Anderson. 1970. Separation of two factors M$_1$ and M$_2$ required for poly U-dependent polypeptide synthesis by rabbit reticulocyte ribosomes at low magnesium ion concentration. *Biochem. Biophys. Res. Comm.* **38**:721.

Stahl, J., H. Welfle and H. Bielka. 1972. Studies on proteins of animal ribosomes. XIV. Analysis of phosphorylated rat liver ribosomal proteins by two-dimensional polyacrylamide gel electrophoresis. *FEBS Letters* **26**:236.

Traugh, J. A. and R. R. Traut. 1974. Characterization of protein kinases from rabbit reticulocytes. *J. Biol. Chem.* **249**:1207.

Traugh, J. A., M. Mumby and R. R. Traut. 1973. Phosphorylation of ribosomal proteins by substrate-specific protein kinases from rabbit reticulocytes. *Proc. Nat. Acad. Sci.* **70**:373.

Trewavas, A. 1973. The phosphorylation of ribosomal protein in *Lemma minor*. *Plant Physiol.* **51**:760.

Vournakis, J. and A. Rich. 1971. Size changes in eukaryotic ribosomes. *Proc. Nat. Acad. Sci.* **68**:3021.

Walton, G. M. and G. N. Gill. 1973. Adenosine 3′,5′-monophosphate and protein kinase-dependent phosphorylation of ribosomal protein. *Biochemistry* **12**:2604.

Weber, M. J. and G. Edlin. 1971. Phosphate transport, nucleotide pools, and ribonucleic acid synthesis in growing and in density-inhibited 3T3 cells. *J. Biol. Chem.* **246**:1828.

Yamamura, H., Y. Inouye, R. Shimomura and Y. Nishizuka. 1972. Similarity and pleiotropic actions of adenosine 3′,5′-monophosphate-dependent protein kinase from mammalian tissues. *Biochem. Biophys. Res. Comm.* **46**:589.

Zahlten, R. N., A. A. Hochberg, F. W. Stratman and H. A. Lardy. 1972. Glucagon-stimulated phosphorylation of mitochondrial and lysosomal membranes of rat liver in vivo. *Proc. Nat. Acad. Sci.* **69**:800.

How Transfer RNA
May Move
Inside the Ribosome

Alexander Rich
Department of Biology
Massachusetts Institute of Technology
Cambridge, Massachusetts 02139

INTRODUCTION

It has been known for many years that transfer RNA (tRNA) is the molecule which carries amino acids into the ribosome and upon which the growing polypeptide chain is assembled. The broad outlines of tRNA and ribosomal participation in protein synthesis are known: tRNA molecules are sequentially lined up in the ribosome through a specific interaction of their anticodon triplets with the codon triplets on the messenger RNA (mRNA) strand and, in this manner, the sequence information encoded in the polynucleotides of mRNA is translated into an individual polypeptide chain. This takes place through a stepwise assembly of amino acids to yield the sequence specificity in proteins. Although this information is generally understood in outline, we do not know in detail how this process occurs. A detailed understanding of a chemical process of necessity requires structural information. The central organelle in this assembly process is the ribosome, and although we have some information concerning its assembly and structure, we do not as yet have a three-dimensional understanding of its structure. However, there has been progress in understanding the three-dimensional structure of tRNA. We have reported the three-dimensional structure of yeast phenylalanine tRNA at a resolution of 5.5 Å (Kim et al. 1972) and 4.0 Å (Kim et al. 1973). More recently we have completed the analysis at a resolution of 3.0 Å (Suddath et al. 1974), which makes it possible to ascertain many details in the folding of the molecule and the positioning of its various components. This naturally prompts us to ask whether the three-dimensional structural knowledge of tRNA can be used at all to infer something of its mode of action within the ribosome. The answer is necessarily incomplete. This information does allow us to speculate within some boundary limits, but it should be understood that these are simply speculations. Here we shall briefly review information concerning the chemistry and structure of tRNA and its positioning within the ribosome. The three-dimensional structure of tRNA points out a few paradoxes that arise when the molecule is combined with messenger RNA. These will be described, and at the same time, a view will be developed which may allow us to resolve some of these paradoxes. We will suggest a specific model for tRNA movement within the ribosome.

The Cloverleaf Sequence

A large number of transfer RNA molecules have been isolated from different species and purified to homogeneity. The nucleotide sequences of these molecules have been determined and they all fall in the well-known cloverleaf arrangement of secondary structure, an example of which is shown in the yeast phenylalanine tRNA sequence in Figure 1 (RajBhandary and Chang 1968). There are four stem regions that are largely composed of complementary base pairs. In addition to this, there are four loop regions indicated in the figure as well as the common CCA sequence at the 3'-OH end. It is significant that over 50 tRNAs have now been sequenced and all of them fit the same cloverleaf pattern. This suggests that the secondary structure expressed in the hydrogen-bonded interactions inferred from the cloverleaf arrangement of nucleotides may also be reflected in the three-dimensional structure of the molecule. It is important to note that all tRNAs that have been sequenced fit in this same general category, irrespective of

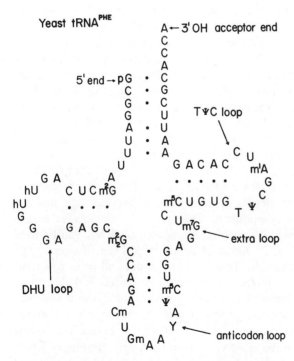

Figure 1   The sequence of nucleotides in yeast phenylalanine-tRNA, shown in the conventional cloverleaf diagram (RajBhandary and Chang 1968).

the species or source (plant, animal or viral). Thus there are significant commonalities among all tRNA molecules. One does not find differences in tRNA molecules comparable to the differences between eukaryotic and prokaryotic ribosomes.

In addition to the common features among different tRNA molecules, it is important to recognize that there are significant differences. Two areas of variability are found when one surveys tRNA sequences. One is found in the extra or variable loop region indicated in Figure 1 and the other is in the dihydrouracil (DHU) stem and loop. The distribution in the number of nucleotides found in the extra loop region of 45 tRNAs is shown diagrammatically in Figure 2 (a recent compilation of tRNA sequences is given in Holmquist, Jukes and Pangburn 1973). Eighty percent of the tRNAs in this group have either four or five nucleotides in the extra loop, whereas the other 20% are much larger and contain 13–21 nucleotides. The large extra-loop regions can often be arranged to include both a stem region, in which there are complementary base pairs, and a terminal loop connecting the two segments of the stem. This degree of variability is significant in thinking about the role of transfer RNA in the ribosome since it means the same ribosomal structure has to accommodate tRNAs with different-sized appendages. Furthermore, it suggests that the extra loop regions must be on the outside of the molecule rather than on the inside.

In contrast to the extreme variability in the extra loop, there is a much more modest distribution in the dihydrouracil (DHU) stem and loop. Figure 3 is a histogram showing the number of nucleotides in the DHU loop regions for the same 45 tRNAs. Among these molecules, 26 had DHU stems containing four base pairs and 19 had stems containing three base pairs. The distribution shown in Figure 3 is modified to yield the heavy solid line by assuming that all DHU stems contain four base pairs. As can be seen in the figure, the resultant size of the DHU loop contains predominantly eight or nine nucleotides, although there are a few containing seven or ten nucleotides. If we assume all tRNA molecules have a common tertiary structure, it is likely that the DHU loop will be on the outside of the molecule.

With these two exceptions, all the other sequence distributions in the tRNA molecule appear to be invariant. The 3'-CCA end is common to all tRNAs. Likewise the number of nucleotides in the acceptor stem, the TψC stem and loop, and the anticodon stem and loop seem to be constant in all species.

In considering the function of tRNA in the ribosome, the commonalities of sequence are as significant as the variabilities. The significant sequence commonalities suggest that there may be a three-dimensional structural framework which is the same for all molecules. Furthermore, it is reasonable to assume that some of the common features of tRNA structure must include elements that interact with the ribosomal machinery, whereas it is likely that the variable regions do not interact with the ribosomal machinery.

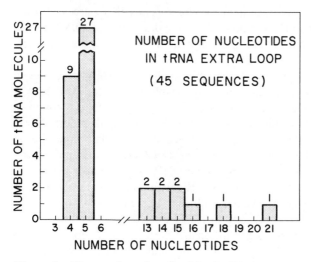

Figure 2   The number of nucleotides in tRNA extra loops. The data are plotted for 45 sequences.

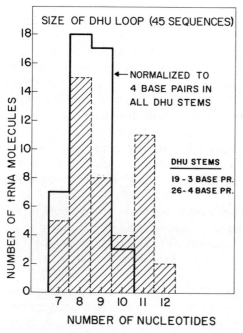

Figure 3   The number of nucleotides in the DHU loop is shown for the same 45 sequences as in Figure 2. If these all had four base pairs in the DHU stem, the solid line distribution would be found.

We make this assumption because many different transfer RNAs are able to work with the same ribosome even though they have significant differences in the number and sequences of their polynucleotides.

Three-dimensional Structure of Transfer RNA

The first transfer RNA to have its structure determined by X-ray crystallographic techniques was yeast phenylalanine tRNA, the sequence of which is shown in Figure 1. At the present time, the electron density map has been computed to 3-Å resolution, which allows the chain to be traced almost in its entirety (Suddath et al. 1974). At 3-Å resolution, the phosphate and ribose residues can be seen as individual clusters of electron density and the hydrogen-bonded bases can be seen when they are stacked, even though they cannot be differentiated one from the other. The molecule has a three-dimensional folding that incorporates the hydrogen bonding implied in the cloverleaf diagram. A solid model of the electron density at 4-Å resolution is shown in Figure 4. The molecule is bent, somewhat L-shaped, consisting of two arms, horizontal and vertical in Figure 4. There are four stem regions containing double helices. One arm of the L is composed of the CCA stem and the TψC stem, both of which are coiled around approxi-

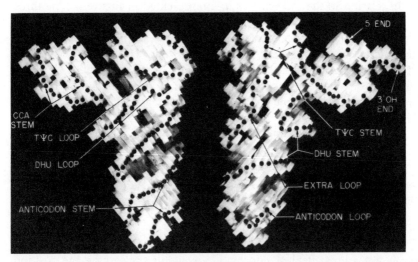

Figure 4 Two views from opposite sides of a solid molecular model of yeast phenylalanine-tRNA as seen at 4.0-Å resolution. The molecule is approximately 20 Å thick in a direction perpendicular to the page. In order to make the tracing of the chains more visible, round-headed pins have been inserted into the molecule. These do *not* represent atoms, but are designed to show the folding of the polynucleotide chain.

mately the same horizontal helical axis in Figure 4. The vertical axis contains both the DHU stem and the anticodon stem, which are organized approximately around the vertical axis in Figure 4, although these two helical segments do not have colinear axes. The corner of the L-shaped molecule is composed of a rather complex coiling together of the TψC and the DHU loops. The extra loop indicated in Figure 4 runs along one edge of the molecule on the exterior. In this position it could accommodate additional residues without interfering with the basic folded structure. Furthermore, it is worth noting that the DHU loop is on the exterior of the molecule and adjacent to the extra loop, so that all the variable regions of the molecule are located close to each other along one edge of the molecule.

The folding in the loop areas is rather complex, and at 3-Å resolution, the details of the loops are not fully revealed. However, the stem regions are fairly clear and have a geometry which is typical of double-helical ribonucleotides. It should be noted that the recent crystal structure determination of two double-helical ribonucleotides (ApU and GpC) is useful in providing the standard bond lengths and angles which are needed to interpret the data found in the 3-Å electron density map (Rosenberg et al. 1973; Day et al. 1973).

The folding in the anticodon loop can be followed in a reasonably clear fashion even though we are not entirely certain about some fine details. The three anticodon bases are oriented so that they project more or less downward in relation to the molecule as it is represented in Figure 4. The three anticodon bases are stacked parallel to one another, although one base does not overlap its neighbors completely. How much of this is due to crystal lattice interactions is not clear at the present time. However, the sense of the folding of the anticodon loop suggests that in this molecule, the anticodon bases are in a position such that there could be hydrogen bonding approximately parallel to the anticodon stem of the molecule.

The distance between the anticodon loop and the 3′-terminal adenosine to which the amino acid is attached during aminoacylation is 77 Å. The distance from the anticodon to the TψC corner of the molecule is 67 Å, and the distance from the TψC corner to the 3′-terminal adenosine is 70 Å. The thickness of the molecule in Figure 4 (approximately perpendicular to the page) is 20 Å, the thickness of a double helix.

Where Is tRNA in the Ribosome?

There is not a great deal of information available concerning the position of transfer RNA within the ribosome. Relatively strong evidence suggests that it is not found near the surface of the ribosome. The best evidence in this regard is the fact that tRNA is completely protected from the action of nucleases once an initiation complex is formed. An example of this is shown in the experiment illustrated in Figure 5 (Kuechler, Bauer and Rich 1972).

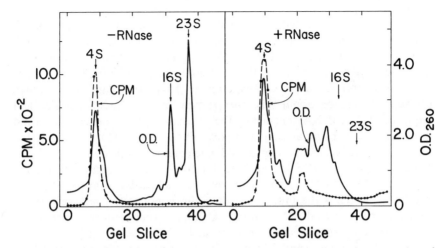

Figure 5 Polyacrylamide gel electrophoresis of RNA from the control and ribonuclease-treated initiation complex. The initiation complex was formed with R17 RNA and [^{35}S]fMet-tRNA^{fMet}; one part was treated with ribonuclease at 0°C for 1 hr and one part was left untreated. The samples were purified by centrifugation through 1 M sucrose. The RNA was extracted, heated for 2 min at 55°C in 0.01 M sodium acetate (pH 5.5), 0.001 M EDTA and 0.5% sodium dodecyl sulfate before layering on 2.8% polyacrylamide gels. Electrophoresis was carried out at 5 mA per tube for 90 min. Gels were scanned at 260 nm (Kuechler, Bauer and Rich 1972).

An initiation complex was formed with *E. coli* ribosomes using the R17 viral messenger RNA and ^{35}S-labeled formylmethionyl-tRNA. After forming the initiation complex, the total RNA was extracted from the ribosome and analyzed on an acrylamide gel. Carrier 4S RNA was added to carry out the gel analysis. In the left side of Figure 5 the RNA extracted from untreated control ribosomes has a typical pattern, showing both 16S and 23S ribosomal RNA, and the ^{35}S radioactivity is seen moving in the 4S region. The right side of Figure 5 shows the gel pattern after these ribosomes have been subjected to digestion with 10 μg of pancreatic ribonuclease for 1 hr at 0°C. Following incubation with ribonuclease, total RNA was extracted and carrier 4S RNA was added; the ribosomal RNA is almost entirely degraded. There is no 23S and only residual amounts of 16S. In contrast to the large destruction of the ribosomal RNA, the radioactivity still migrates with the carrier 4S RNA in the gel. Thus, even though the ribosomal RNA has been subjected to massive digestion by added ribonuclease, the ribosome nonetheless shields the formylmethionyl-tRNA from digestion. The obvious inference from these experiments is that the tRNA is located within the ribosome in a manner which effectively screens it from nuclease in the surrounding media.

Further experiments can be carried out to ask how deeply in the ribosome the tRNA is located. Ribosomes have a diameter of approximately 200–300 Å, depending upon the species of ribosome (Van Holde and Hill, this volume). The tRNA molecule with approximate dimensions of 77 × 20 × 45 Å is considerably smaller and could easily be buried in the ribosomal structure. A number of experiments have been carried out which provide information about this. In one type of experiment, the size and sequence of the mRNA shielded by the ribosome against nuclease digestion is determined for initiation complexes. These experiments show that a segment of mRNA containing 35–45 nucleotides is shielded by the ribosome (Steitz 1969; Hindley and Staples 1969). The reading site for the initiator codon (AUG) is approximately 15 nucleotides from the 3′ end of the fragment. Thus the initiator reading site is not found precisely in the center of the protected tRNA strand. A similar experiment was carried out in which an initiation complex was formed, ribonuclease digestion carried out, and then protein synthesis was allowed to proceed. In this way it was demonstrated that an mRNA piece containing enough information for five amino acids was protected by the ribosome (Kuechler and Rich 1970). Both these experiments suggest that the reading site for the AUG initiator, probably the P site for peptidyl-tRNA, is found approximately 15 nucleotides in from the edge of the ribosome. If the nucleotides are stacked in the normal stable conformation of polynucleotides with a translation per base of 3.4 Å, it suggests that the P site is approximately 50 Å in from one edge of the ribosome.

A similar type of analysis was carried out on the nascent or growing polypeptide chains of reticulocyte polysomes synthesizing hemoglobin (Malkin and Rich 1967). Here the polypeptide chains were digested with a variety of proteolytic enzymes, and it was shown that a resistant segment of polypeptide chain containing 25–30 amino acids remains protected by the ribosome. This segment was shown to be attached to the transfer RNA molecules and as such it provides another dimension concerning the position of the tRNA within the ribosome. If the polypeptide chain is fully extended, the translation per amino acid is 3.7 Å; thus the length of polypeptide chain between the 3′ end of the tRNA and the point at the periphery of the ribosome where the proteolytic enzymes can attack is approximately 100 Å. If the polypeptide is in the form of an alpha helix, this distance would be closer to 50 Å. However, due to the frequent presence of proline residues in the growing polypeptide chain, it is perhaps unreasonable to postulate an alpha-helical conformation of the growing polypeptide chain, and the most probable distance is closer to 100 Å.

These experiments suggest that there may exist a number of channels within the ribosome, one in which the mRNA travels and another in which the growing polypeptide chain travels. To this list one should add an aminoacyl tRNA entrance channel and a tRNA exit channel. These latter two could be the same channel, but they are more likely to be separate.

The familiar model of protein synthesis involves the coordinate activity of two tRNA sites, a peptidyl-tRNA site (P) and an aminoacyl-tRNA site (A). A number of experiments have been carried out that support the concept of a firm binding of two tRNAs to a ribosome when it is active in protein synthesis. Some of the initial experiments carried out on the polysomes of rabbit reticulocytes active in hemoglobin synthesis demonstrated that two tRNA molecules were bound per ribosome in polysomes sedimenting down a sucrose gradient, whereas the single ribosomes inactive in protein synthesis contain approximately one tRNA per ribosome (Warner and Rich 1964). There is good evidence that the requirements for binding to the two sites are different. Binding to the peptidyl site requires acylation of the alpha amino group in aminoacyl-tRNA, and in prokaryotes, this is accomplished by formyl methionine tRNA binding during initiation.

Movement of mRNA in the Ribosome

One can imagine two different mechanisms for the movement of mRNA in the ribosome. One of these involves a "gating" mechanism in which the polynucleotide mRNA strand is advanced three nucleotides at a time. This could be carried out by a component of the ribosomal machinery. A second interpretation of mRNA motion is that it is not carried out directly by the ribosomal machinery, but is instead a sole consequence of the tRNA motion. In this interpretation, the movement of a codon triplet may be governed largely by the movement of the anticodon triplet in the tRNA. Of these two mechanisms, the latter is the most likely to be correct. The reason for this belief stems from the recent sequence analysis of a frameshift suppressor tRNA. Frameshift mutations are those in which the reading frame changes due to a mutation that either deletes or inserts an extra nucleotide in the mRNA strand. A class of suppressor tRNAs is found which acts to correct the reading frame. The nucleotide sequence of a glycine suppressor tRNA has been completed and it has an extra base in its anticodon loop (Riddle and Carbon 1973). The action of the suppressor tRNA strongly suggests that the movement of the suppressor tRNA through the ribosome results in a movement of four bases in the mRNA strand instead of the usual three. This could not occur if a ribosomal gating mechanism solely moved the mRNA three nucleotides at a time. Rather it is more likely that the tRNA movement regulates the mRNA movement.

Flow of tRNA in the Ribosome

The central reaction in protein synthesis involves the formation of consecutive peptide bonds along a polypeptide chain. This is believed to occur through coordinated action involving both the peptidyl- and aminoacyl-tRNA sites. Aminoacyl-tRNA comes into the A site and is positioned next to peptidyl-tRNA in the P site. The ribosomal peptidyl transferase then transfers the peptidyl chain from peptidyl-tRNA to the amino acid of aminoacyl-tRNA with the formation of a peptide bond. The tRNA then leaves the P site once its peptide chain is detached. The A site tRNA, with its newly attached polypeptide chain, is then believed to be moved to the P site. During this translocation process, it is believed that the mRNA moves over at the same time that the tRNA travels from one site to the other. Repetition of this process results in the elongation of the polypeptide chain. This simplified statement is the widely accepted view of protein synthesis. However with our increasing knowledge concerning both the structure of tRNA and the ribosome, we can use these to set new boundary limits on the type of model which we can develop to account for this process in physical terms.

Any model of tRNA movement in the ribosome must of necessity involve a number of assumptions. Before proposing a model, it is worth stating these assumptions explicitly.

We assume that all tRNA molecules have a similar tertiary structure, which is determined by the common features seen in the sequences as discussed above. Thus the yeast phenylalanine-tRNA structure (Figure 4) may illustrate the folding of all tRNAs. Implicit in this is the assumption that the molecular folding seen in the crystal is similar to that seen in solution when it is biologically active. These are substantial assumptions, and even though there is some information supporting them, the evidence is far from complete. Co-crystals of different tRNA molecules have been reported* (Fresco, Blake and Langridge 1968); this and the fact that all tRNAs go through the same ribosomal machinery both suggest common features of tRNA tertiary structure. NMR studies of yeast phenylalanine-tRNA in solution (Wong et al. 1972) support the hydrogen bonding seen in the crystal structure and implied in the cloverleaf configuration.

Another assumption is that the tRNA molecules do not undergo substantial conformational changes within the ribosome. This does not rule out small changes such as an altered folding of the TψC loop, for example, during translocation. However, we assume there are no gross changes in the molecule.

We also assume that the anticodons of the two tRNA molecules in the A and P site both interact with adjacent codons on the mRNA strand at the same time. This is the usual model of tRNA action during protein synthesis. In addition, when the bases of the codon nucleotides are hydrogen bonded with the anticodon bases on tRNA, they remain stacked, i.e., with a separation of 3.4 Å between the bases. This means that the polynucleotide backbone is not stretched out during the codon-anticodon interactions. This is reasonable, since the stability and specificity of the hydrogen bonding is undoubtedly enhanced by the stacking interaction which eliminates rotational freedom of the bases. Finally, we assume that the peptidyl chain is transferred from the 3'-OH of the tRNA in the P site directly to the apha amino group of aminoacyl-tRNA in the A site, i.e., without an intermediate transfer to another structure, such as a ribosomal protein. No one has ever found an intermediate transfer site, even though many experiments have been designed to find it.

A Model of tRNA Movement

These assumptions lead us to an immediate paradox. With the nucleotide bases of the codon stacked, the codon is 3×3.4 Å or about 10 Å along the mRNA strand. If two transfer RNAs are simultaneously interacting with adjacent nucleotide triplets, these tRNA anticodons must also be no further apart than 10 Å. The paradox derives from the fact that the diameter of the transfer RNA molecule near the anticodon end is approximately 20 Å, the diameter of the double-helical anticodon stem.

Can we engage the two tRNA molecules in such a way that they can approach the mRNA strand from different directions? If this were done, two adjacent tRNAs could then form complementary hydrogen bonds with adjacent codon triplets without steric inteference with each other. However, this immediately introduces a problem because if the tRNAs are approaching the mRNA from different directions, this means that the 3'-OH ends of the tRNAs to which the amino acid and the peptidyl chain are attached will be pointing far away from each other. Since the distance between the anticodon and the 3'-OH acceptor end is 77 Å, the separation between the peptidyl chain and the amino acid could be enormous.

This paradox might be resolved by the observation that the tRNA molecule is not, in fact, a linear molecule, but rather one which is bent over in an L-shaped configuration with the anticodon at one end of the L, the 3'-OH acceptor at the other end, and the TψC sequence near the corner of the L. This shape of the molecule suggests that two tRNAs could interact with two adjacent codons and at the same time have their ends close to each other if the molecules were arranged in such a way that two TψC corners were far apart (Figure 6). Here the anticodon stems both converge together, but at an angle from each other, while the 3'-OH acceptor ends are close to each other due to the bent nature of the molecule. These two molecules are related to each other by a rotational operation. The implied suggestion is that the mRNA strand may turn a corner while it is being read. In this manner, the three nucleotide bases in the codon remain in a stacked configuration, but they become unstacked between each triplet of bases. This would then have the consequence that the third codon base (the wobble base) would have a unique environment somewhat different from that of the other two bases in the anticodon when the tRNA is in the peptidyl site.

By this analysis, translocation is largely a rotatory operation on tRNA, during the course of which the tRNA molecule is swung around with the

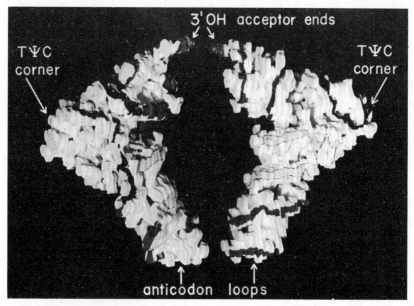

Figure 6   Two molecular models of yeast phenylalanine-tRNA lying on a flat surface with their anticodon loops close together and their 3'-OH acceptor stems close together.

TψC corner moving the maximum distance while the anticodon and 3'-OH ends of the tRNA undergo a minimal translational motion (Figure 7). The tRNA messenger RNA complex would "turn a corner" in order to go from the A site to the P site. The turning angle in Figure 7 is close to 90°; however, that is somewhat arbitrary. The peptidyl transferase could then operate on the two nearby 3'-OH ends of the molecule and transfer the peptidyl chain. A consequence of this model is the relative convenience of the TψC portion of the tRNA molecule as a site to interact with the ribosomal machinery. Recent evidence has suggested that the TψC sequence of tRNA is a binding site to the ribosomal 5S RNA, which may be active during translocation (Erdmann, Sprinzl and Pongs 1973).

There are a number of implications of this model of tRNA movement, some of which can be tested directly. One of these is a statement regarding the position of the anticodon bases in tRNA. They must be in a position such that they can form hydrogen bonds with the nucleotides of two successive codons when such a rotatory operation is carried out. Further high resolution work on the structure of tRNA and, more specifically, on the structure of tRNA-codon complexes should make it possible to answer this question explicitly.

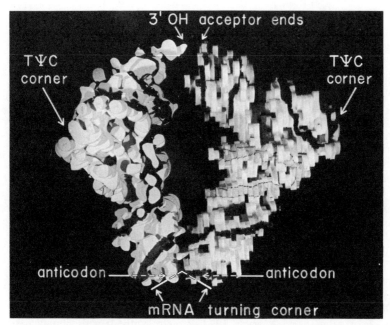

Figure 7 Two tRNA molecular models held approximately at right angles to each other. The model on the left has the TψC corner close to the reader, and the model on the right is lying in the plane of the paper. The mRNA is shown schematically bent around so that two adjacent codons can interact with the two tRNA anticodons at the same time. The two 3'-OH acceptor ends are close enough so that the peptidyl chain can pass from one to the other. In the arrangement shown here, the 3' end of the mRNA strand is at the left, the 5' end is at the right. Aminoacyl-tRNA is at the left and peptidyl-tRNA is at the right. During translocation, the molecule on the left rotates into the position of the molecule on the right after the latter has left the ribosomal site and it drags the mRNA around a corner with this motion.

CONCLUSION

This article presents a frankly speculative description of how tRNA molecules may move within the ribosome during protein synthesis. A large number of models have been proposed for outlining the mechanism of the translocation process. The present model has its genesis in the fact that knowledge of the three-dimensional structure of tRNA provides some boundary conditions for describing these interactions. At the present time our knowledge of tRNA-ribosomal interactions is necessarily limited, so it is difficult to evaluate models of this type. However, such attempts at an explicit formulation have some value in that they serve as a useful basis for asking questions and formulating experiments designed to test the hypothesis.

References

Day, R. O., N. C. Seeman, J. M. Rosenberg and A. Rich. 1973. A crystalline fragment of the double helix: The structure of the dinucleoside phosphate guanylyl-3′,5′-cytidine. *Proc. Nat. Acad. Sci.* **70**:849.

Erdmann, V. A., M. Sprinzl and O. Pongs. 1973. The involvement of 5S RNA in the binding of tRNA to ribosomes. *Biochem. Biophys. Res. Comm.* **54**:942.

Fresco, J. R., R. D. Blake and R. Langridge. 1968. Crystalization of transfer ribonucleic acids from unfractionated mixtures. *Nature* **220**:1285.

Hindley, J. and D. H. Staples. 1969. Sequence of a ribosome binding site in bacteriophage Qβ-RNA. *Nature* **224**:964.

Holmquist, R., T. H. Jukes and S. Pangburn. 1973. Evolution of transfer RNA. *J. Mol. Biol.* **78**:91.

Kim, S. H., G. J. Quigley, F. L. Suddath, A. McPherson, D. Sneden, J. J. Kim, J. Weinzierl and A. Rich. 1973. Three-dimensional structure of yeast phenylalanine transfer RNA: Folding of the polynucleotide chain. *Science* **179**:185.

Kim, S. H., G. Quigley, F. L. Suddath, A. McPherson, D. Sneden, J. J. Kim, J. Weinzierl, P. Blattmann and A. Rich. 1972. The three-dimensional structure of yeast phenylalanine transfer RNA: Shape of the molecule at 5.5 Å resolution. *Proc. Nat. Acad. Sci.* **69**:3746.

Kuechler, E. and A. Rich. 1970. Position of the initiator and peptidyl sites in the *E. coli* ribosome. *Nature* **225**:920.

Kuechler, E., K. Bauer and A. Rich. 1972. Protein synthesis with ribonuclease digested ribosomes. *Biochim. Biophys. Acta* **277**:615.

Malkin, L. I. and A. Rich. 1967. Partial resistance of nascent polypeptide chains to proteolytic digestion due to ribosomal shielding. *J. Mol. Biol.* **26**:329.

RajBhandary, U. L. and S. H. Chang. 1968. Yeast phenylalanine transfer ribonucleic acid: Partial digestion with ribonuclease T and derivation of the total primary structure. *J. Biol. Chem.* **243**:598.

Riddle, D. L. and J. Carbon. 1973. Frameshift suppression: A nucleotide addition in the anticodon of a glycine transfer RNA. *Nature New Biol.* **242**:230.

Rosenberg, J. M., N. C. Seeman, J. J. Kim, F. L. Suddath, H. B. Nicholas and A. Rich. 1973. Double helix at atomic resolution. *Nature* **243**:150.

Steitz, J. A. 1969. Polypeptide chain initiation: Nucleotide sequences of the three ribosomal binding sites in bacteriophage R17 RNA. *Nature* **224**:957.

Suddath, F. L., G. J. Quigley, A. McPrerson, D. Sneden, J. J. Kim, S. H. Kim and A. Rich. 1974. Three-dimensional structure of phenylalanine transfer RNA at 3.0 Å resolution. *Nature* **248**:20.

Warner, J. R. and A. Rich. 1964. The number of soluble RNA molecules on reticulocyte polyribosomes. *Proc. Nat. Acad. Sci.* **51**:1134.

Wong, Y. P., D. R. Kearns, B. R. Reid and R. G. Shulman. 1972. Investigation of exchangeable protons and the extent of base pairings in yeast phenylalanine transfer RNA by high resolution nuclear magnetic resonance. *J. Mol. Biol.* **72**:725.

Reference Index

Numbers in italics refer to pages on which the complete references are listed.

Ab, G., *742*
Abe, H., 151, *161*
Abelson, J., *44*, 355, *362, 363, 802*
Aboderin, A. A., 588, 589, *598*
Aboud, M., 734, *741*
Abrass, I. B., *459*
Acharya, A. S., 116, *135*, 208, *218*, 294, *302*, 559, *570*, 575, *584*, 611, *612*, 642, *660*, 712, 723, *727*
Acs, G., 23, 24, *47*, 474, *487*
Adams, J. M., 171, *187*
Adamson, S., 682, 685, 687, 696, 697, *702, 703*, 859, *868*
Adesnik, M., 15, *42*, 153, *161*, 186, *188*, 394, *412*
Adman, R., 819, *824*
Adoutte, A., 525, *534*
Agsteribbe, E., 519, 520, 521, *536*
Aharonowitz, Y., 340, *361*
Ahmad-Zadeh, C., *91, 114, 307, 330, 392, 666*
Alberghina, F. A. M., 440, 448, *453*
Alberts, B. M., 176, *189*
Albright, L. J., 79, *86*
Aldridge, J., *329*
Alfert, M., 841, *850*
Alföldi, L., 733, 735, *741*
Alix, J. J., 116, *135*
Allaudeen, H. S., 21, *42*
Allen, E. R., 497, *510*
Allende, J. E., *44, 47*
Allet, B., 156, *161*, 245, 255, 263, *264*
Aloni, Y., *161, 168*, 500, 506, *510*, 522, *534*,

538, 539
Altares, C. D., 866, *868*
Altman, S., 18, *42*
Altruda, F., *414*
Amaldi, F., 142, 153, *164*, 346, *362*, 491, 495, 496, 499, 508, *510, 511, 513*, 834, *838*
Amaldi, G., 520, *534*
Amaldi, J., 462, *483*
Amalric, F., 469, *483*
Amelunxen, F., 60, 62, 71, 83, 84, *86, 90*
Ames, B. N., *50*
Amons, R., *138*, 257, 264, 267, *457*, 724, 727, 730
Amoros, M. J., *458*
Amos, H., 430, 432, *457*, 469, *487*, 865, *869*
Anand, N., 774, *785*
Anderegg, J. W., 54, *88*, 274, *304*, 328, 548, 557, 610, *613*
Anderson, J. S., 40, *42*
Anderson, P., *363*, 775, 780, 781, *785, 786*
Anderson, T. F., 6, *11*
Anderson, W. F., 23, *48*, 703, 708, 710, 793, 801, 865, *868, 869*
Anfinsen, C. B., *306*
Apgar, J., *46*
Apirion, D., 129, *136, 137, 139, 140, 190*, 265, *307*, 308, 327, 335, 348, 351, *361, 363, 364, 366, 368*, 408, *414, 415*, 661, 662, 665, 666, 775, 776, 778, 779, 787, 800, *802*, 808, *813*
Apte, B. N., 16, *42*
Arai, K. I., 18, 32, 33, *42*, 52, 734, *742*
Arai, N., *742*

Subject Index

A-site, 33, 299. *See also* Aminoacyl-tRNA:
     P-site: Peptidyl transferase
   Fab inhibitors of, 646–648
   in magic spot synthesis in vitro, 750
AC-DC model of ribosome function, 325
Acetylation of eukaryotic r-proteins, 439
Acheta, gene amplification in, 497
Actinomycin D
   effect on 5S RNA transcription, 151
   effect on ribosome synthesis in HeLa cells,
     463
S-adenosyl methionine as methyl donor to
     rRNA, 184
Affinity chromatography for the study of
     rRNA-protein interactions, 229
Affinity labeling, 573–584
Affinity labels, 295, 298–300, 573
Alanyl-tRNA synthetase mutation suppressed
     by mutant r-proteins, 128, 131
Alkaline phosphatase
   for dephosphorylation of phosphoproteins,
     861
   regulation in *strA* mutants, 347
α_r. *See* Proteins, bacterial and ribosomal, rate
     of synthesis
α-Amanatin, 151, 500
Amino acid composition of eukaryotic r-pro-
     teins, 426
Amino acid sequences of ribosomal proteins
     compared, 117–125
Amino acid starvation, effect of, 376, 473,
     733–736, 747. *See also* Stringent re-
     sponse
Amino acids, modified found in r-proteins, 116
   role in rRNA binding, 256
Aminoacyl-tRNA, 32. *See also* A-site: Elon-

gation factor T
   affinity analogues of, 297–300
   binding inhibited by anti-50S protein anti-
     bodies, 642, 645
   complex with EF-Tu and GTP, 32
   and control of amino acid biosynthesis, 21
   30S proteins involved in binding, 285–287
Aminoacyl-tRNA protein transferases, 20
Aminoacyl-tRNA synthetase, 18–19
   in stringent response, 736
Aminoglycosides, pleiotropic effects on bacte-
     ria, 774. *See also* Streptomycin
Antibiotics
   acting on yeast ribosomes, 821–823
   affinity analogues of, 298–300
   binding to ribosomes inhibited, 627
   genetic loci linked to prokaryotic r-proteins,
     346–351
   macrolide, 130
Antibodies to *E. coli* ribosomal components,
     616–658
   AA-tRNA binding inhibited by, 285
   accessibility of r-proteins to, 275, 628–634
     changed by magnesium concentration,
     658
   association of subunits inhibited by, 284,
     301, 645–655
   to compare r-protein structures, 616–626
     interspecific, 624–626
     in mutants, 622–624
   EF-dependent GTPase inhibited by, 295
   Fmet-tRNA binding inhibited by, 286
   to identify r-proteins, 621, 626
   to locate r-proteins on subunits for EM, 655
   poly(Phe) synthesis inhibited by, 634–637
   protein biosynthesis partial reactions inhib-

911